Contemporary Mathematicians

Gian-Carlo Rota
Editor

Birkhäuser
Boston · Basel · Stuttgart

Kurt Otto Friedrichs

Selecta
Volume 2

Cathleen S. Morawetz
Editor

Birkhäuser
Boston · Basel · Stuttgart
1986

Editor

Cathleen S. Morawetz
Courant Institute of
Mathematical Sciences
New York University
251 Mercer Street
New York
N.Y. 10012
USA

CIP-Kurztitelaufnahme der Deutschen Bibliothek

Friedrichs, Kurt O.:
Selecta / Kurt Otto Friedrichs. Cathleen S.
Morawetz ed. – Boston; Basel; Stuttgart
Birkhäuser
(Contemporary mathematicians)
ISBN 3-7643-3270-0 (Basel ...)
ISBN 0-8176-3270-0 (Boston)
NE: Friedrichs, Kurt O.: [Sammlung]
Vol. 2 (1986).
ISBN 3-7643-3269-7 (Basel ...)
ISBN 0-8176-3269-7 (Boston)

ISBN 0-8176-3270-0 (complete set)
ISBN 0-8176-3268-9 (Vol. 1)
ISBN 0-8176-3269-7 (Vol. 2)

ISBN 3-7643-3270-0 (complete set)
ISBN 3-7643-3268-9 (Vol. 1)
ISBN 3-7643-3269-7 (Vol. 2)

Contents – Volume 2

V Contributions to Magnetohydrodynamics 339
by Harold Weitzner

VI Papers involving Variational Problems.......... 411
Commentaries by Louis Nirenberg

VII Other Papers 444

VIII Contributions to Quantum Mechanics 602
by David Isaacson

**Listed in Reports

Contents – Volume 1

II
Papers in Spectral Theory

Commentaries by Tosio Kato

[34–1] **Spektraltheorie halbbeschränkter Operatoren und Anwendung auf die Spektralzerlegung von Differentialoperatoren, Math. Ann., Teil 1: 109 (1934), 465–487. Teil 2: 109 (1934), 685–713. Berichtigung: 110 (1935), 777–779.**

Commentary:

This massive paper consists of two parts (plus a Berichtigung, which is essentially a supplement). The first and abstract part is a pioneering work on semibounded quadratic forms (written $x\mathbf{G}x$) in a Hilbert space and their relationship with semibounded, self-adjoint operators G. It may be regarded as an abstraction of classical calculus of variations involving quadratic functionals. The decisive notion of closed form is introduced, and the fundamental theorem is given to the effect that there is a one-to-one correspondence between the densely-defined, semibounded closed forms **G** and the semibounded self-adjoint operators G, formally given by $x\mathbf{G}y = x\mathbf{H}Gy$, where **H** is the fundamental form (inner product) in the Hilbert space. The form constructed from a semibounded symmetric operator G_0 is shown to be "closable". From this follows the existence of a distinguished self-adjoint extension G_1 of G_0 (now commonly called the Friedrichs extension), which is the operator associated with the closed form. Criteria for $G_1 = G_0$ to be true (i.e. G_0 is itself self adjoint) are given.

These results are applied in the second part to differential operators $G = -\Delta + v(x)$ in a region Γ, which is R^n, $R^n \setminus \{0\}$, or a ball about $x = 0$ in R^n, $n = 1, 2, 3$, with v real and piecewise continuous ($n = 1$) or continuously differentiable ($n = 2, 3$) (and possibly with some restriction on the negative part of v near the boundary of Γ). G is first defined for twice continuously (or piecewise continuously) differentiable functions with finite Dirichlet integral, satisfying Dirichlet or Neumann boundary condition (in a generalized sense) in case Γ is a ball, so that G becomes a semibounded symmetric operator in $H = L^2(\Gamma)$. The Friedrichs extension G_1 of G is constructed, which is then shown to be identical with the (operator) closure of G. (This means that G is essentially self-adjoint). This method has become one of the standard procedures in problems in which it is required to show that a given operator is essentially self-adjoint. The main body of the proof consists in establishing the regularity of functions belonging to the domain of G_1 and applying the criteria given in the first part. It is interesting to note that the technique of using mollifiers (though not in its fully developed form) and cut-off functions already appears here. There is an article by Kalf [1], in which this paper by Friedrichs is carefully analyzed in relation to later developments.

It may be remarked that in this paper Friedrichs does not work with the L^2-space of Lebesgue-measurable functions, preferring the space of continuous (or piecewise continuous) functions and its completion by ideal elements. Since the domain of G_1 is found to consist of continuous functions, no ideal elements appear in the final results involving eigenfunctions, wave packets, etc.

Another remark is that a nonnegative quadratic form is closed if and only if it is lower semicontinuous (after being extended to the whole space with values $+\infty$ outside of the original domain). This is almost obvious from the important part played by lower semicontinuity in the calculus of variations, but apparently it had not been noticed explicitly until recently (see e.g. Simon [2]).

[1] H. Kalf. Gauss's theorem and the self-adjointness of Schrödinger operators, Ark. Math., 18 (1980), 19–47.

[2] B. Simon. Lower semicontinuity of positive quadratic forms, Proc. Roy. Soc. Edinburgh 79 (1977), 267–273.

Spektraltheorie halbbeschränkter Operatoren und Anwendung auf die Spektralzerlegung von Differentialoperatoren.

Von

Kurt Friedrichs in Braunschweig.

Die vorliegende Arbeit entstand aus dem Versuch, mit den direkten Methoden der Variationsrechnung auch die Eigenwertprobleme solcher linearer partieller Differentialgleichungen zu lösen, die kein gewöhnliches Punktspektrum besitzen und also nicht der Variationsrechnung selbst zugänglich sind. Insbesondere sollte auf diese Weise die Spektraltheorie der quantentheoretischen Energieoperatoren ausgehend von Schrödingers Darstellung gewonnen werden.

Bei der näheren Durchführung dieses Versuchs zeigte sich, daß eine große Reihe gleichartiger Schlüsse und Begriffe bei den verschiedenen Problemen sich einheitlich fassen lassen, wenn man sie der Symbolik der allgemeinen Operatorentheorie des Hilbertschen Raumes unterordnet; und zwar der des „abstrakten" Hilbertschen Raumes, wie sie zuerst konsequent durch v. Neumann [7. 2] ausgearbeitet wurde. Die früher bevorzugte Darstellung des Hilbertschen Raumes durch unendlich viel Veränderliche erweist sich als unhandlich zur Darstellung von Funktionenräumen. Überdies kann, wie v. Neumann [7. 1] erkannt hat, die Darstellung nicht beschränkter Linearoperatoren durch unendliche Matrizen geradezu irreführend sein.

Dagegen erwies es sich als unnötig, die allgemeine Spektraltheorie nicht beschränkter Operatoren von v. Neumann heranzuziehen, da wir von vornherein nur halbbeschränkte Operatoren betrachten; für solche kann die Spektraltheorie unmittelbar auf die der beschränkten Operatoren zurückgeführt werden. In der Tat sind die meisten Energieoperatoren nach unten halbbeschränkt. Auch bei der Behandlung von Eigenwertdifferentialgleichungen durch Variationsrechnung wurde die Halbbeschränktheit wesentlich ausgenutzt.

Eine Theorie halbbeschränkter Operatoren findet sich bei A. Wintner [13]; allerdings ist sie wesentlich auf unendliche Matrizen bezogen.

Man hat zwar bisher schon mehrfach Eigenwertprobleme von Differentialgleichungen auf die Hilbertsche Theorie zurückgeführt (vgl. z. B. [6]);

doch in der Weise, daß man mit Hilfe der explizit bekannten Greenschen Funktion die beschränkte Reziproke des Differentialoperators aufsuchte. Meist handelt es sich um die Fälle, bei denen ein diskretes Punktspektrum auftritt.

Darüber hinaus führt vor allem die Theorie der Differentialgleichungen mit Singularitäten von H. Weyl [12. 1, 12. 2] [1]).

Diese Weylsche Theorie wurde von Stone [10. 2] allgemeiner aufgezogen und ohne Reduktion auf Integraloperatoren anders entwickelt.

Wesentlich wird bei dieser Theorie die bekannte zweiparametrige Schar der Lösungen der Differentialgleichungen benutzt; und von daher rühren die Schwierigkeiten bei dem Versuch einer unmittelbaren Übertragung auf partielle Differentialgleichungen [2]).

Im Teil I dieser Arbeit wird die Spektraltheorie halbbeschränkter symmetrischer Operatoren im abstrakten Hilbertschen Raum entwickelt. Man kann diese Theorie leicht gewinnen, wenn man neben Operatoren noch zugehörige Formen betrachtet. Ohne Einschränkung sei diese Form $\boldsymbol{G}$ positiv halbbeschränkt; d. h. es gebe ein positives γ, so daß mit der Einheitsform $\boldsymbol{H}$ gilt

$$\boldsymbol{G} \geqq \gamma \boldsymbol{H}.$$

Man kann nun die Form $\boldsymbol{G}$ als Maßform eines neuen Hilbertschen Raumes (eines Teilraumes des ursprünglichen) auffassen; dann wird $\boldsymbol{H}$

---

[1]) Weyl behandelt die Eigenwertgleichung

$$-\frac{d}{d x}\left(p \frac{d u}{d x}\right)+q u-\lambda u=0$$

für eine Funktion $u(x)$ in $x \geqq 0$. Dabei sei $p>0$ und p und q stetig in $x \geqq 0$. Bei $x=0$ sei eine Randbedingung $\cos \vartheta\, u+\sin \vartheta \frac{d u}{d x}=0$ gestellt. Weyl zeigt, daß zwei Fälle eintreten können,

1. der Grenzpunktfall, bei dem außer der Existenz von $\int_0^\infty u^2 \, d x$ keine weitere Bedingung für u bei $x=\infty$ zu stellen ist, und

2. der Grenzkreisfall, bei dem noch eine zyklische einparametrige Schar von Randbedingungen bei $x=\infty$ zur Auswahl steht.

Weyl zeigt ferner, wie die Eigenfunktionen des Streckenspektrums aus den nicht im Hilbertschen Raum gelegenen Lösungen der Eigenwertdifferentialgleichung durch Integration nach dem Eigenwert zu gewinnen sind.

[2]) Bei Abfassung dieser Arbeit bemerke ich, daß Carleman [1] angibt, es sei leicht, die Theorie von Weyl auf Gleichungen von mehr Variablen zu übertragen, sei es direkt, sei es vermittels der Theorie der Hermiteschen Integralgleichungen.

eine beschränkte Form. Die für beschränkte Formen bekannte Spektralzerlegung führt so unmittelbar zur Spektralzerlegung der halbbeschränkten Form G [3]).

Solche halbbeschränkten Formen G können stets aus halbbeschränkten Operatoren gewonnen werden. Die Spektralzerlegung solcher Operatoren ist aber nur dann möglich und auf diesem Wege zu gewinnen, wenn sie „selbstadjungiert" (hypermaximal) sind. Diese Bedingung läßt sich nun für halbbeschränkte Operatoren durch wesentlich abgeschwächtere ersetzen, die auch für unsere Differentialoperatoren leichter nachzuweisen sind.

Schließlich werden noch die von Hilbert [5. 2] und Weyl [12. 3] herrührenden Kriterien dafür, daß das Spektrum z. T. diskret ist, auf halbbeschränkte Operatoren übertragen.

Im Teil II wird die vorangehende Theorie auf Differentialoperatoren angewandt. Wir haben uns dabei auf typische Fälle beschränkt. Der Operator sei

$$-\sum_{\nu=1}^{n} \frac{\partial^2}{\partial x_\nu^2} + v(x_1, \ldots, x_n),$$

anzuwenden auf Funktionen $f(x_1, \ldots, x_n)$. Vollständig wird nur der Fall $n = 1, 2, 3$ durchgeführt. Wir behandeln einmal den Fall **(1)** des unendlichen Gebietes mit stetiger nach unten beschränkter Funktion v; zweitens **(2)** wird eine Singularität von v in einem Punkte zugelassen. Sodann wird ein endliches Gebiet zugrunde gelegt, an dessen Rande die Funktion f **(3)** oder ihre normale Ableitung verschwinden soll **(4)**; aus Bequemlichkeit wird dieses Gebiet als Strecke, Kreis, Kugel gewählt. Die Ergebnisse in den Fällen endlichen Gebietes sind nicht neu, ihre Behandlung wurde mit aufgenommen, um zu zeigen, wie weit die Theorie aller Fälle gemeinsam durchgeführt werden kann.

Die erste Aufgabe ist, die Räume der zulässigen Funktionen anzugeben. Diese sind nun zunächst keine Hilbertschen Räume; sie werden aber durch Adjunktion idealer Elemente zu Hilbertschen Räumen fortgesetzt; wir verzichten darauf, diese idealen Elemente durch nach Lebesgue quadratisch integrierbare Funktionen zu realisieren; insbesondere deshalb,

---

3) Übrigens ergibt sich von hier aus auch ein einfacher Zugang zur Spektraltheorie beliebiger unbeschränkter selbstadjungierter Linearoperatoren. Dieser neue Weg bietet gegenüber den bisherigen von v. Neumenn [7. 2], Stone [10. 1], Fr. Riesz [9] den prinzipiellen Vorzug, daß er nicht voraussetzt, daß der zugrunde gelegte Hilbertsche Raum komplex ist; er soll an anderer Stelle durchgeführt werden.

weil gezeigt werden kann, daß die „Eigenelemente", die vor allem interessieren, doch schon den Ausgangsfunktionsräumen angehören[4]).

Auch der Operator ist zunächst nur in einem Raum zweimal differenzierbarer Funktionen erklärt und wird dann formal, aber eindeutig abgeschlossen.

Die Hauptaufgabe ist, nachzuweisen, daß dieser Operator selbstadjungiert ist. Hier liegen überhaupt die eigentlichen Schwierigkeiten der ganzen Theorie. Ihre Überwindung gelingt durch Übertragung der insbesondere von Courant [2. 1, 2. 3, 2. 6, 4. 1] entwickelten Schlußweisen, die bei den direkten Methoden der Variationsrechnung entscheidend sind.

Es zeigt sich so, daß an den für konkrete Differentialoperatoren notwendigen Überlegungen durch die Einordnung in die abstrakte Operatorentheorie kaum etwas gespart wird. Gewonnen ist — abgesehen von der mehr systematischen Anordnung — die Möglichkeit, solche Fälle zugleich mit zu behandeln, bei denen ein diskontinuierliches Spektrum auftritt.

Das *Ergebnis* des zweiten Teiles ist vor allem die Spektralzerlegung des Differentialoperators. Darunter ist die Existenz einer „Spektralschar", einer Schar von Projektionsoperatoren im Sinne der allgemeinen Spektraltheorie verstanden (vgl. Teil I, Nr. 4). Es wird außerdem gezeigt, daß die Eigenelemente dieser Projektionsoperatoren zweimal stetig differenzierbare Funktionen sind. Wir verzichten darauf, diese Projektionsoperatoren und ihre Eigenelemente mit Hilfe der Lösungen der Eigenwertdifferentialgleichungen darzustellen. Wohl folgt unmittelbar für die Eigenfunktionen der Punkteigenwerte, daß sie der Eigenwertdifferentialgleichung genügen. Die Eigenfunktionen des Streckenspektrums könnte man im Falle einer Dimension nach Weyl durch Integration nach dem Eigenwert aus den Lösungen der Eigenwertdifferentialgleichung gewinnen; in den separierbaren Fällen bei mehr Dimensionen ist dasselbe möglich, wie in einer anderen Arbeit gezeigt werden soll; für eine solche Darstellung im allgemeinen Fall bei mehr Dimensionen liegen keine Ansätze vor; hierauf das Hauptschwergewicht der Untersuchung zu legen, scheint mir auch nicht angemessen zu sein.

Ein weiteres Ergebnis bezieht sich auf die Diskussion des Spektrums. Es kann unter einfachen Bedingungen für das „Zusatzpotential" v die Natur des Spektrums genauer bestimmt werden. Es liegt nämlich ein diskretes ins Unendliche anwachsendes Punktspektrum vor, wenn das

---

[4]) Dies Verfahren entspricht ganz dem Vorgang von Hilbert bei seiner Zurückführung der Integralgleichungen auf Gleichungen mit unendlich viel Veränderlichen [5. 1].

Gebiet endlich ist. Dasselbe gilt beim unendlichen Gebiet, wenn das Zusatzpotential v im Unendlichen über alle Schranken wächst. Besitzt dagegen das Zusatzpotential einen endlichen unteren Limes im Unendlichen, so ist das Spektrum unterhalb dieses Wertes diskret.

Diese Kriterien entsprechen einem Teil der Kriterien, die Weyl in seiner Theorie der Differentialgleichungen mit Singularitäten aufgestellt hat; sie können aber in einer von der Anzahl der Variablen unabhängigen Weise bewiesen werden.

Die vorliegende Arbeit enthält den 1. Teil dieser Untersuchungen, der 2. Teil wird in einem folgenden Heft erscheinen.

Literatur.

[1] Carleman, T.
Sur la théorie des équations intégrales et ses applications. Verh. d. intern. Math. Kong. Zürich 1932, Bd. 1, S. 59.

[2] Courant, R.
1. Über die Eigenwerte bei den Differentialgleichungen der math. Physik. Math. Zeitschr. 7. 2. S. 13.
3. Über die Lösungen der Differentialgleichungen der Physik. Math. Annalen 85 (1922). 4. S. 316—317. 5. S. 296.
6. Über direkte Methoden bei Variations- und Eigenwertproblemen. Jahresber. d. d. Math. Vereinig. **34** (1925), S. 105 ff.
7. und Hilbert, D. Methoden der mathematischen Physik 1 (1931), 2. Aufl., Kap. VI, § 5.

[3] Dirac, P. A. M.
Die Prinzipien der Quantenmechanik, § 10 (1930).

[4] Friedrichs, K.
1. Die Rand- und Eigenwertprobleme aus der Theorie der elastischen Platten. Math. Annalen **96** (1927), S. 205 ff. 2. S. 229.

[5] Hilbert, D.
1. Grundzüge einer allgemeinen Theorie der linearen Integralgleichungen. Kap. 13, 14. 2. Kap. 11, Satz 35.

[6] Lichtenstein, L.
Zur Analysis der unendlich vielen Variablen. Math. Zeitschr. **3** (1919), S. 127—160.

[7] Neumann, J. v.
1. Zur Theorie der unbeschränkten Matrizen. J. f. renie u. angew. Math. **161** (1929), S. 208.
2. Allgemeine Eigenwerttheorie Hermitescher Funktionaloperatoren. Math. Annalen **102** (1929), S. 50 ff. 3. S. 70, Def. 5 und Satz 9, S. 72, Def. 9. 4. S. 103, Satz 41. 5. S. 103, [58]).
6. Über adjungierte Funktionaloperatoren. Ann. of Math. **33** (1932), S. 299.

[8] Rellich, F.
Ein Satz über mittlere Konvergenz. Gött. Nachr. 1930.

[9] Riesz, Fr.
1. Les systèmes d'équations linéaires (1913), Nr. 72.
2. Über die linearen Transformationen, Acta Litt. ac Sci. Sectio Math. Szeged 5 (1930). 3. S. 19–54.
[10] Stone, M. H.
1. Linear Transformations in Hilbert Space. New York 1932. Chap. V.
2. Chap. X, § 3.
[11] Toeplitz, O.
Die Jacobische Transformation ..., Gött. Nachr. 1907, S. 101—109.
[12] Weyl, H.
1. Über gewöhnliche lineare Differentialgleichungen mit singulären Stellen. Gött. Nachr. 1909, S. 37.
2. Über gewöhnliche Differentialgleichungen mit Singularitäten. Math. Annalen 68 (1910), S. 220.
3. Über beschränkte quadratische Formen, deren Differenz vollständig ist. Palermo Rend. 27 (1909), S. 373—392.
[13] Wintner, A.
Spektraltheorie der unendlichen Matrizen, § 111.

I. Teil.

Spektraltheorie halbbeschränkter Operatoren.

1. Grundbegriffe.

Es sei erlaubt, zunächst einige wohlbekannte (vgl. [7. 2]) Grundbegriffe und Sätze über Formen und Operatoren zusammenzustellen in einer für unsere Zwecke geeigneten Bezeichnung, um von anderer Literatur unabhängig zu sein.

Unter einem *Raum* $\mathfrak{X}$, $\mathfrak{H}$, $\mathfrak{G}$, $\mathfrak{F}$ von Elementen x, h, g, f sei stets ein reeller *linearer* Raum mit — wenn nicht anders bemerkt — wenigstens abzählbar unendlich vielen linear unabhängigen Elementen verstanden.

Eine Bilinearform ordnet jedem Paar von Elementen x, x_1 eine reelle Zahl zu, linear in x und x_1: wir bezeichnen[5]) sie mit

$$x_1 \boldsymbol{A}\, x.$$

Wir setzen stets

$$x_1 \boldsymbol{A}\, x = x \boldsymbol{A}\, x_1$$

voraus, d. h. $\boldsymbol{A}$ sei *symmetrisch*.

Eine aus einer solchen Bilinearform bestehende quadratische Form $x \boldsymbol{A}\, x$ heiße „nie negativ", wenn

$$x \boldsymbol{A}\, x \geqq 0$$

[5]) Diese Symbolik ist derjenigen von Dirac nachgebildet. Unsere ganze Theorie ist auch im komplexen Raum zu entwickeln, mit nur geringen durchaus geläufigen Abänderungen.

gilt; für solche Formen gilt schon — wie aus der bekannten Schlußweise folgt — die Schwarz-Ungleichung.

$$x_1 \boldsymbol{A} x \leqq \sqrt{(x_1 \boldsymbol{A} x_1)}\sqrt{(x \boldsymbol{A} x)}.$$

Positiv-definit nennen wir sie nur, wenn aus $x \boldsymbol{A} x = 0$ stets $x = 0$ folgt.

Ist in einem Raum $\mathfrak{X}$ durch eine positiv-definite „Maßform" $x \boldsymbol{X} x$ eine Metrik $|x| = \sqrt{x \boldsymbol{X} x}$ eingeführt, läßt sich Dichte einer Menge in $\mathfrak{X}$ und Konvergenz einer Folge x erklären. Wir wollen sagen:

Eine Folge x konvergiert *stark* $(\boldsymbol{X})$

I „in sich", wenn $|x_\nu - x_\mu| \to 0, \quad \nu, \mu \to \infty$

II gegen x, wenn $|x_\nu - x| \to 0, \quad \nu \to \infty$

gilt.

Ist der Raum *separabel* (d. h. er enthält eine abzählbare dichte Menge) und *abgeschlossen* (auch vollständig) (d. h. zu jeder in sich konvergenten Folge gibt es ein Grenzelement), heißt er ein (abstrakter) *Hilbertscher Raum.*

Wir wollen im folgenden einen Hilbertschen Raum $\mathfrak{H}$ von Elementen h zugrunde legen mit der Maßform[6]) $\boldsymbol{H}$.

Wesentlich für unsere Behandlung ist, daß öfter auch Teilräume von $\mathfrak{H}$, z. B. $\mathfrak{G}$ als Hilbertsche Räume mit einer anderen Maßform, z. B. $\boldsymbol{G}$ herangezogen werden, und daß dann Konvergenz und Dichte auf sie bezogen werden. Wir sprechen dann z. B. von Konvergenz $(\boldsymbol{G})$ und von $\boldsymbol{G}$-Dichte.

Teilräume $\mathfrak{F}, \mathfrak{G}, \ldots$ von $\mathfrak{H}$, in denen Formen und Operatoren erklärt sind, seien stets als $\boldsymbol{H}$-dicht in $\mathfrak{H}$ vorausgesetzt.

Ein Operator A in $\mathfrak{F} \leqq \mathfrak{H}$ erklärt, ordnet jedem Element f aus $\mathfrak{F}$ ein h aus $\mathfrak{H}$ zu.

Zu jedem Operator A in $\mathfrak{F}$ „gehört" die Form $\boldsymbol{A}$ in $\mathfrak{F}$:

$$f_1 \boldsymbol{A} f = (f_1 \boldsymbol{H} A f).$$

Der Operator heißt *symmetrisch*, wenn es die zugehörige Form ist; wenn nicht anders bemerkt, seien Operatoren stets als *symmetrisch* und *linear* vorausgesetzt.

Ein Operator B in $\mathfrak{F}$ heißt *beschränkt*[7]), wenn es seine Form $\boldsymbol{B}$ ist: es gibt zwei reelle Zahlen $\underline{\beta}, \bar{\beta}$, so daß $\underline{\beta}\,(f \boldsymbol{H} f) \leqq (f \boldsymbol{B} f) \leqq \bar{\beta}\,(f \boldsymbol{H} f)$ gilt.

[6]) Die übliche Schreibweise entsteht, indem man $\boldsymbol{H}$ durch , ersetzt.

[7]) Der nicht symmetrische Operator S heißt beschränkt, wenn für f, f_1 aus $\mathfrak{F}$ gilt:

$$|f \boldsymbol{H} S f_1| < C\,|f|\,|f_1|.$$

Die nächst einfache Klasse bilden die *halbbeschränkten Formen.* Eine Form G, in $\mathfrak{G} \leqq \mathfrak{H}$ definiert, heißt (nach unten) *positiv-halbbeschränkt*[8]) mit der (unteren) Schranke γ, wenn es ein $\gamma > 0$ gibt, so daß für alle Elemente g aus $\mathfrak{G}$

$$g\,G\,g \geqq \gamma\,(g\,H\,g)$$

ist.

Entsprechend heißt ein Operator *positiv-halbbeschränkt,* wenn es die zugehörige Form ist.

Wichtig ist für das Folgende die Eigenschaft der „Abgeschlossenheit", die positiv-halbbeschränkte Formen besitzen können.

Eine positiv-halbbeschränkte Form G heißt *abgeschlossen* in $\mathfrak{G}$, wenn der Raum $\mathfrak{G}$ abgeschlossen ist mit $g\,G\,g$ als Maßform. $\mathfrak{G}$ ist dann auch ein Hilbertscher Raum mit G als Maßform[9]).

Für (nicht notwendig halbbeschränkte) Operatoren A, in dichten Teilmengen $\mathfrak{F}$ von $\mathfrak{H}$ definiert, hat v. Neumann [7. 3] einen Begriff der Abgeschlossenheit eingeführt, dem der vorangehende nachgebildet ist. Seine Formulierung ist gleichwertig mit der folgenden (ähnlich bei [7. 6]):

Ein Operator A in $\mathfrak{F}$ heißt *abgeschlossen,* wenn der Raum $\mathfrak{F}$ abgeschlossen ist mit

$$A f\,H\,A f + f\,H\,f \quad (\text{kurz } A^2 + H)$$

als Maßform.

2. Fortsetzung durch Abschließen.

Die Differentialoperatoren, die wir vor allem im Auge haben, sind meist nicht in abgeschlossenen Räumen und selbst nicht abgeschlossen gegeben, während diese Eigenschaft bei der allgemeinen Spektraltheorie vorausgesetzt wird. Schon aus diesem Grunde wird man dazu geführt, zu untersuchen, ob sich Räume, Operatoren und Formen in erweiterten Definitionsbereichen zu abgeschlossenen fortsetzen lassen.

Es sei ein Raum $\mathfrak{H}'$ von Elementen h mit der Maßform H gegeben. Den Raum $\mathfrak{H}'$ zu einem Raum $\mathfrak{H}$ fortsetzen heißt: weitere — auch h

---

[8]) Ohne die Annahme $\gamma > 0$ hieße die Form nur halbbeschränkt, eine solche kann aber stets durch Addition von $(1 - \gamma)\,(g\,H\,g)$ zu einer positiv-halbbeschränkten gemacht werden.

[9]) Man konstruiert sich leicht eine abzählbare G-dichte Teilmenge $(\mathfrak{G})$ von $\mathfrak{G}$. Zunächst gibt es — Separabilität von $\mathfrak{H}$ — in jeder Teilmenge $\mathfrak{M}$ von $\mathfrak{H}$ eine abzählbare H-dichte Menge $(\mathfrak{M})$. Man bilde nun die Folge der Teilmengen $\mathfrak{G}_n$ von $\mathfrak{G}$, die durch die Bedingung $g\,G\,g \leqq n\,(g\,H\,g)$ gekennzeichnet sind. Jedes Element g gehört zu einer von ihnen. Die Menge $(\mathfrak{G}_n)$ (d. h. die abzählbare H-dichte Teilmenge von $\mathfrak{G}_n$) ist dann auch G-dicht in $\mathfrak{G}_n$ und die Summe der $(\mathfrak{G}_n)$ hat die von $(\mathfrak{G})$ gewünschte Eigenschaft.

genannte — (ideale) Elemente adjungieren, für sie zusammen mit den Elementen aus $\mathfrak{H}'$ Addition, Multiplikation mit reellen Zahlen und die Form $\boldsymbol{H}$ positiv-definit definieren.

Es gilt dann der Satz:

Satz 1. Ist der Raum $\mathfrak{H}'$ mit der Maßform $\boldsymbol{H}$ separabel, so kann er zu einem abgeschlossenen, also Hilbertschen Raume $\mathfrak{H}$ fortgesetzt werden; und zwar nur auf eine Weise[10]), so, daß $\mathfrak{H}'$ dicht liegt in $\mathfrak{H}$.

Beweis. Jeder Folge h_ν aus $\mathfrak{H}'$, die in sich konvergiert

$$|h_\nu - h_\mu| \to 0,$$

ordne man, wenn nicht schon ein Grenzelement aus $\mathfrak{H}'$ existiert, ein ideales Grenzelement h zu. Zweien solcher Folgen $h_{1\nu}$, $h_{2\nu}$ für die $|h_{1\nu} - h_{2\nu}| \to 0$ strebt, ordne man dasselbe Grenzelement zu. Besitzen zwei Folgen $h_{1\nu}$, $h_{2\nu}$ Grenzelemente h_1, h_2 (aus $\mathfrak{H}'$ oder ideale), so besitzt $h_{1\nu}\boldsymbol{H}h_{2\nu}$ einen Grenzwert; als ihn definiere man $h_1\boldsymbol{H}h_2$. Man zeigt leicht: $\boldsymbol{H}$ ist dann in ganz $\mathfrak{H}$ bilinear, positiv definit, und jede in Satz 1 behauptete Fortsetzung kann auf diese Weise erzeugt werden.

Es sei in $\mathfrak{F}' \leqq \mathfrak{H}$ ein Operator A definiert. Den Operator A in $\mathfrak{F}'$ in einen Teilraum $\mathfrak{F}$ von $\mathfrak{H}$ ($\mathfrak{F}' < \mathfrak{F} \leqq \mathfrak{H}$) fortsetzen, heißt, ihn für die Elemente von $\mathfrak{F}$, die nicht in $\mathfrak{F}'$ liegen, so definieren, daß er in ganz $\mathfrak{F}$ linear und symmetrisch wird.

Entsprechend erklärt sich die Fortsetzung einer Form.

Es gelten dann die Sätze:

Satz 2. Ein beschränkter Operator B bzw. eine beschränkte Form $\boldsymbol{B}$, die nur in einem dichten Teilraum von $\mathfrak{H}$ erklärt sind, können eindeutig mit denselben Schranken β in $\mathfrak{H}$ fortgesetzt werden[11]).

Satz 3. Ein Operator G, im Raum $\mathfrak{F}'$ der Elemente f erklärt, führe zu einer positiv halbbeschränkten Form $\boldsymbol{G}$ mit der Schranke $\gamma > 0$

$$f\boldsymbol{H}Gf = f\boldsymbol{G}f \geqq \gamma(f\boldsymbol{H}f).$$

Dann gibt es einen $\mathfrak{F}'$ enthaltenden Teilraum $\mathfrak{G}$ von $\mathfrak{H}$ ($\mathfrak{F}' \leqq \mathfrak{G} \leqq \mathfrak{H}$), in den sich die Form $\boldsymbol{G}$ zu einer abgeschlossenen mit derselben Schranke γ fortsetzen läßt. $\mathfrak{G}$ und $\boldsymbol{G}$ in $\mathfrak{G}$ sind eindeutig bestimmt, wenn $\mathfrak{F}'$ $\boldsymbol{G}$-dicht liegen soll in $\mathfrak{G}$.

Es heiße $\boldsymbol{G}$ in $\mathfrak{G}$ die zu $\boldsymbol{G}$ in $\mathfrak{F}'$ bzw. G in $\mathfrak{F}'$ gehörige abgeschlossene Form.

---

10) D. h. zwei Fortsetzungen sind so aufeinander abbildbar, daß zugleich Multiplikation mit reellen Zahlen, Addition und Werte der Form $\boldsymbol{H}$ abgebildet werden.

11) Entsprechendes gilt auch für nicht symmetrische Operatoren.

Vor dem Beweis von Satz 3 erwähnen wir noch

Satz 4. Zu jedem Operator A in $\mathfrak{F}'$ gibt es einen $\mathfrak{F}'$ enthaltenden Teilraum $\mathfrak{F}$ von $\mathfrak{H}$, in dem sich eine abgeschlossene Fortsetzung von A definieren läßt. $\mathfrak{F}$ und A in $\mathfrak{F}$ sind eindeutig bestimmt, wenn $\mathfrak{F}'$ in $\mathfrak{F}$ dicht liegen soll mit $\boldsymbol{A}^2 + \boldsymbol{H}$ als Maßform.

A in $\mathfrak{F}$ heißt der abgeschlossene Operator von A in $\mathfrak{F}'$.

Der Satz 4 und sein einfacher Beweis findet sich bei v. Neumann [7. 3].

Bemerkung zu Satz 3 und 4. Der Operator G in $\mathfrak{F}'$ führe zur Form G in $\mathfrak{F}'$. Es seien G in $\mathfrak{F}$ der zugehörige abgeschlossene Operator, G in $\mathfrak{G}$ die zugehörige abgeschlossene Form. Dann liegt $\mathfrak{F}$ in $\mathfrak{G}$.

Wir beweisen zuerst diese Bemerkung, dann Satz 4 und zum Schluß Satz 3.

Beweis der Bemerkung. Liegt f in $\mathfrak{F}$, so kann es $\boldsymbol{H}$-approximiert werden durch eine Folge f_ν, für die auch $G f_\nu$ $\boldsymbol{H}$-konvergiert, also gilt auch

$$(f_\nu - f_\mu)\, \boldsymbol{H} G\, (f_\nu - f_\mu) \to 0 \quad \text{für} \quad \nu, \mu \to \infty,$$

d. h.

$$(f_\nu - f_\mu)\, \boldsymbol{G}\, (f_\nu - f_\mu) \to 0,$$

also konvergiert f_ν auch in bezug auf G; das Grenzelement f muß nach Satz 3 auch in $\mathfrak{G}$ liegen.

Sodann sei der Beweis von Satz 4 nach v. Neumann kurz referiert. Konvergiert für eine Folge f_ν die Form

$$(f_\nu - f_\mu)\, \boldsymbol{H}\, (f_\nu - f_\mu) + A\, (f_\nu - f_\mu)\, \boldsymbol{H} A\, (f_\nu - f_\mu) \to 0 \quad \text{bei} \quad \nu, \mu \to \infty,$$

so gibt es Grenzelemente f_0 und h_0, so daß

$$f_\nu \to f_0, \quad A f_\nu \to h_0\ (\boldsymbol{H}).$$

Alle Elemente f_0, die so entstehen können, bilden den offenbar linearen Raum $\mathfrak{F}$. Wenn nun zwei solche Folgen $f_{1\nu}, f_{2\nu}$ dasselbe Grenzelement $f_{10} = f_{20}$ besitzen, so besitzen auch $A f_{1\nu}$ und $A f_{2\nu}$ dasselbe Grenzelement; denn für alle h aus $\mathfrak{H}$ gilt

$$h\, \boldsymbol{H}\, (f_{1\nu} - f_{2\nu}) \to 0,$$

also auch für

$$h = A f,$$

wo f alle Elemente aus $\mathfrak{F}'$ darstellt, d. h.

$$A f\, \boldsymbol{H}\, (f_{1\nu} - f_{2\nu}) = f\, \boldsymbol{H} A\, (f_{1\nu} - f_{2\nu}) \to 0 \quad \text{bei} \quad \nu, \mu \to \infty$$

und also

$$f\, \boldsymbol{H}\, (h_{10} - h_{20}) = 0;$$

da $\mathfrak{F}'$ dicht liegt, folgt $h_{10} = h_{20}$. Nun kann durch $A f_0 = h_0$ der Operator A in ganz $\mathfrak{F}$ definiert werden; er bleibt offenbar linear und symmetrisch.

Der Beweis von Satz 3 wird ähnlich geführt; übrigens kann nicht jede positiv-halbbeschränkte Form zu einer abgeschlossenen Form fortgesetzt werden; eine Zusatzbedingung wie z. B. die, daß sie zu einem Operator gehört, ist notwendig.

Zum Beweise von Satz 3 werden wir nach Satz 2 den Raum $\mathfrak{F}'$ mit der Maßform G durch Adjunktion von zunächst idealen Elementen abschließen zu einem Raum $\mathfrak{G}$ von Elementen g mit der Maßform G. Jede Folge f_ν von Elementen aus $\mathfrak{F}'$, die mit G als Maßform in sich konvergiert, konvergiert ebenso mit $\boldsymbol{H}$ als Maßform wegen

$$\underline{\gamma}\,(f\boldsymbol{H}f) \leqq f\boldsymbol{G}f,$$

besitzt also ein Grenzelement h aus $\mathfrak{H}$; es ist aber zunächst nicht ausgeschlossen, daß zweien Folgen $f_{1\nu}$ und $f_{2\nu}$ zwei verschiedene Grenzelemente g_1 und g_2 entsprechen, während die Grenzelemente $h_1 = h_2$ identisch sind. Daß das nicht eintritt, soll gezeigt werden. Aus der Voraussetzung, daß zu G ein Operator G gehört, folgern wir nämlich

3, 1. Wenn für eine Folge f_ν aus $\mathfrak{F}'$

$$h\,\boldsymbol{H}\,f_\nu \to 0 \quad \text{für} \quad \text{alle } h \text{ aus } \mathfrak{H}$$

gilt, so folgt

$$f\,G\,f_\nu \to 0 \quad \text{für} \quad \text{alle } f \text{ aus } \mathfrak{F}'\ ^{12)}.$$

Denn es gilt

$$f\,\boldsymbol{G}\,f_\nu = G\,f\,\boldsymbol{H}\,f_\nu \to 0.$$

Aus 3, 1 folgt nun leicht

3, 2. Wenn für eine Folge f_ν aus $\mathfrak{F}'$ gilt

$$(f_\nu - f_\mu)\,G\,(f_\nu - f_\mu) \to 0 \quad \text{für} \quad \nu, \mu \to \infty$$

und

$$f_\nu\,\boldsymbol{H}\,f_\nu \to 0 \quad \text{für} \quad \nu \to \infty,$$

so gilt auch

$$f_\nu\,\boldsymbol{G}\,f_\nu \to 0 \quad \text{für} \quad \nu \to \infty.$$

---

[12]) Übrigens gilt, wenn $f_\nu\,G\,f_\nu$ beschränkt bleibt, $g\,G\,f_\nu \to 0$ auch für alle g aus $\mathfrak{G}$, da $\mathfrak{F}'$ G-dicht liegt in $\mathfrak{G}$.

Es bedeutet dann Hilfssatz 3, 1:

Wenn eine solche Folge f_ν schwach konvergiert in $\mathfrak{H}$ mit $\boldsymbol{H}$ als Maß, so auch in $\mathfrak{G}$ mit G als Maß.

Diese Eigenschaft (3, 1) ist auch gleichwertig mit der Halbstetigkeit von G bei schwacher G-Konvergenz; der folgende Beweis ist nichts als die abstrakte Formulierung häufig angewandter Halbstetigkeitsbeweise; vgl. auch die verwandte Schlußweise bei Courant [2. 4].

Denn für eine solche Folge ist die Voraussetzung von (3, 1) erfüllt; also folgt $f_\mu G f_\nu \to 0$ bei festem μ. Jetzt ziehe man die Schwarz-Ungleichung

$$|f_\mu G f_\mu - f_\mu G f_\nu| \leqq \sqrt{(f_\nu - f_\mu) G (f_\nu - f_\mu)} \sqrt{f_\mu G f_\mu}$$

heran. Die untere Grenze der rechten Seite für $\nu \to \infty$ strebt mit wachsendem μ gegen Null; die untere Grenze der linken Seite für $\nu \to \infty$ ist $f_\mu G f_\mu$; also strebt auch $f_\mu G f_\mu \to 0$ mit wachsendem μ.

Nun erkennt man unmittelbar: Sind $f_{1\nu}$ und $f_{2\nu}$ zwei Folgen aus $\mathfrak{F}'$, die zwei Grenzelemente g_1, g_2 und zwei Grenzelemente h_1, h_2 definieren, und ist $h_1 = h_2$, so sind für die Differenz $f_\nu = f_{1\nu} - f_{2\nu}$ die Voraussetzungen von (3, 2) erfüllt; es folgt also $(f_{1\nu} - f_{2\nu}) G (f_{1\nu} - f_{2\nu}) \to 0$, woraus $(g_1 - g_2) G (g_1 - g_2) = 0$ und also $g_1 = g_2$ folgt.

Ist $g_1 = g_2$ ein reales Element aus $\mathfrak{F}'$, so ist $g_1 = g_2 = h_1 = h_2$; denn die Folge $f_{1\nu}$ konvergiert ja auch in bezug auf $\boldsymbol{H}$ gegen g_1. Ist dagegen $g_1 = g_2$ ein ideales Element aus $\mathfrak{G}$, so identifizieren wir es mit $h_1 = h_2$. Dadurch aber wird $\mathfrak{G}$ zu einem Teilraum von $\mathfrak{H}$.

3. Operator einer Form.

Während unmittelbar jedem Operator eine Form zugehört, ist das Umgekehrte nur in beschränktem Maße der Fall. Es gilt nach F. Riesz (vgl. z. B. [9. 2]).

Satz 5. Zu jeder beschränkten Form $\boldsymbol{B}$ in $\mathfrak{H}$ gehört ein beschränkter Operator B in $\mathfrak{H}$, so daß gilt

$$h \boldsymbol{B} h = h \boldsymbol{H} B h.$$

Beweis. Es sei erlaubt, einen einfachen Beweis anzugeben, der auf die Darstellung durch ein Orthogonalsystem keinen Bezug nimmt.

Wir stellen das *Minimumproblem*: Jedem Element k_1 aus $\mathfrak{H}$ ein solches Element $h = k_0$ zuzuordnen, für das

$$J[h] = h \boldsymbol{H} h - 2 h \boldsymbol{B} k_1$$

möglichst klein wird. Sicher ist $J[h]$ nach unten beschränkt, da $\boldsymbol{B}$ beschränkt ist. Es gibt also eine untere Grenze d und eine Minimalfolge h_ν.

Für sie folgt

$$d_\nu = h_\nu \boldsymbol{H} h_\nu - 2 h_\nu \boldsymbol{B} k_1 \to d$$

und aus

$$(h_\nu + \varepsilon h) \boldsymbol{H} (h_\nu + \varepsilon h) - 2 (h_\nu + \varepsilon h) \boldsymbol{B} k_1 \geqq d,$$

daß

$$(d_\nu - d) + 2 \varepsilon (h \boldsymbol{H} h_\nu - h \boldsymbol{B} k_1) + \varepsilon^2 h \boldsymbol{H} h \geqq 0,$$

also in ε nie negativ ist, so daß

$$\sqrt{(d_\nu - d)\, h \boldsymbol{H} h} \geqq |h \boldsymbol{H} h_\nu - h \boldsymbol{B} k_1|$$

gilt für alle h aus $\mathfrak{H}$. Hieraus folgt zunächst

$$h \boldsymbol{H} h_\nu - h \boldsymbol{B} k_1 \to 0;$$

sodann, indem man $h = h_\nu - h_\mu$ setzt und ν mit μ vertauscht.

$$\begin{aligned}(h_\nu - h_\mu) \boldsymbol{H} (h_\nu - h_\mu) &= |(h_\nu - h_\mu) \boldsymbol{H} h_\nu - (h_\nu - h_\mu) \boldsymbol{B} k_1 \\ &\quad + (h_\mu - h_\nu) \boldsymbol{H} h_\mu - (h_\mu - h_\nu) \boldsymbol{B} k_1| \\ &\leqq (\sqrt{d_\nu - d} + \sqrt{d_\mu - d}) \sqrt{(h_\nu - h_\mu) \boldsymbol{H} (h_\nu - h_\mu)}\end{aligned}$$

und also

$$(h_\nu - h_\mu) \boldsymbol{H} (h_\nu - h_\mu) \leqq (\sqrt{d_\nu - d} + \sqrt{d_\mu - d})^2 \to 0.$$

D. h. aber, h_ν konvergiert gegen ein Grenzelement k_0; für dieses ergibt sich mit jedem h aus $\mathfrak{H}$

$$(*) \qquad h \boldsymbol{H} k_0 - h \boldsymbol{B} k_1 = 0.$$

Die Zuordnung des k_0 zum k_1 bezeichnen wir als Operator B. D. h. wir setzen

$$k_0 = B k_1.$$

Es gilt dann

1. B ist eindeutig; denn die Differenz zweier k_0, die zum selben k_1 gehören, müßte nach (*) auf allen h orthogonal sein.

2. B ist linear; denn die Summe zweier k_0 erfüllt die Relation (*) für die Summe der k_1. Ihr Bestehen ist aber für die Minimumeigenschaft kennzeichnend, da aus ihr

$$\begin{aligned}(k_0 + h) \boldsymbol{H} (k_0 + h) - 2 (k_0 + h) \boldsymbol{B} k_1 \\ \geqq k_0 \boldsymbol{H} k_0 - 2 k_0 \boldsymbol{B} k_1 = d\end{aligned}$$

folgt.

3. B gehört zur Form $\boldsymbol{B}$; denn (*) geht für $k_1 = h$ über in

$$h \boldsymbol{H} B h - h \boldsymbol{B} h = 0.$$

Daraus folgt unmittelbar: 4. B ist symmetrisch und 5. B ist beschränkt.

Als neuen Satz formulieren wir:

S a t z 6. Zu einer abgeschlossenen, durch $\gamma > 0$ halbbeschränkten Form G in $\mathfrak{G} \leqq \mathfrak{H}$ gibt es einen beschränkten Operator B, so daß

$$g G B h = g \boldsymbol{H} h$$

für g aus $\mathfrak{G}$, h aus $\mathfrak{H}$ gilt. Es ist

$$0 \leqq h \boldsymbol{H} B h \leqq \frac{1}{\gamma} (h \boldsymbol{H} h)$$

und der Wertebereich von B liegt in $\mathfrak{G}$.

Dieser Operator B wird sich nachher als Reziproke eines zu G zugehörigen Operators erweisen.

Beweis. Im Hilbertschen Raum $\mathfrak{G}$ ist $\boldsymbol{H}$ eine beschränkte Form

$$g \boldsymbol{H} g \leqq \frac{1}{\gamma} (g G g);$$

also gibt es nach Satz 5 einen in $\mathfrak{G}$ erklärten beschränkten Operator B, für den

$$(**) \qquad g \boldsymbol{H} h = g G B h$$

zunächst für alle h aus $\mathfrak{G}$ gilt; es ist, indem man $g = Bh$ mit h aus $\mathfrak{G}$ setzt,

$$(Bh\, G\, Bh)^2 = (Bh\, \boldsymbol{H}\, h)^2 \leqq (Bh\, \boldsymbol{H}\, Bh)\,(h\, \boldsymbol{H}\, h) \leqq \frac{1}{\gamma}(Bh\, G\, Bh)(h\, \boldsymbol{H}\, h)$$

oder

$$\dagger \qquad (Bh\, G\, Bh) \leqq \frac{1}{\gamma}(h\, \boldsymbol{H}\, h),$$

d. h.

$$(***) \qquad (h\, \boldsymbol{H}\, Bh) \leqq \frac{1}{\gamma}(h\, \boldsymbol{H}\, h).$$

Diese zunächst für h in $\mathfrak{G}$ gültige Relation zeigt, daß B auch $\boldsymbol{H}$-beschränkt und also auf $\mathfrak{H}$ fortsetzbar ist. Aus † und der vorausgesetzten Abgeschlossenheit der Form G folgert man, daß Bh stets in $\mathfrak{G}$ liegt und danach, daß die Relationen (**) und (***) auch für alle h aus $\mathfrak{H}$ bestehen.

Der Satz 6 hätte auch bewiesen werden können, indem man ohne Bezug auf Satz 5 unmittelbar die Minimumforderung anwendet[13]).

Aus dem Satz 6 folgt unmittelbar

Satz 7. Zu einer abgeschlossenen positiv-halbbeschränkten Form G in $\mathfrak{G} \leqq \mathfrak{H}$ gibt es in einem G-dichten Teilraum $\mathfrak{F}_1 \leqq \mathfrak{G}$ einen vermöge

$$g \boldsymbol{H} G f = g G f, \qquad f \text{ aus } \mathfrak{F}_1$$

zugehörigen Operator, dessen Wertebereich $\mathfrak{H}$ ist. Er heiße der *„maximal zugehörige"* Operator zu G.

Beweis. Man konstruiere nach Satz 6 den beschränkten Operator B; der Wertebereich von $f = Bh$ sei $\mathfrak{F}_1$; es liegt dann $\mathfrak{F}_1$ in $\mathfrak{G}$. Jedem f aus $\mathfrak{F}_1$ entspricht nun nur ein h; denn aus $Bh = 0$ folgt wegen

$$g G B h = g \boldsymbol{H} h = 0 \quad \text{für alle } g \text{ aus } \mathfrak{G}$$

---

[13]) Satz 6 ist verwandt mit dem Satz von Toeplitz [11] über die Grenzresolvente positiv-definiter Formen von unendlich viel Veränderlichen, der übrigens ähnlich bewiesen werden kann, bequemer als mit der Jacobischen Transformation. (Siehe Mathem. Annalen **109**, S. 254—256). Der Satz von Toeplitz bildet auch für Wintner [13] den Ausgangspunkt seiner Theorie der halbbeschränkten Matrizen.

auch $h = 0$. Somit ist in $\mathfrak{F}_1$ die Reziproke von B eindeutig definiert; sie heiße G: d. h. $Bh = f$ sei gleichwertig mit $h = Gf$.

Der Raum $\mathfrak{F}_1$ ist G-dicht in $\mathfrak{G}$; denn sonst gäbe es ein Element $g_0 \neq 0$ aus $\mathfrak{G}$, das auf $\mathfrak{F}_1$ G-orthogonal wäre:

$$g_0 \boldsymbol{G} f = 0; \quad \text{also auch} \quad g_0 \boldsymbol{G} B g_0 = g_0 \boldsymbol{H} g_0 = 0, \quad \text{d. h.} \quad g_0 = 0.$$

Nunmehr folgt:

1. G ist linear; weil G eindeutig und B linear ist.

2. G ist symmetrisch; denn B ist symmetrisch.

Zusatz zu Satz 7. Ist G' in $\mathfrak{F}' \leqq \mathfrak{G}$ ein Operator, der zur Form $\boldsymbol{G}$ gehört, d. h. für den mit f' aus $\mathfrak{F}'$, g aus $\mathfrak{G}$ gilt:

$$g \boldsymbol{G} f' = g \boldsymbol{H} G' f',$$

so ist der „maximal zugehörige" Operator G in $\mathfrak{F}_1$ Fortsetzung von G' in $\mathfrak{F}'$.

Beweis. Mit dem Operator B von Satz 6 bilde man für das Element f' aus $\mathfrak{F}'$ das Element $BG'f'$ aus $\mathfrak{F}_1$; für alle g aus $\mathfrak{G}$ gilt dann

$$g \boldsymbol{G} B G' f' = g \boldsymbol{H} G' f' = g \boldsymbol{G} f'; \quad \text{also} \quad B G' f' = f'; \quad \text{d. h.} \quad f' \text{ in } \mathfrak{F}_1.$$

4. Selbstadjungierte (hypermaximale) Operatoren.

Eine Spektralzerlegung ist, wie v. Neumann [7. 2] gezeigt hat, nicht für jeden abgeschlossenen symmetrischen Operator möglich; dazu muß (nach E. Schmidt) die Bedingung der *Hypermaximalität* [7. 3] erfüllt sein; dieselbe Bedingung hat Stone [10. 1] als *Selbstadjungiertheil* seiner Spektraltheorie vorangestellt.

Definition. Es sei ein symmetrischer Operator A in $\mathfrak{F} \leqq \mathfrak{H}$ erklärt. A in $\mathfrak{F}$ heißt selbstadjungiert, wenn folgende Bedingung erfüllt ist:

Ist einem Element h_0 aus $\mathfrak{H}$ ein anderes h_1 aus $\mathfrak{H}$ zugeordnet, so daß für alle f aus $\mathfrak{F}$ gilt

$$h_0 \boldsymbol{H} A f = h_1 \boldsymbol{H} f,$$

so liegt h_0 in $\mathfrak{F}$ und es ist $A h_0 = h_1$.

Ein selbstadjungierten Operator ist offenbar stets abgeschlossen.

Es gilt bekanntlich

Satz 8. Ein beschränkter Operator ist selbstadjungiert.

Dem stellen wir gegenüber

Satz 9[14]). Der einer abgeschlossenen Form G in $\mathfrak{G}$ maximal zugehörige Operator G in $\mathfrak{F}_1 \leqq \mathfrak{G}$ ist selbstadjungiert. Wir sprechen dann vom zugehörigen selbstadjungierten Operator.

Beweis. Es ist nämlich nach v. Neumann [7. 4] jeder symmetrische Operator mit dem Wertebereich $\mathfrak{H}$ selbstadjungiert. Denn dann gibt es zu h_1 ein f_0, so daß $A f_0 = h_1$ ist; also gilt

$$h_0 \boldsymbol{H} A f = A f_0 \boldsymbol{H} f = f_0 \boldsymbol{H} A f,$$

und da die Af ganz $\mathfrak{H}$ durchlaufen, $f_0 = h_0$. Daraus folgt die Behauptung nach Satz 7.

Wir können nun zeigen, daß für positiv halbbeschränkte Operatoren die Selbstadjungiertheit schon aus *schwächeren Bedingungen* folgt, die für unsere Differentialoperatoren leichter nachzuprüfen sind.

Es sei G in $\mathfrak{F}$ ein abgeschlossener positiv-halbbeschränkter Operator, $g\,G\,g$ in $\mathfrak{G}$ die nach Satz 3 zugehörige abgeschlossene Form, G in $\mathfrak{F}_1$ der nach Satz 7 und 9 zu ihr gehörige selbstadjungierte Operator. Dann führen wir die „iterierten" Räume

$$\mathfrak{G}_1, \mathfrak{F}_2, \mathfrak{G}_2, \mathfrak{F}_3, \mathfrak{G}_3, \ldots,$$

ein, die aus allen denjenigen Elementen f von $\mathfrak{F}_1$ bestehen, für die bzw. Gf in

$$\mathfrak{G}, \mathfrak{F}_1, \mathfrak{G}_1, \mathfrak{F}_2, \mathfrak{G}_2, \ldots$$

liegt. Alsdann gilt

Satz 10. Es ist $\mathfrak{F} = \mathfrak{F}_1$, d. h. G in $\mathfrak{F}$ ist selbstadjungiert, wenn für irgendein n ($n = 1, 2, 3, \ldots$) gilt

($\mathfrak{F}_n$) $\qquad$ $\mathfrak{F}_n$ liegt in $\mathfrak{F}$

oder

($\mathfrak{G}_n$) $\qquad$ $\mathfrak{G}_n$ liegt in $\mathfrak{F}$.

Beweis. Da $\mathfrak{F}_{n+1}$ in $\mathfrak{G}_n$ liegt, genügt es, Satz 10 unter der Voraussetzung ($\mathfrak{F}_n$) zu beweisen. Da nach Zusatz zu 7 und ($\mathfrak{F}_n$) die in $\mathfrak{F}_1$, $\mathfrak{F}$, $\mathfrak{F}_n$ erklärten Operatoren G Fortsetzungen voneinander sind und $\mathfrak{F}$ abgeschlossen ist, genügt es zu zeigen:

[14]) In Satz 9 liegt auch der Beweis einer *Vermutung von v. Neumann* [7. 5]. Ein positiv-halbbeschränkter Operator kann zu einem selbstadjungierten Operator mit derselben unteren Schranke fortgesetzt werden.

Denn sei dieser Operator G in $\mathfrak{F}'$; dann werde die zugehörige Form G nach Satz 3 zu einer abgeschlossenen fortgesetzt; der zugehörige selbstadjungierte Operator G in $\mathfrak{F}_1$ ist nach Zusatz zu 7 Fortsetzung von G in $\mathfrak{F}'$; daß er zur selben unteren Schranke gehört, folgt daraus, daß das für die Form G gilt.

Offen bleibt aber noch, ob eine solche Fortsetzung auch auf andere Weise möglich ist.

Der abgeschlossene Operator von G in $\mathfrak{F}_n$ ist G in $\mathfrak{F}_1$. Zum **Beweise** beachte man, daß $\mathfrak{F}_n$ der Wertebereich von $B^n h$ ist, wo B die Reziproke von G in $\mathfrak{F}_1$ nach Satz 6 ist. Daraus folgt, daß $\mathfrak{F}_n$ dicht liegt in $\mathfrak{H}$; anderenfalls gäbe es ein Element $h_0 \neq 0$ mit $0 = h_0 \boldsymbol{H} B^n h = B^n h_0 \boldsymbol{H} h$ für alle h, woraus $B^n h_0 = B^{n-1} h_0 = \ldots = h_0 = 0$ folgte. Sei nun f_1 ein Element von $\mathfrak{F}_1$, so gibt es eine Folge f^ν aus $\mathfrak{F}_{n+1}$, für die $f^\nu \to G f_1$ strebt. Alsdann strebt $B f^\nu \to f_1$ und $G(B f^\nu) \to G f_1$, d. h. der abgeschlossene Operator zu G in $\mathfrak{F}_n$ ist für f_1 erklärt und ergibt $G f_1$.

5. Spektralzerlegung.

Vor Formulierung und Beweis des Spektralsatzes halbbeschränkter Formen sei eine Reihe von bekannten Begriffen und Zusammenhängen über die Spektralschar dargestellt in einer für unsere Zwecke geeigneten Symbolik und Anordnung.

Die Darstellung der Spektralzerlegung ist nach v. Neumann [7.2] auf die „Projektions- oder Einzeloperatoren" zu gründen.

Ein *Projektionsoperator* P ist ein in $\mathfrak{H}$ definierter Operator, der der Relation

$$P^2 = P$$

genügt; offenbar ist $1 - P$ auch ein solcher.

Die Elemente Ph und $(1-P)h$ sollen *Eigen-* und *Gegenelemente* von P heißen; ihre Wertbereiche $\mathfrak{P}$ und $\mathfrak{H} \ominus \mathfrak{P}$ *Eigen-* und *Gegenraum* von P.

Die Forderung der Symmetrie

$$P h_1 \boldsymbol{H} h_2 = h_1 \boldsymbol{H} P h_2$$

für h_1, h_2 aus $\mathfrak{H}$ ist gleichwertig mit

$$P h_1 \boldsymbol{H} (1 - P) h_2 = 0$$

d. h., damit, daß Eigen- und Gegenelemente oder Eigen- und Gegenraum von P *orthogonal* sind.

Gleichwertig ist auch die Identität

$$h \boldsymbol{H} h = P h \boldsymbol{H} P h + (1 - P) h \boldsymbol{H} (1 - P) h.$$

Für die zu symmetrischem P gehörige Form $\boldsymbol{P}$, die „Einzelform" gilt

$$0 \leqq h \boldsymbol{P} h \leqq h \boldsymbol{H} h,$$

wobei Gleichheit (für alle h) nur für $P = 0$ bzw. $P = 1$ besteht. Insbesondere folgt so die Beschränktheit von P aus der Symmetrie.

Die hier auf die Einheitsform $\boldsymbol{H}$ bezogenen Eigenschaften lassen sich analog auch in bezug auf andere Formen erklären. Wir sagen, P ist *symmetrisch* in bezug auf die Form $\boldsymbol{A}$ in $\mathfrak{F}$, kurz: $\boldsymbol{A}$-symmetrisch in $\mathfrak{F}$, wenn mit f auch Pf in $\mathfrak{F}$ liegt und die gleichwertigen Identitäten für f, f_1, f_2 aus $\mathfrak{F}$ bestehen:

$$Pf_1 A f_2 = f_1 A P f_2,$$
$$Pf A (1-P) f = 0,$$
$$f A f = Pf A P f + (1-P) f A (1-P) f.$$

Dann gilt:

$$0 \leqq f A P f \leqq f A f.$$

Ein Projektionsoperator P ist mit einem Operator A in $\mathfrak{F}$ vertauschbar, wenn mit f auch Pf in $\mathfrak{F}$ liegt und

$$APf = PAf$$

gilt. Es ist P vertauschbar mit A in $\mathfrak{F}$, sobald P symmetrisch ist außer in bezug auf $\boldsymbol{H}$ auch in bezug auf die zu A gehörige Form fAf in $\mathfrak{F}$.

Im folgenden sei stets vorausgesetzt, daß P symmetrisch ist (d. h. $\boldsymbol{H}$-symmetrisch in $\mathfrak{H}$).

Daß zwischen zwei Projektionsoperatoren P_1, P_2 für alle h aus $\mathfrak{H}$ die Beziehung

$$h \boldsymbol{P}_1 h \geqq h \boldsymbol{P}_2 h$$

besteht, ist gleichbedeutend damit, daß der Eigenraum von P_1 den von P_2 und der Gegenraum von P_2 den von P_1 enthält. In Formeln:

$$P_2 P_1 = P_2, \quad (1-P_2)(1-P_1) = (1-P_1) \quad \text{oder} \quad P_1 P_2 = P_2.$$

Es ist dann auch $P_1 - P_2$ ein Projektionsoperator.

Die *Spektralschar* ist eine Schar von Projektionsoperatoren, die von einem reellen Parameter α so abhängen, daß die zugehörigen Formen monoton sind. An den Stellen α, an denen diese Formen unstetig sein können, sind zwei Projektionsoperatoren als Grenzwerte von oben und unten zu gewinnen. Es empfiehlt sich, von vornherein jedem Wert von α zwei Projektionsoperatoren zugeordnet zu denken, deren Formen von oben bzw. unten stetig sind. Der bequemeren Ausdrucksweise wegen seien die Projektionsoperatoren Symbolen α^+ und α^- (statt $\alpha + 0$ und $\alpha - 0$) zugeordnet. So motiviert sich folgende Erklärung:

Jeder reellen Zahl α seien die Symbole α^+, α^- zugeordnet; allgemein seien α^+ und α^- mit $\alpha^\cdot$ bezeichnet; für $\alpha = \infty$ und $\alpha = -\infty$ seien $\alpha^\cdot = \infty^-$ und $\alpha^\cdot = -\infty^+$ eingeführt. Wir setzen $\alpha_1^\cdot < \alpha_2^\cdot$ wenn $\alpha_1 < \alpha_2$ ist, ferner $\alpha^- < \alpha^+$.

Das Intervall

$$\Delta\alpha = (\alpha_1^\cdot, \alpha_2^\cdot)$$

enthalte jeden Punkt α, für dessen beide Symbole $\alpha^\cdot$ gilt: $\alpha_1^\cdot \leqq \alpha^\cdot \leqq \alpha_2^\cdot$; so sind zugleich Intervalle mit und ohne Endpunkte erfaßt; (α^-, α^+) enthält nur den Punkt α.

Es sei nun jedem $\alpha^\cdot$ ein Projektionsoperator $P_{\alpha^\cdot}$ zugeordnet, derart, daß für jedes h aus $\mathfrak{H}$ mit den zugehörigen Formen $h\boldsymbol{P}_{\alpha^\cdot}h$ gilt:

1. Ist
$$\alpha_1^\cdot < \alpha_2^\cdot, \quad \text{so} \quad h\boldsymbol{P}_{\alpha_1^\cdot}h \leqq h\boldsymbol{P}_{\alpha_2^\cdot}h.$$

2. Strebt für $\nu \to \infty$
$$\alpha_\nu \downarrow \alpha, \quad \text{so} \quad h\boldsymbol{P}_{\alpha_\nu^\cdot}h \downarrow h\boldsymbol{P}_{\alpha^+}h,$$
$$\alpha_\nu \uparrow \alpha, \quad \text{so} \quad h\boldsymbol{P}_{\alpha_\nu^\cdot}h \uparrow h\boldsymbol{P}_{\alpha^-}h.$$

Dann heißt $P_{\alpha^\cdot}$ eine *Spektralschar*.

Die Spektralschar heißt *vollständig*, wenn gilt
$$P_{-\infty^+} = 0, \quad P_{\infty^-} = 1.$$

Der *Differenzoperator*
$$P_{\Delta\alpha} = P(\alpha_1^\cdot, \alpha_2^\cdot) = P_{\alpha_2^\cdot} - P_{\alpha_1^\cdot}$$
des Intervalles $\Delta\alpha = (\alpha_1^\cdot, \alpha_2^\cdot)$ ist wieder Projektionsoperator. Es ist der *Sprungoperator*
$$P(\alpha^-, \alpha^+) \neq 0$$
nur für eine höchstens abzählbare Menge von Werten α, den *Punkteigenwerten*.

Der *Spektralsatz* für *beschränkte Formen* werde — wie es für uns zweckmäßig ist — in folgender Form ausgesprochen.

Satz 11. Sei $\boldsymbol{B}$ in $\mathfrak{H}$ eine beschränkte Form mit den Schranken $\underline{\beta}, \bar{\beta}$; B in $\mathfrak{H}$ der zugehörige Operator. Dann gibt es genau eine vollständige Spektralschar $Q_{\beta^\cdot}$, so daß gilt

1. $Q_{\beta^\cdot}$ ist symmetrisch in bezug auf die Form $\boldsymbol{B}$ und in bezug auf $\boldsymbol{H}$.

2. Es bestehen die beiden „Eigenwertungleichungen"
$$\beta_2(h\boldsymbol{H}Q_{\Delta\beta}h) \leqq (h\boldsymbol{B}Q_{\Delta\beta}h) \leqq \beta_1(h\boldsymbol{H}Q_{\Delta\beta}h)$$
mit $\Delta\beta = (\beta_2^\cdot, \beta_1^\cdot)$ für alle h aus $\mathfrak{H}$.

Zusatz 11, 1. Es ist $Q_{\underline{\beta}^-} = 0$, $Q_{\bar{\beta}^+} = 1$.

Zusatz 11, 2. Der Operator B in $\mathfrak{H}$ ist mit den $Q_{\beta^\cdot}$ vertauschbar.

Der Beweis von Satz 11 und der Zusätze ist aus den üblichen Formulierungen des Spektralsatzes leicht zu entnehmen (vgl. insbesondere Fr. Riesz [9. 3]).

Der *Spektralsatz* für abgeschlossene positiv *halbbeschränkte Formen* lautet:

Satz 12. Sei G in $\mathfrak{G}$ eine durch $\gamma > 0$ nach unten halbbeschränkte abgeschlossene Form, G in $\mathfrak{F}_1 \leqq \mathfrak{G}^-$ der zugehörige selbstadjungierte

Operator. Dann gibt es genau eine vollständige Spektralschar $R_{\gamma'}$, so daß gilt

1. $R_{\gamma'}$ ist symmetrisch in bezug auf die Form G in $\mathfrak{G}$ und in bezug auf H in $\mathfrak{H}$.

2. Es bestehen die beiden „Eigenwertungleichungen"

$$\gamma_1 (g\, H\, R_{\Delta\gamma} g) \leqq (g\, G\, R_{\Delta\gamma}\, g) \leqq \gamma_2 (g\, H\, R_{\Delta\gamma}\, g)$$

mit $\Delta\gamma = (\gamma_1^{\cdot}, \gamma_2^{\cdot})$ für alle g aus $\mathfrak{G}$.

Zusatz 12, 1. Es ist $R_{\gamma^-} = 0$, $R_{\infty^-} = 1$.

Zusatz 12, 2. Der Operator G in $\mathfrak{F}_1$ ist mit den $R_{\gamma'}$ vertauschbar.

Zusatz 12, 3. Die Eigenelemente der Differentialoperatoren $R_{\Delta\gamma}$ endlicher Intervalle $\Delta\gamma$ liegen in $\mathfrak{F}_1$.

Die Spektralschar $R_{\gamma'}$ liefert — so werden wir sagen — die *Spektralzerlegung* von G in $\mathfrak{G}$ und von G in $\mathfrak{F}_1$.

Beweis. Satz 12 ist eine einfache Folge von Satz 11. Denn: Im Hilbertschen Raum $\mathfrak{G}$ wird H eine in bezug auf G beschränkte Form mit den Schranken 0 und $\frac{1}{\underline{\gamma}} = \bar{\beta}$; es ist Satz 11 anwendbar; er liefert eine Spektralschar $Q_{\beta'}$. Nun werde $\beta = \frac{1}{\gamma}$ für $\beta > 0$ gesetzt; dann wird durch

$$R_{\gamma^+} = 1 - Q_{\beta^-}, \quad R_{\gamma^-} = 1 - Q_{\beta^+}$$

auch eine Spektralschar erklärt. Und zwar ist sie vollständig; denn erstens folgt nach Zusatz 11, 1, daß $Q_{\bar{\beta}^+} = 1$ d. h. $R_{\underline{\gamma}^-} = 0$ gilt. Um zweitens $R_{\infty^-} = 1 - Q_{0^+} = 1$ zu zeigen, ist der Nachweis von $Q_{0^+} = 0$ erforderlich; wäre nicht $Q_{0^+} = 0$, so gäbe es ein Eigenelement $h = Q_{0^+} h$ von Q_{0^+} und nach der zweiten Eigenwertungleichung mit $\beta_1^{\cdot} = 0+$ müßte $(h\, H\, h) \leqq 0$ sein, woraus aber $h = 0$ folgte.

Da $R(\gamma_1^{\pm}, \gamma_2^{\pm}) = Q(\beta_2^{\mp}, \beta_1^{\mp})$ ist, gehen die Eigenwertungleichungen von Satz 11 unmittelbar in die von Satz 12 über. Ebenso folgt unmittelbar die G-Symmetrie der $R_{\gamma'}$ in $\mathfrak{G}$; die Symmetrie in bezug auf H folgt allerdings zunächst nur in $\mathfrak{G}$; die $R_{\gamma'}$ sind ja überhaupt durch obige Definition nur in $\mathfrak{G}$ definiert. Nun folgt aber aus der H-Symmetrie der $R_{\gamma'}$ in $\mathfrak{G}$ ihre H-Beschränktheit in $\mathfrak{G}$; also sind die $R_{\gamma'}$ auf ganz $\mathfrak{H}$ fortsetzbar und zwar — wie man sofort sieht — ohne ihren Charakter als Spektralschar einzubüßen.

Damit ist die Existenz der Spektralschar von $\mathfrak{G}$ gezeigt. Die Eindeutigkeit ergibt sich daraus, daß in umgekehrter Weise die eindeutige Spektralschar von H im Raume $\mathfrak{G}$ aus der von G gewonnen werden kann.

Um Zusatz 12, 2 zu beweisen, ist noch zu zeigen, daß für f aus $\mathfrak{F}_1$ auch $R_{\gamma}\cdot f$ in $\mathfrak{F}_1$ liegt. Es ist aber für alle f, f_1 aus $\mathfrak{F}_1$

$$R_{\gamma}\cdot f\, \boldsymbol{H} G f_1 = R_{\gamma}\cdot f\, \boldsymbol{G} f_1 = f\, \boldsymbol{G} R_{\gamma}\cdot f_1 = G f\, \boldsymbol{H} R_{\gamma}\cdot f_1 = R_{\gamma}\cdot G f\, \boldsymbol{H} f_1.$$

Da aber G in $\mathfrak{F}_1$ selbstadjungiert ist, folgt, daß $R_{\gamma}\cdot f$ in $\mathfrak{F}_1$ liegt und daß $G R_{\gamma}\cdot f = R_{\gamma}\cdot G f$.

Zum Beweise von Zusatz 12, 3 beachten wir, daß nach 12, 2 der Operator $G R_{\Delta\gamma}$ ($\Delta\gamma$ endlich) in $\mathfrak{F}_1$ anwendbar ist. Aus den Eigenwertungleichungen folgt aber, daß $G R_{\Delta\gamma}$ in bezug auf $\boldsymbol{H}$ symmetrisch und beschränkt ist; denn das ist für die zugehörige Form

$$f\, \boldsymbol{H} G R_{\Delta\gamma} f = f\, \boldsymbol{G} R_{\Delta\gamma} f$$

der Fall. Es ist also der Operator $G R \Delta\gamma$ auf ganz $\mathfrak{H}$ fortsetzbar; er heiße dann $(G R \Delta\gamma)$ in $\mathfrak{F}_1$. Nun gilt mit f aus $\mathfrak{F}_1$

$$(G R_{\Delta\gamma})\, h\, \boldsymbol{H} f = R_{\Delta\gamma}\, h\, \boldsymbol{H} G f$$

zunächst für alle h aus $\mathfrak{F}$ und also auch für alle h aus $\mathfrak{H}$. Da G selbstadjungiert ist, folgt, daß $R_{\Delta\gamma} h$ in $\mathfrak{F}_1$ liegt und daß $G R_{\Delta\gamma} h = (G R_{\Delta\gamma})\, h$ ist.

6. Vollstetigkeit und diskretes Spektrum.

Es gibt von Hilbert und Weyl herrührende einfache Kriterien dafür, daß das Spektrum eines Operators in einem Intervall nur aus diskreten Punkteigenwerten endlicher Vielfachheit besteht. Für diese Kriterien soll hier eine einfache Beweisanordnung gegeben werden, die zugleich ihre Übertragung auf nicht beschränkte Operatoren ermöglicht. Zunächst setzen wir fest:

Eine Spektralschar P hat in einem abgeschlossenen Intervall $\Delta\alpha = (\alpha_1-, \alpha_2+)$ ein *diskretes Spektrum*[15]), wenn der Eigenraum von $P_{\Delta\alpha}$ endliche Dimension hat.

Es folgt leicht, daß $P_{\alpha\cdot}$ in diesem Intervall in $\alpha\cdot$ konstant ist bis auf endlich viele Sprungstellen (Punkteigenwerte), deren Eigenräume endliche Dimension besitzen[16]).

Ferner: Das Spektrum heiße diskret in irgendeinem Intervall, wenn es in jedem abgeschlosssenen Teilintervall diskret ist. Das Spektrum heiße überhaupt diskret, wenn es in jedem Intervall diskret ist.

[15]) Auch „diskretes Punktspektrum".

[16]) Die Eigenwerte eines nicht diskreten Punktspektrums können dicht liegen oder unendliche Vielfachheit besitzen.

Wir führen nun den Begriff der *Vollstetigkeit* in bezug auf eine positiv-definite Form ein in Verallgemeinerung des Hilbertschen Begriffs der Vollstetigkeit. Wir kennzeichnen die Vollstetigkeit auch anders als Hilbert durch eine Eigenschaft [17]), die beim Nachweis der folgenden Sätze bequem zu benutzen und bei der Anwendung auf Differentialoperatoren bequem nachzuweisen ist.

Definition. Es sei $g\,G\,g$ eine positiv definite Form in einem Hilbertschen Raum $\mathfrak{G} \leqq \mathfrak{H}$. Dann heiße die Form $g\,V\,g$ in $\mathfrak{G}$ *vollstetig relativ zu* G, (kurz G-vollstetig), wenn es zu jedem ε endlich viele Elemente $h_1, h_2, \ldots, h_n$ aus $\mathfrak{H}$ gibt, so daß gilt:

$$|g\,V\,g| \leqq \sum_{\nu=1}^{n} (h_\nu\,H\,g)^2 + \varepsilon\,(g\,G\,g).$$

Unmittelbar erhält man den

Hilfssatz: Sei ein Teilraum $\mathfrak{Z}$ von $\mathfrak{G}$ von unendlicher Dimension; (d. h. er enthalte unendlich viel linear unabhängige Elemente). Dann enthält $\mathfrak{Z}$ auch ein Element $z \neq 0$, für das

$$|z\,V\,z| \leqq \varepsilon\,(z\,G\,z)$$

gilt, also auch ein Element z mit $z\,G\,z = 1$ und beliebig kleinem $z\,V\,z$.

Denn da der Raum $\mathfrak{Z}$ von unendlicher Dimension ist, enthält er sicher ein Element, das auf beliebig vielen $h_1, h_2, \ldots, h_n$ orthogonal ist.

Das Kriterium von Hilbert [5. 2] lautet:

Satz 16. Eine Form $h\,V\,h$ in $\mathfrak{H}$, die H-vollstetig ist, besitzt ein *diskretes* Punktspektrum in jedem Null nicht enthaltenden Intervall [18]).

Das Weylsche Kriterium [12. 3] bezieht sich auf die Abänderung des Spektrums einer beschränkten Form, wenn man eine vollstetige addiert. Wir begnügen uns hier damit, dies Kriterium auf positiv halbbeschränkte Formen zu übertragen [19]).

Satz 17. Es sei die Form G in $\mathfrak{G} \leqq \mathfrak{H}$ halbbeschränkt mit der unteren Schranke $\underline{\gamma} > 0$ und abgeschlossen; es sei die Form $g\,V\,g$ vollstetig im Hilbertschen Raum $\mathfrak{G}$ mit der Maßform G.

Dann besitzt $G + V$ ein diskretes Punktspektrum unterhalb $\underline{\gamma}$ (d. h. in jedem abgeschlossenen Intervall unterhalb $\underline{\gamma}$).

Zusatz 18. Ist die Einheitsform H vollstetig in bezug auf G, so besitzt G selbst ein überhaupt diskretes Punktspektrum.

[17]) Sie ist verwandt mit der von Hellinger und Toeplitz in ihrem Enzyklopädieartikel bevorzugten Kennzeichnung der Vollstetigkeit.

[18]) Es gilt auch die Umkehrung.

[19]) Die Übertragung des allgemeinen Weylschen Kriteriums auf beliebige nicht beschränkte Operatoren soll an anderer Stelle geschehen.

Die Beweise für die genannten Sätze werden ganz einfach, wenn man sich entsprechend der Methode, die F. Rellich [8] zur Untersuchung des Spektrums von Differentialgleichungen angewandt hat, auf den Hilfssatz stützt.

Diesen Hilfssatz wenden wir zunächst zum Beweise des ersten Kriteriums (Satz 16) an. Angenommen, es gäbe ein Null nicht enthaltendes Intervall $\Delta\,\alpha = (\alpha_1^-, \alpha_2^+)$, in dem das Spektrum von $\boldsymbol{V}$ nicht diskret wäre, so daß der zu $\Delta\,\alpha$ gehörige Eigenraum $\mathfrak{Z}$ der Spektralschar von $\boldsymbol{V}$ von unendlicher Dimension wäre, d. h. unendlich viele lineare unabhängige Eigenelemente z besäße; dann gäbe es nach dem Hilfssatz in $\mathfrak{Z}$ auch ein z, für das $z\boldsymbol{H}z = 1$ und $z\boldsymbol{V}z$ beliebig klein wäre; das aber widerspricht wegen $\alpha_1 > 0$ oder $\alpha_2 < 0$ den Eigenwertungleichungen, die

$$\alpha_1 (z\boldsymbol{H}z) \leqq z\boldsymbol{V}z \leqq \alpha_2 (z\boldsymbol{H}z)$$

verlangen.

Zum Beweise des zweiten Kriteriums, Satz 17, nehmen wir an, zu einem Intervall mit der oberen Grenze $\underline{\gamma}(1-\varepsilon) < \underline{\gamma}$ gehörte ein Eigenraum $\mathfrak{Z}$ von $\boldsymbol{G} + \boldsymbol{V}$ mit unendlich viel linear unahängigen Elementen z; für sie besteht die Eigenwertungleichung

$$z\boldsymbol{G}z + z\boldsymbol{V}z \leqq \underline{\gamma}(1-\varepsilon)(z\boldsymbol{H}z);$$

sie steht aber in Widerspruch mit $z\boldsymbol{G}z \geqq \underline{\gamma}(z\boldsymbol{H}z)$, wenn wir nach dem Hilfssatz 3 aus dem Teilraum $\mathfrak{Z}$ des Hilbertschen Raumes $\mathfrak{G}$ ein Element z herausgreifen, für das $z\boldsymbol{G}z = 1$, aber $z\boldsymbol{V}z$ beliebig klein ist.

Zusatz 18 folgt daraus, daß für die Eigenelemente z von (α_1, α_2) die zweite Eigenwertungleichung

$$z\boldsymbol{G}z \leqq \alpha_2 (z\boldsymbol{H}z)$$

besteht, was für ein z mit $z\boldsymbol{G}z = 1$ und beliebig kleinem $z\boldsymbol{H}z$ zum Widerspruch führt.

(Eingegangen am 14. 7. 1933.)

Spektraltheorie halbbeschränkter Operatoren und Anwendung auf die Spektralzerlegung von Differentialoperatoren.

Zweiter Teil[1]).

Von

Kurt Friedrichs in Braunschweig.

---

1. Spektraltheorie von Differentialoperatoren zweiter Ordnung.

In diesem zweiten Teil soll die Theorie der halbbeschränkten Operatoren auf lineare Differentialoperatoren zweiter Ordnung angewandt werden, um ihre Spektralzerlegung zu gewinnen. Ich habe mich dabei auf einige typische Fälle beschränkt, bei deren Behandlung die Allgemeinheit der Methode wohl schon genügend zum Ausdruck kommt.

Der Operator sei der „Potentialoperator"

$$G = -\sum_{\nu=1}^{n} \frac{\partial^2}{\partial x_\nu^2} + v(x_1, \ldots, x_n);$$

die Elemente, auf die er anzuwenden ist, sind Funktionen f der Variablen $x_1, \ldots, x_n$. Die behandelten Fälle unterscheiden sich einmal durch andere Wahl des Gebietes der Variablen $x_1, \ldots, x_n$ und durch verschiedene Randbedingungen; um zu zeigen, daß sich auch quantentheoretische Energieoperatoren in Schrödingers Darstellung einordnen lassen, habe ich zugleich den Fall behandelt, daß das „Zusatzpotential" v eine Singularität in einem Punkte besitzt. Ferner ist v — wenn nötig, durch Addition einer Konstanten — so groß gewählt, daß G positiv halbbeschränkt wird.

So werde unterschieden:

(1) *Fall des unendlichen Gebietes mit regulärem v.*

Hier ist das Gebiet Γ der gesamte $(x_1, \ldots, x_n)$-Raum. Eine besondere Randbedingung ist nicht zu stellen.

(2) *Fall des unendlichen Gebietes mit singulärem v.* $n > 1$ *).

Hier kann v bei Annäherung an den Punkt $x_1 = \ldots = x_n = 0$ in noch anzugebender Weise unendlich werden.

---

[1]) Teil I dieser Arbeit ist erschienen in Math. Annalen **109**, S. 465—487.

*) Die allgemeine Theorie halbbeschränkter Differentialoperatoren mit Singularitäten bei einer Veränderlichen, $n = 1$, soll an anderer Stelle dargestellt werden.

(3) *Fall des endlichen Kugelgebietes mit regulärem v bei der Randbedingung*

$$f = 0.$$

(4) *Fall des endlichen Kugelgebietes mit regulärem v bei der Randbedingung*

$$\frac{\partial f}{\partial r} = 0.$$

Wir haben uns damit begnügt, als endliches Gebiet die Kugel bzw. Kreis, Strecke anzusetzen, da zur Überwindung der von der Natur des Gebietes herrührenden Schwierigkeiten unsere Theorie keine neuen Gesichtspunkte bietet; dasselbe gilt für die Wahl der Randbedingungen, die übrigens noch etwas abgeschwächt werden. Vollständig durchgeführt haben wir nur die Fälle der Dimensionszahlen $n = 1, 2, 3$.

Für die Dimension $n = 1$ kann die ganze Theorie auch erheblich einfacher dargestellt werden; die Fälle endlichen Gebietes sind auch den bisherigen Methoden (z. B. der Variationsrechnung) ohne weiteres zugänglich. Trotzdem haben wir diese Fälle mitgeführt, um ihre Einordnung unter die allgemeine Theorie klarzustellen.

2. Bezeichnungen.

Ein Punkt des Variablenraumes werde auch mit $x = (x_1, \ldots, x_n)$ bezeichnet. Wir setzen

$$r = |x| = \sqrt{x_1^2 + \ldots + x_n^2}.$$

Das zugrunde gelegte Gebiet Γ sei im Falle

(1) der gesamte x-Raum,

(2) der gesamte x-Raum ohne $x = 0$,

(3) **(4)** die „Kugel" $r < R$.

Wir bezeichnen (für $\varrho \leqq R$) abkürzend mit

Ω_ϱ die Kugelfläche $r = \varrho$,

K_ϱ die Kugel $r < \varrho$,

$K_{\varrho, \mathrm{P}}$ die Kugelschale $\varrho \leqq r < \mathrm{P}$.

Wir verstehen ferner (für $\sigma > 0$) unter Γ_σ im Falle

(1) $\Gamma_\sigma = K_{\frac{1}{\sigma}}$,

(2) $\Gamma_\sigma = K_{\sigma, \frac{1}{\sigma}}$,

(3) **(4)** $\Gamma_\sigma = K_{R-\sigma}$,

so daß für $\sigma \to 0$, Γ_σ ganz Γ ausschöpft.

Als „Quadrat" um $x = \xi$ mit der Seite 2δ bezeichnen wir das Gebiet $|x_1 - \xi_1| < \delta, \ldots, |x_n - \xi_n| < \delta$, kurz $[x - \xi] < \delta$.

Die Integration über den Variablenraum werde mit

$$\int \ldots dx = \iint r^{n-1} \ldots dr\, d\omega$$

bezeichnet. Das uneigentliche Integral über den Gesamtraum $\int\limits_{\Gamma} \ldots dx$ sei als Grenzwert von $\int\limits_{\Gamma_\sigma} \ldots dx$ für $\sigma \to 0$ verstanden.

3. Funktionenräume und Operator.

Zunächst ist der Hilbertsche Raum $\mathfrak{H}$, der Raum $\mathfrak{G}$ der Form G und der Raum $\mathfrak{F}$ des Operators G festzulegen. Zu dem Zweck werden dichte Teilräume, die „Funktionenräume", angegeben, die von Funktionen mit einfachen Differenzierbarkeitseigenschaften gebildet werden; diese werden dann zu Räumen idealer Elemente abgeschlossen.

Um den *Hilbertschen Raum* $\mathfrak{H}$ zu bilden, werden zuerst Teilräume $\mathfrak{H}'$ und $\mathfrak{H}''$ angegeben. $\mathfrak{H}''$ bzw. $\mathfrak{H}'$ besitzen als Elemente h alle in Γ stetigen bzw. stückweise [2]) stetigen Funktionen $h(x)$ der Veränderlichen

$$x = (x_1, \ldots, x_n),$$

für welche

$$h\boldsymbol{H}h = \int\limits_{\Gamma} h^2(x)\, dx$$

existiert. $\boldsymbol{H}$ werde zur Maßform gewählt. $\mathfrak{H}'$ ist bekanntlich [3]) separabel, aber nicht abgeschlossen. Nach Satz 2 (Teil I) kann $\mathfrak{H}'$ eindeutig zu einem abgeschlossenen Hilbertschen Raum $\mathfrak{H}$ fortgesetzt werden; und zwar durch Adjunktion idealer Elemente h, für die auch die Maßform $h\boldsymbol{H}h$ erklärt ist, so daß $\mathfrak{H}'$ dicht liegt in $\mathfrak{H}$. Wohl lassen sich diese idealen Elemente durch quadratisch $\mathfrak{L}$-integrierbare Funktionen realisieren;

---

[2]) Eine Funktion $h(x)$ heiße *stückweise stetig*, wenn sie nicht definiert ist auf einer endlichen Anzahl von Ebenen $x_\nu = \xi_\nu = \text{const}$, Kugelflächen $r = \varrho = \text{const}$, in einer endlichen Anzahl von Punkten $x = \xi$ und sonst stetig ist. Die Integrale $\int\limits_{\Gamma} h^2\, dx$ seien verstanden als Grenzwerte für $\varepsilon \to 0$ der Integrale über die Gebiete, die aus Γ durch Ausschluß der $|x_\nu - \xi_\nu| \leqq \varepsilon$, $|r - \varrho| \leqq \varepsilon$, $|x - \xi| \leqq \varepsilon$ entstehen. Existieren $\int\limits_{\Gamma} h_1^2\, dx$, $\int\limits_{\Gamma} h_2^2\, dx$, so auch $\int\limits_{\Gamma} h_1 h_2\, dx$.

[3]) Die Linearkombinationen der Elemente $e_{\xi,\delta}$ von $\mathfrak{H}'$

$$e_{\xi,\delta} = \frac{1}{(2\delta)^n} \quad \text{in } [x - \xi] < \delta, \qquad e_{\xi,\delta} = 0 \quad \text{außerhalb}$$

liegen bekanntlich $\boldsymbol{H}$-dicht in $\mathfrak{H}'$.

doch scheint es im Falle $n > 1$ keine Vorteile zu bieten, davon Gebrauch zu machen.

Wir verwenden auch für die idealen Elemente die Schreibung

$$h\boldsymbol{H}h = \int_{\Gamma} h^2\,dx;$$

wenn die Elemente h Funktionen $h(x)$ sind, werden wir die Variable x stets sichtbar machen.

Als Teilräume von $\mathfrak{H}''$ seien die *Räume* $\mathfrak{G}''$ *und* $\mathfrak{G}'$ erklärt. $\mathfrak{G}''$ bzw. $\mathfrak{G}'$ bestehe aus allen solchen Funktionen $g(x)$ von $\mathfrak{H}''$, die in Γ stetige bzw. stückweise stetige erste Ableitungen besitzen und für die die Formen

$$g\boldsymbol{D}g = \int_{\Gamma} \sum \left(\frac{\partial}{\partial x} g(x)\right)^2 dx \qquad \left[\sum \left(\frac{\partial}{\partial x} g\right)^2 = \sum_{\nu=1}^{n} \left(\frac{\partial}{\partial x_\nu} g\right)^2\right],$$

$$g\boldsymbol{V}g = \int_{\Gamma} v(x)\,g^2(x)\,dx$$

und also — mit $\boldsymbol{G} = \boldsymbol{D} + \boldsymbol{V}$ —

$$g\boldsymbol{G}g = \int_{\Gamma} \left\{\sum \left(\frac{\partial}{\partial x} g(x)\right)^2 + v(x)\,g^2(x)\right\} dx$$

existieren.

Im Falle **(3)** bei endlichem Gebiet $\Gamma = K_R$ ist schon den Funktionen $g(x)$ von $\mathfrak{G}'$ die Randbedingung des Verschwindens auf Ω_R zu stellen. Wir ersetzen sie durch die abgeschwächte Bedingung:

(1) $$\int_{\Omega_{R-\sigma}} g^2(x)\,dx \underset{\sigma \to 0}{\to} 0 \quad \text{in } \mathfrak{G}'.$$

Im Falle **(4)** wie in den Fällen **(1)**, **(2)** ist keine Randbedingung für $\mathfrak{G}'$ anzusetzen.

Dem Zusatzpotential $v(x)$ seien nun folgende Bedingungen gestellt:

Fall **(1)**, **(3)**, **(4)**.

$v(x)$ sei in Γ stetig[4]) und nach unten beschränkt durch

(2) $$v(x) \geqq 1.$$

Fall **(2)**. $n > 1$.

Es sei mit irgendeinem $\mathsf{P} > 0$

1. für $n \geqq 3$

$$\varphi(r) = \frac{n-2}{2}\,\frac{1}{r} \qquad 0 < r \leqq \mathsf{P},$$

---

[4]) Im Falle $n = 1$ genügt stückweise Stetigkeit; im nicht behandelten Falle $n > 3$ wird von v noch mehr zu verlangen sein.

2. für $n = 2$

$$\varphi(r) = \frac{1}{2}\frac{1}{\ln\frac{A}{r}}\frac{1}{r} \qquad 0 < r \leqq \mathsf{P}$$

für irgendein $A > \mathsf{P}$.

Dann sei $v(x)$ stetig in Γ und es gebe ein $\underline{v}$, ein P und eine Zahl Θ aus $0 \leqq \Theta < 1$, so daß mit $\varphi(r) = 0$ für $r > \mathsf{P}$

$$(2)_\Theta \qquad v(x) \geqq \underline{v} - \Theta\,\varphi^2(r)$$

gilt. Von der Konstanten $\underline{v}$ fordern wir noch, daß sie genügend groß ist, nämlich

$$\underline{v} \geqq 1 + \Theta k,$$

wo die Konstante $k > 0$ nach Anhang (2. 2) zu bestimmen ist [4a]).

Wir bezeichnen noch den Fall **(2)** bei $\Theta = 0$ mit $\mathbf{(2)}_0$, bei $\Theta \neq 0$ mit $\mathbf{(2)}_\Theta$.

Mit der in $\mathfrak{G}'$ erklärten Form $G = D + V$ besteht die Abschätzung

$$(3)_\Theta \qquad g\,G\,g \geqq (1-\Theta)\,g\,D\,g + (g\,H\,g),$$

so daß also die Form G positiv halbbeschränkt ist. Diese Tatsache folgt in den Fällen **(1)**, $\mathbf{(2)}_0$, **(3)**, **(4)** mit $\Theta = 0$ unmittelbar aus $\underline{v} \geqq 1$. Im Falle $\mathbf{(2)}_\Theta$ berufen wir uns darauf, daß nach Anhang (2. 2) und (2. 3) die Abschätzung

$$\int\limits_{K_{\mathsf{P}}} \Big\{ \sum \Big(\frac{\partial}{\partial x} g(x)\Big)^2 - \varphi^2(r)\,g^2(x) \Big\}\,dx + k \int\limits_{K_{\mathsf{P}}} g^2(x)\,dx \geqq 0$$

gilt und also

$$(4) \qquad \int\limits_{K_{\mathsf{P}}} \Big\{ \sum \Big(\frac{\partial}{\partial x} g(x)\Big)^2 + v(x)\,g^2 \Big\}\,dx \geqq (1-\Theta) \int\limits_{K_{\mathsf{P}}} \sum \Big(\frac{\partial}{\partial x} g\Big)^2 dx + \int\limits_{K_{\mathsf{P}}} g^2\,dx \text{ [5]}.$$

Der Raum $\mathfrak{G}'$ läßt sich nun mit G als Maßform zu einem Hilbertschen Raum $\mathfrak{G}$ von Elementen g aus $\mathfrak{H}$ *abschließen.* Das folgt nach Satz 3 (Teil I) aus der sogleich zu beweisenden Tatsache, daß die Form G in einem G-dichten Teilraum $\mathfrak{F}'$ von $\mathfrak{G}''$ zu einem Operator G führt. Für die idealen Elemente aus $\mathfrak{G}$ seien D, V und G wiederum symbolisch durch die entsprechenden Integrale dargestellt. Die Ungleichung $(3)_\Theta$ gilt dann auch in $\mathfrak{G}$.

[4a]) Um das Schrödingersche Problem einzuordnen, wird man $v(x) = -\frac{c}{r} + \text{const}$ setzen.

[5]) Zugleich mit $(3)_\Theta$ folgt, was für spätere Verwendung angemerkt sei, aus (4)

$$g\,G\,g \geqq \int\limits_{\Gamma - K_{\mathsf{P}}} v\,g^2\,dx.$$

Die *Räume* $\mathfrak{F}''$ *und* $\mathfrak{F}'$ bestehen[6]) aus allen Funktionen $f(x)$ aus $\mathfrak{G}''$, die stetige bzw. stückweise stetige zweite Ableitungen in Γ besitzen und für welche die Funktion $-\Delta f(x) + v(x) f(x)$ — zur Abkürzung ist $\Delta = \sum_{\nu=1}^{n} \frac{\partial^2}{\partial x_\nu^2}$ gesetzt — in $\mathfrak{H}''$ bzw. $\mathfrak{H}$ liegt. In $\mathfrak{F}'$ ist der Operator G erklärt durch

$$G = -\Delta + v.$$

Im Falle **(4)** bei endlichem Gebiet $\Gamma = K_P$ ist noch die *Randbedingung* $\frac{\partial}{\partial r} f(x) = 0$ auf Ω_R zu fordern. Wir stellen sie in der abgeschwächten Form: Mit jeder Funktion $g(x)$ von $\mathfrak{G}''$ gelte für $f(x)$ aus $\mathfrak{F}'$ und $\mathfrak{F}''$ im Falle **(4)**

$$\int_{\Omega_\varrho} g(x) \frac{\partial}{\partial r} f(x)\, d\omega \underset{\varrho \to R}{\to} 0. \tag{5}$$

Tatsächlich liegt $\mathfrak{F}'$ G-dicht in $\mathfrak{G}''$ [7]) und es gehört der Operator G in $\mathfrak{F}'$ zur Form $\boldsymbol{G}$; denn es gilt für jedes f aus $\mathfrak{F}'$ und jedes g aus $\mathfrak{G}'$ die „*Greensche Umformung*"

$$f\boldsymbol{G}g = Gf\boldsymbol{H}g; \tag{6}$$

ihren Beweis holen wir im Anhang 3 nach.

Der Operator G in $\mathfrak{F}'$ kann nach Satz 4 (Teil I) eindeutig zu einem abgeschlossenen Operator G in einem Raum $\mathfrak{F}$ von Elementen f fortgesetzt werden. Es liegt $\mathfrak{F}$ in $\mathfrak{G}$ [Bemerkung zu Satz 3 und 4 (Teil I)] und (6) bleibt in $\mathfrak{F}$ und $\mathfrak{G}$ bestehen.

4. Spektralzerlegung. Sätze.

Man kann natürlich nicht erwarten, daß der ursprünglich erklärte Operator G in $\mathfrak{F}'$ eine Spektralzerlegung besitzt, da er nicht abgeschlossen ist; man wird das jedoch verlangen von dem eindeutig zugeordneten abgeschlossenen Operator G in $\mathfrak{F}$. Darüber hinaus gilt aber immerhin, daß die Eigenelemente endlicher Intervalle[8]) schon im Raum $\mathfrak{F}''$ liegen, insbesondere also zweimal stetig differenzierbar sind und die Anwendung von G beliebig oft hintereinander gestatten.

---

[6]) Im Falle **(1)** genügt es, an Stelle der Zugehörigkeit zu $\mathfrak{G}''$ die Existenz stetiger erster Ableitungen zu fordern; die Existenz des Integrals $\boldsymbol{G}$ folgt. Vgl. Anhang 3., Zusatz.

[7]) Durch zweimal stetig differenzierbare Funktionen aus $\dot{\mathfrak{G}}'$ (vgl. Nr. 5. 2) kann man zunächst jede Funktion aus $\mathfrak{F}''$ $\boldsymbol{G}$-approximieren. $\dot{\mathfrak{G}}'$ liegt $\boldsymbol{G}$-dicht in $\mathfrak{G}'$.

[8]) Das heißt die Eigenelemente der zu endlichen Intervallen $\Delta\gamma$ gehörigen Differenzprojektionsoperatoren $R_{\Delta\gamma}$ der Spektralschar R_{γ}. von G.

Die Spektralzerlegbarkeit des Operators G in $\mathfrak{F}$ beruht auf seiner Selbstadjungiertheit. Zu beweisen, daß G in $\mathfrak{F}$ selbstadjungiert ist, ist überhaupt unsere Hauptaufgabe. Wir gehen zu dem Zweck davon aus, daß der Operator G in $\mathfrak{F}'$, der zu G in $\mathfrak{F}$ abgeschlossen wurde, auf andere Weise zu einem selbstadjungierten Operator fortgesetzt werden kann. Nämlich: Man bilde die zu G in $\mathfrak{F}'$ gehörige Form G in $\mathfrak{F}'$, schließe sie ab zu G in $\mathfrak{G}$ nach Satz 3 (I), bilde nach Satz 7 (I) den zu G in $\mathfrak{G}$ gehörigen selbstadjungierten Operator G, in $\mathfrak{F}_1 \leqq \mathfrak{G}$ erklärt [8a]). Es ist dann G in $\mathfrak{F}_1$ Fortsetzung von G in $\mathfrak{F}'$ und also auch von G in $\mathfrak{F}$ (Zusatz zu 7(I)).

Es genügt also, $\mathfrak{F} = \mathfrak{F}_1$ zu zeigen. Hierzu berufen wir uns auf die früher, Satz 10 (I), aufgestellten Kriterien. $(\mathfrak{F}_n)$: $\mathfrak{F}_n$ liegt in $\mathfrak{F}$ oder $(\mathfrak{G}_n)$: $\mathfrak{G}_n$ liegt in $\mathfrak{F}$. Dabei bildeten die Elemente f aus $\mathfrak{F}_1$, für welche Gf in

$$\mathfrak{G}, \mathfrak{F}_1, \mathfrak{G}_1, \mathfrak{F}_2, \ldots$$

liegt, die iterierten Räume

$$\mathfrak{G}_1, \mathfrak{F}_2, \mathfrak{G}_2, \mathfrak{F}_3, \ldots$$

Der erste unserer Sätze werde je nach der Anzahl der Variablen verschieden formuliert.

A. Eine Variable. $n = 1$.

Satz 1 A.

1. $\mathfrak{G}$ liegt in $\mathfrak{H}''$,

2. $\mathfrak{F}_1$ liegt in $\mathfrak{G}''$,

3. $\mathfrak{G}_1$ liegt in $\mathfrak{F}''$.

B. Zwei und drei Variablen. $n = 2, 3$.

Satz 1 B.

1. $\mathfrak{F}_1$ liegt in $\mathfrak{H}''$,

2. $\mathfrak{G}_1$ liegt in $\mathfrak{G}''$,

3. $\mathfrak{F}_2$ liegt in $\mathfrak{F}''$.

Aus dem Satz 1 folgt dann

Satz 2.

$$\mathfrak{F} = \mathfrak{F}_1,$$

d. h. der Operator G in $\mathfrak{F}$ ist selbstadjungiert.

In der Tat ist im Falle A Kriterium $(\mathfrak{G}_1)$, im Falle B Kriterium $(\mathfrak{F}_2)$ erfüllt.

Weiter folgt aus Satz 1 und 2 unmittelbar

Satz 3. Ist auf f aus $\mathfrak{F} = \mathfrak{F}_1$ der Operator G beliebig oft anwendbar, so liegt $f, Gf, G^2 f, \ldots, G^n f, \ldots$ in $\mathfrak{F}''$.

Satz 4. 1. Es existiert zur Form G in $\mathfrak{G}$ und zum Operator G in $\mathfrak{F}$ eine Spektralschar R_γ im Sinne von Satz 12 (I).

[8a]) Es gilt (6) auch für g in $\mathfrak{G}$, f in $\mathfrak{F}_1$.

2. Die Eigenelemente der Differenzoperatoren $R_{\Delta\gamma}$ endlicher Intervalle $\Delta\gamma = (\gamma_1, \gamma_2)$ sind Funktionen $f_{\Delta\gamma}(x)$ aus $\mathfrak{F}''$. Mit ihnen gelten die Eigenwertungleichungen

$$\gamma_1 \int_\Gamma f_{\Delta\gamma}^2(x)\,dx \leqq \int_\Gamma \left\{\sum \left(\frac{\partial}{\partial x} f_{\Delta\gamma}(x)\right)^2 + v(x)\, f_{\Delta\gamma}^2(x)\right\} dx \leqq \gamma_2 \int_\Gamma f_{\Delta\gamma}^2(x)\,dx.$$

Beweis. Satz 4.1 folgt nach Satz 2 aus Satz 12 (I). Satz 4.2 folgt dann nach Zusatz 12.3 (I) aus Satz 3.

5. Vorbereitungen zum Beweis von Satz 1.

5.1. Die Mittelwerte. Ein Haupthilfsmittel werden die Mittelwerte von Elementen h aus $\mathfrak{H}$ über Quadrate $[x-\xi] < \delta$ aus Γ sein. Zu dem Zweck führen wir die Elemente $e_{\xi,\delta}$ aus $\mathfrak{H}'$ ein, die durch

$$e_{\xi,\delta}(x) = \frac{1}{(2\delta)^n} \quad \text{in} \quad [x-\xi] < \delta,$$

$$e_{\xi,\delta}(x) = 0 \quad \text{außer} \quad [x-\xi] < \delta$$

definiert sind. Die Mittelwerte von h aus $\mathfrak{H}$ sind dann durch

$$e_{\xi,\delta}\, \boldsymbol{H}\, h = \int_\Gamma e_{\xi,\delta}\, h\, dx$$

erklärt, wofür wir auch die Schreibung

$$\frac{1}{(2\delta)^n} \int_{[x-\xi]<\delta} h\, dx$$

verwenden.

Die Grenzwerte der Mittelwerte beim Zusammenziehen des Quadrates auf den Mittelpunkt bezeichnen wir — so weit sie existieren — mit h_ξ:

$$\frac{1}{(2\delta)^n} \int_{[x-\xi]<\delta} h\, dx \underset{\delta\to 0}{\to} h_\xi$$

5.2. Räume $\dot{\mathfrak{H}}', \ldots, \dot{\mathfrak{F}}''$.

Oft werden Funktionen der Räume $\mathfrak{H}', \ldots, \mathfrak{F}''$ auftreten, die außerhalb irgend eines Gebietes Γ_σ verschwinden; die von ihnen gebildeten, offenbar in $\mathfrak{H}$ dicht liegenden Teilräume bezeichnen wir beziehungsweise mit

$$\dot{\mathfrak{H}}', \ldots, \dot{\mathfrak{F}}''.$$

Für die Funktionen $h(x)$ aus $\dot{\mathfrak{H}}'$ ist der Operator V durch

$$Vh = v(x)\, h(x)$$

erklärt. Für Elemente f aus $\dot{\mathfrak{F}}'$ ist der Operator

$$\Delta f = \sum_{\nu=1}^{n} \frac{\partial^2}{\partial x_\nu^2} f(x)$$

erklärt und es gilt für f aus $\dot{\mathfrak{F}}'$

$$\int_{\Gamma} \sum \left(\frac{\partial}{\partial x} g \frac{\partial}{\partial x} f\right) dx = -\int_{\Gamma} g \Delta f \, dx \tag{7}$$

ohne weiteres, wenn g zu $\dot{\mathfrak{G}}'$ gehört; diese Umformung überträgt sich auf alle g aus $\mathfrak{G}$, da die Form $\int_{\Gamma} \sum \left(\frac{\partial}{\partial x} g\right)^2 dx$ in $\mathfrak{G}$ beschränkt erklärt ist.

5.3. Der Operator S_σ. Ein weiteres Hilfsmittel wird eine Operation sein, die aus jedem Element ein solches erzeugt, das in genügender Ferne und in Umgebung der singulären Stelle verschwindet. Zu dem Zweck wählen wir eine beliebig oft differenzierbare Funktion $s_\sigma(x)$ mit folgenden Eigenschaften

$$\begin{aligned} s_\sigma(x) &= 1 \quad \text{in} \quad \Gamma_\sigma, \\ 0 < s_\sigma(x) &\leqq 1 \quad \text{in} \quad \Gamma_{\sigma/2} - \Gamma_\sigma, \\ s_\sigma(x) &= 0 \quad \text{außer} \quad \Gamma_{\sigma/2}. \end{aligned} \tag{8}$$

Wir erklären den Operator S_σ in $\mathfrak{H}'$ durch

$$S_\sigma h = s_\sigma(x) h(x);$$

wegen seiner Beschränktheit ist er auf ganz $\mathfrak{H}$ fortzusetzen. Offenbar erzeugt S_σ aus Elementen von $\mathfrak{G}'$ und $\mathfrak{G}''$ wieder solche. Wir zeigen: S_σ ist in $\mathfrak{G}$ mit $\boldsymbol{G}$ als Maßform beschränkt. Zu dem Zweck genügt es offenbar, die Beschränktheit in $\mathfrak{G}'$ nachzuweisen, d. h. — beachte [9]), daß S_σ in bezug auf $\boldsymbol{G}$ nicht symmetrisch ist — daß für zwei Elemente g und g_1 von $\mathfrak{G}'$

$$|S_\sigma g \boldsymbol{G} g_1|^2 \leqq C (g \boldsymbol{G} g)(g_1 \boldsymbol{G} g_1)$$

gilt. Ist $\left|\frac{\partial s_\sigma}{\partial x}\right| \leqq c_1$, $|v(x)| \leqq c_2$ in $\Gamma_{\sigma/2}$, so gilt

$$|S_\sigma g \boldsymbol{D} g_1| = \left| \int_{\Gamma} s_\sigma(x) \sum \frac{\partial}{\partial x} g(x) \frac{\partial}{\partial x} g_1(x) \, dx \right.$$

$$\left. + \int \sum \left(\frac{\partial}{\partial x} s_\sigma(x) \frac{\partial}{\partial x} g_1(x)\right) g(x)\, dx \right| \leqq \sqrt{g \boldsymbol{D} g}\sqrt{g_1 \boldsymbol{D} g_1} + c_1 \sqrt{g \boldsymbol{H} g}\sqrt{g_1 \boldsymbol{D} g_1},$$

$$|S_\sigma g \boldsymbol{V} g_1| \leqq c_2 \sqrt{g \boldsymbol{H} g}\sqrt{g_1 \boldsymbol{H} g_1}.$$

Beachten wir noch Ungleichung $(3)_\Theta$, so ergibt sich mit geeigneter Konstanten C in der Tat $|S_\sigma g \boldsymbol{G} g_1| \leqq C \sqrt{g \boldsymbol{G} g}\sqrt{g_1 \boldsymbol{G} g_1}$.

---

[9]) Vgl. Fußnoten [7]) und [11]) (Teil I).

6. Drei Hilfssätze.

Hilfssatz 1. Wenn für das Element h aus $\mathfrak{H}$ der Grenzwert h_ξ stetig in ξ existiert, derart, daß $\int\limits_\Gamma e_{\xi,\delta} h\,dx \to h_{\xi_0}$ strebt für $\xi = \xi_\delta \to \xi_0$, dann

1. existiert

$$\int\limits_\Gamma h_x^2\,dx,$$

2. ist h eine Funktion $h(x)$ aus $\mathfrak{H}''$ mit $h(x) = h_x$.

Beweis. Zunächst bemerken wir die für stetiges h_x gültige Relation

$$\frac{1}{(2\delta)^n}\int\limits_{[x-\xi]<\delta} h_x\,dx = \frac{1}{(2\delta)^n}\int\limits_{[x-\xi]<\delta} h\,dx, \tag{9}$$

wobei die Integration der linken Seite im gewöhnlichen Sinne zu verstehen ist. In der Tat sei andernfalls $[x-\xi_1] < \delta_1$ ein Quadrat, für welches

$$\frac{1}{(2\delta_1)^n}\left|\int\limits_{[x-\xi_1]<\delta_1} h_x\,dx - \int\limits_{[x-\xi_1]<\delta_1} h\,dx\right| = \alpha > 0$$

gilt. Dann zerlegen wir $[x-\xi_1] < \delta_1$ in 2^n Teilgebiete der Seite $2\delta_2 = \delta_1$. Für eines von ihnen,

$$[x-\xi_2] < \delta_2 = \frac{\delta_1}{2},$$

muß dann auch gelten

$$\frac{1}{(2\delta_2)^n}\left|\int\limits_{[x-\xi_2]<\delta_2} h_x\,dx - \int\limits_{[x-\xi_2]<\delta_2} h\,dx\right| \geqq \alpha.$$

So fortschließend entsteht eine Schachtelfolge von Gebieten

$$[x-\xi_\nu] < \delta_\nu = \frac{\delta_1}{2^\nu},$$

die gegen einen Grenzpunkt ξ_0 konvergieren. Wegen der Stetigkeit von h_ξ gilt aber für $\nu \to \infty$

$$\frac{1}{(2\delta_\nu)^n}\int\limits_{[x-\xi_\nu]<\delta_\nu} h_x\,dx \to h_{\xi_0}, \quad \text{neben} \quad \frac{1}{(2\delta_\nu)^n}\int\limits_{[x-\xi_\nu]<\delta_\nu} h\,dx \to h_{\xi_0},$$

woraus der Widerspruch $|h_{\xi_0} - h_{\xi_0}| \geqq \alpha$ entspringt.

Wir bemerken nun, daß die Funktion $s_\sigma(x)\,h_x$ ein Element

$$h_\sigma^* = s_\sigma(x)\,h_x$$

aus $\mathfrak{H}''$ ist. Die obige Gleichung (9) ist dann für alle Quadrate $[x-\xi] < \delta$ aus Γ_τ gleichwertig mit

$$\int\limits_\Gamma (h_\tau^* - h)\,e_{\xi,\delta}\,dx = 0.$$

Ist nun $k = k(x)$ ein Element aus $\mathfrak{H}''$, so läßt sich sicher die Funktion $S_\sigma k = s_\sigma(x)\, k(x)$ durch Linearkombinationen derjenigen $e_{\xi,\delta}$ approximieren, für die $[x - \xi] < \delta$ in Γ_τ liegt mit $\tau = \frac{\sigma}{3}$, so daß also gilt:

$$\int_\Gamma (h_\tau^* - h)\, S_\sigma k\, d\, x = 0.$$

Beachtet man noch $S_\sigma h_\tau^* = h_\sigma^*$, so folgt

$$\int_\Gamma (h_\sigma^* - S_\sigma\, h)\, k\, d\, x = 0.$$

Es ist also

$$S_\sigma h = h_\sigma^* = s_\sigma(x)\, h_x.$$

Nun gilt für jedes Element h aus $\mathfrak{H}$ die Ungleichung

$$\int_\Gamma (S_\sigma h)^2\, d\, x \leqq \int_\Gamma h^2\, d\, x,$$

weil sie für die h aus $\mathfrak{H}'$ gilt. Insbesondere ist also

$$\int_{\Gamma_\sigma} h_x^2\, d\, x = \int_{\Gamma_\sigma} h_\sigma^{*\,2}(x)\, d\, x \leqq \int_\Gamma h_\sigma^{*\,2}(x)\, d\, x = \int_\Gamma (S_\sigma h)^2\, d\, x \leqq \int_\Gamma h^2\, d\, x.$$

Daraus folgt der erste Teil der Behauptung: die Existenz des Integrals $\int_\Gamma h_x^2\, d\, x$. Überdies ist damit sichergestellt, daß h_x eine Funktion $h^*(x)$ aus $\mathfrak{H}''$ ist, für die $S_\sigma h^* = s_\sigma(x)\, h_x = S_\sigma h$ und also

$$\int_\Gamma S_\sigma k\, (h^* - h)\, d\, x = 0$$

gilt; da aber die $S_\sigma k$ dicht liegen in $\mathfrak{H}$, folgt

$$h^* = h.$$

Darin liegt der zweite Teil der Behauptung.

Ferner gilt der

Hilfssatz 2. Ein Element g aus $\mathfrak{G}$ sei eine Funktion $g(x)$ aus $\mathfrak{H}''$ und $g(x)$ sei stetig differenzierbar.

Dann

1. existiert

$$\int_\Gamma \left\{ \sum \left(\frac{\partial}{\partial\, x}\, g(x)\right)^2 \right\} d\, x \quad \text{und} \quad \int_\Gamma v(x)\, g^2(x)\, d\, x,$$

2. liegt g in $\mathfrak{G}''$.

Zum Beweise ziehen wir die Form G_σ heran, die in $\mathfrak{G}'$ durch

$$g\, G_\sigma\, g = \int_{\Gamma_\sigma} \left\{ \sum \left(\frac{\partial}{\partial\, x}\, g(x)\right)^2 + v(x)\, g^2(x) \right\} d\, x - \Theta\, \psi(\sigma) \int_{\Omega_\sigma} g^2(x)\, d\, \omega$$

erklärt sei mit $\psi(r) = \varphi(r)\, r^{n-1}$ und $\Theta = 0$ außer im Falle $(\mathbf{2})_\Theta$.

Es ist

$$g\,G_\sigma g = (1-\Theta)\int\limits_{\Gamma_\sigma}\sum\left(\frac{\partial}{\partial x}g\right)^2 dx + \int\limits_{\Gamma_\sigma}\{v+\Theta\,\varphi^2(r)\}\,g^2\,dx$$

$$+\,\Theta\Big[\int\limits_{\Gamma_\sigma}\Big\{\sum\left(\frac{\partial}{\partial x}g\right)^2 - \varphi^2(r)\,g^2\Big\}\,dx - \psi(\sigma)\int\limits_{\Omega_\sigma} g^2\,d\omega\Big].$$

Aus den Eigenschaften von $v(x)$ — und im Falle **(2)**$_\Theta$ nach Anhang (2. 1), (2. 3) — erkennen wir: $g\,G_\sigma g$ nimmt bei $\sigma\to 0$ nicht ab und strebt gegen $g\,G\,g$. Insbesondere ist

$$g\,G_\sigma g \leqq g\,G\,g.$$

Es ist somit die Form G_σ in $\mathfrak{G}'$ beschränkt und kann also auf ganz $\mathfrak{G}$ fortgesetzt werden, wobei diese Ungleichung bestehen bleibt.

Da der Operator S_σ in $\mathfrak{G}'$ mit G als Maßform beschränkt ist, ist die in $\mathfrak{G}'$ gültige Identität

$$S_\sigma g\,G_\sigma S_\sigma g = g\,G_\sigma g$$

auch in $\mathfrak{G}$ gültig.

Liegt nun die stetig differenzierbare Funktion $g(x)$ in $\mathfrak{G}$ und $\mathfrak{H}''$, so liegt sicher $S_\sigma g = s_\sigma(x)\,g(x)$ in $\mathfrak{G}''$ und es ist für $\sigma<\sigma_0$ — im Falle **(2)**$_\Theta$ nach Anhang (2. 0) —

$$(1-\Theta)\int\limits_{\Gamma_\sigma}\sum\left(\frac{\partial}{\partial x}g\right)^2 dx + \int\limits_{\Gamma_\sigma}(v+\Theta\,\varphi^2(r))\,g^2\,dx$$

$$\leqq \int\limits_{\Gamma_\sigma}\Big\{\sum\left(\frac{\partial}{\partial x}g(x)\right)^2 + v(x)\,g^2(x)\Big\}\,dx - \Theta\,\psi(\sigma)\int\limits_{\Omega_\sigma} g^2(x)\,d\omega$$

$$= S_\sigma g\,G\,S_\sigma g = g\,G_\sigma g \leqq g\,G\,g.$$

Die linke Seite bleibt also mit wachsendem σ beschränkt. Daraus folgt zunächst die Existenz des Integrals

$$\int\limits_{\Gamma}\sum\left(\frac{\partial}{\partial x}g(x)\right)^2 dx$$

und — im Falle **(2)**$_\Theta$ nach Hilfssatz (2. 3) — die Existenz von

$$\int\limits_{\Gamma} v(x)\,g^2(x)\,dx.$$

Damit ist die erste Behauptung des Hilfssatzes 2 bewiesen. Der zweite Teil: g liegt in $\mathfrak{G}''$ folgt unmittelbar für die Fälle **(1)**, **(2)**, **(4)** nach der Erklärung von $\mathfrak{G}''$.

Im Falle **(3)** muß noch nachgewiesen werden, daß $g(x)$ auch die Randbedingung

(1) $$\int\limits_{\Omega_{R-\sigma}} g^2(x)\,dx \underset{\sigma\to 0}{\to} 0$$

erfüllt. Wir zeigen zu dem Zweck, daß diese Randbedingung gleichwertig ist mit der folgenden:

Es gibt eine Konstante $C > 0$, so daß für $\varrho > \varrho_0 > 0$ gilt:

$$(1^*)\qquad \int\limits_{\Omega_\varrho} g^2\, d\omega \leqq 2\,C\,(R-\varrho) \int\limits_{K_{\varrho,R}} \sum \left(\frac{\partial}{\partial x} g\right)^2 dx.$$

In der Tat: Aus (1*) folgt unmittelbar die Gültigkeit von (1). Umgekehrt: Ist (1) erfüllt, so ziehe man Ungleichung

$$\int\limits_{\Omega_\varrho} g^2\, d\omega \leqq 2 \int\limits_{\Omega_\tau} g^2\, d\omega + 2\,C\,(R-\varrho) \int\limits_{K_{\varrho,R}} \sum \left(\frac{\partial}{\partial x} g\right)^2 dx, \qquad \varrho < \tau < R$$

heran, die aus (1. 1) (Anhang) mit $C = \varrho_0^{1-n}$ folgt; läßt man

$$\tau = R - \sigma \to R$$

streben, so ergibt sich (1*).

Ungleichung (1*) erweist sich unmittelbar als gleichwertig mit der folgenden Bedingung:

Es gibt eine Konstante C, so daß für alle ϱ_1, ϱ_2 aus

$$\varrho_0 \leqq \varrho_1 < \varrho_2 < R$$

gilt:

$$(1^{**})\qquad \int\limits_{K_{\varrho_1,\varrho_2}} g^2\, dx \leqq 2\,C\,(R-\varrho) \int\limits_{\varrho_1}^{\varrho_2} r^{n-1}\, dr \int\limits_{K_{\varrho_1,R}} \sum \left(\frac{\partial}{\partial x} g\right)^2 dx.$$

Denn aus ihr folgt für $\varrho_2 \to \varrho_1 = \varrho$ Ungleichung (1*). Umgekehrt folgert man aus dieser das Bestehen von (1**), wenn man beachtet, daß es einen Zwischenwert ϱ gibt, so daß gilt

$$\int\limits_{K_{\varrho_1,\varrho_2}} g^2\, dx = \int\limits_{\varrho_1}^{\varrho_2} r^{n-1}\, dr \int\limits_{\Omega_\varrho} g^2\, d\omega.$$

Beide Seiten von (1**) stellen in $\mathfrak{G}'$ beschränkte Formen dar, die auf $\mathfrak{G}$ fortzusetzen sind. Es gilt (1**) also auch für die Funktion $g(x)$ aus $\mathfrak{H}''$ und $\mathfrak{G}$, für die soeben die Existenz des Integrals $\int\limits_{\Gamma} \sum \left(\frac{\partial}{\partial x} g\right)^2 dx$ bewiesen wurde; für $g(x)$ werden diese fortgesetzten Formen aber auch durch die Integrale dargestellt. Also ist für diese Funktionen $g(x)$ auch die Bedingung (1) erfüllt; sie liegen also in $\mathfrak{G}''$.

Hilfssatz 3. Ein Element f aus $\mathfrak{F}_1$ sei eine Funktion aus $\mathfrak{G}''$, $Gf = h(x)$ liege in $\mathfrak{H}''$ und $f(x)$ besitze stetige zweite Ableitungen. Dann gilt

1. $Gf = -\Delta f(x) + v(x) f(x)$,
2. f liegt in $\mathfrak{F}''$.

Beweis. Es sei $g(x)$ ein Element aus $\dot{\mathfrak{G}}''$; dann gilt, da g in $\mathfrak{G}$, f in $\mathfrak{F}_1$ liegt, die Formel (6). Da g und f in $\mathfrak{G}''$, Gf in $\mathfrak{H}''$ liegen, geht sie über in

$$\int_\Gamma g(x)\, h(x)\, dx = \int_\Gamma \left\{\sum \frac{\partial}{\partial x} g(x) \frac{\partial}{\partial x} f(x) + v(x)\, g(x)\, f(x)\right\} dx$$

oder, indem man partiell integriert,

$$\int_\Gamma g(x)\, h(x)\, dx = \int_\Gamma g(x) \{-\Delta f(x) + v(x) f(x)\}\, dx.$$

Da $\dot{\mathfrak{G}}''$ dicht liegt in $\mathfrak{H}$, folgt

$$Gf = h(x) = -\Delta f(x) + v(x) f(x)$$

und: $h(x)$ liegt in $\mathfrak{H}''$ geht über in: f liegt in $\mathfrak{F}''$ nach der Erklärung dieses Raumes.

Nur im Falle **(4)** ist noch zu zeigen, daß $f(x)$ die Randbedingung (5) erfüllt. Die Relation

$$\int_\Gamma \left\{g(x)\, G f(x) - \sum \frac{\partial}{\partial x} g(x) \frac{\partial}{\partial x} f(x) - v(x)\, g(x)\, f(x)\right\} dx = 0,$$

die für alle Elemente $g(x)$ aus $\mathfrak{G}'$ gilt, besagt aber, daß dieses Integral über $K_{R-\sigma}$ erstreckt, für $\sigma \to 0$ verschwindet. Nach bekannter Umformung entsteht daraus

$$(5) \qquad \int_{\Omega_{R-\sigma}} g(x) \frac{\partial}{\partial x} f(x)\, dx \underset{\sigma \to 0}{\to} 0$$

für alle Funktionen $g(x)$ aus $\mathfrak{G}'$.

7. Beweis von Satz 1 A.

Die Weiterführung unseres Beweises soll je nach der Anzahl n der unabhängigen Veränderlichen gesondert durchgeführt werden.

A. *Eine unabhängige Veränderliche*, $n = 1$.

Wir ziehen die Grundlösung

$$K(x - \xi) = -\tfrac{1}{2}|x - \xi|$$

des Operators $\Delta = \frac{d^2}{dx^2}$ heran und bilden

$$k_\xi = k_\xi(x) = -\tfrac{1}{2} s_\sigma(x)\, |x - \xi|,$$

wobei $s_\sigma(x)$ nach 5. 3 gewählt sei. $k_\xi(x)$ ist als Funktion von x ein Element von $\dot{\mathfrak{G}}'$, aber nicht von $\mathfrak{F}'$, da die erste Ableitung $\frac{d}{dx} k_\xi(x)$

bei $x = \xi$ unstetig ist. Die zweite Ableitung $\frac{d^2}{dx^2} k_\xi(x) = \Delta k_\xi(x)$ ist aber wieder ein Element von $\mathfrak{H}''$ und verschwindet in Γ_σ und außer $\Gamma_{\sigma/2}$.

Hilfssatz 4 A. Es ist für alle g aus $\mathfrak{G}$ bzw. h aus $\mathfrak{H}$

$$\int_\Gamma k_\xi h \, dx \qquad \text{stetig differenzierbar,}$$

$$\int_\Gamma \Delta k_\xi h \, dx \qquad \text{stetig,}$$

$$\int_\Gamma \frac{d}{dx} k_\xi \frac{d}{dx} g \, dx \text{ [10]} \qquad \text{stetig}$$

in ξ, wenn ξ in Γ_σ liegt.

Beweis. Diese Eigenschaften dürfen als bekannt angesehen werden, vorausgesetzt, daß die Elemente h, g stetige bzw. stetig differenzierbare Funktionen sind. Dann folgen sie aber allgemein, wenn noch gezeigt wird, daß für die drei Funktionen

$$h_\varepsilon = \frac{1}{\varepsilon}(k_{\xi+\varepsilon} - k_\xi), \quad h_\varepsilon = \frac{d}{d\xi} k_{\xi+\varepsilon}, \quad h_\varepsilon = \Delta k_{\xi+\varepsilon}$$

und für $g_\varepsilon = k_{\xi+\varepsilon}$ die Formen

$$\int_\Gamma h_\varepsilon^2 \, dx \quad \text{und} \quad \int_\Gamma \left(\frac{d}{dx} g_\varepsilon\right)^2 dx$$

in ε beschränkt bleiben. Diese Beschränktheiten sind nach der Definition von $k_\xi(x)$ unmittelbar festzustellen.

Für die Funktionen $k_\xi(x)$ bilden wir sodann die ξ-Integrale

$$k_{\xi,\delta} = \frac{1}{2\delta} \int_{\xi-\delta}^{\xi+\delta} k_\eta \, d\eta.$$

Dabei sei stets vorausgesetzt, daß $|x - \xi| < \delta$ in Γ_σ liege. Dann gilt

Hilfssatz 5 A. Die Funktionen $k_{\xi,\delta}(x)$ sind stetig einmal, stückweise stetig zweimal differenzierbar und es ist

$$-\frac{d^2}{dx^2} k_{\xi,\delta}(x) = -\Delta k_{\xi,\delta}(x) = e_{\xi,\delta}(x) \quad \text{in} \quad \Gamma_\sigma.$$

Es liegt also

$$k_{\xi,\delta}(x) \quad \text{in} \quad \dot{\mathfrak{F}}'.$$

Beweis. Man errechnet

$$k_{\xi,\delta} = -\frac{1}{4\delta}\{(x-\xi)^2 + \delta^2\} \quad \text{in} \quad |x - \xi| < \delta$$

$$= -\tfrac{1}{2}|x - \xi|, \qquad \text{wenn } x \text{ sonst in } \Gamma_\sigma \text{ liegt;}$$

[10]) Vgl. Nr. 5. 2.

außer Γ_σ ist $k_{\xi,\delta}$ nach x beliebig oft stetig differenzierbar und außer $\Gamma_{\sigma/2}$ ist $k_{\xi,\delta} = 0$. Daraus folgt die Behauptung.

Hilfssatz 6 A. Es streben bei $\delta \to 0$ und $\xi = \xi_\delta \to \xi_0$ für jedes g aus $\mathfrak{G}$

$$\int\limits_\Gamma \frac{d}{dx} k_{\xi,\delta} \frac{d}{dx} g\, dx \to \int\limits_\Gamma \frac{d}{dx} k_{\xi_0} \frac{d}{dx} g\, dx$$

$$\int\limits_\Gamma (\Delta k_{\xi,\delta} + e_{\xi,\delta})\, g\, dx \to \int\limits_\Gamma \Delta k_\xi g\, dx.$$

Beweis. Die erste Behauptung folgt aus der in x gleichmäßigen Konvergenz

$$\frac{d}{dx} k_{\xi,\delta} \to \frac{d}{dx} k_{\xi_0}$$

außer in einer Umgebung $|x - \xi_0| < \varepsilon$ von $x = \xi_0$. Es ist aber

$$\int\limits_{|x-\xi_0|<\varepsilon} \left(\frac{d}{dx} k_{\xi_0,\delta}\right)^2 dx \leqq \frac{\varepsilon}{2} \underset{\varepsilon \to 0}{\to} 0.$$

Die zweite folgt aus der gleichmäßigen Konvergenz

$$\Delta k_{\xi,\delta} + e_{\xi,\delta} = \Delta k_{\xi,\delta} \to \Delta k_\xi \quad \text{außer } \Gamma_\sigma,$$

$$\Delta k_{\xi,\delta} + e_{\xi,\delta} = 0 = \Delta k_\xi \quad \text{in } \Gamma_\sigma.$$

Nunmehr sind wir in der Lage

Satz 1.1A:

$$\mathfrak{G} \text{ liegt in } \mathfrak{H}''$$

zu beweisen. Es sei g irgendein Element aus $\mathfrak{G}$. Dann ist nach Hilfssatz 5A und, da $k_{\xi,\delta}$ in $\dot{\mathfrak{F}}'$ liegt, nach Nr. 5.2

$$\frac{1}{2\delta} \int\limits_{|x-\xi|<\delta} g\, dx = \int\limits_\Gamma \{e_{\xi,\delta} + \Delta k_{\xi,\delta}\}\, g\, dx + \int\limits_\Gamma \frac{d}{dx} k_{\xi,\delta} \frac{d}{dx} g\, dx;$$

nach Hilfssatz 6A existiert also für $\delta \to 0$ und $\xi = \xi_\delta \to \xi_0$ der Grenzwert g_{ξ_0} und es ist

$$g_\xi = \int\limits_\Gamma \Delta k_\xi g\, dx + \int\limits_\Gamma \frac{d}{dx} k_\xi \frac{d}{dx} g\, dx. \tag{11 A}$$

Nach Hilfssatz 4A ist g_ξ stetig in Γ_σ und da σ willkürlich war, auch in Γ; nach Hilfssatz 1 liegt dann g in $\mathfrak{H}''$ mit $g(x) = g_x$.

Satz 1.2A:

$$\mathfrak{F}_1 \text{ liegt in } \mathfrak{G}''.$$

Beweis. Sei f ein Element von $\mathfrak{F}_1 < \mathfrak{G}$; $Gf = h$ liege in $\mathfrak{H}$. Nach Satz 1.1 liegt $f = f(x)$ in $\mathfrak{H}''$ und es gilt nach (11 A)

$$f(\xi) = \int\limits_\Gamma \Delta k_\xi(x) f(x)\, dx - \int\limits_\Gamma v(x) k_\xi(x) f(x)\, dx + \int\limits_\Gamma \left\{\frac{d}{dx} k_\xi \frac{d}{dx} f + v k_\xi f\right\} dx;$$

ferner, da k_ξ in $\mathfrak{G}$ liegt, nach (6)

$$f(\xi) = \int\limits_{\Gamma} \Delta\, k_\xi(x) f(x)\, dx - \int\limits_{\Gamma} v(x)\, k_\xi(x) f(x)\, dx + \int\limits_{\Gamma} k_\xi\, h\, dx. \tag{12 A}$$

Die ersten beiden Integrale sind, da v und f stetig sind, nach ξ stetig differenzierbar, ebenso das letzte Integral nach Hilfssatz 4 A.

Also ist $f(\xi)$ stetig nach ξ differenzierbar und nach Hilfssatz 2 liegt f in $\mathfrak{G}''$.

Satz 1.3A:

$\mathfrak{G}_1$ liegt in $\mathfrak{F}''$.

Beweis. Sei f ein Element aus $\mathfrak{F}_1$, so daß Gf in $\mathfrak{G}$ liegt. Dann liegt nach Satz 1.1 $Gf = h$ in $\mathfrak{H}''$ und es geht die Darstellung (12 A) über in

$$f(\xi) = \tfrac{1}{2}\int\limits_{\Gamma_\sigma} (x-\xi)\,\{h(x) - v(x) f(x)\}\, dx + \int\limits_{\Gamma_{\sigma/2}-\Gamma_\sigma} \{k_\xi(x)\,(h(x) - v(x) f(x)) + \Delta\, k_\xi(x) f(x)\}\, dx. \tag{13 A}$$

Da nunmehr auch h stetig ist, entsteht durch zweimaliges Differenzieren die stetige Funktion

$$\frac{d^2}{d\xi^2} f(\xi) = -h(\xi) + v(\xi) f(\xi) + \int\limits_{\Gamma_{\sigma/2}-\Gamma_\sigma} \left\{\frac{d^2}{d\xi^2} k_\xi(x)\,(h(x) - v(x) f(x)) + \frac{d^2}{d\xi^2}\Delta\, k_\xi(x) f(x)\right\} dx.$$

Nach Hilfssatz 3 liegt dann f in $\mathfrak{F}''$.

8. Beweis von Satz 1B.

B. *Zwei und drei unabhängige Veränderliche*, $n = 2;\, 3$. An Stelle der Grundlösungen mit $|x-\xi| = \sqrt{\sum_{\nu=1}^{n} (x_\nu - \xi_\nu)^2}$

$$K(x-\xi) = -\frac{1}{2\pi}\ln|x-\xi| \qquad \text{für } n = 2,$$

$$K(x-\xi) = \frac{1}{4\pi}\,\frac{1}{|x-\xi|} \qquad \text{für } n = 3$$

betrachten wir die iterierten Grundlösungen

$$K^2(x-\xi) = -\frac{1}{8\pi}|x-\xi|^2\{\ln|x-\xi| - 1\} \qquad \text{für } n = 2,$$

$$K^2(x-\xi) = \frac{1}{8\pi}|x-\xi| \qquad \text{für } n = 3,$$

die so gewählt sind, daß

$$\Delta K^2 = K$$

ist. Alsdann setzen wir

$$k_\xi = k_\xi(x) = s_\sigma(x) K^2(x-\xi).$$

Es ist $k_\xi(x)$ außer bei $x = \xi$ stetig zweimal differenzierbar; es existiert aber $\int\limits_\Gamma \sum \left(\frac{\partial}{\partial x} k_\xi(x)\right)^2 dx$ und $\int\limits_\Gamma (\Delta k_\xi(x))^2 dx$; $\Delta k_\xi(x)$ ist außer bei $x = \xi$ zweimal stetig differenzierbar und $\Delta^2 k_\xi(x)$ ist stetig; mehr noch $\Delta^2 k_\xi(x) = 0$ in Γ_σ.

Also liegt k_ξ in[11]) $\dot{\mathfrak{G}}'$, Δk_ξ in $\dot{\mathfrak{H}}'$, $\Delta^2 k_\xi$ in $\dot{\mathfrak{H}}''$; mit $v \Delta k_\xi$ sei die in $\dot{\mathfrak{H}}'$ liegende Funktion $v(x) \Delta k_\xi(x)$ bezeichnet.

Hilfssatz 4B. Es ist für alle h aus $\mathfrak{H}$, g aus $\mathfrak{G}$

$\int\limits_\Gamma \Delta^2 k_\xi h\, dx,$	stetig,
$\int\limits_\Gamma \Delta k_\xi h\, dx,$	stetig,
$\int\limits_\Gamma v \Delta k_\xi h\, dx,$	stetig,
$\int\limits_\Gamma k_\xi h\, dx,$	stetig zweimal differenzierbar,
$\int\limits_\Gamma \sum \frac{\partial}{\partial x} k_\xi \frac{\partial}{\partial x} g\, dx$	stetig differenzierbar

in ξ aus Γ_σ.

Die Behauptungen dürfen wieder als bekannt angenommen werden für zweimal stetig differenzierbare Funktionen h und g. Da diese dicht liegen in $\mathfrak{H}$ bzw. $\mathfrak{G}$ mit $\boldsymbol{H}$ bzw. $\boldsymbol{G}$ als Maß, genügt es, mit $\varepsilon = (\varepsilon_1, \ldots, \varepsilon_n)$ zu zeigen:

1. $\Delta^2 k_{\xi+\varepsilon}$, $\Delta k_{\xi+\varepsilon}$, $v \Delta k_{\xi+\varepsilon}$, $\frac{\partial^2}{\partial \xi_\nu \partial \xi_\mu} k_{\xi+\varepsilon}$, $\frac{1}{|\varepsilon|}\left(\frac{\partial}{\partial \xi_\nu} k_{\xi+\varepsilon} - \frac{\partial}{\partial \xi_\nu} k_\xi\right)$ sind Elemente h_ε, für die $\int\limits_\Gamma h_\varepsilon^2\, dx$ in $|\varepsilon|$ beschränkt bleibt;

2. $\frac{\partial}{\partial \xi_\nu} k_{\xi+\varepsilon}$, $\frac{1}{|\varepsilon|}(k_{\xi+\varepsilon} - k_\xi)$ sind Elemente g_ε, für die $\int\limits_\Gamma \sum \left(\frac{\partial}{\partial x} g_\varepsilon\right)^2 dx$ in $|\varepsilon|$ beschränkt bleibt. Diese Beschränktheiten sind aus den Eigenschaften von k_ξ unmittelbar zu entnehmen.

Wieder bilden wir die Mittelwerte

$$\frac{1}{(2\delta)^n} \int\limits_{[\eta - \xi] < \delta} k_\eta\, d\eta = k_{\xi,\delta}.$$

---

11) Es liegt bei $n = 2$ k_ξ in $\mathfrak{F}'$, bei $n = 3$ in $\mathfrak{F}$.

Dann gilt

Hilfssatz 5B. Die Funktion $k_{\xi,\delta}(x)$ ist stetig dreimal und $\Delta k_{\xi,\delta}$ außer bei $x = \xi$ zweimal differenzierbar; es ist

$$-\Delta k_{\xi,\delta}(x) = e_{\xi,\delta}(x) \quad \text{in } \Gamma_\sigma.$$

Es liegt also $\Delta k_{\xi,\delta}$ in $\dot{\mathfrak{F}}'$.

Beweis. Liegt x außer Γ_σ, so ist $k_{\xi,\delta}$ beliebig oft stetig differenzierbar; in Γ_σ aber ist

$$k_{\xi,\delta} = \int\limits_{[\eta-\xi]<\delta} K^2(x-\eta)\,d\eta$$

und von diesem Integral sind die geforderten Eigenschaften bekannt.

Hilfssatz 6B.

Es streben bei $\delta \to 0$ und $\xi = \xi_\delta \to \xi_0$ für jedes h aus $\mathfrak{H}$

$$\int\limits_\Gamma \Delta k_{\xi,\delta} h\,dx \to \int\limits_\Gamma \Delta k_{\xi_0} h\,dx,$$

$$\int\limits_\Gamma v\,\Delta k_{\xi,\delta} h\,dx \to \int\limits_\Gamma v\,\Delta k_{\xi_0} h\,dx,$$

$$\int\limits_\Gamma (\Delta^2 k_{\xi,\delta} + e_{\xi,\delta}) h\,dx \to \int\limits_\Gamma \Delta^2 k_{\xi_0} h\,dx.$$

Beweis. Die ersten beiden Behauptungen folgen daraus, daß gleichmäßig in x

$$\Delta k_{\xi,\delta}(x) \to \Delta k_\xi(x)$$

strebt außer in einer Umgebung $|x-\xi_0| < \varepsilon$, und daß dort

$$\int\limits_{|x-\xi_0|<\varepsilon} (\Delta k_{\xi,\delta})^2\,dx = \int\limits_{|x-\xi_0|<\varepsilon} \left(\frac{1}{(2\delta)^n}\int\limits_{[\eta-\xi]<\delta} K(\eta-x)\,d\eta\right)^2 dx$$

$$\leqq \int\limits_{|x-\xi_0|<\varepsilon} \frac{1}{(2\delta)^n}\int\limits_{[\eta-\xi]<\delta} (K(\eta-x))^2\,d\eta\,dx$$

$$\leqq \frac{1}{(2\delta)^n}\int\limits_{|x-\xi_0|<\varepsilon}\;\int\limits_{|\eta-x|<\varepsilon+2\delta+|\xi-\xi_0|} (K(\eta-x))^2\,dx\,d\eta$$

$$\leqq \int\limits_{|x'|<\varepsilon+2\delta+|\xi-\xi_0|} (K(x'))^2\,dx'$$

zugleich mit ε und δ gegen Null strebt.

Die dritte Behauptung folgt aus der gleichmäßigen Konvergenz

$$\Delta^2 k_{\xi,\delta} + e_{\xi,\delta} = \Delta^2 k_{\xi,\delta} \to \Delta^2 k_{\xi_0} \quad \text{außer } \Gamma_\sigma,$$

$$\Delta^2 k_{\xi,\delta} + e_{\xi,\delta} = 0 \qquad = \Delta^2 k_{\xi_0} \quad \text{in } \Gamma_\sigma.$$

Nunmehr können sofort die drei Sätze 1 bewiesen werden.

Satz 1.1 B

$\mathfrak{F}_1$ liegt in $\mathfrak{H}''$.

Beweis. Für f_1 aus $\mathfrak{F}_1$, $G f_1 = h$ aus $\mathfrak{H}$ gilt — beachte Nr. 5.2 und daß $\Delta k_{\xi,\delta}$ in $\dot{\mathfrak{F}}'$, f_1 in $\mathfrak{G}$ liegt —

$$\frac{1}{(2\,\delta)^n} \int\limits_{|x-\xi|<\delta} f_1\, d x = \int\limits_{\Gamma} (e_{\xi,\delta} + \Delta^2 k_{\xi,\delta}) f_1\, d x + \int\limits_{\Gamma} \sum \frac{\partial}{\partial x} \Delta k_{\xi,\delta} \frac{\partial}{\partial x} f_1\, d x$$

$$= \int\limits_{\Gamma} (e_{\xi,\delta} + \Delta^2 k_{\xi,\delta}) f_1\, d x - \int\limits_{\Gamma} v \Delta k_{\xi,\delta} f_1\, d x + \int\limits_{\Gamma} \Delta k_{\xi,\delta} h\, d x,$$

da f_1 in $\mathfrak{F}_1$, $\Delta k_{\xi,\delta}$ in $\mathfrak{F}'$ liegt und demnach (6) gilt.

Nach Hilfssatz 6 B existiert also für $\delta \to 0$ und $\xi = \xi_\delta \to \xi_0$ der Grenzwert $f_{1\,\xi_0}$ und es ist in Γ_σ

$$\text{(11 B)} \qquad f_{1\xi} = \int\limits_{\Gamma} \Delta^2 k_\xi f_1\, d x - \int\limits_{\Gamma} v \Delta k_\xi f_1\, d x + \int\limits_{\Gamma} \Delta k_\xi h\, d x.$$

Nach Hilfssatz 4 B ist $f_{1\xi}$ stetig in Γ_σ und also in Γ, und nach Hilfssatz 1 liegt f_1 in $\mathfrak{H}''$ mit $f_1(x) = f_{1x}$.

Satz 1.2 B

$\mathfrak{G}_1$ liegt in $\mathfrak{G}''$.

Beweis. Liege g_1 in $\mathfrak{G}_1$, also in $\mathfrak{F}_1$, und $G g_1 = g$ in $\mathfrak{G}$.

Nach Satz 1.1 B liegt $g_1 = g_1(x)$ in $\mathfrak{H}''$ und es gilt, da $k_{\xi,\delta}$ in $\dot{\mathfrak{F}}'$ liegt, nach Nr. 5.2

$$\text{(12 B)} \qquad g_1(\xi) = \int\limits_{\Gamma} \{\Delta^2 k_\xi(x) - v(x) \Delta k_\xi(x)\}\, g_1(x)\, d x - \int\limits_{\Gamma} \sum \frac{\partial}{\partial x} k_\xi \frac{\partial}{\partial x} g\, d x.$$

Das erste Integral ist nach ξ stetig in Γ_σ differenzierbar; ebenso das letzte Glied nach Hilfssatz 4 B; also ist $g_1(\xi)$ in Γ stetig differenzierbar. Nach Hilfssatz 2 liegt dann $g_1(x)$ in $\mathfrak{G}''$.

Satz 1.3 B.

$\mathfrak{F}_2$ liegt in $\mathfrak{F}'$.

Beweis. Es liege f_2 in $\mathfrak{F}_2$ und

$$G f_2 = f_1 \text{ in } \mathfrak{F}_1, \quad G^2 f_2 = h \text{ in } \mathfrak{H}.$$

Nach Satz 1.1 B, 1.2 B liegt $f_1 = f_1(x)$ in $\mathfrak{H}''$, $f_2 = f_2(x)$ in $\mathfrak{G}''$. Die Darstellung (12 B) nimmt — wegen k_ξ in $\mathfrak{G}$, f_1 in $\mathfrak{F}_1$ gilt (6) — die Form an

(13 B)
$$f_2(\xi) = \int_\Gamma \{\Delta^2 k_\xi(x) - v(x)\,\Delta\, k_\xi(x)\}\, f_2(x)\,dx + \int_\Gamma v(x)\,k_\xi(x)\,f_1(x)\,dx - \int_\Gamma k_\xi\, h\, dx.$$

Die zweimalige stetige Differenzierbarkeit der ersten beiden Integrale folgt in bekannter Weise unter Benutzung der stetigen Differenzierbarkeit von $f_2(x)$; nach Hilfssatz 4 B folgt zweimalige stetige Differenzierbarkeit des letzten Gliedes. Nach Hilfssatz 3 B liegt also $f_2(x)$ in $\mathfrak{F}''$.

9. Diskussion des Spektrums.

Wir geben einfache Bedingungen für das Verhalten des Zusatzpotentials v dafür an, daß das Spektrum überhaupt oder unterhalb einer Schranke ein *diskretes* gewöhnliches Punktspektrum ist. Sie entsprechen ganz den Bedingungen von Weyl [12. 2]. In der Methode haben wir uns von der Note von Fr. Rellich [8] leiten lassen. Wir verwenden wesentlich die Sätze aus Teil I, Nr. 7, die an den Begriff der *Vollstetigkeit* anschließen.

Satz 5.

Im Falle **(3)**, **(4)** des endlichen Gebietes $\Gamma = K_R$. Das Spektrum von $G = -\Delta + v$ ist diskret; es gibt eine unendlich anwachsende Folge von Eigenwerten endlicher Vielfachheit.

Satz 6. Im Falle **(1)**, **(2)** des unendlichen Gebietes. Es wachse das Zusatzpotential $v(x)$ gleichmäßig ins Unendliche für $r = |x| \to \infty$.

Dann ist das Spektrum von $G = -\Delta + v$ diskret; es gibt eine unendlich anwachsende Folge von Eigenwerten endlicher Vielfachheit.

Satz 7[12]). Im Falle **(1)**, **(2)** des unendlichen Gebietes. Es sei

$$\underline{\lim}\; v(x) = v_\infty \text{ für } |x| \to \infty.$$

Dann ist das Spektrum von $G = -\Delta + v$ unterhalb v_∞ diskret.

Als wesentliches Hilfsmittel verwenden wir die *Ungleichung von Poincaré* (vgl. [4. 2]) für Gebiete $[x - \xi] < \delta$ aus Γ (die im Fall **(2)** auch $x = 0$ als Ecke besitzen dürfen): Es gibt ein $c > 0$, so daß für alle Funktionen g aus $\mathfrak{G}$ gilt

(14)
$$\int_{[x-\xi]<\delta} g^2\,dx \leqq \frac{1}{(2\delta)^n}\Big\{\int_{[x-\xi]<\delta} g\,dx\Big\}^2 + \delta^2 c \int_{[x-\xi]<\delta} \sum \Big(\frac{\partial}{\partial x} g\Big)^2 dx.$$

---

[12]) Für Schrödingers Wasserstoffproblem vgl. Courant-Hilbert [2. 7].

Beweis. Der für Gebiete $[x - \xi] < \delta$ einfache Beweis sei angemerkt: Es genügt g in $\mathfrak{G}'$ anzunehmen. Sei $[x^1 - \xi] < \delta$, $[x^2 - \xi] < \delta$, so ist

$$|g(x^1) - g(x^2)| \leqq \int\limits_{|x_1 - \xi_1| < \delta} \left| \frac{\partial}{\partial x_1} g(x_1, x_2^1, \ldots, x_n^1) \right| dx_1$$
$$+ \int\limits_{|x_2 - \xi_2| < \delta} \left| \frac{\partial}{\partial x_2} (x_1^2, x_2, x_3^1, \ldots, x_n^1) \right| dx_2$$
$$+ \ldots$$
$$+ \int\limits_{|x_n - \xi_n| < \delta} \left| \frac{\partial}{\partial x_n} (x_1^2, \ldots, x_{n-1}^2, x_n) \right| dx_n.$$

Durch Integration nach x^1 und x^2 entsteht

$$\int\limits_{[x^1 - \xi] < \delta} \int\limits_{[x^2 - \xi] < \delta} |g(x^1) - g(x^2)|^2 \, dx^1 \, dx^2 \leqq (2\delta)^{n+2} n \int\limits_{[x - \xi] < \delta} \sum \left(\frac{\partial}{\partial x} g \right)^2 dx.$$

Die linke Seite ist aber nichts anderes als

$$2(2\delta)^n \int\limits_{[x - \xi] < \delta} g^2 \, dx - 2 \left\{ \int\limits_{[x - \xi] < \delta} g \, dx \right\}^2.$$

So entsteht (14) mit $c = 2n$.

Diese Ungleichung führt uns dann zu dem

Hilfssatz. Die Formen

$$\int\limits_{K_\varrho} g^2 \, dx, \quad \int\limits_{K_\varrho} w(x) g^2 \, dx \qquad \varrho < R$$

mit in Γ stückweise stetiger und *beschränkter* Funktion $w(x)$, die zunächst in $\mathfrak{G}'$ erklärt und dann auf $\mathfrak{G}$ fortgesetzt sind, sind in $\mathfrak{G}$ vollstetig in bezug auf G.

Beweis. Das Gebiet K_ϱ kann man durch eine endliche Summe $\sum\limits_{\xi}$ von fremden Quadraten $[x - \xi] < \delta$ mit beliebig kleinem δ überdecken. Durch Summation der Poincaréschen Ungleichung entsteht

$$\int\limits_{K_\varrho} g^2 dx \leqq (2\delta)^n \sum_{\xi} \left| \int\limits_{[x - \xi] < \delta} g \, dx \right|^2 + 2\delta^2 n \int\limits_{\Gamma} \sum \left(\frac{\partial}{\partial x} g \right)^2 dx.$$

Das besagt unter Beachtung von $(3)_\Theta$ nach der Definition von Teil I gerade die Vollstetigkeit von $\int\limits_{K_\varrho} g^2 \, dx$ in bezug auf G. Die Vollstetigkeit von $\int\limits_{K_\varrho} w g^2 \, dx$ ergibt sich dann unmittelbar daraus, daß mit $|w(x)| \leqq \overline{w}$

$$\left| \int\limits_{K_\varrho} w g^2 \, dx \right| \leqq \overline{w} \int\limits_{K_\varrho} g^2 \, dx$$

abzuschätzen ist, wo die rechte Seite schon als vollstetig erkannt ist.

Zum *Beweise von Satz* 5 berufen wir uns auf Zusatz 18 von Teil I. Es genügt also die Vollstetigkeit von $\int_\Gamma g^2\,dx$ in bezug auf G zu beweisen. Wir zerlegen

$$\int_\Gamma g^2\,dx = \int_{K_{R-\sigma}} g^2\,dx + \int_{K_{R-\sigma,R}} g^2\,dx$$

und verwenden die Ungleichung (1, 1) (Anhang) mit $u = g$.

Nachdem wir dort über ϱ von $R-\sigma$ bis R und über τ von $R-\sigma$ bis $R-\frac{\sigma}{2}$ mitteln, wobei $R-\sigma > \frac{R}{2}$ sei, entsteht aus naheliegenden Abschätzungen

$$\int_{K_{R-\sigma,R}} g^2\,dx \leqq c_1 \int_{K_{R-\sigma/2,\,R-\sigma}} g^2\,dx + \sigma^2 c_2 \int_{K_R} \sum \left(\frac{\partial}{\partial x} g\right)^2 dx$$

z. B. mit $c_1 = 2^{n+1}$, $c_2 = 2R^{n-1}C$.

Durch Einsetzen gewinnen wir

$$\int_\Gamma g^2\,dx \leqq \int_{K_{R-\sigma}} g^2\,dx + c_1 \int_{K_{R-\sigma/2,\,R-\sigma}} g^2\,dx + \sigma^2 c_2 \int_{K_R} \sum \left(\frac{\partial}{\partial x}\right)^2 dx$$

und damit, da die ersten beiden Terme der rechten Seite nach dem Hilfssatz vollstetig sind, die Behauptung.

Zum *Beweise von Satz* 6 genügt es nach Zusatz 18 Teil I zu zeigen, daß $\int_\Gamma g^2\,dx$ G-vollstetig ist. Sei

$$m(\varrho) = \operatorname{Min} v(x)$$

für $x \geqq \varrho$; dann besagt die Voraussetzung von Satz 6

$$m(\varrho) \to \infty \text{ für } \varrho \to \infty.$$

Nun gilt offenbar

$$\int_\Gamma g^2\,dx \leqq \int_{K_\varrho} g^2\,dx + \frac{1}{m(\varrho)} \int_{\Gamma - K_\varrho} v g^2\,dx$$

und ferner, vgl. [5]), S. 689

$$\int_{\Gamma - K_\varrho} v g^2\,dx \leqq g\,G\,g.$$

Darin liegt aber die Behauptung.

Zum *Beweise von Satz* 7 zerlegen wir $v(x)$ in

$$v(x) = v^+(x) - w(x),$$

wo $v^+(x)$ und $w(x)$ in Γ stetige Funktionen sind, so gewählt, daß

$$v^+(x) \geqq v_\infty, \qquad 0 \leqq w(x) \leqq v_\infty$$

wird, und zwar derart, daß gilt

$$w(x) \to 0 \quad \text{für} \quad |x| \to \infty.$$

Im Falle **(2)**$_\theta$ soll dieser Ansatz nur für $|x| \geqq \mathsf{P}$ bestehen; für $0 < |x| \leqq \mathsf{P}$ dagegen

$$v^+(x) = v(x) + v_\infty, \quad w(x) = v_\infty,$$

was durch geeignete Wahl möglich ist, ohne die Stetigkeit bei $|x| = \mathsf{P}$ zu verletzen.

Nach diesen Festsetzungen ist die Form

$$g\,\boldsymbol{W} g = \int\limits_{\Gamma} w(x)\, g^2\, d\,x$$

in $\mathfrak{G}$ beschränkt und es gilt für die in $\mathfrak{G}$ definierte Form

$$g\,G^+ g = \int\limits_{\Gamma} \left\{ \sum \left(\frac{\partial}{\partial x} g\right)^2 + v^+ g^2 \right\} d\,x$$

(*) $$g\,G^+ g \geqq v_\infty \int\limits_{\Gamma} g^2\, d\,x,$$

wobei man sich im Falle **(2)**$_\theta$ noch auf die Abschätzung (4) zu berufen hat. Da ferner offenbar

(⁑) $$v_\infty (g\,G\,g) \geqq g\,G^+ g \geqq g\,G\,g$$

gilt, ist G^+ in $\mathfrak{G}$ abgeschlossen.

Da nun v^+ genau die Bedingungen eines Zusatzpotentials erfüllt, ist auf die Form G^+ die Spektralzerlegung anwendbar. Da ihre untere Schranke offenbar v_∞ ist, besitzt sie unterhalb von v_∞ kein Spektrum. Satz 7: Das Spektrum des Operators G oder der Form $G = G^+ + \boldsymbol{W}$ ist diskret unterhalb v_∞, folgt alsdann unmittelbar aus Satz 17 (I), wenn gezeigt ist, daß $\boldsymbol{W}$ vollstetig ist in bezug auf G^+ in $\mathfrak{G}$ oder — was wegen (⁑) gleichwertig ist — in bezug auf G.

Ist nun

$$m(\varrho) = \operatorname{Max} w(x) \text{ für } |x| \geqq \varrho,$$

so folgt aus $w(x) \to 0$ für $|x| \to \infty$

$$m(\varrho) \to 0, \quad \text{für} \quad \varrho \to \infty.$$

Nun gilt

$$0 \leqq \int\limits_{\Gamma} w\, g^2\, d\,x \leqq v_\infty \int\limits_{K_\varrho} g^2\, d\,x + m(\varrho) \int\limits_{\Gamma - K_\varrho} g^2\, d\,x.$$

Wegen $\int\limits_{\Gamma} g^2\, d\,x \leqq g\,G\,g$ liegt darin nach der Definition von Teil I die Vollstetigkeit von $\boldsymbol{W}$.

10. Anhang.

1. Integralungleichungen.

Sei $u(x)$ eine in $0 < r < R$ stückweise stetig differenzierbare Funktion von $x = (x_1, \ldots, x_n)$. Dann besteht die Ungleichung

$$\left| \sqrt{\int_{\Omega_\varrho} u^2 d\omega} - \sqrt{\int_{\Omega_\tau} u^2 d\omega} \right| \leqq \sqrt{C|\varrho - \tau|} \sqrt{\int_{K_{\tau,\varrho}} \sum \left(\frac{\partial u}{\partial x}\right)^2 dx} \tag{1.0}$$

für

$$C \geqq \frac{1}{\varrho - \tau} \int_\tau^\varrho \frac{dr}{r^{n-1}}$$

mit $0 < \varrho < \mathsf{P}$, $0 < \tau < \mathsf{P}$, $\tau < \varrho$

Beweis. Es sei $x = \xi$ mit $|\xi| = 1$ ein Punkt auf Ω_1; dann ist

$$u(\varrho\xi) - u(\tau\xi) = \int_\tau^\varrho \frac{\partial}{\partial r} u(r\xi)\, dr,$$

$$u^2(\varrho\xi) - 2u(\varrho\xi)u(\tau\xi) + u^2(\tau\xi) \leqq \int_\tau^\varrho \frac{dr}{r^{n-1}} \int_\tau^\varrho r^{n-1} \left(\frac{\partial}{\partial r} u(r\xi)\right)^2 dr.$$

Durch Integration über Ω_1 und Anwendung der Schwarz-Ungleichung auf die linke Seite entsteht

$$\left| \sqrt{\int_{\Omega_\varrho} u^2 d\omega} - \sqrt{\int_{\Omega_\tau} u^2 d\omega} \right|^2 \leqq \int_\tau^\varrho \frac{dr}{r^{n-1}} \int_{K_{\tau,\varrho}} \sum \left(\frac{\partial u}{\partial x}\right)^2 dx$$

und daraus Ungleichung (1.0)

Aus der Ungleichung (1.0) erhält man

$$\int_{\Omega_\varrho} u^2 d\omega \leqq 2 \int_{\Omega_\tau} u^2 d\omega + 2C|\varrho - \tau| \int_{K_{\tau,\varrho}} \sum \left(\frac{\partial u}{\partial x}\right)^2 dx \tag{1.1}$$

mit $K_{\varrho,\tau}$ statt $K_{\tau,\varrho}$ für $\tau > \varrho$.

2. Singularität des Zusatzpotentials.

Es sei $u(x)$ stückweise stetig differenzierbar für $0 < r \leqq \mathsf{P}$,

$$x = x_1, \ldots, x_n; \quad n = 1, 2, 3 \ldots.$$

Dann besteht für $0 < \sigma < \varrho \leqq \mathsf{P}$ die *Ungleichung*

$$(2.0)\quad T_{\sigma,\varrho} \equiv \int\limits_{K_{\sigma,\varrho}} \left\{ \sum \left(\frac{\partial u}{\partial x}\right)^2 - \varphi^2(r)\,u^2 \right\} dx - \varphi(\sigma)\,\sigma^{n-1} \int\limits_{\Omega_\sigma} u^2\,d\omega$$

$$\geqq -\varphi(\varrho)\,\varrho^{n-1} \int\limits_{\Omega_\varrho} u^2\,d\omega$$

mit

$$\varphi(r) = \frac{n-2}{2}\,\frac{1}{r} \qquad \text{für } n = 3, 4, \ldots$$

$$\varphi(r) = \frac{1}{2}\,\frac{1}{\ln\frac{A}{r}}\,\frac{1}{r} \qquad \text{für } n = 2 \quad \text{mit } A > \mathsf{P}.$$

Beweis. Es ist

$$T_{\sigma,\varrho} = \int\limits_{K_{\sigma,\varrho}} \frac{1}{\varphi(r)\,r^{n-1}} \sum \left[\frac{\partial}{\partial x}\left(r^{\frac{n-1}{2}} \sqrt{\varphi(r)}\,u\right)\right]^2 dx - \varphi(\varrho)\,\varrho^{n-1} \int\limits_{\Omega_\varrho} u^2\,d\omega$$

Hilfssätze.

(2.1). Es nimmt

$$T_{\sigma,\varrho}$$

nicht ab für $\sigma \to 0$.

Der Beweis folgt unmittelbar aus der obigen Darstellung von $T_{\sigma,\varrho}$.

(2.2). Es gibt eine positive Konstante, k, so daß für genügend kleines σ ($\sigma \leqq \sigma_0$) gilt

$$T_{\sigma,\mathsf{P}} \geqq -k \int\limits_{K_{\sigma,\mathsf{P}}} u^2\,dx.$$

Beweis: Es sei σ_1 eine solche Zahl zwischen σ_0 und P, daß

$$\sigma_1^{n-1} \int\limits_{\Omega_{\sigma_1}} u^2\,d\omega = \frac{1}{\mathsf{P}-\sigma_0} \int\limits_{K_{\sigma_0,\mathsf{P}}} u^2\,dx$$

wird; dann ist für $\sigma \leqq \sigma_0$ wegen $T_{\sigma,\mathsf{P}} \geqq T_{\sigma_1,\mathsf{P}}$

$$T_{\sigma,\mathsf{P}} \geqq -\frac{\varphi(\sigma_1)}{\mathsf{P}-\sigma_0} \int\limits_{K_{\sigma_0,\mathsf{P}}} u^2\,dx - \int\limits_{K_{\sigma_1,\mathsf{P}}} \varphi^2(r)\,u^2\,dx.$$

Ist nun $k = \operatorname{Max}\left[\frac{\varphi(\sigma_1)}{\mathsf{P}-\sigma_0} + \varphi^2(r)\right]$ in $\sigma_0 \leqq r \leqq \mathsf{P}$, so folgt

$$T_{\sigma,\mathsf{P}} \geqq -k \int\limits_{K_{\sigma_0,\mathsf{P}}} u^2\,dx$$

und damit die Behauptung.

(2.3). Wenn $\int\limits_{K_{\mathsf{P}}} \sum \left(\frac{\partial u}{\partial x}\right)^2 dx$ existiert, so strebt für $n \geqq 2$

$$\varphi(\sigma)\,\sigma^{n-1} \int\limits_{\Omega_\sigma} u^2\,d\omega \underset{\sigma \to 0}{\to} 0$$

und es existiert

$$\int_{K_P} \varphi^2(r)\, u^2\, dx.$$

Beweis. Es folgt aus (2.0) wegen $\varphi(r) > 0$ unmittelbar die Beschränktheit von $\int_{K_{\sigma,\varrho}} \varphi^2(r)\, u^2\, dx$ in σ und damit die Existenz von

$$\int_{K_P} \varphi^2(r)\, u^2\, dx = \int_0^P \varphi(r) \Big[\varphi(r)\, r^{n-1} \int_{\Omega_r} u^2\, d\omega\Big]\, dr.$$

Aus der Existenz dieses Integrals und der Nichtexistenz von

$$\int_0^P \varphi(r)\, dr$$

folgt, daß es eine spezielle Folge $\varrho \to 0$ gibt, für die

$$\varphi(\varrho)\, \varrho^{n-1} \int_{\Omega_\varrho} u^2\, d\omega \underset{\varrho \to 0}{\to} 0$$

strebt. Aus (2.0) entnimmt man dann, daß das für jede Folge $\sigma \to 0$ gilt.

3. Beweis der Greenschen Umformung.

Zum Beweise von

$$\int_\Gamma \Big\{\sum \frac{\partial}{\partial x} f \frac{\partial}{\partial x} g + v f g\Big\}\, dx - \int_\Gamma \{\Delta f g + v f g\}\, dx = 0 \tag{6}$$

für f in $\mathfrak{F}'$, g in $\mathfrak{G}'$ integrieren wir zunächst nur über $K_{\sigma,\varrho}$:

$$\int_{K_{\sigma,\varrho}} \Big\{\sum \frac{\partial}{\partial x} f \frac{\partial}{\partial x} g + v f g + \Delta f g - v f g\Big\}\, dx$$

$$= \varrho^{n-1} \int_{\Omega_\varrho} g \frac{\partial}{\partial r} f\, d\omega - \sigma^{n-1} \int_{\Omega_\sigma} g \frac{\partial}{\partial r} f\, d\omega,$$

wo nur im Falle **(2)** $\sigma \neq 0$ zu wählen ist. Es ist dann zu zeigen, daß es spezielle Folgen $\varrho \to \infty$ bzw. $\varrho \to R$ und $\sigma \to 0$ gibt, für die die rechten Seiten verschwinden.

Fall **(1)**. Aus der Existenz des Integrals

$$\int_0^\infty r^{n-1} \int_{\Omega_r} \Big| g \frac{\partial}{\partial r} f \Big|\, d\omega\, dr \leq \sqrt{\int_\Gamma \sum \Big(\frac{\partial}{\partial x} f\Big)^2 dx \int_\Gamma g^2\, dx}$$

folgt die Existenz einer Folge $\varrho \to \infty$, für die

$$\varrho^n \int_{\Omega_\varrho} g \frac{\partial f}{\partial r}\, d\omega \to 0$$

strebt.

Fall **(2)**. Die Folge $\varrho \to \infty$ wähle wie im Falle **(1)**.

Aus der Existenz von $\int\limits_{K_P} \varphi^2(r)\, g^2\, dx$ und aus der Existenz von

$$\int\limits_{K_P} \sum \left(\frac{\partial}{\partial x} f\right)^2 dx$$

ergibt sich die Existenz von

$$\int\limits_{K_P} \varphi(r) \left| g \frac{\partial f}{\partial r} \right| dx = \int\limits_0^P \varphi(r) \left[r^{n-1} \int\limits_{\Omega_r} \left| g \frac{\partial f}{\partial r} \right| d\omega \right] dr.$$

Aus dieser Existenz und der Nichtexistenz von $\int\limits_0^P \varphi(r)\, dr$ entnimmt man, daß es eine Folge $\sigma \to 0$ gibt, für die

$$\sigma^{n-1} \int\limits_{\Omega_\sigma} \left| g \frac{\partial f}{\partial r} \right| d\omega \to 0$$

strebt wie gewünscht.

Fall **(3)**. Vgl. hierzu Courant [2. 5]. Aus der Existenz des Integrals

$$\int\limits_0^R r^{n-1} \int\limits_\Omega \sum \left(\frac{\partial}{\partial x} f\right)^2 d\omega\, dr$$

folgt die Existenz einer Folge $\varrho \to R$, so daß

$$(R - \varrho) \int\limits_{\Omega_\varrho} \sum \left(\frac{\partial}{\partial x} f\right)^2 d\omega \to 0.$$

Aus der Randbedingung (1*) S. 697

$$\text{(1*)} \qquad \int\limits_{\Omega_\varrho} g^2\, d\omega \leqq 2C(R - \varrho) \int\limits_{K_{\varrho, R}} \sum \left(\frac{\partial}{\partial x} g\right)^2 dx$$

andererseits folgt, daß $\frac{1}{R-\varrho} \int\limits_{\Omega_\varrho} g^2\, d\omega \to 0$ strebt.

Infolgedessen strebt

$$\int\limits_{\Omega_\varrho} \left| g \frac{\partial f}{\partial r} \right| d\omega \leqq \sqrt{\int\limits_{\Omega_\varrho} \sum \left(\frac{\partial f}{\partial x}\right)^2 d\omega \int\limits_{\Omega_\varrho} g^2\, d\omega} \underset{\varrho \to R}{\to} 0.$$

Fall **(4)**. Hier besagt schon die Randbedingung die geforderte Konvergenz

$$\text{(5)} \qquad \varrho^{n-1} \int\limits_{\Omega_\varrho} g \frac{\partial f}{\partial r}\, d\omega \to 0.$$

Zusatz. Im Falle **(1)** folgt für stückweise stetig zweimal differenzierbare Funktionen $f(x)$ aus $\mathfrak{H}''$, für die $-\Delta f(x) + v(x) f(x)$ eine Funktion aus $\mathfrak{H}'$ ist, schon, daß

$$\int_{\Gamma} \left\{ \sum \left(\frac{\partial}{\partial x} f(x)\right)^2 + v(x) f^2(x) \right\} dx$$

existiert; so daß also f in $\mathfrak{F}'$ liegt.

Denn es ist

$$\int_{K_\varrho} \left\{ \sum \left(\frac{\partial}{\partial x} f(x)\right)^2 + v(x) f^2(x) \right\} dx$$

$$= \int_{K_\varrho} f(x)(-\Delta f(x) + v(x) f(x))\, dx + \varrho^{n-1} \int_{\Omega_\varrho} f \frac{\partial}{\partial r} f\, d\omega.$$

Wegen der Existenz von $\int_0^\infty r^{n-1} \int_{\Omega_r} f^2\, d\omega\, dr$ muß es eine Folge $\varrho \to \infty$ geben, in deren Umgebung $\int_{\Omega_r} f^2\, d\omega$ abnimmt, so daß dort also

$$\frac{d}{dr} \int_{\Omega_r} f^2\, d\omega = 2 \int_{\Omega_r} f \frac{\partial}{\partial r} f\, d\omega\ .$$

negativ wird. Für diese Folge bleibt also die linke Seite beschränkt und da der Integrand positiv ist, existiert das Integral über $K_\infty = \Gamma$ erstreckt.

Es erweist sich hierin, daß bei **(1)** der Grenzpunktfall im Sinne von Weyl vorliegt.

(Eingegangen am 14. 7. 1933.)

Sonderabdruck
aus „Mathematische Annalen" ***110****, 777, 1935.*
Verlag von Julius Springer, Berlin W 9.

Berichtigung

zu der Arbeit von Kurt Friedrichs: „Spektraltheorie halbbeschränkter Operatoren I. und II. Teil", Math. Annalen **109**, S. 465–487 und 685–713.

Herr Wecken hat mich auf einige Fehler in der angegebenen Arbeit aufmerksam gemacht, die ich im folgenden berichtigen will.

In Teil I, S. 472, wurde behauptet, daß ein H-dichter Teilraum $\mathfrak{G}$ des Hilbertschen Raumes $\mathfrak{H}$, der mit einer positiv halbbeschränkten Form G als Maßform abgeschlossen ist, auch separabel und also ein Hilbertscher Raum ist mit G als Maßform. Der in Anmerkung [9]) gegebene Beweis ist nicht richtig. Man schließe folgendermaßen.

Nach Satz 6 (S. 477) gibt es einen beschränkten Operator B, dessen Wertbereich $\mathfrak{F}$ in $\mathfrak{G}$ liegt und für den eine Abschätzung $(Bh\ G\ Bh) \leqq \text{const.}\ (h\ H\ h)$ besteht. Er führt offenbar die abzählbare H-dichte Teilmenge $(\mathfrak{H})$ von $\mathfrak{H}$ in eine abzählbare G-dichte Teilmenge $(\mathfrak{F})$ von $\mathfrak{F}$ über. Beim Beweise von Satz 6 wurde in der Tat die G-Separabilität von $\mathfrak{G}$ nicht benutzt, wie Fr. Rellich bemerkt hat; ebenso unabhängig beweist man [1]), daß es zu jeder Teilmenge von $\mathfrak{G}$, die nicht G-dicht in $\mathfrak{G}$ liegt, ein Element $g_0 \neq 0$ gibt, das auf ihr orthogonal ist. Daraus folgt (vgl. Beweis von Satz 7), daß $\mathfrak{F}$ und also auch $(\mathfrak{F})$ G-dicht in $\mathfrak{G}$ liegt.

Teil II. (1). Das Zusatzpotential v soll bei den Dimensionen $n = 2$ und $n = 3$ noch stetig differenzierbar sein statt nur stetig (S. 688, 689).

Beim Beweise von Satz 1.3 B (S. 705) wird anschließend an Formel (13 B) in der Tat die stetige Differenzierbarkeit von v benutzt statt nur die von $f_2(x)$.

Im Falle (4) soll ferner (S. 688) vorausgesetzt werden, daß v in Γ nach oben beschränkt bleibt.

(2). Für die Tatsache, daß die Räume $\mathfrak{F}''$, $\mathfrak{F}'$, $\mathfrak{G}''$, $\mathfrak{G}'$ ineinander dicht liegen mit G als Maßform (S. 689—690) wurde in Fußnote [7]) ein Beweis angedeutet, der für den Fall (4) nicht stichhaltig ist. Er soll hier für alle Fälle näher ausgeführt werden.

Es ist also zu jeder Funktion $g(x)$ aus $\mathfrak{G}'$ eine Funktion $f(x)$ aus $\mathfrak{F}''$ anzugeben, so daß für die Differenz die Form G beliebig klein wird.

Man wird sich zu dem Zwecke auf die bekannte Tatsache berufen, daß jede stetige, (im Sinne von S. 687) stückweise stetig differenzierbare Funktion $g(x)$ in jedem Teilgebiet Γ_σ derart durch viermal stetig differenzierbare Funktionen $f(x)$ approximiert werden kann, daß

$$\int_{\Gamma_\sigma} \left\{ \sum \left(\frac{d}{dx}(f-g) \right)^2 + (f-g)^2 \right\} dx$$

beliebig klein wird. Ist $g(x)$ außerhalb eines Teilgebietes Γ_τ von Γ_σ $(\tau > \sigma)$ viermal stetig differenzierbar, so läßt sich f in Γ zweimal stetig differenzierbar so wählen, daß $f = g$ ist außer Γ_σ.

Nun genügt es, zur Funktion $g(x)$ eine Funktion $\tilde{g}(x)$ aus $\mathfrak{G}'$ mit beliebig kleinem $g - \tilde{g}\ G\ g - \tilde{g}$ zu konstruieren, die außerhalb eines geeigneten Teil-

[1]) Vgl. H. Löwig, Fr. Riesz, Acta Litt. Sci. Szeged 7 (1934); Fr. Rellich, Math. Annalen 110 (1934), S. 342.

gebietes Γ_τ in den Fällen (1), (2), (3) identisch verschwindet, im Falle (4) in radialer Richtung konstant und viermal stetig differenzierbar ist. Denn wählen wir zu dieser Funktion $\tilde{g}(x)$ nach Vorangehendem eine zweimal stetig differenzierbare Funktion $f(x)$ mit beliebig kleinem $\tilde{g} - f\, G\, \tilde{g} - f$, die außerhalb Γ_σ mit $g(x)$ übereinstimmt, so liegt $f(x)$ in $\mathfrak{F}''$ und es ist auch $g - f\, G\, g - f$ beliebig klein.

Zur Konstruktion der Funktion $\tilde{g}(x)$ im Falle (4) gehen wir aus von einer in $K_{R-\tau_0}$ mit $\tau_0 > 0$ erklärten viermal stetig differenzierbaren Funktion $g^*(x)$ und setzen sie über $K_{R-\tau}$ in K_R hinein konstant in radialer Richtung fort, wo τ aus $\tau_0 \leqq \tau \leqq 2\tau_0$ noch zu bestimmen ist; d. h. wir bilden $\tilde{g}(x) = g^*(x)$ in $K_{R-\tau}$, $= g^*\left(x\,\frac{R-\tau}{r}\right)$ für $R - \tau \leqq r < R$. Diese Funktion $\tilde{g}(x)$ liegt offenbar in $\mathfrak{G}'$; sie ist außerhalb Γ_τ viermal stetig differenzierbar und es ist $\frac{d\,g}{d\,r} = 0$ in $R - \tau < r < R$. Für die Differenz $g - \tilde{g}$ läßt sich die Form G abschätzen durch

$$g\text{-}\tilde{g}\,G\,g\text{-}\tilde{g} \leqq \int\limits_{K_{R-\tau_0}} \Phi(g-g^*)\,dx + 4R^{n-1}\tau \int\limits_{\Omega_{R-\tau}} \Phi(g^*-g)\,d\omega + 4R^{n-1}\tau \int\limits_{\Omega_{R-\tau}} \Phi(g)\,d\omega + 8\left(\frac{R}{R-2\tau_0}\right)^2 \int\limits_{K_{R-2\tau_0,\,R}} \Phi(g)\,dx,$$

wenn wir $\Phi(g) = \sum \left(\frac{d}{d\,x}\,g\right)^2 + v_0\,g^2$ setzen, wo v_0 die obere Schranke für v in Γ ist.

Nun kann man zunächst τ_0 und τ so wählen, daß die beiden letzten Terme, und g^* so bestimmen, daß die beiden ersten Glieder beliebig klein werden.

In den Fällen (1), (2), (3) haben wir zur Funktion $g(x)$ eine Funktion $\tilde{g}(x)$ aus $\mathfrak{G}'$ zu konstruieren, die außerhalb eines geeigneten Γ_τ verschwindet. Zu positiven σ und $\tau < \sigma$ wählen wir eine Funktion $s(r)$, die in Γ_σ verschwindet, außerhalb Γ_τ den Wert 1 hat. In den Fällen (3) und (1) steige sie im Zwischengebiet $\Gamma_\sigma - \Gamma_\tau$ in radialer Richtung linear von 0 zu 1 an; dann ist $\Delta\, s \geqq 0$ und die Ableitung s_r bleibt für $\tau \to 0$ bei festem σ beschränkt.

Nunmehr ersetze man die Funktion $g(x)$ durch die Funktion

$$\tilde{g}(x) = (1 - s(r))\,g(x),$$

die offenbar in $\mathfrak{G}'$ liegt und außerhalb Γ_τ verschwindet. Alsdann besteht die Identität

$$(*)\qquad g - \tilde{g}\,G\,g - \tilde{g} = \int\limits_{\Gamma-\Gamma_\sigma} \left\{ s^2 \sum \left(\frac{\partial}{\partial\,x}\,g\right)^2 + (s^2\,v - s\,\Delta\,s)\,g^2 \right\} d\,x + \int\limits_{\Omega_{(\tau)}} r^{n-1}\,s_r(r)\,g^2\,d\,\omega,$$

wenn $\Omega_{(\tau)}$ den Rand von Γ_τ bedeutet. Das Integral über $\Gamma - \Gamma_\sigma$, das wegen $0 \leqq s \leqq 1$ und $v \geqq 1$ nach oben abzuschätzen ist durch

$$(\ast\ast)\qquad \int\limits_{\Gamma-\Gamma_\sigma} \left\{ \left(\frac{d}{d\,x}\,g\right)^2 + v\,g^2 \right\} d\,x,$$

kann durch Wahl von σ beliebig klein gemacht werden; ebenso das Randglied durch Wahl von τ bei festem σ, und zwar im Falle (3) wegen der Randbedingung (1) S. 688 im Falle (1) wegen der Existenz des Integrals $\int\limits_\Gamma g^2\,d\,x$.

Im Falle (2) wähle man $s(r)$ ebenso wie bei (1) für $\frac{1}{\tau} \leqq r \leqq \frac{1}{\sigma}$ linear in r, in $\tau \leqq r \leqq \sigma$ dagegen

$$s(r) = \left(r^{n-1}\,\varphi(r)\right)^{-\frac{1}{2}} \int\limits_r^\sigma \varphi(r')\,d\,r' : \left(\tau^{n-1}\,\varphi(\tau)\right)^{-\frac{1}{2}} \int\limits_\tau^\sigma \varphi(r')\,d\,r',$$

wo $\varphi(r)$ die Bedeutung von S. 688, 689 hat. Dann ist, wie man nachrechnet, $-\Delta s = \varphi^2 s$ in $\tau \leqq r < \sigma$, und bei festem σ strebt $-s_r(\tau) : \varphi(\tau) \underset{\tau \to 0}{\to} 1$. Zur Identität (*) tritt noch das Glied $-\int\limits_{\Omega_\tau} r^{n-1} s_r(r)\, g^2\, d\,\omega$, das nach Anhang 2. 3 (S. 711) bei festem σ mit τ gegen Null strebt; zur Abschätzung (⁑) tritt wegen $s^2 v - s\Delta s = s^2(v + \varphi^2) \leqq v + \varphi^2$ [vgl. $(2)_\theta$ S. 689] noch das Glied $\int\limits_{K_{\tau,\sigma}} \varphi^2 g^2\, d\,x$, das nach Anhang 2. 3 ebenfalls mit σ beliebig klein wird.

(3). Die Abschätzung von $\int\limits_{|x-\xi_0| \leqq \varepsilon} (\Delta k_{\xi,\delta})^2\, d\,x$ (S. 703) ist unrichtig; sie werde durch die folgende ersetzt. Es gilt

$$\Delta k_{\xi,\delta} = \frac{1}{(2\delta)^n} \int\limits_{[\eta-\xi]<\delta} K(\eta - x)\, d\,\eta \leqq \frac{1}{(2\delta)^n} \int\limits_{|\eta-\xi|<\delta\sqrt{n}} K(\eta - x)\, d\,\eta \leqq C_n\, K(x - \xi)$$

mit $C_n = \left(\frac{\sqrt{n}}{2}\right)^n \int\limits_{|x| \leqq 1} d\,x$, wenn δ und ε so klein sind, daß $K(\eta - x) \geqq 0$ für $|x - \xi_0| \leqq \varepsilon$, $|\eta - \xi| \leqq \delta\sqrt{n}$ ist, auch für $n = 2$. Nun folgt in der Tat, daß

$$\int\limits_{|x-\xi_0| \leqq \varepsilon} (\Delta k_{\xi,\delta})^2\, d\,x \leqq C_n^2 \int\limits_{|x'-(\xi_0-\xi)| \leqq \varepsilon} (K(x'))^2\, d\,x' \leqq C_n^2 \int\limits_{|x'| \leqq \varepsilon + |\xi_0 - \xi|} (K(x'))^2\, d\,x',$$

zugleich mit ε und δ gegen Null strebt.

(Eingegangen am 19. 10. 1934.)

[34–2] **Beiträge zur Theorie der Spektralschar, Math. Ann. 110 (1934), 54–62.**

Commentary:

This is one of the rare abstract papers by Friedrichs, in which a new proof of the spectral theorem for unitary operators is given. Here he takes the view that the spectral family is primarily an interval function, rather than a point function as was the case with Hilbert and von Neumann. A plan is indicated to go further and introduce the spectral measure on Borel-measurable sets.

Correction:

Line 10 on Page 56 should read:

$$\begin{aligned}&= (\Sigma\,|\varphi(\alpha_\nu^2) - \varphi(\alpha_\mu^2)|^2\,|E\Delta_\nu^1\,\Delta_\mu^2\,\alpha h\,|^2)^{1/2}\\ &\leqq (\delta_1+\delta_2)\,(\Sigma\,|E\Delta_\nu^1\,\Delta_\mu^2\,\alpha h\,|^2)^{1/2}\\ &\leqq (\delta_1+\delta_2)\,|h|.\end{aligned}$$

Beiträge zur Theorie der Spektralschar.

I. Mitteilung.

Spektralschar auf Intervallen und Spektralzerlegung unitärer Operatoren.

Von

Kurt Friedrichs in Braunschweig.

In dieser Mitteilung soll gezeigt werden, daß es mancherlei Vorteile bietet, wenn man die Projektionsoperatoren einer Spektralschar von vornherein den Intervallen zugeordnet denkt. So kann man dann sehr bequem die stetige Funktion $\varphi(A)$ eines spektralzerlegten Operators A durch einen unmittelbaren stark konvergenten Grenzprozeß gewinnen. Ferner läßt sich das Prinzip, nach dem die Spektralschar von $\varphi(A)$ aus der von A zu gewinnen ist, besonders handlich formulieren.

Diese Vorteile sollen dann an der Spektraltheorie der unitären Operatoren geprüft werden: Es gelingt, die Spektralzerlegung der unitären Operatoren auf die der beschränkten zurückzuführen, ohne neuen Grenzübergang.

In einer späteren Mitteilung soll die Spektralschar auf alle B-meßbaren Mengen übertragen und der Operator $\varphi(A)$ für alle B-meßbaren Funktionen $\varphi(\alpha)$ definiert werden. Anschließend wird sich ein vollständiges System unitärer Invarianten einer Spektralschar besonders einfach erklären lassen. Ferner soll dort der Satz von der Eindeutigkeit der kanonisch Konjugierten neu bewiesen werden.

§ 1.

Spektralschar auf Intervallen.

Die Elemente des komplexen Hilbertschen Raumes heißen h; die Bilinearform

$$(h^1, h^2)$$

sei im hinteren Element h^2 linear; es sei

$$(h, h) = |h|^2.$$

Ein Projektionsoperator, kurz „Projektor", ist ein symmetrischer (d. h. Hermitescher) Operator E, für den $E^2 = E$ gilt; die Elemente Eh bilden seinen „Eigenraum".

Intervalle $\alpha_1 \underset{(=)}{<} \alpha \underset{(=)}{<} \alpha_2$ einer reellen α-Achse bezeichnen wir mit $\Delta\alpha = (\alpha_1^{\mp}, \alpha_2^{\pm})$, wobei das obere oder untere der Zeichen $+$ oder $-$

steht, wenn der betreffende Endpunkt hinzugehört oder nicht[1]. Insbesondere ist ein Punkt α auch ein Intervall (α^-, α^+).

Eine *Spektralschar* bestehe aus einer Schar von Projektoren $E\,\Delta\,\alpha$, die den Intervallen $\Delta\,\alpha$ so zugeordnet sind, daß sie die folgenden Eigenschaften besitzen.

1. Ist $\Delta^1\alpha\,\Delta^2\alpha = \Delta^1\,\Delta^2\alpha$ der Durchschnitt der Intervalle $\Delta^1\alpha$, $\Delta^2\alpha$, so ist

$$E\,\Delta^1\alpha\,E\,\Delta^2\alpha = E\,\Delta^1\,\Delta^2\alpha;$$

diese Eigenschaft ist gleichwertig mit

$$\begin{cases} E\,\Delta\,\alpha\,E\,\Delta\,\alpha = E\,\Delta\,\alpha, & \\ E\,\Delta^1\alpha\,E\,\Delta^2\alpha = 0, & \text{wenn } \Delta^1\alpha,\ \Delta^2\alpha \text{ fremd sind.}\end{cases}$$

2. Ist die endliche oder unendliche[2] Summe der fremden Intervalle $\Delta^1\alpha$, $\Delta^2\alpha$, ... wieder ein Intervall $\Delta\,\alpha$,

$$\Delta^1\alpha + \Delta^2\alpha + \ldots = \Delta\,\alpha,$$

so gilt[3]

$$E\,\Delta^1\alpha + E\,\Delta^2\alpha + \ldots = E\,\Delta\,\alpha.$$

3. Es gebe ein Intervall $(\underset{\sim}{\alpha}^-, \widehat{\alpha}^+)$ so daß wird

$$E\,(\underset{\sim}{\alpha}^-, \widehat{\alpha}^+) = 1.$$

Für alle Intervalle $\Delta\,\alpha$ außerhalb von $(\underset{\sim}{\alpha}^-, \widehat{\alpha}^+)$ muß dann $E\,\Delta\,\alpha = 0$ sein.

In dieser Fassung der Vollständigkeit ist zugleich gefordert, daß das Spektrum beschränkt ist, was für das Folgende ausreicht.

In den Darstellungen der Spektraltheorie werden im allgemeinen nur die Spektralprojektoren angegeben, die zu den Intervallen $(\underset{\sim}{\alpha}^-, \alpha^+)$ gehören: $E\alpha^+ = E\,(\underset{\sim}{\alpha}^-, \alpha^+)$, und die also von oben stetig sind; aus ihnen gewinnt man die Projektoren $E\alpha^- = E\,(\underset{\sim}{\alpha}^-, \alpha^-)$, die von unten stetig sind und in Stetigkeitspunkten mit $E\,\alpha^+$ übereinstimmen. Alsdann kann man die Projektoren beliebiger Intervalle gewinnen aus

$$E\,(\alpha_1^{\mp}, \alpha_2^{\pm}) = E\,\alpha_2^{\pm} - E\,\alpha_1^{\pm};$$

leicht überzeugt man sich dann von den obigen Eigenschaften.

Es sei nun $\varphi(\alpha)$ eine stückweise stetige beschränkte komplexwertige Funktion; es genügt, sie in $(\underset{\sim}{\alpha}^-, \widehat{\alpha}^+)$ zu definieren. Dann soll der Operator

$$\int \varphi(\alpha)\,E\,d\,\alpha$$

erklärt werden.

---

[1] α^+ und α^- sind an Stelle von $\alpha + 0$ und $\alpha - 0$ gesetzt.

[2] Dabei sollen in der unendlichen Reihe $\sum_{\nu=1}^{\infty} \Delta^\nu\alpha$ alle Partialsummen auch Intervalle sein.

[3] Die Summe der unendlichen Reihe $\sum_{\nu=1}^{\infty} E\,\Delta^\nu\alpha$ ist im Sinne starker Konvergenz zu verstehen.

Es seien $\Delta_1 \alpha, \Delta_2 \alpha, \ldots$ endlich viele fremde Intervalle, deren Summe $\sum_\nu \Delta_\nu \alpha = (\underline{\alpha}^-, \widehat{\alpha}^+)$ wird; α_ν sei eine Stelle in $\Delta_\nu \alpha$. Dann bilde man den Operator

$$\sum \varphi(\alpha_\nu) E \Delta_\nu \alpha.$$

Für eine Folge von Einteilungen, für die die Maximalschwankung δ von φ in den $\Delta_\nu \alpha$ gegen Null strebt, konvergiert diese Summe gegen einen Grenzoperator, unabhängig von der Wahl dieser Folge. Denn: Seien $\sum_\nu \Delta_\nu^1 \alpha$, $\sum_\mu \Delta_\mu^2 \alpha$ zwei solche Einteilungen mit den Maximalschwankungen δ_1, δ_2, so wird

$$\left|\left(\sum_\nu \varphi(\alpha_\nu^1) E \Delta_\nu^1 \alpha - \sum_\mu \varphi(\alpha_\mu^2) E \Delta_\mu^2 \alpha\right) h\right| = \left|\sum_{\nu,\mu} (\varphi(\alpha_\nu^1) - \varphi(\alpha_\mu^2)) E \Delta_\nu^1 \Delta_\mu^2 \alpha h\right|$$

$$\leqq \sum_{\nu,\mu} |\varphi(\alpha_\nu^1) - \varphi(\alpha_\mu^2)| \, |E \Delta_\nu^1 \Delta_\mu^2 \alpha h| \leqq (\delta_1 + \delta_2) \sum_{\nu,\mu} |E \Delta_\nu^1 \Delta_\mu^2 h|$$

$$= (\delta_1 + \delta_2) |h| \to 0 \qquad \text{für} \qquad \delta_1, \delta_2 \to 0.$$

Nach einem bekannten Satz von Fr. Riesz gibt es also einen beschränkten Grenzoperator, gegen den die Folge $\sum_\nu \varphi(\alpha_\nu) E \Delta_\nu \alpha$ bei $\delta \to 0$ konvergiert. Er heiße

$$\int \varphi(\alpha) E \, d\alpha.$$

Man zeigt unmittelbar

$$1. \quad c_1 \int \varphi_1(\alpha) E \, d\alpha + c_2 \int \varphi_2(\alpha) E \, d\alpha = \int (c_1 \varphi_1(\alpha) + c_2 \varphi_2(\alpha)) E \, d\alpha.$$

Ferner gilt

$$2. \quad \int \varphi(\alpha) E \, d\alpha \int \psi(\alpha) E \, d\alpha = \int \varphi(\alpha) \psi(\alpha) E \, d\alpha.$$

Diese Relation (Hilberts Vollständigkeitsrelation) läßt sich, bei der gegebenen Erklärung von $\int \varphi(\alpha) E \, d\alpha$ durch einen *stark* konvergenten Grenzprozeß, in besonders einfacher Weise ableiten. Sie folgt nämlich unmittelbar aus:

$$\sum_\nu \varphi(\alpha_\nu) E \Delta_\nu \alpha \sum_\nu \psi(\alpha_\nu) E \Delta_\nu \alpha = \sum_\nu \varphi(\alpha_\nu) \psi(\alpha_\nu) E \Delta_\nu \alpha,$$

wenn man beachtet, daß nach Fr. Riesz das Produkt von zwei stark konvergenten Operatoren stark gegen das Produkt der Grenzoperatoren strebt.

Insbesondere werde

$$A = \int \alpha E \, d\alpha$$

gesetzt. Dann rechtfertigen obige Relationen die Definition

$$\varphi(A) = \int \varphi(\alpha) E \, d\alpha;$$

denn es gilt:

1. Ist
$$c_1\,\varphi_1(\alpha)+c_2\,\varphi_2(\alpha) = (c_1\,\varphi_1+c_2\,\varphi_2)(\alpha),$$
so
$$c_1\,\varphi_1(A)+c_2\,\varphi_2(A) = (c_1\,\varphi_1+c_2\,\varphi_2)(A).$$
2. Ist
$$\varphi(\alpha)\,\psi(\alpha) = \varphi\,\psi(\alpha),$$
so
$$\varphi(A)\,\psi(A) = \varphi\,\psi(A).$$

Wir formulieren nun, weil wir im nächsten Paragraphen von ihm Gebrauch machen, den

Spektralsatz: Zu jedem beschränkten symmetrischen Operator A gibt es eine und nur eine Spektralschar $E\,\Delta\,\alpha$, so daß wird
$$A = \int \alpha\, E\, d\alpha.$$
Ferner den

Vertauschungssatz: Ist der beschränkte Operator B mit A vertauschbar, so auch mit dessen Spektralprojektoren $E\,\Delta\,\alpha$.

Ist $\varphi(\alpha)$ reell, so ist $\varphi(A)$ auch symmetrisch. Man kann leicht die Spektralschar von $\varphi(A)$ aus der von A gewinnen. Wir wollen uns hier damit begnügen, $\varphi(\alpha)$ als stückweise[4]) monoton und stetig vorauszusetzen, so daß die α-Menge, in der $\varphi(\alpha)$ in einem Intervall der φ-Achse liegt, eine endliche Summe von Intervallen ist. Wir wollen allgemein der Summe von Intervallen auch einen Projektionsoperator zuordnen, nämlich: die Summe der Projektoren der fremden Summandenintervalle. Es gilt dann das

Lemma: Sei $\varphi(\alpha)$ reell, stückweise monoton und stetig in $(\underset{\smile}{\alpha}^-, \widehat{\alpha}^+)$; sei jedem Intervall $\Delta\,\varphi$ der φ-Achse die endliche Summe fremder Intervalle $\sum\limits_{\varkappa} \Delta_\varkappa\,\alpha$ zugeordnet, in denen $\varphi(\alpha)$ monoton ist und Werte aus $\Delta\,\varphi$ annimmt.

Ist dann $E_A\,\Delta\,\alpha$ die Spektralschar des symmetrischen beschränkten Operators A, so ist
$$E_{\varphi(A)}\,\Delta\,\varphi = E\sum_{\varkappa}\Delta_\varkappa\,\alpha = \sum_{\varkappa} E\,\Delta_\varkappa\,\alpha$$
die Spektralschar des Operators $\varphi(A)$; dabei wähle man $\underset{\smile}{\varphi}, \widehat{\varphi}$ als untere und obere Grenze von $\varphi(\alpha)$ in $(\underset{\smile}{\alpha}^-, \widehat{\alpha}^+)$.

Beweis. Zunächst stellt man unmittelbar fest, daß die so definierten $E_{\varphi(A)}\,\Delta\,\varphi$ wieder eine Spektralschar bilden.

Sodann identifiziere man $\int \varphi\, E_{\varphi(A)}\, d\,\varphi$ mit $\int \varphi(\alpha)\, E_A\, d\,\alpha$. Sei zu dem Zweck $\sum\limits_{\nu} \Delta_\nu\,\varphi = (\underset{\smile}{\varphi}^-, \widehat{\varphi}^+)$ eine Intervalleinteilung der φ-Achse, $\sum\limits_{\nu,\varkappa} \Delta_{\nu\varkappa}\,\alpha$ die zugehörige Einteilung der α-Achse; $\alpha_{\nu\varkappa}$ liege in $\Delta_{\nu\varkappa}\alpha$, φ_ν in $\Delta_\nu\,\varphi$;

[4]) Es sei $(\underset{\smile}{\alpha}^-, \widehat{\alpha}^+)$ eine Summe von endlich vielen Intervallen, in denen $\varphi(\alpha)$ monoton und stetig ist.

δ sei die Maximalschwankung von $\varphi(\alpha)$ in den $\Delta_{\nu\varkappa}\alpha$, also nicht größer als die Maximallänge der $\Delta_\nu \varphi$. Dann ist

$$\sum_\nu \varphi_\nu E_{\varphi(A)} \Delta_\nu \varphi = \sum_{\nu,\varkappa} \varphi_\nu E_A \Delta_{\nu\varkappa}\alpha$$

und

$$\left|\left(\sum_\nu \varphi_\nu E_{\varphi(A)} \Delta_\nu \varphi - \sum_{\nu,\varkappa} \varphi(\alpha_{\nu\varkappa}) E_A \Delta_{\nu\varkappa}\alpha\right) h\right| = \left|\sum_{\nu,\varkappa} (\varphi_\nu - \varphi(\alpha_{\nu\varkappa})) E_A \Delta_{\nu\varkappa} h\right)\right|$$

$$\leqq \delta \sum_{\nu,\varkappa} |E_A \Delta_{\nu\varkappa}\alpha h| = \delta |h| \to 0 \quad \text{für} \quad \delta \to 0.$$

Daraus folgt für $\delta \to 0$

$$\int \varphi E_{\varphi(A)} d\varphi = \int \varphi(\alpha) E_A d\alpha = \varphi(A).$$

§ 2.

Unitäre Operatoren.

Ein Operator U heißt unitär, wenn er intakt und längentreu ist; d. h. wenn er beschränkt ist, eine beschränkte Reziproke besitzt und $|Uh| = |h|$ erfüllt.

Aus jedem symmetrischen beschränkten Operator B kann man in

$$U = e^{iB}$$

einen unitären Operator gewinnen; Wintner hat gezeigt, daß jeder unitäre Operator sich so darstellen läßt. Dabei ist B dann eindeutig bestimmt, wenn das Spektrum von B z. B. auf $(-\pi^+, \pi^+)$ eingeschränkt wird. Es ist also zu beweisen:

Satz 2.1[5]). Zu jedem unitären Operator U gibt es genau eine Spektralschar $E_B \Delta\beta$ mit $E_B(-\pi^+, \pi^+) = 1$, so daß wird

$$U = \int e^{i\beta} E_B d\beta = e^{iB},$$

$E_B \Delta\beta$ heiße dann auch die Spektralschar von U.

Nun läßt sich jeder unitäre Operator U darstellen als

$$U = V + iW,$$

wo V und W beschränkt, symmetrisch, miteinander vertauschbar sind und der Relation

$$V^2 + W^2 = 1$$

genügen; man setze nämlich $V = \frac{1}{2}(U + U^{-1})$, $W = \frac{1}{2i}(U - U^{-1})$ und beachte, daß $U^{-1} = U^*$ zu U adjungiert ist.

Wir wollen zeigen, wie sich die Spektralschar von U unmittelbar aus denen von V und W ergibt.

---

[5]) Vgl. Wintner, v. Neumann, Bochner.

Zunächst werde angenommen, U sei als $U = e^{iB}$ dargestellt. Dann wird

$$V = \cos B = \int \cos\beta\, E_B\, d\beta, \quad W = \sin B = \int \sin\beta\, E_B\, d\beta.$$

Die Spektralscharen $E_V\, \Delta\mu$, $E_W\, \Delta\nu$ von V und W lassen sich dann nach dem Lemma aus $E_B\, \Delta\beta$ berechnen.

Um uns bequem ausdrücken zu können, wollen wir vorbereitend einige Bezeichnungen einführen.

Zwei Werte β werden als gleich angesehen, wenn sie sich um 2π unterscheiden; damit fällt die Einschränkung $-\pi < \beta \leqq \pi$ fort. Jedem Intervall $\Delta\beta$, $\Delta\mu$, $\Delta\nu$ ordnen wir zu als $\Delta f(\beta)$, $\Delta f(\mu)$, $\Delta f(\nu)$ die Wertbereiche der Funktion f in $\Delta\beta$, $\Delta\mu$, $\Delta\nu$. Insbesondere sei stets $\Delta \cos\beta = \Delta\mu$, $\Delta \sin\beta = \Delta\nu$ gesetzt.

Nunmehr stellen wir auf:

Satz 2.2. Sei ein unitärer Operator dargestellt als

$$U = e^{iB} = \int e^{i\beta} E_B\, d\beta, \qquad E(-\pi^+, \pi^+) = 1.$$

Seien $E_V\, \Delta\mu$, $E_W\, \Delta\nu$ die Spektralscharen von

$$V = \cos B = \int \cos\beta\, E_B\, d\beta, \quad W = \sin B = \int \sin\beta\, E_B\, d\beta.$$

Sei $\Delta\beta$ ein solches Intervall, daß zwei Punkte aus ihm einen Abstand kleiner als π besitzen; seien $\Delta\mu = \Delta \cos\beta$, $\Delta\nu = \Delta \sin\beta$ die Wertebereiche von $\mu = \cos\beta$, $\nu = \sin\beta$ in ihm. Dann ist

$$E_B\, \Delta\beta = E_V\, \Delta\mu\, E_W\, \Delta\nu.$$

Beweis. Die β-Bereiche, in denen $\mu = \cos\beta$, $\nu = \sin\beta$ Werte aus $\Delta\mu$, $\Delta\nu$ annehmen, sind $\Delta\beta + \Delta(-\beta)$, $\Delta\beta + \Delta(\pi - \beta)$. Nach dem Lemma ist also

$$E_V\, \Delta\mu = E_B(\Delta\beta + \Delta - \beta),$$
$$E_W\, \Delta\nu = E_B(\Delta\beta + \Delta\,\pi - \beta).$$

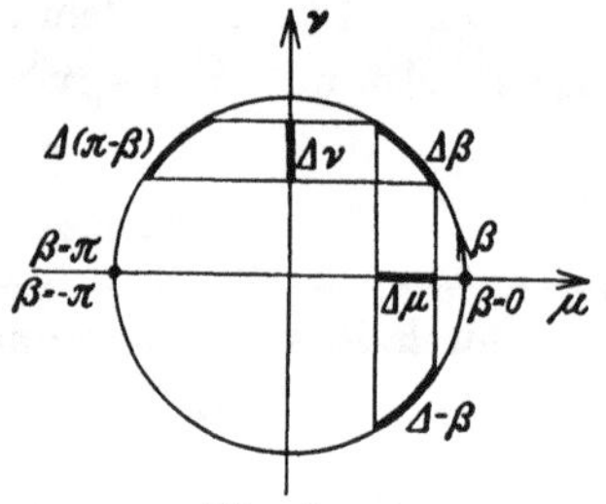

Fig. 1.

Nach der Annahme über $\Delta\beta$ ist der Durchschnitt von

$$\Delta\beta + \Delta - \beta \quad \text{und} \quad \Delta\beta + \Delta\,\pi - \beta$$

gerade $\Delta\beta$ selbst. Damit folgt die Behauptung.

Beweis von Satz 2.1.

Der Satz 2.2 legt nun nahe, wie die Spektralzerlegung eines beliebigen unitären Operators $U = V + iW$ auf den Spektralsatz für beschränkte symmetrische Operatoren zurückzuführen ist. Es sei also angenommen, daß zu dem Operator V, W Spektralscharen $E_V\, \Delta\mu$, $E_W\, \Delta\nu$ bekannt sind mit[6]) $E_V(-1^-, 1^+) = 1$, $E_W(-1^-, 1^+) = 1$, so daß wird

$$V = \int \mu\, E_V\, d\mu, \quad W = \int \nu\, E_W\, d\nu.$$

[6]) Die Schranken ± 1 folgen aus $V^2 + W^2 = 1$.

Wir führen nun die „Teilbereiche"

$$\Delta^+\beta = (0^+, \pi^-), \quad \Delta^-\beta = (-\pi^+, 0^-) \quad \text{und} \quad \Delta^0\beta$$

bestehend aus $\beta = 0$, $\beta = \pi$ ein. Liegt dann $\Delta\beta$ in einem Teilbereich, so setzen wir

$$E\,\Delta\beta = E_V\,\Delta\mu\,E_W\,\Delta\nu;$$

sonst zerlegen wir

$$\Delta\beta = \Delta^+\Delta\beta + \Delta^-\Delta\beta + \Delta^0\Delta\beta$$

und setzen

$$E\,\Delta\beta = E\,\Delta^+\Delta\beta + E\,\Delta^-\Delta\beta + E\,\Delta^0\Delta\beta.$$

Wir behaupten nun, daß die so definierten Projektoren $E\,\Delta\beta$ eine Spektralschar bilden, derart, daß wird

$$U = \int e^{i\beta}\,E\,d\beta.$$

1. Man beachte zunächst: Da V, W vertauschbar sind, sind es auch die $E_V\,\Delta\mu$, $E_W\,\Delta\nu$. (Vertauschungssatz).

2. Es gilt

$$E\,\Delta\beta\;E\,\Delta\beta = E\,\Delta\beta, \tag{1}$$

$$E\,\Delta_1\beta\;E\,\Delta_2\beta = 0, \qquad \text{wenn } \Delta_1\beta,\ \Delta_2\beta \text{ fremd sind.} \tag{2}$$

Denn: Liegen zunächst $\Delta\beta$, $\Delta_1\beta$, $\Delta_2\beta$ im selben Teilbereich, so folgt (1) unmittelbar, (2) daraus, daß mit $\Delta_1\beta$, $\Delta_2\beta$ auch $\Delta_1\mu$, $\Delta_2\mu$ fremd sind. Liegen $\Delta_1\beta$, $\Delta_2\beta$ in verschiedenen Teilbereichen, so folgt (2) daraus, daß $\Delta_1\nu$, $\Delta_2\nu$ fremd sind. Danach ergibt sich (1), (2) allgemein.

3. Bevor wir die Additivität von $E\,\Delta\beta$ nachprüfen, beweisen wir die Relation

$$E_V\,\Delta\mu = E_V\,\Delta\mu\;E_W(\Delta\nu + \Delta - \nu). \tag{3}$$

Es ist nämlich nach dem Lemma $E_V(\Delta\mu + \Delta - \mu)$ der Spektralprojektor von V^2 auf dem Intervall $\Delta\mu^2$, $E_W(\Delta\nu + \Delta - \nu)$ der Spektralprojektor von W^2 auf dem Intervall $\Delta\nu^2$ oder von $1 - W^2$ auf dem Intervall $\Delta(1 - \nu^2)$. Nun ist $1 - W^2 = V^2$ und $\Delta(1 - \nu^2) = \Delta(1 - \sin^2\beta)$ $= \Delta\cos^2\beta = \Delta\mu^2$; und da die Spektralschar eines Operators, hier V^2, eindeutig bestimmt ist, folgt

$$E_V(\Delta\mu + \Delta - \mu) = E_W(\Delta\nu + \Delta - \nu). \tag{4}$$

Durch Multiplikation mit $E_V\,\Delta\mu$ entsteht (3).

4. Additivität. Es genügt, die Additivität für Intervalle desselben Teilbereiches zu beweisen. Es seien $\Delta_1\beta$, $\Delta_2\beta$ zwei fremde Intervalle im selben Teilbereich; dann sind auch $\Delta_1\mu$ und $\Delta_2\mu$ fremd und es ist

$$(\Delta_1\nu + \Delta_2\nu)(\Delta_1\nu + \Delta_1 - \nu) = \Delta_1\nu,\ (\Delta_1\nu + \Delta_2\nu)(\Delta_2\nu + \Delta_2 - \nu) = \Delta_2\nu.$$

Es folgt dann

$$E(\Delta_1\beta+\Delta_2\beta)=E_V(\Delta_1\mu+\Delta_2\mu)\,E_W(\Delta_1\nu+\Delta_2\nu)$$
$$=E_V\,\Delta_1\mu\,E_W(\Delta_1\nu+\Delta_2\nu)+E_V\,\Delta_2\mu\,E_W(\Delta_1\nu+\Delta_2\nu)$$

und nach (3)

$$=E_V\,\Delta_1\mu\,E_W(\Delta_1\nu+\Delta_1-\nu)\,E_W(\Delta_1\nu+\Delta_2\nu)$$
$$+E_V\,\Delta_2\mu\,E_W(\Delta_2\nu+\Delta_2-\nu)\,E_W(\Delta_1\nu+\Delta_2\nu)$$
$$=E_V\,\Delta_1\mu\,E_W\,\Delta_1\nu+E_V\,\Delta_2\mu\,E_W\,\Delta_2\nu$$
$$=E\,\Delta_1\beta+E\,\Delta_2\beta.$$

Damit folgt die Additivität für endliche viele fremde $\Delta\beta$ aus demselben Teilbereich

$$\sum_\sigma E\,\Delta_\sigma\beta=E\sum_\sigma\Delta_\sigma\beta=E_V\sum_\sigma\Delta_\sigma\mu\,E_W\sum_\sigma\Delta_\sigma\nu=\Big(\sum_\sigma E_V\,\Delta_\sigma\mu\Big)\Big(\sum_\sigma E_W\,\Delta_\sigma\nu\Big).$$

Nach dem oben erwähnten Satz von Riesz folgt dann in der Grenze die Additivität für unendliche Summen.

5. Vollständigkeit. Es ist nach Definition

$$E(-\pi^+,\pi^+)=E_V\,\Delta^-\mu\,E_W\,\Delta^-\nu+E_V\,\Delta^+\mu\,E_W\,\Delta^+\nu+E_V\,\Delta^0\mu\,E_W\,\Delta^0\nu;$$

dabei ist z. B. $\Delta^+\mu=\Delta^+\cos\beta$ der Wertebereich von $\mu=\cos\beta$ in $\Delta^+\beta$. Es ist aber

$$\Delta^-\mu=\Delta^+\mu=(-1^+,1^-),\quad \Delta^0\mu=(-1^-,-1^+)+(1^-,1^+),$$
$$\Delta^-\nu=\Delta^+-\nu,\quad \Delta^0\nu=\Delta^0-\nu;$$

so entsteht nach (3), da $\Delta^+-\nu$ fremd zu $\Delta^+\nu$ ist,

$$E(-\pi^+,\pi^+)=E_V\Delta^+\mu+E_V\Delta^0\mu=E_V(-1^-,1^+)=1.$$

6. Mit dieser so gewonnenen Spektralschar $E\,\Delta\,\beta$ bilde man $B=\int\beta\,E\,d\beta$, $\cos B$ und $\sin B$. Dann ist zu beweisen

$$V=\cos B,\quad W=\sin B.$$

Zu dem Zweck genügt es zu zeigen, daß $E_V\,\Delta\,\mu$ und $E_W\,\Delta\,\nu$ die Spektralscharen von $\cos B$ und $\sin B$ sind. Es sei $\Delta\,\mu$ irgendein Intervall aus $-1\leqq\mu\leqq 1$; wir bilden den Wertebereich

$$\Delta^+\Delta\beta+\Delta^-\Delta\beta+\Delta^0\Delta\beta,$$

in dem $\mu=\cos\beta$ Werte aus $\Delta\,\mu$ annimmt, und beachten:

$$\Delta^+\Delta\mu=\Delta^-\Delta\mu,\quad \Delta^+\Delta\mu+\Delta^0\Delta\mu=\Delta\mu,$$
$$\Delta^-\Delta\nu=\Delta^+\Delta-\nu,\quad \Delta^0\Delta\nu=\Delta^0\Delta-\nu.$$

Dann wird der Spektralprojektor von $\cos B$ auf $\Delta\,\mu$ nach dem Lemma

$$E\,\Delta^+\Delta\beta+E\,\Delta^-\Delta\beta+E\,\Delta^0\Delta\beta$$
$$=E_V\,\Delta^+\Delta\mu\,E_W\,\Delta^+\Delta\nu+E_V\,\Delta^-\Delta\mu\,E_W\,\Delta^-\Delta\nu+E_V\,\Delta^0\Delta\mu\,E_W\,\Delta^0\Delta\nu$$
$$=E_V\,\Delta^+\Delta\mu\,E_W(\Delta^+\Delta\nu+\Delta^+\Delta-\nu)+E_V\,\Delta^0\Delta\mu\,E_W(\Delta^0\Delta\nu+\Delta^0\Delta-\nu)$$

und nach (3)

$$= E_V \varDelta^+ \varDelta \mu + E_V \varDelta^0 \varDelta \mu = E_V \varDelta \mu.$$

Ebenso zeigt man, indem man (4) benutzt, daß $E_W \varDelta \nu$ Spektralprojektor von $\sin B$ ist. Damit ist der Beweis geschlossen.

Literatur.

Bochner, S. Spektralzerlegung linearer Scharen unitärer Operatoren. S.-B. Pr. Akad. Wiss. 1933, S. 376.

v. Neumann, J. Allgemeine Eigenwerttheorie, Math. Annalen 102 (1929), III, S. 74ff.

Riesz, Fr. Über die linearen Transformationen des komplexen Hilbertschen Raumes. Acta Litt. Sci. Math. Szeged. 5 (1930), S. 19ff.

Stone, M. H. Linear Transformations in Hilbert Space, New York 1932, Chap. V, VI.

Wintner, A. Spektraltheorie unendlicher Matrizen, Leipzig 1929, § 82, 99.

(Eingegangen am 24. 10. 1933.)

[35–1] **Über die ausgezeichnete Randbedingung in der Spektraltheorie der halbbeschränkten gewöhnlichen Differentialoperatoren zweiter Ordnung, Math. Ann., 112 (1935), 1–23.**

Commentary:

Here Friedrichs deals with singular Sturm-Liouville operators

$$L = r^{-1}(-DpD + q), \qquad D = d/dx, \; X^- < x < X^+$$

where p, q, r are real-valued, continuous functions on (X^-, X^+) with $p, r > 0$. (In [35–1] it is assumed that $r = 1$. We include r for convenience in referencing paper [48–1]). According to the general theory due to Weyl and Stone, L has a spectral decomposition (i.e. L is self-adjoint) in the Hilbert space $L^2(X^-, X^+; rdx)$ if and only if it is restricted by certain boundary conditions, the number of which may be 2, 1, or 0 (Weyl's classification). Which case occurs depends on (X^-, X^+) and on p, q, r. In this paper Friedrichs constructs a distinguished self-adjoint operator, and describes its domain precisely, independently of which case of Weyl's classification occurs, under the assumption that the minimal operator for L (restriction of L to functions vanishing identically near the end points) is semibounded. The distinguished operator is exactly the Friedrichs extension of the minimal operator, and is characterized, among other possible self-adjoint realizations of L, by the condition that the associated quadratic form has the smallest domain. The results are extremely useful since the distinguished operator is the most important one in most applications (Legendre, Bessel, Laguerre, Schrödinger equations).

It may be remarked that in this paper Friedrichs does use the complete L^2-space of Lebesgue measurable functions, in contrast to [34–1], part 2.

Corrections:

Page 8 Satz 2.2 replace $g(x)$ by $g(x)^2$

Page 16, lines 10 and 11; the two formulas should read:

$$\varkappa(x) = \frac{1}{4y^2 p}, \quad \text{wenn } Y^+ = \infty$$

$$= \frac{1}{4(Y^+ - y)^2 p}, \quad \text{wenn } Y^+ < \infty \text{ ist.}$$

Über die ausgezeichnete Randbedingung in der Spektraltheorie der halbbeschränkten gewöhnlichen Differentialoperatoren zweiter Ordnung.

Von

Kurt Friedrichs in Braunschweig.

In der Spektraltheorie von Differentialoperatoren ist es eine Hauptaufgabe, für die Funktionen, auf die der Operator anwendbar sein soll, die zulässigen Randbedingungen richtig zu formulieren. Zulässig ist eine Randbedingung dann, wenn sie den Anwendungsbereich des Operators so einschränkt, daß er genau eine Spektralzerlegung besitzt.

Der Operator L sei auf Funktionen $f(x)$ anwendbar, die in einem Grundgebiet $X^- < x < X^+$ erklärt sind; wo auch $X^+ = \infty$ und $X^- = -\infty$ sein kann. Er werde durch zwei Koeffizienten $p(x) > 0$ und $q(x)$ gegeben und erzeuge aus $f(x)$ die Funktion

$$L f(x) = -\frac{d}{dx} p(x) \frac{d}{dx} f(x) + q(x) f(x).$$

Die entscheidenden Erkenntnisse über die zulässigen Randbedingungen verdankt man H. Weyl [1], [2], [3]. Weyl zeigt (allerdings nur für den Fall der Grundgebiete $0 < x < \infty$ und $-\infty < x < \infty$), daß man von den Funktionen $f(x)$ vorerst verlangen muß, daß

$$\int_{X^-}^{X^+} f^2(x)\, dx \quad \text{und} \quad \int_{X^-}^{X^+} (L f(x))^2\, dx$$

existieren. Über die zulässigen Randbedingungen gibt alsdann die Alternative Auskunft, die Weyl zunächst für den Fall formuliert, daß X^- regulärer Endpunkt ist, d. h. daß $X^- > -\infty$, p, q bei $x = X^-$ stetig, $p(X^-) \neq 0$ ist. Wird dann bei $x = X^-$ irgendeine Randbedingung aus der linearen Schar $a p \frac{df}{dx} + b f = 0$ gestellt, so können je nach dem Verlauf von p und q für $x \to X^+$ zwei Fälle eintreten.

I. Entweder: Der Operator L ist spektralzerlegbar, ohne daß eine zusätzliche Randbedingung bei $x = X^+$ zu stellen ist (Grenzpunktfall).

II. Oder: Es gibt eine lineare Schar von Randbedingungen für die Funktionen $f(x)$ bei $x = X^+$, von denen jede zulässig ist (Grenzkreisfall).

Bevor man die Spektralzerlegung des Operators L untersuchen kann, wird man also zu entscheiden haben, welcher Fall der Alternative vorliegt.

Bei den klassischen Differentialoperatoren von Legendre, Bessel, Laguerre, Schrödinger zeigt sich nun, daß beim „Grundproblem" Fall II, bei den „Oberproblemen" Fall I vorliegt. Beim Grundproblem ist also unter der Schar zulässiger Randbedingungen noch eine auszuzeichnen. Diese Tatsache legt es nahe zu fragen[1]), ob es nicht möglich ist, von vornherein eine Randbedingung anzugeben, die im Falle I von selbst erfüllt ist, im Falle II mit der ausgezeichneten zusammenfällt. Dann könnte man sich die Mühe sparen, die Alternative zu entscheiden.

Das ist in der Tat möglich, wenn der Operator nach unten **halbbeschränkt** ist. Wir werden in diesem Fall ohne Bezug auf die Alternative eine zulässige Randbedingung angeben, die, wie sich zeigt, im Falle II der Alternative gegen die anderen zulässigen Randbedingungen in mehrfacher Hinsicht ausgezeichnet ist. Diese Randbedingung lautet besonders einfach, wenn der Koeffizient $q(x)$ nach unten beschränkt ist[2]). Dann fordert sie nämlich, daß neben

$$\int_{X^-}^{X^+} f^2\,dx, \qquad \int_{X^-}^{X^+} (Lf)^2\,dx$$

noch das Integral

$$\int_{X^-}^{X^+} \left\{p\left(\frac{df}{dx}\right)^2 + q f^2\right\} dx$$

existiert; überdies soll noch $f(x)$ am Endpunkte X^+ bzw. X^- verschwinden, wenn $\int_{x_0}^{x} \frac{dx}{p(x)}$ bei $x \to X^+$ bzw. $x \to X^-$ endlich bleibt ($X^- < x_0 < X^+$). In dieser Fassung läßt sich in der Tat auch für die genannten klassischen Differentialoperatoren, da bei ihnen q nach unten beschränkt ist[2]), das Spektralproblem in einheitlicher Weise richtig ansetzen.

Weyl stützt sich bei seiner Untersuchung auf die Hilbertsche Theorie der beschränkten Operatoren und muß zunächst, da L nicht beschränkt ist, explicite den beschränkten Operator $\frac{1}{L+i}$ konstruieren. Dagegen hat M. H. Stone [1] Ch. X den Differentialoperator unmittelbar der Theorie der nicht beschränkten Operatoren eingeordnet, die v. Neumann [1] und er entwickelt haben; zugleich hat er die Alternative von Weyl auf neue und allgemeinere Weise formuliert und bewiesen.

---

1) Diesen Hinweis verdanke ich Fr. Hund.

2) Oder am Rande nicht zu stark negativ unendlich wird. Vgl. § 5.

Hier schließe ich mich zunächst (§ 1) dem Ansatz von Stone an und untersuche sodann (§§ 2, 3, 5), ohne Bezug auf die Alternative, die ausgezeichnete Randbedingung. Um sie als zulässig zu erweisen (§ 4) und ihre ausgezeichnete Rolle zu klären (§ 6), berufe ich mich auf Begriffe und Sätze über halbbeschränkte Operatoren aus dem 1. Teil einer früheren Arbeit [1]. Wesentlich stütze ich mich noch auf eine Reihe von Integralungleichungen, die im Anhang zusammengestellt sind.

Die Schlußweise, die zur Theorie der ausgezeichneten Randbedingung herangezogen wird, macht keinen wesentlichen Gebrauch davon, daß die Funktionen nur von einer Veränderlichen abhängen; sie ließe sich weitgehend auf partielle Differentialoperatoren übertragen. Für einige Sonderfälle wurde das in Teil II der genannten Arbeit [2] [3] durchgeführt; ihr gegenüber besitzt die vorliegende Arbeit noch einen vereinfachten Aufbau der Beweisführung.

Literatur.

Carleman, T.

[1] Sur les équations intégrales singulières à noyau réel et symétrique (1923).

Friedrichs, K.

Spektraltheorie halbbeschränkter Operatoren und Anwendung auf die Spektralzerlegung von Differentialoperatoren. Math. Annalen (1934).

[1] Teil I 109, S. 466. — [2] Teil II 109, S. 688. — [3] Berichtigung 110, S. 777.

v. Neumann

[1] Allgemeine Eigenwerttheorie Hermitescher Funktionaloperatoren. Math. Annalen 102 (1929), S. 50.

Stone, M. H.

[1] Linear Transformations in Hilbert Space. Am. Math. Soc. Coll. Publ. Vol. XV (1932).

Weyl, H.

Über gewöhnliche lineare Differentialgleichungen mit singulären Stellen und ihre Eigenfunktionen. Gött. Nachr.

[1] 1. Note (1909), S. 37. — [2] 2. Note (1910), S. 442.

[3] Über gewöhnliche Differentialgleichungen mit Singularitäten und die zugehörigen Entwicklungen willkürlicher Funktionen. Math. Annalen 68 (1910), S. 220.

§ 1.

Der Funktionenraum.

Die auftretenden Funktionen seien stets in einem Grundgebiet $X^- < x < X^+$ erklärt; x_0 sei eine feste Stelle im Innern.

Mit $h(x)$ bezeichnen wir Funktionen, die über jedes innere Intervall[3]) quadratisch nach x L-integrierbar sind; mit $[h]$ ihre Gesamtheit. Mit $\mathfrak{H}$ bezeichnen wir den Raum der über das gesamte Grundgebiet

[3]) Das Intervall (x^-, x^+) heiße ein inneres, wenn $x^- > X^-$, $x^+ < X^+$ ist.

quadratisch L-integrierbaren Funktionen h; bekanntlich bildet $\mathfrak{H}$ einen reellen Hilbertschen Raum (v. Neumann [1]) mit dem inneren Produkt

$$(h_1, h_2) = \int_{X^-}^{X^+} h_1 h_2 \, dx.$$

Es sei $p(x)$ eine positive stetige Funktion; wir verwenden die Abkürzung

$$y = \int_{x_0}^{x} \frac{dx'}{p(x')};$$

Werte von y an gekennzeichneten Stellen x kennzeichnen wir entsprechend; z. B.

$$Y^+ = \int_{x_0}^{X^+} \frac{dx}{p}, \qquad Y^- = \int_{x_0}^{X^-} \frac{dx}{p}.$$

Mit $g(x)$ bezeichnen wir stetige Funktionen von x, die eine Derivierte $\frac{dg}{dx}$ besitzen, für welche über jedes innere Intervall (x^-, x^+)

$$\int_{x^-}^{x^+} p \left(\frac{dg}{dx}\right)^2 dx = \int_{y^-}^{y^+} \left(\frac{dg}{dy}\right)^2 dy$$

existiert; die Gesamtheit dieser $g(x)$ heiße $[g]$.

Mit $f(x)$ bezeichnen wir die mit stetigen ersten Ableitungen versehenen Funktionen, die eine zweite Derivierte $\frac{d}{dx} p \frac{df}{dx} = \frac{d}{dx} \frac{df}{dy}$ besitzen, die über innere Intervalle quadratisch L-integrierbar ist, also zu $[h]$ gehört; ihre Gesamtheit heiße $[f]$.

Es sei $q(x)$ eine stetige Funktion. Dann definieren wir den Operator L durch

$$L = -\frac{d}{dx} p \frac{d}{dx} + q = -\frac{d}{dx} \frac{d}{dy} + q.$$

Er erzeugt aus der Funktion f von $[f]$ die Funktion

$$Lf = -\frac{d}{dx} p \frac{df}{dx} + qf = -\frac{d}{dx} \frac{df}{dy} + qf$$

aus $[h]$.

Es gilt mit innerem Intervall (x^-, x^+) für f aus $[f]$, g aus $[g]$:

$$\int_{x^-}^{x^+} g L f \, dx = \int_{y^-}^{y^+} \frac{dg}{dy} \frac{df}{dy} dy - g \frac{df}{dy}\Big|_{x^-}^{x^+}, \tag{1.1}$$

und für f_1, f_2 aus $[f]$:

$$\int_{x^-}^{x^+} \{f_1 L f_2 - L f_1 \cdot f_2\} \, dx = f_1 \frac{df_2}{dy} - \frac{df_1}{dy} f_2 \Big|_{x^-}^{x^+}. \tag{1.2}$$

Mit $\mathring{\mathfrak{F}}$ bezeichnen wir den Raum aller Funktionen f aus $[f]$, die in $\mathfrak{H}$ liegen und aus denen der Operator L eine Funktion aus $\mathfrak{H}$ erzeugt; für f aus $\mathring{\mathfrak{F}}$ gilt also

$$\Omega \qquad \int_{X^-}^{X^+} f^2\,dx < \infty \quad \text{und} \quad \int_{X^-}^{X^+} (Lf)^2\,dx < \infty.$$

Liegen f_1, f_2 in $\mathring{\mathfrak{F}}$, so existiert in (1.2) der Grenzwert der linken, also auch der rechten Seite für $x^+ \to X^+$ und für $x^- \to X^-$. Die Grenzwerte von $f_1 \left.\frac{d f_2}{d y}\right|^x$ für $x \to X^+$ bzw. X^- und ihre Differenz bezeichnen wir mit $f_1 \left.\frac{d f_2}{d y}\right|^{X^+}$ bzw. $f_1 \left.\frac{d f_2}{d y}\right|^{X^-}$ und $f_1 \left.\frac{d f_2}{d y}\right|_{X^-}^{X^+}$.

Die Aufgabe ist nun, aus $\mathring{\mathfrak{F}}$ solche Teilräume $\mathfrak{F}$ auszuwählen, in denen der Operator L selbstadjungiert (hypermaximal)[4]) ist; denn dann ist er nach der Theorie von v. Neumann [1] und Stone [1] Ch. V. spektralzerlegbar. Der Operator L heißt in einem Raume $\mathfrak{F}$ **selbstadjungiert** (der Deutlichkeit halber auch „überhaupt selbstadjungiert"), wenn er die folgenden zwei Eigenschaften besitzt.

1. Für alle f_1, f_2 aus $\mathfrak{F}$ gilt

$$(1.4) \qquad \int_{X^-}^{X^+} \{f_1 L f_2 - L f_1 \cdot f_2\}\,dx = 0. \qquad \text{(Symmetrie).}$$

1*. Besteht für ein Paar h_0, h_1 aus $\mathfrak{H}$ die Relation

$$(1.4)^* \qquad \int_{X^-}^{X^+} \{f h_1 - Lf \cdot h_0\}\,dx = 0$$

mit allen f aus $\mathfrak{F}$, so liegt h_0 in $\mathfrak{F}$ und es ist $L h_0 = h_1$.

Es empfiehlt sich, diese Forderungen zu zerspalten in solche, die sich auf das Innere und solche, die sich auf den Rand des Grundgebietes beziehen.

Wir führen den Raum $\dot{\mathfrak{F}}$ aller Funktionen f ein, zu denen es ein inneres Intervall gibt, außerhalb dessen sie identisch verschwinden[5]). Dann gilt

Satz 1.1. *Die Relation* (1.4) *ist für* f_1, f_2 *aus* $\dot{\mathfrak{F}}$ *erfüllt.*

Satz 1.1*. *Besteht für ein Paar* h_0, h_1 *aus* $\mathfrak{H}$ *Relation* (1.4)* *mit allen* f *aus* $\dot{\mathfrak{F}}$, *so liegt* h_0 *in* $\mathring{\mathfrak{F}}$ *und es ist* $L h_0 = h_1$.

Die beiden Eigenschaften, die in diesen beiden Sätzen behauptet sind, wollen wir auch mit „selbstadjungiert im Innern" bezeichnen. Satz 1.1 folgt sofort aus (1.2); zum Beweise von Satz 1.1* vgl. Stone[6]).

[4]) Vgl. v. Neumann [1], Stone [1] Def. 2.11, p. 50.

[5]) Stone führt nicht $\dot{\mathfrak{F}}$, sondern den $\dot{\mathfrak{F}}$ enthaltenden Raum $\mathring{\mathfrak{F}}^*$ aller f aus $\mathring{\mathfrak{F}}$ ein, für die (1.5)* mit allen f_0 aus $\mathring{\mathfrak{F}}$ gilt. [1] Th. 10.11, p. 458.

[6]) [1] Th. 10.11, p. 458. Stone beweist mehr: Der Operator L in $\mathring{\mathfrak{F}}$ ist adjungiert zu L in $\mathring{\mathfrak{F}}^*$. Vgl. [5]).

Damit der Operator L in einem Raume $\mathfrak{F}$, der $\dot{\mathfrak{F}}$ enthalte, überhaupt selbstadjungiert ist, muß er noch die folgenden beiden Eigenschaften besitzen, die wir auch mit „selbstadjungiert am Rande" bezeichnen.

2. Für f_1, f_2 aus $\mathfrak{F}$ ist

(1.5) $$f_1 \frac{d f_2}{d y} - \frac{d f_1}{d y} f_2 \Big|_{X-}^{X+} = 0.$$ („Symmetrie am Rande").

2*. Gilt für ein f_0 aus $\mathring{\mathfrak{F}}$ Relation

(1.5)* $$f \frac{d f_0}{d y} - \frac{d f}{d y} f_0 \Big|_{X-}^{X+} = 0$$

mit allen f aus $\mathfrak{F}$, so liegt f_0 in $\mathfrak{F}$.

Man prüft sofort nach

Satz 1.2. *Ist der Operator L in $\mathfrak{F}$ selbstadjungiert am Rande, so überhaupt.*

Denn aus (1.5) und (1.2) folgt (1.4). Aus (1.4)* folgt nach Satz 1.1* zunächst, daß h_0 in $\mathring{\mathfrak{F}}$ liegt mit $L h_0 = h_1$; aus (1.2) folgt dann (1.5)* für $f_0 = h_0$, so daß also h_0 auch in $\mathfrak{F}$ liegt.

Von dem Raume $\mathfrak{F}$, den wir im folgenden angeben, werden wir zunächst nachweisen, daß in ihm der Operator L symmetrisch ist; (d. h., daß 2. und also 1. gilt). Sodann werden wir unter Benutzung von Satz 1.1* die Eigenschaft 1* und also die Selbstadjungiertheit von L in $\mathfrak{F}$ unmittelbar nachprüfen können[6a]).

§ 2.

Der Raum $\mathfrak{G}$.

Die ausgezeichnete Randbedingung werden wir zunächst explicite angeben unter der Voraussetzung, daß der Koeffizient q vermöge

(2.0) $$q(x) \geqq \underline{q}$$

durch eine Konstante $\underline{q}$ nach unten beschränkt ist. Wir führen dann vorbereitend einen Raum $\mathfrak{G}$ von Funktionen g ein und stellen seine Eigenschaften zusammen. Es ist bequem, die Eigenschaften, die $\mathfrak{G}$ kennzeichnen, aufzuspalten in solche, die sich auf je einen Endpunkt beziehen. Wir nennen $\mathfrak{G}^+$ die Mannigfaltigkeit der Funktionen $g(x)$ aus $[g]$, die den folgenden beiden Bedingungen genügen.

A$^+$: Es existieren die Integrale

$$\int_{x_0}^{X+} p \left(\frac{d g}{d x}\right)^2 d x = \int_0^{Y+} \left(\frac{d g}{d y}\right)^2 d y, \quad \int_{x_0}^{X+} q g^2 d x, \quad \int_{x_0}^{X+} g^2 d x.$$

6a) Satz 1.2 und Eigenschaft 2* werden also im folgenden nicht ausdrücklich herangezogen; sie dienen hier nur der Klärung des Begriffes selbstadjungiert.

B⁺: Wenn $\int\limits_{x_0}^{X^+} \frac{dx}{p(x)} = Y^+ < \infty$ ist, so strebt

$$g(x) \to 0$$

wenigstens auf einer speziellen Folge $x \to X^+$, was wir hier und analog im folgenden durch

$$g(x) \underset{x \dot{\to} X^+}{\to} 0$$

ausdrücken wollen.

Aus Anhang Satz A 1 entnimmt man, daß dann auch $g(x) \to 0$ strebt für jede Folge $x \to X^+$. Man überzeugt sich alsdann davon, daß $\mathfrak{G}^+$ ein linearer Raum ist.

Entsprechend definieren wir den linearen Raum $\mathfrak{G}^-$ von Funktionen $g(x)$ durch

A⁻: $$\int\limits_{Y^-}^{x_0} \left(\frac{dg}{dy}\right)^2 dy < \infty, \quad \int\limits_{X^-}^{x_0} q g^2 dx < \infty, \quad \int\limits_{X^-}^{x_0} g^2 dx < \infty,$$

B⁻: $$g(x) \underset{x \dot{\to} X^-}{\to} 0, \quad \text{wenn} \quad Y^- > -\infty \text{ ist.}$$

Unter $\mathfrak{G}$ wird dann der Raum aller $g(x)$ aus $[g]$ verstanden, die den Forderungen A⁺, B⁺, A⁻, B⁻ genügen, d. h. der Durchschnitt

$$\mathfrak{G} = \mathfrak{G}^+ \mathfrak{G}^-;$$

$\mathfrak{G}$ liegt offenbar in $\mathfrak{H}$.

Wir wählen eine Zahl $k \geqq 1 - q$, die dann festgehalten wird. Für Funktionen g_1, g_2 aus $\mathfrak{G}$ ist die Form

$$g_1 G g_2 = \int\limits_{Y^-}^{Y^+} \frac{dg_1}{dy} \frac{dg_2}{dy} dy + \int\limits_{X^-}^{X^+} (q+k) g_1 g_2 dx \tag{2.1}$$

erklärt, und es gilt für g aus $\mathfrak{G}$

$$g G g \geqq \int\limits_{Y^-}^{Y^+} \left(\frac{dg}{dy}\right)^2 dy + \int\limits_{X^-}^{X^+} g^2 dx. \tag{2.2}$$

Es seien zunächst zwei Eigenschaften des Raumes $\mathfrak{G}^+$ angegeben, die sich auf $\mathfrak{G}^-$ analog übertragen. Ihre Beweise finden sich im Anhang: Sätze A 1. 2. 3.

Satz 2.1. *Im Raume $\mathfrak{G}^+$ existieren die Integrale*

$$\begin{aligned} &\int\limits_{0}^{Y^+} \frac{g^2}{(Y^+ - y)^2} dy, && \textit{wenn} \quad Y^+ < \infty, \\ &\int\limits_{y_0'}^{Y^+} \frac{g^2}{y^2} dy, && \textit{wenn} \quad Y^+ = \infty \textit{ ist mit } y_0' > 0. \end{aligned} \tag{2.3}$$

Satz 2.2. *Im Raume $\mathfrak{G}^+$ gilt für $x \to X^+$*

(2.4)
$$\frac{g(x)}{Y^+ - y} \to 0, \quad \textit{wenn} \quad Y^+ < \infty,$$
$$\frac{g(x)}{y} \to 0, \quad \textit{wenn} \quad Y^+ = \infty \ \textit{ist}.$$

Ferner besteht

Satz 2.3. *Der Raum $\dot{\mathfrak{G}}$ aller Funktionen g aus $[g]$, zu denen es innere Intervalle gibt, außerhalb deren sie verschwinden, liegt G-dicht in $\mathfrak{G}$. D. h. zu jedem g aus $\mathfrak{G}$ gibt es ein $\dot g$ aus $\dot{\mathfrak{G}}$, so daß*

(2.5)
$$(g - \dot g)\, G\, (g - \dot g)$$

beliebig klein ist.

Beweis[7]: Mit Werten $x_2 > x_1 > x_0$, $x_{-2} < x_{-1} < x_0$ setzen wir

$$\begin{aligned} \eta(x) &= 0 && \text{in} \quad x_{-1} \leqq x \leqq x_1, \\ &= \frac{y - y_1}{y_2 - y_1} && \text{in} \quad x_1 \leqq x \leqq x_2, \\ &= \frac{y - y_{-1}}{y_{-2} - y_{-1}} && \text{in} \quad x_{-2} \leqq x \leqq x_{-1}, \\ &= 1 && \text{in} \quad x_2 \leqq x < X^+ \quad \text{und} \quad X^- < x \leqq x_{-2}. \end{aligned}$$

Dann genügt

$$\dot g = (1 - \eta)\, g$$

den Forderungen des Satzes. Es ist nämlich, wie man nachrechnet,

$$\begin{aligned} &\int_0^{Y^+} \left(\frac{d\,\eta g}{d\,y}\right)^2 d\,y + \int_{x_0}^{X^+} (q + k)\,\eta^2 g^2\, d\,x \\ &= \int_{y_1}^{y_2} \left(\frac{d\,\eta g}{d\,y}\right)^2 d\,y + \int_{y_2}^{Y^+} \left(\frac{d\,g}{d\,y}\right)^2 d\,y + \int_{x_1}^{X^+} (q + k)\,\eta^2 g^2\, d\,x \\ &= \frac{g^2(x_2)}{y_2 - y_1} + \int_{y_1}^{y_2} \eta^2 \left(\frac{d\,g}{d\,y}\right)^2 d\,y + \int_{y_2}^{Y^+} \left(\frac{d\,g}{d\,y}\right)^2 d\,y + \int_{x_1}^{X^+} (q + k)\,\eta^2 g^2\, d\,x \\ &\leqq \frac{g^2(x_2)}{y_2 - y_1} + \int_{y_1}^{Y^+} \left(\frac{d\,g}{d\,y}\right)^2 d\,y + \int_{x_1}^{X^+} (q + k)\, g^2\, d\,x. \end{aligned}$$

Man bestimme nun zunächst x_1 so nahe an X^+, daß die beiden Integrale rechts beliebig klein werden. Sodann kann man x_2 so wählen, daß $\frac{g^2(x_2)}{y_2 - y_1}$ beliebig klein wird; ist $Y^+ < \infty$, so lasse man zu dem Zwecke $x_2 \to X^+$ die spezielle Folge durchlaufen, auf der $g(x_2) \to 0$ strebt; ist

[7]) Vgl. Friedrichs [3].

$Y^+ = \infty$, so stütze man sich auf Satz 2. 2. Also läßt sich auch die linke Seite beliebig klein machen; ebenso macht man

$$\int_{Y^-}^{0} \left(\frac{d\,\eta\, g}{d\,y}\right)^2 d\,y + \int_{X^-}^{x_0} (q+k)\,\eta^2 g^2\, d\,x$$

beliebig klein. Damit ist Satz 2. 3 bewiesen.

Satz 2. 4. *Der Raum* $\mathfrak{G}$ *ist G-abgeschlossen. D. h. zu einer Folge* $g^\nu(x)$ *von Funktionen aus* $\mathfrak{G}$ *für die*

(2. 6) $$(g^\nu - g^\mu)\, G\, (g^\nu - g^\mu) \underset{\nu,\,\mu\to\infty}{\to} 0$$

strebt, gibt es eine Funktion $g_0(x)$ *aus* $\mathfrak{G}$, *so daß gilt*

(2. 7) $$(g^\nu - g_0)\, G\, (g^\nu - g_0) \underset{\nu\to\infty}{\to} 0 \,.$$

Beweis. Wegen der Konvergenz von $\int_{Y^-}^{Y^-} \left(\frac{d\,(g^\nu - g^\mu)}{d\,y}\right)^2 d\,y \underset{\nu,\,\mu\to\infty}{\to} 0$ gibt es, dem Satze von Fischer gemäß, eine Funktion $g_0(x)$, für die $\int_{Y^-}^{Y^-}\left(\frac{d\,g_0}{d\,y}\right)^2 d\,y$ existiert, so daß $\int_{Y^-}^{Y^+}\left(\frac{d\,(g^\nu - g_0)}{d\,y}\right)^2 d\,y \to 0$ strebt; und zwar kann $g_0(x)$ so gewählt werden, daß $g(x)$ in jedem Teilgebiet gleichmäßig gegen $g_0(x)$ konvergiert, wie man aus der Abschätzung

$$|g(x_1) - g(x_2)|^2 \leqq (y_2 - y_1) \int_{y_1}^{y_2} \left(\frac{d\,g}{d\,y}\right)^2 d\,y,$$

auf $g = g^\nu - g^\mu$ angewandt, erschließt. Aus der Konvergenz von $\int_{X^-}^{X^+} (q+k)\,(g^\nu - g^\mu)^2\, d\,x \underset{\nu,\,\mu\to\infty}{\to} 0$ folgt, daß $\int_{X^-}^{X^+} q\, g_0^2\, d\,x$ und $\int_{X^-}^{X^+} g_0^2\, d\,x$ existieren (beachte $q + k \geqq 1$), daß $\int_{X^-}^{X^+} q\,(g^\nu - g_0)^2\, d\,x \underset{\nu\to\infty}{\to} 0$, $\int_{X^-}^{X^+} (g^\nu - g_0)^2\, d\,x \underset{\nu\to\infty}{\to} 0$ streben und daß also (2. 7) besteht. Ist $Y^+ < \infty$, so beachte man, daß die Relation

$$\frac{g^2(x^+)}{Y^+ - y^+} \leqq 2 \int_{y^+}^{Y^+} \left(\frac{d\,g}{d\,y}\right)^2 d\,y, \quad 0 \leqq y^+ < Y^+, \quad Y^+ < \infty,$$

die nach Anhang Satz 1 für alle g aus $\mathfrak{G}$ und also für die g^ν besteht, sich durch Grenzübergang auf $g_0(x)$ überträgt; somit folgt $\frac{g_0(x)}{Y^+ - y} \to 0$, also erst recht $g_0(x) \to 0$ für $x \to X^+$. Ebenso zeigt man $g_0(x) \underset{x \to X^-}{\to} 0$, wenn $Y^+ > -\infty$ ist. Es liegt also $g_0(x)$ in $\mathfrak{G}$. Damit ist Satz 2.4 bewiesen.

§ 3.

Symmetrie im Raume $\mathring{\mathfrak{F}}\mathfrak{G}$ und Folgerungen.

Die ausgezeichnete Randbedingung besteht nun in der Forderung, daß die Funktion f aus $\mathring{\mathfrak{F}}$ auch noch im Raume $\mathfrak{G}$ liegen soll; wir werden in § 4 zeigen, daß der Operator L im Raume aller solchen Funktionen f, also im Durchschnitt

$$\mathfrak{F} = \mathring{\mathfrak{F}}\mathfrak{G}$$

selbstadjungiert ist. In diesem § 3 beschäftigen wir uns mit der Symmetrie von L in $\mathfrak{F}$ und ziehen aus ihr einige Konsequenzen. Wir zeigen zunächst, daß die Relation

$$(3.0)^+ \quad \int_{x^+}^{X^+} g\,L f\,dx - g\,\frac{df}{dx}\Big|^{X^+} = \int_{y^+}^{Y^+} \frac{dg}{dy}\frac{df}{dy}\,dy + \int_{x^+}^{X^+} q\,g\,f\,dx,\quad x_0 \leqq x^+ < X^+,$$

für alle Funktionen f aus $\mathring{\mathfrak{F}}\mathfrak{G}^+$ und g aus $\mathfrak{G}^+$ besteht. Zu dem Zwecke stützen wir uns auf den

Hilfssatz: Für ein Paar von Funktionen g aus $\mathfrak{G}^+$, f aus $\mathring{\mathfrak{F}}\mathfrak{G}^+$ gilt

$$g\,\frac{df}{dy}\Big|^{x} \underset{x \to X^+}{\to} 0.$$

Beweis: Wir setzen

$$\mu(x) = \frac{1}{y} \qquad \text{für } Y^+ = \infty$$

$$\mu(x) = \frac{1}{Y^+ - y} \quad \text{für } Y^+ < \infty.$$

Da $\int_{y_0'}^{Y^+} \mu\,dy = \infty$ wird, $(y_0' > 0)$, gibt es ein $M > 0$ und zwei Folgen $x' \to X^+$, $x'' \to X^+$ mit $x'' > x'$, so daß $\int_{y'}^{y''} \mu\,dy \geqq M$ bleibt. Nun bilde man den Ausdruck

$$\int_{y'}^{y''} \mu\,g\,\frac{df}{dy}\,dy : \int_{y'}^{y''} \mu\,dy.$$

Er stellt offenbar den Wert $g\,\frac{df}{dy}\Big|^{\bar{x}}$ von $g\,\frac{df}{dy}$ an einer Stelle $\bar{x}$ zwischen x' und x'' dar und läßt sich absolut abschätzen durch

$$\left|g\,\frac{df}{dy}\right|^{\bar{x}} \leqq \frac{1}{M}\sqrt{\int_{y'}^{y''} \mu^2 g^2\,dy}\;\sqrt{\int_{y'}^{y''} \left(\frac{df}{dy}\right)^2 dy}.$$

Hier strebt nun die rechte Seite gegen Null mit x', $x'' \to X^+$. Denn es existiert $\int\limits_{y_0}^{Y+} \left(\frac{df}{dy}\right)^2 dy$ und nach Satz 2.1 auch $\int\limits_{y_0}^{Y+} \mu^2 g^2\, dy$. Somit folgt für die spezielle Folge $\bar{x} \to X^+$ die Behauptung des Hilfssatzes: $g \frac{df}{dy}\Big|^{\bar{x}} \to 0$.

Wir ziehen nun zum Beweise der Formel $(3.0)^+$ die Identität

$$\int\limits_{x+}^{\bar{x}} g\, L f\, dx + g \frac{df}{dy}\Big|_{x+}^{\bar{x}} = \int\limits_{y+}^{\bar{y}} \frac{dg}{dy}\frac{df}{dy}\, dy + \int\limits_{x+}^{\bar{x}} q g f\, dx$$

heran, die nach (1.2) sicher für g aus $\mathfrak{G}^+$ und f aus $\mathring{\mathfrak{F}}\mathfrak{G}^+$ besteht. Läßt man hier $\bar{x} \to X^+$ nach dem Hilfssatze eine solche Folge durchlaufen, daß $g \frac{df}{dy}\Big|^{\bar{x}} \to 0$ strebt, und beachtet man, daß alle drei Integrale dieser Identität für $\bar{x} \to X^+$ existieren, so entsteht die zu beweisende Greensche Formel $(3.0)^+$.

Aus dieser Formel folgt nun, daß für jede Folge $x^+ \to X^+$ auch $g \frac{df}{dy}\Big|^{x+} \to 0$ strebt, kurz, daß $g \frac{df}{dy}\Big|^{X+} = 0$ gilt. Setzt man hier für f und g zwei Funktionen f_1, f_2 aus $\mathring{\mathfrak{F}}\mathfrak{G}^+$, vertauscht und subtrahiert, so folgt

Satz 3.1$^+$. *Für zwei Funktionen f_1, f_2 aus $\mathfrak{G}^+ \mathring{\mathfrak{F}}$ gilt*

$$(3.1)^+ \qquad f_1 \frac{df_2}{dy} - \frac{df_1}{dy} f_2 \Big|^{X+} = 0.$$

Entsprechendes ergibt sich im Raume $\mathring{\mathfrak{F}}\mathfrak{G}^-$, und da $\mathfrak{F} = \mathring{\mathfrak{F}}\mathfrak{G}$ der Durchschnitt von $\mathring{\mathfrak{F}}\mathfrak{G}^+$ mit $\mathring{\mathfrak{F}}\mathfrak{G}^-$ ist, entspringt die Formel

$$(3.0) \qquad \int\limits_{X-}^{X+} g\, L f\, dx = \int\limits_{Y-}^{Y+} \frac{dg}{dy}\frac{dg}{dy}\, dy + \int\limits_{X-}^{X+} q g f\, dx$$

für g aus $\mathfrak{G}$, f aus $\mathring{\mathfrak{F}}\mathfrak{G}$ und der

Satz 3.1. *Für zwei Funktionen f_1, f_2 aus $\mathfrak{G}\mathring{\mathfrak{F}}$ gilt*

$$(3.1) \qquad f_1 \frac{df_2}{dy} - \frac{df_1}{dy} f_2 \Big|_{X-}^{X+} = 0.$$

D. h. der Operator L ist in $\mathring{\mathfrak{F}}\mathfrak{G}$ symmetrisch am Rande und also überhaupt. (Vgl. Satz 1.2 oder (3.0)).

Es ist nun bemerkenswert, daß die Bedingung, in $\mathfrak{G}$ zu liegen, die die Forderung A enthält, daß gewisse Integrale existieren, sich für Funktionen f aus $\mathring{\mathfrak{F}}$ auch durch eine Kennzeichnung des Grenzverhaltens der Funktion f für $x \to X^+$ bzw. X^- ersetzen läßt. Es gilt nämlich

Satz 3.2$^+$. *Der Raum* $\mathring{\mathfrak{F}}\mathfrak{G}^+$ *besteht aus allen Funktionen* $f(x)$ *aus* $\mathring{\mathfrak{F}}$, *für die gilt*

$$(3.2)^+ \qquad \begin{aligned} \frac{f(x)}{y} &\underset{x \dot{\to} X^+}{\to} 0 \quad \textit{wenn} \quad Y^+ = \infty \\ f(x) &\underset{x \dot{\to} X^+}{\to} 0 \quad \textit{wenn} \quad Y^+ < \infty \textit{ ist.} \end{aligned}$$

Beweis[8]): Diese Randbedingungen sind im Raume $\mathring{\mathfrak{F}}\mathfrak{G}^+$ nach Satz 2.2 (sogar mit $\longrightarrow$ statt $\dot{\to}$) erfüllt. Wir haben nur die Umkehrung zu zeigen: Genügt f aus $\mathring{\mathfrak{F}}$ diesen Bedingungen, so liegt f in $\mathfrak{G}^+$.

Wir nehmen zunächst den Fall $Y^+ < \infty$. Wir gehen aus von der Formel (vgl. (1.1))

$$(\dagger) \qquad \int_{x_0}^{x^+} f(L+k)f\,dx + f\frac{df}{dy}\bigg|_{x_0}^{x^+} = \int_{y_0}^{y^+}\left(\frac{df}{dy}\right)^2 dy + \int_{x_0}^{x^+}(q+k)f^2\,dx.$$

Aus der Voraussetzung $f(x) \underset{x \longrightarrow X^+}{\to} 0$ für die Funktion f folgt, daß es auch eine spezielle Folge von Werten $x^+ \to X^+$ gibt, für welche die Ableitung $\frac{df^2}{dy}\Big|^{x^+} = 2f\frac{df}{dy}\Big|^{x^+}$ nicht positiv ist. Für diese spezielle Folge bleibt die linke Seite der Formel ($\dagger$) nach oben beschränkt, denn da f und Lf in $\mathfrak{H}$ liegen, existiert $\int_{x_0}^{X^+} f(L+k)f\,dx$. Infolgedessen bleiben für diese Folge auch die Integrale der rechten Seite nach oben beschränkt; damit ist aber (beachte $q+k \geqq 1$) gesichert, daß $\int_0^{Y^+}\left(\frac{df}{dy}\right)^2 dy$ und $\int_{x_0}^{X^+} qf^2\,dx$ existieren; $\int_{x_0}^{X^+} f^2\,dx$ existiert, da f in $\mathring{\mathfrak{F}}$ liegt; die Bedingung B^+ (vgl. F. 7.) ist in $(3.2)^+$ vorausgesetzt. Also liegt f in $\mathfrak{G}^+$.

Im Falle $Y^+ = \infty$ würde man durch dieselbe Schlußweise unter der Voraussetzung $f(x) \underset{x \dot{\to} X^+}{\to} 0$ zum Ziele kommen. Um mit der schwächeren Voraussetzung $\frac{f(x)}{y} \underset{x \dot{\to} X^+}{\to} 0$ auszukommen, gehen wir von der Identität

$$(\ddagger) \qquad \int_{y_0'}^{y^+}\left(\frac{df}{dy}\right)^2 dy = \int_{y_0'}^{y^+} y^2\left(\frac{d}{dy}\frac{f}{y}\right)^2 dy + \frac{f^2}{y}\bigg|_{y_0'}^{y^+}, \quad (y_0' > 0),$$

8) Zur Schlußweise vgl. Carleman [1], S. 177.

aus; in Formel (†) eingesetzt entsteht

$$(\S)\quad \int\limits_{x'_0}^{x^+} f(L+k)f\,dx + \left(f\frac{df}{dy}-\frac{f^2}{y}\right)\Bigg|_{x'_0}^{x^+} = \int\limits_{y'_0}^{y^+} y^2\left(\frac{d}{dy}\frac{f}{y}\right)^2 dy + \int\limits_{x'_0}^{y^+}(q+k)f^2\,dx.$$

Aus der Voraussetzung, daß $\frac{f}{y}\to 0$ strebt auf wenigstens einer Folge $x\to X^+$, folgt, daß es eine spezielle Folge $x^+\to X^+$ gibt, wo

$$\frac{d}{dy}\frac{f^2}{y^2}=\frac{2}{y^2}\left[f\frac{df}{dy}-\frac{f^2}{y}\right],$$

also auch $f\frac{df}{dy}-\frac{f^2}{y}$ nicht positiv ist. Infolgedessen bleibt die linke, also auch die rechte Seite von (§) nach oben beschränkt. Es existiert somit $\int\limits_{x'_0}^{X^+} q f^2\,dx$ und $\int\limits_{y'_0}^{\infty} y^2\left(\frac{d}{dy}\frac{f}{y}\right)^2 dy$. Führen wir hier $y_*=-\frac{1}{y}$ mit $Y_*^+=0$ als neue Variable, $\frac{f}{y}=f_*$ als neue Funktion ein, so existiert also $\int\limits_{y'_{*0}}^{0}\left(\frac{df_*}{dy_*}\right)^2 dy_*$ und es gilt $f_*\underset{y_*\to 0}{\longrightarrow} 0$. Daraus folgt nach Anhang Satz A 1: $\frac{f_*}{\sqrt{-y_*}}\underset{x\to X^+}{\longrightarrow} 0$, das heißt $\frac{f^2}{y}\underset{x\to X^+}{\longrightarrow} 0$. Infolgedessen bleibt in (‡) die rechte Seite beschränkt für $x^+\to X^+$, also auch die linke. Es existiert somit $\int\limits_{y'_0}^{\infty}\left(\frac{df}{dy}\right)^2 dy$. Damit ist gezeigt, daß f in $\mathfrak{G}^+$ liegt, und also ist Satz 3. 2⁺ auch für $Y^+=\infty$ bewiesen.

Als Folge von Satz 3. 2⁺ gewinnen wir

Satz 3. 3⁺. *Ist*

$$(3.3)^+\quad \begin{cases} X^+=\infty, & \textit{wenn}\quad Y^+<\infty,\\ \int\limits_{x'_0}^{X^+} y^2\,dx=\infty, & \textit{wenn}\quad Y^+=\infty,\end{cases}$$

so sind schon im Raume $\mathring{\mathfrak{F}}$ *selbst die Forderungen* A⁺, B⁺ *aus* § 2 *erfüllt; d. h. es liegt* $\mathring{\mathfrak{F}}$ *in* $\mathfrak{G}^+$ *und es ist* $\mathring{\mathfrak{F}}\mathfrak{G}^+=\mathring{\mathfrak{F}}$.

Man beachte, daß die Bedingungen $(3.3)^+$ nicht vom Verlauf von q abhängen, sondern sich nur auf das Verhalten von x und y bzw. p am rechten Endpunkte beziehen.

Aus Satz 3. 3⁺ und 3. 1⁺ folgt dann der

Zusatz 3. 4⁺. Für f_1, f_2 aus $\mathring{\mathfrak{F}}$ besteht, wenn $(3.3)^+$ gilt, die Symmetrierelation am Endpunkte X^+

$$(3.4)^+\quad f_1\frac{df_2}{dy}-\frac{df_1}{dy}f_2\Bigg|^{X^+}=0.$$

Beweis von Satz 3.3+. Wir brauchen nur zu zeigen, daß unter der Bedingung $(3.3)^+$ jede Funktion $f(x)$ aus $\mathring{\mathfrak{F}}$ den Randbedingungen $(3.2)^+$: $\frac{f(x)}{y} \to 0$ bzw. $f(x) \to 0$ für $x \to X^+$ genügt; denn dann folgt $\mathring{\mathfrak{F}}$ in $\mathfrak{G}^+$ aus Satz 3.2.

1. Ist $Y^+ = \infty$ und $\int\limits_{x_0'}^{X^+} y^2\,dx = \infty$, so gibt es zur Funktion f aus $\mathring{\mathfrak{F}}$, da $\int\limits_{x_0'}^{X^+} f^2\,dx = \int\limits_{x_0'}^{X^+} \frac{f^2}{y^2}\,y^2\,dx < \infty$ wird, eine spezielle Folge $x \to X^+$, für die $\frac{f^2(x)}{y^2} \to 0$ strebt.

2. Ist $Y^+ < \infty$ und $X^+ = \infty$, so gibt es zur Funktion f aus $\mathring{\mathfrak{F}}$, da $\int\limits_{x^0}^{\infty} f^2\,dx < \infty$ wird, eine spezielle Folge $x \to X^+$, für die $f^2(x) \to 0$ strebt.

Die Sätze 3.3+ und 3.4+ besagen, daß es unnötig war, am Ende X^+ die Randbedingungen A^+, B^+ oder $(3.2)^+$ zu stellen, vorausgesetzt, daß $(3.3)^+$ gilt. Damit gewinnen wir sofort ein von Weyl[9]) angegebenes Kriterium dafür, daß bei X^+ der Grenzpunktfall I der Alternative vorliegt. Um es zu formulieren, werde angenommen, daß durch eine Randbedingung bei X^-, die so beschaffen ist, daß ihr jedenfalls alle Funktionen f aus $\mathring{\mathfrak{F}}$ genügen, die in Umgebung von X^- verschwinden, der Raum $\mathring{\mathfrak{F}}$ zu einem Raume $\mathfrak{F}$ so eingeschränkt sei, daß folgendes gilt:

3. Für f_1, f_2 aus $\mathfrak{F}$ ist

$$(3.5) \qquad f_1 \frac{d f_2}{d y} - \frac{d f_1}{d y} f_2 \Big|^{X^-} = 0.$$

3*. Besteht für ein f_0 aus $\mathring{\mathfrak{F}}$ die Relation

$$(3.5)^* \qquad f \frac{d f_0}{d y} - \frac{d f}{d y} f_0 \Big|^{X^-} = 0$$

mit allen f aus $\mathfrak{F}$, so liegt f_0 in $\mathfrak{F}$.[9a])

Dann lautet das Kriterium von Weyl[9]): *Der Operator L ist im Raume $\mathfrak{F}$ selbstadjungiert, wenn* $(3.3)^+$ *gilt.*

Sein Beweis entspringt sofort aus Zusatz 3.4+. Aus (3.5) und $(3.4)^+$ folgt nämlich (1.5) für f_1, f_2 aus $\mathfrak{F}$. Aus $(1.5)^*$ für f_0 aus $\mathring{\mathfrak{F}}$, alle f aus $\mathfrak{F}$ und $(3.4)^+$ folgt $(3.5)^*$, also: f_0 liegt in $\mathfrak{F}$. D. h. der Operator L ist im Raume $\mathfrak{F}$ am Rande, also überhaupt, selbstadjungiert.

---

[9]) [3] Satz 5, S. 238. Bei Weyl kommt statt $(3.3)^+$ nur $X^+ = \infty$ vor.

[9a]) Die Eigenschaft, daß 3. und 3*. gilt, könnten wir auch mit „Selbstadjungiertheit am linken Endpunkte X^-" bezeichnen.

§ 4.

Selbstadjungiertheit im Raume $\mathring{\mathfrak{F}}\mathfrak{G}$.

Wir formulieren nun den Hauptsatz unserer Theorie, nämlich daß der Operator L selbstadjungiert ist im Raume aller Funktionen f, die in $\mathring{\mathfrak{F}}$ und $\mathfrak{G}$ liegen, also den Bedingungen Ω, A, B (vgl. §§ 1 und 2) genügen.

Satz 4. 1. *Der Operator L ist im Raume $\mathring{\mathfrak{F}}\mathfrak{G}$ selbstadjungiert.*

Beweis. Da die Form G, wie in Satz 2. 5 ausgesprochen war, im Raume $\mathfrak{G}$ abgeschlossen ist, gehört zu ihr ein selbstadjungierter Operator. Das folgt aus einem an anderer Stelle[10]) bewiesenen Satze, der für unseren Fall folgendes besagt: Es gibt einen Teilraum $\hat{\mathfrak{F}}$ von Funktionen $\hat{f}$ aus $\mathfrak{G}$ (der G-dicht in $\mathfrak{G}$ liegt), in dem ein selbstadjungierter Operator $\hat{G}$ erklärt ist, so daß die Relation

$$(4.1)\qquad \int_{x^-}^{X^+} g\hat{G}\hat{f}\,dx = \int_{Y^-}^{Y^+} \frac{dg}{dy}\frac{d\hat{f}}{dy}\,dy + \int_{X^-}^{X^+} (q+k)\,g\hat{f}\,dx = g\,G\,\hat{f}$$

für g aus $\mathfrak{G}$ und $\hat{f}$ aus $\hat{\mathfrak{F}}$ besteht.

Nun folgt sofort, daß $\hat{\mathfrak{F}}$ in $\mathring{\mathfrak{F}}$ liegt. Denn für alle $\hat{f}$ aus $\hat{\mathfrak{F}}$ und alle $\dot{f}$ aus $\dot{\mathfrak{F}}$, die ja auch in $\mathfrak{G}$ liegen, folgt aus (4. 1) und aus der Symmetrierelation (3. 0), die in § 3 bewiesen wurde,

$$\int_{X^-}^{X^+} \dot{f}\cdot\hat{G}\,\hat{f}\,dx = \int_{X^-}^{X^+} (L+k)\dot{f}\cdot\hat{f}\,dx.$$

Satz 1. 1* lehrt dann, daß $\hat{f}$ zu $\mathring{\mathfrak{F}}$ gehört und daß $\hat{G}\,\hat{f} = (L+k)\hat{f}$ ist.

Der Raum $\hat{\mathfrak{F}}$ liegt also auch im Durchschnitt $\mathfrak{F} = \mathring{\mathfrak{F}}\mathfrak{G}$. Es ist somit der in $\mathfrak{F}$ erklärte Operator $L+k$ Fortsetzung des Operators $\hat{G}$ in $\hat{\mathfrak{F}}$; da dieser selbstadjungiert ist, muß $\hat{\mathfrak{F}} = \mathfrak{F}$ sein. Zugleich mit $\hat{G} = L + k$ ist auch L im Raume $\mathfrak{F}$ selbstadjungiert. So ist Satz 4. 1 bewiesen.

§ 5.

Abschwächung der Bedingung für q.

Die Voraussetzung, die seit § 3 gemacht wurde, daß der Koeffizient q bei X^+ nach unten beschränkt ist, läßt sich noch etwas abschwächen, was für manche Anwendungen bequem zu benutzen ist. Wir formulieren die Abschwächung nur für die Umgebung des Endpunktes X^+.

---

[10]) Vgl. Friedrichs [1], Sätze 7, 9, S. 478, 480.

Es sei $\omega(x)$ eine positive Funktion, die mit Ableitungen $\frac{d\omega}{dy}, \frac{d}{dx}\frac{d\omega}{dy}$ stetig ist, die für $x \leqq x_0$ identisch 1 ist und von einer Stelle $x_0' > x_0$ an gegeben ist durch

$$\omega(x) = \sqrt{y}, \qquad \text{wenn} \quad Y^+ = \infty$$
$$= \sqrt{Y^+ - y}, \quad \text{wenn} \quad Y^+ < \infty \text{ ist.}$$

Wir setzen dann

$$\varkappa(x) = -\frac{1}{\omega}\frac{d}{dx}\frac{d\omega}{dy},$$

also für

$$x_0' \leqq x < X^+$$
$$\varkappa(x) = \frac{p}{4y^2}, \qquad \text{wenn} \quad Y^+ = \infty$$
$$= \frac{p}{4(Y^+ - y)^2}, \quad \text{wenn} \quad Y^+ < \infty \text{ ist;}$$

es sei $\underline{\varkappa}$ die untere Grenze von $\varkappa(x)$ im Grundgebiet; $\underline{\varkappa} > -\infty$. Die abgeschwächte Bedingung für q lautet dann:

Es gibt zu q eine Zahl $\tau > 0$ und eine Konstante $\underline{q}$, so daß

$$q(x) \geqq \underline{q} - (1 - \tau^2)\,\varkappa(x)$$

im Grundgebiet gilt.

Alsdann werde der Raum $\mathfrak{G}^+$ wie in § 2 durch die Forderungen A^+, B^+ erklärt. Dann gilt:

Es gibt eine Konstante k, so daß die durch (2. 1) definierte Form G der Ungleichung (2. 2) genügt, und es übertragen sich die Sätze 2. 1, 2. 2, 2. 3, 2. 4, 3. 1^+ und 4. 1 wörtlich.

Zum Beweise dieser Behauptungen führe man neue Variable, Koeffizienten und Funktionen ein durch die Transformation

$$d\bar{x} = \omega^2 dx$$
$$\bar{p} = \omega^4 p, \quad \text{also} \quad \omega^2 d\bar{y} = dy$$
$$\bar{q} = q + \varkappa \quad \text{und} \quad \bar{L} = -\frac{d}{d\bar{x}}\frac{d}{d\bar{y}} + \bar{q}$$
$$h = \omega\bar{h}.$$

Dabei wird, wie man nachrechnet,

$$Lf = \omega\bar{L}\bar{f}.$$

Wegen $\qquad \bar{q} = q + \varkappa \geqq \underline{q} + \tau^2\varkappa \geqq \underline{q} + \tau^2\underline{\varkappa}$

ist die Theorie von §§ 2. 3. 4 anwendbar.

Sie führt zwangsläufig zu den behaupteten Sätzen, wenn man noch beachtet, daß durch die Transformation die Räume $\mathring{\mathfrak{F}}$, $\mathfrak{G}^-$, $\mathfrak{G}^+$ in die Räume $\mathring{\bar{\mathfrak{F}}}$, $\bar{\mathfrak{G}}^-$, $\bar{\mathfrak{G}}^+$ übergehen, die in bezug auf die neuen Variablen und Funktionen im Sinne von § 1 und § 2 erklärt sind. Für $\mathring{\mathfrak{F}}$ und $\mathfrak{G}^-$ ergibt sich das unmittelbar; für $\mathfrak{G}^+$ stütze man sich auf die Sätze A 2 und 3 im Anhang

Es sei noch erwähnt, daß sich auch die Sätze 3.2+, 3.3+, 3.4+ übertragen, wenn in ihnen ersetzt werden die Bedingungen

$(3.2)^+$ durch

$$\frac{f(x)}{y^{1+\tau}} \underset{x \to X^+}{\to} 0, \quad \text{wenn } Y^+ = \infty,$$

$$\frac{f(x)}{(Y^+ - y)^{1-\tau}} \underset{x \to X^+}{\to} 0, \quad \text{wenn } Y^+ < \infty \text{ ist},$$

$(3.3)^+$ durch

$$\int_{x_0'}^{X^+} y^{1+\tau} dx = \infty, \quad \text{wenn } Y^+ = \infty,$$

$$\int_{x_0}^{X^+} (Y^+ - y)^{1-\tau} dx = \infty, \quad \text{wenn } Y^+ < \infty \text{ ist}.$$

Zum Beweise wird man in derselben Weise wie oben mit ω, mit der Funktion $\omega^{1+\tau}$, wenn $Y^+ = \infty$, mit $\omega^{1-\tau}$, wenn $Y^+ < \infty$ ist, transformieren, wobei, wie man nachrechnet, $\bar{q} = q + (1 - \tau^2)\varkappa \geqq \underline{q}$ wird. Unmittelbar gehen die obigen Ersatzbedingungen in die Bedingungen $(3.2)^+$, $(3.3)^+$ für die transformierten Größen über.

§ 6.

Der halbbeschränkte Operator und die ausgezeichnete Rolle der Randbedingung.

Der Operator L heißt halbbeschränkt, wenn es eine „untere Schranke" $\underline{\lambda}$ gibt, so daß für alle f aus $\mathfrak{F}$ gilt

$$\int_{X^-}^{X^+} f L f \, dx = \int_{Y^-}^{Y^+} \left(\frac{df}{dy}\right)^2 dy + \int_{X^-}^{X^+} q f^2 dx \geqq \underline{\lambda} \int_{X^-}^{X^+} f^2 dx.$$

Offenbar ist L halbbeschränkt, wenn q nach unten beschränkt ist.

Ein solcher halbbeschränkter Operator L läßt sich nun stets von $\dot{\mathfrak{F}}$ aus auf einen weiteren Raum fortsetzen, in dem er selbstadjungiert wird. Wir wollen nun (unter den möglicherweise verschiedenen) eine selbstadjungierte Fortsetzung aufzeigen, die in mehrfacher Weise grundsätzlich ausgezeichnet ist; sie wird, wenn q nach unten beschränkt ist, gerade durch unsere Randbedingung von § 2, § 3 gekennzeichnet.

Unter selbstadjungierter Fortsetzung von L in $\dot{\mathfrak{F}}$ ist ein selbstadjungierter Operator $\tilde{L}$ zu verstehen, in einem Teilraume $\tilde{\mathfrak{F}}$ von $\mathfrak{H}$, der $\dot{\mathfrak{F}}$ enthält, derart erklärt, daß er in $\dot{\mathfrak{F}}$ mit L übereinstimmt. Man überzeugt sich sofort davon, daß ein solcher Raum $\tilde{\mathfrak{F}}$ in $\mathfrak{F}$ liegt und daß in $\tilde{\mathfrak{F}}$ der Operator $\tilde{L}$ mit L übereinstimmt, so daß man von L in $\tilde{\mathfrak{F}}$ sprechen kann.

Denn für $\tilde f$ aus $\widetilde{\mathfrak F}$, $\dot f$ aus $\dot{\mathfrak F}$ (das also auch zu $\widetilde{\mathfrak F}$ gehört) gilt wegen der Symmetrie von L in $\widetilde{\mathfrak F}$ und wegen $\widetilde L \dot f = L \dot f$

$$\int_{X-}^{X+} (\dot f \widetilde L \tilde f - L \dot f \cdot \tilde f)\, d x = 0 ;$$

daraus folgt nach Satz 1.1* in der Tat, daß $\tilde f$ zu $\mathring{\mathfrak F}$ gehört und $\widetilde L \tilde f = L \tilde f$.

Es sei L in $\widetilde{\mathfrak F}$ eine solche selbstadjungierte Fortsetzung von L in $\dot{\mathfrak F}$; wir nehmen an, daß sie halbbeschränkt ist mit der unteren Schranke $\tilde\lambda$:

$$\int_{X-}^{X+} \tilde f L \tilde f\, d x \geqq \tilde\lambda \int_{X-}^{X+} \tilde f^2\, d x \qquad \text{für } \tilde f \text{ aus } \widetilde{\mathfrak F}.$$

Dann gibt es[11]) genau einen Teilraum $\widetilde{\mathfrak G}$ aus Elementen $\tilde g$ von $\mathfrak H$ und in ihm eine bilineare Form $\tilde g_1 \widetilde G \tilde g_2$ mit folgenden Eigenschaften. 1. Es besteht die Ungleichung

$$\tilde g \widetilde G \tilde g \geqq \tilde\lambda \int_{X-}^{X+} \tilde g^2\, d x$$

und $\widetilde G$ ist in $\widetilde{\mathfrak G}$ abgeschlossen. 2. $\widetilde{\mathfrak G}$ enthält den Raum $\widetilde{\mathfrak F}$, und zwar dicht mit $\widetilde G$ als Maßform. 3. $\widetilde G$ hängt mit dem Operator L in $\widetilde{\mathfrak F}$ vermöge

$$\tilde g \widetilde G \tilde f = \int_{X-}^{X+} \tilde g L \tilde f\, d x$$

für alle $\tilde g$ aus $\widetilde{\mathfrak G}$, $\tilde f$ aus $\widetilde{\mathfrak F}$ zusammen. Wir nennen $\widetilde{\mathfrak G}$ den zu L in $\widetilde{\mathfrak F}$ gehörigen $\mathfrak G$-Raum und $\widetilde G$ die zugehörige Form. Umgekehrt gibt es zu einer abgeschlossenen Form $\widetilde G$ in $\widetilde{\mathfrak G}$ auch nur einen zugehörigen Raum $\widetilde{\mathfrak F}$ mit selbstadjungiertem Operator L in $\widetilde{\mathfrak F}$.

Es existiert nun ein selbstadjungierter Operator L, in einem Raume $\mathfrak F_\infty$ erklärt, der Fortsetzung von L in $\dot{\mathfrak F}$ und durch jede der drei folgenden Eigenschaften gekennzeichnet ist; (d. h. er besitzt alle drei Eigenschaften und ist der einzige, der irgendeine dieser Eigenschaften besitzt). Dabei sei $\mathfrak G_\infty$ der zu L in $\mathfrak F_\infty$ gehörige $\mathfrak G$-Raum und G_∞ die zugehörige Form.

1. Der Raum $\dot{\mathfrak F}$ liegt dicht in $\mathfrak G_\infty$ mit G_∞ als Maßform.

2. Ist L in $\widetilde{\mathfrak F}$ eine andere halbbeschränkte selbstadjungierte Fortsetzung von L in $\dot{\mathfrak F}$, so ist $\mathfrak G_\infty$ echter Teilraum des zugehörigen $\mathfrak G$-Raumes $\widetilde{\mathfrak G}$; in $\mathfrak G_\infty$ stimmt $\widetilde G$ mit G_∞ überein.

3. Es ist $\mathfrak F_\infty$ genau der Durchschnitt

$$\mathfrak F_\infty = \mathring{\mathfrak F}\, \mathfrak G_\infty$$

von $\mathring{\mathfrak F}$ mit dem zugehörigen $\mathfrak G$-Raume $\mathfrak G_\infty$.

[11]) Friedrichs [1], Satz 3, S. 473.

Wir nennen L in $\mathfrak{F}_\infty$ die „ausgezeichnete Fortsetzung" von L in $\dot{\mathfrak{F}}$.

Daß es eine Fortsetzung gibt, die Eigenschaft 1. besitzt, wurde an anderer Stelle bewiesen[12]). Aus 1. aber folgt 2. Denn ist L in $\tilde{\mathfrak{F}}$ eine andere selbstadjungierte Fortsetzung, so stimmt $\tilde{G}$ mit $\mathfrak{G}_\infty$ in $\dot{\mathfrak{F}}$ überein; daraus, daß $\dot{\mathfrak{F}}$ in $\mathfrak{G}_\infty$ dicht liegt mit G_∞ als Maßform, folgt, daß $\mathfrak{G}_\infty$ in $\tilde{\mathfrak{G}}$ liegt und daß $G_\infty = \tilde{G}$ ist in $\mathfrak{G}_\infty$. Es ist aber $\mathfrak{G}_\infty \neq \tilde{\mathfrak{G}}$; denn sonst müßten auch die Räume $\mathfrak{F}_\infty$ und $\tilde{\mathfrak{F}}$ übereinstimmen, da sie den Räumen $\mathfrak{G}_\infty$ und $\tilde{\mathfrak{G}}$ samt Form G, wie oben bemerkt, eindeutig zugeordnet sind. Offenbar ist L in $\mathfrak{F}_\infty$ die einzige selbstadjungierte Fortsetzung von L in $\dot{\mathfrak{F}}$, die 2., also auch die einzige, die 1. erfüllt.

Ebenfalls folgt 3. aus 1. Denn für f_0 aus $\mathring{\mathfrak{F}}\mathfrak{G}_\infty$ gilt die Relation

$$\dot{f}\, G_\infty f_0 = \int_{X-}^{X+} L\dot{f}\cdot f_0\, dx = \int_{X-}^{X+} \dot{f}\, L f_0\, dx$$

mit allen $\dot{f}$ aus $\dot{\mathfrak{F}}$; aus ihr folgt, da $\dot{\mathfrak{F}}$ in $\mathfrak{G}_\infty$, also in $\mathfrak{F}_\infty$ ja G_∞-dicht liegt, für alle f aus $\mathfrak{F}_\infty$ die Relation

$$f\, G_\infty f_0 = \int_{X-}^{X+} f\, L f_0\, dx,$$

somit

$$\int_{X-}^{X+} (Lf\cdot f_0 - f\, L f_0)\, dx = 0.$$

Da L in $\mathfrak{F}_\infty$ selbstadjungiert ist, folgt, daß f_0 in $\mathfrak{F}_\infty$ liegt und damit Eigenschaft 3. Für jede andere selbstadjungierte Fortsetzung L in $\tilde{\mathfrak{F}}$ liegt $\mathfrak{G}_\infty$ in $\tilde{\mathfrak{G}}$ wegen 2., also $\mathfrak{F}_\infty = \mathring{\mathfrak{F}}\mathfrak{G}_\infty$ in $\mathring{\mathfrak{F}}\tilde{\mathfrak{G}}$; da aber $\mathfrak{F}_\infty$ nicht in $\tilde{\mathfrak{F}}$ enthalten ist, ist $\tilde{\mathfrak{F}} \neq \mathring{\mathfrak{F}}\tilde{\mathfrak{G}}$.

In § 2 wurde unter der Voraussetzung, daß q nach unten beschränkt ist (wonach ja L halbbeschränkt wird), ein Raum $\mathfrak{G}$ eingeführt, und in § 3.4 wurde bewiesen, daß L in $\mathfrak{F} = \mathring{\mathfrak{F}}\mathfrak{G}$ selbstadjungiert ist. Dieser Raum $\mathfrak{G}$ stimmt mit dem hier abstrakt gekennzeichneten Raume $\mathfrak{G}_\infty$ überein. Das folgt, da die Eigenschaft 1. den Raum $\mathfrak{G}_\infty$ eindeutig kennzeichnet, aus der Tatsache, daß der Raum $\dot{\mathfrak{F}}$ in $\mathfrak{G}$ dicht liegt mit G als Maßform. Diese Tatsache ergibt sich[13]) aus Satz 2. 2. Aus diesem Satze entnimmt man nämlich, daß jede Funktion g aus $\mathfrak{G}$ durch eine Funktion $\dot{g}$ aus $\mathfrak{G}$, die in Umgebung der Enden identisch verschwindet, so angenähert

12) Friedrichs [1. 2].

13) Vgl. Friedrichs [3].

werden kann, daß $(g - \dot{g})\, G\, (g - \dot{g})$ beliebig klein wird. Die Funktion $\dot{g}$ läßt sich aber samt Ableitungen $\frac{d\dot{g}}{dy}$ gleichmäßig durch Funktionen f aus $\mathring{\mathfrak{F}}$ approximieren.

Um die hier eingeführten Begriffe zu erläutern, sei an den regulären Fall erinnert, wo X^+, X^-, Y^+, Y^- endlich sind und $|q|$ beschränkt ist. Der Raum $\mathfrak{G}_\infty$ besteht aus den Funktionen g aus $[g]$ und $\mathfrak{H}$, für welche $\int\limits_{Y^-}^{Y^+}\left(\frac{dg}{dy}\right)^2 dy$ existiert und die an den Endpunkten verschwinden. Der Operator L ist selbstadjungiert im Raume $\mathfrak{F} = \mathfrak{F}_\infty = \mathring{\mathfrak{F}}\,\mathfrak{G}_\infty$ aller Funktionen f_0 aus $\mathring{\mathfrak{F}}$, die in $\mathfrak{G}_\infty$ liegen (nach Satz 3.2 genügt es zu fordern, daß f_0 an den Enden verschwindet). Es gibt andere selbstadjungierte Fortsetzungen L in $\tilde{\mathfrak{F}}$ von L in $\mathring{\mathfrak{F}}$; man erhält solche Räume $\tilde{\mathfrak{F}} = \mathfrak{F}_\sigma$ aus $\mathring{\mathfrak{F}}$ z. B., indem man die Randbedingungen $\frac{df}{dy} + \sigma^+ f = 0$, bzw. $\frac{df}{dy} + \sigma^- f = 0$ an den Enden X^+ bzw. X^- stellt. Der zugehörige $\mathfrak{G}$-Raum $\mathfrak{G}_\sigma$ besteht aus allen Funktionen g aus $[g]$ und $\mathfrak{H}$, für welche $\int\limits_{Y^-}^{Y^+}\left(\frac{dg}{dy}\right)^2 dy$ existiert; denn jede solche Funktion g läßt sich mit $\int\limits_{Y^-}^{Y^+}\left(\frac{dg}{dy}\right)^2 dy + \sigma g^2\Big|_{X^-}^{X^+} + \int\limits_{X^-}^{X^+}(q+k)\,g^2\,dx$ (k genügend groß) als Maßform approximieren durch solche Funktionen f aus $\mathring{\mathfrak{F}}$, die den obigen Randbedingungen genügen; d. h. durch Funktionen aus $\mathfrak{F}_\sigma$. Der Raum $\mathfrak{G}_\sigma$ enthält, wie in Eigenschaft gefordert, den Raum $\mathfrak{G}_\infty$; er ist von den Parametern σ der Randbedingungen unabhängig und der Raum $\mathfrak{F}_\sigma$ ist nicht mit dem Durchschnitt $\mathring{\mathfrak{F}}\,\mathfrak{G}_\sigma$ identisch.

Diese Eigenschaft der Randbedingungen $\frac{df}{dy} + \sigma f = 0$, daß sie noch nicht beim zugehörigen Raume $\mathfrak{G}_\sigma$, sondern erst beim Raume $\mathfrak{F}_\sigma$ zu stellen sind, kennzeichnet sie als natürliche[14]) Randbedingungen, zu denen also die hier behandelte ausgezeichnete Randbedingung im Gegensatz steht.

---

[14]) Diese Bezeichnung wird motiviert durch die folgende aus der Selbstadjungiertheit entspringende Eigenschaft: Genügt eine Funktion g_0 aus $\mathfrak{G}_\sigma$ mit allen g aus $\mathfrak{G}_\sigma$ (die also den Randbedingungen noch nicht zu genügen brauchen) mit einer Funktion h_1 aus $\mathfrak{H}$ der Relation

$$g\,G\,g_0 = (g, h_1),$$

so liegt g_0 in $\mathfrak{F}_\sigma$, erfüllt also neben der Differentialgleichung $L g_0 = h_1$ die Randbedingung.

Anhang.

Integralungleichungen.

Wir stellen eine Reihe von Integralungleichungen für Funktionen g aus $[g]$ zusammen, von denen wir Gebrauch gemacht haben.

Mit $\mathfrak{D}^+$ bezeichnen wir den Raum aller Funktionen g aus $[g]$, für die $\int\limits_0^{Y^+} \left(\frac{d\,g}{d\,y}\right)^2 d\,y < \infty$ ist.

Satz A. 1. *Sei* $Y^+ < \infty$. *Dann besteht für die Funktionen* g *aus* $\mathfrak{D}^+$, *für die* $g(x) \underset{x \to X^+}{\to} 0$ *strebt, die Ungleichung*

$$\frac{g^2(x^+)}{Y^+ - y^+} \leqq \int\limits_{y^+}^{Y^+} \left(\frac{d\,g}{d\,y}\right)^2 d\,y \quad \textit{für alle } x^+ \geqq x_0,$$

und infolgedessen strebt

$$\frac{g(x)}{\sqrt{Y^+ - y}} \underset{x \to X^+}{\to} 0\,.$$

Beweis. Man lasse in der Abschätzung

$$|\,g(x^+) - g(x^*)\,|^2 \leqq (y^* - y^+) \int\limits_{y^+}^{y^*} \left(\frac{d\,g}{d\,y}\right)^2 d\,y$$

x^* die spezielle Folge durchlaufen, auf der $g(x^*) \to 0$ strebt; dann folgt die Behauptung.

Satz A. 2. *Sei* $Y^+ < \infty$.

1. *Dann existieren für jede Funktion* g *aus* $\mathfrak{D}^+$, *für die* $g(x) \underset{x \to X^+}{\to} 0$ *strebt, die Integrale*

$$\int\limits_0^{Y^+} Y^+ - y) \left(\frac{d}{d\,y} \frac{g}{\sqrt{Y^+ - y}}\right)^2 d\,y < \infty,$$

$$\int\limits_0^{Y^+} \frac{g^2}{(Y^+ - y)^2}\, d\,y < \infty.$$

2. *Existieren für eine Funktion* g *aus* $[g]$ *diese Integrale, so liegt* g *in* $\mathfrak{D}^+$ *und es strebt* $g(x) \underset{x \to X^+}{\to} 0$.

3. *Es besteht für* g *in* $\mathfrak{D}^+$ *und* $x^+ \geqq x_0$ *die Identität*

$$\int\limits_{y^+}^{Y^+} \left(\frac{d\,g}{d\,y}\right)^2 d\,y = \int\limits_{y^+}^{Y^+} (Y^+ - y) \left(\frac{d}{d\,y} \frac{g}{\sqrt{Y^+ - y}}\right)^2 d\,y + \frac{1}{4} \int\limits_{y^+}^{Y_+} \frac{g^2}{(Y^+ - y)^2}\, d\,y + \frac{1}{2}\, \frac{g^2(x^+)}{Y^+ - y^+}.$$

Beweis. Wir gehen aus von der in $[g]$ gültigen Identität

$$(\times)\quad \int_{y^+}^{\bar y} (Y^+-y)\left(\frac{d}{d\,y}\frac{g}{\sqrt{Y^+-y}}\right)^2 d\,y + \frac{1}{4}\int_{y^+}^{\bar y} \frac{g^2}{(Y^+-y)^2}\,d\,y + \frac{1}{2}\,\frac{g^2(x^+)}{Y^+-y^+}$$
$$= \int_{y^+}^{\bar y} \left(\frac{d\,g}{d\,y}\right)^2 d\,y + \frac{1}{2}\,\frac{g^2(\bar x)}{Y^+-\bar y}.$$

1. Liegt g in $\mathfrak{D}^+$ mit $g(x) \underset{x \to X^+}{\to} 0$, so folgt $\frac{g^2(\bar x)}{Y^+-\bar y}$ nach A. 1; also existieren die Integrale der linken Seite bis Y^+ erstreckt und es gilt die Identität 3.

2. Existieren diese Integrale für ein g aus $[g]$, so bleibt die rechte Seite der Identität beschränkt; es existiert also $\int_0^{Y^+}\left(\frac{d\,g}{d\,y}\right)^2 d\,y$ und g liegt in $\mathfrak{D}^+$; aus der Existenz von $\int_0^{Y^+} \frac{g^2}{Y^+-y}\,\frac{d\,y}{Y^+-y}$ folgt sodann, daß es eine spezielle Folge $\bar x \to X^+$ gibt, für die $\frac{g(\bar x)}{Y^+-\bar y} \to 0$ strebt; für diese Folge gilt erst recht $g(\bar x) \to 0$.

Satz A. 3. *Sei* $Y^+ = \infty$; $y'_0 > 0$.

1. *Dann existieren für Funktionen g aus $\mathfrak{D}^+$ die Integrale*

$$\int_{y'_0}^{Y^+} y\left(\frac{d}{d\,y}\frac{g}{\sqrt{y}}\right)^2 d\,y,\qquad \int_{y'_0}^{Y^+} \frac{g^2}{y^2}\,d\,y.$$

2. *Existieren diese Integrale für eine Funktion g aus $[g]$, so liegt sie in $\mathfrak{D}^+$.*

3. *Für g aus $\mathfrak{D}^+$ und $x^+ \geqq x'_0$ besteht die Identität*

$$\int_{y^+}^{Y^+} y\left(\frac{d}{d\,y}\frac{g}{\sqrt{y}}\right)^2 d\,y + \frac{1}{4}\int_{y^+}^{Y^+}\frac{g^2}{y^2}\,d\,y = \int_{y^+}^{Y^+}\left(\frac{d\,g}{d\,y}\right)^2 d\,y + \frac{1}{2}\,\frac{g^2(x^+)}{y^+}.$$

Beweis. Wir gehen aus von der in $[g]$ gültigen Identität

$$(\mp)\quad \int_{y^+}^{\bar y} y\left(\frac{d}{d\,y}\frac{g}{\sqrt{y}}\right)^2 d\,y + \frac{1}{4}\int_{y^+}^{\bar y}\frac{g^2}{y^2}\,d\,y + \frac{1}{2}\,\frac{g^2(\bar x)}{\bar y} = \int_{y^+}^{\bar y}\left(\frac{d\,g}{d\,y}\right)^2 d\,y + \frac{1}{2}\,\frac{g^2(x^+)}{y^+}.$$

1. Aus ihr entnimmt man für $\bar x \to X^+$, daß für g aus $\mathfrak{D}^+$ die Integrale der linken Seite bis Y^+ erstreckt existieren.

2. Existieren diese Integrale, also insbesondere $\int_{y'_0}^{Y^+}\frac{g^2}{y}\,\frac{d\,y}{y}$, so muß es

eine spezielle Folge $\bar{x} \to X^+$ geben, für welche $\frac{g^2(\bar{x})}{\bar{y}} \to 0$ strebt; mit ihr bleibt die linke, also die rechte Seite beschränkt; d. h. es ist

$$\int_{y^+}^{Y^+} \left(\frac{d g}{d y}\right)^2 d y < \infty.$$

Zugleich ergibt sich die Identität 3.

Lassen wir in $(\mp)$ $x^+ \to X^+$ die spezielle Folge durchlaufen, auf der $\frac{g^2(x^+)}{y^+} \to 0$ strebt, so erhalten wir unmittelbar den

Satz A. 4. *Sei $Y^+ = \infty$. Dann strebt für Funktionen g aus $\mathfrak{D}^+$*

$$\frac{g(x)}{\sqrt{y}} \underset{x \to X^+}{\to} 0.$$

(Eingegangen am 23. 6. 1935.)

[38–1] **Über die Spektralzerlegung eines Integraloperators, Math. Ann., 115 (1938), 249–273.**

Commentary:

This paper on perturbation theory for continuous spectra deals with a rather special class of operators $T + K$, where T is an operator of multiplication: $Tx(t) = tx(t)$ for complex-valued functions x on $[-1, 1]$, and where K is an integral operator with kernel $k(t, s)$ which is doubly Hölder continuous with exponent μ, $0 < \mu < 1$. It is shown, among other things, that if $k(t, s) = 0$ for $s = \pm 1$ and k is sufficiently small, $T + K$ is similar to T (so that the spectrum of $T + K$ is identical with $[-1, 1]$). The underlying function space is $L^2(-1, 1)$ or the space of Hölder continuous functions on $[-1, 1]$. The proof consists in constructing the transformation operator U with $U^{-1}TU = T + K$ in the form $U = 1 + i\check{R}$, where $\check{R}$ is a singular integral operator given by

$$\check{R}x(t) = \pi r(t, t)\, x(t) - i \int_{-1}^{1} (s - t)^{-1}\, r(t, s)\, x(s)\, ds$$

with a certain kernel $r(t, s)$ with properties similar to those of $k(t, s)$. The proof is elementary but depends on extremely delicate estimates of Privalov type involving singular integrals with Hölder continuous kernels.

No motivation for the study of this problem is given in the paper except that it is a typical perturbation problem involving continuous spectra, in contrast to the perturbation theory for discrete spectra that had been developed by Rellich [1]. It turned out that this was another case of "dramatic anticipations in the history of mathematics." The paper contains fundamental results indispensable in mathematical scattering theory, which essentially started a decade later with another paper [48–8] of Friedrichs and was followed by numerous investigations. In fact the operator U given above is the prototype of the wave operator in scattering theory.

[1] F. Rellich. Störungstheorie der Spektralzerlegung, Parts I/II, Math. Ann. 113 (1937), 600–619, 677–685.

Über die Spektralzerlegung eines Integraloperators.

Von

Kurt Friedrichs in New York.

Die vorliegende Arbeit behandelt die Spektralzerlegung einer speziellen Klasse von Operatoren der Form $T+K$, wobei T ein bestimmter Operator mit Streckenspektrum und K ein Integraloperator ist. Wir geben eine Klasse von Integraloperatoren K an, derart daß der Operator $T+K$ dasselbe oder nahezu dasselbe Spektrum besitzt wie T.

T ist dabei der Operator, der aus Funktionen $x(t)$ die Funktion $t\,x(t)$ erzeugt. Die zugelassenen Funktionen $x(t)$ sind komplexwertig, für $|t| \leqq 1$ definiert und bilden einen linearen Raum, der in bezug auf eine Norm $\|x\|$ vollständig ist. Wir legen zwei verschiedene Funktionenräume zugrunde: einmal den Hilbertschen Raum aller L^2-integrablen Funktionen, andererseits den Raum aller stetigen Funktionen, die einer Lipschitz-Hölderbedingung mit einem Exponenten $\mu < 1$ genügen,

In jedem Falle erzeugt der Operator T aus $x(t)$ eine Funktion

$$T\,x(t) = t\,x(t)$$ [1]

desselben Raumes.

[1]) Um zu klären, in welchem Maße unser Operator T speziell ist, sei darauf hingewiesen, daß auf Grund der Hellingerschen Theorie jeder selbstadjungierte Operator des Hilbertschen Raumes einem Operator T äquivalent ist, der durch $T\,x(t) = t\,x(t)$ gegeben ist; dabei kann x außer von t noch von weiteren Parametern abhängen. T heißt Spektraldarstellung des Operators. Jeder abgeschlossenen t-Menge kann man eine Vielfachheit zuordnen. Kommen nur die Vielfachheiten 0 und 1 vor, so spricht man von einfachem Spektrum; in diesem Falle besitzt der Operator eine Spektraldarstellung, in der die Funktionen x außer von t nicht von weiteren Parametern abhängen. Das Spektrum heißt regulär, wenn jede Nullmenge die Vielfachheit Null hat (also auch keine Punkteigenwerte auftreten). Man sagt, das Spektrum besteht aus einer Strecke, wenn jede Menge außerhalb und jede Nullmenge innerhalb der Strecke die Vielfachheit Null hat.

Unser Operator T repräsentiert somit den Operator, der ein einfaches reguläres Spektrum besitzt, das aus der Strecke $|t| \leqq 1$ besteht. Wesentlich ist für uns übrigens nur die Regularität.

Von der allgemeinen Spektraltheorie (vgl. z. B. M. H. Stone, Linear Transformations in Hilbert Space, 1932) machen wir in dieser Arbeit nur hinsichtlich der Fragestellung Gebrauch, nicht hinsichtlich der Methoden und Sätze.

Die behandelten Operatoren K sind Integraloperatoren mit einem Kerne $k(t, s)$, die aus der Funktion $x(t)$ die Funktion

$$K x(t) = \int_{-1}^{1} k(t, s) x(s) d s$$

erzeugen. Der (nicht notwendig symmetrische) Kern $k(t, s)$ soll stetig sein und gewissen Hölderbedingungen mit dem Exponenten $\mu < 1$ genügen (es ist zweckmäßig, eine Norm $\|K\|$ einzuführen, so daß diese Bedingungen mit $\|K\| < \infty$ gleichbedeutend sind). Ferner ist $k(t, s)$ der Randbedingung

(0) $$k(t, s) = 0 \quad \text{für} \quad |s| = 1$$

unterworfen.

Dann ist das Hauptergebnis, daß der Operator $T + K$ dasselbe Spektrum hat wie T, nämlich die Strecke $|t| \leqq 1$, vorausgesetzt, daß $\|K\|$ genügend klein ist. Auch ohne diese Einschränkung für $\|K\|$ kann (unter gewissen Zusatzbedingungen) die Spektralzerlegung von $T + K$ gewonnen werden; dabei zeigt sich, daß neben demselben Streckenspektrum nur noch endlich viele Punkteigenwerte auftreten.

Um die Rolle der Randbedingung (0) zu klären, sei das einfachste Beispiel erwähnt, in dem sie nicht erfüllt ist:

$$k(t, s) = \varepsilon = \text{const.}$$

Hier hat der Operator $T + K$ für $\varepsilon \neq 0$ einen Punkteigenwert[2])

$$\tau = \frac{e^{\frac{1}{\varepsilon}} + 1}{e^{\frac{1}{\varepsilon}} - 1}$$

mit der Eigenfunktion

$$x(t) = \frac{1}{t - \tau};$$

denn es ist, wie man nachrechnet,

$$t x(t) + \varepsilon \int_{-1}^{1} x(s) d s = \tau x(t).$$

Offenbar gilt $\tau \downarrow 1$ für $\varepsilon \downarrow 0$ und $\tau \uparrow -1$ für $\varepsilon \uparrow 0$[3]).

Ein solches Vorkommnis wird durch die Randbedingung (0) ausgeschlossen.

---

[2]) Es ließe sich leicht zeigen, daß $T + K$ außerdem das Streckenspektrum $|t| \leqq 1$ hat.

[3]) Es liegt hier ein einfaches Beispiel dafür vor, daß die Qualität des Spektrums auch bei beliebig kleiner analytischer Störung sich ändern kann. Vgl. die Arbeiten von Fr. Rellich zur Störungstheorie der Spektralzerlegung, Math. Annalen 113 (1936), S. 600 u. 677, wo das Problem der Abhängigkeit der Spektralzerlegung von einem Störungsparameter grundsätzlich aufgerollt wird.

Um die Spektraldarstellung des Operators $T + K$ zu gewinnen[4]), werden wir zwei Transformationen U und $\mathbf{U}$ einführen, die jede Funktion $x(t)$ des zugrundegelegten Raumes in Funktionen $U x(t)$ bzw. $\mathbf{U} x(t)$ desselben Raumes transformieren und den Relationen

$$(1) \qquad (T + K)\, U = U\, T$$

$$(\mathbf{1}) \qquad \mathbf{U}(T + K) = T\, \mathbf{U}$$

genügen. Wir wollen solche Transformationen Rechts- bzw. Links-*Normatoren* von $T + K$ nennen. Wir sagen, sie bilden ein *Paar*, wenn noch

$$(2) \qquad \mathbf{U}\, U = 1$$

gilt.

Ein „Linksnormator“ $\mathbf{U}$ transformiert jede Funktion $x(t)$ in eine Funktion

$$\mathbf{U} x(t) = y(t)$$

derart, daß $(T + K)\, x(t)$ in

$$\mathbf{U}(T + K)\, x(t) = t\, y(t)$$

übergeht. Umgekehrt vermittelt der „Rechtsnormator“ U, daß jeder Funktion $y(t)$ eine Funktion $x(t)$ entspricht, nämlich

$$x(t) = U\, y(t)$$

derart, daß

$$(T + K)\, x(t) = U\, t\, y(t)$$

wird, also der Funktion $t\, y(t)$ entspricht.

Existiert ein solches Paar von Normatoren, so ist gesichert, daß die Strecke $|t| \leqq 1$ zum Spektrum von $T + K$ gehört; und der zugehörige Anteil der Spektralzerlegung von $T + K$ wird durch die Normatoren geliefert. Genügt das Paar von Normatoren noch der Relation

$$(3) \qquad U\, \mathbf{U} = 1$$

(wir nennen es dann *intakt*), so liefert es die volle Spektralzerlegung von $T + K$ und das Spektrum von $T + K$ besteht genau aus der Strecke $|t| \leqq 1$.

Ist der Hilbertsche Raum zugrunde gelegt, und ist der Operator K symmetrisch, so kann das intakte Paar von Normatoren unitär gewählt werden. Doch die gegebene Formulierung der Spektralzerlegung nimmt weder Bezug auf die spezielle Metrik des Raumes, noch auf die Symmetrie des Operators.

Die Hauptaufgabe ist nun, den funktionalen Charakter der Normatoren aufzufinden. Es zeigt sich, daß sie aus zwei Operatoren R, $\mathbf{R}$ gebildet

---

[4]) Bzw. zu definieren im Falle, daß $T + K$ nicht symmetrisch ist oder nicht im Hilbertschen Raume operiert.

werden, deren Kerne $r(t,s)$ und $\boldsymbol{r}(t,s)$ denselben Bedingungen wie $k(t,s)$ genügen, und zwar in folgender Weise:

$$U\,y(t) = \big(1 + i\pi\, r(t,t)\big)\, y(t) + \int_{-1}^{1} \frac{r(t,s)}{s-t}\, y(s)\, ds$$

$$\boldsymbol{U} x(t) = \big(1 - i\pi\, \boldsymbol{r}(t,t)\big) x(t) - \int_{-1}^{1} \frac{\boldsymbol{r}(t,s)}{s-t}\, x(s)\, ds.$$

Der erste zu beweisende Satz besagt dann genauer, daß $T + K$ für genügend kleines $\|K\|$ genau ein intaktes Paar von Normatoren dieser Form besitzt.

Die Konstruktion von Normatoren hängt somit wesentlich am Studium von Integraloperatoren, deren Kerne eine Singularität wie $\frac{1}{s-t}$ besitzen.

Jedem Integraloperator G mit dem Kern $g(t,s)$, der denselben Bedingungen wie $k(t,s)$ genügt, ordnen wir einen Operator $\breve{G}$ zu, vermöge

$$\breve{G} x(t) = \pi g(t,t)\, x(t) - i \int_{-1}^{1} g(t,s)\, x(s) \frac{ds}{s-t},$$

wo das Integral als Cauchys Hauptwert zu verstehen ist. Dann lassen sich die Normatoren der angegebenen Gestalt kurz als

$$U = 1 + i\breve{R}, \qquad \boldsymbol{U} = 1 - i\breve{\boldsymbol{R}}$$

schreiben.

Die Operatoren $\breve{G}$ haben nun folgende Eigenschaften:

A: $\breve{G} x(t)$ liegt im selben Raume wie $x(t)$ und es gibt eine von μ abhängige Konstante N, so daß

$$\|\breve{G} x\| \leqq \mathsf{N}\, \|G\|\, \|x\|.$$

B: Mit G und H gehören auch $\breve{G}H$ und $G\breve{H}$ der genannten Klasse an und es ist

$$\|\breve{G}H\| \leqq \mathsf{N}\, \|G\|\, \|H\|, \quad \|G\breve{H}\| \leqq \mathsf{N}\, \|G\|\, \|H\|.$$

C: Es besteht die Identität

$$\breve{G}\breve{H} = \breve{\breve{G}H} + \breve{G\breve{H}},$$

und schließlich

D: $$T\breve{G} - \breve{G}T = iG.$$

Eigenschaft A hängt eng mit einem Satze von Priwaloff[5]) zusammen. Sie gilt nicht für $\mu = 1$ und sie beruht wesentlich auf der Randbedingung (0) für G.

[5]) Sur les fonctions conjuguées. Bull. de la soc. math. de France **44** (1916), S. 100; im Anschluß an Fatou, Acta math. **30** (1916), S. 361.

Die Identität C ist verwandt mit einer Identität, die Hilbert 1904 aufgestellt hat und die seitdem mehrfach untersucht worden ist[6]).

Die Kenntnis dieser vier Eigenschaften reicht völlig aus, um die Existenz eines intakten Paares von Normatoren $1 + i\breve{R}$ und $1 - i\breve{R}$ für genügend kleines $\|K\|$ zu beweisen.

Zur Behandlung der Spektralzerlegung von $T + K$ ohne Einschränkung für $\|K\|$ sind noch weitere Identitäten heranzuziehen. Alsdann gelangt man zum Ziele, indem man der Idee des Abspaltungsverfahrens von Erhardt Schmidt folgt.

§ 1.

Lipschitz-Hölder-Funktionen.

In diesem § 1 formulieren und beweisen wir vorbereitend eine Reihe von grundsätzlich bekannten Sätzen über Funktionen, die einer Lipschitz-Hölder-Bedingung genügen.

1. Es sei μ eine Zahl aus

$$0 < \mu < 1,$$

die stets festgehalten werde; wir setzen

$$\mathsf{M} = \frac{1}{\mu(1-\mu)}.$$

2. Die im folgenden behandelten Funktionen der reellen Variablen t seien komplexwertig und für $|t| \leqq 1$ definiert.

Ist $p(t)$ eine solche Funktion, so setzen wir

$$\Delta p = \frac{p(t'') - p(t')}{|t'' - t'|^{\mu}}.$$

---

[6]) D. Hilbert, Verh. d. 3. intern. Math. Kongr., Heidelberg (1904), S. 233—240; O. D. Kellogg, Math. Annalen 58 (1904), S. 441—456; F. Noether, Math. Annalen 82 (1921), S. 42. Weitere Literatur (Poincaré, Villat u. a.) in der Enz. d. math. Wissensch.; II C. 3. L. Lichtenstein Nr. 13, S. 233 [146]); II C. 13. E. Hellinger-O. Toeplitz, Nr. 21 b), S. 1452. Bei allen diesen Arbeiten liegt das Hauptinteresse in der Umkehrung von Operationen des Typus

$$y(t) = v(t)\,x(t) + \lambda \int_{-1}^{1} \frac{w(t,s)}{s-t}\,x(s)\,ds.$$

Das entsprechende Problem für Funktionen von mehr Variablen mit zugehörigen Identitäten behandelt G. Giraud, Ann. scient. Éc. norm. sup. 51 (1934), S. 251—372; 53 (1936), S. 1—40, J. math. pures appl. (9) 15 (1936), S. 193—205 im Anschluß an Poincaré, Picard, Bertrand. Ein ähnliches Problem behandelt Tricomi. Nachweise siehe bei Giraud.

Wir bezeichnen die obere Grenze von $|p(t)|$ für $|t| \leqq 1$ mit $\sup |p|$ und mit $\sup |\varDelta p|$ die obere Grenze von $|\varDelta p|$ für $-1 \leqq t' < t'' \leqq 1$. Dann setzen wir

$$\|p\| = \sup |p| + \sup |\varDelta p|.$$

Die Gesamtheit aller Funktionen $p(t)$, für die $\|p\| < \infty$ ist, heiße $\mathfrak{P}$. Die Gesamtheit der $p(t)$ aus $\mathfrak{P}$, die für $|t| = 1$ verschwinden, heiße $\dot{\mathfrak{P}}$; diese Funktionen $p(t)$ denken wir uns für $|t| > 1$ identisch Null fortgesetzt.

Es sei $p(t_1, \ldots, t_n)$ eine für $|t_1| \leqq 1, \ldots, |t_n| \leqq 1$ erklärte komplexwertige Funktion. Die Operation $\varDelta$, hinsichtlich der Variablen t_ν angewandt, heiße $\varDelta_{t_\nu}$. Mit ∇ bezeichnen wir jede der 2^n Operationen, die entstehen, wenn wir auf jede der n Variablen t_ν die Operation $\varDelta_{t_\nu}$ einmal oder keinmal anwenden[7]. $\sup |\nabla p|$ ist die obere Grenze von $|\nabla p|$ für $-1 \leqq t'_\nu < t''_\nu \leqq 1$ bzw. $|t_\nu| \leqq 1$ $(\nu = 1, \ldots, n)$, je nachdem ob $\varDelta_{t_\nu}$ angewandt wurde oder nicht. Wir setzen

$$\|p\| = \sum_{\nabla} \sup |\nabla p|.$$

Die Gesamtheit aller Funktionen $p(t_1, \ldots, t_n)$, für die $\|p\| \leqq \infty$ ist, heiße $\mathfrak{P}^n$. Die Mannigfaltigkeit aller p aus $\mathfrak{P}^n$, die hinsichtlich einiger Variablen am Ende des Intervalls, d. h. für $|t| = 1$ verschwinden, kennzeichnen wir, indem wir in $\mathfrak{P}^n = \overbrace{\mathfrak{P} \ldots \mathfrak{P}}$ an entsprechenden Stellen $\mathfrak{P}$ durch $\dot{\mathfrak{P}}$ ersetzen. Auch diese Funktionen denken wir uns für die betreffenden Variablen t_ν in $|t_\nu| \geqq 1$ identisch Null fortgesetzt.

$\mathfrak{P}^n$ und jede der genannten Teilmannigfaltigkeiten ist ein linearer Raum[8] und hinsichtlich der Norm $\|p\|$ vollständig.

Satz 1.1. *Gehört* $p(t_1, \ldots, t_n)$ *zu* $\mathfrak{P}^n$, *so liegt* $p'(t_1, \ldots, t_{n-1}) = p(t_1, \ldots, t_{n-1}, t_{n-1})$ *in* $\mathfrak{P}^{n-1}$ *und es ist*

$$\|p'\| \leqq \|p\|.$$

Beweis. Es enthalte ∇_0 nicht die Operationen $\varDelta_{t_{n-1}}$ und $\varDelta_{t_n}$. Dann entsteht $\nabla_0 p'$ aus $\nabla_0 p$, indem man die Nebenbedingung $t_n = t_{n-1}$ stellt; es ist also $\sup |\nabla_0 p'| \leqq \sup |\nabla_0 p|$. Ist $\nabla_1 = \nabla_0 \varDelta_{t_{n-1}}$ so entsteht $\nabla_1 p'$ aus $\nabla_0 \varDelta_{t_{n-1}} p + \nabla_0 \varDelta_{t_n} p$, indem man im ersten Summanden $t_n = t''_n$, im zweiten $t_{n-1} = t'_{n-1}$ und schließlich in beiden $t'_n = t'_{n-1}$, $t''_n = t''_{n-1}$ setzt. Somit ist $\sup |\nabla_1 p'| \leqq \sup |\nabla_0 \varDelta_{t_{n-1}} p| + \sup |\nabla_0 \varDelta_{t_n} p|$. Hieraus folgt $\sum_{\nabla} \sup |\nabla p'| \leqq \|p\|$.

Satz 1.2. *Gehört*

$$p(t_1, \ldots, t_m;\ r_1, \ldots, r_l) \ \ zu \ \ \mathfrak{P}^{m+l},$$
$$q(r_1, \ldots, r_l;\ s_1, \ldots, s_n) \ \ zu \ \ \mathfrak{P}^{l+n}, \qquad (l = 0, 1, \ldots),$$

[7]) Für $n = 2$ bezeichnet also ∇p jeden der Ausdrücke p, $\varDelta_{t_1} p$, $\varDelta_{t_2} p$, $\varDelta_{t_1} \varDelta_{t_2} p$.

[8]) $p = 0$ soll stets $p(t) \equiv 0$ bedeuten.

so liegt

$$pq(t_1, \ldots, t_m; r_1, \ldots, r_l; s_1, \ldots, s_n)$$
$$= p(t_1, \ldots, t_m; r_1, \ldots, r_l)\, q(r_1, \ldots, r_l; s_1, \ldots, s_n) \quad \textit{in} \quad \mathfrak{P}^{m+l+n}$$

und es ist

$$\|pq\| \leqq \|p\| \; \|q\|.$$

Beweis. Sei zunächst $l = 0$. Dann ist $\nabla pq = \nabla_t p \nabla_s q$, wo ∇_t und ∇_s bzw. auf die Variablen t und s operieren. Aus

$$\sup |\nabla pq| = \sup |\nabla_t p| \sup |\nabla_s p|$$

folgt

$$\sum_{\nabla} \sup |\nabla pq| = \sum_{\nabla_t} \sup |\nabla_t p| \sum_{\nabla_s} \sup |\nabla_s q|;$$

d. h.

$$\|pq\| = \|p\| \; \|q\|.$$

Ist nun $l > 0$, so folgt nach vorangehendem, daß die Funktion $pq' = p(t_1, \ldots, t_m; r_1, \ldots, r_l)\, q(r_1', \ldots, r_l'; s_1, \ldots, s_n)$ in $\mathfrak{P}^{m+2l+n}$ liegt mit $\|pq'\| = \|p\| \; \|q\|$. Identifizieren wir nun die Variablen r und r', so ergibt sich die Behauptung aus Satz 1. 1.

3. *Integrale* ohne Grenzen sollen sich stets über den Definitionsbereich des Integranden erstrecken, also, wenn der Integrand hinsichtlich der Integrationsvariablen zu $\mathfrak{P}$ bzw. $\dot{\mathfrak{P}}$ gehört, über $(-1, +1)$ bzw. $(-\infty, +\infty)$.

Satz 1. 3. Gehört $p(t_1, \ldots, t_n)$ zu $\mathfrak{P}^n$, so liegt

$$p'(t_1, \ldots, t_{n-1}) = \int p(t_1, \ldots, t_{n-1}, \tau)\, d\tau$$

in $\mathfrak{P}^{n-1}$ und es gilt

$$\|p'\| \leqq 2\|p\|.$$

Beweis. Enthält ∇ nicht Δ_{t_n}, so ist $\nabla p' = \int \nabla p\, d\tau$, also $\sup |\nabla p'| \leqq 2 \sup |\nabla p|$, woraus $\sum\limits_{\nabla} \sup |\nabla p'| \leqq 2\|p\|$ folgt.

4. Entscheidend ist das Studium der folgenden Operation. Der Funktion $q(r) = q(t_1, \ldots, t_{n-1}, r)$ aus $\mathfrak{P}^{n-1}\dot{\mathfrak{P}}$ wird die Funktion

$$p(r) = p(t_1, \ldots, t_{n-1}, r) = \int q(\varrho) \frac{d\varrho}{\varrho - r}$$

als „Cauchys Hauptwert" zugeordnet. D. h. wir definieren für $|r| \leqq 1$

$$p(r) = \int\limits_{|\varrho - r| \geqq a} q(\varrho) \frac{d\varrho}{\varrho - r} + \int\limits_0^a [q(r+\alpha) - q(r-\alpha)] \frac{d\alpha}{\alpha},$$

offenbar unabhängig von a, wobei man beachte, daß $q(r) \equiv 0$ für $|r| \geqq 1$ erklärt ist.

Hilfssatz 1. 1. Für $q(r)$ aus $\dot{\mathfrak{P}}$, $p(r) = \int q(\varrho) \frac{d\varrho}{\varrho - r}$ gilt

$$|p(r)| \leqq 4\,\mathsf{M} \sup |\Delta q|.$$

Beweis. Ohne Beschränkung sei $\sup |\Delta q| = 1$. Wir schließen

$$|p(r)| \leqq \left| \int_0^2 [q(r+\alpha) - q(r-\alpha)] \frac{d\alpha}{\alpha} \right| \leqq \int_0^2 |2\alpha|^{\mu} \frac{d\alpha}{\alpha} < 4\,\mathrm{M}.$$

Hilfssatz 1.2. Für $q(r)$ aus $\dot{\mathfrak{P}}$, $p(r) = \int q(\varrho) \frac{d\varrho}{\varrho - r}$ gilt

$$|\Delta p(r)| \leqq 8\,\mathrm{M} \sup |\Delta q|.$$

Dieser Satz entspricht dem von Priwaloff[9]).

Beweis. Ohne Beschränkung sei $\sup |\Delta q| = 1$. Wir setzen $r'' = r + b$, $r' = r - b$, $b > 0$; dann wird

$$p(r'') - p(r') = 2b \int q(\varrho) \frac{d\varrho}{(\varrho - r'')(\varrho - r')}$$

$$= 2b \int_{-\infty}^{r} [q(\varrho) - q(r')] \frac{d\varrho}{(\varrho - r'')(\varrho - r')} + 2b \int_{r}^{\infty} [q(\varrho) - q(r'')] \frac{d\varrho}{(\varrho - r'')(\varrho - r')};$$

denn es ist, da für $\varrho = r - b$ bzw. für $\varrho = r + b$ Cauchys Hauptwert zu nehmen ist,

$$\int_{-\infty}^{r} \frac{d\varrho}{(\varrho - r'')(\varrho - r')} = \int_{r}^{\infty} \frac{d\varrho}{(\varrho - r'')(\varrho - r')} = 0.$$

Wir gewinnen

$$|\Delta p| = \frac{|p(r'') - p(r')|}{(2b)^{\mu}} \leqq (2b)^{1-\mu} \int_{-\infty}^{r} \frac{d\varrho}{|\varrho - r''|\,|\varrho - r'|^{1-\mu}}$$

$$+ (2b)^{1-\mu} \int_{r}^{\infty} \frac{d\varrho}{|\varrho - r''|^{1-\mu}\,|\varrho - r'|} \leqq 2(2b)^{1-\mu} \int_{-b}^{\infty} \frac{d\alpha}{|\alpha|^{1-\mu}(\alpha + 2b)} < 8\,\mathrm{M}.$$

Satz 1.4. *Für* $q(r) = q(t_1, \ldots, t_{n-1}, r)$ *aus* $\mathfrak{P}^{n-1}\dot{\mathfrak{P}}$ *liegt*

$$p(t_1, \ldots, t_{n-1}, r) = \int q(\varrho) \frac{d\varrho}{\varrho - r}$$

in $\mathfrak{P}^n$ *und es ist*

$$\|p\| \leqq 12\,\mathrm{M}\,\|q\|.$$

Beweis. Für $n = 1$ entspringt die Behauptung aus den beiden Hilfssätzen. Ist $n > 1$, so wenden wir die Hilfssätze statt auf q auf $V_0 q$ an, wo V_0 die Operation Δ_r nicht enthält; dann gewinnen wir $|V_0 p| \leqq 4\,\mathrm{M} \sup |\Delta_r V_0 q|$, $|\Delta_r V_0 p| \leqq 8\,\mathrm{M} \sup |\Delta_r V_0 q|$ und damit $\sum_{V} \sup |V p| \leqq 12\,\mathrm{M}\,\|q\|$.

---

9) l. c. Für $p(r) = \int q(\varrho) \operatorname{cotg} \frac{\pi}{2}(\varrho - r)\, d\varrho$ und periodisches $q(r)$.

5. Unter $\mathfrak{Z}$ verstehen wir die Gesamtheit der komplexen Zahlen $\zeta = \xi + i\eta$, inklusive $z = \infty$, mit Ausnahme der Strecke $\eta = 0$, $|\xi| \leqq 1$. Für $q(r) = q(t_1, \ldots, t_{n-1}, r)$ aus $\mathfrak{P}^{n-1}\dot{\mathfrak{P}}$ ist dann

$$Q(\zeta) = \int q(\varrho) \frac{d\varrho}{\varrho - \zeta}$$

eine analytische Funktion von ζ in $\mathfrak{Z}$. Wir setzen mit $\varepsilon = \pm 1$, $|r| \leqq 1$,

$$Q^\varepsilon(r) = \int q(\varrho) \frac{d\varrho}{\varrho - r} + \varepsilon i \pi q(r);$$

nach Satz 1.4 gehört $Q^\varepsilon(r)$ zu $\mathfrak{P}^n$.

Hilfssatz 1.3. Mit $q(r)$ aus $\dot{\mathfrak{P}}$ gilt

$$|Q(r + \varepsilon i\varkappa) - Q^\varepsilon(r)| \leqq 2\,\mathsf{M}\,\varkappa^\mu \sup |\Delta q|$$

für $\varkappa > 0$ und $|r| \leqq 1$. Es konvergiert also $Q(r + \varepsilon i\varkappa)$ für $\varkappa \downarrow 0$ gleichmäßig in r gegen $Q^\varepsilon(r)$. (Das entspricht einer in der Potentialtheorie geläufigen Tatsache.)

Beweis. Ohne Beschränkung sei $\sup |\Delta q| = 1$. Es ist für $A > 2$

$$Q(r + \varepsilon i\varkappa) = \int_{-A}^{A} q(r + \alpha) \frac{d\alpha}{\alpha - \varepsilon i\varkappa}$$

$$= \int_{-A}^{A} [q(r + \alpha) - q(r)] \frac{d\alpha}{\alpha - \varepsilon i\varkappa} + \left[\log \frac{A - \varepsilon i\varkappa}{A + \varepsilon i\varkappa} + \varepsilon i\pi\right] q(r),$$

$$Q^\varepsilon(r) = \int_{-A}^{A} q(r + \alpha) \frac{d\alpha}{\alpha} + \varepsilon i\pi q(r)$$

$$= \int_{-A}^{A} [q(r + \alpha) - q(r)] \frac{d\alpha}{\alpha} + \varepsilon i\pi q(r),$$

$$Q(r + \varepsilon i\varkappa) - Q^\varepsilon(r) = \varepsilon i\varkappa \int_{-A}^{A} [q(r + \alpha) - q(r)] \frac{d\alpha}{(\alpha - \varepsilon i\varkappa)\alpha} + \log\left[\frac{A - \varepsilon i\varkappa}{A + \varepsilon i\varkappa}\right] q(r)$$

$$= \varepsilon i\varkappa \int_{-\infty}^{\infty} [q(r + \alpha) - q(r)] \frac{d\alpha}{(\alpha - \varepsilon i\varkappa)\alpha},$$

da $A > 2$ beliebig war. Also

$$|Q(r + \varepsilon i\varkappa) - Q^\varepsilon(r)| \leqq \varkappa \int_{-\infty}^{\infty} \frac{d\alpha}{|\alpha - \varepsilon i\varkappa|\,|\alpha|^{1-\mu}} \leqq 2\,\mathsf{M}\,\varkappa^\mu.$$

6. Wir stellen einige Sätze über die Vertauschung von Integrationen zusammen. Die Reihenfolge der Integrationen soll stets durch die Reihen-

folge der Differentiale festgelegt sein. Ist $q(r, s)$ eine Funktion aus $\dot{\mathfrak{P}}\mathfrak{P}$ bzw. $\dot{\mathfrak{P}}^2$, so liegt $\int q(\varrho, s)\frac{d\varrho}{\varrho - t}$ nach Satz 1.4 in $\mathfrak{P}^2$ bzw. $\mathfrak{P}\dot{\mathfrak{P}}$; also liegt nach Satz 1.1 $\int q(\varrho, s)\frac{d\varrho}{\varrho - s}$ in $\mathfrak{P}$ bzw. $\dot{\mathfrak{P}}$.

Hilfssatz 1.4. Für $q(r, s)$ aus $\dot{\mathfrak{P}}\mathfrak{P}$ gilt

$$\iint q(\varrho, \sigma)\frac{d\varrho}{\varrho - t}\,d\sigma = \iint q(\varrho, \sigma)\,d\sigma\,\frac{d\varrho}{\varrho - t}\,.$$

Beweis. Wenden wir Hilfssatz 1.3 auf $q(r) = q(r, \sigma)$ an und integrieren nach σ, so entsteht

$$\iint q(\varrho, \sigma)\frac{d\varrho}{\varrho - t - \varepsilon i \varkappa}\,d\sigma \to \iint q(\varrho, \sigma)\frac{d\varrho}{\varrho - t}\,d\sigma + \varepsilon i \pi \int q(t, \sigma)\,d\sigma;$$

wenden wir dagegen Hilfssatz 1.3 auf $q(r) = \int q(r, \sigma)\,d\sigma$ an, so kommt

$$\iint q(\varrho, \sigma)\,d\sigma\,\frac{d\varrho}{\varrho - t - \varepsilon i \varkappa} \to \iint q(\varrho, \sigma)\,d\sigma\,\frac{d\varrho}{\varrho - t} + \varepsilon i \pi \int q(t, \sigma)\,d\sigma.$$

Die linken Seiten stimmen überein, also auch die rechten.

Hilfssatz 1.5. Für $q(r, s)$ aus $\dot{\mathfrak{P}}^2$ gilt

$$\iint q(\varrho, \sigma)\frac{d\varrho}{\varrho - \sigma}\,d\sigma = \iint q(\varrho, \sigma)\frac{d\sigma}{\varrho - \sigma}\,d\varrho.$$

Beweis. Wir gehen aus von der Identität

$$\int\Big\{\int\limits_{|\sigma - \varrho| \geqq a} q(\varrho, \sigma)\frac{d\varrho}{\varrho - \sigma}\Big\}\,d\sigma = \int\Big\{\int\limits_{|\sigma - \varrho| \geqq a} q(\varrho, \sigma)\frac{d\varrho}{\varrho - \sigma}\Big\}\,d\varrho$$

und beachten, daß nach Definition von Cauchys Hauptwert die Integranden $\{\ \}$ für $a \to 0$ gleichmäßig in σ bzw. ϱ gegen $\int q(\varrho, \sigma)\frac{d\varrho}{\varrho - \sigma}$ bzw. $\int q(\varrho, \sigma)\frac{d\sigma}{\varrho - \sigma}$ streben.

7. Mit einer Funktion $q(r, s)$ aus $\dot{\mathfrak{P}}^2$ bilden wir die Funktion

$$Q(\zeta', \zeta) = \iint q(\varrho, \sigma)\frac{d\sigma}{\sigma - \zeta'}\,\frac{d\varrho}{\varrho - \zeta},$$

die für ζ, ζ' in $\mathfrak{Z}$ definiert ist. Mit $\varepsilon = \pm 1$, $\delta = \pm 1$ setzen wir

$$\begin{aligned} Q^{\varepsilon\delta}(r, s) = & \iint q(\varrho, \sigma)\frac{d\sigma}{\sigma - s}\,\frac{d\varrho}{\varrho - r} + \varepsilon i \pi \int q(t, \sigma)\frac{d\sigma}{\sigma - s} \\ & + \delta i \pi \int q(\varrho, s)\frac{d\varrho}{\varrho - r} - \delta\varepsilon\pi^2 q(r, s). \end{aligned}$$

Da $\int q(r, \sigma)\frac{d\sigma}{\sigma - s}$ in $\dot{\mathfrak{P}}\mathfrak{P}$ liegt, ist das erste Integral definiert. $Q^{\varepsilon\delta}(r, s)$ liegt somit in $\mathfrak{P}^2$.

Hilfssatz 1.6. Mit $q(r, s)$ aus $\dot{\mathfrak{P}}^2$ gilt

$$|Q(r + \varepsilon i \varkappa, s + \delta i \lambda) - Q^{\varepsilon\delta}(r, s)| \leqq (\varkappa^{\mu}\lambda^{\mu} + \varkappa^{\mu} + \lambda^{\mu})\, 16\, \mathsf{M}^2 \|q\|.$$

Es strebt also für $\varkappa \downarrow 0$, $\lambda \downarrow 0$, $Q(r + \varepsilon i \varkappa, s + \delta i \lambda)$ gleichmäßig in s, r gegen $Q^{\varepsilon\delta}(r, s)$.

Beweis. Wir setzen mit $\zeta = r + \varepsilon i \varkappa$, $\zeta' = s + \delta i \lambda$

$$p(r) = \int q(r,\sigma)\frac{d\sigma}{\sigma-\zeta'}, \qquad p_\delta(r) = \int q(r,\sigma)\frac{d\sigma}{\sigma-s} + \delta i \pi q(r,s),$$

$$P(\zeta) = \int p(\varrho)\frac{d\varrho}{\varrho-\zeta}, \qquad P_\delta(\zeta) = \int p_\delta(\varrho)\frac{d\varrho}{\varrho-\zeta},$$

$$P^\varepsilon(r) = \int p(\varrho)\frac{d\varrho}{\varrho-r} + \varepsilon i \pi p(r), \quad P_\delta^\varepsilon(r) = \int p_\delta(\varrho)\frac{d\varrho}{\varrho-r} + \varepsilon i \pi p_\delta(r).$$

Dann wird $Q(\zeta,\zeta') = P(\zeta)$, $Q^{\varepsilon\delta}(r,s) = P_\delta^\varepsilon(r)$,

Wir haben also $|P(\zeta) - P_\delta^\varepsilon(r)|$ abzuschätzen. Wir gewinnen nun der Reihe nach folgende Ungleichungen:

(1) $$|p(r) - p_\delta(r)| \leqq 2\,\mathsf{M}\,\lambda^\mu \sup|\Delta_s q|,$$

(2) $$|\Delta_r p(r)| - \Delta_r p_\delta(r)| \leqq 2\,\mathsf{M}\,\lambda^\mu \sup|\Delta_s \Delta_r q|,$$

indem wir Hilfssatz 1. 3 auf $q(s) = q(r,s)$ und $q(s) = \Delta_r q(r,s)$ anwenden.

(I) $$|P^\varepsilon(r) - P_\delta^\varepsilon(r)| \leqq 16\,\mathsf{M}^2\lambda^\mu \sup|\Delta_s\Delta_r q| + 2\pi\,\mathsf{M}\,\lambda^\mu \sup|\Delta_s q|,$$

indem wir Hilfssatz 1. 2 auf $p(r) - p_\delta(r)$ anwenden und Formeln (2) und (1) benutzen.

(II) $$|(P(\zeta) - P_\delta(\zeta)) - (P^\varepsilon(r) - P_\delta^\varepsilon(r))| \leqq 4\,\mathsf{M}^2\varkappa^\mu\lambda^\mu \sup|\Delta_s\Delta_r q|;$$

hierzu wenden wir Hilfssatz 1. 3 auf $q(r) = p(r) - p_\delta(r)$ an und benutzen (2).

(III) $$|P_\delta(\zeta) - P_\delta^\varepsilon(r)| \leqq 8\,\mathsf{M}^2\varkappa^\mu \sup|\Delta_s\Delta_r q| + 2\pi\,\mathsf{M}\,\varkappa^\mu \sup|\Delta_r q|;$$

hierzu wenden wir Hilfssatz 1. 3 auf $q(r) = p_\delta(r)$ an und schätzen $|\Delta_r p_\delta(r)|$ nach Hilfssatz 1. 1 ab durch

(3) $$|\Delta_r p_\delta(r)| \leqq 4\,\mathsf{M} \sup|\Delta_s\Delta_r q| + \pi \sup|\Delta_r q|.$$

Addieren wir die Abschätzungen (I), (II), (III), so entsteht

$$\begin{aligned}|P(\zeta) - P_\delta^\varepsilon(r)| &\leqq (4\varkappa^\mu\lambda^\mu + 16\lambda^\mu + 8\varkappa^\mu)\,4\,\mathsf{M}^2 \sup|\Delta_s\Delta_r q| \\ &\quad + 2\pi\,\mathsf{M}\,\varkappa^\mu \sup|\Delta_r q| + 2\pi\,\mathsf{M}\,\lambda^\mu \sup|\Delta_s q| \\ &\leqq (\varkappa^\mu\lambda^\mu + \varkappa^\mu + \lambda^\mu)\,16\,\mathsf{M}^2\{\sup|\Delta_s\Delta_r q| + \sup|\Delta_s q| + \sup|\Delta_s q|\};\end{aligned}$$

(denn es ist $\mathsf{M} \geqq 4$, also $2\pi < 16\,\mathsf{M}$), und damit die Behauptung von Hilfssatz 1. 6.

Satz 1. 5. *Für $q(r,s)$ aus $\mathfrak{P}^2$ besteht die Identität*[10])

$$\iint q(\varrho,\sigma)\frac{d\varrho}{\sigma-\varrho}\frac{d\varrho}{\varrho-t}$$
$$-\iint q(\varrho,\sigma)\frac{d\varrho}{\sigma-\varrho}\frac{d\sigma}{\sigma-t} - \iint q(\varrho,\sigma)\frac{d\varrho}{\varrho-t}\frac{d\sigma}{\varrho-t} + \pi^2 q(t,t) = 0.$$

---

[10]) Sie entspricht der von Hilbert, Kellogg, F. Noether l. c. verwandten Identität:

$$\iint \frac{\pi^2}{4} b(\varrho)\,\varphi(\sigma)\cot g\frac{\pi}{2}(\sigma-\varrho)\cot g\frac{\pi}{2}(\varrho-t)\,d\sigma\,d\varrho$$
$$-\iint \frac{\pi^2}{4} b(\varrho)\,\varphi(\sigma)\cot g(\sigma-\varrho)\cot g\frac{\pi}{2}(\varrho-t)\,d\varrho\,d\sigma + \pi^2 b(t)\,\varphi(t) = 0,$$

wobei $b(r)$ periodisch und bei F. Noether sonst nur stetig sein soll.

Beweis. Wir gehen aus von der Identität

$$\frac{1}{\sigma-\varrho}\,\frac{1}{\varrho-\zeta}-\frac{1}{\sigma-\varrho}\,\frac{1}{\sigma-\zeta}-\frac{1}{\sigma-\zeta}\,\frac{1}{\varrho-\zeta}=0.$$

Aus ihr gewinnen wir, für ζ aus $\mathfrak{Z}$,

$$\iint q(\varrho,\sigma)\frac{d\sigma}{\sigma-\varrho}\,\frac{d\varrho}{\varrho-\zeta}-\iint q(\varrho,\sigma)\frac{d\varrho}{\sigma-\varrho}\,\frac{d\sigma}{\sigma-\zeta}-\iint q(\varrho,\sigma)\frac{d\sigma}{\sigma-\zeta}\,\frac{d\varrho}{\varrho-\zeta}=0,$$

indem wir erst nach σ und dann nach ϱ integrieren und beim zweiten Integral auf Grund von Hilfssatz 1.5 die Reihenfolge vertauschen. Wir setzen $\zeta = t + i\varkappa$ und lassen $\varkappa \downarrow 0$ streben. Wir wenden Hilfssatz 1.3 auf die ersten beiden, Hilfssatz 1.6 mit $\zeta' = \zeta$ auf das dritte Integral an. Dann entsteht in der Grenze

$$\iint q(\varrho,\sigma)\frac{d\sigma}{\sigma-\varrho}\,\frac{d\varrho}{\varrho-t}+i\pi\int q(t,\sigma)\frac{d\sigma}{\sigma-t}$$
$$-\iint q(\varrho,\sigma)\frac{d\varrho}{\sigma-\varrho}\,\frac{d\sigma}{\sigma-t}-i\pi\int q(\varrho,t)\frac{d\varrho}{t-\varrho}$$
$$-\iint q(\varrho,\sigma)\frac{d\sigma}{\sigma-t}\,\frac{d\varrho}{\varrho-t}-i\pi\int q(t,\sigma)\frac{d\sigma}{\sigma-t}-i\pi\int q(\varrho,t)\frac{d\varrho}{\varrho-t}$$
$$+\pi^2 q(t,t)=0.$$

Diese Relation ergibt die Behauptung von Satz 1.5.

§ 2.

Operatoren.

1. Wir untersuchen einige Klassen von Operatoren, die auf Funktionen $x(t)$ des Raumes $\mathfrak{P}$ operieren.

Jede Funktion $f(t)$ von $\mathfrak{P}$ führt zu einem Operator $f(T)$, der aus $x(t)$ von $\mathfrak{P}$ die Funktion

$$f(T)\,x(t)=f(t)\,x(t)$$

erzeugt; diese liegt (nach Satz 1.2) auch in $\mathfrak{P}$ und es ist

$$\|f(T)\,x\| \leqq \|f\|\,\|x\|.$$

Jede Funktion $g(t,s)$ aus $\mathfrak{P}^2$ ist Kern eines *Integraloperators G*, der aus $x(t)$ von $\mathfrak{P}$ die Funktion

$$G\,x(t)=\int g(t,\sigma)\,x(\sigma)\,d\sigma$$

erzeugt, die (nach Satz 1.2 und 1.3) auch in $\mathfrak{P}$ liegt. Wir setzen

$$\|G\| = \|g\|;$$

dann gilt (nach Satz 1.3)

$$\|G\,x\| \leqq 2\,\|G\|\,\|x\|.$$

Die Klassen aller solchen Operatoren G, deren Kerne in $\mathfrak{P}^2$, $\mathfrak{P}\dot{\mathfrak{P}}$, $\dot{\mathfrak{P}}\mathfrak{P}$, $\dot{\mathfrak{P}}^2$ liegen, bezeichnen wir bzw. mit $\mathfrak{P}^2$, $\mathfrak{P}\dot{\mathfrak{P}}$, $\dot{\mathfrak{P}}\mathfrak{P}$, $\dot{\mathfrak{P}}^2$.

Operatoren G, G^* mit den Kernen $g(t,s)$, $g^*(t,s)$ heißen zueinander „formal-adjungiert", wenn

$$g^*(t,s) = \overline{g(s,t)}$$

ist; gehört G zu $\dot{\mathfrak{P}}^2$, so auch G^*.

2. Dem Operator G von $\mathfrak{P}\dot{\mathfrak{P}}$ mit dem Kern $g(t,s)$ ordnen wir den Operator $\tilde{G}$ mit dem Kern $\tilde{g}(t,s) = g(t,s)\frac{1}{s-t}$ zu; er erzeugt aus der Funktion $x(t)$ von $\mathfrak{P}$ die Funktion

$$\tilde{G}x(t) = \int g(t,\varrho)\, x(\varrho)\frac{d\varrho}{\varrho - t},$$

die (nach Satz 1.2, 1.4 und 1.1) in $\mathfrak{P}$ liegt mit

$$\|\tilde{G}x\| \leqq 12\,\mathsf{M}\,\|G\|\,\|x\|.$$

Die Rolle dieses Operators $\tilde{G}$ kommt zum Ausdruck in der *Identität*

$$\tilde{G}T - T\tilde{G} = G,$$

genauer, $\tilde{G}T\,x(t) - T\tilde{G}\,x(t) = G\,x(t)$, die offenbar für $x(t)$ aus $\mathfrak{P}$ besteht.

Eine erste wichtige Eigenschaft dieser Operatoren ist die folgende: Sind G und H Operatoren aus $\mathfrak{P}\dot{\mathfrak{P}}$, so gehören auch $\tilde{G}H$ und $G\tilde{H}$ zu $\mathfrak{P}\dot{\mathfrak{P}}$ und es ist

$$(*)\qquad \|\tilde{G}H\| \leqq 12\,\mathsf{M}\,\|G\|\,\|H\|;\quad \|G\tilde{H}\| \leqq 12\,\mathsf{M}\,\|G\|\,\|H\|;$$

sind $g(t,s)$, $h(t,s)$ die Kerne von G und H, so sind die Kerne von $\tilde{G}H$ und $G\tilde{H}$ bzw.

$$\tilde{g}h(t,s) = \int g(t,\varrho)\,h(\varrho,s)\frac{d\varrho}{\varrho - t};\qquad g\tilde{h}(t,s) = \int g(t,\varrho)\,h(\varrho,s)\frac{d\varrho}{s-\varrho}.$$

Da nach Satz 1.2 $g(t,r)\,h(r,s)$ in $\mathfrak{P}\dot{\mathfrak{P}}^2$ liegt, folgt aus Satz 1.4, daß $g\tilde{h}(t,s)$ und $\tilde{g}h(t,s)$ in $\mathfrak{P}\dot{\mathfrak{P}}$ liegen mit $\|g\tilde{h}\| \leqq 12\,\mathsf{M}\,\|g\|\,\|h\|$, $\|\tilde{g}h\| \leqq 12\,\mathsf{M}\,\|g\|\,\|h\|$. Ferner liegt nach Satz 1.2 auch $g(t,r)\,h(r,s)\,x(s)$ in $\mathfrak{P}\dot{\mathfrak{P}}^2$. Wir können die Hilfssätze 1.4 bzw. 1.5 auf $q(\varrho,\sigma) = g(t,\varrho)\,h(\varrho,\sigma)\,x(\sigma)$ anwenden und erhalten in der Tat

$$\begin{aligned}\tilde{G}H\,x(t) &= \iint g(t,\varrho)\,h(\varrho,\sigma)\,x(\sigma)\frac{d\sigma}{\varrho-t}\,d\varrho\\ &= \iint g(t,\varrho)\,h(\varrho,\sigma)\,x(\sigma)\frac{d\varrho}{\varrho-t}\,d\sigma = \int \tilde{g}h(t,\sigma)\,x(\sigma)\,d\sigma\end{aligned}$$

bzw.

$$\begin{aligned}G\tilde{H}\,x(t) &= \iint g(t,\varrho)\,h(\varrho,\sigma)\,x(\sigma)\,d\sigma\frac{d\varrho}{\sigma-\varrho}\\ &= \iint g(t,\varrho)\,h(\varrho,\sigma)\,x(\sigma)\frac{d\varrho}{\sigma-\varrho}\,d\sigma = \int g\tilde{h}(t,\sigma)\,x(\sigma)\,d\sigma.\end{aligned}$$

3. Dem Operator G mit dem Kern $g(t,s)$ von $\mathfrak{P}\dot{\mathfrak{P}}$ ordnen wir ferner den Operator

$$\dot{G} = \pi g(T,T)$$

zu; er erzeugt aus $x(t)$ von $\mathfrak{P}$ die Funktion

$$\mathring{G}\, x(t) = \pi\, g(t,t)\, x(t)$$

aus $\mathfrak{P}$ und es gilt (nach Satz 1. 2)

$$(\substack{*\\*}) \qquad \|\mathring{G}x\| \leqq \pi \|G\|\, \|x\|.$$

Mit G und H gehören auch $\mathring{G}H$ und $G\mathring{H}$ zu $\mathfrak{P}\dot{\mathfrak{P}}$ und es ist offenbar

$$(\substack{\circ\\\circ}) \qquad \overset{\circ}{\mathring{G}H} = \overset{\circ}{G\mathring{H}} = \mathring{G}\mathring{H},$$

$$(\substack{\sim\\\circ}) \qquad \widetilde{\mathring{G}H} = \mathring{G}\widetilde{H}, \qquad \widetilde{G\mathring{H}} = \widetilde{G}\mathring{H}.$$

Grundlegend sind die folgenden beiden *Identitäten* für G, H aus $\mathfrak{P}\dot{\mathfrak{P}}$:

$$(\substack{\circ\\\sim}) \qquad \overset{\circ}{\widetilde{G}H} + \overset{\circ}{G\widetilde{H}} = 0,$$

$$(\substack{\sim\\\sim}) \qquad \widetilde{\widetilde{G}H} + \widetilde{G\widetilde{H}} = \widetilde{G}\widetilde{H} - \mathring{G}\mathring{H}.$$

Die erste $(\substack{\circ\\\sim})$ folgt daraus, daß der Kern von $\widetilde{G}H + G\widetilde{H}$:

$$\widetilde{g}h(t,s) + g\widetilde{h}(t,s) = \int g(t,\varrho)\, h(\varrho,s)\left[\frac{1}{\varrho - t} + \frac{1}{s-\varrho}\right] d\varrho$$

für $s = t$ verschwindet. Die zweite Identität $(\substack{\sim\\\sim})$ entspringt unmittelbar aus dem Satz 1. 5, wenn man ihn auf die Funktion $q(r,s) = g(t,r)\, h(r,s)\, x(s)$ anwendet, die (nach Satz 1. 2) in $\mathfrak{P}\dot{\mathfrak{P}}^2$ liegt.

4. Dem Operator G aus $\mathfrak{P}\dot{\mathfrak{P}}$ ordnen wir noch den Operator

$$\breve{G} = \mathring{G} - i\widetilde{G}$$

zu. Aus dem Vorhergehenden gewinnen wir dann die Sätze A, B, D der Einleitung mit $\mathfrak{P}$ als Raum der $x(t)$, $\dot{\mathfrak{P}}\mathfrak{P}$ als Klasse der G, H und

$$\mathsf{N} = \pi + 12\,\mathsf{M};$$

an Stelle von Satz C erhalten wir allgemeiner die Identitäten:

$$(\substack{\circ\\\smile}) \qquad \overset{\circ}{\breve{G}H} + \overset{\circ}{G\breve{H}} = 2\,\mathring{G}\mathring{H},$$

$$(\substack{\smile\\\smile}) \qquad \breve{\breve{G}H} + \breve{G\breve{H}} = \breve{G}\breve{H}.$$

Die erste folgt aus $(\substack{\circ\\\circ})$ und $(\substack{\circ\\\sim})$; die zweite aus $(\substack{\circ\\\circ})$, $(\substack{\sim\\\sim})$, $(\substack{\circ\\\sim})$, $(\substack{\sim\\\circ})$:

$$\begin{aligned}\breve{\breve{G}H} + \breve{G\breve{H}} &= \overset{\circ}{\mathring{G}H} + \overset{\circ}{G\mathring{H}} - (\widetilde{\widetilde{G}H} + \widetilde{G\widetilde{H}}) \\ &\quad - i(\overset{\circ}{\widetilde{G}H} + \overset{\circ}{G\widetilde{H}}) - i(\widetilde{\mathring{G}H} + \widetilde{G\mathring{H}}) \\ &= \mathring{G}\mathring{H} - \widetilde{G}\widetilde{H} - i(\mathring{G}\widetilde{H} + \widetilde{G}\mathring{H}) = (\mathring{G} - i\widetilde{G})(\mathring{H} - i\widetilde{H}).\end{aligned}$$

5. Schließlich sei noch für G, H aus $\dot{\mathfrak{P}}^2$

$$(\breve{G}H)^* = H^*\breve{G}^*, \quad (G\breve{H})^* = \breve{H}^* G^*$$

angemerkt.

§ 3.

Spektralzerlegung des Operators $T + K$.

1. Es sei im folgenden K stets ein Operator aus $\mathfrak{P}\dot{\mathfrak{P}}$ mit dem Kern $k(t, s)$, der also die Randbedingung (0) der Einleitung erfüllt. Unsere Absicht ist, die Spektraleigenschaften des Operators

$$T + K$$

zu untersuchen. Zu dem Zweck suchen wir, wie in der Einleitung bemerkt, lineare Transformationen U, $\boldsymbol{U}$, die aus Funktionen von $\mathfrak{P}$ wieder solche erzeugen, derart, daß

$$(T + K)\,U = U\,T, \tag{1}$$

$$\boldsymbol{U}(T + K) = T\,\boldsymbol{U} \tag{1}$$

gilt; solche Transformationen nennen wir „Normatoren".

Zwei Normatoren U, $\boldsymbol{U}$ nennen wir ein „Paar", wenn

$$\boldsymbol{U}\,U = 1 \tag{2}$$

gilt; diese Relation hat zur Folge, daß $U\,y(t) \neq 0$ ist für $y(t) \neq 0$. Wir nennen das Paar von Normatoren „intakt", wenn noch die Relation

$$U\,\boldsymbol{U} = 1 \tag{3}$$

besteht; sie hat zur Folge, daß $\boldsymbol{U}\,x(t) \neq 0$ ist für $x(t) \neq 0$. Umgekehrt gilt:

Ist U, $\boldsymbol{U}$ ein Paar von Normatoren, und folgt $x(t) = 0$ aus $\boldsymbol{U}\,x(t) = 0$, so ist das Paar intakt. Denn es ist nach (2) $\boldsymbol{U}(U\,\boldsymbol{U} - 1)\,x(t) = 0$.

Beiläufig bemerken wir: Besitzt der Operator $T + K$ ein Paar intakter Normatoren U, $\boldsymbol{U}$ so läßt er sich darstellen als

$$T + K = U\,T\,\boldsymbol{U};$$

diese Darstellung können wir auch als die Spektralzerlegung von $T + K$ bezeichnen.

Wir wollen zeigen, daß man solche Normatoren unter den Operatoren

$$U = 1 + i\breve{R}$$

$$\boldsymbol{U} = 1 - i\breve{\boldsymbol{R}}$$

zu suchen hat, wo R und $\boldsymbol{R}$ Operatoren aus $\mathfrak{P}\dot{\mathfrak{P}}$ sind.

Liegt K in $\dot{\mathfrak{P}}^2$ und ist $U = (1 + i\breve{R})$ Rechtsnormator von $T + K$, so ist $U^* = (1 - i\breve{R}^*)$ Linksnormator von $T + K^*$. Ist $K = K^*$ formal-selbstadjungiert, so gewinnt man also aus dem Rechtsnormator $U = (1 + i\breve{R})$ in $U^* = (1 - i\breve{R}^*)$ einen Linksnormator.

2. Für Normatoren $U = 1 + i\breve{R}$, $\boldsymbol{U} = 1 - i\breve{\boldsymbol{R}}$ sind die Relationen (1), (1) gleichwertig mit

$$K(1 + i\breve{R}) = R, \tag{1)'$$

$$(1 - i\breve{\boldsymbol{R}})K = \boldsymbol{R}, \tag{1)'$$

wie sich nach D aus $\breve{R}\, T - T \breve{R} = i R$, $\breve{\mathbf{R}}\, T - T \breve{\mathbf{R}} = i \mathbf{R}$ ergibt. Aus den Relationen (1)', (1)' folgert man

$$(1 - i \breve{\mathbf{R}})\, R = (1 - i \breve{\mathbf{R}})\, K\, (1 + i \breve{R}) = \mathbf{R}\,(1 + i \breve{R}),$$

also

$$(\dagger) \qquad R - \mathbf{R} = i\,(\breve{\mathbf{R}} R + \mathbf{R} \breve{R}).$$

Nach § 2 $(\smile)$ wird somit $\breve{R} - \breve{\mathbf{R}} = i \breve{\mathbf{R}} \breve{R}$, also

$$\begin{aligned} \mathbf{U} U &= (1 - i \breve{\mathbf{R}})\,(1 + i \breve{R}) \\ &= 1 + i\,(\breve{R} - \breve{\mathbf{R}}) + \breve{\mathbf{R}} \breve{R} = 1. \end{aligned}$$

So gewinnen wir

$$(2) \qquad \mathbf{U} U = 1,$$

d. h.

Satz 3.1. *Zwei Normatoren der Gestalt $U = 1 + i \breve{R}$, $\mathbf{U} = 1 - i \breve{\mathbf{R}}$ des Operators $T + K$ bilden ein Paar.*

Der Operator $T + K$ besitzt höchstens *ein* Paar von intakten Normatoren der Form $U = (1 + i \breve{R})$, $\mathbf{U} = (1 - i \breve{\mathbf{R}})$. Denn ist $U' = (1 + i \breve{R}')$ noch ein Rechtsnormator, so folgt aus Satz 3.1 $U' = U \mathbf{U} U' = U$.

3. Wir nehmen nun an, daß der Operator $T + K$ ein Paar von Normatoren $U = 1 + i \breve{R}$, $\mathbf{U} = 1 - i \breve{\mathbf{R}}$ besitzt. Wir können dann die Spektralzerlegung gewinnen, auch wenn die Normatoren nicht intakt sind. Es zeigt sich nämlich, daß zum Spektrum $|t| \leqq 1$ nur noch endlich viele Punkteigenwerte hinzutreten.

Punkteigenwert von $T + K$ ist jede komplexe Zahl τ, zu der es eine Funktion $z(t) \neq 0$ von $\mathfrak{P}$ gibt, mit der

$$(T + K)\, z(t) = \tau\, z(t)$$

gilt; Hauptfunktion zu λ ist jede Funktion $z(t) \neq 0$ aus $\mathfrak{P}$, zu der es ein ganzes $m > 0$ gibt, so daß

$$(T + K - \tau)^m\, z(t) = 0$$

ist. Für jede solche Hauptfunktion $z(t)$ gilt $\mathbf{U} z(t) = 0$. Denn es ist

$$(t - \lambda)^m\, \mathbf{U} z(t) = (T - \lambda)^m\, \mathbf{U} z(t) = \mathbf{U}(T + K - \lambda)^m z(t) = 0;$$

also $\mathbf{U} z(t) = 0$.

4. Wir untersuchen zunächst den Raum $\mathfrak{O}$ aller Funktionen $z(t)$ aus $\mathfrak{P}$, für die

$$\mathbf{U} z(t) = 0$$

ist, und wir werden zeigen, daß $\mathfrak{O}$ von den Hauptfunktionen aller Punkteigenwerte von $T + K$ aufgespannt wird. Das ist sehr einfach auf Grund der folgenden entscheidenden Eigenschaft des Raumes $\mathfrak{O}$.

Satz 3.2[11]. *Der Raum* $\mathfrak{O}$ *aller* $z(t)$ *von* $\mathfrak{P}$ *mit* $Uz(t) = 0$ *hat endliche Dimension.*

(Insbesondere hat also $T + K$ höchstens endlich viele Punkteigenwerte und diese sind von endlicher Vielfachheit.)

Zum Beweise führen wir den Operator

$$O = i(\boldsymbol{R} - R) - \breve{R}\boldsymbol{R} - R\breve{\boldsymbol{R}}$$

aus $\mathfrak{P}\dot{\mathfrak{P}}$ ein; wie man nach § 2 ($\smile$) errechnet, wird

$$UU = 1 - \breve{O}.$$

Für jede Funktion $z(t)$ von $\mathfrak{O}$ ist also

$$z(t) = \breve{O}z(t).$$

Der Operator O hat nun die Eigenschaft, daß

$$\mathring{O} = 0$$

ist; denn nach § 2 ($\circ$) ist zunächst $\mathring{O} = i(\mathring{\boldsymbol{R}} - \mathring{R}) - 2\mathring{R}\mathring{\boldsymbol{R}}$; andererseits gilt nach (†)

$$i(\boldsymbol{R} - R) = \breve{\boldsymbol{R}}R + \boldsymbol{R}\breve{R},$$

also nach § 2 ($\circ$)

$$i(\mathring{\boldsymbol{R}} - \mathring{R}) = 2\mathring{\boldsymbol{R}}\mathring{R};$$

wegen

$$\mathring{\boldsymbol{R}}\mathring{R} = \mathring{R}\mathring{\boldsymbol{R}}$$

folgt daher $\mathring{O} = 0$.

Ist $o(t, s)$ der Kern von O, so ist also $o(t, t) = 0$; folglich

$$|o(t, s)| = |o(t, t) - o(t, s)| \leqq |t - s|^{\mu}\,\|o\|.$$

Für den Kern $\tilde{o}(t, s) = o(t, s)\frac{1}{s-t}$ von $\tilde{O}$ gilt also

$$|\tilde{o}(t, s)| \leqq |t - s|^{\mu-1}\,\|o\|;$$

d. h. der Integraloperator $\tilde{O}$ wird bei $s = t$ höchstens von der Ordnung $1 - \mu < 1$ unendlich. Infolgedessen gibt es ein ganzes $k > 0$ und $2k$ stetige Funktionen $p_1(t), \ldots, p_k(t)$; $q_1(s), \ldots, q_k(s)$, so daß mit

$$\tilde{o}_k(t, s) = p_1(t)\overline{q_1(s)} + \ldots + p_k(t)\overline{q_k(s)}$$

$$\sup_t \int |\tilde{o}(t, s) - \tilde{o}_k(t, s)|\, ds < 1$$

ist.

---

[11] Dieser Satz ist verwandt mit dem Satze von F. Noether (loc. cit.): Es sei $v(t)$, $w(t)$ stetig, $\int\int A(s, t)^2\, ds\, dt < \infty$. Dann hat der Raum aller $z(t)$ mit

$$v(t)\,z(t) + \int w(s) \operatorname{cotg} \frac{\pi}{2}(t - s)\, z(s)\, ds + \int A(s, t)\, z(s)\, ds = 0$$

endliche Dimension, wenn in $|t| \leqq 1$

$$v^2(t) + w^2(t) \neq 0$$

ist. (An Stelle dieser Bedingung nutzen wir $UU = 1$ aus.) F. Noether beweist übrigens ferner, daß diese Dimension nicht von $A(s, t)$ abhängt.

Wäre nun die Dimension von $\mathfrak{D}$ unendlich (oder auch nur $> k$), so enthielte $\mathfrak{D}$ eine Funktion $z(t)$ mit $\sup_t |z(t)| = 1$ und

$$\int \overline{q_1(s)}\, z(s)\, ds = \ldots = \int \overline{q_k(s)}\, z(s)\, ds = 0;$$

also wäre

$$z(t) = \breve{O} z(t) = -i \breve{O} z(t) = -i \int \tilde{o}(t,s)\, z(s)\, ds = -i \int [\tilde{o}(t,s) - \tilde{o}_k(t,s)]\, z(s)\, ds,$$

und somit

$$|z(t)| \leqq \sup_t \int |\tilde{o}(t,s) - \tilde{o}_k(t,s)|\, ds \, \sup_s |z(s)| < 1,$$

im Widerspruch zu $\sup_t |z(t)| = 1$.

(Kurz gesagt: der Operator $\breve{O}$ ist vollstetig in bezug auf die Norm $\sup_t |z(t)|$ und infolgedessen hat sein Eigenwert 1 endliche Vielfachheit.)

5. Der Operator $T + K$ erzeugt aus jeder Funktion $z(t)$ von $\mathfrak{D}$ wieder eine Funktion aus $\mathfrak{D}$. Denn es ist für $z(t)$ aus $\mathfrak{D}$

$$U(T + K)\, z(t) = T\, U z(t) = 0.$$

Da der Raum $\mathfrak{D}$ endliche Dimension hat, läßt sich die Elementarteilertheorie auf $T + K$ in $\mathfrak{D}$ anwenden. Es gibt also endlich viele Eigenwerte τ_ν, $(\nu = 1, \ldots, n)$, (die nicht verschieden zu sein brauchen), und zu jedem von ihnen endlich viele Hauptfunktionen $u_{\nu,\mu}(t)$, $(\mu = 1, \ldots, m_\nu)$, die der Relation

$$(T + K - \tau_\nu)\, u_{\nu,\mu}(t) = u_{\nu,\mu-1}(t), \quad \text{mit } u_{\nu,0}(t) = 0,$$

genügen. Zu jeder Funktion $z(t)$ von $\mathfrak{D}$ gibt es Zahlen $y_{\nu,\mu}$, so daß

$$z(t) = \sum_{\nu,\mu} y_{\nu,\mu}\, u_{\nu,\mu}(t)$$

ist; diese Zuordnungen bezeichnen wir mit $y_{\nu,\mu} = \mathbf{U}_{\nu,\mu}\, z(t)$. Eine Funktion $x(t)$ von $\mathfrak{P}$ zerlegen wir in

$$x(t) = U\, \mathbf{U}\, x(t) + (1 - U\, \mathbf{U})\, x(t),$$

dann liegt $(1 - U\,\mathbf{U})\, x(t)$ in $\mathfrak{D}$; wir setzen $\mathbf{U}_{\nu,\mu}\, x(t) = \mathbf{U}_{\nu,\mu}(1 - U\mathbf{U})\, x(t)$. So gewinnen wir

Satz 3. 3. *Der Operator $T + K$ mit K aus $\mathfrak{P}\dot{\mathfrak{P}}$ besitze ein Paar von Normatoren $U = (1 + i\breve{R})$, $\mathbf{U} = (1 - i\breve{\mathbf{R}})$; dann gibt es endlich viele Punkteigenwerte τ_ν, $(\nu = 1, \ldots, n)$, mit Hauptfunktionen $u_{\nu,\mu}(t)$, $(\mu = 1, \ldots, m_\nu)$.*

Jede Funktion $x(t)$ von $\mathfrak{P}$ läßt sich darstellen mit Hilfe von Zahlen

$$y_{\nu,\mu} = \mathbf{U}_{\nu,\mu}\, x(t)$$

und von

$$y(t) = \mathbf{U}\, x(t)$$

in der Form

$$x(t) = U\, y(t) + \sum_{\nu,\mu} u_{\nu,\mu}(t)\, y_{\nu,\mu},$$

so daß

$$(T + K)\, x(t) = U\, t\, y(t) + \sum_{\nu, \mu} u_{\nu,\mu}(t)\, (\tau_\nu\, y_{\nu,\mu} + y_{\nu,\mu+1})$$

wird, wo $y_{\nu, m_\nu + 1} = 0$ *zu setzen ist.*

Damit ist die Spektralzerlegung von $T + K$ geleistet.

§ 4.

Existenz von Normatoren.

1. Unter einfachen Bedingungen für den Operator $T + K$ läßt sich die Existenz eines Paares von Normatoren sicherstellen.

Satz 4.1. *Der Operator* $T + K$ *(*K *aus* $\mathfrak{P}\dot{\mathfrak{P}}$*) besitzt ein Paar von intakten Normatoren* $U = 1 + i\breve{R}$, $\boldsymbol{U} = 1 - i\breve{\boldsymbol{R}}$ *(*R, $\boldsymbol{R}$ *aus* $\mathfrak{P}\dot{\mathfrak{P}}$*), wenn* $\|K\|$ *genügend klein ist.*

Wir nehmen an

$$\odot \qquad \|K\| < \frac{1}{2\,\mathsf{N}}.$$

I. Dann zeigen wir zunächst, daß ein Paar von Normatoren $U = (1 + i\breve{R})$, $\boldsymbol{U} = (1 - i\breve{\boldsymbol{R}})$ intakt ist, wenn $\odot$ gilt.

a) Aus $(\mathbf{1})'$, d. h. $(1 - i\breve{\boldsymbol{R}})K = \boldsymbol{R}$ folgt nach B

$$\|\boldsymbol{R}\| \leqq \{1 + \mathsf{N}\,\|\boldsymbol{R}\|\}\,\|K\|,$$

also

$$2\,\mathsf{N}\,\|\boldsymbol{R}\| < 1 + \mathsf{N}\,\|\boldsymbol{R}\|;$$

d. h.

$$\mathsf{N}\,\|\boldsymbol{R}\| < 1.$$

b) Ist $(1 - i\breve{\boldsymbol{R}})\,x(t) = 0$, für ein $x(t)$ von $\mathfrak{P}$, so ist $x(t) = 0$. Denn aus $x(t) = i\breve{\boldsymbol{R}}\,x(t)$ folgt $\|x\| \leqq \mathsf{N}\,\|\boldsymbol{R}\|\;\|x\| < \|x\|$, wenn $x(t) \neq 0$. Also ist das Paar $U, \boldsymbol{U}$ intakt.

II. Wir zeigen, daß unter $\odot$ ein Paar von Normatoren zu $T + iK$ existiert. Wir lösen die Gleichung $(1)'$ $R = K(1 + i\breve{R})$ durch Iteration. Wir setzen $R_0 = 0$, $R_{n+1} = K(1 + i\breve{R}_n)$. Nach B wird dann

$$\|R_{n+1} - R_n\| \leqq \|K\|\;\mathsf{N}\,\|R_n - R_{n-1}\| < \tfrac{1}{2}\,\|R_n - R_{n-1}\|.$$

Also konvergiert die Folge R_n in sich; da der Raum $\mathfrak{P}\dot{\mathfrak{P}}$ abgeschlossen ist, enthält er einen Operator R mit $\|R_n - R\| \to 0$. Aus

$$R - K(1 + i\breve{R}) = (R - R_{n+1}) - iK(\breve{R} - \breve{R}_n)$$

folgt

$$\|R - K(1 - i\breve{R})\| \leqq \|R - R_{n+1}\| + \|K\|\;\|\mathsf{N}\;\|R - R_n\| \to 0,$$

also $R = K(1 + i\breve{R})$.

Ebenso ergibt sich die Existenz eines Operators $\boldsymbol{R}$ aus $\mathfrak{P}\dot{\mathfrak{P}}$, der $(\mathbf{1})'$ erfüllt.

2. Wir geben nun Bedingungen dafür an, daß $T + K$ wenigstens ein Paar von (nicht notwendig intakten) Normatoren $U = (1 + i\,R)$, $\overline{U} = (1 - i\,R)$ besitzt.

Jeder reellen Zahl τ aus $|\tau| \leqq 1$ ordnen wir den Raum $\mathfrak{P}_\tau$ der Funktionen

$$\frac{x(t)}{t-\tau}$$

zu, wo $x(t)$ die Funktionen von $\mathfrak{P}$ durchläuft. Ist G ein Operator aus $\mathfrak{P}\dot{\mathfrak{P}}$, so erklären wir $G\frac{x(t)}{t-\tau}$ durch

$$G\frac{x(t)}{t-\tau} = \int g(t,s)\,x(s)\,\frac{ds}{s-\tau} - i\,\pi\,g(t,\tau)\,x(\tau);$$

diese Funktion liegt nach Satz 1.2, 1.4 in $\mathfrak{P}$. Ferner setzen wir

$$\mathring{G}\frac{x(t)}{t-\tau} = g(t,t)\frac{x(t)}{t-\tau}$$

und erklären $\widetilde{G}\frac{x(t)}{t-\tau}$ durch

$$\widetilde{G}\frac{x(t)}{t-\tau} = \frac{1}{t-\tau}\Big[\int g(t,s)\,x(s)\,\frac{ds}{s-t} - \int g(t,s)\,x(s)\,\frac{ds}{s-\tau} + i\,\pi\,g(t,\tau)\,x(\tau)\Big];$$

diese Funktionen liegen wieder in $\mathfrak{P}_\tau$.

Man überzeuge sich davon, daß diese Definitionen, wenn $\frac{x(t)}{t-\tau}$ selbst in $\mathfrak{P}$ liegt, in die von § 2 übergehen.

Eine reelle Zahl τ aus $|\tau| \leqq 1$ heiße „Punkteigenwert im weiteren Sinne" von $T + K$, wenn es eine Funktion $\frac{x(t)}{t-\tau} \neq 0$ von $\mathfrak{P}_\tau$ gibt mit

$$(T+K)\frac{x(t)}{t-\tau} = \tau\frac{x(t)}{t-\tau},$$

was gleichbedeutend ist mit

$$x(t) + \int k(t,s)\,x(s)\,\frac{ds}{s-\tau} - i\pi\,k(t,\tau)\,x(\tau) = 0.$$

Der Operator K soll im folgenden zu $\dot{\mathfrak{P}}^2$ gehören.

3. Ein Operator K' aus $\dot{\mathfrak{P}}^2$ heiße von endlichem Range, wenn es ein ganzes $k > 0$ und $2\,k$ Funktionen $p_\alpha(t)$, $q_\alpha(s)$ $(\alpha = 1, \ldots, k)$ aus $\dot{\mathfrak{P}}$ gibt, so daß [12])

$$k'(t,s) = p_\alpha(t)\,\overline{q_\alpha(s)} = p_1(t)\,\overline{q_1(s)} + \ldots + p_k(t)\,\overline{q_k(s)}$$

ist. Wir verlangen nun:

(!) Der Operator K aus $\dot{\mathfrak{P}}^2$ läßt sich durch Operatoren K' aus $\dot{\mathfrak{P}}^2$ von endlichem Range so approximieren, daß $\|K' - K\|$ beliebig klein ist.

Diese Bedingung ist jedenfalls erfüllt, wenn $k(t,s)$ stetige erste und gemischt-zweite Ableitungen besitzt.

---

[12]) Wir unterdrücken hier und im folgenden Summenzeichen, wenn der Index α oder β zweimal auftritt.

Satz 4.2. *Es sei K ein Operator aus $\dot{\mathfrak{P}}^2$, der (!) erfüllt. Hat $T + K$ keinen reellen Eigenwert τ aus $|\tau| \leqq 1$ im weiteren Sinne, so besitzt er einen Rechtsnormator der Gestalt $U = 1 + i\breve{R}$.*

Hat $T + K^*$ keinen reellen Eigenwert τ aus $|\tau| \leqq 1$ im weiteren Sinne, so gibt es zu $T + K$ einen Linksnormator $\boldsymbol{U} = 1 - i\breve{\boldsymbol{R}}$.

Die zweite Behauptung folgt aus der ersten; denn ist U_* Rechtsnormator von $T + K^*$, so ist U_*^* Linksnormator von $T + K$.

Die erste Behauptung beweisen wir nach dem Muster des Abspaltungsverfahrens von E. Schmidt. Wir approximieren K durch einen Operator K' endlichen Ranges so, daß $\|K' - K\| < \frac{1}{2\,\mathrm{N}}$ ist. Es sei $k'(t, s) = p_\alpha(t)\overline{q_\alpha(s)}$ der Kern von K'. Wir dürfen $K \neq 0$ voraussetzen und annehmen, daß die Funktionen $p_\alpha(t)$ linear unabhängig sind. Nach Satz 4.1 gibt es ein Paar von intakten Normatoren $U_0 = 1 + i\breve{R}_0$, $\boldsymbol{U}_0 = 1 - i\breve{\boldsymbol{R}}_0$, so daß $\boldsymbol{U}_0(T + K - K')U_0 = T$, also $\boldsymbol{U}_0(T + K)U_0 = T + \boldsymbol{U}_0 K' U_0$ ist. Der Operator $T + \boldsymbol{U}_0 K' U_0$ hat dieselben Eigenwerte wie $T + K$, auch im weiteren Sinne, wie man nachprüfe; ist U' zu ihm ein Rechtsnormator, so ist $U_0 U'$ ein Rechtsnormator von $T + K$.

Der Operator $\boldsymbol{U}_0 K' U_0$ ist aber ebenso von endlichem Range wie K'; sein Kern lautet nämlich

$$\boldsymbol{U}_0\, p_\alpha(t)\, \overline{U_0^*\, q_\alpha(s)}$$

und es sind die $\boldsymbol{U}_0 p_\alpha(t)$ linear unabhängig, da es die $p_\alpha(t)$ sind.

4. Infolgedessen genügt es, den Satz unter der Annahme zu beweisen, daß $K \neq 0$ von endlichem Range ist. Es sei

$$k(t, s) = p_\alpha(t)\,\overline{q_\alpha(s)},$$

$p_\alpha(t)$, $q_\alpha(s)$ aus $\dot{\mathfrak{P}}$; die $p_\alpha(t)$ seien linear unabhängig. Wir gewinnen den Normator $U = 1 + i\breve{R}$ durch den Ansatz

$$r(t, s) = p_\alpha(t)\, j_\alpha(s)$$

und bestimmen $j_\alpha(t)$ nach (1)': $K(1 + i\breve{R}) = R$ aus

$$\overline{q_\alpha(s)} + i\pi\,\overline{q_\alpha(s)}\, p_\beta(s)\, j_\beta(s) + \int \overline{q_\alpha(\sigma)}\, p_\beta(\sigma)\, \frac{d\sigma}{s - \sigma}\, j_\beta(s) = j_\alpha(s).$$

Dies System von Gleichungen für die Unbekannten $j_1(s), \ldots, j_k(s)$ ist eindeutig lösbar, wenn die Derminante $D(s)$ der Matrix

$$\delta_\alpha^\beta - i\pi\,\overline{q_\alpha(s)}\, p_\beta(s) - \int \overline{q_\alpha(\sigma)}\, p_\beta(\sigma)\, \frac{d\sigma}{s - \sigma}$$

für $|s| \leqq 1$ nicht verschwindet. Angenommen, es gäbe ein reelles τ aus $|\tau| \leqq 1$ mit $D(\tau) = 0$; dann gäbe es ein System von Zahlen ι_α, ($\alpha = 1, \ldots, \varkappa$), die nicht alle verschwinden, so daß

$$\iota_\alpha - i\pi\,\overline{q_\alpha(\lambda)}\, p_\beta(\lambda)\, \iota_\beta - \int \overline{q_\alpha(\sigma)}\, p_\beta(\sigma)\, \frac{d\sigma}{\tau - \sigma}\, \iota_\beta = 0$$

ist; wir multiplizieren mit $p_\alpha(t)$ und setzen

$$\iota_\alpha p_\alpha(t) = p(t);$$

es ist $p(t) \neq 0$, da die $p_\alpha(t)$ linear unabhängig sind. Wir gewinnen so:

$$p(t) - i\pi k(t,\tau)\, p(\tau) - \int k(t,\sigma)\, p(\sigma) \frac{d\sigma}{\tau - \sigma} = 0;$$

d. h. die Funktion $\frac{p(t)}{t-\tau}$ aus $\mathfrak{P}_\lambda$ ist Eigenfunktion zum Eigenwert τ im weiteren Sinne. Das aber war ausgeschlossen.

Es ist also $D(s) \neq 0$ für $|s| \leqq 1$. Das obige Gleichungssystem besitzt also eine Lösung $j_\alpha(s)$. Da $D(s)$ in $\mathfrak{P}$ liegt, gehört $\frac{1}{D(s)}$, also auch $j_\alpha(s)$ zu $\mathfrak{P}$. $j_\alpha(s)$ gehört sogar zu $\dot{\mathfrak{P}}$; denn für $|s| = 1$ ist $q_\alpha(s) = 0$; das Gleichungssystem hat somit für $|s| = 1$ die Lösung $j_\alpha(s) = 0$, und die ist eindeutig bestimmt. Es liegt $r(t,s) = p_\alpha(t) j_\alpha(s)$ in $\dot{\mathfrak{P}}^2$ und $(1 + i\breve{R})$ ist in der Tat ein Rechtsnormator von $T + K$.

§ 5.

Spektralzerlegung von $T + K$ im Hilbert-Raume.

1. Als Hilbert-Raum $\mathfrak{H}$ haben wir den Raum aller in $|t| \leqq 1$ erklärten quadratisch integrierbaren Funktionen $x(t)$ zu wählen. Wir setzen

$$(x, y) = \int \overline{x(t)}\, y(t)\, dt, \quad |x| = \sqrt{(x, x)}.$$

Der Raum $\mathfrak{P}$ liegt dicht in $\mathfrak{H}$ (in bezug auf die Norm $|x|$); es ist für x in $\mathfrak{P}$

$$|x| \leqq 2\,\|x\|.$$

Jeder in $\mathfrak{P}$ erklärte lineare Operator, der (in bezug auf $|x|$) beschränkt ist, läßt sich eindeutig zu einem linearen in $\mathfrak{H}$ erklärten Operator mit derselben Schranke fortsetzen. So sind die Operatoren $f(T)$ aus $\mathfrak{P}$ und G aus $\mathfrak{P}\dot{\mathfrak{P}}$ auf $\mathfrak{H}$ zu übertragen; denn es gilt für x in $\mathfrak{P}$

$$|f(T)x| \leqq \|f\|\,|x|; \quad |Gx| \leqq 2\,\|G\|\,|x|.$$

Der Nachweis, daß dasselbe auch für Operatoren $\breve{G}$ mit G aus $\dot{\mathfrak{P}}^2$ gilt, bedarf einiger Vorbereitungen.

Wir betrachten zunächst Operatoren G aus $\mathfrak{P}\dot{\mathfrak{P}}$, für die $\mathring{G} = 0$ ist, deren Kerne $g(t,s)$ also auf der Diagonalen verschwinden:

$$g(t,t) = 0.$$

Für sie besteht die Abschätzung

$$(+) \qquad g(t,s) \leqq \tfrac{1}{2}\,\|g\|\,|t-s|^\mu;$$

denn es ist

$$\begin{aligned} 2|g(t,s)| &= |g(t,s)-g(t,t)+g(t,s)-g(s,s)| \\ &\leqq |t-s|^{\mu}\sup|\Delta_s g| + |t-s|^{\mu}\sup|\Delta_t g| \\ &\leqq |t-s|^{\mu}\,\|g\|. \end{aligned}$$

Hilfssatz 5.1. Ist G ein Operator aus $\mathfrak{P}\dot{\mathfrak{P}}$ mit $\mathring{G}=0$, so gilt für $x(t)$ aus P

$$|(x, \breve{G}y)| \leqq 2\,\mathsf{M}\,\|G\|\,|x|\,|y|.$$

Der Beweis beruht auf der Relation (+). Mit ihrer Hilfe schätzen wir ab:

$$\left|\int\limits_{|t-s|\geqq 2\varepsilon}\int \overline{x(t)}\,g(t,s)\,y(s)\,\frac{ds}{s-t}\,dt\right|$$

$$= \left|\int\limits_{\varepsilon\leqq\alpha\leqq 1}\;\int\limits_{-1-|\alpha|}^{-1+|\alpha|} g(\beta-\alpha,\beta+\alpha)\,\overline{x(\beta-\alpha)}\,y(\beta+\alpha)\,d\beta\,\frac{d\alpha}{\alpha}\right|$$

$$\leqq \frac{1}{2}\|G\|\,|x|\,|y| \int\limits_{|\alpha|\leqq 1} |2\alpha|^{\mu}\frac{d\alpha}{\alpha} \leqq 2\,\mathsf{M}\,\|G\|\,|x|\,|y|.$$

Die linke Seite geht aber für $\varepsilon\to 0$ über in

$$\left|\iint \overline{x(t)}\,g(t,s)\,y(s)\,\frac{ds}{s-t}\,dt\right| = |(x,\widetilde{G}y)| = |(x,\breve{G}y)|.$$

Hilfssatz 5.2. Ist G ein Operator aus $\dot{\mathfrak{P}}^2$, G^* seine Formal-Adjungierte, so gilt für $x(t)$, $y(t)$ aus $\mathfrak{P}$

$$(\breve{G}x, y) = (x, \breve{G}^*y).$$

Beweis. Da $g(t,s)$ in $\dot{\mathfrak{P}}^2$ liegt, ist Hilfssatz 1.5 auf

$$q(r,s) = \overline{x(s)}\,\overline{g(r,s)}\,y(r)$$

anwendbar:

$$(\widetilde{G}x, y) = \int\overline{\widetilde{G}x(\varrho)}\,y(\varrho)\,d\varrho = \iint\overline{x(\sigma)}\,\overline{g(\varrho,\sigma)}\,y(\varrho)\,\frac{d\sigma}{\sigma-\varrho}\,d\varrho$$

$$= -\iint\overline{x(\sigma)}\,\overline{g(\varrho,\sigma)}\,y(\varrho)\,\frac{d\varrho}{\varrho-\sigma}\,d\sigma = -\int\overline{x(\sigma)}\,\widetilde{G}^*y(\sigma)\,d\sigma = -(x,\widetilde{G}^*y);$$

und daraus folgt die Behauptung.

Satz 5.1. *Liegt der Operator G in $\dot{\mathfrak{P}}^2$, so ist $\breve{G}$ auf $\mathfrak{H}$ beschränkt fortsetzbar und es gilt für x in $\mathfrak{H}$*

$$|\breve{G}x| \leqq 2\,\mathsf{N}\,\|G\|\,|x|.$$

Ein entsprechender Satz wurde von Lichtenstein[13]) angegeben. Unser Beweis benutzt im Prinzip einen von Hilbert (l. c.) verwandten Kunstgriff.

---

[13]) l. c. Mit $x(t)$ ist auch $\int_{-1}^{1}[x(\tau)-x(t)]\cot g\,\frac{\pi}{2}(\tau-t)\,d\tau$ quadratisch integrierbar.

Beweis. Es ist für x aus $\mathfrak{P}$ nach Hilfssatz 5.2 und § 2 ($\breve{\smile}$), ($\overset{\circ}{\circ}$), ($\tilde{\circ}$), ($\overset{\circ}{\sim}$)

$$|\breve{G}x|^2 = (\breve{G}x, \breve{G}x) = (x, \breve{G}^* \breve{G} x) = (x, (\breve{G}^{\breve{*}} G + G^{\breve{*}} \breve{G}) x)$$

$$= -i(x, (\widetilde{G}^{\breve{*}} G + G^{\breve{*}} \widetilde{G}) x) + 2(x, \mathring{G}^* \mathring{G} x).$$

Nach § 2 ($\overset{\circ}{\sim}$) ist $\widetilde{G}^{\mathring{*}} G + G^{\mathring{*}} \widetilde{G} = 0$, also folgt aus Hilfssatz 5.1 und ($*$), ($\overset{*}{*}$) § 2

$$|\breve{G}x|^2 \leqq (48\,\mathsf{M}^2 + 2\pi^2)\,\|g^2\|\,|x|^2$$

und damit die Behauptung.

2. Ist K ein Operator aus $\dot{\mathfrak{P}}^2$, so ist $T + K$ ein in $\mathfrak{H}$ beschränkter Operator. Besitzt $T + K$ ein Paar von Normatoren $U = 1 + i\breve{R}$, $\boldsymbol{U} = 1 - i\breve{\boldsymbol{R}}$ mit R, $\boldsymbol{R}$ aus $\dot{\mathfrak{P}}^2$, so besitzen U, $\boldsymbol{U}$ nach Satz 5.1 in $\mathfrak{H}$ beschränkte Fortsetzungen, die auch U, $\boldsymbol{U}$ heißen sollen. Mit ihnen bestehen die Relationen (1), (**1**), (2) und, wenn sie intakt sind, auch (3). Der Operator $P = U\boldsymbol{U}$ ist wegen $P^2 = P$ ein (nicht notwendig orthogonaler) Projektor.

Es sei Δt ein Teilintervall von $|t| \leqq 1$ mit den Endpunkten t_1, t_2 und ΔE_t der Operator, der aus der Funktion $x(t)$ von $\mathfrak{H}$ die auch zu $\mathfrak{H}$ gehörige Funktion

$$\Delta E_t\, x(t) = \begin{cases} x(t) & \text{für } t \text{ in } \Delta t \\ 0 & \text{für } t \text{ außer } \Delta t \end{cases}$$

erzeugt. Dann ist

$$P_{t_2} - P_{t_1} = \Delta P_t = U \Delta E_t \boldsymbol{U}$$

auch ein Projektor und es bestehen die Relationen

$$\int_{-1}^{1} d P_t = P, \qquad \int_{-1}^{1} t\, d P_t = (T + K) P,$$

wobei die Integration in bekannter Weise nach dem Muster des Stieltjes-Integrals zu definieren ist. Die erste Relation folgt unmittelbar aus

$$\int_{-1}^{1} d P_t = U \int_{-1}^{1} d E_t \boldsymbol{U} = U \boldsymbol{U} = P,$$

die zweite aus

$$\int_{-1}^{1} t\, d P_t = U \int_{-1}^{1} t\, d E_t \boldsymbol{U} = U T \boldsymbol{U} = (T + K) P.$$

Es ist also ΔP_t der zum Intervall Δt gehörige Projektor des Operators $T + K$.

Ist $U \boldsymbol{U} = 1$, also $P = 1$, so gibt ΔP_t die vollständige Spektralschar von $T + K$.

Ist $K = K^*$ symmetrisch, so ist $\boldsymbol{U} = U^*$ Linksnormator. Dann sind die Projektoren $P = U U^*$ und ΔP_t orthogonal; U ist längentreu, und, wenn $U U^* = 1$ ist, auch unitär.

(Eingegangen am 5. 2. 1937.)

[48–1] **Criteria for the discrete character of the spectra of ordinary differential operators, R. Courant Ann'y. Volume, (1948), 145–160.**

Commentary:

This paper deals, as did [35–1], with Sturm-Liouville operators L. Rather general criteria for the operator to have a discrete spectrum, or a discrete spectrum below a certain value, are given. If L has many self-adjoint realizations, the distinguished one, introduced in [35–1], is meant. (Incidentally, if the whole spectrum is discrete for one realization, the same is true for all others.) A sufficient condition for the whole spectrum to be discrete is given by

$$r^{-1}[q+(1/4ph^2)] \to +\infty \quad \text{as} \quad x \to X^{\pm}$$

where h is, for x close to X^+, any indefinite integral of $1/p$ except that it should be normalized by $h(X^+) = 0$ if $1/p$ is integrable near X^+, and similarly for x close to X^-.

The proof depends on an ingenious device by which the problem is converted into a standard form on $(-\infty, \infty)$ with no boundary conditions, no matter which case of Weyl's classification occurs in the original problem.

In the special case $(X^-, X^+) = (-\infty, \infty), p = r = 1$, a definitive result in this problem was subsequently given by Molčanov [1].

[1] A.M. Molčanov. On conditions for discreteness of the spectrum of self-adjoint differential equations of second order, Trudy Moskov. Mat. Obsc. 1 (1953), 169–199, (Russian).

COURANT
Anniversary Volume
1948

CRITERIA FOR THE DISCRETE CHARACTER OF THE SPECTRA OF ORDINARY DIFFERENTIAL OPERATORS

K. O. Friedrichs

Self-adjoint differential operators with a discrete spectrum deserve particular attention, certainly more attention than they have generally received in the literature. Not the least reason for this is the fact that such differential operators can be completely treated independently of the general theory of spectral resolution or spectral representation. The most natural approach to the spectral theory of differential operators with a discrete spectrum employs the minimum property of their eigenvalues or, more generally, the beautiful maximum-minimum property discovered by Courant. The arguments which Courant has based on this property in order to reveal the behavior of eigenvalues of regular differential operators remain valid for differential operators which have singularities at the boundary of the domain, provided that their spectra are discrete.

In the present note we shall formulate and prove simple sufficient conditions that the spectrum of an ordinary differential operator is discrete, discrete below a finite value, or non-discrete. These criteria refer to the behavior of the coefficients of the operator at the end points of the interval, which may extend to infinity. Criteria for discreteness have been given in the literature, by Weyl [1], Stone [2], Titchmarsh [3], who made specific use of the theory of ordinary differential equations.

The advantage of the method of the present note, which was partly used in earlier papers, see [4, 5], is that it can be used in a similar manner for differential operators referring to functions of several variables, see [6].

1. The problem. The independent variable will be denoted by ξ, its domain by $\mathcal{J}$: $\xi_- < \xi < \xi_+$. Here ξ_- may be $-\infty$ and ξ_+ may be ∞. The functions defined in this domain are denoted by $x = x(\xi)$. Differentiation with respect to ξ is denoted by D. The differential operator considered is

(1) $$L = -r^{-1}[DpD - q],$$

so that the eigenvalue equation $Lx = \lambda x$ can also be written in the form

(2) $$DpD\,x - q\,x + \lambda r\,x = 0.$$

The coefficients are of course also defined in the interval $\mathcal{J}$ as continuous functions. We require p and r to be positive,

(3) $$p(\xi) > 0,\ r(\xi) > 0 \text{ in } \mathcal{J},$$

but do not restrict the behavior of these functions at the endpoints. Conditions for q will be formulated later on.

Two tasks should be carried out in the treatment of the differential operator L. Firstly, appropriate conditions for the function x, in particular boundary conditions should be formulated so that the operator L, applicable on functions satisfying these conditions, is self-adjoint. We shall dispose of this task in a simple manner, imposing a condition on the function x which in the case of regular behavior at the endpoints is equivalent to requiring the function to vanish there.

Secondly, the nature of the spectrum of the operator L should be characterized in as much detail as can be deduced by simple general criteria from the nature of the coefficients p,q,r without determining the spectral resolution explicitly. In particular, conditions for these coefficients should be formulated which insure that the spectrum of the operator L is discrete. We call a spectrum discrete in the present context if there is an increasing sequence of eigenvalues, each of finite multiplicity, whose eigenfunctions form a complete system, spanning the whole space of admissible functions x. More generally, we shall call the spectrum discrete definitely below a value λ_* if there is such an increasing sequence of eigenvalues $\lambda < \lambda_*$ which exhausts the part of the spectrum below λ_*; for a precise definition of this notion see [7].

2. The criteria. We postpone the discussion of the first task and proceed to formulate the main objective of this note, our criterion for completeness. This criterion covers all the common eigenvalue problems with a discrete spectrum for regular or singular differential equations of the type (2). We choose any points ξ_0, ξ_1, ξ_{-1} in $\mathcal{I}$ such that $\xi_- < \xi_{-1} < \xi_0 < \xi_1 < \xi_+$. We then introduce the variable

$$\tau = \int_{\xi_0}^{\xi} \frac{d\xi}{p} \tag{4}$$

and set

$$\tau_+ = \int_{\xi_0}^{\xi_+} \frac{d\xi}{p}, \qquad \tau_- = -\int_{\xi_-}^{\xi_0} \frac{d\xi}{p}. \tag{5}$$

Finally we introduce a function h such that

$$\begin{aligned} h &= \tau && \text{for } \xi_1 \le \xi < \xi_+ , \text{ if } \tau_+ = \infty, \\ &= \tau_+ - \tau && \text{for } \xi_1 \le \xi < \xi_+ , \text{ if } \tau_+ < \infty, \\ &= -\tau && \text{for } \xi_- < \xi \le \xi_{-1} , \text{ if } \tau_- = -\infty, \\ &= \tau - \tau_- && \text{for } \xi_- < \xi \le \xi_{-1} , \text{ if } \tau_- > -\infty, \end{aligned} \tag{6}$$

Obviously, we can construct a function h which is positive in $\mathcal{I}$, possesses continuous second derivatives with respect to τ, and satisfies relations (6).

Our criterion now is that <u>the spectrum of the operator L given by (1)</u> is discrete if

$$r^{-1}[q + \frac{1}{4ph^2}] \to \infty \text{ as } \xi \to \xi_+ \text{ and as } \xi \to \xi_-. \tag{7}$$

In addition to this criterion for discreteness we formulate as criterion for the discreteness of the spectrum definitely "below" a value" λ_* the condition

$$\text{lim. inf. } r^{-1}[q + \frac{1}{4ph^2}] \geq \lambda_*, \text{ as } \xi \to \xi_+ \text{ and as } \xi \to \xi_-. \tag{8}$$

Finally we shall show that the spectrum of the operator L is <u>not discrete</u> if there is a positive constant c_o such that

$$r^{-1}[|q| + \frac{1}{4ph^2}] \leq c_o \tag{9}$$

either for $\xi_1 \leq \xi < \xi_+$ or for $\xi_- < \xi \leq \xi_{-1}$, and if in addition there is a constant $\gamma \geq 0$ such that

$$r^{-1}[q + \frac{1}{4ph^2}] \geq -\gamma \text{ for } \xi_- < \xi < \xi_+ . \tag{10}$$

3. <u>Applications</u>. We first want to show that these criteria are easily applied to the common eigenvalue problems.

1. The regular operator

$$L = - D^2 + q \tag{11}$$

for functions $x(\xi)$ in a bounded interval is discrete if q has a lower bound. For $p = r = 1$, $\tau = \xi - \xi_o$, and hence

$q + \dfrac{1}{4(\xi_+ - \xi)^2} \to \infty$ as $\xi \to \xi_+$, $q + \dfrac{1}{4(\xi - \xi_-)^2} \to \infty$ as $\xi \to \xi_-$.

Thus (7) is satisfied.

2. If the same operator (11) is to be applied on functions $x(\xi)$ defined in an unbounded interval $0 < \xi < \infty$, for example, the spectrum is not discrete if the coefficient q has, in addition to a lower bound, an upper bound for $\xi_1 \leq \xi < \infty$. For since $h = \tau = \xi$ in this interval, conditions (9) and (10) are then satisfied.

3. If, again with reference to an interval extending to $+\infty$, the function q approaches infinity with $|\xi|$, relation (7) holds once more. The operator L has then a discrete spectrum. This applies in particular to the operator $-D^2 + \frac{1}{4}\xi^2$, whose eigenfunctions are the <u>Hermite functions</u> $x = \exp(-\frac{1}{4}\xi^2)\, H_n(\xi)$.

4. The <u>Legendre polynomials</u> are eigenfunctions of the operator

(12) $$L = -D(1 - \xi^2)D$$

with reference to the interval $-1 < \xi < 1$. Since $r = 1$, $p = 1 - \xi^2$, we have

(13) $$\tau = \frac{1}{2} \ln \frac{1 + \xi}{1 - \xi},$$

and $\tau_+ = -\tau_- = \infty$. Clearly $\frac{1}{4} p \tau^2 \to \infty$ as $|\xi| \to 1$ and, consequently, the spectrum of the operator (12) is discrete.

5. The <u>Laguerre polynomials</u> are eigenfunctions of the operator

(14) $$L = -e^{\xi} D \xi e^{-\xi} D$$

with reference to the interval $0 < \xi < \infty$. Here $r = e^{-\xi}$,

$p = \xi e^{-\xi}$, and

(15) $$\tau = \int_1^{\xi} e^{\xi} \frac{d\xi}{\xi}$$

becomes infinite at the endpoints. Since $\tau \simeq e^{\xi}/\xi$ for large ξ and $\tau \approx \log \xi$ for $\xi \approx 0$, it is clear that $rp\tau^2 \to 0$ as $\xi \to \infty$ and as $\xi \to 0$. Thus condition (7) is satisfied and, consequently, the spectrum is discrete.

6. The eigenfunctions of the operator

(16) $$L = -\xi^{-1} D \xi D + m^2 \xi^{-2}$$

are Bessel functions J_m of an appropriate argument. Suppose we restrict the domain to $0 < \xi < 1$. Then condition (7) is satisfied as regards the endpoint $\xi = 1$. Since $r = p = \xi$, $q = m^2\xi^{-1}$, we find

(17) $$\tau = \log (\xi - \xi_0),$$

and $r^{-1}[q + (4p\tau^2)^{-1}] = m^2 \xi^{-2} + \frac{1}{4}\xi^{-2} \log^{-2}(\xi - \xi_0)$ becomes infinite as ξ approaches zero. Consequently the spectrum is discrete.

7. The <u>potential operator</u>, for functions of $n > 0$ variables, specialized to functions depending only on the distance ξ from the origin, is

(18) $$L = -\xi^{-(n-1)} D \xi^{n-1} D.$$

We have $r = p = \xi^{n-1}$ and hence

(19) $$\tau = \frac{1}{n-2} \left(\frac{1}{\xi^{n-2}} - \frac{1}{\xi_0^{n-2}}\right) \text{ for } n > 2,$$

$$\tau = \log \xi/\xi_0 \text{ for } n = 2.$$

Clearly, $rp\tau^2 \to 0$ as $\xi \to 0$. Hence the part of condition (7) which refers to the endpoint $\xi = 0$ is satisfied. Consequently the spectrum of the operator is discrete if it is referred to a finite interval such as $0 < \xi < 1$. Since, on the other hand, $rp\tau^2 \to \infty$ as $\xi \to \infty$, condition (9) is satisfied. Consequently the spectrum is not discrete if it is referred to the infinite interval $0 < \xi < \infty$.

8. The Schroedinger operator for functions depending only on the distance ξ from the origin, in the interval $0 < \xi < \infty$, is

(20) $$L = -\xi^2 D\xi^2 D - 2Z\xi^{-1}$$

(in Hartree units). Here, $r = p = \xi^2$, $q = -2Z\xi$, and

(21) $$\tau = \xi^{-1} - \xi_o^{-1}.$$

Hence $rp\tau^2 = \xi^2(1 - \xi_o^{-1}\xi)^2$ and

(22) $$r^{-1}[q + (4p\tau^2)^{-1}] = -2Z\xi^{-1} + \frac{1}{4}\xi^{-2}(1 - \xi_o^{-1}\xi)^{-2}.$$

This expression approaches infinity as ξ approaches zero, since $\frac{1}{4}\xi^{-2}$ dominates $-2Z\xi^{-1}$; it evidently approaches zero as ξ approaches infinity. Hence conditions (9) and (10) are satisfied with $\lambda_* = 0$. Consequently the spectrum of the Schroedinger operator is discrete below zero, see [11].

4. Definition of the operator. Before justifying our criterion we must come back to the first task mentioned above and define precisely the operator L and the manifold of functions on which it operates.

We introduce for convenience the new independent variable

(24) $$\eta = \int_{\xi_0}^{\xi} \frac{d\xi}{ph} .$$

From (4) and (6), ξ_1 and ξ_{-1} can clearly be chosen so close to ξ_+, ξ_-, respectively, that

(25) $$|\eta| = |\log h| + \text{const. for } \xi_1 \leq \xi < \xi_+ \text{ and } \xi_- < \xi \leq \xi_{-1}.$$

The range of η corresponding to the interval $\xi_- < \xi < \xi_+$ is consequently

(26) $$-\infty < \eta < \infty .$$

Instead of x we introduce the quantity

(27) $$y = x/\sqrt{h} ,$$

which we consider a function of η. Differentiation with respect to η will be denoted by D_η. We further introduce the quantity

$$\sigma = (pDh)^2 - 2hpD(pDh),$$

which by (6) reduces to

(28) $$\sigma = 1 \qquad \text{for } \eta_1 < \eta \text{ and } \eta < \eta_{-1} .$$

The operator L given by (1) is then formally equivalent to the operator

(29) $$M = - r^{-1}p^{-1}h^{-2}[D_\eta^2 - qph^2 - \frac{1}{4}\sigma]$$

in the sense that

(30) $$My = \frac{Lx}{\sqrt{h}}$$

for differentiable functions x for which pDx is again differentiable. The eigenvalue equation (2) is equivalent to

(31) $$D_\eta^2 y - [qph^2 + \frac{1}{4}\sigma] y + \lambda rph^2 y = 0.$$

The Hilbert space of all functions y for which

(32) $$(y\mathcal{H}y) = \int_{-\infty}^{\infty} rph^2y^2 \, d\eta < \infty$$

will be denoted by $\mathfrak{H}$. Further we introduce the manifold $\mathfrak{G}$ of functions y which possess derivatives (almost everywhere) and for which

(33) $$\int_{-\infty}^{\infty} [(D_\eta y)^2 + \{qph^2 + \frac{1}{4}\sigma + \gamma rph^2\}y^2] \, d\eta < \infty .$$

Here γ is a constant such that

(34) $$qph^2 + \frac{1}{4}\sigma + \gamma rph^2 \geq 0 \qquad \text{for } -\infty < \eta < \infty .$$

The existence of such a constant is implied by either (7), (8), or (10), and (28). We set

(35) $$(y\mathcal{G}y) = \int_{-\infty}^{\infty} [(D_\eta y)^2 + \{qph^2 + \frac{1}{4}\sigma\}y^2] \, d\eta .$$

Finally we introduce the manifold $\mathfrak{F}$ of all functions y in $\mathfrak{G}$ for which $D_\eta y$ is differentiable (almost everywhere) and $My = -D_\eta^2 y + [qph^2 + \frac{1}{4}\sigma]$ belongs to $\mathfrak{H}$. We then maintain: Firstly,

(36) $$(y_1\mathcal{H}My_2) = (y_1\mathcal{G}y_2) \text{ for } y_1 \text{ in } \mathfrak{G} \text{ and } y_2 \text{ in } \mathfrak{F}$$

(in obvious notation). This property implies the symmetry of the operator M:

(37) $$(y_1\mathcal{H}My_2) = (My_1\mathcal{H}y_2) \text{ for } y_1, y_2 \text{ in } \mathfrak{F} .$$

Secondly, if y and y' are functions in $\mathfrak{H}$ such that

(38) $$(y\mathcal{H}My_*) = (y'\mathcal{H}y_*)$$

holds for all y_* in $\mathfrak{F}$, then y' is in $\mathfrak{F}$ and $My = y'$. This last property combined with the symmetry (37) is the "self-adjointness" in the sense of v. Neumann [8] and Stone [2]. For the proof of these two statements we refer to earlier papers [5, 9].

We now adopt the following definition of the operator L. It is applicable in a subspace of the Hilbert space of all functions $x(\xi)$ for which

$$(40) \qquad (x,x) = \int_{\xi_-}^{\xi_+} rx^2 \, d\xi$$

is finite. (We denote this space also by $\mathfrak{H}$, considering x and $y = x/\sqrt{h}$ as different representations of the same element in $\mathfrak{H}$). The subspace then consists of all functions x for which $y = x/\sqrt{h}$ belongs to $\mathfrak{F}$; this subspace is again denoted by $\mathfrak{F}$. The operator L is then so defined that it transforms every function x in $\mathfrak{F}$ into the function

$$(41) \qquad Lx = h^{1/2}M(h^{-1/2}x),$$

in accordance with (30). Evidently, Lx lies in $\mathfrak{H}$. Relations (37) and (38) then become

$$(x_1, Lx_2) = (Lx_1, x_2),$$

$$(x, Lx_*) = (x', x_*).$$

It is thus clear that the operator L now defined is self-adjoint with reference to the unit form (40).

Our present definition of the operator L has the advantage that no boundary condition need be imposed explicitly. Condition (34) here takes the place of a boundary condition. Also no cases need be distinguished, and the derivation of our criteria becomes extremely simple, as we shall see. These are the reasons why we have not worked with the forms (40) and

$$(42) \qquad \int_{\xi_-}^{\xi_+} [p(Dx)^2 + qx^2] \, d\xi,$$

which are naturally associated with the operator L as given by (1), instead of (33) and (34), respectively, although that would be possible provided that, instead of (34), condition

$$(43) \qquad qph^2 + \frac{\theta}{4}\sigma + \gamma rph^2 \geq 0 \quad \text{for } -\infty < \eta < \infty$$

is satisfied with any positive constant $\theta < 1$. In that case the form (42) has the same value as the form $(y, \mathcal{B}y)$, see (35), if appropriate boundary conditions are imposed on the functions x.

By way of illustration we consider the regular case, in which ξ_+ and ξ_- are finite, say $\xi_+ = -\xi_- = 1$, and $r = p = 1$, $q = 0$. The form (42) would be simply

$$(44) \qquad \int_{-1}^{1} (Dx)^2 \, d\xi \, .$$

It is known that it is not sufficient to require this form to be finite in order to insure that the operator $-D^2$ be self-adjoint. Boundary conditions must be imposed. The form $\mathcal{B}$, as defined by (35), is

$$(45) \qquad \int_{-\infty}^{\infty} [(D_\eta y)^2 + \tfrac{1}{4} y^2] \, d\eta ,$$

when we set

$$(46) \qquad h = \frac{1 - \xi^2}{2}, \qquad \sigma = 1, \qquad \eta = \log \frac{1 + \xi}{1 - \xi} ,$$

(slightly deviating from the properties (6) in the definition of h). The fact that $\int_{-\infty}^{\infty} y^2 \, d\eta$ is finite implies that there are sequences of values $\eta' \to \infty$, $\eta'' \to -\infty$ for which $y^2 \to 0$ or $x^2/(1 - \xi^2) \to 0$. Now

$$\int_{\eta''}^{\eta'} [(D_\eta y)^2 + \tfrac{1}{4}y^2] \, d\eta = \int_{\xi''}^{\xi'} [(1-\xi^2)(D[\frac{x}{\sqrt{1-\xi^2}}])^2 + \frac{x^2}{\sqrt{1-\xi^2}}] \, d\xi$$

$$= \int_{\xi''}^{\xi'} (Dx)^2 \, d\xi + [\frac{\xi x^2}{1 - \xi^2}]_{\xi''}^{\xi'} \, .$$

Hence

$$\int_{-\infty}^{\infty} [(D_\eta y)^2 + \frac{1}{4} y^2]\, d\eta = \int_{-1}^{1} (Dx)^2\, d\xi . \tag{47}$$

From the fact that the latter integral is finite and that x $\to 0$ for $\xi' \to 1$, $\xi'' \to -1$ it can easily be derived that x is a continuous function which vanishes at $\xi = 1$ and $\xi = -1$. Thus it is seen that the finiteness of $(y\mathfrak{G}y)$ entails that the function $x(\xi)$ satisfies a boundary condition.

According to the theory of H. Weyl [1] there are four possibilities, two for each endpoint. Either the condition that (x,x) and (Lx,Lx), see (40), are finite determines a class of functions x for which the operator L is self-adjoint, or there is a variety of boundary conditions available to be imposed at $\xi = \xi_+$ or $\xi = \xi_-$ or at both endpoints, to achieve the same effect.*) Our condition (33) is redundant in the first case. In the other cases it selects a particular boundary condition, distinguished by various properties, see [5]. In regular cases, such as the one considered above, this particular condition requires vanishing of the function at the endpoint in question.

6. <u>Proof of the criteria</u>. We now proceed to justify our criteria. We use the following fact: The spectrum of the operator L is discrete definitely below the value $\lambda = \lambda_*$ if for every $\lambda < \lambda_*$ there is a finite number of functions $y^{(1)}, \ldots, y^{(n)}$ in $\mathfrak{H}$ such that

$$(y\mathfrak{G}y) + \sum_{\nu=1}^{n} (y\mathcal{H}y^{(\nu)})^2 \geq \lambda(y\mathcal{H}y) \tag{48}$$

for all functions y in $\mathfrak{G}$, see [7, 12]. If this is true for every

*) For a way of determining these possible boundary conditions, see Rellich [10].

(arbitrarily large) value of λ with a value of n depending on λ, the spectrum of M is discrete.

Suppose now condition (8) is satisfied, or, what is equivalent,

(49) $$\liminf\ (qph^2 + \frac{1}{4}\sigma - \lambda_* rph^2) \geq 0$$

as $|\eta| \to \infty$. Then for any $\lambda < \lambda_*$ we can find a value $H > 0$ such that

(50) $$qph^2 + \frac{1}{4}\sigma - \lambda rph^2 \geq 0 \qquad \text{for } |\eta| \geq H,$$

whence

(51) $$\int_{|\eta| \geq H} [(D_\eta y)^2 + (qph^2 + \frac{1}{4}\sigma)y^2]\, d\eta \geq \lambda \int_{|\eta| \geq H} rph^2 y^2\, d\eta.$$

Next we use Poincaré's inequality, see [13],

(52) $$2\int_{\Delta\eta} y^2\, d\eta \leq 2\,|\Delta\eta|^{-1} (\int_{\Delta\eta} y\, d\eta)^2 + |\Delta\eta| \int_{\Delta\eta} (D_\eta y)^2\, d\eta$$

for any interval $\Delta\eta$ of length $|\Delta\eta|$. Choosing a number n so large that

(53) $$n \geq (\lambda + \gamma)\, rph^2 H \qquad \text{for } |\eta| \leq H$$

(for the definition of γ see (34)), and applying (52) to the n subintervals $\Delta_\nu\eta$ of $|\eta| \leq H$ with the length $|\Delta\eta| = 2H/n$, we find

(54) $$(\lambda + \gamma) \int_{\Delta_\nu\eta} rph^2 y^2\, d\eta \leq \frac{2}{|\Delta\eta|} \int_{\Delta_\nu\eta} y^2 d\eta$$

$$\leq \frac{2}{|\Delta\eta|^2} (\int_{\Delta_\nu\eta} y\, d\eta)^2 + \int_{\Delta_\nu\eta} (D_\eta y)^2\, d\eta.$$

Adding the n inequalities (54) to (51) and using (34), we obtain the inequality

$$(\lambda + \gamma) \int_{-\infty}^{\infty} rph^2y^2 \, d\eta \leq \sum_{\gamma=1}^{n} \frac{2}{|\Delta\eta|^2} \Big(\int_{\Delta_\gamma \eta} y \, d\eta \Big)^2$$
$$+ \int_{-\infty}^{\infty} [(D_\eta y)^2 + (qph^2 + \frac{1}{4}\sigma + \gamma rph^2)y^2] \, d\eta,$$

which is equivalent to (48). Thus we have proved that condition (8) insures that the spectrum of the operator L is discrete definitely below $\lambda_{\#}$. If condition (7) is satisfied we may choose $\lambda_{\#}$ arbitrarily large. It follows that the spectrum of L is then discrete.

To justify our criterion (9), (10) that the spectrum of the operator L is not discrete we use Hilbert's original definition of complete continuity instead of condition (48). Accordingly, the spectrum of the operator L is not discrete if there exists a sequence of functions $y = y^{(\alpha)}$ in $\mathfrak{G}$ for which, with an appropriate constant C,

(55) $\qquad (y \mathcal{Y} y) \qquad C,$

(56) $\qquad (y \mathcal{H} y) = 1,$

(57) $\qquad (y_{\#} \mathcal{H} y) \to 0 \qquad$ for every $y_{\#}$ in $\mathfrak{H}$,

see [14]. To construct such a sequence $y = y^{(\alpha)}$ we set, with $\alpha > \eta > 0$,

(58)
$$\begin{aligned} y &= 0, & \eta &\leq \alpha, \\ y &= A(\eta - \alpha), & \alpha \leq \eta &\leq 2\alpha, \\ y &= A\alpha, & 2\alpha \leq \eta &\leq 3\alpha, \\ y &= A(4\alpha - \eta), & 3\alpha \leq \eta &\leq 4\alpha, \\ y &= 0, & 4\alpha &\leq \eta, \end{aligned}$$

and assume that relation (9) holds for $\xi \to \xi_+$. The constant A is to be so determined that (56) holds. Hence, by (32) and (9),

$$1 \geq \int_{2\alpha}^{3\alpha} rph^2y^2 \, d\eta \geq \frac{1}{4c_0} \int_{2\alpha}^{3\alpha} y^2 \, d\eta = \frac{A^2\alpha^3}{4c_0}$$

or

(59) $$A^2 \leq 4c_0\alpha^{-3}.$$

Next we have by (35), (28), and (9),

$$(y\mathcal{G}y) \leq \int_{\alpha}^{4\alpha} (D_\eta y)^2 \, d\eta + c_0 \int_{\alpha}^{4\alpha} rph^2y^2 \, d\eta$$

$$= 2\alpha A^2 + c_0 \leq (8\alpha^{-2} + 1) c_0.$$

Hence (55) is satisfied. To verify (57) we need only observe that it is sufficient to establish this relation for every function y_* to which corresponds a value β such that $y_*(\eta) = 0$ for $\eta \geq \beta$; for the manifold of all such functions y_* is dense in $\mathfrak{H}$. For such functions y_* obviously $(y_*\mathcal{N}y^{(\alpha)}) = 0$ as soon as $\alpha \geq \beta$.

Thus our criterion for the non-discreteness of the spectrum is also justified.

REFERENCES

1 Weyl, H., "Ueber gewoehnliche Differentialgleichungen mit singulaeren Stellen," Goett. Nachr., 1909, p. 37; "Ueber gewoehnliche Differentialgleichungen mit Singularitaeten," Math. Annalen, v.68, 1909, p. 373-392.

2 Stone, M. H.: Linear Transformations in Hilbert Space, New York, 1932, Ch. X, 484, Theorem 10.19.

2A Ibid., Ch. V, p. 155-197.

3 Titchmarsh, E. C.: Eigenfunction Expansions associated with Second-Order Differential Equations, Oxford, 1946, Ch. V, p. 97-117.

4 Friedrichs, K.: "Spektraltheorie linearer Differentialoperatoren," Jahresberichte d. Deutschen Math.-Vereinigung, v.45, 1935, p. 181-193.

5 Friedrichs, K.: Ueber die ausgezeichnete Randbedingung in der Spektraltheorie der halbbeschränkten gewöhnlichen Differentialoperatoren zweiter Ordnung, Mathem. Annalen, Vol. 112,

1935, p. 1-23.

6 Friedrichs, K.: "Spektraltheorie halbbeschraenkter Operatoren und Anwendung auf die Spektralzerlegung von Differentialoperatoren. Zweiter Teil,"" Math. Annalen, v. 109, 1934, p. 685-713.

7 Friedrichs, K.: "Spektraltheorie halbbeschraenkter Operatoren und Anwendung auf die Spektralzerlegung von Differentialoperatoren. I," Math. Annalen, v. 109, 1934, p. 465-487.

8 v. Neumann, J.: "Allgemeine Eigenwerttheorie Hermitescher Funktionaloperatoren," Math. Annalen, v. 102, 1929, p. 50-103.

9 Friedrichs, K.: On Differential Operators in Hilbert Spaces. Amer. Journ. of Math. Vol. 61, 1939, p. 523-544.

10 Rellich, F.: Die zulässigen Randbedingungen bei den singulären Eigenwertproblemen der mathematischen Physik. Mathematische Zeitschrift, vol. 49, 1944, p. 702-723.

11 Courant, R. and Hilbert, D.: Methoden der Mathematischen Physik, v. I, Berlin, 1931, Ch. V, p. 294-296.

12 Courant, R. and Hilbert, D.: Methoden der Mathematischen Physik, v. II, Berlin, 1937, Ch. VII, p 489.

13 Ibid., p. 488.

14 Hilbert, D.: Grundzuege einer allgemeinen Theorie der Linearen Integralgleichungen, Leipzig, 1924, Ch. XI, p. 147-148, Ch. XIV, p. 174-175.

Conditions of a different kind for the behavior of eigenfunctions of a discrete spectrum were given by:

15 Kemble, I.: Note on the Sturm-Liouville Eigenvalue-Eigenfunction Problem with Singular End-Points. Proc. Nat. Acad. Sci. 19, 1933, p. 710.

[48–8] **On the perturbation of continuous spectra, Comm. Pure Appl. Math., I (1948), 361–406.**

Commentary:

This is a partially expository paper, apparently motivated by the progress of quantum mechanics made during the decade with little mathematical justification. It deals with several typical perturbation phenomena involving continuous spectra. First, the results of [38–1] are summarized, with partial generalization and with new notation, to the effect that a continuous spectrum is stable in certain situations. Then an example is given in which an additional point spectrum is created by a small perturbation of rank one, and another example, in which an eigenvalue immersed in a continuous spectrum disappears under a small perturbation of rank two. Finally, it is shown that in the first case, the solutions to the time dependent Schrödinger equation have definite asymptotic form so that Heisenberg's scattering operator can be defined rigorously. In the example of the immersed eigenvalue, a similar analysis leads to a mathematical foundation of the theory of the Auger effect and the width of spectral lines. As mentioned above, all these problems have since been extensively studied.

[Reprinted from *Communication on* APPLIED MATHEMATICS
Vol. 1, Number 4, December, 1948]

On the Perturbation of Continuous Spectra

By **K. O. FRIEDRICHS**

This paper is concerned with the perturbation of operators which have a continuous spectrum, or a mixed continuous and discrete spectrum. The aim is to develop a mathematically consistent formalism for such perturbations and to put this perturbation calculus on a rigorous mathematical foundation. Naturally, parts of these perturbation procedures are substantially equivalent with perturbation procedures developed for the purposes of the quantum theory. It seems, though, that the approach presented in this paper is more suitable to clarify the mathematical structure of the problem of the perturbation of operators with continuous spectra.

The present paper is to a large extent of an expository character. For detailed proofs of various statements reference is made to an earlier paper of the author—see [2].

The investigations reported in this paper were prompted by the rather complete investigations of F. Rellich on the perturbation of operators with discrete spectra [1]. Most of the results of Rellich concern (bounded and non-bounded) self-adjoint operators $L = L_0$ with a discrete spectrum which are varied so as to depend analytically on a parameter ϵ, $L = L_\epsilon$. Rellich derives conditions under which eigen-elements u_ϵ of L_ϵ can be so selected that they, together with the corresponding eigen-values λ_ϵ, depend analytically on the parameter ϵ in a neighborhood of $\epsilon = 0$. He shows that the analyticity of the dependence of L_ϵ on the parameter ϵ is essential in the neighborhood of multiple eigen-values: he gives examples of operators L_ϵ, depending continuously on ϵ, for which no eigen-functions u_ϵ exist which depend continuously on ϵ and reduce, for $\epsilon = 0$, to eigen-functions associated with a multiple eigen-value of the undisturbed operator L_0.

Naturally, it is desirable to investigate whether or not a similar theory of perturbations can be developed for operators which possess a continuous, or a mixed continuous and discrete, spectrum. There are a great variety of possibilities of operators with non-discrete spectra and of perturbations of such operators. Only a few simple, but in many respects typical, cases will be treated in this paper.

Firstly we shall consider an operator with a simple continuous spectrum and its perturbation through an integral operator. We shall show that the character of the spectrum is not changed if a certain simple condition is imposed on the kernel of the perturbing operator. The spectral representation can then easily be established. The mathematical tools needed for this purpose were developed in an earlier paper [2], to which we shall frequently refer.

Next we shall discuss typical examples of perturbations by integral operators which do not satisfy the condition mentioned and in which a point eigen-value appears through an arbitrarily small perturbation. We shall indicate how one could derive the spectral theory of operators disturbed in this manner.

Further, we shall consider an operator whose spectrum consists of a simple continuous spectrum and a single point eigen-value immersed in it. We shall consider the simplest possible perturbation of such an operator involving an "interaction" between the eigen-elements of the continuous spectrum and those of the point eigen-values. This problem, which is closely related to the theory of the Auger effect—see e.g. [11, page 736]—and to Weisskopf and Wigner's theory of the width of spectral lines [4] (see also [3, page 473]), is of particular interest. The formal perturbation procedure breaks down in this case. As a matter of fact, the term of first order in the formal perturbation procedure gives a well determined change of the value of the point eigen-value, a result which is misleading since the disturbed operator has no point eigen-value. The term of second order in the formal perturbation procedure becomes infinite. We shall give explicitly the correct perturbation procedure and reveal the significance of the apparent shift of the point eigen-value.[1] To this end we shall use the spectral representation of the disturbed operator. We shall formulate this representation, given by Dirac [10], in a precise way, determine its inverse, and prove its completeness.

The spectral representations of the disturbed operators which we shall derive may be applied to a discussion of the solutions of the Schroedinger equations connected with these operators. Of particular interest is the asymptotic behavior of these solutions for infinite times $|t|$. We shall show that the limits for $t \to \infty$ and $t \to -\infty$ are connected with each other by a certain unitary operator, which corresponds to the "scattering operator" introduced by Heisenberg [7]. We shall formulate conditions under which this scattering operator is not only length-preserving but in addition unitary. We shall also give an exact solution of Weisskopf and Wigner's equation and show that their formulas give a correct asymptotic approximation.

1. *Spectral Representation*

As starting point for *perturbation investigations* we shall use the spectral representation and not the spectral resolution of the operator. The spectral resolution consists in giving to every interval of the spectral variable λ the appropriate projection operator—see [8]. If the projection operator assigned to a certain open interval vanishes, this interval is said to lie outside of the spectrum. If the projection operator assigned to a single point (which is to be considered a closed point set) does not vanish, the point is said to be a point eigen-value.

[1]The situation is analogous to that in the theory of the Stark effect and other problems of "weak quantization" (see, e.g. [5, page 178]), in which the point spectrum of the undisturbed operator also disappears through perturbation.

The aim of the *spectral representation* goes farther. It can simply be formulated thus: To represent the elements of the space on which the operator is permitted to act, by one or several functions of a "spectral variable" λ, defined in appropriate domains, in such a way that the operator is represented by the multiplication by the spectral variable λ. In other words, if the element x is represented through a set of functions $x_\nu(\lambda)$,

$$x = \{x_1(\lambda), x_2(\lambda), \cdots\},$$

then the element Lx produced by applying the operator L on x is represented through the functions $\lambda x_\nu(\lambda)$

$$Lx = \{\lambda x_1(\lambda), \lambda x_2(\lambda), \cdots\}.$$

The spectrum of the operator is called simple if the spectral representation involves only one representing function,

$$x = \{x(\lambda)\}.$$

The domain of values of λ in which $x(\lambda)$ is defined is then called the spectrum of the operator.

Sometimes it is preferable to employ representations through functions $x(\lambda)$ whose values are not numbers but elements in a finite dimensional or Hilbert space, although it would be possible to represent such a "function" $x(\lambda)$ by a sequence of functions whose values are numbers. We shall employ this generalized notion in connection with our first problem in Sections 2 to 4.

The operator occurring in our third problem, in Sections 6 to 8, will be represented by two functions having different domains of definition. The first function $x(\lambda)$ has an interval of the λ-axis as domain. The second function ξ has as domain just one particular value λ_0 ; it is therefore nothing but a single variable. The elements x are then represented by the pair $x(\lambda)$, ξ,

$$x = \{x(\lambda), \xi\}$$

and the operator L transforms x into the element represented by the pair $\lambda x(\lambda)$, $\lambda_0\xi$,

$$Lx = \{\lambda x(\lambda), \lambda_0\xi\}.$$

The value λ_0 is clearly a point eigen-value. The corresponding eigen-elements are those elements that are represented as

$$u = \{0, \xi\},$$

i.e. those for which the function $x(\lambda)$ vanishes; for, evidently $Lu = \lambda_0 u$ for these elements.

2. *Regular Perturbation of a Continuous Spectrum*

We consider an operator which has a simple continuous spectrum consisting of an interval $\mathfrak{s}$ of the λ-axis. This interval $\mathfrak{s}$ may either be the whole λ-axis or a semi-axis $\Lambda_- \leq \lambda < \infty$, or a finite interval $\Lambda_- \leq \lambda \leq \Lambda_+$. We assume from the outset that each element on which the operator may act is represented by a complex-valued function $x(\lambda)$ defined in $\mathfrak{s}$ so that the operator consists in transforming the function $x(\lambda)$ into the function $\lambda x(\lambda)$. This representation of the operator we denote by L (while in Section 1 we denoted by L the operator to be represented). Thus we can write

$$Lx(\lambda) = \lambda x(\lambda). \tag{2.01}$$

The "values" of the functions $x(\lambda)$ should be elements of a Hilbert space $\mathfrak{P}$, which may also be a space of a finite number of dimensions. The manifold of functions $x(\lambda)$ considered shall be the Hilbert space $\mathfrak{H}$ of all functions quadratically integrable over $\mathfrak{s}$. To explain this notion we denote the inner product of two elements $x(\lambda)$ and $y(\lambda)$ in $\mathfrak{P}$ by $\overline{x(\lambda)}y(\lambda)$ and the norm of $x(\lambda)$ by $| x(\lambda) | = (\overline{x(\lambda)}x(\lambda))^{1/2}$. We then can define

$$(x, y) = \int_{\mathfrak{s}} \overline{x(\lambda)}y(\lambda)\, d\lambda \tag{2.02}$$

and

$$|| x || = (x, x)^{1/2}. \tag{2.03}$$

The space $\mathfrak{H}$ then consists of all functions $x(\lambda)$ for which $|| x ||$ is finite. If $\mathfrak{s}$ extends to infinity the operator L is not bounded; it is then applicable only to the subspace $\mathfrak{F}$ of functions x in $\mathfrak{H}$ for which

$$|| Lx ||^2 = \int_{\mathfrak{s}} \lambda^2 \, | x(\lambda) |^2 \, d\lambda$$

is finite.

By K we denote the integral operator which transforms $x(\lambda)$ into

$$Kx(\lambda) = \int_{\mathfrak{s}} k(\lambda, \mu)x(\mu)\, d\mu. \tag{2.04}$$

The "kernel" $k(\lambda, \mu)$ is a bounded operator in $\mathfrak{P}$ which transforms the element $x(\mu)$ into the element $k(\lambda, \mu)x(\mu)$. Its adjoint operator is denoted by $\overline{k(\lambda, \mu)}$ We require that $k(\lambda, \mu)$ be Hermitian,

$$k(\mu, \lambda) = \overline{k(\lambda, \mu)}. \tag{2.05}$$

The disturbed operator to be considered is then represented by

$$L_\epsilon = L + \epsilon K. \tag{2.06}$$

Under appropriate conditions for the "kernel" $k(\lambda, \mu)$, to be formulated below, *the operators* L_ϵ *possess also a continuous spectrum, covering the same interval* $\mathcal{J}$, *provided that* $|\epsilon|$ *is sufficiently small.*

To find the *spectral representation* of the operator the functions $x(\lambda)$ should be represented by functions $y(\lambda)$ in $\mathfrak{H}$ such that $L_\epsilon x(\lambda)$ is represented by $\lambda y(\lambda)$ or, what is the same thing, by $Ly(\lambda)$. The transformation of the functions $y(\lambda)$ into the functions $x(\lambda)$ may be denoted by U_ϵ so that

(2.07) $$x(\lambda) = U_\epsilon y(\lambda)$$

should imply

(2.08) $$L_\epsilon x(\lambda) = U_\epsilon \lambda y(\lambda) = U_\epsilon L y(\lambda).$$

Combining these two relations, which should hold for all $y(\lambda)$ in $\mathfrak{H}$, we see that the pair of relations (2.07) and (2.08) is equivalent with the one relation

(2.09) $$L_\epsilon U_\epsilon = U_\epsilon L.$$

The functions $y(\lambda)$ representing the functions $x(\lambda)$ through (2.07) should be obtainable by a transformation

(2.10) $$y(\lambda) = U_\epsilon^* x(\lambda),$$

so that

(2.11) $$U_\epsilon^* U_\epsilon = 1$$

holds (we denote the unit operator by 1). Since every function $x(\lambda)$ should be representable through a function $y(\lambda)$ by (2.07), the inverse relation to (2.11) should also hold,

(2.12) $$U_\epsilon U_\epsilon^* = 1.$$

In other words the operators U_ϵ and U_ϵ^* should be inverse to each other. Relation (2.09) is then equivalent with $U_\epsilon^* L_\epsilon = L U_\epsilon^*$.

Finally the norm of the function $x(\lambda)$ should not be changed by the transformation into $y(\lambda)$;

2.13) $$\int_{\mathcal{J}} |x(\lambda)|^2 \, d\lambda = \int_{\mathcal{J}} |y(\lambda)|^2 \, d\lambda;$$

in other words, the operators U_ϵ and U_ϵ^* should be unitary and hence U_ϵ^* should be adjoint to U_ϵ.

Transformations U_ϵ and U_ϵ^* for which $L_\epsilon U_\epsilon = U_\epsilon L$ and $U_\epsilon^* L_\epsilon = L U_\epsilon^*$ hold were called "normators" in [2]; they were said to form a "pair" if $U_\epsilon^* U_\epsilon = 1$ and such a pair was said to be "intact" if $U_\epsilon U_\epsilon^* = 1$.

It is natural to try to determine the transformations U_ϵ by a *formal perturbation procedure* setting

(2.14) $$U_\epsilon = 1 + \epsilon U^{(1)} + \epsilon^2 U^{(2)} + \cdots.$$

Insertion into (2.09) and use of (2.06) gives successively

$$U^{(1)}L - LU^{(1)} = K,$$

$$(2.15)\qquad U^{(2)}L - LU^{(2)} = KU^{(1)},$$

$$\cdots .$$

Suppose now $U^{(1)}$ were an operator of the form

$$(2.16)\qquad U^{(1)}y(\lambda) = \omega^{(1)}(\lambda)y(\lambda) + \int_s u^{(1)}(\lambda, \mu)y(\mu)\, d\mu,$$

involving a "factor" $\omega^{(1)}(\lambda)$, a bounded operator in $\mathfrak{P}$, and an integral operator with the kernel $u^{(1)}(\lambda, \mu)$; then equation (2.15.1) would assume the form

$$\int_s u^{(1)}(\lambda, \mu)\mu y(\mu)\, d\mu - \lambda \int_s u^{(1)}(\lambda, \mu)y(\mu)\, d\mu = \int_s k(\lambda, \mu)y(\mu)\, d\mu$$

or

$$(2.17)\qquad \int_s (\mu - \lambda)u^{(1)}(\lambda, \mu)y(\mu)\, d\mu = \int_s k(\lambda, \mu)y(\mu)\, d\mu.$$

Since this relations were to hold for all $y(\mu)$ we would conclude

$$(2.18)\qquad u^{(1)}(\lambda, \mu) = \frac{k(\lambda, \mu)}{\mu - \lambda}$$

and

$$(2.19)\qquad U^{(1)}y(\lambda) = \omega^{(1)}(\lambda)y(\lambda) + \int_s \frac{k(\lambda, \mu)}{\mu - \lambda}\, y(\mu)\, d\mu;$$

the factor $\omega^{(1)}$ is so far not determined. The condition that the operator U_ϵ be unitary requires that the operator

$$(2.20)\qquad U_\epsilon^* = 1 + \epsilon U^{(1)*} + \epsilon^2 U^{(2)*} + \cdots$$

be the adjoint to U_ϵ, whence

$$U^{(1)*}x(\lambda) = \bar{\omega}^{(1)}(\lambda)x(\lambda) + \int_s \overline{u^{(1)}(\mu, \lambda)}x(\mu)\, d\mu$$

or, in view of (2.18) and (2.05),

$$(2.21)\qquad U^{(1)*}x(\lambda) = \bar{\omega}^{(1)}(\lambda)x(\lambda) - \int_s \frac{k(\lambda, \mu)}{\mu - \lambda}\, x(\mu)\, d\mu.$$

Insertion of (2.14), (2.20) in (2.11) yields the relation

$$(2.22)\qquad U^{(1)} + U^{(1)*} = 0,$$

whence by (2.19) and (2.20)

$$(2.23)\qquad \omega(\lambda) + \bar{\omega}(\lambda) = 0;$$

in other words the factor ω is such that $i\omega$ is symmetric. If $\mathfrak{P}$ is one-dimensional this means that ω is purely imaginary.

The value of the factor ω, except for the sign which remains arbitrary, is to be determined from the terms of second order. The terms of second order in relation (2.11) evidently involve the iterated operation

$$\int_{\mathfrak{g}} \frac{k(\lambda, \mu)}{\mu - \lambda} \int_{\mathfrak{g}} \frac{k(\mu, \nu)}{\nu - \mu} y(\nu) \, d\nu \, d\mu.$$

Before continuing the formal perturbation procedure we shall investigate the nature of operators of the type

$$\int_{\mathfrak{g}} \frac{k(\lambda, \mu)}{\mu - \lambda} y(\mu) \, d\mu$$

and of the successive application of such operators. Definite conditions on the kernel k will then have to be imposed. Next we shall establish the functional nature of the transformations U_ϵ, U_ϵ^* and prove their existence, for sufficiently small values of ϵ, under proper conditions for the kernel k. Only then shall we come back to the formal perturbation procedure, continue it, and establish its validity.

3. *Integral Operators of Class* (*R*)

Several conditions may be imposed on kernels $r(\lambda, \mu)$ of integral operators

$$Rx(\lambda) = \int_{\mathfrak{g}} r(\lambda, \mu) x(\mu) \, d\mu \tag{3.01}$$

in order that the singular integral operators

$$ARx(\lambda) = \int_{\mathfrak{g}} \frac{r(\lambda, \mu)}{\mu - \lambda} x(\mu) \, d\mu \tag{3.02}$$

can be defined and possess various desired properties.

For the sake of simplicity we shall confine ourselves here to the case that the interval $\mathfrak{g}$ is bounded, $\mathfrak{g} : \Lambda_- \leq \lambda \leq \Lambda_+$. Of the kernel $r(\lambda, \mu)$ we require firstly that it vanish when λ or μ is at one of the end points

$$\begin{aligned} r(\Lambda_-, \mu) &= r(\Lambda_+, \mu) = 0, \\ r(\lambda, \Lambda_-) &= r(\lambda, \Lambda_+) = 0; \end{aligned} \tag{3.03}$$

secondly, that it satisfy a Hoelder condition with respect to λ and μ in a sense to be explained. We define the norm of an operator $s(\lambda, \mu)$ in $\mathfrak{P}$ as the least upper bound of all numbers σ for which

$$| s(\lambda, \mu) x(\mu) | \leq \sigma | x(\mu) |$$

holds for all $x(\mu)$ in $\mathfrak{P}$. Then we require that the ratio

$$| r(\lambda_1, \mu_1) - r(\lambda_2, \mu_1) - r(\lambda_1, \mu_2) + r(\lambda_2, \mu_2) | / | \lambda_1 - \lambda_2 |^\theta | \mu_1 - \mu_2 |^\theta$$

remain bounded for all $\lambda_1 \neq \lambda_2$, $\mu_1 \neq \mu_2$ in $\mathfrak{s}$; here θ is a positive number less than one, kept fixed in the following. Denoting by ρ the least upper bound for this ratio we have

$$\begin{aligned} &| r(\lambda_1, \mu_1) - r(\lambda_2, \mu_1) - r(\lambda_1, \mu_2) + r(\lambda_2, \mu_2) | \\ &\qquad \leq \rho | \lambda_1 - \lambda_2 |^\theta | \mu_1 - \mu_2 |^\theta. \end{aligned} \tag{3.04}$$

(We could combine the two conditions (3.03) and (3.04) in one by defining $r(\lambda, \mu)$ to be zero if λ or μ lies outside of $\mathfrak{s}$ and then requiring (3.04) to hold for all $\lambda_1 \neq \lambda_2$, $\mu_1 \neq \mu_2$.) Integral operators R whose kernels satisfy these two conditions will be called of class (R).

It would be possible to extend the conditions imposed on $r(\lambda, \mu)$ so that the case in which the domain $\mathfrak{s}$ extends to infinity is also covered. To this end it would be sufficient to modify condition (3.04) by giving the right member of (3.04) the denominator

$$(1 + | \lambda_1 |^\theta + | \lambda_2 |^\theta + | \lambda_1\lambda_2 |^\theta)(1 + | \mu_1 |^\theta + | \mu_2 |^\theta + | \mu_1\mu_2 |^\theta),$$

to require that $r(\lambda, \mu)$ vanish if λ or μ becomes infinite, and in addition to impose conditions which insure that the functions $Rx(\lambda)$ lie in $\mathfrak{H}$ when $x(\lambda)$ lies in an appropriate subspace of $\mathfrak{H}$. In the appendix, Section 12, a condition of quite a different kind will be given; it will permit us to handle the case that the domain $\mathfrak{s}$ consists of the whole λ-axis in a particularly simple manner.

We now formulate various properties of the integral operators of class (R). These properties were proved in [2] for the case that the "values" of the function are numbers or, what is equivalent, that the dimension of the space $\mathfrak{P}$ is one; the proofs given there can be carried over literally to the case that $\mathfrak{H}$ is a Hilbert space.

An operator R of class (R) transforms every function $x(\lambda)$ in $\mathfrak{H}$ into a function $Rx(\lambda)$ which also lies in $\mathfrak{H}$. There is a constant ι, depending on the interval $\mathfrak{s}$ (and on θ), such that with

$$\| R \| = \iota\rho \tag{3.05}$$

the inequality

$$\| Rx \| \leq \| R \| \, \| x \| \tag{3.06}$$

holds.

The manifold of operators of class (R) form a complete space so that the relation $\| R^{(n)} - R^{(m)} \| \to 0$ implies the existence of an operator R such that $\| R^{(n)} - R \| \to 0$.

The operation ARx, given by (3.02), is first defined for functions x which satisfy a Hoelder condition of the same type as (3.04), viz.

$$\mathrm{A}Rx(\lambda) = \int_{\mathfrak{s}, |\lambda-\mu| \geq \epsilon} \frac{k(\lambda, \mu)}{\mu - \lambda} x(\mu)\, d\mu + \int_{\lambda-\epsilon}^{\lambda+\epsilon} \frac{k(\lambda, \mu)}{\mu - \lambda} x(\mu)\, d\mu,$$

the second contribution being defined as Cauchy's main value. It can then be shown that $\mathrm{A}Rx(\lambda)$ again satisfies a Hoelder condition of the type (3.04) and that further the inequality

$$\| \mathrm{A}Rx \| \leq \alpha \| R \| \| x \| \tag{3.07}$$

holds with a factor α depending on $\mathfrak{s}$ (and θ). By virtue of inequality (3.07) the definition of the operator $\mathrm{A}R$ can then be extended to all of $\mathfrak{H}$ so that (3.07) remains valid—see [2, §5].

In addition to the operator $\mathrm{A}R$ we introduce the operator

$$\Gamma R = \mathrm{A}R + i\mathrm{B}R, \tag{3.08}$$

with

$$\mathrm{B}Rx(\lambda) = \pi r(\lambda, \lambda) x(\lambda). \tag{3.09}$$

Thus

$$\Gamma Rx(\lambda) = i\pi r(\lambda, \lambda) x(\lambda) + \int_{\mathfrak{s}} \frac{r(\lambda, \mu)}{\mu - \lambda} x(\mu)\, d\mu. \tag{3.10}$$

On occasion we shall use the operator

$$\Gamma^* R = \mathrm{A}R - i\mathrm{B}R, \tag{3.08*}$$

which has essentially the same properties as the operator ΓR. (The present notation seems more appropriate than that used in [2].)

Evidently, an inequality

$$\| \Gamma Rx \| \leq \gamma \| R \| \| x \| \tag{3.11}$$

holds with a constant γ depending on $\mathfrak{s}$ (and θ).

The fundamental identity

$$(\Gamma R)L - L(\Gamma R) = R \tag{3.12}$$

holds, as is easily verified from the definitions (2.01) and (3.08) of the operators L and ΓR.

The product of an operator R_1 and an operator ΓR_2, R_1 and R_2 being of class (R), is also of the class (R), and inequalities

$$\| R_1(\Gamma R_2) \| \leq \gamma \| R_1 \| \| R_2 \|, \quad \| (\Gamma R_1) R_2 \| \leq \gamma \| R_1 \| \| R_2 \|, \tag{3.13}$$

hold. A second fundamental identity is

$$(\Gamma R_1)(\Gamma R_2) = \Gamma\{(\Gamma R_1) R_2 + R_1(\Gamma R_2)\}; \tag{3.14}$$

its proof requires a more detailed analysis; see [2, §2]. Easily, however, one verifies from (3.10) the identity

$$2i(BR_1)(BR_2) = B\{(\Gamma R_1)R_2 + R_1(\Gamma R_2)\}. \tag{3.15}$$

Incidentally, from identity (3.14) one could evaluate the product (AR_1) (AR_2), which appeared to be needed when determining the terms of second order in the formal perturbation procedure as discussed at the end of Section 2. The result is

$$(AR_1)(AR_2) = A\{(AR_1)R_2 + R_1(AR_2)\} + (BR_1)(BR_2).$$

Actually, we shall not need this identity; we shall find it more suitable to use identity (3.14).

Finally we note: The adjoint of the operator R is the integral operator R^* with the kernel

$$r^*(\lambda, \mu) = \overline{r(\mu, \lambda)}. \tag{3.16}$$

The adjoint of the operator ΓR is then

$$(\Gamma R)^* = -\Gamma R^*. \tag{3.17}$$

4. *Spectral Representation of the Regularly Disturbed Operator*

We now assume that the Hermitian operator K is of the class (K), i.e. that the kernel $k(\lambda, \mu) = \overline{k(\mu, \lambda)}$ of the operator K given by (2.04) is Hermitian and of class (R). In addition to the Hoelder condition (3.04), the condition (3.03) is imposed requiring the kernel $k(\lambda, \mu)$ to vanish if λ or μ lies at an end point of the interval $\mathcal{I}$. We want to emphasize this condition. We shall show later on, in Section 7, that the character of the spectrum of the operator $L + \epsilon K$ differs in general from that of L if this condition is violated.

We now maintain that there are spectral transformations U_ϵ and U_ϵ^* of the form

$$U_\epsilon = 1 + \epsilon\Gamma R_\epsilon\,, \qquad U_\epsilon^* = 1 - \epsilon\Gamma R_\epsilon^*\,, \tag{4.01}$$

in which R_ϵ and R_ϵ^* are appropriate integral operators of the class (R) with kernels $r_\epsilon(\lambda, \mu)$ and $r_\epsilon^*(\lambda, \mu) = \overline{r_\epsilon(\mu, \lambda)}$, which depend analytically on ϵ in the neighborhood of $\epsilon = 0$. The meaning of the operators ΓR_ϵ and ΓR_ϵ^* is defined through (3.08) or (3.10). Note, by (3.17), that U_ϵ^* is the adjoint of U_ϵ if ϵ is real.

Relation (2.09) now takes the form

$$(L + \epsilon K)(1 + \epsilon\Gamma R_\epsilon) = (1 + \epsilon\Gamma R_\epsilon)L; \tag{4.02}$$

the adjoint to this relation is

$$(1 - \epsilon\Gamma R_\epsilon^*)(L + \epsilon K) = L(1 - \epsilon\Gamma R_\epsilon^*), \tag{4.02*}$$

since $K^* = K$ is assumed. By virtue of (3.12), these relations are equivalent with

$$R_\epsilon = K(1 + \epsilon\Gamma R_\epsilon) = KU_\epsilon \tag{4.03}$$

and

$$R_\epsilon^* = (1 - \epsilon\Gamma R_\epsilon^*)K = U_\epsilon^* K. \tag{4.03*}$$

It is easy to prove that to a given operator K of class (R) operators R_ϵ of class (R) exist such that (4.03) holds, provided that $|\epsilon|$ is small enough, see [2]. (It is just as easy to prove that to any operator R an operator K_ϵ exists which satisfies the equation $R = K_\epsilon(1 + \epsilon\Gamma R)$.) The value of ϵ may even be permitted to be complex. One need only set up iterations

$$R_\epsilon^{(n+1)} = K(1 + \epsilon\Gamma R_\epsilon^{(n)}), \tag{4.04}$$

beginning with $R_\epsilon^{(0)} = K$, and take note of the fact that $K(\Gamma R_\epsilon^{(n)})$ is an integral operator of the class (R) if $R_\epsilon^{(n)}$ is such an operator. From the inequality

$$\| R_\epsilon^{(n+1)} - R_\epsilon^{(n)} \| \leq \gamma \, |\epsilon| \, \| K \| \, \| R_\epsilon^{(n)} - R_\epsilon^{(n-1)} \|,$$

which follows from (3.11), we deduce $\| R_\epsilon^{(m)} - R_\epsilon^{(n)} \| \to 0$ as $m, n \to \infty$, provided that

$$|\epsilon| < 1/\gamma \, \| K \| . \tag{4.05}$$

Consequently, according to a remark made in Section 3, after (3.06), the operators $R_\epsilon^{(n)}$ converge to a limit operator R_ϵ, which then satisfies (4.03).

It is not difficult to show that operators R_ϵ are continuously differentiable with respect to ϵ. Assuming that $R_\epsilon^{(n)}$ is continuously differentiable, one proves the same of $\Gamma R_\epsilon^{(n)}$, $K(\Gamma R_\epsilon^{(n)})$, and hence of $R_\epsilon^{(n+1)}$, finding

$$\frac{\partial}{\partial\epsilon} R_\epsilon^{(n+1)} = K(\Gamma R_\epsilon^{(n)}) + \epsilon K\left(\Gamma \frac{\partial}{\partial\epsilon} R_\epsilon^{(n)}\right).$$

The convergence of these derivatives to a limit operator of class (R) is easily established and it is also easily shown that this limit operator is the derivative of R_ϵ, satisfying the relation

$$\frac{\partial}{\partial\epsilon} R_\epsilon = K(\Gamma R_\epsilon) + \epsilon K\left(\Gamma \frac{\partial}{\partial\epsilon} R_\epsilon\right), \tag{4.06}$$

and depending continuously on ϵ, provided (4.05) holds.

Since the preceding arguments are valid even for complex values of ϵ, continuous differentiability with respect to ϵ implies analytic dependence on ϵ. Therefore, the operator R_ϵ admits a convergent expansion with respect to powers of ϵ,

$$R_\epsilon = R^{(1)} + \epsilon R^{(2)} + \cdots . \tag{4.07}$$

Insertion into (4.03) gives successively

$$R^{(1)} = K,$$

(4.08)
$$R^{(2)} = K(\Gamma K),$$

$$R^{(3)} = K(\Gamma(K(\Gamma K))),$$

and so on. Thus we have established the *formal perturbation procedure* rigorously. It assumes a form much simpler than would be expected from the first steps of such a procedure discussed at the end of Section 2. The term of first order, in particular, assumes now the form

(4.09)
$$U^{(1)} = \Gamma K = \mathrm{A}K + i\mathrm{B}K;$$

comparing this formula with relation (2.19) we see from (3.09) that the factor $\omega^{(1)}$ is now specified to be

(4.10)
$$\omega^{(1)}(\lambda) = i\pi k(\lambda, \lambda).$$

Our choice (4.01) of the operator U_ϵ is not the only one that gives a solution of the equation (4.02); we might also have chosen the expression

$$U_\epsilon = (1 + \epsilon\Gamma R_\epsilon)(1 + \epsilon f_\epsilon(L)),$$

in which the operator $f(L)$ is defined by

$$f(L)x(\lambda) = f(\lambda)x(\lambda).$$

In that case we would have

$$\omega^{(1)}(\lambda) = i\pi k(\lambda, \lambda) + f^{(1)}(\lambda),$$

while the contribution AK to $U^{(1)}$ would not change.

In Section 2 we attempted to determine the factor $\omega^{(1)}$ from the condition that the transformation U_ϵ be unitary. We want to show that just the form (4.01) of the transformation U_ϵ we had chosen implies that U_ϵ is unitary.

From now on we again assume ϵ to be real. If the operator R_ϵ with the kernel $r_\epsilon(\lambda, \mu)$ satisfies relation (4.03), the adjoint operator R_ϵ^* with the kernel $\overline{r_\epsilon(\mu, \lambda)}$ satisfies the relation (4.03*). From these two relations we obtain immediately

$$U_\epsilon^* R_\epsilon = U_\epsilon^* K U_\epsilon = R_\epsilon^* U_\epsilon\,.$$

Thus we have

(4.11)
$$U_\epsilon^* R_\epsilon - R_\epsilon^* U_\epsilon = 0,$$

or in detail, by (4.01),

(4.12)
$$R_\epsilon - R_\epsilon^* - \epsilon(\Gamma R_\epsilon^*)R_\epsilon - \epsilon R_\epsilon^* \Gamma R_\epsilon = 0.$$

Important conclusions can be drawn from this relation. First, from the fundamental identity (3.14), we derive the identity

$$(1 - \epsilon\Gamma R_\epsilon^*)(1 + \epsilon\Gamma R_\epsilon) - 1 = \epsilon\Gamma R_\epsilon - \epsilon\Gamma R_\epsilon^* - \epsilon^2(\Gamma R_\epsilon^*)(\Gamma R_\epsilon)$$

$$= \epsilon\Gamma\{R_\epsilon - R_\epsilon^* - \epsilon(\Gamma R_\epsilon^*)R_\epsilon - \epsilon R_\epsilon^*(\Gamma R_\epsilon)\}$$

or

(4.13) $$U_\epsilon^* U_\epsilon - 1 = \epsilon\Gamma\{U_\epsilon^* R_\epsilon - R_\epsilon^* U_\epsilon\}$$

for any two transformations $U_\epsilon = 1 + \epsilon\Gamma R_\epsilon$, $U_\epsilon^* = 1 - \epsilon\Gamma R_\epsilon^*$. For our transformations which satisfy (4.03) and (4.03*) we conclude from (4.11) that relation

(4.14) $$U_\epsilon^* U_\epsilon = 1$$

holds, which expresses the fact that the transformations U_ϵ and U_ϵ^* form a pair—see (2.11).

Secondly, we employ identity (3.14) and derive from (4.12) the relation (see [2, page 265]),

(4.15) $$\mathrm{B}R_\epsilon - \mathrm{B}R_\epsilon^* - 2i\epsilon(\mathrm{B}R_\epsilon^*)(\mathrm{B}R_\epsilon) = 0,$$

which is equivalent with

(4.16) $$(1 - 2i\epsilon\mathrm{B}R_\epsilon^*)(1 + 2i\epsilon\mathrm{B}R_\epsilon) = 1.$$

This relation can be written in the form

(4.17) $$S_\epsilon^* S_\epsilon = 1$$

when we introduce the operator

(4.18) $$S_\epsilon = 1 + 2i\epsilon\mathrm{B}R_\epsilon$$

with the adjoint

(4.18*) $$S_\epsilon^* = 1 - 2i\epsilon\mathrm{B}R_\epsilon\,.$$

The operator S_ϵ transforms the function $x(\lambda)$ into the function

(4.19) $$S_\epsilon x(\lambda) = (1 + 2i\pi\epsilon r_\epsilon(\lambda, \lambda))x(\lambda).$$

If the space $\mathfrak{P}$ is one-dimensional, i.e. if the "values" of the functions $x(\lambda)$ are numbers, the operator S_ϵ simply consists in multiplying $x(\lambda)$ by the number

(4.20) $$S_\epsilon(\lambda) = 1 + 2i\pi\epsilon r_\epsilon(\lambda, \lambda).$$

Relation (4.17) then expresses the fact that the absolute value of this number is one:

(4.21) $$| S_\epsilon(\lambda) | = 1.$$

If $\mathfrak{P}$ is a general Hilbert space, however, the operator $S_\epsilon(\lambda) = 1 + 2i\pi\epsilon r_\epsilon(\lambda, \lambda)$ is an operator in this Hilbert space $\mathfrak{P}$. Relation (4.21) or rather relation

$$S_\epsilon^*(\lambda)S_\epsilon(\lambda) = 1 \tag{4.22}$$

then expresses the fact that this operator is length preserving. This operator will be shown to be unitary in $\mathfrak{P}$ under the same conditions under which U_ϵ will be shown to be unitary in $\mathfrak{H}$.

The operator S_ϵ corresponds to the "scattering operator" introduced by Wheeler [6] and Heisenberg [7]; our proof of its "length preserving" character is related to the derivation of the corresponding character for the scattering operator given by Pauli and Bargmann (quoted in [9, page 48]).

The inverse relation to (4.11), the "intactness,"

$$U_\epsilon U_\epsilon^* = 1, \tag{4.23}$$

does not follow in a purely formal manner as (4.14) did. As a matter of fact relation (4.23) need not hold if $|\epsilon|$ is too large; but we are able to establish relation (4.23) if $|\epsilon|$ is bounded by

$$|\epsilon| < 1/2\gamma \, \| K \|. \tag{4.24}$$

Relations (4.03*) and (3.12) then entail

$$\| R_\epsilon^* \| \leq \| K \| + \gamma \, |\epsilon| \, \| K \| \, \| R_\epsilon^* \|,$$

whence by (4.24)

$$\epsilon\gamma \, \| R_\epsilon^* \| < 1. \tag{4.25}$$

From this relation we conclude: *Whenever* $U_\epsilon^* z(\lambda) = 0$ for a function $z(\lambda)$ in $\mathfrak{H}$, then $z(\lambda) = 0$.

For, $z(\lambda) = \epsilon\Gamma R_\epsilon^* z(\lambda)$ implies $\| z \| \leq |\epsilon| \gamma \, \| R_\epsilon^* \| \, \| z \|$ or

$$\{1 - |\epsilon| \gamma \, \| R_\epsilon^* \|\} \, \| z \| \leq 0,$$

whence $\| z \| = 0$ by (4.25).

Let now $x(\lambda)$ be any function $x(\lambda)$ in $\mathfrak{H}$; then by (4.14)

$$U_\epsilon^*(U_\epsilon U_\epsilon^* - 1)x(\lambda) = 0;$$

consequently, $(U_\epsilon U_\epsilon^* - 1)x(\lambda) = 0$. In other words, (4.23) holds.

We can derive a somewhat stronger result. From (4.14), (4.23), and (4.11) we deduce for any $x(\lambda)$ in $\mathfrak{H}$

$$U_\epsilon^*(R_\epsilon U_\epsilon^* - U_\epsilon R_\epsilon^*)x(\lambda) = (U_\epsilon^* R_\epsilon - R_\epsilon^* U_\epsilon)U_\epsilon^* x(\lambda) = 0.$$

Consequently,

$$R_\epsilon U_\epsilon^* - U_\epsilon R_\epsilon^* = 0, \tag{4.26}$$

or, which is equivalent,

$$R_\epsilon - R_\epsilon^* - \epsilon R_\epsilon(\Gamma R_\epsilon^*) - \epsilon(\Gamma R_\epsilon^*)R_\epsilon = 0. \tag{4.27}$$

From this relation we could re-derive (4.23) by using the fundamental identity (3.14). Using identity (3.15) we obtain from (4.27) the relation

$$\mathrm{B}R_\epsilon - \mathrm{B}R_\epsilon^* - 2i\epsilon(\mathrm{B}R_\epsilon)(\mathrm{B}R_\epsilon^*) \tag{4.28}$$

which is equivalent with

$$(1 + 2i\epsilon\mathrm{B}R_\epsilon)(1 - 2i\epsilon\mathrm{B}R_\epsilon^*) = 1, \tag{4.29}$$

or by (4.18), (4.18*)

$$S_\epsilon S_\epsilon^* = 1. \tag{4.30}$$

Thus the intactness of the pair of operators U_ϵ, U_ϵ^* implies that the operators $S_\epsilon(\lambda) = (1 + 2i\pi\epsilon r_\epsilon(\lambda, \lambda))$ and $S_\epsilon^*(\lambda) = (1 - 2i\pi\epsilon r_\epsilon^*(\lambda, \lambda))$ in $\mathfrak{P}$ are not only length preserving but, in addition, unitary,

$$S_\epsilon(\lambda)S_\epsilon^*(\lambda) = 1. \tag{4.31}$$

We should mention that the spectral transformation is not uniquely determined. Instead of choosing $U_\epsilon = 1 + \epsilon\Gamma R_\epsilon$ we might just as well have chosen $U_\epsilon = 1 + \epsilon\Gamma^* R_\epsilon$—see (3.08*); this would even have been of advantage in certain applications—see Section 9.

We should also admit that our insistence on unitary transformations U_ϵ, U_ϵ^* was more than necessary for a spectral representation. Instead of the form (4.01) of the transformations, as mentioned above, we might have taken any transformations of the form

$$\begin{aligned} U_\epsilon &= (1 + \epsilon\Gamma R_\epsilon)(1 + \epsilon f_\epsilon(L)), \\ U_\epsilon^* &= (1 + \epsilon f_\epsilon(L))^{-1}(1 - \epsilon\Gamma R_\epsilon^*); \end{aligned} \tag{4.32}$$

relations (4.11) and (4.23) are then also satisfied. Unless $| 1 + \epsilon f_\epsilon(\lambda) | = 1$, U_ϵ^* is no longer the adjoint of U_ϵ and these transformations are then not unitary.

By choosing

$$1 + \epsilon f_\epsilon(L) = \{1 + i\epsilon\pi r_\epsilon(L, L)\}^{-1} = \{1 + i\epsilon\mathrm{B}R_\epsilon\}^{-1}, \tag{4.33}$$

for example, we could achieve that the transformation U_ϵ is of the form

$$U_\epsilon = 1 + \epsilon\mathrm{A}R_\epsilon' \tag{4.34}$$

with

$$R_\epsilon' = R_\epsilon\{1 + i\epsilon\pi r_\epsilon(L, L)\}^{-1} = R_\epsilon\{1 + i\epsilon\mathrm{B}R_\epsilon\}^{-1},$$

which is simpler than the form $U_\epsilon = 1 + \epsilon\Gamma R_\epsilon = 1 + i\epsilon\mathrm{B}R_\epsilon + \epsilon\mathrm{A}R_\epsilon$ assumed in (4.01). The inverse operator of (4.34), however, would then not be of the same form.

Suppose the spaces $\mathfrak{P}$ and $\mathfrak{H}$ are real, and not complex—i.e., suppose these spaces admit multiplication of their elements only by real numbers—and let $k(\lambda, \mu)$ be a symmetric operator in $\mathfrak{P}$. Then the kernel $r'_\epsilon(\lambda, \mu)$ is also defined in $\mathfrak{P}$. This can be seen from the fact that R'_ϵ satisfies the equation

$$R'_\epsilon = K(1 + \epsilon A R'_\epsilon),$$

which can also be solved by iterations. By choosing

$$(4.35) \qquad 1 + \epsilon f_\epsilon(\lambda) = ((1 - i\epsilon\pi r_\epsilon(\lambda, \lambda))/(1 + i\epsilon\pi r_\epsilon(\lambda, \lambda)))^{1/2}$$

in (4.32) we obtain again a unitary operator since the absolute value of this function is unity. Evidently, the transformation so obtained can be written in the form

$$(4.36) \qquad U_\epsilon = (1 + \epsilon A R'_\epsilon)(1 + \epsilon^2\pi^2 r_\epsilon^2(L, L))^{1/2}$$

and, consequently, this transformation is defined in a real space $\mathfrak{H}$. The same is of course, true for its adjoint.

For some purposes this transformation (4.36) is more suitable, but the transformation (4.01) on which we have based our theory is easier to characterize.

5. *Perturbation Through Which a Point Eigen-value Appears*

It was stated at the beginning of Section 4 that the boundary condition that the kernel $k(\lambda, \mu)$ of the disturbing operator K vanish at the end points of the interval was essential. We shall show here that the perturbed operator $L + \epsilon K$ may possess a point eigen-value for arbitrarily small values of the parameter ϵ if this boundary condition is not satisfied. We then shall call the integral operator K "abnormal." We shall consider special abnormal operators $K = J$ with the kernel

$$(5.01) \qquad j(\lambda, \mu) = j(\lambda)j(\mu)$$

assuming that the function

$$(5.02) \qquad k(\lambda) = (\lambda - \Lambda_-)^{\alpha_-}(\lambda - \Lambda_+)^{\alpha_+}j(\lambda)$$

vanishes at the end points of $\mathfrak{J}$ and satisfies a Hoelder condition

$$(5.03) \qquad | k(\lambda_1) - k(\lambda_2) | \leq a | \lambda_1 - \lambda_2 |^\theta, \qquad \lambda_1 \neq \lambda_2 \text{ in } \mathfrak{J}$$

with appropriate non-negative constants α_-, $\alpha_+ < 1/2$.

This case is not quite as special as it might appear since every kernel $j(\lambda, \mu)$ with

$$j(\Lambda_+, \Lambda_+) \neq 0, \qquad j(\Lambda_-, \Lambda_-) \neq 0$$

can be represented as the sum or difference of two kernels $j^{(1)}$, $j^{(2)}$ of the type (5.01) and one kernel $k(\lambda, \mu)$ of the class (K) considered in Section 4 in the form $J = J^{(1)} \pm J^{(2)} + K$. By applying the transformation U_ϵ, U_ϵ^*, associated with the operator $L + \epsilon K$ as set up in Section 4, the operator $L + \epsilon K$ can be reduced to the operator

$$L = U_\epsilon^*(L + \epsilon K)U_\epsilon .$$

The total operator $L + \epsilon J = L + \epsilon J^{(1)} \pm \epsilon J^{(2)} + \epsilon K$ is then reduced to

$$L + \epsilon U_\epsilon^* J^{(1)} U_\epsilon \pm \epsilon U_\epsilon^* J^{(2)} U_\epsilon . \tag{5.04}$$

The last two operators are again of the type (5.01) as easily seen. Operators of the type (5.04) can be treated similarly to the way we shall treat the operator

$$L + \epsilon J \tag{5.05}$$

with an operator J having a kernel as given by (5.01).

Note that the operator J transforms the function $x(\lambda)$ into the function

$$Jx(\lambda) = j(\lambda)(j, \lambda) \tag{5.06}$$

We proceed to show that the operator (5.05) possesses a point eigen-value for $\epsilon \neq 0$. If there were such an eigen-value λ_ϵ with the eigen-function $x_\epsilon(\lambda)$ we would have

$$\lambda x_\epsilon(\lambda) + \epsilon j(\lambda)(j, x_\epsilon) = \lambda_\epsilon x_\epsilon(\lambda), \tag{5.07}$$

whence

$$x_\epsilon(\lambda) = -\epsilon(j, x_\epsilon)\frac{j(\lambda)}{\lambda - \lambda_\epsilon}; \tag{5.08}$$

Insertion into (5.07) yields

$$(j, x_\epsilon) = -\epsilon(j, x_\epsilon)[j; \lambda_\epsilon] \tag{5.09}$$

with

$$[j; \lambda_\epsilon] = \int_{\mathfrak{s}} \frac{|j(\mu)|^2}{\mu - \lambda_\epsilon}\, d\mu. \tag{5.10}$$

Since $(j, x_\epsilon) = 0$ would entail $x_\epsilon = 0$, see (5.08), we have $(j, x_\epsilon) \neq 0$; hence we have from (5.09)

$$1 = -\epsilon[j; \lambda_\epsilon]. \tag{5.11}$$

We maintain that this equation (5.11) possesses a real root $\lambda_\epsilon > \Lambda_+$ for any $\epsilon > 0$ approaching Λ_+ as $\epsilon \to 0$, if $j(\Lambda_+) \neq 0$, and further a root $\lambda_\epsilon < \Lambda_-$ for any $\epsilon < 0$, approaching Λ_- as $\epsilon \to 0$ if $j(\Lambda_-) \neq 0$. This statement is obvious from the following evident facts: The function $[j; \lambda]$, defined by (5.10), approaches zero as $|\lambda|$ approaches infinity; as λ approaches an end point of $\mathfrak{s}$, the function $[j; \lambda]$ becomes infinite if the function $j(\lambda)$ does not vanish at that end point. Further, $[j; \lambda] < 0$ for $\lambda > \Lambda_+$, > 0 for $\lambda < \Lambda_-$.

Suppose we choose such a root λ_ϵ and define $x_\epsilon(\lambda)$ by

(5.12) $$x_\epsilon(\lambda) = -\epsilon \frac{j(\lambda)}{\lambda - \lambda_\epsilon};$$

in accordance with (5.08); then by (5.10)

(5.13) $$(j, x_\epsilon) = 1.$$

Consequently, relation (5.08) holds, and hence relation (5.06), which expresses the fact that $x_\epsilon(\lambda)$ is eigen-function with the eigen-value λ_ϵ.

The spectral transformation of the operator $L + \epsilon J$ will be given in the appendix, Section 11.

6. *A Perturbation Through Which a Point Eigen-value Disappears*

We shall treat in detail the spectral representation of disturbed operators which possess no point eigen-value while the undisturbed operator possesses one. As in Section 2 we assume that the undisturbed operator is given in spectral representation. The Hilbert space of elements $x(\lambda)$ is then represented by a pair $\{x(\lambda), \xi\}$, $x(\lambda)$ being a quadratically integrable function defibed in an interval $\mathcal{J}$ of the type considered in Section 2, ξ being a complex number. As inner product we assume

(6.01) $$(\{x, \xi\}, \{y, \eta\}) = (x, y) + \bar{\xi}\eta = \int_{\mathcal{J}} \overline{x(\lambda)}y(\lambda)\, d\lambda + \bar{\xi}\eta,$$

and as norm

(6.02) $$(\| x \|^2 + | \xi |^2)^{1/2}.$$

The undisturbed operator L transforms the pair $\{x(\lambda), \xi\}$ into the pair

(6.03) $$L\{x(\lambda), \xi\} = \{\lambda x(\lambda), \lambda_0 \xi\}.$$

The point eigen-value λ_0 is assumed to lie in the interior of the domain $\mathcal{J}$. The corresponding eigen-functions are $\{0, \xi\}$.

With the aid of a function $f(\lambda)$, which should satisfy conditions formulated below, we form the operation

(6.04) $$F\{x(\lambda), \xi\} = \{f(\lambda)\xi, (f, x)\}.$$

The operator F is evidently symmetric:

(6.05) $$\begin{aligned}(\{x, \xi\}, F\{x', \xi'\}) &= \xi' \int_{\mathcal{J}} \overline{x(\lambda)}f(\lambda)\, d\lambda + \bar{\xi} \int_{\mathcal{J}} \overline{f(\lambda)}x'(\lambda)\, d\lambda \\ &= (F\{x, \xi\}, \{x', \xi'\}).\end{aligned}$$

The disturbed operator is then assumed of the form $L + \epsilon F$, transforming every element $\{x(\lambda), \xi\}$ into the element

(6.06) $$(L + \epsilon F)\{x(\lambda), \xi\} = \{\lambda x(\lambda) + \epsilon f(\lambda)\xi, \epsilon(f, x) + \lambda_0 \xi\}.$$

The disturbing operator transforms the eigen-functions $\{x(\lambda),\ 0\}$ of the continuous spectrum into eigen-functions $\{0,\ \xi\}$ of the point eigen-value and vice versa. This form of the operator is less restricted than it might appear. Suppose we had added a term $\{\epsilon Kx,\ 0\}$ to the operator (6.06) with an operator K as treated in Sections 2 and 4, then we could have reduced the disturbed operator to one of the form (6.05) by using the transformation U_ϵ set up in Section 4. Similarly, if we had added a term $\{0,\ \epsilon\kappa\xi\}$ to (6.05) we could have reduced the operator so disturbed to the form (6.06).

We first attempt to carry out a *formal perturbation procedure* assuming that the character of the spectrum is not changed; in particular we assume that the operator $L + \epsilon F$ has a point eigen-value λ_ϵ which depends analytically on ϵ. We set

$$\lambda_\epsilon = \lambda_0 + \epsilon\lambda_1 + \epsilon^2\lambda_2 + \cdots \tag{6.07}$$

and assume that the eigen-element $\{x_\epsilon(\lambda),\ \xi_\epsilon\}$ corresponding to it admits an expansion

$$x_\epsilon(\lambda) = \epsilon x_1(\lambda) + \epsilon^2 x_2(\lambda) + \cdots, \qquad \xi_\epsilon = 1. \tag{6.08}$$

The assumption $\xi_\epsilon = 1$ is no restriction since the eigen-functions are only determined within a factor, and we can choose the factor such that $\xi_\epsilon = 1$.

We insert the expansions (6.07), (6.08) in the relation

$$(L + \epsilon F)\{x_\epsilon(\lambda),\ \xi_\epsilon\} = \lambda_\epsilon\{x_\epsilon(\lambda),\ \xi_\epsilon\} \tag{6.09}$$

or

$$(\lambda - \lambda_\epsilon)x_\epsilon(\lambda) + \epsilon f(\lambda)\xi_\epsilon = 0, \tag{6.10}$$

$$\epsilon(f,\ x_\epsilon) + (\lambda_0 - \lambda_\epsilon)\xi_\epsilon = 0. \tag{6.11}$$

We then obtain in first order

$$x_1(\lambda) = -\frac{f(\lambda)}{\lambda - \lambda_0}, \qquad \lambda_1 = 0; \tag{6.12}$$

in second order

$$x_2(\lambda) = 0, \qquad \lambda_2 = -\int_s \frac{|f(\lambda)|^2}{\lambda - \lambda_0}\, d\lambda; \tag{6.13}$$

in third order

$$x_3(\lambda) = -\lambda_2 \frac{f(\lambda)}{(\lambda - \lambda_0)^2}, \qquad \lambda_3 = 0; \tag{6.14}$$

in fourth order

$$x_4(\lambda) = 0, \qquad \lambda_4 = -\lambda_2 \int_s \frac{|f(\lambda)|^2}{(\lambda - \lambda_0)^2}\, d\lambda. \tag{6.15}$$

Relation (6.13.2) gives a well-determined shift $\epsilon^2\lambda_2$ (in second order) of the eigen-value, since the integral occurring in (6.13), regarded as a Cauchy main value, is finite for functions $f(\lambda)$ of the type that we are going to consider below. The fourth order term given by relation (6.15.2), however, is infinite. Thus it is seen that *the formal perturbation procedure breaks down.* As a matter of fact, it already breaks down at the first step since the function $x_1(\lambda)$ given by (6.12.1) is not quadratically integrable—unless $f(\lambda_0) = 0$.

We proceed to show that *the operator* $L + \epsilon F$ *has no point eigen-value at all* if the function $f(\lambda)$ does not vanish in the interior of $\mathfrak{s}$ and if ϵ is small enough, certainly if ϵ satisfies the conditions

$$\epsilon^{-2} > \frac{1}{\Lambda_+ - \lambda_0} \int_{\mathfrak{s}} \frac{|f(\mu)|^2}{\Lambda_+ - \mu} d\mu,$$

6.16)

$$\epsilon^{-2} > \frac{1}{\lambda_0 - \Lambda_+} \int_{\mathfrak{s}} \frac{|f(\mu)|^2}{\mu - \Lambda_-} d\mu,$$

in which $\Lambda_- < \Lambda_+$ are the end points of the interval $\mathfrak{s}$.

To prove our statement we proceed indirectly. Let λ_ϵ be a point eigen-value of the operator $L + \epsilon F$ with the eigen-function $\{x_\epsilon(\lambda), \xi_\epsilon\}$.

We first show that λ_ϵ is real (assuming ϵ to be real). This follows in the standard manner. Multiplying the equations (6.10), (6.11) by $\overline{x_\epsilon(\lambda)}$ and $\overline{\xi_\epsilon}$ respectively, integrating, and adding, we find

$$(x_\epsilon, Lx_\epsilon) + \lambda_0 |\xi_\epsilon|^2 + 2\epsilon \Re e \xi_\epsilon (x_\epsilon, f) = \lambda_\epsilon \{(x_\epsilon, x_\epsilon) + |\xi_\epsilon|^2\}.$$

Since the left member is real the same is true for the right member and hence for λ_ϵ.

Next we show that condition (6.16) is violated if a point eigen-value λ_ϵ exists. Without restriction we set $\xi_\epsilon = 1$. From equation (6.10) we have

$$x_\epsilon(\lambda) = -\epsilon \frac{f(\lambda)}{\lambda - \lambda_\epsilon}. \tag{6.17}$$

Since $x_\epsilon(\lambda)$ is assumed quadratically integrable and $f(\lambda)$ is assumed not to vanish in the interior of $\mathfrak{s}$ it follows that the real number λ_ϵ does not lie in the interior of $\mathfrak{s}$. Inserting (6.17) into (6.11) we find

$$\lambda_0 - \lambda_\epsilon = \epsilon^2 AF(\lambda_\epsilon) \tag{6.18}$$

when we set

$$AF(\lambda) = \int_{\mathfrak{s}} \frac{|f(\mu)|^2}{\mu - \lambda} d\mu. \tag{6.19}$$

The function $AF(\lambda)$ is evidently monotone in λ for real λ outside of $\mathfrak{s}$. The same is then evidently true for the positive function $(\lambda_0 - \lambda)^{-1} AF(\lambda)$ for $\lambda \geq \Lambda_+$ and $\lambda \leq \Lambda_-$. It is clear that this function assumes its maximum either at $\lambda = \Lambda_-$ or at $\lambda = \Lambda_+$. In particular, either

$$(\lambda_0 - \lambda_\epsilon)^{-1} AF(\lambda_\epsilon) \leq (\lambda_0 - \Lambda_-)^{-1} AF(\Lambda_-)$$

or

$$(\lambda_0 - \lambda_\epsilon)^{-1} AF(\lambda_\epsilon) \leq (\lambda_0 - \Lambda_+)^{-1} AF(\Lambda_+).$$

This fact is by (6.18) not compatible with (6.16) if $\epsilon \neq 0$. Thus we have proved that the operator $L + \epsilon F$ has no point eigen-value for $\epsilon \neq 0$ if condition (6.16) is satisfied.

7. *Spectral Representation of the Disturbed Operator Without Point Eigen-value*

The spectral representation of the operator $L + \epsilon F$ defined by (6.06) requires representing the pair $\{x(\lambda), \xi\}$ by one function $y(\lambda)$ defined in $\mathfrak{s}$. The transformation of $y(\lambda)$ into $\{x(\lambda), \xi\}$ involves an operator

(7.01) $$U_\epsilon = 1 + \epsilon\Gamma R_\epsilon$$

and a function g_ϵ in $\mathfrak{H}$, which will be given explicitly below. The transformation then is

(7.02) $$x(\lambda) = U_\epsilon y(\lambda), \qquad \xi = (g_\epsilon , y).$$

The inverse transformation is

(7.03) $$y(\lambda) = U_\epsilon^* x(\lambda) + g_\epsilon(\lambda)\xi.$$

The norms are preserved under these transformations in the sense that

(7.04) $$\| x \|^2 + | \xi |^2 = \| y \|^2.$$

The requirement that the transformation (7.02) furnish a spectral representation of the operator $L + \epsilon F$ is expressed through the relations

(7.05) $$\begin{aligned} Lx(\lambda) + \epsilon f(\lambda)\xi &= U_\epsilon Ly(\lambda), \\ \epsilon(f, x) + \lambda_0\xi &= (g_\epsilon , Ly). \end{aligned}$$

The spectral transformations (7.02), (7.03) can now be given explicitly. The function $g_\epsilon(\lambda)$ is given[2] by

(7.06) $$g_\epsilon(\lambda) = \frac{\epsilon f(\lambda)}{\lambda - \lambda_0 + \epsilon^2[f; \lambda]}$$

with

(7.07) $$[f; \lambda] = \int_{\mathfrak{s}} \frac{| f(\mu) |^2}{\mu - \lambda} d\mu + i\pi \mid f(\lambda) \mid^2.$$

The integral here is defined as Cauchy's main value; since $f(\mu)$ satisfies a Hoelder condition and vanishes at the end points of $\mathfrak{s}$, the function $[f; \lambda]$ also satisfies a Hoelder condition (but it does not vanish at the end points of $\mathfrak{s}$). Since we have assumed that $f(\lambda)$ does not vanish at λ_0, the modulus $| \lambda - \lambda_0 + \epsilon[f; \lambda] |$ for λ in $\mathfrak{s}$ has a positive lower bound; hence $g_\epsilon(\lambda)$ does also satisfy a Hoelder condition and vanishes at the end points of $\mathfrak{s}$.

[2]Formula (7.06) corresponds to a formula given by Dirac [10, §69, eq. (52), p. 206].

The integral operators R_ϵ and R_ϵ^* are given through the kernels

$$r_\epsilon(\lambda, \mu) = f(\lambda)\overline{g_\epsilon(\mu)}, \qquad r_\epsilon^*(\lambda, \mu) = g_\epsilon(\lambda)\overline{f(\mu)}; \tag{7.08}$$

hence they are of the class (R) described in Section 3.

We state that the transformations (7.02), (7.03) with g_ϵ and R_ϵ given by (7.06), (7.08) are the spectral transformations for the operator $L = \epsilon F$, satisfying (7.05) and (7.04).

It is clear that the transformations so described do not depend analytically on the parameter ϵ. Evidently, the formal perturbation procedure must fail. Nevertheless, a certain significance can be given to the well-determined shift (in second order)

$$\epsilon^2\lambda_2 = -\epsilon^2 \int_g \frac{|f(\mu)|^2}{\mu - \lambda_0}\, d\mu \tag{7.09}$$

of the eigen-value found by the formal perturbation procedure. We maintain that the function $g_\epsilon(\lambda)$ has approximately its maximum at $\lambda_0 + \epsilon^2\lambda_2$. The reason is that the absolute value of the denominator $\lambda - \lambda_0 + \epsilon^2[f; \lambda]$ has its minimum approximately where its real part vanishes; for at any other place the real part dominates if ϵ is sufficiently small. The real part vanishes when

$$\lambda - \lambda_0 + \epsilon^2 \int_g \frac{|f(\mu)|^2}{\mu - \lambda}\, d\mu = 0$$

and this is approximately the case for

$$\lambda = \lambda_0 - \epsilon^2 \int_g \frac{|f(\mu)|^2}{\mu - \lambda_0}\, d\mu = \lambda_0 + \epsilon^2\lambda_2 .$$

The eigen-element of the point eigen-value λ_0 of the undisturbed operator L, represented by $x(\lambda) = 0$, $\xi = 1$, is represented by the function $y(\lambda) = g_\epsilon(\lambda)$, which vanishes approximately except near $\lambda = \lambda_0 + \epsilon^2\lambda_2$ if ϵ is sufficiently small.

On the other hand, an element for which $y(\lambda)$ vanishes exactly outside of a neighborhood of $\lambda = \lambda_0 + \epsilon^2\lambda_2$ with $\|y\| = 1$ is represented by a pair $\{x(\lambda), \xi\}$ for which nearly $|\xi| = 1$ and hence nearly $\|x\| = 0$.

Suppose the operator $L + \epsilon F$ corresponds to the energy of a system in the sense of quantum physics, and suppose the system is further subjected to an additional perturbation of its energy represented by an operator which interacts only with the eigen-elements $\{x(\lambda), 0\}$ of the continuous spectrum of L and not with the eigen-elements $\{0, \xi\}$ of the point eigen-value of L. We then ask for approximately stationary states of the system which are not affected by the additional perturbation. Such states are represented by functions $y(\lambda)$ which vanish outside of a small neighborhood of a value $\lambda = \lambda'$, the approximate value of the energy of the system in that state. Now, only if approximately $\lambda' = \lambda_0 + \epsilon^2\lambda_2$, is the system approximately stable because only then is the state represented by a pair $\{x(\lambda), \xi\}$ with $x(\lambda) = 0$, $|\xi| = 1$ approximately.

Since, however, such a function $x(\lambda)$ does not vanish exactly there is a certain probability of interaction with the physical entity, for example a measuring device, producing the additional perturbation.

Such is essentially the situation in the Auger effect and other occurrences of "spontaneous emission" due to "weak quantization."

Other problems of considerable quantum-physical significance will be treated in Sections 9 and 10.

8. *Justification of the Spectral Transformation Formulas Given for the Operator* $L + \epsilon F$

We first show that relation (7.04) follows from (7.02) and (7.03). Multiplying (7.03) by $\overline{y(\lambda)}$ and integrating over $\mathcal{g}$ we find

$$\begin{aligned}(y, y) &= (y, U_\epsilon^* x) + (y, g_\epsilon)\xi \\ &= (U_\epsilon y, x) + \overline{(g_\epsilon\,, y)}\xi \\ &= (x, x) + \bar{\xi}\xi.\end{aligned}$$

Next we show that the representation of $\{x(\lambda),\ \xi\}$ by $y(\lambda)$ through (7.02) entails relations (7.05), which express that the operator $L + \epsilon F$ is represented by L.

To prove (7.05.1) we derive from (7.02) and (7.08)

$$\begin{aligned}Lx(\lambda) + \epsilon f(\lambda)\xi - U_\epsilon Ly &= (LU_\epsilon - U_\epsilon L + \epsilon R_\epsilon)y \\ &= \epsilon(L(\Gamma R_\epsilon) - (\Gamma R_\epsilon)L + R_\epsilon)y \\ &= 0,\end{aligned}$$

by (7.01) and (3.12). Further we have from (7.02) and (7.06)

$$\begin{aligned}\epsilon(f, x) + \lambda_0\xi - (g_\epsilon\,, Ly) &= \epsilon(f, U_\epsilon y) + (g_\epsilon\,, (\lambda_0 - L)y) \\ &= \epsilon(U_\epsilon^* f, y) - ((L - \lambda_0)g_\epsilon\,, y) \\ &= 0\end{aligned}$$

since by (7.08)

$$\Gamma R_\epsilon^* f(\lambda) = g_\epsilon(\lambda)[f; \lambda]$$

and hence by (7.06)

$$U_\epsilon^* \epsilon f(\lambda) = (\lambda - \lambda_0) g_\epsilon(\lambda). \tag{8.01}$$

Thus (7.05) is derived from (7.02).

Next we show that the transformation (7.03) follows from the transformation (7.02). Introducing the symmetrical operator G_ϵ with the kernel

$$g_\epsilon(\lambda)\overline{g_\epsilon(\mu)}, \tag{8.02}$$

we have

$$y(\lambda) - U_\epsilon^* x(\lambda) - g_\epsilon(\lambda)\xi = (1 - U_\epsilon^* U_\epsilon - G_\epsilon)y(\lambda). \tag{8.03}$$

By (7.01) and (4.13) we have

$$1 - U_\epsilon^* U_\epsilon = \epsilon\Gamma\{R_\epsilon^* U_\epsilon - U_\epsilon^* R_\epsilon\}. \tag{8.04}$$

The operator in the braces can be evaluated explicitly; we have by (7.08)

$$\epsilon R_\epsilon y(\lambda) = \epsilon f(\lambda)(g_\epsilon, y),$$

hence by (8.01)

$$U_\epsilon^* R_\epsilon y(\lambda) = (\lambda - \lambda_0)g_\epsilon(\lambda)(g_\epsilon, y).$$

Using the operator G_ϵ defined by (8.02) we may write the last equation in the form

$$U_\epsilon^* R_\epsilon = (L - \lambda_0)G_\epsilon . \tag{8.05}$$

The adjoint of this equation gives

$$R_\epsilon^* U_\epsilon = G_\epsilon(L - \lambda_0). \tag{8.05*}$$

The difference of these last equations gives

$$\begin{aligned} R_\epsilon^* U_\epsilon - U_\epsilon^* R_\epsilon &= G_\epsilon(L - \lambda_0) - (L - \lambda_0)G_\epsilon \\ &= G_\epsilon L - LG_\epsilon ; \end{aligned} \tag{8.06}$$

hence we have from (8.04)

$$1 - U_\epsilon^* U_\epsilon = \epsilon\Gamma\{G_\epsilon(L - \lambda_0) - (L - \lambda_0)G_\epsilon\}$$

or

$$1 - U_\epsilon^* U_\epsilon = \epsilon G_\epsilon \tag{8.07}$$

by (3.12). Thus (8.03) gives

$$y(\lambda) - U_\epsilon^* x(\lambda) - g_\epsilon(\lambda)\xi = 0,$$

i.e. transformation (7.03).

As a by-product of relation (8.06) we have relation

$$(1 + \overline{2i\pi\epsilon r_\epsilon(\lambda, \lambda)})(1 + 2i\pi\epsilon r_\epsilon(\lambda, \lambda)) = 1, \tag{8.08}$$

which follows in the same way from (8.06) as (4.16) followed from (4.12) but can of course be directly verified from (7.08), (7.06), and (7.07).

Transformation (7.02) does not follow in a purely formal manner from transformation (7.03). Explicit use will be made of the fact that the operator $L + \epsilon F$ does not have a point eigen-value. From this fact we shall derive the relations

$$(8.09) \qquad (g_\epsilon , g_\epsilon) = 1,$$

$$(8.10) \qquad [g_\epsilon ; \lambda] = -\frac{1}{\lambda - \lambda_0 + \epsilon^2[f; \lambda]}.$$

Before doing so we shall show that (7.03) entails (7.02) if (8.09) and (8.10) hold. We have

$$x(\lambda) - U_\epsilon y(\lambda) = (1 - U_\epsilon U_\epsilon^*)x(\lambda) - U_\epsilon g_\epsilon(\lambda)\xi,$$

$$\xi - (g_\epsilon , y) = -(g_\epsilon , U_\epsilon^* x) + (1 - (g_\epsilon , g_\epsilon))\xi.$$

We are supposed to derive the relations

$$(8.11) \qquad 1 - U_\epsilon U_\epsilon^* = 0,$$

$$(8.12) \qquad U_\epsilon g_\epsilon(\lambda) = 0,$$

which then imply

$$(g_\epsilon , U_\epsilon^* x) = (U_\epsilon g_\epsilon , x) = 0,$$

and

$$1 - (g_\epsilon , g_\epsilon) = 0.$$

The latter relation is (8.09). To derive (8.12) we calculate by (7.01), (7.08)

$$U_\epsilon g_\epsilon(\lambda) = g_\epsilon(\lambda) + \epsilon f(\lambda)[g_\epsilon ; \lambda].$$

By (8.10) and (7.06) relation (8.12) then follows.

To derive (8.11) from (8.10) we start off with the relation

$$(8.13) \qquad 1 - U_\epsilon U_\epsilon^* = \Gamma\{U_\epsilon R_\epsilon^* - R_\epsilon U_\epsilon^*\},$$

see (4.13). Now, by (7.08), (8.10), and (7.06)

$$\epsilon(\Gamma R_\epsilon)R_\epsilon^* y(\lambda) = \epsilon f(\lambda)[g_\epsilon ; \lambda](f, y) = -g_\epsilon(\lambda)(f, y)$$

$$= -R_\epsilon^* y(\lambda),$$

or

$$(8.14) \qquad U_\epsilon R_\epsilon^* = R_\epsilon^* + (\Gamma R_\epsilon)R_\epsilon^* = 0.$$

The adjoint of this relation is

$$(8.14^*) \qquad R_\epsilon^* U_\epsilon = R_\epsilon - R_\epsilon(\Gamma R_\epsilon^*) = 0.$$

Subtraction of the last two relations and insertion into (8.13) yields (8.11).

$$U_\epsilon R_\epsilon^* - R_\epsilon U_\epsilon^* = 0 \tag{8.15}$$

and hence (8.11) by (8.13).

From (8.15) the relation

$$(1 + 2i\pi\epsilon r_\epsilon(\lambda, \lambda))(1 + \overline{2i\pi\epsilon r_\epsilon(\lambda, \lambda)}) = 1 \tag{8.16}$$

follows, which together with (8.12) expresses the fact that the operator $(1 + 2i\pi\epsilon r_\epsilon(\lambda, \lambda))$ is unitary.

In order to establish relations (8.09) and (8.10) we introduce the function

$$[f\,; z] = \int_{\mathfrak{s}} \frac{|f(\mu)|^2}{\mu - z}\, d\mu \tag{8.17}$$

of the complex variable $z = \lambda + i\lambda'$, and note that this function approaches $[f; \lambda]$ when λ' approaches zero through positive values, and approaches $\overline{[f; \lambda]}$ when λ' approaches zero through negative values. For a derivation of this well known fact see, e.g., [2, page 257].

Further we note

$$2i\pi\, |g_\epsilon(\mu)|^2 = \frac{1}{\mu - \lambda_0 + \epsilon\,\overline{[f\,;\mu]}} - \frac{1}{\mu - \lambda_0 + \epsilon^2[f\,;\mu]} \tag{8.18}$$

as follows from (7.06) and (7.07). Consequently we have

$$\begin{aligned}(g_\epsilon, g_\epsilon) &= \int_{\mathfrak{s}} |g_\epsilon(\mu)|^2\, d\mu \\ &= \frac{1}{2\pi i}\int_{\mathfrak{s}} \left\{\frac{1}{\mu - \lambda_0 + \epsilon\,\overline{[f\,;\mu]}} - \frac{1}{\mu - \lambda_0 + \epsilon^2[f\,;\mu]}\right\} d\mu \\ &= \frac{1}{2\pi i}\oint \frac{d\zeta}{\zeta - \lambda_0 + \epsilon^2[f\,;\zeta]};\end{aligned} \tag{8.19}$$

defining $[g_\epsilon\,; z]$ similarly as $[f; z]$ through (8.17) we further have

$$\begin{aligned}[g_\epsilon\,; z] &= \int_{\mathfrak{s}} \frac{|g_\epsilon(\mu)|^2}{\mu - z}\, d\mu \\ &= \frac{1}{2\pi i}\oint \frac{1}{\zeta - \lambda_0 + \epsilon^2[f\,;\zeta]}\,\frac{d\zeta}{\zeta - z}\end{aligned} \tag{8.20}$$

The complex integrals (8.19) and (8.20) are easily evaluated. In the discussion at the end of Section 6 it was shown that $\zeta - \lambda_0 + \epsilon^2[f; \zeta]$ does not vanish for any complex value of ζ outside of $\mathfrak{s}$ provided that $|\epsilon|$ is so small that condition (6.16) is satisfied. Since $[f; \zeta] \to 0$ as $\zeta \to \infty$ we find by shifting the path of integration to infinity

(8.21) $$(g_\epsilon, g_\epsilon) = \frac{1}{2\pi i} \oint \frac{d\zeta}{\zeta} = 1,$$

(8.22) $$[g_\epsilon ; z] = -\frac{1}{z - \lambda_0 + \epsilon^2[f ; z]}$$

Relation (8.21) is the same as (8.09). In relation (8.22) we let $z = \lambda + i\lambda'$ approach a value λ in $\mathfrak{s}$ by letting λ' approach zero through positive values. Then

(8.23) $$[g_\epsilon ; z] \to [g_\epsilon ; \lambda].$$

Consequently,

(8.24) $$[g_\epsilon ; \lambda] = -\frac{1}{\lambda - \lambda_0 + \epsilon^2[f ; \lambda]};$$

this is relation (8.10). Thus the formulas given in Section 7 for the spectral representation of the operator $L + \epsilon F$ are established.

9. *On the Solution of the Schroedinger Equation for Regularly Disturbed Operators*

The Schroedinger equation for "states" $\psi(t)$ depending on the time t is

(9.01) $$\partial\psi/\partial t = -iH\psi$$

when an appropriate unit of the time is chosen; H is the "energy operator." The solution $\psi(t)$ is simply given by

(9.02) $$\psi(t) = e^{-itH}\psi(0)$$

in terms of the initial state $\psi(0)$; the unitary operator $\exp\{-iH\}$ is defined whenever H is a self-adjoint operator in the Hilbert space $\mathfrak{H}$.

If a representation $\psi = y(\lambda)$ is chosen in which the energy operator is spectrally represented, $H\psi = \lambda\psi$, the operator $\exp\{-i\lambda\}$ is nothing but multiplication by the number $\exp\{-iH\}$, and accordingly

(9.03) $$y(\lambda, t) = e^{-it\lambda}y(\lambda, 0).$$

It is, however, of considerable interest to discuss different representations of the states $\psi(t)$, and in particular their asymptotic behavior as t increases indefinitely. It is also important to investigate how such a representation varies if the energy operator is disturbed.

In this section we shall identify the energy operator with the operator $L + \epsilon K$ studied in Section 4; in Section 10 we shall identify it with the operator $L + \epsilon F$ studied in Sections 6 to 8.

The Schroedinger equation for the operator $L + \epsilon K$ is

(9.04) $$\frac{\partial}{\partial t} x(\lambda, t) = -i(L + \epsilon K)x(\lambda, t)$$

with the initial conditions

(9.05) $$x(\lambda, 0) = x^0(\lambda),$$

$x^0(\lambda)$ being a given function in $\mathfrak{H}$. The transform

(9.06) $$y(\lambda, t) = U_\epsilon^* x(\lambda, t),$$

see (2.10), (4.01), satisfies by (4.02*) the equation

(9.07) $$\frac{\partial}{\partial t} y(\lambda, t) = -iLy(\lambda, t)$$

with the initial condition

(9.08) $$y(\lambda, 0) = y^0(\lambda)$$

when we set

(9.09) $$y^0(\lambda) = U_\epsilon^* x^0(\lambda).$$

The solution of (9.07), (9.08) is evidently

(9.10) $$y(\lambda, t) = e^{-itL} y^0(\lambda) = e^{-it\lambda} y^0(\lambda)$$

and hence the solution of (9.04), (9.05) is

(9.11) $$x(\lambda, t) = U_\epsilon e^{-itL} y^0(\lambda),$$

see (4.14) and (4.23).

The most interesting property of this solution is its asymptotic behavior as $t \to \infty$ or $\to -\infty$. We maintain that the function $e^{it\lambda} x(\lambda, t)$ approaches definite limits as $t \to \pm\infty$.

(9.12⁺) $$e^{it\lambda} x(\lambda, t) \to x^{\infty}(\lambda) \quad \text{as} \quad t \to \infty,$$

(9.12⁻) $$e^{it\lambda} x(\lambda, t) \to x^{-\infty}(\lambda) \quad \text{as} \quad t \to -\infty.$$

Specifically

(9.13⁺) $$x^{\infty}(\lambda) = y^0(\lambda)$$

as $t \to -\infty$,

(9.13⁻) $$x^{-\infty}(\lambda) = S_\epsilon(\lambda) y^0(\lambda)$$

with $S_\epsilon(\lambda) = 1 + 2i\pi\epsilon r_\epsilon(\lambda, \lambda)$—see (4.20).

If we assume here that $x^0(\lambda)$ and hence $y^0(\lambda)$ satisfy a Hoelder condition the convergence is uniform. Otherwise the statement $e^{it\lambda} x(\lambda, t) \to x^{\pm\infty}(\lambda)$ means

$$\| e^{itL} x(t) - x^{\pm\infty} \| \to 0.$$

The proof of these statements is immediate. By (9.11), (4.01), (4.11), and (3.10) we have

$$e^{it\lambda}x(\lambda, t) = y^0(\lambda) + i\pi\epsilon r_\epsilon(\lambda, \lambda)y^0(\lambda)$$

(9.13)

$$+ \epsilon \int_{\mathfrak{s}} \frac{r_\epsilon(\lambda, \mu)}{\mu - \lambda} \exp \{-it(\mu - \lambda)\} y^0(\mu)\, d\mu.$$

Clearly, as $| t | \to \infty$,

$$\int_{\mathfrak{s}} \frac{r_\epsilon(\lambda, \mu)y^0(\mu) - r_\epsilon(\lambda, \lambda)y^0(\lambda)}{\mu - \lambda} \exp \{-it(\mu - \lambda)\}\, d\mu \to 0$$

since the function $y^0(\lambda)$ is assumed to satisfy a Hoelder condition; for then the fraction in the integral is absolutely integrable and the Fourier transform of such a function is known to approach zero. Further we have

$$\int_{\mathfrak{s}} \exp \{-it(\mu - \lambda)\} \frac{d\mu}{\mu - \lambda} \to \int_{-\infty}^{\infty} \exp \{\mp i\sigma\} \frac{d\sigma}{\sigma} = \mp i\pi \qquad \text{as} \qquad t \to \pm \infty.$$

Insertion in (9.13) then yields (9.12) and ($9.13^{\pm}$).

Formula (9.13^+) shows that *the effect of the spectral transformation* U_ϵ *on the solution of the Schroedinger equation is asymptotically, as* $t \to \infty$, *equivalent to the identity*; in other words, the effect of the disturbance ϵK is wiped out asymptotically, while as $t \to -\infty$ the effect is expressed through the operator

$$1 + 2\pi i\epsilon r_\epsilon(\lambda, \lambda),$$

which by (4.17) and (4.23) is unitary provided that the transformations U_ϵ and U_ϵ^* form an intact pair.

The results expressed in (9.13) depend, however, on the fact that we had chosen $U_\epsilon = 1 + \epsilon\Gamma R_\epsilon$ as spectral transformation. Had we chosen a transformation of the form $\tilde{U}_\epsilon = 1 + \epsilon\Gamma^*\tilde{R}_\epsilon$—see (3.08*) and Section 4 (page 375)—we would have found

$$x^{-\infty}(\lambda) = \tilde{y}^0(\lambda) \text{ and } x^{\infty}(\lambda) = \tilde{S}_\epsilon^*(\lambda)\tilde{y}^0(\lambda) = (1 - 2i\pi\tilde{r}_\epsilon(\lambda, \lambda))\tilde{y}^0(\lambda).$$

The latter choice is therefore more suitable for representations in connection with scattering problems; the transformation $\tilde{U}_\epsilon$ corresponds to the one used by Dirac [10, §50, eq. (35)].

Independently of the choice of the form of the operator U_ϵ we can state: There exists, to every value of λ in $\mathfrak{s}$, a unitary operator $S_\epsilon^*(\lambda)$, operating in the Hilbert space $\mathfrak{P}$ such that

$$x^{\infty}(\lambda) = S_\epsilon^*(\lambda)x^{-\infty}(\lambda). \tag{9.14}$$

If the spectral transformation $U_\epsilon = 1 + \epsilon\Gamma R_\epsilon$ is chosen, the operator $S_\epsilon^*(\lambda)$ is the inverse of $1 + 2\pi i r_\epsilon(\lambda, \lambda)$, i.e.

$$S_\epsilon^*(\lambda) = 1 - 2i\pi\epsilon r_\epsilon^*(\lambda, \lambda) = 1 + \overline{2i\pi\epsilon r_\epsilon(\lambda, \lambda)}. \tag{9.15}$$

That the elements $x^{\infty}(\lambda)$ and $x^{-\infty}(\lambda)$ of the Hilbert space $\mathfrak{H}$ are connected through a unitary operator in $\mathfrak{H}$ is clear since as well known

$$\| x(\lambda, t) \|^2 = \int_s | x(\lambda, t) |^2 \, d\lambda$$

is constant for a solution of the Schroedinger equation. The remarkable fact here revealed is that this unitary operator is a function, $S_\epsilon^*(L)$, of the undisturbed energy operator.

This operator $S_\epsilon^*(\lambda)$ corresponds to the "scattering operator" of Heisenberg [7]—see also Wheeler [6]. It embodies the over-all effect of the disturbance expressed through ϵK. As a matter of fact, it is this operator $S_\epsilon^*(\lambda)$ which is amenable to measurement in a scattering experiment.

To illustrate these remarks we may first consider the case that the space $\mathfrak{P}$ is two-dimensional so that every element x in $\mathfrak{P}$ is represented by two numbers x_1, x_2 with $| x |^2 = | x_1 |^2 + | x_2 |^2$. Accordingly, every "function" $x(\lambda)$ in $\mathfrak{H}$ is represented by two functions $x_1(\lambda)$, $x_2(\lambda)$ whose values are numbers. This case arises in one-dimensional motion of a particle when the functions $x_1(\lambda)$, $x_2(\lambda)$ represent states with positive or negative momentum respectively, λ being the kinetic energy.

The operator $r_\epsilon(\lambda, \lambda)$ is a matrix in this case. We set

$$S_\epsilon^*(\lambda) = 1 - 2i\pi\epsilon r_\epsilon^*(\lambda, \lambda) = \begin{pmatrix} \sigma_{11}(\lambda) & \sigma_{12}(\lambda) \\ \sigma_{21}(\lambda) & \sigma_{22}(\lambda) \end{pmatrix}. \tag{9.16}$$

The pair of functions $\{x_1^{-\infty}(\lambda), x_2^{-\infty}(\lambda)\}$ representing the "state" at the time $t = -\infty$ is transformed into the pair of functions

$$\begin{aligned} x_1^{\infty}(\lambda) &= \sigma_{11}(\lambda)x_1^{-\infty}(\lambda) + \sigma_{12}(\lambda)x_2^{-\infty}(\lambda) \\ x_2^{\infty}(\lambda) &= \sigma_{21}(\lambda)x_1^{-\infty}(\lambda) + \sigma_{22}(\lambda)x_2^{-\infty}(\lambda) \end{aligned} \tag{9.17}$$

representing the "state" at the time $t = \infty$.

If, for example, the momentum of the particle had definitely a forward direction in the original state, $x_2^{-\infty}(\lambda) = 0$, it has in the final state a forward direction with the probability $| \sigma_{11}(\lambda)x_1^{-\infty}(\lambda) |^2$ and a backward direction with the probability $| \sigma_{21}(\lambda)x_1^{-\infty}(\lambda) |^2$. Its energy λ is however not changed.

The unitary character of the operator $S_\epsilon^*(\lambda)$ is expressed by relations

$$\begin{aligned} &| \sigma_{11}(\lambda) |^2 + | \sigma_{21} |^2 = 1, \quad | \sigma_{12}(\lambda) |^2 + | \sigma_{22} |^2 = 1, \\ &\sigma_{11}(\lambda)\sigma_{12}(\lambda) + \sigma_{21}(\lambda)\sigma_{22}(\lambda) = 0, \end{aligned} \tag{9.18}$$

the well-known relations for transmission and reflection coefficients.

To represent a three-dimensional scattering process one should take as space $\mathfrak{P}$ the space of all quadratically integrable functions $x(\alpha)$ of the point

$\alpha = \{\alpha_1, \alpha_2, \alpha_3\}$ on the unit sphere $\mathcal{S}$. Denoting the differential of the area on the unit sphere by $d\alpha$ we set

$$| x |^2 = \int_{\mathcal{S}} | x(\alpha) |^2 \, d\alpha.$$

The functions $x(\lambda)$ in $\mathfrak{H}$ are then actually functions $x(\lambda, \alpha)$ of the energy λ and the direction α. Assuming the disturbing operator K to be an integral operator transforming $x(\lambda, \alpha)$ into

$$\int_{\mathcal{J}} \int_{\mathcal{S}} k(\lambda, \mu; \alpha, \beta) x(\mu, \beta) \, d\mu \, d\beta$$

with a sufficiently smooth kernel, the operators $r_\epsilon(\lambda, \mu)$ are also integral operators, transforming $x(\alpha)$ into

$$\int_{\mathcal{S}} r_\epsilon(\lambda, \mu; \alpha, \beta) x(\beta) \, d\beta.$$

The unitary operator $S_\epsilon^*(\lambda)$ then differs from the identity by an integral operator

$$S_\epsilon^*(\lambda) x(\lambda, \alpha) = x(\lambda, \alpha) - 2i\pi\epsilon \int_{\mathcal{S}} \overline{r_\epsilon(\lambda, \lambda; \alpha, \beta)} x(\lambda, \beta) \, d\beta.$$

This form of the scattering operator expresses the fact that the over-all effect of scattering consists in a re-arranging of directions without any change in energy. The quantity $| \epsilon r_\epsilon(\lambda, \lambda; \alpha, \beta) |^2$ is proportional to the "effective cross section" per unit angle of direction for a wave arriving in the direction β and scattered in the direction α.

10. *On the Solution of the Schroedinger Equation for the Operator* $L + \epsilon F$

We proceed to study the solution of the Schroedinger equation for the operator $L + \epsilon F$ studied in Sections 6 to 8. We remember that the undisturbed operator L here had a continuous spectrum covering an interval $\mathcal{J}$ and also, a point eigen-value λ_0 within $\mathcal{J}$, while the disturbed operator $L + \epsilon F$ had no such point eigen-value. In this case we are not only interested in the asymptotic behavior of the solutions as $| t | \to \infty$, but also in the dependence on the parameter ϵ. We shall show that *the solution of the Schroedinger equation depends analytically on the parameter* ϵ, *although the spectral transformation does not*, as we have seen in Section 8. As to the asymptotic behavior we shall show that the point eigen-value component of the solution, i.e. its projection into the eigen-element of the point eigen-value, vanishes asymptotically, as $t \to \infty$ and $t \to -\infty$. In addition we shall show that this vanishing takes place exponentially in a sense to be specified.

An operator of the type $L + \epsilon F$ occurs as the energy operator in a simplified model for a system consisting of an atom and an electromagnetic field. An approximate solution of the Schroedinger equation for such an energy operator was given by Weisskopf and Wigner in their theory of the natural width of spectral lines. From our explicit exact solution we shall derive the correct asymptotic behavior and shall show in which sense Weisskopf and Wigner's solution is an approximation.

The operator $L + \epsilon F$ applies on "states" which are represented by a pair $\{x(\lambda, t), \xi(t)\}$ satisfying the equation

$$\frac{\partial}{\partial t}\{x(\lambda, t), \xi(t)\} = -i(L + \epsilon F)\{x(\lambda, t), \xi(t)\} \tag{10.01}$$

with given initial values

$$\{x(\lambda, 0), \xi(0)\} = \{x^0(\lambda), \xi^0\}. \tag{10.02}$$

The transform

$$y(\lambda, t) = U_\epsilon^* x(\lambda, t) + g_\epsilon(\lambda)\xi(t) \tag{10.03}$$

of such a solution, see (7.03), then satisfies the equation

$$\frac{\partial}{\partial t} y(\lambda, t) = -iLy(\lambda, t) \tag{10.04}$$

with the solution

$$y(\lambda, t) = e^{-it\lambda}y^0(\lambda) \tag{10.05}$$

when we set

$$y^0(\lambda) = U_\epsilon^* x^0(\lambda) + g_\epsilon(\lambda)\xi^0. \tag{10.06}$$

The solution of (10.01), (10.02) is then by (7.02) given as

$$\begin{aligned} x(\lambda, t) &= U_\epsilon e^{-itL}y^0(\lambda), \\ \xi(t) &= (g_\epsilon, e^{-itL}y^0). \end{aligned} \tag{10.07}$$

The asymptotic behavior for $t \to \infty$ of $x(\lambda, t)$ is the same as that of the function $x(\lambda, t)$ given by (9.11), hence the functions $e^{it\lambda}x(\lambda, t)$ possess limits $x^{\infty}(\lambda)$, $x^{-\infty}(\lambda)$, as $t \to \pm\infty$, and these limits are given by (9.13).

Since $\xi(t)$ is simply the Fourier transform of the function $g_\epsilon(\lambda)y^0(\lambda)$, it approaches zero as $t \to \infty$. Thus we have

$$e^{it\lambda}x(\lambda, t) \to x^{\infty}(\lambda) = y^0(\lambda) \quad \text{as} \quad t \to \infty, \tag{10.08$^+$}$$

$$e^{it\lambda}x(\lambda, t) \to x^{-\infty}(\lambda) = S_\epsilon(\lambda)y^0(\lambda) \quad \text{as} \quad t \to -\infty, \tag{10.08$^-$}$$

$$\xi(t) \to 0 \quad \text{as} \quad |t| \to \infty. \tag{10.09}$$

Thus *the disturbance ϵF has asymptotically the effect of wiping out the component* *of the point eigen-state of the undisturbed operator; otherwise the spectral transformation is asymptotically the identity for $t = \infty$ and equal to multiplying by a number of absolute value one for $t = -\infty$.*

The refined asymptotic representation which we shall derive is valid only for small values of the parameter ϵ; it agrees with the approximate solution, given by Weisskopf and Wigner—(see [4])—and takes in our notation the form

$$x(\lambda, t) \sim \epsilon f(\lambda_0) \frac{\exp\{-it\lambda\} - \exp\{-it\lambda_0 + it\epsilon^2[f; \lambda_0]\}}{\lambda - \lambda_0 + \epsilon^2[f; \lambda_0]} \xi^0, \tag{10.10}$$

$$\xi(t) \sim \exp\{-it\lambda_0 + it\epsilon^2[f; \lambda_0]\}\xi^0, \qquad \text{for} \quad t \sim \infty. \tag{10.11}$$

Note that the imaginary part of $[f; \lambda_0]$ is $\pi| f(\lambda_0) |^2 > 0$. Therefore the real part in the exponent in (10.11) is negative. Consequently $\xi(t) \to 0$ as $| t | \to \infty$, in agreement with (10.09). Similarly one obtains (10.08^+) as $t \to \infty$ from (10.10).

The precise sense in which (10.10) and (10.11) are asymptotic representations is the following. One must let ϵ approach zero as $t \to \infty$ in such a way that $t\epsilon^2$ remains fixed. Also one must let λ approach λ_0 in such a way that $t(\lambda - \lambda_0)$ remains fixed. Then the precise form of (10.10) and (10.11) is

$$\begin{aligned} &t^{-1/2}x(\lambda,t) \exp\{it\lambda\} \\ &\qquad \to t^{1/2}\epsilon f(\lambda_0) \frac{1 - \exp\{it(\lambda - \lambda_0) + it\epsilon^2[f; \lambda_0]\}}{t(\lambda - \lambda_0) + t\epsilon^2[f; \lambda_0]} \xi^0, \end{aligned} \tag{10.12}$$

$$\xi(t) \exp\{it\lambda_0\} \to \exp\{it\epsilon^2[f; \lambda_0]\}\xi^0, \tag{10.13}$$

as $t \to \infty$. The contributions to $t^{-1/2}x(\lambda, t)$ and $\xi(t)$ resulting from the initial function $x^0(\lambda)$ approach zero of the order $t^{-3/2}$.

To derive all these statements the representation (10.07) of the solution of Schroedinger's equation is not suitable. It is necessary to express the function $y^0(\lambda)$ in (10.07) in terms of $x^0(\lambda)$, ξ^0 through the inverse transformation (7.03). Thus we obtain

$$x(\lambda, t) = W_\epsilon(t)x^0(\lambda) + w_\epsilon(\lambda, t)\xi^0, \tag{10.14}$$

$$\xi(t) = (w_\epsilon(-t), x^0) + q_\epsilon(t)\xi^0, \tag{10.15}$$

with

$$q_\epsilon(t) = (g_\epsilon, e^{-it\lambda}g_\epsilon), \tag{10.16}$$

$$w_\epsilon(t) = w_\epsilon(\lambda, t) = U_\epsilon e^{-itL}g_\epsilon(\lambda), \tag{10.17}$$

$$W_\epsilon(t) = U_\epsilon e^{-itL}U_\epsilon^* . \tag{10.18}$$

The function w_ϵ can by (7.01), (7.08), and (7.06) be written in the form

$$w_\epsilon(\lambda, t) = \epsilon f(\lambda) v_\epsilon(\lambda, t) e^{-it\lambda} \tag{10.19}$$

with

$$\begin{aligned} v_\epsilon(\lambda, t) &= \frac{1}{\lambda - \lambda_0 + \epsilon^2[f; \lambda]} + i\pi \mid g_\epsilon(\lambda) \mid^2 \\ &\quad + \int_s \frac{\mid g_\epsilon(\mu) \mid^2 \exp\{it(\lambda - \mu)\}}{\mu - \lambda} \, d\mu. \end{aligned} \tag{10.20}$$

Note that the imaginary parts of the first two terms here cancel each other so that also

$$\begin{aligned} v_\epsilon(\lambda, t) &= \frac{1}{\lambda - \lambda_0 + \epsilon^2\overline{[f; \lambda]}} - i\pi \mid g_\epsilon(\lambda) \mid^2 \\ &\quad + \int_s \frac{\mid g_\epsilon(\mu) \mid^2 \exp\{it(\lambda - \mu)\}}{\mu - \lambda} \, d\mu. \end{aligned} \tag{10.21}$$

The operator $W_\epsilon(t)$ given by (10.18) can be written in the form

$$W_\epsilon(t) = e^{-itL}\{1 + \epsilon \Gamma S_\epsilon(t)\} \tag{10.22}$$

with

$$\begin{aligned} S_\epsilon(t) &= e^{itL} R_\epsilon e^{-itL} - R_\epsilon^* - \epsilon e^{itL} R_\epsilon e^{-itL}(\Gamma R_\epsilon^*) - \epsilon e^{itL}(\Gamma R_\epsilon) e^{-itL} R_\epsilon^* \\ &= e^{itL} R_\epsilon e^{-itL} U_\epsilon^* - e^{itL} U_\epsilon e^{-itL} R_\epsilon^* . \end{aligned}$$

Making use of (7.01), (7.08), (7.06) and (10.21), (10.22) we find for the kernel $s_\epsilon(\lambda, \mu, t)$ of $S_\epsilon(t)$,

$$s_\epsilon(\lambda, \mu, t) = \epsilon f(\lambda)(\exp\{it(\lambda - \mu)\} v_\epsilon(\mu, t) - v_\epsilon(\lambda, t)) \epsilon \overline{f(\mu)}. \tag{10.23}$$

It is now easy to prove our statement that *the solution of the Schroedinger equation depends analytically on the parameter* ϵ. It is clear from the representations (10.14) to (10.18) that we need only prove that $q_\epsilon(t)$ and $v_\epsilon(\mu, t)$ depend analytically on ϵ in order to be sure that this is the case of $x(\lambda, t)$ and $\xi(t)$. To this end we represent the quantity $q_\epsilon(t)$ given by (10.16) in the form

$$\begin{aligned} q_\epsilon(t) &= \int_s \mid g_\epsilon(\mu) \mid^2 e^{-it\mu} \, d\mu \\ &= \frac{1}{2\pi i} \int_s \left\{ \frac{1}{\mu - \lambda_0 + \epsilon^2\overline{[f; \mu]}} - \frac{1}{\mu - \lambda_0 + \epsilon^2[f; \mu]} \right\} e^{-it\mu} \, d\mu \end{aligned}$$

using (8.18). In the same way that (8.19) was derived we now obtain the expression

$$(10.24)\qquad q_\epsilon(t) = \frac{1}{2\pi i}\oint \frac{e^{-it\zeta}}{\zeta - \lambda_0 + \epsilon^2[f;\zeta]}\,d\zeta$$

in which the variable ζ should run outside of the interval $\mathcal{J}$ encircling it. The function $[f; z]$ occurring here is defined by (8.17). Since the denominator here does not vanish for $\epsilon = 0$ it is clear that the right member in (10.24) can be expanded in powers of ϵ.

To express the function $v_\epsilon(\lambda, t)$ defined by (10.20) as a complex integral we first note that $v_\epsilon(\lambda, t)$ is the limit of the function

$$(10.25)\qquad v_\epsilon(z, t) = \frac{1}{z - \lambda_0 + \epsilon^2[f; z]} + \int_{\mathcal{J}} \frac{|\, g_\epsilon(\mu)\,|^2 \exp\{-it(\mu - z)\}}{\mu - z}\,d\mu$$

as $z = \lambda + i\lambda'$ approaches the real value λ with $\lambda' > 0$. Using (8.22) and (8.20) we find

$$(10.26)\qquad v_\epsilon(z, t) = \int_{\mathcal{J}} \frac{1}{\zeta - \lambda_0 + \epsilon^2[f;\zeta]}\,\frac{\exp\{-it(\zeta - z)\} - 1}{\zeta - z}\,d\zeta.$$

Since the integrand here is continuous even for $z = \mu$ we may let z approach any value λ on $\mathcal{J}$ and obtain

$$v_\epsilon(\lambda, t) = \int_{\mathcal{J}} \frac{1}{\zeta - \lambda_0 + \epsilon^2[f;\zeta]}\,\frac{\exp\{-it(\zeta - \lambda)\} - 1}{\zeta - \lambda}\,d\zeta$$

or, since the denominator here does not vanish outside of $\mathcal{J}$ and is of the order ζ^2 at infinity,

$$(10.27)\qquad v_\epsilon(\lambda, t) = \int_{\mathcal{J}} \frac{1}{\zeta - \lambda_0 + \epsilon^2[f;\zeta]}\,\frac{\exp\{-it(\zeta - \lambda)\}}{\zeta - \lambda}\,d\zeta.$$

Evidently the right member here admits an expansion with respect to powers of ϵ.

It then follows from the remarks made above that the functions $x(\lambda, t)$ and $\xi(t)$ admit expansions with respect to powers of ϵ.

To derive the refined asymptotic expressions (10.10) and (10.11) we introduce the new variable

$$(10.28)\qquad \beta = t(\zeta - \lambda_0)$$

and write the representation (10.24) of the quantity $q_\epsilon(t)$ in the form

$$(10.29)\qquad \begin{aligned} q_\epsilon(t)e^{it\lambda_0} &= \frac{1}{2\pi i}\int_{-\infty - ic}^{\infty - ic} \frac{e^{-i\beta}\,d\beta}{\beta + t\epsilon^2[f; \lambda_0 + \beta/t]} \\ &\quad - \frac{1}{2\pi i}\int_{-\infty + ic}^{\infty + ic} \frac{e^{-i\beta}\,d\beta}{\beta + t\epsilon^2[f; \lambda_0 + \beta/t]} \end{aligned}$$

by extending the path surrounding the interval $\mathcal{I}$ into two parallel straight lines $\mathcal{I}m\,\beta = c$, $\mathcal{I}m\,\beta = -c$, $c > 0$. The two integrals exist since $\int_{-\infty}^{\infty} e^{-i\beta}\,d\beta/\beta$ does, as is well known. We now let t approach ∞, keeping $t\epsilon^2$ fixed. We note that $[f;\lambda_0 + \beta/t] \to [f;\lambda_0]$ if $\mathcal{I}m\,\beta > 0$ and $\to \overline{[f;\lambda_0]}$ if $\mathcal{I}m\,\beta < 0$. Therefore we obtain

$$q_\epsilon(t)e^{it\lambda_0} \to \frac{1}{2\pi i}\int_{-\infty-ic}^{\infty-ic} \frac{e^{-i\beta}\,d\beta}{\beta + t\epsilon^2\overline{[f;\lambda_0]}} - \frac{1}{2\pi i}\int_{-\infty+c}^{\infty+c} \frac{e^{-i\beta}\,d\beta}{\beta + t\epsilon^2[f;\lambda_0]} \tag{10.30}$$

We evaluate the integrals by pushing the path of integration toward $\beta = -i\infty$. The denominator of the first integral does not vanish in the lower half-plane; therefore the first integral approaches zero and hence is zero. We note that the denominator of the second integral has a zero at $\beta = -t\epsilon^2[f;\lambda_0]$; hence its value is given by the residue at that point. Thus formula (10.30) simplifies itself to

$$q_\epsilon(t)\exp\{it\lambda_0\} \to \exp\{it\epsilon^2[f;\lambda_0]\}. \tag{10.31}$$

To derive an asymptotic expression for the function $v_\epsilon(\lambda, t)$ given by (10.20) we set

$$\alpha = t(\lambda - \lambda_0) \tag{10.32}$$

and write the representation (10.20) in the form

$$t^{-1}v_\epsilon(\lambda, t) = \frac{1}{2\pi i}\int_{-\infty-ic}^{\infty-ic} \frac{e^{-i\beta}}{\beta + \epsilon^2 t\overline{[f;\lambda_0 + \beta/t]}}\frac{d\beta}{\beta - \alpha} - \frac{1}{2\pi i}\int_{-\infty+ic}^{\infty+ic} \frac{e^{-i\beta}}{\beta + \epsilon^2 t[f;\lambda_0 + \beta/t]}\frac{d\beta}{\beta - \alpha}$$

Letting $t \to \infty$, keeping $\epsilon^2 t$ and α fixed, we obtain

$$t^{-1}v_\epsilon(\lambda, t) \to \frac{1}{2\pi i}\int_{-\infty-ic}^{\infty-ic} \frac{e^{-i\beta}}{\beta + \epsilon^2 t\overline{[f;\lambda_0]}}\frac{d\beta}{\beta - \alpha} - \frac{1}{2\pi i}\int_{-\infty+ic}^{\infty+ic} \frac{e^{-i\beta}}{\beta + \epsilon^2 t[f;\lambda_0]}\frac{d\beta}{\beta - \alpha} \tag{10.33}$$

Evaluating the integrals here by pushing down the path of integration to $\beta = -i\infty$, and taking the residues at $\beta = \alpha$ and $\beta = -\epsilon^2 t[f;\lambda_0]$ into account, we can write formula (10.33) in the simple form

$$t^{-1}v_\epsilon(\lambda, t) \to \frac{1 - \exp\{i(\alpha + \epsilon^2 t[f;\lambda_0])\}}{\alpha + \epsilon^2 t[f;\lambda_0]} \tag{10.34}$$

We further have, as $t \to \infty$ and $\epsilon^2 t$ is fixed,

(10.35) $$f(\lambda) \to f(\lambda_0)$$

and hence by (10.34) and (10.19)

(10.36) $$t^{-1/2} e^{it\lambda} w_\epsilon(\lambda, t) \to \epsilon\, t^{1/2} f(\lambda_0) \frac{1 - \exp\{i(\alpha + \epsilon^2 t[f; \lambda_0])\}}{\alpha + \epsilon^2 t[f; \lambda_0]}$$

We now insert (10.36) and (10.31) into (10.14) and (10.15). Making use of (10.22) and (10.23) we see that $t^{-1/2} W_\epsilon(t) x^0(\lambda)$ and $(w_\epsilon(-t), x^0)$ vanish of the order $t^{-3/2}$. Consequently, formulas (10.12) and (10.13) result.

Thus all stated properties of the solution of the Schroedinger equation for the operator $L + \epsilon F$ are proved.

Appendix

11. *Spectral Transformation of the Operator* $L + \epsilon J$

We shall now formulate a spectral transformation of the operator $L + \epsilon J$ considered in Section 5, without so far having established its validity rigorously. The functions $x(\lambda)$ on which the operator $L + \epsilon J$ acts are represented by a pair consisting of a function $y(\lambda)$ defined in $\mathcal{I}$ and a number η through the transformation

(11.01) $$x(\lambda) = U_\epsilon y(\lambda) + g_\epsilon(\lambda)\eta,$$

in which

(11.02) $$U_\epsilon = 1 + \epsilon \Gamma R_\epsilon\,,$$

(11.03) $$r_\epsilon(\lambda, \mu) = j(\lambda)\overline{h_\epsilon(\mu)},$$

(11.04) $$h_\epsilon(\lambda) = \frac{j(\lambda)}{1 + \epsilon[j; \lambda]},$$

(11.05) $$[j] = [j; \lambda] = \int_{\mathcal{I}} \frac{|\, j(\mu)\,|^2}{\mu - \lambda}\, d\mu + i\pi\, |\, j(\lambda)\,|^2 \qquad \text{for } \lambda \text{ in } \mathcal{I}.$$

Further,

(11.06) $$g_\epsilon(\lambda) = \frac{a_\epsilon j(\lambda)}{\lambda_\epsilon - \lambda};$$

here λ_ϵ is defined as root of equation (5.10), and a_ϵ is so chosen that

(11.07) $$(g_\epsilon\,,\, g_\epsilon) = 1.$$

The operator R_ϵ here is not of the class (R) considered in Section 3 since its kernel $r_\epsilon(\lambda, \mu)$ does not vanish if λ is at an end point of $\mathcal{I}$. It is true that $r_\epsilon(\lambda, \mu)$

vanishes if μ is at one end point of $\mathfrak{s}$ since the denominator in the expression (11.04) of h_ϵ becomes infinite there; however, this denominator becomes only logarithmically infinite so that h_ϵ approaches zero at that end point only like the reciprocal logarithm while of the kernels r in Section 3 vanishing like a power was required. Nevertheless it seems that some of the properties of the operators of class (R) remain valid. In particular R and ΓR remain bounded operators and the identity (3.14) remains true for $S = R^*$ and also when R and R^* are interchanged.

The inverse transformation to (11.01) is

$$y(\lambda) = U_\epsilon^* x(\lambda), \qquad \eta = (g_\epsilon\,, x). \tag{11.08}$$

The fact that the representation of $x(\lambda)$ by $\{y(\lambda), \eta\}$ is a spectral representation of the operator $L + \epsilon J$ is expressed through the relation

$$(L + \epsilon J)x(\lambda) = U_\epsilon \lambda y(\lambda) + \lambda_\epsilon g_\epsilon(\lambda)\eta. \tag{11.09}$$

Finally we have for the norms

$$\| x \|^2 = \| y \|^2 + | \eta |^2, \tag{11.10}$$

as follows immediately from (11.01) and (11.08).

Relation (11.08) follows formally from (11.01). Relation (11.01) does not follow formally from (11.08); the derivation must make material use of the fact that the operator $L + \epsilon J$ has only one eigen-value, which is real and lies outside of $\mathfrak{s}$. We shall not present these derivations here; firstly because arguments similar to those needed were given in detail in Section 8 for a different problem; secondly, because these arguments have not yet been established rigorously.

We shall, however, show how relation (11.09) follows from (11.01). To this end we observe

$$LU_\epsilon y(\lambda) = U_\epsilon Ly(\lambda) - \epsilon R_\epsilon y(\lambda)$$

$$= U_\epsilon \lambda y(\lambda) - \epsilon j(\lambda)(h_\epsilon\,, y),$$

$$JU_\epsilon y(\lambda) = j(\lambda)\{(j, y) - \epsilon([j; \lambda]h_\epsilon\,, y)\},$$

as easily verified. Now, by (6.04)

$$j(\lambda) - \epsilon[j; \lambda]h_\epsilon(\lambda) = h_\epsilon(\lambda);$$

hence

$$JU_\epsilon y(\lambda) = j(\lambda)(h_\epsilon\,, y).$$

Consequently,

$$LU_\epsilon y(\lambda) + JU_\epsilon y(\lambda) = U_\epsilon \lambda y(\lambda). \tag{11.11}$$

Further, by (11.06), (5.01), (5.09), and (5.10)

$$(L + \epsilon J)g_\epsilon(\lambda) = a_\epsilon \frac{\lambda j(\lambda)}{\lambda_\epsilon - \lambda} - \epsilon a_\epsilon j(\lambda)[j; \lambda_\epsilon]$$

$$= a_\epsilon j(\lambda)\left\{\frac{\lambda}{\lambda_\epsilon - \lambda} + 1\right\} = a_\epsilon j(\lambda) \frac{\lambda_\epsilon}{\lambda_\epsilon - \lambda}$$

or

(11.12) $$(L + \epsilon J)g_\epsilon(\lambda) = \lambda_\epsilon g_\epsilon(\lambda).$$

Relations (11.11) and (11.12) establish (11.09).

Next we derive the representation (11.08) from (11.01) or, what is equivalent, the identities

(11.13) $$U_\epsilon^* U_\epsilon = 1,$$

(11.14) $$U_\epsilon^* g_\epsilon = 0,$$

(11.15) $$(g_\epsilon\,,\, U_\epsilon y) = 0,$$

(11.16) $$(g_\epsilon\,,\, g_\epsilon) = 1.$$

Using identity (3.14) for R^* and R (instead of R and S), we have from (11.02)

(11.17) $$U_\epsilon^* U_\epsilon - 1 = \epsilon\Gamma\{R_\epsilon - R_\epsilon^* - \epsilon(\Gamma R_\epsilon^*)R_\epsilon - \epsilon R_\epsilon^*(\Gamma R_\epsilon)\}.$$

Introducing the integral operator H_ϵ with the kernel $\overline{h_\epsilon(\lambda)}h_\epsilon(\mu)$ we find from (11.03) and (11.05)

$$\epsilon R_\epsilon^*(\Gamma R_\epsilon)x = \epsilon\overline{h_\epsilon(\lambda)}([j]h_\epsilon\,,\, x)$$

$$= \overline{h_\epsilon(\lambda)}\{(j, x) - (h, x)\}$$

by (11.04). Or,

$$\epsilon R_\epsilon^*(\Gamma R_\epsilon) = R^* - H.$$

The adjoint of this relation is

$$-\epsilon(\Gamma R_\epsilon^*)R_\epsilon = R - H.$$

Subtraction gives

$$\epsilon R_\epsilon^*(\Gamma R_\epsilon) + \epsilon(\Gamma R_\epsilon^*)R_\epsilon = R^* - R.$$

Consequently, the right member in (11.17) vanishes and (11.13) is proved.

To derive relation (11.14) we note that by (11.06)

$$\frac{1}{\mu - \lambda}\overline{j(\mu)}g(\mu) = \frac{a_\epsilon}{\lambda_\epsilon - \lambda}\left\{\frac{\overline{j(\mu)}j(\mu)}{\mu - \lambda} - \frac{\overline{j(\mu)}j(\mu)}{\mu - \lambda_\epsilon}\right\}.$$

Hence

$$\Gamma R_\epsilon^* g_\epsilon(\lambda) = \frac{a_\epsilon h_\epsilon(\lambda)}{\lambda_\epsilon - \lambda} \{[j; \lambda] - [j; \lambda_\epsilon]\};$$

further, by (11.04)

$$g_\epsilon(\lambda) = \frac{a_\epsilon h_\epsilon(\lambda)}{\lambda_\epsilon - \lambda} \{1 + \epsilon[j; \lambda]\}.$$

Thus

$$U^* g_\epsilon = \frac{a_\epsilon h_\epsilon(\lambda)}{\lambda_\epsilon - \lambda} \{1 + [j; \lambda_\epsilon]\} = 0,$$

since λ_ϵ satisfies equation (5.10). Relation (11.15) follows from (11.14),

$$(g_\epsilon , Uy) = (U_\epsilon^* g_\epsilon , y) = 0.$$

Relation (11.16) was already stipulated in (11.07).

Finally we show that, conversely, representation (11.01) follows from (11.08). To this end we need only establish the identity

$$U_\epsilon U_\epsilon^* + G_\epsilon = 1, \tag{11.18}$$

when G_ϵ is the integral operator with the kernel $\overline{g_\epsilon(\lambda) g_\epsilon(\mu)}$. By virtue of the fundamental identity (3.14) used for $S = R^*$ relation (11.18) is equivalent with

$$\epsilon\Gamma\{R_\epsilon^* - R_\epsilon + \epsilon R_\epsilon(\Gamma R_\epsilon^*) + \epsilon(\Gamma R_\epsilon) R_\epsilon^*\} = G_\epsilon . \tag{11.19}$$

We first have from (11.03)

$$R_\epsilon(\Gamma R_\epsilon^*)x = -j([h_\epsilon]j, x) \tag{11.20}$$

with

$$[h_\epsilon] = [h_\epsilon ; \lambda] = \int_s \frac{| h_\epsilon(\mu) |^2}{\mu - \lambda} d\mu + i\pi | h_\epsilon(\lambda) |^2. \tag{11.21}$$

We now introduce the function

$$[h_\epsilon ; z] = \int_s | h_\epsilon(\mu) |^2 \frac{d\mu}{\mu - z}$$

for complex numbers $z = \lambda + i\lambda'$ not in s. We have $[h_\epsilon ; z] \to [h_\epsilon ; \lambda]$ if $\lambda' \to 0$ through values $\lambda' > 0$. On the other hand we can, by (11.04), write $[h_\epsilon ; z]$ in the form

$$[h_\epsilon ; z] = -\frac{1}{2\pi i\epsilon} \int_s \left\{\frac{1}{1 + \epsilon[j; \mu]} - \frac{1}{1 + \epsilon\overline{[j; \mu]}}\right\} \frac{d\mu}{\mu - z} .$$

Since the function

$$[j; z] = \int_s | j(\mu) |^2 \frac{d\mu}{\mu - z}$$

approaches $[j; \lambda]$ or $\overline{[j; \lambda]}$ as $z = \lambda + i\lambda' \to \lambda$ depending on whether $\lambda' > 0$ or $\lambda' < 0$ we can write

$$[h_\epsilon ; z] = -\frac{1}{2\pi i\epsilon} \oint \frac{d\zeta}{\{1 + \epsilon[j; \zeta]\}(\zeta - z)}$$

when the path of integration is so chosen that it encircles the interval $\mathfrak{g}$, but not the points z and λ_ϵ . This integral can be evaluated by residues. The denominator vanishes at $\zeta = \lambda_\epsilon$ according to the definition of λ_ϵ as root of equation (5.10); it further vanishes at $\zeta = z$, and behaves like ζ at infinity. To evaluate the residue at $\zeta = \lambda_\epsilon$ we must evaluate the derivative of $[j; \zeta]$ with respect to ζ at $\zeta = \lambda_\epsilon$. This derivative evidently equals

$$\int_{\mathfrak{g}} \frac{|j(\mu)|^2}{|\mu - \lambda_\epsilon|^2} d\mu = a_\epsilon^{-2}(g_\epsilon , g_\epsilon) = a_\epsilon^{-2}$$

by (11.06) and (11.07). Thus we obtain

$$\epsilon[h_\epsilon ; z] = \frac{a_\epsilon^2}{\epsilon(\lambda_\epsilon - z)} + \frac{1}{1 + \epsilon[j; z]} - 1.$$

Letting $z = \lambda + i\lambda'$ approach λ with $\lambda' > 0$ we find

$$\epsilon[h_\epsilon ; \lambda] = \frac{a_\epsilon^2}{\epsilon(\lambda_\epsilon - \lambda)} + \frac{1}{1 + \epsilon[j; \lambda]} - 1.$$

By (11.04) and (11.06) this relation gives

$$\epsilon j(\lambda)[h_\epsilon ; \lambda] = \frac{a_\epsilon}{\epsilon} g_\epsilon(\lambda) + h_\epsilon(\lambda) - j(\lambda).$$

Insertion of this relation into (11.20) yields

$$\epsilon R_\epsilon(\Gamma R_\epsilon^*)x = j(\lambda)\left\{\frac{a_\epsilon}{\epsilon} (g_\epsilon , x) + (h_\epsilon , x) - (j, x)\right\}$$

or, since $a_\epsilon j(\lambda) = (\lambda_\epsilon - \lambda)g_\epsilon(\lambda)$,

$$\epsilon R_\epsilon(\Gamma R_\epsilon^*) = \frac{\lambda_\epsilon - L}{\epsilon} G_\epsilon + R_\epsilon - J.$$

Subtracting the adjoint relation

$$-\epsilon(\Gamma R_\epsilon)R_\epsilon^* = G_\epsilon \frac{\lambda_\epsilon - L}{\epsilon} + R_\epsilon^* - J$$

we obtain

$$R_\epsilon^* - R_\epsilon + \epsilon R_\epsilon(\Gamma R_\epsilon^*) + \epsilon(\Gamma R_\epsilon)R_\epsilon^*$$

$$= \frac{1}{\epsilon} \{(\lambda_\epsilon - L)G_\epsilon - G_\epsilon(\lambda_\epsilon - L)\} = \frac{1}{\epsilon} \{G_\epsilon L - LG_\epsilon\}.$$

Using the fundamental relation (3.12) we obtain just relation (11.19). Thus relation (11.18) is established.

12. *Regular Perturbation Represented through Fourier Transforms*[3]

It was mentioned in Section 3 that conditions for the kernel of the integral operators can be given which are different from those imposed in Section 3 and which are particularly suitable for treating perturbation problems in which the spectrum of the undisturbed operator covers the whole λ-axis.

The conditions for the symmetric kernel $k(\lambda, \mu) = \overline{k(\mu, \lambda)}$ of the disturbing operator K, which we shall adopt in the present section, are firstly that k is the Fourier transform

$$(12.01) \qquad k(\lambda, \mu) = \frac{1}{2\pi} \int_{-\infty}^{\infty} \int_{-\infty}^{\infty} m(\sigma, \tau) \exp \{i(\lambda\sigma - \mu\tau)\}\, d\sigma\, d\tau$$

of an absolutely integrable function, i.e. of a function $m(\sigma, \tau)$ for which

$$(12.02) \qquad \int_{-\infty}^{\infty} \int_{-\infty}^{\infty} | m(\sigma, \tau) | \, d\sigma\, d\tau < \infty .$$

Secondly, k should be such that the operator K transforms quadratically integrable functions into such functions.

Under these conditions again transformations U_ϵ and U_ϵ^* exist satisfying the relations (2.09), (2.11), and (2.12) and thus furnishing the spectral transformation of the operator $L_\epsilon = L + \epsilon K$, provided $| \epsilon |$ is small enough.

The transformations U_ϵ , U_ϵ^* assume a particularly simple form, simpler than the form given by (4.01), if one works with the Fourier transforms of the operators L and K and of the functions $x(\lambda)$ on which they operate.

The operator L is applicable on functions $x(\lambda)$ for which

$$(12.03) \qquad \int_{-\infty}^{\infty} \{1 + \lambda^2\} \, | x(\lambda) |^2 \, d\lambda < \infty$$

and it transforms such functions into functions

$$Lx(\lambda) = \lambda x(\lambda)$$

for which

$$\int_{-\infty}^{\infty} | Lx(\lambda) |^2 \, d\lambda < \infty .$$

The Fourier transforms

$$(12.04) \qquad u(\sigma) = \int_{-\infty}^{\infty} e^{-i\sigma\mu} x(\mu) \, d\mu$$

[3]The approach to the perturbation problem as given in this section and the proofs of the statements made will be worked out in detail in a future publication.

of quadratically integrable functions $x(\lambda)$ are again quadratically integrable, and vice versa. The functions $x(\lambda)$ which satisfy (12.03), to which thus L is applicable, are transformed into continuous functions $u(\sigma)$ which possess almost everywhere a derivative

$$Du(\sigma) = \frac{d}{d\sigma} u(\sigma) \tag{12.05}$$

which is quadratically integrable,

$$\int_{-\infty}^{\infty} | Du(\sigma) |^2 \, d\sigma < \infty . \tag{12.06}$$

The operator L is transformed into the operator iD; that means $\lambda x(\lambda)$ is transformed into $iDu(\sigma)$. The integral operator K, which transforms $x(\lambda)$ into

$$Kx(\lambda) = \int_{-\infty}^{\infty} k(\lambda, \mu) x(\mu) \, d\mu, \tag{12.07}$$

is transformed into the integral operator M with the kernel $m(\sigma, \tau)$ which transforms the function $u(\sigma)$ into

$$Mu(\sigma) = \int_{-\infty}^{\infty} m(\sigma, \tau) u(\tau) \, d\tau. \tag{12.08}$$

The operator $L + \epsilon K$ is therefore represented by the operator $iD + \epsilon M$.

In the following we shall deal only with the operator $iD + \epsilon M$ and shall establish its transformation into the operator iD. Since the operator iD can be transformed into the operator L we may consider the spectral transformation of $iD + \epsilon M$ and hence of $L + \epsilon K$ as established.

We denote by $\mathfrak{H}$ the space of all quadratically integrable complex-valued functions, i.e. of all functions $u(\sigma)$ for which

$$|| u || = \left[\int_{-\infty}^{\infty} | u(\sigma) |^2 \, d\sigma \right]^{1/2} \tag{12.09}$$

is finite; by $\mathfrak{F}$ we denote the subspace of all functions $u(\sigma)$ which possess a derivative in $\mathfrak{H}$.

We introduce the class (N) of integral operators whose kernels $n(\sigma, \tau)$ satisfy firstly the condition

$$\int_{-\infty}^{\infty} \int_{-\infty}^{\infty} | n(\sigma, \tau) | \, d\sigma \, d\tau < \infty \tag{12.10}$$

and which secondly are such that the operator N transforms a function $x(\sigma)$ in $\mathfrak{H}$ into the function

$$Nu(\sigma) = \int_{-\infty}^{\infty} n(\sigma, \tau) u(\tau) \, d\tau, \tag{12.11}$$

which also lies in $\mathfrak{H}$. The latter condition is satisfied either if

$$\int_{-\infty}^{\infty}\int_{-\infty}^{\infty} | n(\sigma, \tau) |^2 \, d\sigma \, d\tau < \infty,$$

or if

$$\int_{-\infty}^{\infty} | n(\sigma, \tau) | \, d\sigma + \int_{-\infty}^{\infty} | n(\sigma, \tau) | \, d\tau < \infty,$$

or if there is a function $n_0(\sigma) \geq 0$ with

$$\int_{-\infty}^{\infty} n_0(\sigma) \, d\sigma < \infty$$

such that $| n(\sigma, \tau) | \leq n_0(\sigma - \tau)$.

To operators of the class (N) we assign the integral operator ZN with the kernel

$$(12.12) \qquad i \int_{-\infty}^{\sigma} n(\sigma + \rho, \tau + \rho) \, d\rho = i \int_{-\infty}^{\sigma} n(\rho, \tau - \sigma + \rho) \, d\rho.$$

The operators ZN transform functions $u(\sigma)$ in $\mathfrak{H}$ into functions $ZNu(\sigma)$ in $\mathfrak{H}$ and there is a constant, $\| ZN \|$, such that

$$(12.13) \qquad \| ZNu \| \leq \| ZN \| \, \| u \|.$$

If the functions $u(\sigma)$ are in $\mathfrak{F}$ the functions $ZNu(\sigma)$ are also in $\mathfrak{F}$ and the first fundamental identity

$$(12.14) \qquad ZNiD - iDZN = N$$

holds. The products $(ZP)N$ and $P(ZN)$ formed with two operators P and N of class (N) are again of class (N) and the second fundamental identity,

$$(12.15) \qquad (ZP)(ZN) = Z\{(ZP)N + P(ZN)\},$$

holds.

Evidently, the relations (12.13), (12.14), and (12.15) correspond to the relations (3.11), (3.12), and (3.14) for the operators ΓR. As a matter of fact the operator ZN is the Fourier transform of the operator ΓR if N is the transform of R. The relations (12.13), (12.14), and (12.15) for the operator ZN are, however, much more easily verified than the corresponding relations for the operator ΓR. The reason is that the operation Z is by far more regular in form than the operation Γ.

The disturbing operator M is required to be of class (N) and, in addition, to be symmetric (in the Hermitian sense),

$$(12.16) \qquad m(\sigma, \tau) = \overline{m(\tau, \sigma)}.$$

The statement now is that operators N_ϵ of class (N) exist such that the transformations

$$V_\epsilon = 1 + \epsilon\Gamma N_\epsilon, \qquad V_\epsilon^* = 1 - \epsilon\Gamma N_\epsilon^*, \tag{12.17}$$

satisfy the relations

$$(iD + \epsilon M)V_\epsilon = V_\epsilon iD, \tag{12.18}$$

$$V_\epsilon^* V_\epsilon = 1, \tag{12.19}$$

$$V_\epsilon V_\epsilon^* = 1. \tag{12.20}$$

If now functions $u(\sigma)$ are represented through functions $v(\sigma)$ in $\mathfrak{F}$ by

$$u(\sigma) = V_\epsilon v(\sigma) \tag{12.21}$$

then, by (12.18)

$$(iD + \epsilon M)u(\sigma) = V_\epsilon iDv(\sigma) \tag{12.22}$$

and thus the operator $iD + \epsilon M$ is represented by iD. According to (12.19) the function $v(\sigma)$ is found by

$$v(\sigma) = V_\epsilon^* u(\sigma) \tag{12.23}$$

and, according to (12.20), every function $u(\sigma)$ is represented by (12.21).

The analytic dependence of N_ϵ and N_ϵ^* on ϵ near $\epsilon = 0$ is easily established.

The formal perturbation procedure, i.e. the expansion of V_ϵ in powers of ϵ, is immediately set up exactly as in Section 4. Inserting in the relation

$$N_\epsilon = M(1 + \epsilon Z N_\epsilon), \tag{12.24}$$

which is equivalent with (12.18) by (12.14), the expansion

$$N_\epsilon = N^{(1)} + \epsilon N^{(2)} + \cdots \tag{12.25}$$

we find successively

$$\begin{aligned} N^{(1)} &= M, \\ N^{(2)} &= M(ZM), \\ N^{(3)} &= M(ZM(ZM)), \cdots, \end{aligned} \tag{12.26}$$

in perfect analogy with (4.08).

Bibliography

1. Rellich, F., *Störungstheorie der Spektralzerlegung.* Mathematische Annalen, Volume 113, pp. 600–619, 1936; Volume 113, pp. 677–685, 1937; Volume 116, pp. 555–570, 1939; Volume 117, pp. 355–382, 1939; Volume 118, pp. 462–484, 1942.
2. Friedrichs, K., *Über die Spektralzerlegung eines Integraloperators.* Mathematische Annalen, Volume 115, No. 2, pp. 249–272, 1938.

3. Kramers, H. A., *Theorien des Aufbaues der Materie.* Hand- und Jahrbuch der chemischen Physik, edited by A. Eucken and K. L. Wolf, Volume 1, 1933.
4. Weisskopf, V., and E. Wigner, *Über die natürliche Linienbreite in der Strahlung des harmoninchen Oszillators*, Zeitschrift für Physik, Volume 65, p. 18, 1930. *Berechnung der naturlichen Linienbreite auf Grund der Diracschen Lichttheorie*, Zeitschrift für Physik, Volume 63, p. 54, 1930.
5. Kemble, E. C., *The Fundamental Principles of Quantum Mechanics, with Elementary Applications.* McGraw-Hill, 1937.
6. Wheeler, J. A., *On the Mathematical Description of Light Nuclei by the Method of Resonating Group Structure.* Physical Review, Volume 52, p. 1107, 1937.
7. Heisenberg, W., *Die "beobachtbaren Grössen" in der Theorie der Elementarteilchen*, Zeitschrift für Physik, Volume 120, pp. 513–538, 1943. *Die "beobachtbaren Grössen" in der Theorie der Elementarteilchen. II*, pp. 673–702, 1943.
8. Neumann, J. v., *Mathematische Grundlagen der Quantenmechanik.* Springer, 1932; Dover Publishing Co., New York, 1943.
9. Pauli, W., *Meson Theory of Nuclear Forces.* Interscience Publishers, Inc., New York, second edition, 1948.
10. Dirac, P. A. M., *Principles of Quantum Mechanics.* Oxford University Press, London, Third Edition, 1947.
11. Wentzel, G., *Wellenmechanik der Stoss- und Strahlungsvorgänge.* Handbuch der Physik, 2nd ed., Volume XXIV, Part I, Chapter 5, Berlin, 1933.

[66–2] **Spectral Perturbation Phenomena, Proc. of a Seminar in Perturbation Theory and Its Application in Quantum Mechanics, ed. C.H. Wilcox, John Wiley & Sons, Inc., New York, (1966), 35–47.**

Commentary:

This is another expository paper on perturbation theory and its application to quantum mechanics. Here Friedrichs discusses wider topics than in [48–2], ranging from the classical perturbation theory for discrete eigenvalues to various aspects of perturbation of continuous spectra, including, in addition to the results in [48–2], problems of spectral concentration, more recent results on scattering theory, and some problems in quantum field theory. This paper gives not only an extensive outlook over the state of things at that time, but remains an excellent introduction to the subject.

KURT O. FRIEDRICHS

Spectral Perturbation Phenomena

Offprint from Perturbation Theory
and Its Application in Quantum Mechanics
Edited by Calvin H. Wilcox
Published by John Wiley & Sons, Inc., 1966.

One of the great attractions of the theory of perturbation of spectra lies in the great variety of phenomena that the spectrum of an operator may show if this operator is disturbed. Spectra are very touchy objects; they are bound to react irrationally under the slighest provocation. This nervousness offers quite some attraction to theoretical mathematicians. The computing mathematician may feel differently, though.

The computation of disturbed spectra will naturally involve various stages. I am inclined to formulate such states in the following manner:

1. Just compute. Run into trouble.
2. Investigate the qualitative character of the change of spectrum.
3. Taking this into account, redesign the approximation procedure. Compute, run into trouble again.
4. Discover the faults of your computing scheme.
5. Redesign the computing scheme. Compute successfully.

The present lecture is concerned with item 2. I shall describe a number of peculiar phenomena that may happen to the spectrum of an operator if this operator is disturbed. If these phenomena do occur, but are not taken into account, trouble will inevitably result.

Naturally, one likes to have criteria for the absence of irregular happenings. Some such criteria will be mentioned in this lecture.

These criteria may not always be directly applicable to a perturbation problem arising in physics or chemistry; nevertheless they may indicate which properties of the operator will be responsible for this or that phenomenon.

The operators considered in this lecture will be assumed to act on vectors Φ of a (complex) Hilbert space $\mathcal{H}$ endowed with a unit form (Φ, Φ). The undisturbed operator, H_o, will always be assumed self-adjoint in the strict sense, so that it admits a spectral resolution. The disturbed operator, H, need not always share this property. The difference $H - H_o$ will be referred to as the disturbing operator and denoted by V. The disturbed operator is to depend on a parameter ϵ in such a way that it reduces to H_o for $\epsilon = 0$. Thus

$$H = H(\epsilon) = H_o + V(\epsilon), \quad V(0) = 0 .$$

The rigorous mathematical theory of perturbation of spectra was initiated by Rellich in a series of fundamental papers [2]. Rellich treated operators with discrete spectra. He proved that the expansion procedure is valid under certain circumstances; He also discovered a number of peculiar phenomena which may occur under other circumstances and cause the expansion procedure to break down. I intend to describe some such phenomena.

In doing this I shall at first assume that the undisturbed operator H_o is completely continuous, so that its spectrum is bounded and essentially discrete. That is to say, in each closed interval of real numbers excluding zero the spectrum of H_o consists at most of a finite number of point eigenvalues of finite multiplicity. For simplicity I shall also assume that zero itself is not eigenvalue.

Suppose now such an operator H_o is disturbed by a bounded operator $V = V(\epsilon)$ which depends continuously on ϵ. Then one may expect that each eigenvalue of H_o will go over into an eigenvalue of $H(\epsilon) = H_o + V(\epsilon)$ which also moves continuously and that the same is true for properly chosen associated eigenvectors. This is indeed the case if the eigenvalue of H_o is simple, but not necessarily so if it is a multiple one. Rellich has given a simple example for this irregular behavior: a rotating ellipse which approaches a circle. The disturbing operator here may just as well depend infinitely differentiably on the parameter ϵ. It could not depend analytically on ϵ, though. For, in that case there are always branches of eigenvalues $\lambda = \lambda(\epsilon)$ and eigenvectors $\Phi = \Phi(\epsilon)$ of $H(\epsilon)$ which are regular analytic in ϵ near $\epsilon = 0$. This need not be the case if V depends on two parameters ϵ_1, ϵ_2 in a regular analytic manner [Rellich [2], also [19, Ch. I]. These facts illustrate the sensitivity of the spectrum and motivate the search for conditions of good behavior.

Rellich's basic result concerning regular behavior of eigenvalues and eigenvectors in case of regular dependence of the disturbance on a single parameter implies that perturbation expansions converge in this case. Such expansions may be derived by various procedures. Roughly speaking there are two classes of such procedures, the "explicit" and the "implicit" ones. The explicit procedures are the Rayleigh-Schrödinger procedure and its variants;

they can be derived from an equation for the eigenvector obtained by eliminating the eigenvalue. The implicit procedures are the Brillouin-Wigner procedure and its variants; they can be derived from an equation for the eigenvalue obtained by eliminating the eigenvector. These equations [see e.g. 19, Ch. I] are essentially nonlinear, although they are concerned with the behavior of linear operators. This nonlinearity is connected with the fact that the point spectrum of an operator in general varies when disturbed. This need not be the case for continuous spectra as will be discussed later on.

The regular branches of eigenvalues and eigenvectors may come to an end if the parameter ϵ reaches a certain limit. This limit will depend on which eigenvalue is pursued; i.e. it will depend on the label n of the nth eigenvalue $\lambda_n(\epsilon)$. Naturally, one should hope that there is a common bound on the values of ϵ such that for all n the eigenvalue $\lambda_n(\epsilon)$ and the associated eigenvector can be continued until ϵ reaches this bound. Rellich has shown that this need not be so and he also gave conditions under which it is.

Suppose now there is such a common bound $\bar{\epsilon}$ such that the spectrum of $H_o + V(\epsilon)$ behaves regularly for $|\epsilon| < \bar{\epsilon}$; then another question arises. Since, by assumption, H_o is completely continuous and does not have zero as eigenvalue, the sequence of eigenvectors associated with the eigenvalues $\lambda_n(0)$ is complete. Naturally, one will expect that the same is true of the eigenvectors associated with the values $\lambda_n(\epsilon)$. Rellich showed that this need not be so and he described the circumstances under which it is. More extensive results on this question and related matters were recently obtained by Turner [20].

What will happen if the system of disturbed eigenvectors is not complete?

Rellich has given an example of a very harmless looking disturbance which produces an extra eigenvalue, that leaks out of zero as ϵ increases; but that is only one possibility.

In other cases a whole sequence of extraneous eigenvalues may appear; even a continuous section of the spectrum may come out of zero. As a matter of fact, it may happen that such an extraneous part of the spectrum eventually takes over by gradually absorbing more and more of the eigenvalues $\lambda_n(\epsilon)$. The same phenomenon may appear as soon as ϵ moves away from zero in case there is no common bound. Of course, a continuous spectrum cannot appear if the disturbing operator V is completely continuous; but we have not made the assumption that this be so. We only have assumed that V is bounded.

In case V is unbounded, a new possibility arises: it may happen that the spectrum becomes continuous instantaneously, as soon as $\epsilon \neq 0$. One may be inclined to regard such an occurrence as just one of those pathologies some mathematicians like to indulge in. And in a way, it is. Still, it is just this phenomenon that has a definite significance for concrete problems in physics; for it is this phenomenon that occurs in the theory of the Stark effect. Here is an example where the naive original computation was misleading in first order and failed when pushed to 2nd order. Clarification of this failure resulted from an analysis of the phenomenon.

A correct description of this phenomenon requires an appropriate asymptotic analysis. Several papers have been devoted to this question, papers by Titchmarsh, Rejto and myself, and others [8,13,14]. In these Proceedings, the papers on spectral concentration by McLeod and Conley are concerned with related questions.

To motivate the term " spectral concentration" one may look at the variation of the parameter in the opposite direction, letting $H(\epsilon)$ stand for a family of operators which have continuous spectra (in some interval) when $\epsilon \neq 0$ while for $\epsilon = 0$ the spectrum there consists of an isolated point.

This phenomenon may be called "isolated" spectral concentration in contrast to another phenomenon, which may be called "embedded" spectral concentration. In this phenomenon a point eigenvalue is embedded in a continuous spectrum, for $\epsilon = 0$ while no point eigenvalue was present (in its neighborhood) for $|\epsilon| \neq 0$ (and small). Thus, in the latter case, the undisturbed operator H_0 has embedded in a continuous spectrum a point eigenvalue, which disappears when H_0 is disturbed. (For this concentration phenomenon to occur, the disturbance V need neither be bounded nor have itself a continuous spectrum). The analytic treatment of such embedded concentration is simpler than that of isolated concentration [3].

Embedded spectral concentration was already described by Dirac [see 1] in connection with problems of quantum mechanics. To describe the role of this phenomenon in quantum mechanics we shall refer to the Auger effect. We consider a Helium atom, first disregarding the interaction between the two electrons. Among the bound eigenstates of the atom in which both electrons are excited there will be some in which they have the same energy as in states in which one electron is in the ground state while the other electron has its energy in the continuous spectrum (not to say: is out of bounds), and is ready to move away. If now the interaction between the two electrons is taken into consideration and regarded as a disturbance, one will find that there is no bound eigenstate for the disturbed energy that would correspond to the doubly excited eigenstate for the undisturbed energy. The latter state would reappear by "concentration" if the interaction were made to disappear. If now the atom at some time is in the doubly excited eigenstate for H_0, (the energy without interaction of the two electrons) it will, with high probability, move over into an eigenstate for H_o (with nearly the same energy) in which one electron is in the groundstate of H_o while the other one has its energy in the continuous spectrum of H_o. The latter electron will move away and the description in terms of eigenstates for H_o will eventually become appropriate.

New phenomena appear if the undisturbed operator has a continuous spectrum to begin with (even without having point eigenvalues embedded in it). To describe them it is necessary to say in specific terms what it means to say than an operator has a continuous spectrum. This can be done conveniently by first describing the motion of functional representation.

We say that a Hilbert space is functionally represented if the vectors Φ are linearly represented by a function ψ of a variable α,

$$\Phi \Longleftrightarrow \psi(\alpha) ,$$

such that the unit form of Φ is given as

$$(\Phi, \Phi) = \int |\psi(\alpha)|^2 \, dm(\alpha) ,$$

in terms of the representer ψ. Here $dm(\alpha)$ is an appropriate measure differential. To make the assignment of Φ and ψ one-to-one, we simply stipulate that $\psi = 0$ if and only if $\int |\psi(\alpha)|^2 \, dm(\alpha) = 0$.

Such a functional representation gives a spectral representation of an operator H if there is a function $\lambda(\alpha)$ such that application of the operator H on the vector Φ is represented by

$$H\Phi \Longleftrightarrow \lambda(\alpha)\, \psi(\alpha) .$$

The closed support of the measure function $m(\alpha)$ is then the spectrum of H. If $m(\alpha)$ jumps a point α_o the value $\lambda(\alpha_o)$ is a point eigenvalue of H. If any change of $m(\alpha)$ is composed solely of jumps the spectrum of H is called a (pure) point spectrum. If there are no point eigenvalues in the spectrum it is called continuous. In general the measure function $m(\alpha)$ will be the sum of two parts, one associated with a pure point spectrum, the other one with a continuous spectrum, $m(\alpha) = m_{pt}(\alpha) + m_{ct}(\alpha)$.

Let us first assume that H_o has only a continuous spectrum; for simplicity we assume that it consists of the ray $0 \le \lambda$ and that we may take $m(\lambda) = \lambda$ as measure function for $0 \le \lambda$. With $\alpha = \lambda$ we then write

$$\Phi \Longleftrightarrow \psi(\lambda) \quad \text{implies} \quad H_o \Phi \underset{o}{\Longleftrightarrow} \lambda \psi(\lambda) .$$

The values of the functions ψ here may be complex numbers or they may be complex valued functions of discrete or continuous "accessory variables" (such as labels, spin variables, or angular variables); more generally, they maybe themselves vectors of an "accessory" Hilbert space.

A continuous spectrum may undergo drastic changes under rather mild disturbances; as von Neumann has shown, it may readily go over into a pure (though dense) point spectrum. (It has been suggested to call this phenomenon "curdling".) Instead of investigating all the peculair things that may happen with a continuous spectrum we shall start from the other end. We shall formulate the mildest variety of changes that a continuous spectrum may undergo and ask under which circumstances will that happen.

The mild change of spectrum we have in mind involves no change of the spectrum itself; it involves only a one-to-one correspondence of the spectral representations with the same measure function. That is to say there should be a spectral representation for $H(\epsilon)$ through function $\varphi_\epsilon(\lambda)$ such that

$$\Phi_\epsilon \underset{\epsilon}{\Longleftrightarrow} \varphi_\epsilon(\lambda) \quad \text{implies} \quad H(\epsilon)\Phi \underset{\epsilon}{\Longleftrightarrow} \lambda\, \varphi_\epsilon(\lambda) \ ;$$

moreover

$$(\Phi, \Phi) = \int_0^\infty |\varphi_\epsilon(\lambda)|^2 \, d\lambda \ .$$

The two spectral representations can then be connected through a unitary operator $U(\epsilon)$ in such a way that $H(\epsilon)$ can be written as

$$H(\epsilon) = U(\epsilon)\, H_0 U^*(\epsilon) \ .$$

The operator $U(\epsilon)$ will be referred to as a "spectral transformation". If such an operator exists the two operators H and H_0 are unitarily equivalent.

It should be mentioned that the problem of finding such a transformation is a linear one, which is consistent with the assumed property of the perturbation that it does not change the spectrum.

Whether or not the operator $H_0 + V$ is unitarily equivalent with H_0 certainly depends on the nature of the disturbing operator V. It is not smallness of V that matters primarily; it is a certain type of smoothness of V. I myself gave such smoothness conditions quite some time ago [3]. They involve the assumption that V can be written as an integral operator with respect to the spectral representation of H_0; i.e. $V\Phi$ is represented as

$$V\Phi \underset{o}{\Longleftrightarrow} \int_0^\infty v(\lambda, \lambda') \psi(\lambda') d\lambda' .$$

If the kernel v is sufficiently smooth the desired representation of $H(\epsilon)$ could be established by constructing a spectral transformation $U(\epsilon)$. Disturbing operators V satisfying an appropriate class of such smoothness conditions have been called "gentle".

These smoothness conditions include the requirement that the kernel $v(\lambda, \lambda')$ vanish for $\lambda = 0$, and for $\lambda' = 0$; i.e. if either λ or λ' is at an end point of the spectrum. If this condition is abandoned it may happen that one or more point eigenvalues come out of the end point $\lambda = 0$ of the spectrum for arbitrarily small values of ϵ. On the other hand, if this condition, together with the other gentleness conditions, is satisfied, point eigenvalues cannot develop for arbitrarily small ϵ. Still such point eigenvalues can develop if ϵ exceeds a certain threshold. But that is the worst that can happen under wide circumstances; see [9, 7,

The specific gentleness requirements I have referred to are stronger than they should be. It is therefore desirable to have a theory of perturbation of continuous spectra under less restricted requirements. Such a theory was developed by Rosenbloom, Kato, and others, [5, 7, 12, 15, 16]. The restrictions imposed in this theory on the disturbing operator V are rather unspecific: this operator should be of the trace class or satisfy even weaker related conditions. On the other hand, the results derived under these conditions are somewhat weaker than those derived from gentleness of V inasmuch as the spectrum of the disturbed operator H may differ from that of H_0 in its "singular spectrum". This notion refers to the decomposition of the measure differential dm into two contributions, $dm = dm_{ac} + dm_{sg}$, where the first, associated with the "absolutely continuous" part of the spectrum is such that $\int_{\mathfrak{s}} dm_{ac}(x) = 0$ on every set $\mathfrak{s}$ having Lebesgue measure zero, $\int_{\mathfrak{s}} d\lambda = 0$, while the second contribution associated with the "singular" part of the spectrum is such that $d\lambda \neq 0$ implies $\int_{\mathfrak{s}} dm_{sg}(\lambda) = 0$. The special undisturbed operator H_0 that we have considered above has only an absolutely continuous spectrum, $0 \leq \lambda < \infty$ with $dm(\lambda) = d\lambda$; if it is disturbed by an operator V of the trace class, an additional singular spectrum may appear. That this actually can happen has been shown by simple examples (Aronszajn, Donoghue [4, 22]. Here then we meet a new spectral perturbation phenomenon that may appear when an operator is modified by a relatively well behaved disturbance.

According to what was said before no singular spectrum can appear under a small gentle disturbance. Thus there is a gap between the results of the two approaches, the one involving disturbances of trace class and related ones, and the one involving gentle disturbances. Naturally, one should like to be able to close this gap.

Special progress has been made on this question quite recently. Kato as well as Kuroda have formulated conditions on the disturbance, related to trace conditions, under which no singular spectrum appears. On the other hand Rejto has characterized another class of disturbances of comparable degree of generality which comprise gentle and small disturbances and which produce disturbed operators H which are unitarily equivalent to H_0. These various characterizations cover various classes of potential disturbances of the Laplacian. Reports on these matters are referred to in other papers in these Proceedings [Kuroda, Rejto].

The phenomena that appear if the disturbed operator is not unitarily equivalent to the undisturbed one affect the mathematical description of scattering; the standard description of scattering may not be valid if such phenomena occur.

The mathematical description of scattering employs an asymptotic description of the Schrödinger operator e^{-itH}. Under favorable circumstances this description can be given with the aid of a pair of operators $W_\pm$ in the form

$$e^{-itH} \sim e^{-itH_0} W_\pm^* \quad \text{as} \quad t \sim \pm\infty .$$

The operators $W_\pm$, the "wave operators", can then be described as the limits

$$W_\pm = \lim e^{itH} e^{-itH_0} \quad \text{as} \quad t \to \pm\infty .$$

These wave operators can be given explicitly in terms of the spectral transformation $U(\epsilon)$ constructed for gentle disturbances, [3]. On the other hand, in the earlier approach to treating trace class disturbances the wave operators were established first and then used as spectral transformations, which is possible if they are unitary.

Scattering can be described by identifying any state of the form $e^{-itH}\Phi$ asymptotically, as $t \to \pm\infty$, with states of the form $e^{-itH_0}\Phi_\pm$, the states $\Phi_\pm$ being given by $W_\pm^*\Phi$. The assignment of Φ_+ to Φ_- is then given by

$$\Phi_+ = W_+^* W_- \Phi_- \quad .$$

The operator $W_+^* W_-$ is the "scattering operator".

There are cases, however, in which this description is impossible or incomplete. For it can happen that either the limits defining the wave operators $W_\pm$ do not exist or that they do exist and satisfy the relation $W_{\pm}^{*} W_{\pm} = 1$, but not the relation $W_{\pm} W_{\pm}^{*} = 1$.

An example of the latter occurrence was described by Kato and Kuroda and later on recognized by Kuroda [6, 7], as a case involving multiple channel scattering in the sense of Jauch [10]. The relevant situation can be described in somewhat more general terms [19, App. Ch. II]. Its main feature is that the disturbance produces an additional continuous spectrum (involving fewer accessory variables than the main part of the spectrum). In this case a state $e^{-itH}\Phi$ is asymptotically of the form $e^{-itH_o}\Phi_{\pm}$ only if $\Phi = \Phi^o$ belongs to the main part of the spectrum of H. If $\Phi = \Phi^a$ belongs to the "new" part of this spectrum its asymptotic description is of the form $e^{-itH_a}\Phi_{\pm}^a$ in which H_a is an operator of the form $H_a = H_o + V_a$ where V_a stands for a contribution to V which is responsible for the new spectrum. Writing $\Phi = \Phi + \Phi^a$ one may describe scattering as an assignment of the pair of associated states $(\Phi_{+}^{o}, \Phi_{+}^{a})$ to the pair of states $(\Phi_{-}^{o}, \Phi_{-}^{a})$. The qualifiers ^{o} and ^{a} then are said to indicate different "channels".

The last perturbation phenomenon I should like to mention appears in an area which is, if not explicitly then at least implicitly, excluded from this conference, quantum theory of fields. I should like to describe this phenomenon nevertheless because it throws some light on what I have discussed before.

In the perturbation treatment of field problems one deals with undisturbed operators which have a compound spectrum, consisting of a point eigenvalue and a sequence of continuous spectra which can be described as being $1, 2, 3, \ldots, n, \ldots$ parametric. If the state of the field is given by a vector belonging to the nth component of the spectrum one says that exactly n particles (of some sort) are present in the field. The lower end of the nth part of the spectrum is at the place nm where m is the mass of a single particle of the kind involved.

The various types of interaction that occur in the quantum theory of fields and in the quantum theory of elementary particles have analytical and combinatorial properties. If the interaction operators are written as integral operators, with reference to the spectral representation of H_o, their kernels will be highly singular. Let us suppose that all these singularities have been smeared out so that the perturbation phenomena result only from the combinatorial properties. These properties result from the fact that the interaction operators can be expressed in terms of particle annihilation and creation operators; these are operators characterized by canonical commutation laws.

This fact has a remarkable consequence: the spectrum is shifted under the influence of the perturbation and all parts of the spectrum are shifted by the same amount, which thus is independent of the number n of particles present; [see e.g. 19, Ch. III].

In the original perturbation approach to constructing a spectral transformation (or rather to constructing the associated scattering operator) the occurrence of this shift was not recognized (at least not explicitly) and a formal perturbation procedure was adopted which could not be valid unless there was no shift. Later on the existence of the shift was recognized (still later an explicit expression of it, the Goldstone formula, was given). If now the shift is eliminated by addition of an appropriate constant to the disturbed (or undisturbed) operator, the terms of the perturbation expansion will be meaningful; certain meaningless terms that appeared before the shift was incorporated in the disturbance now do not occur anymore.

That this is so depends essentially on the assumption that the singularities of the interaction kernel have been smeared out. Among these singularities there is one delta function factor which implies that the total momentum of the particles involved is conserved when they interact. Suppose now we allow this factor, but no other singularities, to be present. Then the disturbed spectrum will undergo another shift. This new shift will be different for the different parts of the spectrum of H_0; it will be proportional to the number of particles; i.e. it will be a multiple $n\delta m$ of a number δm which may be interpreted as a shift of the mass of a single particle. It is possible to modify the disturbed (or undisturbed) operator in such a way that such a shift no longer occurs; (one then speaks of mass renormalization). If this is done certain (but not all) meaningless terms that otherwise would appear will not occur in a perturbation expansion.

It should be mentioned that the two adjustments mentioned are just the two simplest renormalizations needed in quantum field theory. The real difficulties only start from here.

In conclusion let me say that in some way the situation is analogous in perturbation problems of quantum mechanics. To be sure it is a necessary, and at the same time attractive, task to clarify all the various more or less qualitative phenomena involved in the perturbation of spectra (item 2). But there remains still quite a way to go to the end of item 5: compute successfully.

REFERENCES

1. P. A. M. Dirac, Principles of Quantum Mechanics, Oxford University Press, London, Third Edition, 1947.

2. F. Rellich, Störungstheorie der Spektralzerlegung, Math. Ann. I, II, 113 (1936), III, 116 (1939), IV, 117 (1940), V, 118 (1942).

F. Rellich, Störungstheorie der Spektralzerlegung, Proc. Internat. Congress Mathematicians (Cambridge, Mass. 1950), Vol. 1, pp. 606-613, Amer. Math. Soc., Providence, R. I., 1952.

3. K. O. Friedrichs, Ueber die Spektralzerlegung eines Integral-operators, Math. Ann., 115 (1938), pp. 249-272.

__________, On the perturbation of continuous spectra, Comm. Pure Appl. Math., 1 (1948), pp. 361-406.

4. N. Aronszajn, On a problem of Weyl in the theory of singular Sturm-Liousville equations, Amer. J. Math., Vol. 79, 1947.

5. T. Kato, On the convergence of the perturbation method, J. Fac. Sci. Tokyo Univ., 6 (1951), pp. 198-205.

__________, Perturbation of continuous spectra by trace class operators, Proc. Japan Acad., 33 (1957), pp. 260-264.

__________, Wave operators and unitary equivalence, Tech. Rept. No. 11, Univ. of California, Berkeley, Calif., 1963.

6. T. Kato and S. T. Kuroda, A remark on the unitary property of the scattering operator, Nuovo Cimento, 14 (1959), pp. 1102-1107.

7. S. T. Kuroda, On the existence and unitary property of the scattering operator, Nuovo Cimento, 12 (1959), pp. 431-454.

__________, Perturbation of continuous spectra by unbounded operators, I, II, J. Math. Soc. Japan, 11 (1959), pp. 247-262; 12 (1960), 243-257.

__________, An example of a scattering system in Jauch's sense, Progr. Theoret. Phys., 24 (1960), p. 461.

__________, On a stationary approach to scattering problems, Bull. Amer. Math. Soc., 70 (1964), pp. 556-560.

8. E. C. Titchmarsh, Some theorems on perturbation theory, V, J. Analyse Math., 4 (1954/55), pp. 187-208.

9. O. A. Ladyzhenskaya and L. D. Faddeev, On perturbation theory of a continuous spectrum, Dokl. Akad. Nauk, SSSR, 120 (1958), pp. 1187-1190 (Russian).

10. J. M. Jauch, Theory of the scattering operator, Helv. Phys. Acta 31 (1958), pp. 127-158.

__________, Theory of the scattering operator, II. Multichannel scattering, Helv. Phys. Acta 31 (1958) pp. 661-684.

11. J. Schwartz, Some non-selfadjoint operators, I, II, Comm. Pure Appl. Math. 13 (1960), pp. 609-639; ibid. 14 (1961), pp.619-626.

12. L. de Branges, Perturbation of self-adjoint transformation, Amer. J. Math. 84 (1962), pp. 543-560.

13. K. O. Friedrichs and P. A. Rejto, On a perturbation through which the discrete spectrum becomes continuous, Comm. Pure Appl. Math. 15 (1962), pp. 218-235.

14. P. A. Rejto and C. C. Conley, On spectral concentration, IMM-NYU Report No. 293, 1962.

15. M. Sh. Birman, Conditions for the existence of wave operators, Dokl. Akad. Nauk SSSR, 143 (1962), pp. 506-509 = Soviet Math. Dokl. 3 (1962) pp. 408-411.

16. M. Sh. Birman and M. G. Krein, On the theory of wave operators, Dokl. Akad. Nauk, SSSR 144 (1962), pp. 475-478 = Soviet Math. Dokl. 3 (1962), 740-743.

17. L. D. Faddeev, Mathematical questions in the quantum theory of scattering for a system of three particles, Trudy Mat. Inst. Steklov, 69 (1963) (Russian).

__________, About Friedrichs' model in the theory of perturbations of continuous spectra, Trudy Mat. Inst. Steklov, 73 (1964) (Russian).

18. P. A. Rejto, On gentle perturbations, I, II, Comm. Pure Appl. Math. 16 (1963), pp. 279-303; 17 (1964), pp. 257-292.

19. K. O. Friedrichs, Perturbation of Spectra in Hilbert Space, Lectures in Applied Mathematics, Proceedings of the Summer Seminar, Boulder, Colo. (1960), Amer. Math. Soc., 1965.

20. R. E. L. Turner, Perturbation of compact spectral operators, Comm. Pure Appl. Math. 18, No. 3, (1965), pp. 519-541.

21. J. Pincus, Commutators, generalized eigenfunction expansions and singular integral operators. To appear, Trans. Amer. Math. Soc.

22. W. Donoghue, On the perturbation of spectra, to appear: Comm. Pure Appl. Math., 18, No. 4, 1965.

23. T. Kato, Wave operators and similarity for some non-selfadjoint operators. Preprint, Berkeley, June 1965.

III
Papers in Elasticity

Commentaries by Fritz John

[41–2] **The non-linear boundary value problem of the buckled plate (with J.J. Stoker), Amer. J. Math., LXIII (1941), 839–888.**

[42–2] **Buckling of the circular plate beyond the critical thrust (with J.J. Stoker), J. Appl. Mech., 9 (1942), 7–14.**

Commentary:

The rigorous mathematical treatment is given in the first paper; the second discusses numerical approximation methods. (The results had been announced in [39–3].) Both papers study the buckling of a thin circular plate under compressive forces at the edge with circular symmetry. For sufficiently large edge thrust the plate buckles. A boundary layer forms near the edge, while in the interior of the plate a uniform state of tension develops. These papers present a novel application of Prandtl's boundary layer concept to elasticity. Appropriate "stretching" near the edge yields a description of the boundary layer, which can be fitted to the asymptotic state in the interior obtained by direct expansion. The general underlying philosophy is given in Friedrichs' Gibbs Lecture [55–2].

That compression of a circular plate at the edge can produce tension in the interior somehow runs counter to intuition. However, for people unconvinced by mathematical arguments the Department of Mechanical Engineering at the Technical University of Delft has on exhibit a model showing this phenomenon very clearly.

THE NON-LINEAR BOUNDARY VALUE PROBLEM OF THE BUCKLED PLATE.*

By K. O. FRIEDRICHS and J. J. STOKER.

Introduction. In this paper we develop methods for solving a non-linear boundary value problem which has its origin in a problem in elasticity. The methods yield (Part I) a complete numerical solution of the problem; we obtain also (Part II) the relevant uniqueness, existence, and convergence theorems. Although we treat a specific problem, the basic principles of our methods could be applied to a considerable variety of non-linear problems.

Our problem concerns the buckling of a thin elastic plate under forces acting in the plane of the plate. While it is true that the lowest "critical" load at which buckling *begins* can be determined by solving a linear eigenvalue problem, the treatment of the buckling for loads beyond the critical load requires of necessity the description of the situation by a boundary value problem for non-linear differential equations.

Non-linear differential equations for the case of the thin plate have been derived by v. Kármán,[1] cf. [13]. They contain the linear biharmonic differential operator (of order four) and quadratic terms in the second derivatives.

We confine our discussion here to the special case of the circular plate under compressive forces at the boundary with *radial symmetry* assumed. In this case the v. Kármán equations reduce to a pair of ordinary non-linear differential equations, each of the second order. The boundary value problem considered in this paper concerns the latter pair of equations. It is found that the problem depends essentially upon one parameter N—the ratio of the pressure $\bar{p}$ applied at the edge to the lowest critical pressure p^0 at which buckling just begins.

In Part I we explain a procedure for complete *numerical* solution of our problem. We have carried out the numerical solution and will report here on some of the results.[2] Our procedure yields solutions for the entire range

* Received August 31, 1940.

[1] For general theories of non-linear elasticity see Trefftz [21], Biot [1], and Murnaghan [16]. The theory of Biot, which assumes small strains, appears to include that of v. Kármán as a special case; this theory has been applied to the buckling of thick plates. Murnaghan does not even make the assumption of small strains, and is successful in applying his theory to two special problems involving very large strains.

[2] A complete numerical and graphical discussion of these results, particularly those

$0 \leq N \leq \infty$; to cover this range we employ three methods which are different but which interlock. Each of the three methods is suitable for a particular range of values of N: 1. perturbation method (sec. 2) for low values of N, 2. power series method (sec. 3) for intermediate N, 3. asymptotic solution (sec. 4) for $N \to \infty$.

The perturbation method consists in developing the essential quantities with respect to a parameter and solving the sequence of linear boundary value problems which arise from the v. Kármán equations. The first such linear problem is nothing but the above mentioned eigenvalue problem, the lowest eigenvalue of which yields the critical load where bifurcation begins. This method is used for a rather low range of the ratio N, say $1 \leq N \leq 2.5$; beyond this range the amount of numerical computation involved is excessive.

The power series method consists in developing the essential quantities with respect to powers of the distance r from the center and satisfying the boundary conditions by solving a transcendental equation. Certain peculiar difficulties encountered in solving the latter equation can be overcome easily by using results furnished by the perturbation method. In this way one can obtain solutions for a higher range of values of N, say $N \leq 15$, but hardly for larger values of N since the labor of calculating again becomes excessive.

The asymptotic solution (sec. 4) gives the limit situation for $N \to \infty$. It is also characterized by a non-linear boundary value problem. The formulation of this problem is based on the occurrence of a boundary layer effect which in our computations had already become apparent for $N < 15$. The boundary layer effect can be roughly described as follows: While with increasing N all quantities tend to become constant in the interior of the plate, they change rapidly in a narrow strip at the boundary.[3] Once the asymptotic solution has been found it is possible by a perturbation method to develop the solution in the neighborhood of $N = \infty$. In sec. 4 we carry out the first step in such a development.

In Part I (sec. 1 to 4) of this paper our problem is treated from the point of view of applied mathematics. In Part II, which is quite independent[4] of Part I, we investigate our problem with regard to existence and uniqueness of the solutions and their continuous dependence on N for $N = \infty$. The

---

of practical significance, will appear, cf. [6]. A short note discussing our results has appeared, cf. [5].

[3] Such an edge effect presents some analogy to Prandtl's boundary layer phenomena encountered in connection with the flow of viscous fluids around obstacles (cf. e.g. [19]). This analogy aided us materially in finding the proper mode of attack for the asymptotic solution.

[4] However, some of the assumptions made in Part I receive justification in Part II.

discussion is based on a minimum problem and its relation to the boundary value problem.[5]

In sec. 5 these problems are completely formulated for the case of finite N. In sec. 7 the same problems are formulated in terms of new variables in such a way as to make possible simultaneous treatment of the finite ($N < \infty$) and asymptotic ($N = \infty$) cases. We prove in sec. 8 that the minimum problem has at most one solution, apart from sign, and that such a solution never vanishes; in addition we show that a solution of the boundary value problem which never vanishes solves the minimum problem. To prove the latter statement we show that the solution of our non-linear minimum problem, which is of the fourth degree, is at the same time the solution of a certain quadratic minimum problem to which Jacobi's transformation of the second variation can be applied. The existence of the solutions is treated in sec. 9 by direct methods; in addition, we prove the continuous dependence of the solution on N for $N = \infty$, that is, we show that the solutions for finite N converge to the asymptotic solution as $N \to \infty$. This asymptotic treatment, however, refers only to the boundary layer effect. In sec. 10 we give a rigorous treatment of the limit state in the *interior* of the plate and its connection with the boundary layer.

While it is true that the minimum problem has only one solution, apart from sign, the boundary value problem will have more and more solutions as N increases. We show, however, in sec. 6 that the boundary value problem has at most three solutions, one identically zero, the others differing only in sign, provided that N is not too large. The method used combines well-known facts concerning linear eigenvalue problems with geometrical reasoning and could be applied to a more general class of non-linear problems.

Appendix I is devoted to an investigation of the onset of buckling from the point of view of E. Schmidt's bifurcation theory (cf. e. g. [14]); this leads to a justification of some assumptions made in working with the perturbation method.

PART I.

1. Formulation of the problem. We first introduce the v. Kármán equations (cf. 13), reference to which has been made in the introduction. They are the following pair of non-linear differential equations for two functions ϕ and w of the variable x and y:

$$(1.01) \quad \nabla^4\phi = w_{xx}w_{yy} - w^2{}_{xy}, \quad \nabla^4 = \nabla^2\nabla^2, \quad \nabla^2 = \partial^2/\partial x^2 + \partial^2/\partial y^2,$$

$$(1.02) \quad (\gamma h)^2\nabla^4 w + (\phi_{yy}w_{xx} - 2\phi_{xy}w_{xy} + \phi_{xx}w_{yy}) = 0.$$

[5] For a different class of non-linear boundary value problems and their treatment by minimum problems, see Hammerstein [10].

The quantity w is the deflection of the middle surface of the plate and ϕ is the Airy stress function, from which the stresses in the middle surface (the "membrane" stresses) are derived from the relations

$$\sigma_{xx} = \phi_{yy}, \qquad \sigma_{xy} = -\phi_{xy}, \qquad \sigma_{yy} = \phi_{xx}, \tag{1.03}$$

implying that the equations of equilibrium are identically satisfied. Subscripts x, y, and r on all quantities but stresses σ and strains ϵ denote differentiation with respect to these variables. (What we call stresses here are to be understood as stresses per modulus of elasticity E). The stresses in the middle surface are also the averages of the stresses over the thickness of the plate. As a consequence of the hypothesis that the squares of the slopes of the middle surface are of the same order of magnitude as the strains in the middle surface, the following non-linear relations hold between the strains ϵ, stresses σ, and displacements u, v, w of the middle surface

$$\begin{aligned} \epsilon_{xx} &= u_x + \tfrac{1}{2}w_x^2 = -(1+\nu)\sigma_{xx} + \nu(\sigma_{xx}+\sigma_{yy}), \\ \epsilon_{xy} &= \tfrac{1}{2}(u_y + v_x) + \tfrac{1}{2}w_x w_y = -(1+\nu)\sigma_{xy}, \\ \epsilon_{yy} &= v_y + \tfrac{1}{2}w_y^2 = -(1+\nu)\sigma_{yy} + \nu(\sigma_{xx}+\sigma_{yy}). \end{aligned} \tag{1.04}$$

The expressions for the strains here differ by the quadratic terms in w_x, w_y from the expressions of the linear theory. The quantity ν is Poisson's ratio. Equation (1.01), the compatibility equation, is obtained from the second and third members of (1.04) as the result of eliminating u and v. Equation (1.02) is the same as that obtained from the usual linear bending theory of plates; in (1.02) h is the thickness of the plate and $\gamma^2 = 1/12(1-\nu)$.

We shall take as boundary conditions on ϕ and w those corresponding to the following physical situation: Only one external force is applied—a uniform normal pressure at the edge and in the plane of the plate. The plate is simply supported, i. e., the deflection at the edge is zero and no bending moment is applied there. Our methods can, however, be extended with obvious modifications to other cases, e. g., the case of the clamped plate.

The essential restriction of this paper is that we consider only the circular plate with radial symmetry. As domain of the variables x and y we take therefore the circle $r^2 = x^2 + y^2 \leq R^2$ and assume that ϕ and w *depend upon* r *alone*; in this case (1.01) and (1.02) become ordinary differential equations. We introduce the new functions $p = r^{-1}\phi_r$, $q = -Rr^{-1}w_r$, and the linear differential operator

$$G = R^2 r^{-3} \frac{d}{dr} r^3 \frac{d}{dr}. \tag{1.05}$$

The quantity p represents physically the membrane stress on an element perpendicular to the radius. The differential equations (1.01) and (1.02) become

$$\frac{d}{dr}r^2[Gp-\tfrac{1}{2}q^2]=0;\qquad \frac{d}{dr}r^2[\eta^2Gq+pq]=0,$$

where $\eta=\gamma h/R$; integration of these equations yields

$$(1.06)\qquad Gp-\tfrac{1}{2}q^2=c_1/r^2,\qquad\qquad (1.07)\qquad \eta^2Gq+qp=c_2/r^2.$$

Unless the constants c_1 and c_2 are zero, the original functions ϕ and w could not have continuous fourth derivatives, since continuous fourth derivatives of ϕ and w at $r=0$ imply, on account of radial symmetry, continuous second derivatives of p and q with respect to r and hence continuity of the left members of (1.06) and (1.07) at $r=0$. We therefore assume $c_1=c_2=0$ and take as *the fundamental differential equations*[6] for our investigation

$$(1.08)\qquad Gp=\tfrac{1}{2}q^2,\qquad\qquad (1.09)\qquad \eta^2Gq+pq=0.$$

As boundary conditions at the center we take

$$(1.10)\qquad \frac{dp}{dr}=p_r=0,\qquad\qquad (1.11)\qquad \frac{dq}{dr}=q_r=0\quad\text{for}\quad r=0$$

which must be satisfied, again because of symmetry, if ϕ and w are to possess continuous fourth derivatives. The physical situation which we consider leads to the following conditions at the boundary:

$$(1.12)\qquad\qquad p=\bar{p}\quad\text{for}\quad r=R,$$

where $\bar{p}$ is a prescribed positive constant;

$$(1.13)\qquad B_\nu q(R)\equiv Rq_r(R)+(1+\nu)q(R)=0\qquad (r=R),$$

which implies vanishing of the bending moment at the edge.

The equations (1.08) to (1.13) constitute the formulation of the boundary value problem to be discussed in what follows.

Once the functions p and q have been determined, all quantities of interest from the physical point of view are easily obtained. In particular, the most important stresses, the circumferential membrane stress p_c and the radial bending stress p_b at the lower surface of the plate are given by

$$(1.15)\qquad p_c=rp_r+p=B_0p,\quad\text{and}\quad (1.16)\qquad p_b=6\gamma\eta B_\nu q,$$

where the operator B_ν is defined in (1.13).

Our problem would seem to depend upon several parameters; but only

[6] These equations are simpler in form than those used by others, for example Nádai [17, p. 288] and Way [22], who work with quantities other than p and q.

two are really essential, namely, Poisson's ratio ν and what might be called the "thrust ratio"

$$\lambda^2 = \bar{p}/\eta^2. \tag{1.17}$$

This becomes evident upon introduction of the quantities

$$p^* = p/\eta^2, \qquad q^* = q/\eta, \qquad r_* = r/R \tag{1.18}$$

into equations (1.08) to (1.14) as well as (1.05), which then assume the form

(1.05)* $$G = r_*^{-3}\frac{d}{dr_*}\left(r_*^3\frac{d}{dr_*}\right),$$

(1.08)* and (1.09)* $$Gp^* = \tfrac{1}{2}q^{*2}, \qquad Gq^* + p^*q^* = 0,$$

(1.10)* and (1.11)* $$p^*_{r_*} = q^*_{r_*} = 0 \quad \text{for} \quad r_* = 0,$$

(1.12)* and (1.13)* $$\bar{p}^* = \lambda^2, \qquad B_\nu q^* = 0 \quad \text{for} \quad r_* = 1.$$

(Note that the parameters λ and ν occur only in the boundary conditions). It follows that two plates with the same values of λ and ν possess solutions differing only by constant multipliers.

The material constant ν may have any value between zero and one; usually it is about 0.3. In general, variations in ν affect the solutions of problems in elasticity very little. We therefore took a convenient fixed value (0.318) for ν in our numerical calculations. Hence only λ remains as the essential parameter (cf. [8]).

2. Perturbation method. Our boundary value problem has the obvious solution $q \equiv 0$, $p \equiv$ const. This is also the *only* solution for sufficiently small values of the parameter λ—a fact which we prove in sec. 6 (cf. Theorem 6.1). At a certain value λ_0 of λ there will be a bifurcation; for $\lambda > \lambda_0$ two solutions for which $q \not\equiv 0$ will exist; these solutions differ only in the sign of q. The onset of such a "buckling" is usually treated by linearizing the v. Kármán equations (i. e., by setting the right member of (1.01), or, for us, of (1.08) equal to zero); the result is an eigenvalue problem for which the lowest eigenvalue is λ_0. For information respecting what occurs for $\lambda > \lambda_0$, it is necessary to solve the non-linear equations. Our problem can be solved for λ beyond λ_0 by the perturbation method. One finds that the above linearization is nothing but the result of the first step of this method.

The perturbation method consists in developing $\bar{p}$, p, and q with respect to a parameter ϵ. Since solutions (p, q) must appear in pairs (p, q), $(p, -q)$, it is natural to assume developments of the following type:[7]

[7] That there exist solutions given by expansions of this type can be shown by the methods of the bifurcation theory of E. Schmidt (cf. App. I).

$$\bar{p} = \bar{p}^0 + \epsilon^2\bar{p}^{(2)} + \epsilon^4\bar{p}^{(4)} + \cdots$$
$$p(r) = p^0(r) + \epsilon^2 p^{(2)}(r) + \epsilon^4 p^{(4)}(r) + \cdots$$
$$q(r) = \epsilon q^{(1)}(r) + \epsilon^3 q^{(3)}(r) + \epsilon^5 q^{(5)}(r) + \cdots.$$

From $\bar{p}(\epsilon)$ we then obtain ϵ as a function of $\bar{p}$.

We substitute the above expansions into the differential equations (1. 08) and (1. 09) and the corresponding boundary conditions. By collecting terms of the same order in ϵ we obtain a sequence of linear differential equations for $p^0, q^{(1)}, p^{(2)}, q^{(3)}, \cdots$ accompanied by boundary conditions. For p^0 we find

$$L_0: \qquad Gp^0 = 0, \qquad p^0(R) = \bar{p}^0, \qquad p_r{}^0(0) = 0;$$

the only solution of L_0 is $p^0 \equiv$ const., i. e., $p^0 = \bar{p}^0$. The value of $\bar{p}^0$ is, however, not yet determined. The next equations are

$$L_1: \qquad \eta^2 Gq^{(1)} + p^0 q^{(1)} = 0, \qquad B_\nu q^{(1)}(R) = 0, \qquad q_r{}^{(1)}(0) = 0.$$

This homogeneous differential equation with homogeneous boundary conditions will have as sole solution $q^{(1)} \equiv 0$ except when p^0 is an eigenvalue. The system serves as an eigenvalue problem for fixing the value of $p^0 = \bar{p}^0$; it is the same problem as would result from the customary linear treatment of stability, in accordance with the above remark. The lowest eigenvalue

$$p^0 = \lambda_0{}^2\eta^2 \tag{2. 1}$$

characterizes the thrust at which buckling begins and the solution bifurcates.

We are now in a position to introduce the parameter N by

$$N = \bar{p}/p^0 = \lambda^2/\lambda_0{}^2; \tag{2. 2}$$

N will be used frequently as the essential parameter (cf. end of sec. 1) in place of λ.

The solution $q^{(1)}$ of the differential equation in L_1 satisfying $q_r{}^{(1)}(0) = 0$ is easily found in terms of the Bessel function of order one:

$$q^{(1)} = aRr^{-1}J_1(\lambda_0 R^{-1} r), \tag{2. 3}$$

where a is a constant which we may choose arbitrarily. The lowest value of λ_0 for which the boundary condition $B_\nu q^{(1)}(R) = 0$ is satisfied, is

$$\lambda_0 = 2.06, \text{ when } \nu = .318.^8 \tag{2. 4}$$

In this manner the lowest critical value $\lambda_0{}^2 = 4.2436$ of the thrust ratio is determined.

---

[8] This value of ν is a convenient and reasonable one; we have used it throughout our numerical calculations (cf. end of sec. 1).

The equations for $p^{(2)}$ are

$$L_2: \qquad Gp^{(2)} = \tfrac{1}{2}(q^{(1)})^2, \qquad p^{(2)}(R) = \bar{p}^{(2)}, \qquad p_r^{(2)}(0) = 0.$$

Using (1.08) and (1.10) we obtain

$$\frac{dp^{(2)}}{dr} = \tfrac{1}{2}r^{-3}R^{-2}\int_0^r (q^{(1)})^2 r^3 dr.$$[9]

By a second integration one obtains $p^{(2)}$, except that $\bar{p}^{(2)}$ has not yet been determined. It will, however, be fixed in the next step.

The problem for $q^{(3)}$ is:

$$L_3: \qquad Gq^{(3)} + p^0 q^{(3)} = -p^{(2)}q^{(1)}, \qquad B_\nu q^{(3)}(R) = 0, \qquad q_r(0) = 0.$$

The differential equation here is non-homogeneous, but the corresponding homogeneous problem, which is the same as L_1, has a solution not identically zero; thus L_3 presents the exceptional case in which the non-homogeneous problem need not have a solution and, if it has, the solution is not unique. In order that L_3 possess a solution, the right-hand side $-p^{(2)}q^{(1)}$ must be orthogonal to the solution $q^{(1)}$ of the homogeneous problem. That is $p^{(2)}$ must satisfy the orthogonality condition

$$\int_0^R p^{(2)}(q^{(1)})^2 r^3 dr = 0.$$

This relation serves to determine the boundary value $\bar{p}^{(2)}$.

Having satisfied the preceding condition we are sure that L_3 has a solution. Let $\hat{q}^{(3)}$ be one such solution; it follows that all functions of the form $q^{(3)} = \hat{q}^{(3)} + \alpha q^{(1)}$ are also solutions. The constant α is, in principle, undetermined.[10] We may choose α arbitrarily, but we wish to choose it in such a way as to obtain a good approximation to the complete solution.

---

[9] While this integral is expressible in terms of Bessel functions:

$$\frac{dp^{(2)}}{dr} = (4\lambda_0)^{-1}a^2r^{-1}J_1{}^2(\lambda_0 R^{-1}r) - J_0(\lambda_0 R^{-1}r)J_2(\lambda_0 R^{-1}r),$$

$p^{(2)}$ itself, unfortunately, appears not to be expressible in terms of known functions.

[10] This does not mean that the final solution is not uniquely determined, but refers to the fact that the parameter ϵ can be chosen in different ways. Instead of ϵ one might just as well have taken for parameter any *function* of ϵ of the form

$$\hat{\epsilon} = \epsilon + a\epsilon^3 + \cdots.$$

The development

$$q = \hat{\epsilon}q^{(1)} + \hat{\epsilon}^3\hat{q}^{(3)} + \cdots$$

is then equivalent to

$$q = \epsilon q^{(1)} + \epsilon^3(\hat{q}^{(3)} + aq^{(1)}) + \cdots.$$

The Rayleigh-Ritz method is a natural procedure for this purpose. By inserting $q = (\epsilon + \alpha\epsilon^3)q^{(1)} + \epsilon^3\hat{q}^{(3)}$ into the strain energy functional[11] $V(q)$ for fixed p we obtain a function of α and ϵ. Since q can be characterized (cf. sec. 5) as the function minimizing $V(q)$, a suitable relation between α and ϵ can be obtained by minimizing V considered as a function of α. The resulting condition for α, however, is a rather complicated cubic equation with coefficients depending on ϵ; instead of solving this equation, we develop it in powers of ϵ and retain the term of lowest order, which incidentally is linear in α. We choose for α the value obtained from this linear equation.[12]

Having fixed the constant α, and thus $q^{(3)}$, we proceed to determine $p^{(4)}$ and $p^{(5)}$ from the equations:

$$L_4:\quad Gp^{(4)} = q^{(1)}q^{(3)},\quad p^{(4)}(R) = \bar{p}^{(4)},\quad p_r^{(4)}(0) = 0.$$

$$L_5:\quad \eta^2 Gq^{(5)} + p^0 q^{(5)} = -p^{(2)}q^{(3)} - p^{(4)}q^{(1)},\ B_\nu q^{(5)}(R) = 0,\ \ q_r^{(5)}(0) = 0.$$

Here we face the same problem as that encountered above: the boundary value $\bar{p}^{(4)}$ must be determined from the condition that L_5 have a solution, while $q^{(5)}$ is determined only within a multiple of $q^{(1)}$, which multiple may be chosen arbitrarily, or fixed by a Rayleigh-Ritz procedure as above.

The general type of equations to be solved is:

$$L_n:\qquad Gp^{(n)} = f^{(n)}(r),\qquad (n \text{ even}),$$

$$L_{n+1}:\qquad \eta^2 Gq^{(n+1)} + p^0 q^{(n+1)} = f^{(n+1)}(r),$$

where the expressions $f^{(n)}$ and $f^{(n+1)}$ are quadratic in previously determined functions $q^{(i)}$ and $p^{(j)}$ with $i < n$ and $j \leq n$. Equations of the type L_n can always be solved by two quadratures, which is obviously not the case with

---

[11] The functional V is defined and discussed in sec. 5.

[12] The linear equation is

$$\Big[2\int_0^R p_r^{(11)}p_r^{(33)}r^3dr + 4\int_0^R (p_r^{(13)})^2 r^3 dr - R^{-2}p^{(2)}\int_0^R q^{(3)})^2 r^3 dr\Big]\int_0^R (q^{(1)})^2 r^3 dr$$

$$-\Big[6\int_0^R p_r^{(11)}p_r^{(13)}r^3dr - R^{-2}\bar{p}^{(2)}\int_0^R q^{(1)}q^{(3)}r^3dr\Big]\int_0^R q^{(1)}q^{(3)}r^3dr = 0,$$

where

$$p_r^{(mn)} = (R^{-2}/2)\int_0^r q^{(m)}(r_1)q^{(n)}(r_1)r_1^3dr_1;\qquad m, n = 1, 3$$

and

$$q^{(3)} = \hat{q}^{(3)} + \alpha q^{(1)}.$$

We used this equation in our calculations with results which seemed to justify the labor involved in such a procedure.

L_{n+1}. However, L_{n+1} can be transformed by introducing the new dependent variable $u^{(n)} = q^{(n+1)}/q^{(1)}$ and it then takes the form

$$r^{-3}\frac{d}{dr}r^3(q^{(1)})^2\frac{d}{dr}u^{(n)} = q^{(1)}f^{(n+1)}.$$

The first integration yields

$$\frac{du^{(n)}}{dr} = r^{-3}\int_0^r q^{(1)}f^{(n+1)}r^3dr/(q^{(1)})^2.$$

A second integration determines $u^{(n)}$ and consequently $q^{(n+1)}$. The boundary condition $B_\nu q^{(n+1)}(R) = 0$ is transformed into $u_r^{(n)} = 0$ for $r = R$, i. e., $\int_0^R q^{(1)}f^{(n+1)}r^3dr = 0$, and this relation holds automatically since it expresses the condition that L_{n+1} have a solution satisfying the boundary conditions. We carried the solutions so far as to calculate $q^{(5)}$ and $p^{(6)}$.

Since all quadratures, after the first, appear not to be expressible in terms of known functions, we found it necessary to operate with power series. Although the integration of power series presents no difficulty, the obvious necessity for multiplication and division of one power series by another makes the numerical calculations very laborious.[13]

The rapidity of convergence of perturbation series is a matter of general significance, in view of the wide use of such methods in problems similar to ours [cf. 9, 15, 18].[14] In our problem it is possible to check the accuracy of the perturbation method by comparison with the results of a different method (cf. sec. 3). Our calculations show that the perturbation series converge satisfactorily only for a rather small range of values of the ratio $N = \bar{p}/p^0 = \lambda^2/\lambda_0^2$. If errors up to about 4% are permitted, q may be calculated for $N = \bar{p}/p^0 = 1.15, 1.8, 2.5$ by using terms up to and including those of first, third, and fifth order respectively; p may be calculated for $N = 1.4, 2.2, 2.8$ by using terms up to and including those of second, fourth, and sixth order respectively. For the stresses p_b and p_c (cf. (1.16) and (1.15)) the convergence is not quite so good. Beyond these values of N the approximations become inaccurate rather quickly.

---

[13] We retained eight terms of each power series and found this just accurate enough to compute $q^{(5)}$ and $p^{(6)}$, with which our calculations ended.

[14] Marguerre [15], for instance, mentions that his calculations for the rectangular plate, which implied two perturbations, may perhaps be valid up to $N = 20$. In our simpler case we find that values of N of such magnitudes cannot be treated with three perturbations. In fact, such values of N lead to solutions already in the asymptotic range (cf. sec. 4).

3. Power series method. The solutions of equations (1.08) and (1.09) can be obtained as power series in r. All coefficients of the series could be fixed in terms of the first coefficient in each series, the latter being determined by the boundary conditions. This would involve solving very complicated transcendental equations. It is possible, however, to transform the problem in such a way that only one transcendental equation of relatively simple type need be solved.[15]

We introduce the new independent variable

$$\alpha = \mathrm{A}R^{-1}r, \qquad 0 \leq \alpha \leq \mathrm{A};$$

and the new functions

$$\begin{aligned} \pi &= \mathrm{A}^{-2}p^* = \mathrm{A}^{-2}p/\eta^2 \\ \kappa &= \mathrm{A}^{-2}q^* = \mathrm{A}^{-2}q/\eta, \quad \text{(cf. (1.18))}, \end{aligned} \tag{3.1}$$

where A is a parameter at our disposal. The differential equations (1.08) and (1.09) assume the form

$$\begin{aligned} \alpha^{-3}\frac{d}{d\alpha}\alpha^3\frac{d}{d\alpha}\pi &= \tfrac{1}{2}\kappa^2, \\ \alpha^{-3}\frac{d}{d\alpha}\alpha^3\frac{d}{d\alpha}\kappa + \pi\kappa &= 0 \end{aligned} \tag{3.2}$$

with boundary conditions

$$\frac{d\pi}{d\alpha}\bigg|^0 = 0, \qquad \frac{d\kappa}{d\alpha}\bigg|^0 = 0, \tag{3.3}$$

$$B_\nu\kappa = 0 \quad \text{for} \quad \alpha = \mathrm{A}, \tag{3.4}$$

$$\pi(\mathrm{A}) = \bar{\pi}. \tag{3.5}$$

Instead of prescribing $\bar{\pi}$ and A we proceed as follows: $\pi(0)$ and $\kappa(0)$ are chosen, A is determined by solving (3.4), and $\bar{\pi}$ is calculated from (3.5).

The power series for π and κ contain only even powers of α, a conclusion that can be drawn from (3.2) and (3.3); we write, therefore,

$$\pi = \sum_{k=0}^{\infty}\pi_k\alpha^{2k}, \qquad \kappa = \sum_{k=0}^{\infty}\kappa_k\alpha^{2k}.$$

Upon insertion of these power series into (3.2) we obtain the following recursion formulas for π_k and κ_k:

$$2k(2k+2)\pi_k = \tfrac{1}{2}\sum_{m+n=k-1}\kappa_m\kappa_n, \qquad 2k(2k+2)\kappa_k = -\sum_{m+n=k-1}\pi_m\kappa_n.$$

[15] The power series method (with a different arrangement) was used by Hencky [11] and Way [22] for the problem of the bending of circular plates under lateral loads or edge moments.

After having chosen $\pi(0) = \pi_0$ and $\kappa(0) = \kappa_0$ the successive coefficients are calculated by these formulas. Equation (3. 4) becomes

$$\sum_{k=0}^{\infty} (2k + 1 + \nu)\kappa_k\alpha^{2k} = 0, \tag{3.6}$$

and must be solved for its lowest root[16] $\alpha = \mathrm{A}$; p^* and q^* are then calculated from (3. 1).

By this method we obtained solutions for a much higher range of values of N with very much less labor in numerical computation than would be required by the perturbation method. However, it was necessary to calculate a rather large number of coefficients in order to obtain sufficient accuracy for the higher values of N, e. g., for $N = 14.7$, the most extreme case calculated, we found thirty terms in each series barely sufficient.

In applying the power series method for the higher values of N, the following considerations are essential. We wish to obtain a fairly even distribution for the values of N and at the same time we want to know roughly where the solutions $\alpha = \mathrm{A}$ of equations (3. 6) will be. The amount of labor in computation reduces to a minimum if $\mathrm{A} \sim 1$: for then we have a means of knowing the accuracy with which the coefficients need be calculated and also the number of terms to be computed in the series. In order to obtain $\mathrm{A} \sim 1$ and a pre-determined distribution of the values of N, it is necessary to make in advance fairly accurate estimates of the values of $p^* = p/\eta^2$ and $q^* = q/\eta$ at the center ($r = 0$). How to do this is not obvious, for the values of p and q at the center change with increasing N in rather surprising and unforeseen ways, as our computations showed. This is one disadvantage of the power series method which, however, can be overcome by beginning with the perturbation method (which requires no estimates in advance), and pursuing the latter until the trend of the solutions with increasing N becomes apparent. The solutions obtained above by the perturbation method for $N \leqq 2.5$ proved in fact to be amply sufficient for this purpose, although not all of the distinctive qualitative features of the solutions, in their dependence upon N, had yet appeared in this range.

In comparing the perturbation and power series methods, it should be mentioned that the former is applicable in principle to similar problems without radial symmetry; this appears not to be the case with the power series method.

4. Asymptotic solution. Although the power series method can be

---

[16] If we were to take the second root of (3. 6) for A, we would find the prolongation into the non-linear range of the *second* eigenfunction of the linearized problem L_1 (sec. 2).

applied to solve our problem for rather high values of N, it will not serve to determine what occurs when N tends to infinity, or, what is the same thing, when the thrust ratio $\lambda^2 = \bar{p}/\eta^2$ tends to infinity. Such a passage to the limit may be achieved physically in various ways which are mathematically equivalent: for example, one might take a fixed plate (η fixed) and allow $\bar{p}$ to increase indefinitely, or hold $\bar{p}$ fixed and consider plates with slenderness ratios η tending to zero.

In order to determine the asymptotic behavior of the solutions, it is necessary to formulate a limit boundary value problem by a passage to the limit in the original differential equations and boundary conditions. A simple and rather natural procedure would be to hold $\bar{p}$ fixed and let η tend to 0 in equations (1. 08) and (1. 09) which then take the form

$$Gp = \tfrac{1}{2}q^2, \qquad pq = 0. \tag{4.01}$$

The only solution of these equations which satisfies the regularity conditions (1. 10) and (1. 11) is $q \equiv 0$, $p \equiv$ constant. One is then tempted to fix this constant by setting it equal to the prescribed value $\bar{p}$ at the edge. This means that in the limit there would be a hydrostatic compression ($p > 0$) throughout the plate. Such a procedure corresponds to the treatment of laterally loaded clamped sheets by Hencky [11, 12], cf. also [4, 2], and a similar method may well be legitimate in cases where no edge *compression* is prescribed. Applied to our case, however, wrong results would be obtained. The correct limit procedure can be found only by a deeper analysis of the nature of the solutions.

In our case we have found by numerical calculation (cf. [5] and [6]) that with increasing thrust ratio the membrane stress p approaches a state of constant tension ($p < 0$) over an increasingly large part of the interior of the plate and that the transition from tension in the interior to the prescribed compression $\bar{p} > 0$ at the edge takes place in a narrow strip, the breadth of which decreases with increasing λ. (We shall use the parameter λ instead of N from now on).

These results of the numerical calculation indicate strongly the nature of the limit situation. In the interior of the plate, the above solution $q \equiv 0$, $p \equiv$ const. of (4. 01) seems valid, but the constant should not be determined by setting it equal to $\bar{p}$. The constant can be fixed only by an investigation of the transition phenomena in the "boundary layer." The boundary layer phenomena are coupled with the fact that the order of the system of differential equations (1. 08) and (1. 09) has been reduced from four to two on passing to (4. 01). It is not to be expected that the solutions of the limit system (4. 01) can satisfy four boundary conditions. The above discussion indicates that the lost boundary conditions are those at the edge.

A treatment of such an edge effect requires that the scale be stretched with increasing λ in such a manner that the width of the edge strip, or boundary layer (as measured in the new scale) does not shrink to zero. This will be accomplished, as we shall see, by introducing the new independent variable

(4.02) $$\beta = \lambda(1 - r/R), \qquad 0 \leq \beta \leq \lambda,$$

where

(4.03) $$\lambda = \bar{p}^{\frac{1}{2}}/\eta = \lambda_0 \sqrt{N}.$$

Subscripts β in what follows denote differentiation with respect to β. Upon introduction of

(4.04) $$P = p/\bar{p}, \qquad Q = \eta q/\bar{p}$$

as new dependent variables, the original differential equations (1.08) and (1.09) take the forms:

(4.05) $$\begin{aligned} P_{\beta\beta} - [3/(\lambda - \beta)]P_\beta &= Q^2/2 \\ Q_{\beta\beta} - [3/(\lambda - \beta)]Q_\beta + PQ &= 0. \end{aligned}$$

The boundary conditions (1.12) and (1.13) become

(4.06) $$P(0) = 1, \qquad Q_\beta(0) - [(1 + \nu)/\lambda]Q(0) = 0,$$

while the regularity conditions (1.10) and (1.11) take the form

(4.07) $$P_\beta(\lambda) = Q_\beta(\lambda) = 0.$$

We now let λ tend to infinity and obtain the limit differential equations

(4.08) $$P_{\beta\beta} = Q^2/2$$

(4.09) $$Q_{\beta\beta} + PQ = 0;$$

the boundary conditions (4.06) become

(4.10) $Q_\beta(0) = 0$, (4.11) $P(0) = 1$, (4.12) $Q_\beta(\infty) = P(\infty) = 0$.

The equations (4.08) to (4.12) constitute the formulation of the limit boundary value problem concerned with the "boundary layer." (This is not to be confused with the limit problem partially formulated in (4.01), which is concerned with the interior of the plate.) The present problem has the trivial solution $P = 1$, $Q = 0$; if it has another solution P, $Q \not\equiv 0$, then also P, $-Q$ is a solution. Hence the sign of Q is arbitrary.

The equations (4.08) and (4.09) possess the first integral:

(4.13) $$Q_\beta^2 - P_\beta^2 + PQ^2 = \text{const.}$$

It is plausible to take zero as the value of the constant. This, in view of the boundary conditions (4.12), is equivalent to assuming $PQ^2 = 0$ at $\beta = \infty$. (For a justification of this assumption see [34]). We take, therefore,

$$Q_\beta{}^2 - P_\beta{}^2 + PQ^2 = 0. \tag{4.14}$$

In view of (4.10) and (4.11) we find $P_\beta{}^2 = Q^2$ for $\beta = 0$, or $P_\beta = \pm Q$. Since we may choose either sign for Q, we set

$$P_\beta(0) = -Q(0). \tag{4.15}$$

We introduce new variables x, y, z (not to be confused with the space variables used earlier) as follows:

$$x = \xi e^{-\omega\beta}, \qquad 0 \leq x \leq \xi, \tag{4.16}$$

$$y = -\omega^{-2}P, \tag{4.17}$$

$$z = \tfrac{1}{2}2^{\frac{1}{2}}\omega^{-2}Q, \tag{4.18}$$

where ξ and ω are numbers to be determined. The differential equations (4.08) and (4.09) in these variables are:

$$x(xy_x)_x + z^2 = 0, \tag{4.19}$$

$$x(xz_x)_x - yz = 0. \tag{4.20}$$

The introduction of the new variable x has the effect that the resulting differential equations (4.19) and (4.20) possess solutions expressible as power series in x:

$$y = \sum_{k=0}^{\infty} (-1)^k y_k x^{2k} \tag{4.21}$$

$$z = \sum_{m=0}^{\infty} (-1)^m z_m x^{2m+1}. \tag{4.22}$$

Upon substituting (4.21) and (4.22) into (4.19) and (4.20) we find the following formulas for y_k and z_m:

$$(2k)^2 y_k = \sum_{m+n=k-1} z_m z_n, \tag{4.23}$$

$$(2m+1)^2 z_m = \sum_{n+k=m} z_n y_k. \tag{4.24}$$

There is only one arbitrary coefficient, namely z_0, to which we assigned the numerical value $z_0 = 4$. The coefficient y_0 is determined from (4.24) for $m = 0$ and has obviously the value $y_0 = 1$. This fact makes it possible to re-write (4.24) as a proper recursion formula:

$$(4.25) \qquad 2m(2m+2)z_m = \sum_{n=0}^{m-1} z_n y_{m-n}.$$

We found it amply sufficient to compute ten terms in each series.

We turn now to consideration of the boundary condition associated with (4. 19) and (4. 20). The regularity conditions (4. 12) are satisfied automatically in view of (4. 16), (4. 17), (4. 18), and the assumed development into the power series (4. 21) and (4. 22). The boundary conditions (4. 10) and (4. 11) become in the new variables

$$(4.26) \qquad z_x(\xi) = 0, \qquad (4.27) \qquad y(\xi) = -\omega^{-2}.$$

The first is a transcendental equation in ξ, to be solved for its lowest root, which is found to be $\xi = .98618$.[17] This value inserted in (4. 27) determines ω, which is found to be $\omega = .68754$.

Once ξ and ω are determined, the limit boundary value problem is solved in principle. The function $P(\beta)$ starts with the prescribed value $P(0) = 1$, decreases monotonically, assumes the value zero at $\beta = .941$, and approaches the value $P(\infty) = -\omega^2 = -.47271$ as $\beta \to \infty$, the latter value resulting from (4. 17). The function $Q(\beta)$ decreases monotonically and approaches zero as β tends to ∞. For $Q(0)$ and $P(0)$ we find $Q(0) = -P_\beta(0) = 1.61436$, thus checking (4. 15).

We can now discuss the connection[18] between our results from the boundary layer theory and the limit procedure for the interior of the plate described above. The inner edge of the boundary layer is to be identified in the limit with the outer edge $(r \to R)$ of the interior region. Since the value $P(\infty) = -\omega^2$ is the limit value of $p/\bar{p}$ at the inner edge of the boundary layer, $-\omega^2\bar{p}$ is the proper value to be taken for fixing the essential constant for the limit problem in the interior of the plate.[19] Thus we see that the limit membrane stress p in the interior is a tension. The value $Q(\infty) = 0$ is the limit value of $q/\bar{p}$ at the inner edge of the boundary layer and this result is also consistent with the solution $q = 0$ of the equations (4. 01).

This solution of the "asymptotic" boundary value problem furnishes limit values for all quantities. From the physical standpoint it is of especial interest to discuss limit values of those quantities which give information on the ultimate stress distribution. The value $Q(0)$ is the limit value of $\eta\bar{q}/\bar{p}$

---

[17] The reason for assuming $z_0 = 4$ was to make $\xi \sim 1$.

[18] A rigorous proof of the validity of the limit process for the *interior* of the plate is given in sec. 10.

[19] This procedure differs from that in Prandtl's boundary layer theory: there the limit state in the interior furnishes a quantity which must be used to determine the solution of the boundary layer equations.

as one sees from the definition (4.04) of Q. The value $P_\beta(0)$ is the limit value of $\eta \bar{p}_c/\bar{p}^{3/2} = \bar{p}_c/\bar{p}\lambda$, where $\bar{p}_c = R\bar{p}_r + \bar{p}$ (cf. 1.15) is the circumferential membrane stress at the edge of the plate; this follows for $\lambda \to \infty$ from the identity

(4.28) $$P_c/\bar{p}\lambda = -(1-\beta/\lambda)P_\beta + P/\lambda,$$

where P is defined in (4.04). For the radial bending stress p_b (cf. 1.16) we obtain from (4.04) the formula

(4.29) $$p_b/\bar{p}\lambda = -6\gamma[(1-\beta/\lambda)Q_\beta - (1+\nu)Q/\lambda].$$

The stress p_b attains its maximum at the point where β satisfies

$$(1-\beta/\lambda)Q_{\beta\beta} - (2+\nu)Q/\lambda = 0.$$

In the limit ($\lambda \to \infty$) this equation reduces to $Q_{\beta\beta} = 0$; in view of (4.09) it is satisfied at the point $\beta = .941$ where $P(\beta) = 0$. It follows that in the limit the point of maximum bending stress lies in the boundary layer. As a result, we find from (1.28) that the limiting value of $p_{b\,\max}/\bar{p}\lambda$ is $-6\gamma Q_\beta(.941) = 1.1123$.

Asymptotic development. We proceed to explain a method of developing our solutions P and Q in the neighborhood of $\lambda = \infty$. This leads to approximate formulas for rather large values of λ. Although we have not proved the validity of such a development, our numerical results indicate strongly that it is justified.

We confine ourselves to the first step of such a development; it could be carried further. We assume that $P(\beta)$ and $Q(\beta)$ considered as functions of $\kappa = \lambda^{-1}$ possess first derivatives with respect to κ at $\kappa = 0$, namely

(4.30) $$\delta P(\beta) = \frac{\partial}{\partial\kappa} P(\beta)|^{\kappa=0}, \qquad \delta Q(\beta) = \frac{\partial}{\partial\kappa} Q(\beta)|^{\kappa=0}.$$

Differentiation of equations (4.05) and (4.06) with respect to κ for $\kappa = 0$ leads to the linear differential equations

(4.31) $$\delta P_{\beta\beta} - Q\delta Q = 3P_\beta,$$

(4.32) $$\delta Q_{\beta\beta} + P\delta Q + Q\delta P = 3Q_\beta$$

for δP and δQ, and the boundary conditions

(4.33) $$\delta P(0) = 0,$$

(4.34) $$\delta Q_\beta(0) = (1+\nu)Q_0, \quad \text{where} \quad Q_0 = Q(0).$$

Differentiation of (4.07) with respect to κ yields

$$\delta P_\beta(\kappa^{-1}) - \kappa^{-2}P_{\beta\beta}(\kappa^{-1}) = 0 \quad \text{and} \quad \delta Q_\beta(\kappa^{-1}) - \kappa^{-2}Q_{\beta\beta}(\kappa^{-1}) = 0.$$

Our numerical calculations indicate that the second terms in each of the latter equations tend to zero as $\kappa \rightarrow 0$. We assume, then, as boundary conditions

$$(4.35) \qquad \delta P_\beta(\infty) = 0, \qquad\qquad (4.36) \qquad \delta Q_\beta(\infty) = 0:$$

this assumption seems justified by the results. In equations (4.31) to (4.34) the functions $P(\beta)$ and $Q(\beta)$ refer to the solutions of the limit boundary value problem treated earlier in this section.

Before solving the differential equations we note the relation

$$(4.37) \qquad \delta P_\beta + \delta Q = 6/5 \quad \text{for} \quad \beta = 0,$$

which will serve as a useful check on the solution. To prove (4.37) we start with the identity

$$(4.38) \qquad [P_\beta \delta P_\beta - Q_\beta \delta Q_\beta - PQ\delta Q - Q^2 \delta P/2]_\beta = 3(P_\beta^2 - Q_\beta^2)$$

which follows from (4.31), (4.32), (4.08), (4.09). Integration yields

$$(4.39) \qquad P_\beta \delta P_\beta - Q_\beta \delta Q_\beta - PQ\delta Q - Q^2 \delta P/2 |^0 = -3 \int_0^\infty (P_\beta^2 - Q_\beta^2)\, d\beta,$$

where the boundary conditions for $\beta = \infty$ have been used. The right member can be evaluated: we multiply (4.08) by P, (4.09) by $-Q$, add, and integrate from 0 to ∞. The result, after integration by parts, and use of the boundary conditions for $\beta = \infty$, can be written

$$\int_0^\infty (P_\beta^2 - Q_\beta^2 + 3Q^2P/2)\, d\beta = -PP_\beta + QQ_\beta|^0 = Q(0) = Q_0,$$

where (4.10), (4.11), and (4.15) have been used. In view of (4.14) this yields

$$(4.40) \qquad (5/2) \int_0^\infty (P_\beta^2 - Q_\beta^2)\, d\beta = Q_0.$$

Insertion of this in (4.39), use of (4.10), (4.11), (4.15), and division by $-Q_0$ gives

$$(4.41) \qquad \delta P_\beta + \delta Q + \tfrac{1}{2} Q \delta P |^0 = 6/5$$

and (4.37) follows from (4.33).

It is possible to obtain solutions of the homogeneous differential equations $(4.31)_0$ and $(4.32)_0$ derived from (4.31) and (4.32). One considers a one-parameter set of solutions $P(\beta, \alpha)$, $Q(\beta, \alpha)$ of the asymptotic differential equations (4.08) and (4.09); the derivatives of P and Q with respect to α satisfy the homogeneous equations $(4.31)_0$ and $(4.32)_0$, as one easily verifies.

One sees readily that $P(\beta+\alpha)$, $Q(\beta+\alpha)$ is such a set; the derivatives of these with respect to α for $\alpha=0$, i. e.,

$$\delta P = P_\beta, \qquad \delta Q = Q_\beta \tag{4.42}$$

constitute, therefore, solutions of $(4.31)_0$ and $(4.32)_0$. We note the following values[20]

$$\delta P = -Q_0, \quad \delta Q = 0, \quad \delta P_\beta = Q_0{}^2/2, \quad \delta Q_\beta = -Q_0 \text{ for } \beta = 0, \tag{4.43}$$

which follow from (4.10) and (4.15).

A second one-parameter set is formed by $\alpha^2 P(\alpha\beta)$, $\alpha^2 Q(\alpha\beta)$, as can be readily verified; the derivatives of these with respect to α for $\alpha=1$, i. e.,

$$\delta P = \beta P_\beta + 2P, \qquad \delta Q = \beta Q_\beta + 2Q \tag{4.44}$$

form a second solution of $(4.31)_0$ and $(4.32)_0$. We note the following values[20] for this solution:

$$P = 2, \quad Q = 2Q_0, \quad P = -3Q_0, \quad Q = 0 \text{ for } \beta = 0, \tag{4.45}$$

which follow from (4.10) and (4.15).

We proceed now to construct a solution of the non-homogeneous equations (4.31) and (4.32) satisfying the conditions (4.35) and (4.36). Since these conditions are satisfied by the above solutions of the homogeneous equations $(4.31)_0$ and $(4.32)_0$ we may satisfy the remaining boundary conditions by adding an appropriate linear combination of the latter solutions.

We introduce the new variable $x=\xi e^{-\omega\beta}$ as in (4.16) and the new functions δy and δz by

$$\delta P = 3\omega\delta y, \qquad \delta Q = -3\sqrt{2}\omega\delta z. \tag{4.46}$$

With the notation of (4.17) and (4.18) equations (4.31) and (4.32) become

$$x(x\delta y_x)_x + 2z\delta z = xy_x, \tag{4.47}$$

$$x(x\delta z_x)_x - y\delta z - z\delta y = xz_x. \tag{4.48}$$

Assuming for δy and δz the power series

$$\delta y = \sum_{k=0}^{\infty} (-1)^k \delta y_k x^{2k}, \tag{4.49}$$

$$\delta z = \sum_{m=0}^{\infty} (-1)^m \delta z_m x^{2m+1}, \tag{4.50}$$

and the series (4.21) and (4.22) for y and z we obtain by insertion in (4.47) and (4.48) the relation

---

[20] We observe that the left member of (4.41) vanishes, as it should, for these values.

(4. 51) $$(2k)^2 y_k = \sum_{n+m=k-1} z_m \delta z_n + k y_k, \qquad (k = 1, 2, \cdots).$$

(4. 52) $$(2m+1)^2 \delta z_m = \sum_{k+n=m} (y_k \delta z_n + z_n \delta y_k) + (2m+1) z_m, \qquad (m = 0, 1, \cdots).$$

The coefficient of δz_m in the right member of (4. 52) is $y_0 = 1$; hence we may replace (4. 52) by

(4. 53) $$2m(2m+2)\delta z_m = \sum_{n=0}^{m-1} y_{m-n} \delta z_n + \sum_{k+m=n} z_n \delta y_k + (2m+1) z_m, \qquad (m = 0, 1, 2, \cdots).$$

This and (4. 51) constitute proper recursion formulas for the series (4. 49) and (4. 50). The coefficient δy_0 is fixed by (4. 53) for $m = 0$; its value is -1. The coefficient δz_0 can be chosen arbitrarily; it is convenient to take $\delta z_0 = 0$. The values y_k and z_m have been previously calculated from (4. 23) and (4. 24); the coefficients δy_k and δz_m can then be determined from (4. 51) and (4. 53). The convergence of the resulting series for δy and δz appears to be very satisfactory.

In this way we find a particular solution δP, δQ of the non-homogeneous equations (4. 31), (4. 32); by adding a proper linear combination of the two previously found solutions of $(4.31)_0$ and $(4.32)_0$ we obtain a solution of (4. 31) and (4. 32) satisfying the conditions (4. 33), (4. 34). We give a few numerical results:[21]

$$\delta P = 0, \quad \delta Q = -.74, \quad \delta P_\beta = 1.94, \quad \delta Q_\beta = 2.13 \text{ for } \beta = 0,$$
$$\delta P = -1.03, \quad \delta Q = 0, \quad \delta P_\beta = 0, \quad \delta Q_\beta = 0 \text{ for } \beta = \infty.$$

Hence we have the following approximation formulas for large λ (that is, for large N): for $\beta = 0$

$$P = 1, \quad Q = 1.61 - .74/\lambda, \quad P_\beta = -1.61 + 1.94/\lambda, \quad Q = 2.13/\lambda;$$

for $\beta = \infty$ we have

$$P = -.47 - 1.03/\lambda, \quad Q = 0, \quad P_\beta = 0, \quad Q_\beta = 0.$$

It is clear that the preceding calculations yield also approximate formulas for all physical quantities such as stresses and deflections.

PART II.

In Part II of this paper we consider the purely mathematical aspects of the boundary value problems which were discussed in Part I from the point of view of explicit numerical solution.

[21] Note that our results check with formula (4. 37).

5. Minimum and boundary value problems for the finite case. Our subsequent discussion will be based upon the possibility of formulating minimum problems equivalent to our boundary value problems.[22] In this section we consider the problems for finite λ. The functional to be minimized, essentially the total potential energy,[23] is

$$V = \eta^2 \int_0^R q_r^2(r) r^3 dr + (1+\nu) R^2 q(R) - \bar{p} R^{-2} \int_0^R q^2(r) r^3 dr + \int_0^R p_r^2(r) r^3 dr. \tag{5.01}$$

We do not vary p and q independently. Rather, q must be varied while p is considered a functional in q through the differential equation (1.08) and the boundary conditions (1.10) and (1.12).

It is convenient to work with the invariant quantities p^*, q^*, and r_* defined at the end of sec. 1, but we omit the * from now on. We introduce the functionals

$$qHq = \int_0^1 q^2(r) r^3 dr, \tag{5.02}$$

$$qDq = \int_0^1 q_r^2(r) r^3 dr + (1+\nu) q^2(1), \tag{5.03}$$

$$p_r K p_r = \int_0^1 p_r^2(r) r^3 dr, \tag{5.04}$$

and

$$V_\lambda[q] = qDq - \lambda^2 qHq + p_r K p_r. \tag{5.05}$$

The new functional V_λ to be minimized is related to (5.01) through $V = \eta^4 R^2 V_\lambda$.

We define $p(r)$ and $p_r(r)$ as functionals in $q(r)$ by means of the formulas

$$p_r(r) = \tfrac{1}{2} r^{-3} \int_0^r q^2(r_1) r_1^3 dr_1. \tag{5.06}$$

$$p(r) = \lambda^2 - \int_r^1 p_r(r_1) dr_1. \tag{5.07}$$

We are now in a position to formulate *admissibility conditions.* By admissible functions we mean functions $q(r)$ continuous in $0 < r \leq 1$ which possess L^2-integrable[24] derivatives in every interval $0 < \epsilon \leq r \leq 1$ and for

---

[22] Incidental use of this fact was made in sec. 2.

[23] The total potential energy is given by

$$\pi h E [V - (1-\nu) R^2 \bar{p}^2].$$

[24] For a method avoiding Lebesgue derivatives which could be applied here, see [8].

which the integrals H, D, and K are finite, p_r being considered a functional in $q(r)$ through (5.06).

The *minimum problem* M_λ is that of minimizing $V_\lambda[q]$ with respect to admissible functions $q(r)$.

The first variation δV_λ of V_λ is

$$\delta V_\lambda[q] = qD\delta q - \lambda^2 qH\delta q + p_r K\delta p_r, \tag{5.08}$$

where the bilinear forms corresponding to (5.02), (5.03), and (5.04) have been used; δq is any admissible function and

$$\delta p_r = r^{-3} \int_0^r q\delta q r_1{}^3 dr_1. \tag{5.09}$$

V_λ is said to be stationary for a function q if $\delta V_\lambda[q] = 0$, for all admissible δq. We refer to the problem of making V_λ stationary in this sense as the problem S_λ. A solution of M_λ is also a solution of S_λ; this conclusion results from the standard reasoning.

We formulate the *boundary value problem* B_λ as follows; an admissible function q is to be found which possesses a continuous second derivative and satisfies the differential equation

$$Gq + pq = 0, \qquad G = r^{-3}\frac{d}{dr}r^3\frac{d}{dr}, \tag{5.10}$$

and the boundary condition

$$B_\nu q = rq_r + (1+\nu)q = 0 \quad \text{for} \quad r = 1. \tag{5.11}$$

The function p in (5.10) is defined by (5.07) and (5.06). These conditions imply that p satisfies the differential equation

$$Gp = q^2/2, \tag{5.12}$$

and the boundary condition

$$p = \lambda^2 \quad \text{for} \quad r = 1. \tag{5.13}$$

The conditions $p_r = q_r = 0$ at $r = 0$ which were used in calculating the solutions in sec. 1 are not required; we have sufficiently characterized the behavior of p and q at $r = 0$ through the admissibility conditions [25] that the integrals H, D, and K be finite.

The problems S_λ and B_λ are related as follows: 1) A solution of B_λ solves S_λ. 2) If an admissible function q solves S_λ it possesses a continuous second derivative and solves B_λ. These statements are proved in sec. 7 (cf.

[25] It could be shown that a solution of the boundary value problem which satisfies the admissibility conditions would also satisfy the regularity conditions at $r = 0$.

Theorems 7.1 and 7.2) in terms of convenient new variables, which at the same time permit proofs of the corresponding statements for the asymptotic case. We may mention here that the connection between the two problems is based on the Green's formula

$$\delta V_\lambda(q) = -\int_0^1 (Gq + pq)\delta q r^3 dr + B_\nu q \delta q|^1, \tag{5.14}$$

which, in terms of new variabes, is justified in sec. 7 (cf. formula 7.21).

6. Multiplicity of solutions of the boundary value problem. In this section we are interested in the *number* of possible *distinct* solutions q of the boundary value problem B_λ in their dependence upon the parameter λ. We proceed first to give a general description of what can be expected to occur with increasing λ. For sufficiently small λ there will be only one solution, i. e., $q \equiv 0$. This holds up to $\lambda = \lambda_0$, where λ_0 is the lowest eigenvalue of the linearized buckling problem. Beyond λ_0 two new solutions will appear which differ only in sign. We shall prove in this section that these solutions together with $q \equiv 0$ are the only solutions as long as $\lambda < \lambda_1$, where λ_1 is the second eigenvalue of the linear buckling problem. It could be proved, for example by the perturbation method, that two additional solutions will appear when λ passes the next critical value λ_1. It seems likely that there would be exactly $2n+1$ solutions when $\lambda_{n-1} < \lambda < \lambda_n$, one the solution $q \equiv 0$, and n other pairs differing only in sign. In this section we prove less, namely:

THEOREM 6.1. *For $0 \leq \lambda \leq \lambda_0$ the sole solution of B_λ is $q = 0$.*

THEOREM 6.2.[26] *For $\lambda_0 < \lambda < \lambda_1$ there are at most three solutions of B_λ; one of these is $q \equiv 0$ and the other two differ only in sign.*

We shall make use of the following properties[27] of the eigenvalues λ_0^2 and λ_1^2. The inequality

$$qD_\lambda q \equiv qDq - \lambda^2 qHq > 0 \tag{6.1}$$

holds for admissible functions $q \not\equiv 0$ (cf. sec. 5) if $\lambda < \lambda_0$, equality holding for the eigenfunction $q \equiv q_0$; the inequality (6.1) holds also for $\lambda_0 < \lambda < \lambda_1$ if the additional restriction

[26] This theorem could be easily generalized. The essential properties used are readily seen to be: $D + H$ is a positive definite quadratic form, H is a completely continuous quadratic form with respect to $D + H$ as unit form, and K is a positive definite, centro-symmetric, and convex functional of degree greater than one. The reasoning given here could, with slight adaptations, be taken over for such a generalization.

[27] Cf. Courant-Hilbert [3] Volume I, Chapter VI.

$$(6.2) \qquad qHq_0 = 0, \qquad q \not\equiv 0,$$

is imposed, where q_0 is the eigenfunction belonging to λ_0.

In sec. 5 it was stated that solutions q of the boundary value problem B_λ also solve S_λ, i. e., they are completely characterized by the statement that $\delta V_\lambda[q] = 0$ for all admissible δq.

If q renders $V_\lambda[q]$ stationary, the identity

$$(6.3) \qquad qD_\lambda q + 2p_rKp_r = 0$$

holds. It results when δq is replaced by q in (5.08) and (5.09).

For $\lambda \leq \lambda_0$ we may apply (6.1), and (6.3) leads to $p_rKp_r = 0$. In view of (5.04) and (5.06) this is possible only for $q \equiv 0$. Thus Theorem 6.1 is proved.

In the proof of Theorem 6.2 we make use of the fact that the positive definite form p_rKp_r is a *convex* functional (of fourth degree) in q. By convexity we mean that the second variation $\delta p_rK\delta p_r + \delta^2 p_rKp_r$ of p_rKp_r is positive. This is the case since the second term is positive because $p_r > 0$ and $\delta^2 p_r = r^{-3}\int_0^r (\delta q)^2 r_1^3 dr_1 > 0$, while the first term is obviously positive.

The proof of Theorem 6.2 will be given indirectly. Assume, then, that there are *two* linearly independent solutions $q^{(1)} \not\equiv 0$, $q^{(2)} \not\equiv 0$. We introduce the linear combinations $q = \alpha_1 q^{(1)} + \alpha_2 q^{(2)}$ and consider the homogeneous forms

$$Q(\alpha_1, \alpha_2) = qD_\lambda q, \qquad P(\alpha_1, \alpha_2) = p_rKp_r.$$

Since $q^{(1)}$ and $q^{(2)}$ make V_λ stationary, the sum $V_\lambda = Q(\alpha_1, \alpha_2) + P(\alpha_1, \alpha_2)$ is stationary for $(1, 0)$, $(-1, 0)$ and for $(0, 1)$, $(0, -1)$. Our Theorem 6.2 is proved if we can show that all stationary points of $V_\lambda = P + Q$ lie on the same straight line through the origin in the (α_1, α_2)-plane, in contradiction with the assumption that $(1, 0)$ and $(0, 1)$ are stationary points.

The quadratic form $Q(\alpha_1, \alpha_2)$ is indefinite since 1. $Q(1, 0) = q^{(1)}D_\lambda q^{(1)} < 0$ as we see from (6.3); 2. the linear set q contains a function $\hat{q} = \hat{\alpha}_1 q^{(1)} + \hat{\alpha}_2 q^{(2)} \not\equiv 0$ satisfying $\hat{q}Hq_0 = 0$ and since $\lambda < \lambda_1$ we know that (6.1) holds for $\hat{q}$; thus $Q(\hat{\alpha}_1, \hat{\alpha}_2) = \hat{q}D_\lambda\hat{q} > 0$. The form $P(\alpha_1, \alpha_2)$ is of the fourth degree, and is positive definite and convex since, as we have seen, this is the case for p_rKp_r. Both forms are clearly centro-symmetric and stationary points occur in symmetrically located pairs.

Let $\bar{\alpha}_1, \bar{\alpha}_2$ be any values of α_1 and α_2 $(\alpha_1^2 + \alpha_2^2 \not\equiv 0)$ for which $V_\lambda = Q(\alpha_1, \alpha_2) + P(\alpha_1, \alpha_2)$ is stationary, i. e., for which $\delta V_\lambda = \delta Q + \delta P = 0$. As a consequence we have

$$(6.4) \qquad -Q_{\alpha_1}d\alpha_1 - Q_{\alpha_2}d\alpha_2 = P_{\alpha_1}d\alpha_1 + P_{\alpha_2}d\alpha_2$$

for $\alpha_1 = \bar{\alpha}_1$, $\alpha_2 = \bar{\alpha}_2$. Let $\bar{P}$ and $\bar{Q}$ be the values of P and Q for $\alpha_1 = \bar{\alpha}_1$, $\alpha_2 = \bar{\alpha}_2$. In view of (6.4) the curves $P(\alpha_1, \alpha_2) = \bar{P}$ and $Q(\alpha_1, \alpha_2) = \bar{Q}$ have a common tangent at $(\bar{\alpha}_1, \bar{\alpha}_2)$.

Consider a second point where $V_\lambda = P + Q$ is stationary and where, consequently, two curves $P =$ const., $Q =$ const. have a common tangent. By a similarity transformation in the (α_1, α_2)-plane we can transform $P =$ const. into the original curve $P = \bar{P}$ while the equation for the new curve obtained from $Q =$ const. by the transformation may be written in the form $Q = \epsilon\bar{Q}$ with $\epsilon > 0$. In order to show that such a second stationary point is on the same ray as $(\bar{\alpha}_1, \bar{\alpha}_2)$ it is obviously sufficient to show that it concides with $(\bar{\alpha}_1, \bar{\alpha}_2)$ after the similarity transformation. This we accomplish by proving that the set of curves $Q = \epsilon\bar{Q}$ contains only one curve, namely $Q = \bar{Q}$, tangent to $P = \bar{P}$ and that $Q = \bar{Q}$ has only two symmetrically located points of contact with $P = \bar{P}$.

The set of points $P \leq \bar{P}$ is convex since its boundary is a level line of a convex surface. The curves $Q = \epsilon\bar{Q}$ constitute a set of hyperbolas since $Q(\alpha_1, \alpha_2)$ is an indefinite quadratic form; we need evidently consider only one branch of these hyperbolas. Each such branch is the boundary of an unbounded convex set S_ϵ. In addition it is to be noted that the origin is not contained in S_ϵ, but is an inner point of the set $P \leq \bar{P}$.

When the curves $P = \bar{P}$ and $Q = \epsilon\bar{Q}$ have a common tangent, the two convex point sets $P \leq \bar{P}$ and S_ϵ lie entirely on the same side of the tangent line, or they lie entirely on opposite sides. The first case is excluded since the tangent to the hyperbola $Q = \epsilon\bar{Q}$ separates the set S_ϵ from the origin and this point lies in the interior of the set $P \leq \bar{P}$. In the second case the two convex sets have only one point in common since S_ϵ is bounded by a hyperbola and $P \leq \bar{P}$ is convex.

We have assumed that a common tangent line exists for $\epsilon = 1$; such a tangent line separates the sets $P \leq \bar{P}$ and S_1. For every $\epsilon > 1$ it is obvious that the sets $P \leq \bar{P}$ and S_ϵ have no common points. For every $\epsilon < 1$ it is obvious that the two sets have inner points in common; hence a possible new tangency would fall under the first case which has already been excluded. With this we have established the contradiction which proves our theorem.

It appears that the type of reasoning used above would not be sufficient to give a corresponding theorem for $\lambda > \lambda_1$.

7. Simultaneous formulation of finite and asymptotic problems. For the investigations of sections 8 and 9 it is convenient and useful to formulate our problems in terms of certain new variables. At the same time the new formulations make possible simultaneous treatment of the finite and the asymptotic problems.

The new variables are (cf. sec. 5)

(7.01) $P = p/\bar{p}$ and (7.02) $Q = q/\bar{p}$;

the new independent variable is

(7.03) $\beta = \lambda(r^{-2} - 1)/2$ with the domain $0 \leq \beta < \infty$.

With the notation

(7.04) $\kappa = \lambda^{-1}$ we have[28] (7.05) $r = (1 + 2\kappa\beta)^{-1/2}$.

The parameter κ occurs essentially only in the combination

(7.06) $$\rho^{\kappa}(\beta) = (1 + 2\kappa\beta)^{-6}.$$

In the following formulations we admit $\kappa \geq 0$; the cases $\kappa > 0$ are equivalent to the cases for finite λ (cf. sec. 5), while the case $\kappa = 0$ corresponds to the transition $\lambda \to \infty$ and will be seen to represent the asymptotic case.

We introduce the forms

(7.07) $$H^{\kappa}[Q] = \int_0^{\infty} Q^2(\beta)\rho^{\kappa}(\beta)\,d\beta,$$

(7.08) $$D^{\kappa}[Q] = \int_0^{\infty} Q_{\beta}^2(\beta)\,d\beta + \kappa(1 + \nu)Q^2(0),$$

and

(7.09) $$K^{\kappa}[Q] = \int_0^{\infty} P_{\beta}^2(\beta)\,d\beta,$$

where P_{β} is a functional in Q through

(7.10) $$P_{\beta} = P_{\beta}^{\kappa}[Q] = -(1/2)\int_{\beta}^{\infty} Q^2(\beta_1)\rho^{\kappa}(\beta_1)\,d\beta_1.$$

The functional[29] to be minimized is

(7.11) $$W^{\kappa}[Q] = D^{\kappa}[Q] - H^{\kappa}[Q] + K^{\kappa}[Q].$$

By *admissible functions* we mean functions $Q(\beta)$ continuous in $0 \leq \beta < \infty$ with L^2-integrable derivatives in every interval $0 \leq \beta \leq b < \infty$ and for which the integrals in (7.07), (7.08), and (7.09) are finite, P_{β} being defined by (7.10).

The *minimum problem* M^{κ} is that of minimizing $W^{\kappa}[Q]$, the problem S^{κ} is that of making $W^{\kappa}[Q]$ stationary, in each case with respect to admissible functions Q.

---

[28] The variable β here is not the same as that of sec. 4 but it yields the same transition to the asymptotic case. The transformation $r = (1 - \kappa\beta)$ of sec. 4 could be obtained as the term of first order in the development of (7.05) with respect to κ.

[29] The relation between V_{λ} (see (5.5)) and W^{κ} is $W^{\kappa} = \kappa^5 V_{\lambda}$.

The boundary value problem B^κ requires the determination of an admissible function $Q(\beta)$ possessing a continuous second derivative and satisfying the differential equation

$$Q_{\beta\beta} + \rho^\kappa PQ = 0 \tag{7.12}$$

and the boundary condition

$$Q_\beta(0) - \kappa(1+\nu)Q(0) = 0. \tag{7.13}$$

The function $P(\beta)$ in (7.12) is defined by

$$P(\beta) = 1 + \int_0^\infty P_\beta(\beta_1)\,d\beta_1, \tag{7.14}$$

where P_β is given by (7.10). The function P therefore satisfies the differential equation

$$P_{\beta\beta} = \rho^\kappa Q^2/2 \tag{7.15}$$

and the boundary condition

$$P(0) = 1. \tag{7.16}$$

The problems M^κ, S^κ, and B^κ ($\kappa > 0$) are the equivalents, in the new variables, of the problems M_λ, S_λ, and B_λ of sec. 5.

To obtain the formulation of the asymptotic problems M^0, S^0, and B^0 we need only set $\kappa = 0$ in the preceding. Since $\kappa = 0$ implies $\rho^0 = 1$ (cf. (7.06)), this formulation yields the same differential equations and boundary conditions as in sec. 4 except that the conditions used in sec. 4 at $\beta = \infty$ are here replaced by the admissibility conditions.

The connection between the problems S^κ and B^κ (including $\kappa = 0$) is based on two " Green's " formulas. They refer to the first variation

$$\delta W^\kappa[Q] = \int_0^\infty [Q_\beta \delta Q_\beta - \rho^\kappa Q\delta Q + P_\beta \delta P_\beta]\,d\beta + \kappa(1+\nu)Q(0)\delta Q(0), \tag{7.17}$$

where δP_β is defined, in accordance with (7.10), by

$$\delta P_\beta = -\int_\beta^\infty Q(\beta_1)\delta Q(\beta_1)\rho^\kappa(\beta_1)\,d\beta_1. \tag{7.18}$$

It is convenient to state first the following lemma, the proof of which will be given later:

LEMMA 7.1. *If $f(\beta)$ and $g(\beta)$ are continuous functions with L^2-integrable derivatives for which $\int_0^\infty f^2 d\beta < \infty$, $\int_0^\infty g^2 d\beta < \infty$, then there is a special sequence $\beta \to \infty$ on which*

$$f(\beta)g_\beta(\beta) \to 0. \tag{7.19}$$

COROLLARY. *If f_1, g_1 and f_2, g_2 are pairs of functions with the properties of f and g in Lemma 7.1 there exists a special sequence $\beta \to \infty$ on which (7.19) holds for both pairs.*

We consider (for finite $b > 0$) the relation

$$\int_0^b P_\beta \delta P_\beta d\beta = -\int_0^b (P-1)\delta P_{\beta\beta} d\beta + [(P-1)\delta P_\beta]_0^b,$$

obtained by product integration. Lemma 7.1 with $f = (P-1)$ and $g_\beta = \delta P_\beta$ [30] shows that $[(P-1)\delta P_\beta]^b \to 0$ if $b \to \infty$ on a special sequence. Hence we have

$$\int_0^\infty P_\beta \delta P_\beta d\beta = -\int_0^\infty (P-1)\delta P_{\beta\beta} d\beta + [(P-1)\delta P_\beta]_0.$$

If this and the relations $P(0) = 1$, $\delta P_{\beta\beta} = Q\delta Q$ (cf. 7.18) are used one obtains from (7.17) the first Green's formula:

$$(7.20) \qquad \delta W^\kappa[Q] = \int_0^\infty [Q_\beta \delta Q_\beta - \rho^\kappa P Q \delta Q] d\beta + \kappa(1+\nu) Q(0) \delta Q(0).$$

It holds for all admissible Q and δQ and $\int_0^\infty$ means $\lim \int_0^b$ when $b \to \infty$ on a special sequence.

If Q possesses a second derivative we may write

$$\int_0^b Q_\beta \delta Q_\beta d\beta = -\int_0^b Q_{\beta\beta} \delta Q_\beta d\beta + [Q_\beta \delta Q]_0^b.$$

The corollary to Lemma 7.1 applied to $[(P-1)\delta P_\beta]^b$ and $[Q_\beta \delta Q]^b$ yields the existence of a common special sequence $b \to \infty$ on which both expressions tend to zero. If this sequence is used and if $\int_0^\infty Q_\beta \delta Q_\beta d\beta$ is replaced by $-\int_0^\infty Q_{\beta\beta} \delta Q_\beta d\beta + [Q_\beta \delta Q]_0$ in (7.20) we obtain the second Green's formula

$$(7.21) \qquad \delta W^\kappa[Q] = -\int_0^\infty [Q_{\beta\beta} + \rho^\kappa P Q] \delta Q d\beta - [Q_\beta(0) + \kappa(1+\nu) Q(0)] \delta Q(0),$$

which holds for admissible functions Q possessing second derivatives, the integral again being "conditional." Formula (7.21) yields immediately

---

[30] $\int_0^\infty \delta P_\beta{}^2 d\beta < \infty$ follows from the admissibility conditions on Q and δQ and the Schwarz inequality applied to (7.18).

THEOREM 7.1.[31] *A solution of B^κ solves S^κ.*

The converse also holds:

THEOREM 7.2.[31] *A solution of S^κ (hence also a solution of M^κ) possesses a continuous second derivative and solves B^κ.*

To prove Theorem 7.2 we make use of the following well known

LEMMA 7.2.[32] *Let $R(\beta)$ be an L^2-integrable function and $S(\beta)$ be a continuous function such that $\int_0^\infty RT_\beta d\beta = \int_0^\infty ST d\beta$ holds for all continuous functions $T(\beta)$ with L^2-integrable derivatives T_β which vanish identically in the neighborhood of $\beta = 0$ and of $\beta = \infty$. Then $R(\beta)$ coincides (almost everywhere) with a function $\bar{R}$ which possesses the continuous derivative $-S$. We note in addition that $\bar{R} = R$ in case R is the derivative of a continuous function.*

We apply this lemma to $R = Q_\beta$, $S = \rho^\kappa PQ$, where Q is a solution of S^κ. Since $\delta W^\kappa[Q] = 0$ for the admissible variations $\delta Q = T$ it follows from (7.20) that Q possesses the continuous second derivative $-\rho^\kappa PQ$; thus Q satisfies the differential equation (7.12). We may now apply (7.21); it yields $0 = [Q_\beta(0) + \kappa(1+\nu)Q(0)]\delta Q(0)$ and since $\delta Q(0)$ is arbitrary the boundary condition (7.13) is satisfied. Hence Theorem 7.2 is proved.

We have to supply the proof of Lemma 7.1 and its corollary. This will be done with the aid of two additional lemmas of which we shall make frequent use in later sections.

LEMMA 7.3. *"Jacobi's identity." If $f(\beta)$ is a function with L^2-integrable derivative and the positive function $\omega(\beta)$ possesses continuous second derivatives, then*

$$\int_a^b [f_\beta^2 + \omega^{-1}\omega_{\beta\beta} f^2]\, d\beta = \int_a^b \omega^2[(\omega^{-1}f)_\beta]^2 d\beta + \omega^{-1}\omega_\beta f^2\Big|_a^b. \tag{7.22}$$

The identity (7.22) follows from

$$\begin{aligned}\int_a^b \omega^2[(\omega^{-1}f)_\beta]^2 d\beta &= \int_a^b (f_\beta^2 - 2\omega^{-1}\omega_\beta f f_\beta + \omega^{-2}\omega_\beta^2 f^2)\, d\beta \\ &= \int_a^b [f_\beta^2 - (\omega^{-1}\omega_\beta f^2)_\beta + \omega^{-1}\omega_{\beta\beta} f^2]\, d\beta \\ &= \int_a^b (f_\beta^2 + \omega^{-1}\omega_{\beta\beta} f^2)\, d\beta - \omega^{-1}\omega_\beta f^2\Big|_a^b.\end{aligned}$$

[31] These theorems are equivalent (for $\kappa > 0$) to the statements made at the end of sec. 5.

[32] The lemma can be proved in much the same way as the lemma of du Bois-Reymond to which it reduces if $R(\beta)$ is assumed continuous.

LEMMA 7.4. *If $f(\beta)$ is a continuous function with L^2-integrable derivative for which $\int_0^\infty f^2(\beta)\,d\beta < \infty$, then*

$$\beta^{-1/2}f(\beta) \to 0 \quad \text{as } \beta \to \infty. \tag{7.23}$$

To prove the lemma we start with Jacobi's identity for $\omega = \beta^{1/2}$:

$$\int_a^b [f_\beta^2 - \tfrac{1}{4}\beta^{-2}f^2]\,d\beta = \int_a^b \beta(\beta^{-1/2}f)_\beta^2\,d\beta + \tfrac{1}{2}[\beta^{-1}f^2]_a^b,$$

or

$$\tfrac{1}{4}\int_a^b \beta^{-2}f^2\,d\beta + \int_a^b \beta(\beta^{-1/2}f)_\beta^2\,d\beta + \tfrac{1}{2}b^{-1}f^2(b) = \int_a^b f_\beta^2\,d\beta + \tfrac{1}{2}a^{-1}f^2(a).$$

For $b \to \infty$ the right member is bounded; hence the terms of the left member are bounded since all are positive. From

$$\int_a^\infty \beta^{-1}f^2\frac{d\beta}{\beta} = \int_a^\infty \beta^{-2}f^2\,d\beta < \infty$$

we see that $\beta^{-1}f^2(\beta) \to 0$ at least on a special sequence $\beta \to 0$. If b is taken on that special sequence we obtain

$$\tfrac{1}{4}\int_a^\infty \beta^{-2}f^2\,d\beta + \int_a^\infty (\beta^{-1/2}f)^2\,d\beta = \int_a^\infty f_\beta^2\,d\beta + \tfrac{1}{2}a^{-1}f^2(a),$$

which, for $a \to \infty$, yields $a^{-1}f^2(a) \to 0$ and the Lemma 7.4 is proved.

We turn now to the proof of Lemma 7.1. From $\int_0^\infty \beta g_\beta^2 \beta^{-1}\,d\beta < \infty$ we infer that a special sequence exists on which $\beta^{1/2}g_\beta \to 0$; in view of Lemma 7.4 we have $\beta^{-1/2}f(\beta) \to 0$ and (7.19) follows immediately.

For the proof of the corollary it is sufficient to observe that from $\int_0^\infty (\beta[g_{1\beta}]^2 + \beta[g_{2\beta}]^2)(\beta^{-1}d\beta) < \infty$ a special sequence $\beta \to \infty$ exists on which $\beta[g_{1\beta}]^2 + \beta[g_{2\beta}]^2 \to 0$ and that therefore also $\beta^{1/2}g_{1\beta} \to 0$ and $\beta^{1/2}g_{2\beta} \to 0$.

8. Uniqueness theorems. In sec. 6 we proved that the boundary value problem B_λ has, apart from sign, only one solution $q \not\equiv 0$ for a certain range of values of λ $(\lambda < \lambda_1)$. It was also indicated there that the problem B_λ will possess more and more distinct solutions as λ increases beyond λ_1. The situation is quite different as regards the minimum problem M_λ. We shall prove in this section (with the notation of sec. 7):

THEOREM 8.1. *There is at most one solution Q of the minimum problem M^κ, apart from the sign of Q.*

We first dispose of the case in which $Q \equiv 0$ is a solution of M^κ. In this case the functional W^κ is non-negative; otherwise there would be a function Q with $W^\kappa[Q] < 0$ and $W^\kappa[Q] = 0$ would not be the minimum. If Q is *any* solution of M^κ with

$$W^\kappa[Q] = D^\kappa[Q] - H^\kappa[Q] + K^\kappa[Q] = 0,$$

then, for constant ϵ

$$W^\kappa[\epsilon Q] = \epsilon^2\{D[Q] - H[Q]\} + \epsilon^4 K[Q] = (\epsilon^4 - \epsilon^2)K[Q]$$

would be negative for $\epsilon < 1$ unless $K[Q] = 0$, which implies $P_\beta = 0$, and $Q^2 = 2P_{\beta\beta} = 0$. Hence $Q \equiv 0$ is the only solution of M^κ if W^κ is non-negative. From now on we may thus leave aside the case in which $Q \equiv 0$ solves M^κ.

We have noted in the preceding section that a solution of the minimum problem M^κ also solves S^κ and the boundary value problem B^κ (cf. Theorem 7.2). Hence Theorem 8.1 (for $Q \not\equiv 0$) results from the following two theorems:

THEOREM 8.2. *A solution $Q(\beta) \not\equiv 0$ of M^κ is nowhere zero in the interval $0 \leq \beta < \infty$.*

THEOREM 8.3. *A solution $Q(\beta)$ of B^κ which is nowhere zero in the interval $0 \leq \beta < \infty$ is, apart from the sign of Q, the sole solution of M^κ.*

An immediate consequence of Theorem 8.3 is the following

COROLLARY.[33] *The problem B^κ has only one solution, apart from the sign of Q, which is nowhere zero in the interval $0 \leq \beta < \infty$.*

We prove Theorem 8.2 indirectly. Assuming then that the solution $Q(\beta)$ of M^κ vanishes for some value $\beta = \beta_0$. We distinguish two cases:

1. $P(\beta) \leq 0$ for $\beta \geq \beta_0$, and 2. $P(\beta) > 0$ for $\beta \leq \beta_0$.

That Case 1 or Case 2 occurs follows from the fact that $P(\beta)$ decreases monotonically from the value $P(0) = 1$.

Case 1: We make use of the fact that $Q(\beta)$ solves B^κ and hence satisfies the differential equation $Q_{\beta\beta} + \rho^\kappa PQ = 0$. If a solution $Q(\beta)$ of such a differential equation vanishes at a point $\beta = \beta_0$, then, as is well known,

[33] This corollary shows that the solutions for the finite and the asymptotic cases, as treated in sec. 3 and sec. 4, are the sole solutions of B^κ provided that the power series used there converge and the transcendental equations are solved. The admissibility conditions are satisfied if the series converge and the non-vanishing of the solutions is guaranteed by taking the lowest root of the transcendental equations, as one easily shows.

$Q_\beta(\beta_0) \neq 0$ unless $Q(\beta) \equiv 0$. Hence, in our case, $Q_\beta(\beta_0) \neq 0$. We may assume $Q_\beta(\beta_0) > 0$. Let b be the greatest value such that $Q(\beta) \geqq 0$ for $\beta_0 \leqq \beta < b$. In this range we then have $\rho^\kappa PQ \leqq 0$, $Q_{\beta\beta} \geqq 0$ and, consequently, $Q_\beta(\beta) \geqq Q_\beta(\beta_0)$ and $Q(\beta) \geqq (\beta - \beta_0)Q_\beta(\beta_0)$. If b were finite the latter inequality would yield $Q(b) > 0$, in contradiction with the definition of b. Hence $b = \infty$. The inequality $Q_\beta(\beta) \geqq Q_\beta(\beta_0) > 0$ for $\beta_0 \leqq \beta < \infty$ is, however, in contradiction with $\int_0^\infty Q^2 d\beta < \infty$. This rules out Case 1.

Case 2: We shall dispose of this case by constructing an admissible function Q^ϵ for which $W[Q^\epsilon] < W[Q]$ in contradiction with the minimum property of Q. We first replace Q by $|Q|$ and observe that $W[|Q|] = W[Q]$. We choose $\beta_1 \leqq \beta_0$ and $\beta_2 > \beta_0$ such that

$$|Q|_\beta = -|Q_\beta| \text{ for } \beta_1 \leqq \beta < \beta_0, \quad |Q|_\beta = |Q_\beta| \text{ for } \beta_0 < \beta \leqq \beta_2,$$
$$P > 0 \text{ for } \beta \leqq \beta_2.$$

We then introduce a non-negative function $\eta(\beta) \not\equiv 0$ with continuous derivative which has the following properties:

$$\eta(\beta) \geqq 0,\ \eta(\beta) \equiv 0 \quad \text{for} \quad \beta \geqq \beta_2 \quad \text{and} \quad \beta \leqq \beta_1,$$
$$\eta_\beta(\beta) \geqq 0 \quad \text{for} \quad \beta \leqq \beta_0, \quad \leqq 0 \quad \text{for} \quad \beta \geqq \beta_0.$$

Upon introduction of the function $Q^\epsilon = |Q| + \epsilon\eta$, we notice that

$$\int_\beta^\infty [Q^\epsilon(\beta_1)]^2 d\beta_1 - \int_\beta^\infty Q^2(\beta_1)\, d\beta_1 = \epsilon\delta^\epsilon(\beta),$$

where

$$\delta^\epsilon(\beta) = 2\int_\beta^\infty \eta(\beta_1)\,|\,Q(\beta_1)\,|\, d\beta_1 + \epsilon\int_\beta^\infty \eta^2(\beta_1)\, d\beta_1.$$

Hence

$$H[Q^\epsilon] - H[Q] = \epsilon\delta^\epsilon(0)\,; \quad P_\beta[Q^\epsilon] - P_\beta[Q] = -\epsilon\delta^\epsilon(\beta)/2,$$

$$\begin{aligned} K[Q^\epsilon] - K[Q] &= -\int_0^\infty \delta^\epsilon(\beta) P_\beta d\beta + (\epsilon^2/4)\int_0^\infty \delta^\epsilon(\beta)\, d\beta \\ &= \epsilon\delta^\epsilon(0) + \epsilon\int_0^\infty \frac{d}{d\beta}\,\delta^\epsilon P d\beta + (\epsilon^2/4)\int_0^\infty (\delta^\epsilon)^2 d\beta \qquad (8.01) \\ &= \epsilon\delta^\epsilon(0) + \epsilon[-2\int_0^\infty \eta\,|\,Q\,|\,P d\beta \\ &\quad - \epsilon\int_0^\infty \eta^2 P d\beta + (\epsilon/4)\int_0^\infty (\delta^\epsilon)^2 d\beta]. \end{aligned}$$

For the latter equation the definition of $\delta^\epsilon(\beta)$ was used. Since $\eta \equiv 0$ for $\beta \geqq \beta_2$ and $P > 0$ for $\beta \leqq \beta_2$, we have $\int_0^\infty \eta\,|\,Q\,|\,P d\beta = \int_0^{\beta_2} \eta\,|\,Q\,|\,P d\beta > 0$.

It is clear, therefore, that ϵ can be made so small that the quantity in the brackets is negative. Hence

$$K[Q^\epsilon]-K[Q]<\epsilon\delta^\epsilon(0). \tag{8.02}$$

Further we have, from the definition of $\eta(\beta)$,

$$\begin{aligned}\int_0^\infty [Q^\epsilon{}_\beta]^2 d\beta-\int_0^\infty Q_\beta{}^2 d\beta &=\epsilon[-2\int_{\beta_1}^{\beta_0}\eta_\beta\,|\,Q_\beta\,|\,d\beta+2\int_{\beta_0}^{\beta_2}\eta_\beta\,|\,Q_\beta\,|\,d\beta+\epsilon\int_{\beta_1}^{\beta_2}\eta_\beta{}^2 d\beta]\\ &=\epsilon[-2\int_{\beta_1}^{\beta_2}|\,\eta_\beta\,|\,|\,Q_\beta\,|\,d\beta+\epsilon\int_{\beta_1}^{\beta_2}\eta_\beta{}^2 d\beta].\end{aligned}$$

Evidently we can choose ϵ so small that also here the quantity in the brackets is negative; i. e.,

$$\int_0^\infty [Q^\epsilon{}_\beta]^2 d\beta-\int_0^\infty Q_\beta{}^2 d\beta<0. \tag{8.03}$$

By combining (8.01), (8.02), and (8.03) we obtain

$$W[Q^\epsilon]-W[Q]<0$$

in contradiction with the minimum property of Q. This contradiction establishes Theorem 8.2.

Theorem 8.3 is equivalent to the statement: if Q^0 is any solution of B^κ which does not vanish for $0\leq\beta<\infty$, then

$$W^\kappa[Q]\geq W^\kappa[Q^0]$$

for every admissible function Q and the equality holds only for $Q=\pm Q^0$.

Let P and P^0 be the functions corresponding to Q and Q^0. We derive the identity

$$\int_0^\infty (1-P^0)Q^2\rho^\kappa d\beta=2\int_0^\infty P_\beta{}^0 P_\beta d\beta. \tag{8.04}$$

The integral on the right-hand side exists in view of $\int_0^\infty [P_\beta{}^0]^2 d\beta<\infty$ and $\int_0^\infty P_\beta{}^2 d\beta<\infty$. To prove (8.04) we introduce the identity

$$\int_0^a (1-P^0)Q^2\rho^\kappa d\beta=2\int_0^a P_\beta{}^0 P_\beta d\beta+2(1-P^0)P_\beta|^a.$$

From Lemma 7.1 applied to $f=1-P^0$, $g=P$ it follows that the last term tends to zero when $a\to\infty$ on a special sequence and (8.04) is established. Formula (8.04) implies that $\int_0^\infty (1-P^0)Q^2\rho^\kappa d\beta$ exists for every admissible

Q; in view of $(1-P^0)\geq 0$, $|P^0|\leq 1+(1-P^0)$ and $\int_0^\infty Q^2\rho^\kappa d\beta<\infty$ we can infer

$$(8.05)\text{[34]}\qquad \int_0^\infty |P^0|\, Q^2\rho^\kappa d\beta\ <\infty.$$

We now introduce the quadratic functional

$$(8.06)\qquad T[Q]=\int_0^\infty Q_\beta{}^2 d\beta+\kappa(1+\nu)Q^2(0)-\int_0^\infty P^0Q^2\rho^\kappa d\beta,$$

which, in view of (8.05), is defined for admissible Q. From (8.04) we have

$$(8.07)\qquad \begin{aligned} W[Q]&=T[Q]+\int_0^\infty P_\beta{}^2 d\beta-\int_0^\infty (1-P^0)Q^2\rho^\kappa d\beta\\ &=T[Q]+\int_0^\infty P_\beta{}^2 d\beta-2\int_a^\infty P^0{}_\beta P_\beta d\beta,\end{aligned}$$

and for $Q=Q^0$

$$W[Q^0]=T[Q^0]-\int_0^\infty (P^0{}_\beta)^2 d\beta.$$

Subtraction yields the identity

$$(8.08)\qquad W[Q]-W[Q^0]=T[Q]-T[Q^0]+\int_0^\infty (P_\beta-P^0{}_\beta)^2 d\beta.$$

Theorem 8.3 is a consequence of (8.08) and

LEMMA 8.1. *For admissible* Q

$$(8.09)\qquad T[Q]\geq 0,$$

where the equality holds only for $Q=cQ^0$, $c=$ *const.*

Lemma 8.1 implies

$$(8.09)_0\qquad T[Q^0]=0.$$

If Lemma 8.1 holds, (8.08) yields $W[Q]\geq W[Q^0]$, the equality holding only for $Q=cQ^0$, $P_\beta=P^0{}_\beta$ or $\int_0^\infty [Q^2-(Q^0)^2]\rho^\kappa d\beta=0$. Since $\rho^\kappa>0$, the last equality holds only when $Q=\pm Q^0$. Hence Theorem 8.3 is proved once the inequality (8.09) is established. This inequality states that Q^0 minimizes the quadratic functional $T[Q]$.

We proceed to establish Lemma 8.1. Our proof follows the line of

---

[34] From (8.05) and the admissibility conditions for $\kappa=0$, $\rho^0=1$, it follows that $\int_0^\infty [Q_\beta{}^2-P_\beta{}^2+PQ^2]d\beta$ is finite. For the solution Q, P the integrand is constant according to (4.13). This constant is therefore zero, confirming the assumption made in (4.14).

Jacobi's treatment of the second variation; however, the infinite range of the independent variable and the singularity occurring in our problem require considerable modifications of the standard reasoning.[35] In view of $Q^0 > 0$ it is possible to introduce the function

(8.10) $$\theta = Q/Q^0.$$

With this function θ the identity

(8.11) $$T[Q] = \int_0^\infty \theta_\beta^2 (Q^0)^2 \, d\beta$$

holds, as we shall prove. The identity (8.11) implies that $T[Q] \geq 0$ and that the equality holds only if $\theta_\beta = 0$ (since $Q^0 \not\equiv 0$), that is, if $\theta \equiv c = \text{const.}$, or $Q = cQ^0$. Thus Lemma 8.1 follows from (8.11).

To prove (8.11) we apply Jacobi's identity (7.22) to $\omega = Q^0$ and obtain

$$\int_0^a [Q_\beta^2 + (Q^0)^{-1} Q^0_{\beta\beta} Q^2] d\beta = \int_0^a (Q^0)^2 \theta_\beta^2 d\beta + (Q^0)^{-1} Q^0_\beta Q^2 \big|_0^a,$$

or, in view of $Q^0_{\beta\beta} = -\rho^\kappa P^0 Q^0$ and $Q^0_\beta - \kappa(1+\nu) Q^0 |^0 = 0$,

$$\int_0^a [Q_\beta^2 - P^0 Q^2 \rho^\kappa] d\beta + \kappa(1+\nu) Q^2(0) = \int_0^a (Q^0)^2 \theta_\beta^2 d\beta + (Q^0)^{-1} Q^0_\beta Q^2 |^a.$$

If it could be shown that the last term tends to zero as $a \to \infty$ on at least a special sequence, (8.11) would result in view of (8.06). It is doubtful whether such a direct attack is possible.

Instead, we proceed as follows: Since $\beta^{-1}[Q^0(\beta)]^2 \to 0$, according to Lemma 7.4, there is a special sequence $\beta \to \infty$ on which $(\beta^{-1/2} Q^0)_\beta \leq 0$, or $Q^0_\beta \leq \frac{1}{2}\beta^{-1} Q$, or $(Q^0)^{-1} Q^0_\beta Q^2 \leq \frac{1}{2}\beta^{-1} Q^2$. But $\beta^{-1} Q^2(\beta) \to 0$. If a is taken on this sequence, the inequality

(8.11)⁻ $$T[Q] \leq \int_0^\infty (Q^0)^2 \theta_\beta^2 d\beta \text{ results.}$$

Identity (8.11) will be established once the reverse inequality

(8.11)⁺ $$T[Q] \geq \int_0^\infty (Q^0)^2 \theta_\beta^2 d\beta \text{ is proved.}$$

To this end we introduce the function $Q^\epsilon = Q^0 + \epsilon\beta$ and set

(8.10) $$\theta^\epsilon = Q/Q^\epsilon \quad \text{for} \quad \epsilon > 0.$$

We then apply once more Jacobi's identity (7.22) with $\omega = Q^\epsilon$ and obtain

[35] Our reasoning could be made the basis of a general theory extending Jacobi's treatment of the second variation to the case of singular or infinite end points.

$$\int_0^a [Q_\beta^2 + (Q^\epsilon)^{-1}(Q^\epsilon_{\beta\beta})^2]d\beta = \int_0^a (Q^\epsilon)^2(\theta^\epsilon_\beta)^2 d\beta + (Q^\epsilon)^{-1}(Q^\epsilon_\beta)^2 Q^2|_0^a;$$

using the relation

$$-Q^\epsilon_{\beta\beta} = \rho^\kappa P^0 Q^0, \quad (Q^\epsilon)^{-1}Q^0 = 1 - \epsilon\beta(Q^\epsilon)^{-1}, \text{ and } Q^\epsilon_\beta - (1+\nu)Q^\epsilon|^0 = \epsilon,$$

we have

$$\int_0^a [Q_\beta^2 - \rho^\kappa P^0 Q^2]d\beta + (1+\nu)Q^2(0)$$
$$= \int_0^a [(Q^\epsilon)^2(\theta^\epsilon_\beta)^2 - \epsilon\beta Q^{-1}\rho^\kappa P^0 Q^2]d\beta - \epsilon(Q^0)^{-1}Q^2|^0 + (Q^\epsilon)^{-1}Q^\epsilon_\beta Q^2|^a.$$

Since $\int_0^\infty Q_\beta^2 d\beta < \infty$, there is a special sequence $\beta \to \infty$ on which $\beta^{1/2}Q^0{}_\beta(\beta) \to 0$. On this sequence we have

$$\beta^{-1}Q^\epsilon = \beta^{-1}Q^0 + \epsilon \to \epsilon, \qquad Q^\epsilon{}_\beta = Q^0{}_\beta + \epsilon \to \epsilon, \qquad \beta^{-1}Q^2 \to 0$$

in view of Lemma 7.4; hence $(Q^\epsilon)^{-1}Q^\epsilon{}_\beta Q^2 \to 0$. If we let $a \to \infty$ on this special sequence we obtain the identity

$$(8.11)_\epsilon \quad T[Q] = \int_0^\infty [(Q^\epsilon)^2(\theta^\epsilon_\beta)^2 - \epsilon\beta(Q^\epsilon)^{-1}\rho^\kappa P^0 Q^2]d\beta - \epsilon(Q^0)^{-1}Q^2|^0.$$

Allowing ϵ to tend to zero in $(8.11)_\epsilon$ we have

$$\epsilon(Q^0)^{-1}Q^2|^0 \to 0, \qquad \int_0^a \epsilon\beta(Q^\epsilon)^{-1}\rho^\kappa P^0 Q^2 d\beta \to 0,$$

and

$$\left|\int_a^\infty \epsilon\beta(Q^\epsilon)^{-1}\rho^\kappa P^0 Q^2 d\beta\right| \leq \int_a^\infty \rho^\kappa \,|\, P^0 \,|\, Q^2 d\beta.$$

The last quantity can be made arbitrarily small by choice of a. Further we have

$$\int_0^b (Q^\epsilon)^2(\theta^\epsilon_\beta)^2 d\beta \to \int_0^b (Q^0)^2\theta_\beta^2 d\beta, \qquad \int_b^\infty (Q^\epsilon)^2(\theta^\epsilon_\beta)^2 d\beta \geq 0.$$

Hence $(8.11)_\epsilon$ yields in the limit

$$T[Q] \geq \int_0^b (Q^0)^2\theta_\beta^2 d\beta$$

and $(8.11)^+$ is established since b is arbitrary. Hence (8.11) holds, Lemma 8.1 is proved and, with it, Theorem 8.3.

9. Existence and asymptotic convergence. In this section we prove the existence of the solutions of the minimum problem M^κ, including the asymptotic case M^0, and establish the convergence of the solutions for $\kappa > 0$ to the asymptotic solution ($\kappa = 0$) as $\kappa \to 0$. We apply direct methods similar to those used for linear boundary value problems (cf. Courant-Hilbert [3], Volume II, Chapter VII).

We use the same formulation of the minimum problem M^κ as in sec. 7. Functions Q admissible with respect to the problem M^κ in the sense of sec. 7 are here referred to as κ-admissible functions.

Our theorems are

THEOREM 9.1. *To every $\kappa \geq 0$ there exists at least one κ-admissible function $Q(\beta)$ for which $W^\kappa[Q]$ attains its minimum.*

Such a minimizing function will be denoted henceforth by $Q^\kappa(\beta)$; it is uniquely determined (Theorem 8.1) once the condition $Q^\kappa(0) \geq 0$ has been imposed.

THEOREM 9.2. *The minimizing functions $Q^\kappa(\beta)$ with $Q^\kappa(0) \geq 0$ tend as $\kappa \rightarrow 0$ to the minimizing function $Q^0(\beta)$ with $Q^0(0) \geq 0$ uniformly in each finite interval $0 \leq \beta \leq b < \infty$, and $W^\kappa[Q]$ tends to $W^0[Q^0]$.*

The proofs of the theorems are based upon a number of preliminary lemmas and inequalities.

We prove first

$$\left\{ \int_b^\infty Q^2 \rho^\kappa d\beta \right\}^2 \leq 4b^{-1} K^\kappa[Q] \tag{9.01}$$

for any κ-admissible Q and $b > 0$. Due to $P_{\beta\beta} \geq 0$ (cf. (7.10)) the non-negative quantity $-P_\beta$ does not increase with increasing β. Hence

$$K^\kappa[Q] = \int_0^\infty P_\beta^2 d\beta \geq \int_0^b P_\beta^2 d\beta \geq b P_\beta^2(b) = (b/4) \left\{ \int_b^\infty Q^2 \rho^\kappa d\beta \right\}^2,$$

establishing (9.01).

Next we show that a constant $c > 0$ exists such that

$$H^\kappa[Q] \leq \tfrac{1}{2} D^\kappa[Q] + 2c\{K^\kappa[Q]\}^{1/2} \tag{9.02}$$

for any κ-admissible[36] Q and all $\kappa \geq 0$.

To prove (9.02) we distinguish two cases: $\kappa \leq 1$, $\kappa \geq 1$.

Case 1 ($\kappa \leq 1$). We consider the successive inequalities

$$\begin{aligned} |Q(\beta_1) - Q(\beta_1 + \tfrac{1}{2})|^2 &\leq \left[\int_{\beta_1}^{\beta_1 + 1/2} Q_\beta d\beta \right]^2 \leq \tfrac{1}{2} \int_0^\infty Q_\beta^2 d\beta, \\ Q^2(\beta_1) &\leq 2Q^2(\beta_1 + \tfrac{1}{2}) + 2\,|Q(\beta_1) - Q(\beta_1 + \tfrac{1}{2})|^2 \\ &\leq 2Q^2(\beta_1 + \tfrac{1}{2}) + \int_0^\infty Q_\beta^2 d\beta. \end{aligned}$$

---

[36] The argument used to prove (9.02) could be so modified as to show that the admissibility condition on H^κ is a consequence of the others and could therefore be omitted.

Integration with respect to β_1 yields

$$\int_0^{1/2} Q^2 d\beta \leq 2 \int_{1/2}^1 Q^2 d\beta + \tfrac{1}{2} \int_0^\infty Q_\beta^2 d\beta.$$

Since $\rho^\kappa \leq 1$ and $\rho^\kappa = (1 + 2\kappa\beta)^{-6} \geq 3^{-6}$ for $\beta \leq 1$, $\kappa \leq 1$, we have

$$\int_0^{1/2} \rho^\kappa Q^2 d\beta \leq 2 \cdot 3^6 \int_{1/2}^1 \rho^\kappa Q^2 d\beta + \tfrac{1}{2} \int_0^\infty Q_\beta^2 d\beta.$$

By adding $\int_{1/2}^\infty \rho^\kappa Q^2 d\beta$ to both sides we obtain

$$\int_0^\infty \rho^\kappa Q^2 d\beta \leq (1 + 2 \cdot 3^6) \int_{1/2}^\infty \rho^\kappa Q^2 d\beta + \tfrac{1}{2} \int_0^\infty Q_\beta^2 d\beta.$$

We now use (9. 01) with $b = \tfrac{1}{2}$, that is

$$\int_0^\infty \rho^\kappa Q^2 d\beta \leq 2\sqrt{2} \left\{ \int_0^\infty P_\beta^2 d\beta \right\}^{1/2}$$

and thus obtain

$$\int_0^\infty \rho^\kappa Q^2 d\beta \leq 2c \left\{ \int_0^\infty P_\beta^2 d\beta \right\} + \tfrac{1}{2} \int_0^\infty Q_\beta^2 d\beta$$

where $c = \sqrt{2}(1 + 2 \cdot 3^6)$. This establishes (9. 02) for $\kappa \leq 1$.

Case 2 ($\kappa \geq 1$). We set

$$\omega = (1 + \beta)^{-1}(1 + 2\beta)$$

in Jacobi's identity (7. 22) noting that

$$\omega_\beta = (1 + \beta)^{-2}, \quad \omega^{-1}\omega_\beta = (1 + \beta)^{-1}(1 + 2\beta)^{-1}, \quad \omega_{\beta\beta} = -2(1 + \beta)^{-3},$$

and

$$\omega^{-1}\omega_{\beta\beta} = 2(1 + \beta)^{-2}(1 + 2\beta)^{-1} \geq 2(1 + 2\kappa\beta)^{-6} = 2\rho^\kappa(\beta).$$

The result is

$$\int_0^\infty (Q^2 - 2\rho^\kappa Q^2)\, d\beta \geq \int_0^\infty \omega^2 (\omega^{-1} Q)_\beta^2 d\beta + \omega^{-1}\omega_\beta Q^2 \big|_0^\infty.$$

Since

$$\omega^{-1}\omega_\beta \big|^0 = 1, \quad \beta\omega^{-1}\omega_\beta \big|^\infty = 0, \quad \text{and} \quad \beta^{-1} Q^2 \big|^\infty = 0,$$

according to Lemma 7. 4, the right member of the inequality is non-negative, i. e.,

$$\int_0^\infty \rho^\kappa Q^2 d\beta \leq \tfrac{1}{2} \int_0^\infty Q_\beta^2 d\beta.$$

This establishes (9. 02) for $\kappa \geq 1$ and hence generally.

From (9. 02) we deduce the inequality

(9. 03) $$W^\kappa[Q] \geq D^\kappa[Q] + \{(K^\kappa[Q])^{1/2} - c\}^2 - c^2,$$

which implies

LEMMA 9.1. *For κ-admissible functions Q, $W^\kappa[Q]$ has a lower bound $(-c^2)$ independent of κ.*

From (9.03) we obtain

$$\begin{aligned} &D^\kappa[Q] \leq W^\kappa[Q] + c^2, \\ (9.04)\qquad &K^\kappa[Q] \leq \{c + (W^\kappa[Q] + c^2)^{1/2}\}^2, \text{ and from (9.02)} \\ &H^\kappa[Q] \leq 3c^2 + W^\kappa[Q] + 2c(W^\kappa[Q] + c^2)^{1/2}. \end{aligned}$$

Consider a set of κ-admissible functions Q (with κ not necessarily fixed) for which W^κ has an upper bound M. We conclude from (9.04)

LEMMA 9.2. *An upper bound M for W^κ implies upper bounds for D^κ, H^κ, and K^κ which depend upon M but not on κ.*

In what follows we shall make frequent use of the following well-known lemmas:

LEMMA A. *If $f_s(\beta)$ is a sequence of non-negative continuous functions, defined for $0 \leq \beta < \infty$, which converge to a limit function $f_0(\beta)$ uniformly in every finite interval, then $f_0(\beta)$ is continuous, non-negative, and*

$$\int_0^\infty f_0(\beta)\,d\beta \leq \underline{\lim} \int_0^\infty f_s(\beta)\,d\beta.$$

LEMMA B. *If, in addition, to every $\epsilon > 0$ a quantity $b = b(\epsilon)$ exists such that*

$$\int_b^\infty f_s(\beta)\,d\beta \leq \epsilon \text{ for all } s, \text{ then}$$

$$\int_0^\infty f_s(\beta)\,d\beta \to \int_0^\infty f_0(\beta)\,d\beta.$$

We proceed to state the following basic

LEMMA 9.3. *Let κ_m be a sequence of values of κ tending to a finite value κ_*, and let Q_m be a sequence of κ_m-admissible functions for which $W^{\kappa_m}[Q_m]$ is bounded. Then there exists a subsequence Q_s converging uniformly in every finite interval to a κ_*-admissible function Q such that*

$$(9.05)\qquad \underline{\lim}\, D^{\kappa_s}[Q_s] \geq D^{\kappa_*}[Q_*],$$

$$(9.06)\qquad \underline{\lim}\, H^{\kappa_s}[Q_s] = H^{\kappa_*}[Q_*],$$

$$(9.07)\qquad \underline{\lim}\, K^{\kappa_s}[Q_s] \geq K^{\kappa_*}[Q_*],$$

$$(9.08)\qquad \underline{\lim}\, W^{\kappa_s}[Q_s] \geq W^{\kappa_*}[Q_*].$$

Proof. Lemma 9.2 shows that $D^{\kappa_m}[Q_m]$, $H^{\kappa_m}[Q_m]$, and $K^{\kappa_m}[Q_m]$ are bounded. A bound for $D^{\kappa_m}[Q_m]$ (cf. (7.08)) implies a bound for the sequence $\int_0^\infty Q^2_{m\beta}(\beta)\,d\beta$. Hence, as is well known, there exists a subsequence converging uniformly in every finite interval $0 \leq \beta \leq b$ to a continuous limit function Q_* with L^2-integrable derivative and such that

$$\underline{\lim} \int_0^\infty [Q_{s\beta}]^2(\beta)\, d\beta \geq \int_0^\infty [Q_{*\beta}]^2 d\beta.$$

This inequality yields (9.05), in view of (7.08) and $Q_s(0) \to Q_*(0)$. In order to establish (9.06) we use (9.01) in the form

$$\int_0^\infty \rho^\kappa Q_s^2 d\beta \leq 2b^{-1/2} K^{\kappa_s}[Q_s].$$

Since K^{κ_s} is bounded, the right member can be made arbitrarily small by proper choice of b. Hence we may apply Lemma B with $f_s = \rho^{\kappa_s} Q_s^2$ and thus obtain (9.06). By virtue of the definition (7.10) of P_β we have

$$| P_\beta^{\kappa_s}[Q_s] - P_\beta^*[Q_*] | \leq 2 \, | H^{\kappa_s}[Q_s] - H^{\kappa_*}[Q_*] |$$

and (9.06) yields the uniform convergence of $P_\beta^{\kappa_s}[Q_s]$ to $P_\beta^{\kappa_*}[Q_*]$. Hence we may apply Lemma A to K^{κ_s}; the result is (9.07).

As a consequence of (9.05), (9.06), and (9.07) Q_* is κ_*-admissible and, in addition, (9.08) holds.

We are now in a position to prove Theorem 9.1. We turn, therefore, to the problem of minimizing $W^\kappa[Q]$ by a κ-admissible function Q. From Lemma 9.1 we know that the g. l. b. ω^κ of $W^\kappa[Q]$ is finite; hence there exists a minimizing sequence, i. e., a sequence of κ-admissible functions Q_m for which $W^\kappa[Q_m]$ has as limit ω^κ. We now apply Lemma 9.3 with $\kappa_m = \kappa_* = \kappa$; it yields the existence of a subsequence Q_s and a κ-admissible function[37] $Q = Q_*$ for which (cf. (9.08))

$$W^\kappa[Q] \leq \underline{\lim}\, W^\kappa[Q_s].$$

Since the right member here is the g. l. b. ω^κ of W^κ, the equality must hold. Hence Q solves the minimum problem M^κ. This proves Theorem 9.1.

We add the remark that the minima ω^κ of W^κ have the common *upper* bound zero, i. e.,

(9.09) $$\omega^\kappa \leq 0,$$

which follows immediately from $W^\kappa[0] = 0$ since $Q \equiv 0$ is an admissible function.

Before proving Theorem 9.2 we establish two lemmas:

[37] If we had required the existence of continuous second derivatives for admissibility it would have been necessary at this point to prove that the limit function Q has this property—otherwise Q could not be identified as the solution of M^κ. Our procedure, which requires only L^2-integrable derivatives for admissibility, makes it possible to separate the problems of existence and continuous differentiability of the solution of M^κ.

LEMMA 9.4. *A 0-admissible function $Q(\beta)$ is also κ-admissible for every $\kappa > 0$.*

This is an immediate consequence of $\rho^\kappa \leq \rho^0 = 1$ and

$$P_\beta^\kappa[Q] \leq P_\beta^0[Q], \tag{9.10}$$

in view of definitions (7.07) to (7.10).

LEMMA 9.5. *For any 0-admissible function $Q(\beta)$:*

$$W^\kappa[Q] \to W^0[Q] \text{ as } \kappa \to 0. \tag{9.11}$$

Proof. From (7.08) it is obvious that

$$D^\kappa[Q] \to D^0[Q] \text{ as } \kappa \to 0. \tag{9.12}$$

We apply Lemma B to

$$H^0[Q] - H^\kappa[Q] = \int_0^\infty Q^2(\beta)(1-\rho^\kappa)\,d\beta,$$

observing that

$$\int_b^\infty Q^2(\beta)(1-\rho^\kappa)\,d\beta \leq \int_b^\infty Q^2(\beta)\,d\beta$$

can be made arbitrarily small by proper choice of b. It follows that

$$H^\kappa[Q] \to H^0[Q] \text{ as } \kappa \to 0. \tag{9.13}$$

The latter implies, in view of (7.10), that

$$P_\beta^\kappa[Q] \to P_\beta^0[Q] \text{ as } \kappa \to 0. \tag{9.14}$$

We now consider

$$K^0[Q] - K^\kappa[Q] = \int_0^\infty \{(P_\beta^0)^2 - (P_\beta^\kappa)^2\}\,d\beta. \tag{9.15}$$

The right member, in view of (9.10), has a positive integrand and

$$\int_b^\infty \{(P_\beta^0)^2 - (P_\beta^\kappa)^2\}\,d\beta < \int_b^\infty (P_\beta^0)^2\,d\beta$$

can be made small. This fact and (9.14) permit us to apply Lemma B to (9.15) with the result

$$K^\kappa[Q] \to K^0[Q] \text{ as } \kappa \to 0. \tag{9.16}$$

Relations (9.12), (9.13), and (9.16) lead immediately to (9.11).

We now take *any* sequence of positive values κ tending to zero and solutions Q^κ of the corresponding minimum problems M^κ (which exist according to Theorem 9.1). We may assume $Q^\kappa(0) \geq 0$. As remarked above (cf. (9.09)), the values $W^\kappa[Q^\kappa] = \omega^\kappa$ have an upper bound. We can therefore apply Lemma 9.3 to the sequence Q^κ with $\kappa_* = 0$. The lemma insures the existence of a subsequence converging in the sense of the lemma to a 0-admissible limit function Q_0. From now on Q^κ refers to such a subsequence. From (9.08) we have, in particular,

$$\underline{\lim}\, W^\kappa[Q^\kappa] \geq W^0[Q_0]. \tag{9.17}$$

We proceed to show that Q_0 *solves* the minimum problem M^0. This minimum problem, according to Theorem 9.1, has a solution Q^0 for which (Lemma 9.5)

$$\lim W^\kappa[Q^0] = W^0[Q^0]. \tag{9.18}$$

As a consequence of the minimum properties of Q^0 and Q^κ we have

$$\omega^0 = W^0[Q^0] \leq W^0[Q_0] \quad \text{and} \tag{9.19}$$

$$\omega^\kappa = W^\kappa[Q^\kappa] \leq W^\kappa[Q^0]. \tag{9.20}$$

Here we have made use of the fact that Q^0 is κ-admissible (Lemma 9.4). From (9.20) we find

$$\overline{\lim}\, W^\kappa[Q^\kappa] \leq \overline{\lim}\, W^\kappa[Q^0]. \tag{9.21}$$

Successive consideration of (9.21), (9.18), (9.19), and (9.17) yields

$$\overline{\lim}\, W^\kappa[Q^\kappa] \leq W^0[Q^0] \leq W^0[Q_0] \leq \underline{\lim}\, W^\kappa[Q^\kappa] \tag{9.22}$$

and this obviously implies the equality

$$W^0[Q^0] = W^0[Q_0]. \tag{9.23}$$

Since $W^0[Q^0]$ is the g. l. b. of W^0 it follows that the function $Q = Q_0$ is a solution of the minimum problem M^κ for $\kappa = 0$ with $Q(0) \geq 0$.

From Theorem 8.1 we know that the minimum problem M^0 has at most one solution Q with $Q(0) \geq 0$. Therefore $Q^0 \equiv Q_0$, i.e., all convergent sequences Q^κ converge to the same limit function. If a sequence has the property that every subsequence contains a convergent subsequence with limit L and if L is the same for all such convergent subsequences, then the original sequence itself converges to L. Therefore we conclude in our case that the solutions $Q^\kappa(\beta)$ with $Q^\kappa(0) \geq 0$ of the minimum problems M^κ converge, as $\kappa \to 0$, to the unique solution $Q^0(\beta)$ with $Q^0(0) \geq 0$ of the minimum problem M^0. This completes the proof of Theorem 9.2.

10. Limit state in the interior of plate. While the limit procedure of sec. 9 concerns the *boundary layer*, we deal in this section with the limit procedure in the *interior* of the plate as $\lambda \to \infty$. These two limit procedures have already been discussed and contrasted with each other in sec. 4.

We summarize the discussion of sec. 4 concerning the interior limit process, but we use the notation of sec. 5. By multiplying (5.10) by κ^3 and (5.12), (5.13) by κ^2 ($\kappa = \lambda^{-1}$) we obtain

$$\kappa^2 G(\kappa q) + (\kappa^2 p)(\kappa q) = 0, \tag{10.01}$$

(10.02) $$G(\kappa^2 p) = (\kappa q)^2/2, \quad \text{and}$$

(10.03) $$\kappa^2 p = 1 \quad \text{for} \quad r = 1.$$

When $\kappa \to 0$, the following equations (cf. 4.01) result

(10.04) $$\hat{p}\hat{q} = 0,$$

(10.05) $$G(\hat{p}) = \hat{q}^2/2, \quad \text{and}$$

(10.06) $$\hat{p} = 1 \quad \text{for} \quad r = 1, \quad \text{where}$$

(10.07) $$\hat{p} = \lim (\kappa^2 p), \quad \hat{q} = \lim (\kappa q) \quad \text{for} \quad \kappa \to 0.$$

One expects that the limit equations (10.04) and (10.05) will be satisfied. The only admissible solution of them is $\hat{p} =$ const., $\hat{q} = 0$. It was pointed out in sec. 4 that the proper constant value for $\hat{p}$ is not to be taken from (10.06) but rather from the asymptotic solution of the boundary layer problem, i.e., $\hat{p} = P(\infty)$. Since $P(\infty)$ is negative, this means that the radial membrane stress tends to a negative constant in every interior region.

Precisely, we prove the following

THEOREM 10.1. *As $\kappa \to 0$ the limit relations*

(10.08) $\kappa^2 p_\lambda(r) \to P^0(\infty)$, (10.09) $\kappa q_\lambda(r) \to 0$

hold uniformly in every interval $0 \leq r \leq r_0 < 1$. Here $q_\lambda(r)$ is the solution of the minimum problem M_λ (cf. sec. 5) and p_λ is given by (5.07) and (5.06).

Once Theorem 10.1 has been established, it is clear that the limit procedure discussed above leads to correct results for those solutions of B_λ which do not vanish (cf. Theorem 8.3).

With the notation of sec. 7 the convergence relations (10.08) and (10.09) become

(10.10) $P^\kappa(\kappa^{-1}\zeta) \to P^0(\infty)$, (10.11) $\kappa^{-1}Q^\kappa(\kappa^{-1}\zeta) \to 0$;

where

(10.12) $$\zeta = \tfrac{1}{2}(r^{-2} - 1) = \kappa\beta.$$

Theorem 10.1 is evidently proved once these relations are shown to hold uniformly in every interval $0 < \zeta_0 \leq \zeta$. $Q^\kappa(\beta)$ is the solution of the minimum problem M^κ. In what follows the superscript κ on the quantities P and Q is omitted where no confusion could result.

We make use of various properties of the functions Q and P. From Theorem 8.2 we have

(10.13) $$Q^\kappa(\beta) \geq 0.$$

We show next that there exists a value β_0 of β such that

(10.14) $$P^0(\beta_0) < 0.$$

This can be seen indirectly as follows. If $P^0(\beta) \geq 0$ for all β, (10.13) and (7.12) for $\kappa = 0$ show that $Q^0{}_{\beta\beta}$ would be ≤ 0. Consequently the function $Q^0(\beta) \not\equiv 0$ would be non-decreasing or would become negative infinite. But both cases are impossible in view of the admissibility condition

$$\int_0^\infty (Q^0)^2 d\beta < \infty.$$

Since $P^\kappa(\beta_0)$ and $Q^\kappa(\beta_0)$ converge to $P^0(\beta_0)$ and $Q^0(\beta_0)$ as $\kappa \to 0$ (Theorem 9.2) there exist positive values κ_0, c, c_1, and d such that for $\kappa \leq \kappa_0$

(10.15) $$0 < c \leq -P^\kappa(\beta_0) \leq c_1, \quad \text{and}$$

(10.16) $$Q^\kappa(\beta_0) \leq d$$

hold. From now on we assume always $\kappa \leq \kappa_0$ and $\beta \geq \beta_0$.

Our subsequent discussion is based on the following formulas, which hold for all $\kappa \geq 0$:

(10.17) $$-P_\beta = \tfrac{1}{2} \int_\beta^\infty \bar{Q}^2 \bar{\rho}^\kappa d\bar{\beta},$$

(10.18) $$P(\beta_1) - P(\beta_2) = \tfrac{1}{2} \int_{\beta_1}^{\beta_2} \bar{Q}^2 (\bar{\beta} - \beta_1) \bar{\rho}^\kappa d\bar{\beta} - (\beta_2 - \beta_1) P_\beta(\beta_2),$$

(10.19) $$-Q_\beta(\beta) = \int_\beta^\infty (-\bar{P}) \bar{Q} \bar{\rho}^\kappa d\bar{\beta}, \quad \text{and}$$

(10.20) $$Q(\beta_1) - Q(\beta_2) = \int_{\beta_1}^{\beta_2} (-\bar{P}) \bar{Q} (\bar{\beta} - \beta_1) \bar{\rho}^\kappa d\bar{\beta} - (\beta_2 - \beta_1) Q_\beta(\beta_2),$$

where the bar indicates that β has been replaced by $\bar{\beta}$.

Formula (10.17) is the definition of P_β (sec. (7.10)). Formula (10.18) is obtained by integrating (10.17) from β_1 to β_2 and applying integration by parts to the right-hand side. Formula (10.19) results from the differential equation (7.12) by integration from β to ∞ and use of the fact that $Q_\beta(\beta) \to 0$ if $\beta \to \infty$ at least on a special sequence, the latter following from the admissibility condition $D^\kappa[Q] < \infty$ (cf. (7.08)). Formula (10.20) follows from (10.19) in the same way as (10.18) was obtained from (10.17).

From (10.17) we have

(10.21) $-P_\beta \geq 0$; from (10.18) and (10.21)

(10.22) $$-P(\beta_1) \leq -P(\beta_2) \quad \text{for} \quad \beta_1 \leq \beta_2,$$

and, in addition, in view of (10.15),

(10.23) $$-P(\beta) \geq c > 0, \qquad \beta \geq \beta_0.$$

Correspondingly we obtain from (10.19), (10.20), and (10.16), in view of (10.23) and (10.13):

(10.24) $$Q_\beta \leq 0,$$

(10.25) $$Q(\beta_1) \geq Q(\beta_2) \quad \text{for} \quad \beta_1 \leq \beta_2, \quad \text{and}$$

(10.26) $$Q(\beta) \leq d, \qquad \beta \geq \beta_0.$$

In view of (10.23), (10.25), and (10.24) we decrease the right member of (10.20) when we replace $-\bar{P}$ by c, $\bar{Q}$ by $Q(\beta_2)$ and omit the second term. We have thus

$$Q(\beta_1) - Q(\beta_2) \geq cQ(\beta_2) \int_{\beta_1}^{\beta_2} (\bar{\beta} - \beta_1)\bar{\rho}^\kappa d\bar{\beta}.$$

The left member of this inequality, in view of (10.26) and (10.13), is $\leq d$. Hence we obtain

(10.27) $$Q(\beta_2) \leq dc^{-1}\left[\int_{\beta_1}^{\beta_2} (\bar{\beta} - \beta_1)\bar{\rho}^\kappa d\bar{\beta}\right]^{-1}.$$

For $\kappa = 0$, $(\rho^0 = 1)$, (10.27) yields, with $\beta_1 = \beta_0$, $\beta_2 = \beta$,

(10.28) $$Q^0(\beta) \leq 2dc^{-1}(\beta - \beta_0)^{-2}.$$

For $\kappa > 0$ we estimate the bracket on the right side of (10.27). We introduce the function $\sigma(\beta_1, \beta_2)$ by the formula

(10.29) $$\sigma(\beta_1, \beta_2) = \int_{\beta_1}^{\beta_2} (\bar{\beta} - \beta_1)\bar{\rho}^1 d\bar{\beta},$$

from which, in view of $\rho^\kappa(\beta) = \rho^1(\kappa\beta)$ (cf. (7.06)), we have

(10.30) $$\int_{\beta_1}^{\beta_2} (\bar{\beta} - \beta_1)\bar{\rho}^\kappa d\bar{\beta} = \kappa^{-2}\sigma(\kappa\beta_1, \kappa\beta_2).$$

We insert (10.30) in (10.27) with $\beta_1 = \beta_0$, replace β_2 by $\kappa^{-1}\zeta$ (cf. (10.12)), and obtain for $\kappa \leq \zeta/\beta_0$

(10.31) $$\kappa^{-2}Q(\kappa^{-1}\zeta) \leq dc^{-1}/\sigma(\kappa\beta_0, \zeta).$$

We have already restricted ζ to the interval $0 < \zeta_0 \leq \zeta$, and, if we restrict κ by $\kappa \leq \zeta_0/2\beta$ (as we may do), we have $\sigma(\kappa\beta_0, \zeta) \geq \sigma(\zeta_0/2, \zeta_0)$. Hence we derive from (10.31) the inequality

(10.32) $$\kappa^{-2}Q(\kappa^{-1}\zeta) \leq dc^{-1}/\sigma(\zeta_0/2, \zeta_0).$$

The left member is bounded as κ tends to zero for every $\zeta_0 > 0$. This obviously proves the convergence relation (10.11).

We turn now to the proof of the relation (10.10). An immediate consequence of (10.26) and (10.23) is

$$Q^2 \leq dc^{-1}(-PQ);$$

insertion of this in (10.18) after replacing P_β in accordance with (10.17), and comparison with (10.20) and (10.19) yields for $\beta_2 > \beta_1$

$$P(\beta_1) - P(\beta_2) \leq \tfrac{1}{2}dc^{-1}[Q(\beta_1) - Q(\beta_2)]. \tag{10.33}$$

The right member is $\leq \frac{1}{2}dc^{-1}Q(\beta_1)$ in view of (10.25). Hence, when $\beta_2 \to \infty$, $-P(\beta_2)$ tends to a finite limit $-P(\infty)$ since it increases monotonically (cf. (10.22)).

We consider the inequality

$$|P^\kappa(\kappa^{-1}\zeta) - P^0(\infty)| \leq |P^\kappa(\kappa^{-1}\zeta) - P^\kappa(\beta_1)| + |P^\kappa(\beta_1) - P^0(\beta_1)| + |P^0(\beta_1) - P^0(\infty)|.$$

The first term on the right is to be estimated by (10.33) and

$$|Q^\kappa(\kappa^{-1}\zeta) - Q^\kappa(\beta_1)| \leq |Q^\kappa(\kappa^{-1}\zeta)| + |Q^\kappa(\beta_1) - Q^0(\beta_1)| + |Q^0(\beta_1)|.$$

The resulting inequality is

$$|P^\kappa(\kappa^{-1}\zeta) - P^0(\infty)| \leq \tfrac{1}{2}dc^{-1}|Q^\kappa(\kappa^{-1}\zeta)| + \tfrac{1}{2}dc^{-1}|Q^\kappa(\beta_1) - Q^0(\beta_1)| + \tfrac{1}{2}dc^{-1}|Q^0(\beta_1)| + |P^\kappa(\beta_1) - P^0(\beta_1)| + |P^0(\beta_1) - P^0(\infty)|. \tag{10.34}$$

Since $P^0(\beta_1) \to P^0(\infty)$ and $Q^0(\beta_1) \to 0$ as $\beta_1 \to \infty$ (cf. (10.28)) we can choose β_1 so large that the third and fifth terms on the right side of (10.34) are arbitrarily small. Since $P^\kappa(\beta_1) \to P^0(\beta_1)$ and $Q^\kappa(\beta_1) \to Q^0(\beta_1)$ (Theorem 9.2), and since $Q^\kappa(\kappa^{-1}\zeta) \to 0$ (cf. (10.32)) as $\kappa \to 0$, we can choose κ so small that the remaining terms on the right side of (10.34) become arbitrarily small. This establishes the convergence relation (10.10) and the proof of Theorem 10.1 is completed.

APPENDIX.

Perturbation method and E. Schmidt's bifurcation theory. The perturbation method of sec. 2 finds its theoretical justification in its close relation to the bifurcation theory developed by E. Schmidt for a class of non-linear integral equations (cf. [14] and the literature given there). In this appendix we shall reduce our problem to an integral equation of a similar type. A certain singularity occurring in our problem does not interfere with the applicability of E. Schmidt's procedure, of which we give a brief indication.

We begin by considering the linear boundary value problem

$$(G + \lambda_0^2)q = -f, \quad rq_r + (1+\nu)q|^1 = 0, \quad qDq < \infty, \quad qHq < \infty, \tag{B$_f$}$$

where the operator G and the forms D and H are defined in sec. 5 and $f = f(r)$ is any given function. The corresponding homogeneous problem

($f \equiv 0$) is denoted by $(B)_0$; we assume λ_0^2 to be the lowest eigenvalue of $(B)_0$ (cf. sec. 5).

Our non-linear boundary value problem can also be written in the form $(B)_f$ by taking for f the functional

(I. 01) $$f = \mu q - O(q^2)q,$$

where the functional O is defined by

(I. 02) $$O(q^2) = \tfrac{1}{2}\int_r^1 r_1^{-3} \int_0^{r_1} q^2(r_2)r_2^3 dr_2 dr,$$

while μ is given by

(I. 03) $$\mu = \bar{p} - \lambda_0^2 = \bar{p} - p^0$$

and represents the excess of the applied load $\bar{p}$ over the critical load p^0. The problem thus defined will be referred to as (B) from now on. That (B) is our original problem can be seen as follows: instead of (5. 07) we have $\bar{p} - p = O(q^2)$, hence (I. 01), in view of (I. 03), is the same as $f = (p - \lambda_0^2)q$ and the differential equation in (B) becomes $Gq + pq = 0$, i. e., equation (5. 10).

We proceed to construct the solutions of the linear problem $(B)_f$. If λ_0^2 were not an eigenvalue of $(B)_0$, the problem $(B)_f$ could be solved by means of the Green's function; in our case, however, λ_0^2 being an eigenvalue, an improper Green's function must be used to represent the solutions. Let $j(r) = ar^{-1}J_1(\lambda_0 r)$ (cf. (2. 3)) be the first eigenfunction of $(B)_0$ where the constant a is determined by

(I. 04) $$jHj = 1.$$

The improper Green's function $s(r, \rho)$ (cf. Courant-Hilbert [3] Bd. I, Kap. V, § 14, 2) is then characterized as the solution of the boundary value problem $(B)_f$ with $f = -j(r)j(\rho)$, which is continuous but has for $r = \rho$ the jump singularity

$$\left[\frac{d}{dr}s(r,\rho)\right]_{\rho-}^{\rho+} = -1.$$

We note that $s(r, \rho)$ is symmetric in r and ρ:

$$s(r,\rho) = s(\rho, r).$$

Such a function can be determined explicitly:

$$s(r,\rho) = -(\pi/2a^2)n(r)j(\rho) + \tfrac{1}{2}[m(r)j(\rho) + j(r)m(\rho)] + bj(r)j(\rho)$$

for $r \geq \rho$, and, by symmetry, for $r \leq \rho$. Here $j(r)$ is the function defined above, and $n(r) = ar^{-1}Y_1(\lambda_0 r)$, $m(r) = a\lambda_0^{-1}J_2(\lambda_0 r)$, where Y_1 is the Bessel function of first order and second kind and J_2 that of second order and first kind. The constant b is arbitrary and indicates that $s(r,\rho)$ is determined only within a multiple of $j(r)j(\rho)$.

With $s(r,\rho)$ we construct the operation

$$Sh = \int_0^1 s(r,\rho)h(\rho)\rho^3 d\rho$$

for any function $h(r)$ with $hHh < \infty$. The function Sh satisfies the differential equation

$$(G+\lambda^2)Sh = -h + j(jHh), \tag{I. 05}$$

as can be inferred from the above properties of the Green's function $s(r,\rho)$; in addition, $q = Sh$ satisfies the other conditions of problem $(B)_f$. For a function f orthogonal to j, i. e., satisfying

$$jHf = 0, \tag{I. 06}$$

the function $q = Sf$ is then a solution of $(B)_f$. Since such a solution of $(B)_f$ is unique within a multiple of j, we obtain *all* solutions of $(B)_f$ in the form

$$q = \epsilon j + Sf \text{ for constant } \epsilon. \tag{I. 07}$$

Upon applying the operation S to $h = j$ we obtain $(G+\lambda_0{}^2)Sj = 0$, indicating that Sj is an eigenfunction of $(B)_0$ and hence that Sj is a multiple of j, say $Sj = cj$. After replacing $s(r,\rho)$ by $s(r,\rho) - cj(r)j(\rho)$ (which amounts to fixing the constant b in the explicit expression for $s(r,\rho)$), the relation

$$Sj = 0, \tag{I. 08}$$

follows from the definition of the operation S and (I. 04). We assume (I. 08) to hold from now on. An immediate consequence of (I. 08) and the symmetry of $s(r,\rho)$ is

$$jHSh = 0, \tag{I. 09}$$

indicating that the function Sh is orthogonal to j for arbitrary h.

We now turn to problem (B) where f is given as a functional in q through (I. 01). Upon inserting f as given by (I. 01) into (I. 07) we obtain

$$q = \epsilon j + \mu Sq - S\{qO(q^2)\}, \tag{I. 10}$$

an integral equation which must be satisfied by every solution q of (B). However, a function satisfying (I. 10) is a solution of (B) only if the condition (I. 06) is satisfied, which, upon using (I. 01), takes the form

$$jHq - jH\{qO(q^2)\} = 0.$$

This "bifurcation equation" is to be considered as a relation between μ and ϵ. Bv virtue of (I. 07), (I. 04), and (I. 09) it reduces to

(I. 11) $$\mu\epsilon - jH\{qO(q^2)\} = 0.$$

As we are interested in solutions in the neighborhood of $\epsilon = 0$, $q \equiv 0$, it is appropriate to introduce a new function t through

(I. 12) $$q = \epsilon j + \epsilon t, \text{ where } t, \text{ of course, depends on } \epsilon.$$

In view of (I. 06) equation (I. 10) becomes

(I. 13) $$t = \mu St - \epsilon^2 S\{(j+t)O(j+t)^2\},$$

while the bifurcation equation (I. 11) assumes the form

(I. 14) $$\epsilon[\mu - \epsilon^2 jH\{(j+t)O(j+t)^2\}] = 0.$$

We exclude the trivial solution $\epsilon = 0$, $t \equiv 0$ valid for any μ. Then (I. 14) reduces to

(I. 15) $$\mu = \epsilon^2 jH\{(j+t)O(j+t)^2\}.$$

Once the relations (I. 13) and (I. 15) have been obtained, the procedure of E. Schmidt can be applied: 1) ϵ and μ are considered as independent parameters and the integral equation (I. 13) is solved[38] on this basis; 2) the solution $t(r;\epsilon^2,\mu)$ of (I. 13) is inserted in the relation (I. 15); μ is then considered a function $\mu(\epsilon^2)$ of ϵ^2 through the resulting transcendental equation, the bifurcation equation; 3) $t(r;\epsilon^2,\mu(\epsilon^2))$ inserted in (I. 12) determines finally the solution of problem (B).

Following the procedure of E. Schmidt, it is possible to show that (I. 13) has a solution which is a *power* series in μ and ϵ^2 for small enough μ and ϵ. The relation between μ and ϵ derived from (I. 15) will then be such that μ can be expressed as a power series in ϵ^2, starting with the term $\epsilon^2 jHjO(j^2)$ which does not vanish.[39] That is: $\mu = \bar{p} - p^0$ is a power series in ϵ^2 and q a series in odd powers of ϵ. This was *assumed* in working with the perturbation method (sec. 2).

NEW YORK UNIVERSITY.

---

[38] Note that $O(q^2)$ and Sq are continuous even at $r = 0$ if q has this property. Hence iterations on (I. 13) are applicable if t is assumed to be continuous for $0 \leq r \leq 1$.

[39] From the definition of $O(q^2)$ we may write

$$jHjO(j^2) = \tfrac{1}{4}\int_0^1\int_0^1 \left(\frac{1}{r_m} - 1\right) j^2(r_1) j^2(r_2) r_1^3 r_2^3 dr_1 dr_2$$

where $r_m = \max(r_1, r_2)$.

REFERENCES

[1]. M. Biot, "Non-linear Theory of Elasticity," *Phil. Mag.*, Ser. 7, vol. 27 (1939), pp. 468-489; "Elastizitätstheorie zweiter Ordnung," *Z. f. ang. Math. u. Mech.*, vol. 20 (1940), pp. 89-99.

[2]. D. G. Bourgin, "The Clamped Square Sheet," *American Journal of Mathematics*, vol. 61 (1939), pp. 417-439.

[3]. R. Courant und D. Hilbert, *Methoden der mathematischen Physik*, vol. 1, 2nd ed. (1931), vol. 2 (1938).

[4]. A. Föppl, *Vorlesungen über technische Mechanik*, vol. 5, § 24 (1907).

[5]. K. O. Friedrichs and J. J. Stoker, "The Non-linear Boundary Value Problem of the Buckled Plate," *Proceedings of the National Academy of Sciences*, vol. 25 (1939), pp. 535-540.

[6]. ——— "Buckling of the Circular Plate beyond the Critical Thrust." To appear in the *Journal of Applied Mechanics*.

[7]. K. O. Friedrichs, "Über die ausgezeichnete Randbedingung in der Spektraltheorie," *Mathematische Annalen*, vol. 112 (1936), pp. 1-23.

[8]. ——— "On Differential Operators in Hilbert Spaces," *American Journal of Mathematics*, vol. 61 (1939), pp. 523-544.

[9]. A. Grzedzielski and W. Billewicz, "Sur la rigidité de la tôle flambée," *Sprawozdania Inst. Techn. Lotnictwa*, vol. 10 (1937), pp. 5-22.

[10]. A. Hammerstein, "Nichtlineare Integralgleichungen," *Acta math.*, vol. 54 (1930), pp. 117-176.

[11]. H. Hencky, "Über den Spannungszustand in kreisrunden Platten," *Z. f. Math. u. Phys.*, vol. 63 (1915), pp. 311-317.

[12]. ——— "Die Berechnung dünner rechteckiger Platten," *Z. f. ang. Math. u. Mech.*, vol. 1 (1921), pp. 81-89, 423-424.

[13]. Th. v. Kármán, "Festigkeitsproblem im Maschinenbau," *Enz. d. math. Wiss.*, Bd. IV⁴ (1910), pp. 438-352.

[14]. L. Lichtenstein, *Vorlesungen über einige Klassen nichtlinearer Integralgleichungen* (1931).

[15]. K. Marguerre und A. Kromm, "Verhalten eines Plattenstreifens oberhalb der Beulgrenze," *Luftfahrtf.*, vol. 14 (1937).

[16]. F. D. Murnaghan, "Finite Deformations of an Elastic Solid," *American Journal of Mathematics*, vol. 59 (1937), pp. 235-260.

[17]. A. Nádai, *Elastische Platten* (1925).

[18]. P. Polubarinova-Kotschina, "Zum Problem der Plattenstabilität," *Appl. Math. a. Mech.*, vol. 3 (1936).

[19]. L. Prandtl, "The Mechanics of Viscous Fluids," in Durand, *Aerodynamic Theory*, vol. 3, Sections 13, 14 (1935).

[20]. S. Timoshenko, *Theory of Elastic Stability* (1936).

[21]. E. Trefftz, "Über die Ableitung der Stabilitätskriterien," *Third International Congress of Mechanics*, Stockholm (1930).

[22]. S. Way, "Bending of Circular Plates," *Transactions of the American Society of Mechanical Engineers*, vol. 56 (1934).

Buckling of the Circular Plate Beyond the Critical Thrust

By K. O. FRIEDRICHS[1] and J. J. STOKER,[2] NEW YORK, N. Y.

In this paper a complete mathematical solution of the problem of the buckling with large deflections of a thin circular plate under uniform radial pressure is given, assuming radial symmetry. Buckling takes place as soon as the prescribed thrust p_e at the edge reaches a certain critical value p_E. Thin plates do not, however, fail if the thrust p_e is increased beyond p_E. It is of importance, then, to determine the stresses when the ratio $\Lambda = p_e/p_E$ becomes greater than unity. This problem, which is a nonlinear one, is solved by two methods for finite values of Λ; also an asymptotic solution is given for the limit state when Λ tends to infinity. The most notable single result is that the membrane stresses for large values of Λ become tensions in the interior of the plate and change abruptly to compressions in a narrow "boundary layer" at the edge of the plate. Curves showing the behavior of the deflection and the stresses are given for a series of values of Λ. Such curves show clearly, in particular, that the limit state is approached quite closely for relatively small values of Λ, i.e., $\Lambda > 5$. In closing, the authors discuss the relation between their results and von Karmán's theory of effective width for the buckling of rectangular plates.

Notation

x, y = rectangular coordinates
r = distance from center
R = radius of plate
h = thickness of plate
E = Young's modulus
ν = Poisson's ratio
$\gamma = [12(1 - \nu^2)]^{-1/2}$
$\eta = \gamma h/R$ slenderness ratio
w = deflection of the middle surface
$q = -Rr^{-1}dw/dr$
q_e = slope of the middle surface at the edge
φ = Airy's stress function
$-p$ = radial membrane stress
$-p_c$ = circumferential membrane stress
σ_b = radial bending stress at the upper surface
σ_{cb} = circumferential bending stress at the upper surface
τ_b = shearing stress due to bending at the upper surface
p_0, p_{c0}, σ_{b0} = values of stresses at the center of the plate
p_e, p_{ce}, σ_{be} = values of stresses at the edge of the plate
p_E = critical value of p_e
$\Lambda = p_e/p_E$

[1] Associate Professor of Applied Mathematics, Graduate School, New York University.

[2] Assistant Professor of Mathematics, College of Engineering, New York University.

Presented at the National Meeting of the Applied Mechanics Division, Philadelphia, Pa., June 20–21, 1941, of The American Society of Mechanical Engineers.

Discussion of this paper should be addressed to the Secretary, A.S.M.E., 29 West 39th Street, New York, N. Y., and will be accepted until one month after final publication of the paper itself in the Journal of Applied Mechanics.

Note: Statements and opinions advanced in papers are to be understood as individual expressions of their authors and not those of the Society.

If a straight slender column is subjected to gradually increasing thrusts applied at its ends, it will become instable and buckle when a certain critical value of the thrust is reached. If the thrust is increased a small amount beyond the critical value, the column will generally collapse completely. The behavior of a thin flat plate subjected to compressive forces at its edges and in its plane is quite different. The plate will become instable when the compressive forces reach certain critical values, but the forces can be increased considerably above such values without causing a collapse of the plate. The fundamental reason for this experimentally well-established difference in the behavior of column and plate is that a straight line (the axis of the column) can be bent into a curve without changing its length, while a plane surface, the edge of which is restrained, cannot be deformed without stretching the surface. Thus, when the plate buckles, the stretch which results relieves the central part of the plate of a portion of the compressive stresses and as a consequence the plate is relatively much stiffer than the column.

It is a matter of practical importance to determine the strength of thin plates when subjected to loads greater than those at which buckling begins, because such thin-walled structures are used extensively, for example, in aircraft, where lightness combined with maximum strength is desired and where buckling is permissible as long as it does not result in failure.

A formula for the design of rectangular plates for thrusts beyond the lowest buckling thrust has been developed by von Kármán. It is based on the assumption that the compressive stresses tend to become concentrated in a strip near the edge of the plate. Such an edge effect has been observed, in fact, by a number of experimenters (2, 3, 9, 12, 13, 18, 19).[3]

The purpose of this investigation is to give a complete mathematical solution for the case of the circular plate with simply supported edges, buckled with radial symmetry under the action of a uniformly distributed edge thrust. The results include, in particular, an explanation and discussion of an edge effect which is found here as a natural part of the solution. Important and interesting conclusions can be drawn which retain at least qualitative significance for the case of the rectangular plate—the case of more immediate practical importance.

The mathematical formulation of the problem of the buckled plate involves a pair of nonlinear differential equations derived by von Kármán for the bending of thin plates with large deflections (8). Various writers (6, 10, 11, 12, 13, 14, 16, 20, 24) have solved these equations using the perturbation and energy methods in order to obtain solutions valid at least for a limited range of the ratio Λ of applied pressure p_e to the lowest buckling pressure p_E. Such solutions deal in the main with the rectangular plate. An exact solution for the rectangular plate valid for an unlimited range of the ratio Λ presents seemingly insurmountable difficulties; for the circular plate,[4] however, the authors obtain

[3] Numbers in parentheses refer to the Bibliography.

[4] The problem of the bending of a circular plate under lateral pressure and edge moment has been solved by Way (22) by an exact method for a rather large range of applied load.

Federhofer (24) has solved the problem of the buckling of the circular plate for simply supported and clamped edges. In the former case, which is the same as that treated here, the method of Federhofer yields accurate results up to about $\Lambda = 1.25$.

rigorous solutions valid for an unlimited range of Λ. The existence of an edge effect can then be derived by an asymptotic treatment for $\Lambda \to \infty$, which presents some mathematical analogy with the Prandtl boundary-layer theory for the flow of a viscous fluid around an obstacle (15). It is possible, as will be shown, to design plates for arbitrarily large values of Λ without exceeding given limits for the maximum stresses and without violating the assumptions upon which the von Kármán equations rest.

In the present paper are given the results of an investigation together with an outline of the mathematical methods. A detailed explanation, with proofs, of the mathematical methods will be given in a forthcoming paper (26).[5]

1—Mathematical Formulation

Consider a plane circular plate with radius R and thickness h subjected to a constant radial thrust at the edge; assume the plate simply supported, i.e., that deflection and radial bending moment vanish at the edge. Further, assume that the deflection w of the middle surface depends only upon the distance r from the center of the plate; a consequence of this assumption is that the radial and circumferential stresses also depend only upon r. Quantities which carry the subscript e or 0 refer to their values at the edge ($r = R$) or center ($r = 0$) respectively.

The von Kármán equations in Cartesian coordinates x and y are

$$E^{-1}\nabla^4\varphi = \left(\frac{\partial^2 w}{\partial x \partial y}\right)^2 - \frac{\partial^2 w}{\partial x^2}\frac{\partial^2 w}{\partial y^2} \qquad [1]$$

$$\gamma^2 h^2 E \nabla^4 w = \frac{\partial^2\varphi}{\partial y^2}\frac{\partial^2 w}{\partial x^2} - 2\frac{\partial^2\varphi}{\partial x \partial y}\frac{\partial^2 w}{\partial x \partial y} + \frac{\partial^2\varphi}{\partial x^2}\frac{\partial^2 w}{\partial y^2} \qquad [2]$$

where φ is the Airy stress function, E the modulus of elasticity, and $\gamma = [12(1 - \nu^2)]^{-1/2}$, where ν is Poisson's ratio. These equations imply that the stresses in the plate are given by combination of two stress systems, the bending stresses and the "membrane" stresses.

For the case of the circular plate with radial symmetry these nonlinear equations can be greatly simplified. Since φ and w depend only on r, the operation ∇^2 reduces to

$$\nabla^2 = \frac{d^2}{dr^2} + r^{-1}\frac{d}{dr} = r^{-1}\frac{d}{dr}r\frac{d}{dr} \qquad [3]^6$$

and the von Kármán equations assume the form

$$E^{-1}r^{-1}\frac{d}{dr}r\frac{d}{dr}r^{-1}\frac{d}{dr}r\frac{dw}{dr} = -\frac{d^2w}{dr}r^{-1}\frac{dw}{dr} \qquad [4]$$

$$\gamma^2 h^2 E r^{-1}\frac{d}{dr}r\frac{d}{dr}r^{-1}\frac{d}{dr}r\frac{d\varphi}{dr} = \frac{d^2\varphi}{dr^2}r^{-1}\frac{dw}{dr} + r^{-1}\frac{d\varphi}{dr}\frac{d^2w}{dr^2} \qquad [5]$$

The right members here can be written in the form

$$-\frac{1}{2}r^{-1}\frac{d}{dr}\left(\frac{dw}{dr}\right)^2 \quad \text{and} \quad r^{-1}\frac{d}{dr}\left(\frac{d\varphi}{dr}\frac{dw}{dr}\right)0;$$

hence the equations can be integrated

$$E^{-1}r\frac{d}{dr}r^{-1}\frac{d}{dr}r\frac{dw}{dr} = -\frac{1}{2}\left(\frac{dw}{dr}\right)^2 + \text{const} \qquad [6]$$

$$h^2 E r\frac{d}{dr}r^{-1}\frac{d}{dr}r\frac{d\varphi}{dr} = \frac{d\varphi}{dr}\frac{dw}{dr} + \text{const} \qquad [7]$$

Since the left members and also dw/dr vanish for $r = 0$, the constants are zero. The equations are further simplified by introducing the quantities

$$p = -r^{-1}\,d\varphi/dr, \quad q = -Rr^{-1}\,dw/dr$$

the operation

$$\Gamma = R^{-2}r^{-1}\frac{d}{dr}r^{-1}\frac{d}{dr}r^{-2} = R^{-2}r^{-3}\frac{d}{dr}r^3\frac{d}{dr}$$

and the constant

$$\eta = \gamma h/R$$

which may be called the "slenderness ratio." Then the equations assume the form

$$E^{-1}\Gamma p = q^2/2 \qquad [8]$$

$$E\Gamma q + pq = 0 \qquad [9]$$

The quantity $-p$ is the radial membrane stress; the authors prefer to denote a membrane stress by $-p$ rather than by σ since it is a compression when buckling begins, i.e., p is then positive. The significance of the quantity q is that $(r/R)q$ is the slope of the deflected plate; in particular, q_e, the value of q at the edge, is the slope at the edge; the quantity $-q/R = r^{-1}dw/dr$ could also be interpreted as the circumferential curvature of the deflected plate.

The boundary conditions to be satisfied at the edge ($r = R$) in terms of p and q are

$$rdq/dr + (1 + \nu)q = 0 \qquad [10]$$

and

$$p = p_e \qquad [11]$$

The important quantity p_e is the prescribed radial compression at the edge. Equation [10] states that the bending moment at the edge vanishes. The conditions for $r = 0$ are

$$dp/dr = 0, \quad dq/dr = 0 \qquad [12, 13]$$

which, on account of symmetry, imply regularity of p and q at the center of the plate.

Of the constants R, h, E, ν, p_e, which characterize the problem, only two, ν and $p_e/E\eta^2$, are essential in the sense that deflection and stresses in two plates having the same values of these ratios differ only by constant multipliers. This can be seen by introducing r/R, $p/E\eta^2$, q/η as new variables; the formulation of the problem obtained in this way contains only ν and p_e/η^2E as parameters.

Once p and q have been determined, the circumferential membrane stress $-p_c$ and the radial bending stress σ_b at the upper surface of the plate are given by

$$p_c = rdp/dr + p \qquad [14]$$

and

$$\sigma_b = 6\gamma\eta[(rdq/dr) + (1 + \nu)q] \qquad [15]$$

The deflection w is given by

$$w = R^{-1}\int^R qr^{-1}dr \qquad [16]$$

thus $w(R) = 0$.

For sufficiently small values of p_e/η^2E there will be no buckling of the plate—corresponding to the fact that the mathematical problem has then as sole solution $q \equiv 0$, $p \equiv$ const. At a certain value p_E of p_e buckling will begin; for $p_e > p_E$ there will

[5] For a preliminary report, see reference (5).
[6] Cf., e.g., reference (21).

be two stable buckled states[7] differing only in the sign of w, while the state $w \equiv 0$, which continues to be a possible state of equilibrium, is instable. The critical value p_E of p_e is given by $p_E/\eta^2 E = 4.24$ when the value 0.318 is taken[8] for Poisson's ratio ν. This value of the critical or "Euler" pressure p_E is obtained from Equation [9] by assuming p to be constant, that is, by the standard linear treatment of the buckling problem. Cf. reference (26).

Instead of using $p_e/\eta^2 E$ as essential parameter, the authors prefer to use the ratio $\Lambda = p_e/p_E = p_e/4.24\eta^2 E$ in what follows.

2—Methods of Solution

There are at least two methods available for solving this problem: the perturbation and the power-series methods. Both methods are presented. The perturbation method is found to be manageable only for a rather low range of values of the ratio Λ. This fact is in itself of interest since the perturbation method has been used extensively in the case of the rectangular plate, where an analogue of the power-series method is not available. The power-series method can be used successfully for a much higher range of values of Λ if solutions for a low range of values of Λ have been obtained by some other means, for example by the perturbation method. In general, the authors recommend the power-series method for problems similar to the present one, and for this reason it will be explained in some detail. However, the power-series method also becomes cumbersome as Λ increases, and eventually one is forced to turn to an asymptotic treatment.

3—Perturbation Method

This method consists in a development of p and q with respect to a parameter ϵ, which may be chosen, for example, as $\epsilon = \sqrt{(\Lambda - 1)}$:

$$p = p^{(0)} + \epsilon^2 p^{(2)} + \epsilon^4 p^{(4)} + \ldots \quad [17]$$

$$q = \epsilon q^{(1)} + \epsilon^3 q^{(3)} + \ldots \quad [18]$$

Upon substituting in the original differential equations and equating coefficients of like powers of ϵ, one obtains a sequence of

[7] That the equations have only two solutions is a consequence of their nonlinearity; linear theories of buckling always yield infinitely many buckled states differing only by constant factors.

[8] This value for ν, a typical value for aluminum alloy, was taken to facilitate calculation of p_E and was also used throughout the calculations.

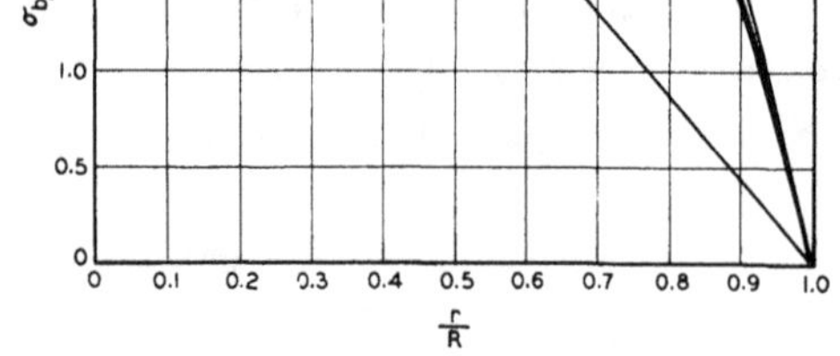

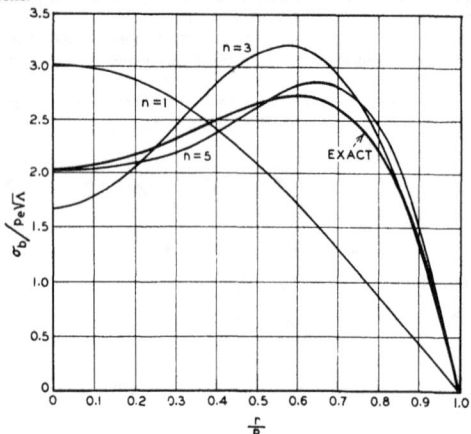

Fig. 1 Convergence of Perturbation Method for $\Lambda = 2.5$

linear differential equations for $p^{(0)}, q^{(1)}, p^{(2)}, q^{(3)}, \ldots$, which can be solved successively, though with great labor. The quantity $p^{(0)}$ proves to be a constant identical with the lowest value p_E of p_e for which buckling begins and is the same as that obtained by the linear theory (26); its value, as stated previously, is $p^{(0)} = p_E = 4.24\eta^2 E$.

The authors have computed $q^{(1)}$ to $q^{(5)}$ and $p^{(0)}$ to $p^{(6)}$. The convergence is satisfactory only for a small range of values of $\Lambda = p_e/p_E$, namely, for q up to $\Lambda = 1.2, 1.8, 2.5$ if terms of the first, third, and fifth orders respectively are included; and for p up to $\Lambda = 1.4, 2.2, 2.8$ for terms of the second, fourth, and sixth orders respectively. For the other stresses the convergence is not quite as good. Nevertheless, even for these low values of Λ, some distinctive and rather surprising qualitative phenomena of the buckling are found. See section 5.

In Fig. 1 curves are shown which indicate the accuracy of the successive approximations. For this purpose it was chosen to plot $\sigma_b/p_e\sqrt{\Lambda}$ as a function of r/R for $\Lambda = 2.5$. This appears to be the least accurately obtained stress. The figure shows the curves obtained by taking successively terms up to first, third, and fifth orders in ϵ, together with the curve obtained by an exact solution to be discussed in the next section. The curve obtained by taking terms up to fifth order would give for the stress σ_b values in error by as much as 10 per cent.[9]

4—Power-Series Method

The power-series method was used by Way (22) for solving the von Kármán equations as applied to the bending of circular plates under lateral loads and edge moment. See also reference (7). The method consists in setting up series for p and q in powers of r, with assumed values for the first coefficient of each series, i.e., for the values of p and q at the center of the plate.

Here are introduced a new independent variable

$$\alpha = AR^{-1}r, \quad 0 \leq \alpha \leq A \quad [19]$$

and the new functions to replace p and q

$$\pi = A^{-2}p/\eta^2 E, \quad \kappa = A^{-2}q/\eta \quad [20]$$

Differential Equations [8] and [9] assume the form

$$\alpha^{-3}\frac{d}{d\alpha}\left(\alpha^3 \frac{d\pi}{d\alpha}\right) = \frac{1}{2}\kappa^2 \quad [21]$$

$$\alpha^{-3}\frac{d}{d\alpha}\left(\alpha^3 \frac{d\kappa}{d\alpha}\right) + \pi\kappa = 0 \quad [22]$$

with boundary conditions

$$\frac{d\pi}{d\alpha} = 0, \quad \frac{d\kappa}{d\alpha} = 0 \quad \text{for} \quad \alpha = 0 \quad [23]$$

$$B(\alpha) \equiv \alpha\frac{d\kappa}{d\alpha} + (1 + \nu)\kappa = 0 \quad \text{for} \quad \alpha = A \quad [24]$$

$$\pi = \pi_e \quad \text{for} \quad \alpha = A \quad [25]$$

Instead of prescribing π_e and A, which would be equivalent to the

[9] Marguerre (13) mentions that his solutions for the rectangular plate, which implied two perturbations, may perhaps be valid up to $\Lambda = 20$; Timoshenko (20) applies his Ritz method with three constants for $\Lambda = 59.7$ and mentions that an additional constant would be necessary only for $\Lambda > 50$. In the simpler case of the circular plate it is found that values of Λ of these magnitudes cannot be treated with three perturbations. In fact such values of Λ lead to solutions already well in the asymptotic range. See section 6.

The method of Federhofer (24), see also footnote,[4] corresponds to the determination of $q^{(1)}$ and $p^{(2)}$ by the Galerkin method. Federhofer's numerical results coincide with the authors' quite well up to $\Lambda = 1.2$; they deviate from them by 10 per cent for $\Lambda = 1.5$.

original formulation, one proceeds as follows: $\pi(0)$ and $\kappa(0)$ are chosen, A is determined by solving Equation [24], and π_e is calculated from Equation [25]; p and q are then determined from Equation [20].

It is easily seen that power series for π and κ contain only even powers of α

$$\pi = \sum_{k=0}^{\infty} \pi_k \alpha^{2k}, \quad \kappa = \sum_{k=0}^{\infty} \kappa_k \alpha^{2k} \quad \ldots\ldots [26]$$

Upon insertion of these power series into Equations [21] and [22] one obtains the following recursion formulas for π_k and κ_k

$$\left.\begin{aligned} 2k(2k+2)\pi_k &= \frac{1}{2}\sum_{n+m=k-1} \kappa_m \kappa_n \\ 2k(2k+2)\kappa_k &= -\sum_{m+n=k-1} \pi_m \kappa_n \end{aligned}\right\} \ldots\ldots [27]$$

After having chosen $\pi(0) = \pi_0$ and $\kappa(0) = \kappa_0$ the successive coefficients are calculated from these formulas. Equation [24] becomes

$$B(\alpha) \equiv \sum_{k=0}^{\infty} (2k + 1 + \nu)\kappa_k \alpha^{2k} = 0 \ldots\ldots [28]$$

It must be solved for its lowest root $\alpha = A$.

One of the principal difficulties encountered in working with the power-series method consists in solving Equation [28]. Unless $A \sim 1$, the amount of labor involved in its solution is excessive. One reason for this is that otherwise one has no means of knowing the accuracy with which π_k and κ_k need be calculated and also the number of such coefficients necessary for reasonable accuracy. It might also be noted that the left side of Equation [28] changes rapidly with changes in α, and hence it is difficult to find the root A if no advance estimate of its value is available. What is evidently required for this purpose is a good estimate for π_0 and κ_0. This is, of course, also necessary if the solution is desired for a given value of Λ. One of the main purposes of this paper is to give solutions over a considerable range of values of Λ and with a fairly even distribution of such values.[10] For this objective it was found that solutions by the perturbation method were very helpful as a starting point.

As an example, the calculations are presented which yielded the solution for $\Lambda = 14.7$ (the highest value of Λ for which the power-series method was used). What was desired was the solution for $\Lambda = 15$; by extrapolation from previously calculated solutions,[11] which included values of Λ up to 7.3, it was found that $\kappa_0 = 7.0$ and $\pi_0 = -44.0$ were reasonable values

TABLE 1

k	π_k	κ_k	k	π_k	κ_k
0	—44.00	7.00	16	2.71	0.86
1	3.06	38.50	17	1.99	—0.96
2	11.23	69.69	18	1.02	—1.75
3	25.60	59.79	19	0.18	—1.71
4	38.77	22.57	20	—0.35	—1.21
5	40.74	—10.25	21	—0.55	—0.59
6	29.55	—28.29	22	—0.51	—0.08
7	12.36	—33.37	23	—0.35	0.23
8	— 2.39	—28.56	24	—0.16	0.34
9	—10.60	—17.79	25	—0.01	0.30
10	—12.44	— 5.86	26	0.08	0.20
11	—10.03	3.30	27	0.10	0.09
12	— 5.79	8.00	28	0.09	0.00
13	— 1.64	8.62	29	0.06	—0.05
14	1.29	6.61	30	0.02	—0.06
15	2.65	3.60			

[10] It is impractical for two reasons to choose values of π_0 and κ_0 more or less at random: After lengthy computations it can appear that two different choices for π_0 and κ_0 lead to nearly the same solution; the values of π_0 and κ_0 change with increasing Λ in rather unexpected ways, for example, π_0 changes sign, and κ_0 starts at zero for $\Lambda = 1$ and tends again to zero for large values of Λ.

[11] Figs. 2 and 6 give values of p_0/p_e and $\sigma_{b0}/p_e\sqrt{\Lambda}$. From these one derives the values $\pi_0 = (p_E/\eta^2{}_E)\Lambda p_0/p_e$ and $\kappa_0 = C\Lambda^{3/2}\sigma_{b0}/p_e\sqrt{\Lambda}$, where $C = (p_E/\eta^2 E)/6\gamma(1+\nu) = 1.76$ and $(p_E/\eta^2{}_E) = 4.24$.

for the first coefficients in the series. Equations [27] then lead to the coefficients π_k and κ_k given in Table 1.

In order to solve Equation [28] $B(1)$ was calculated and found to be —155.0 instead of zero. To apply Newton's method $\alpha dB/d\alpha$ was calculated for $\alpha = 1$ and found to be —7800. The resulting new value for A, replacing the value 1, is then $A = 0.98$. Whence $B(0.98) = -9.14$. Repeating this process once more, $A = 0.978$, $B(A) = +2.16$. The value 2.16 for $B(A)$ is of the same order as the values of the last terms in the series for $B(\alpha)$; therefore further improvement in the value of A is not possible using these coefficients. With this value of A, the quantity π_e was calculated from Equation [25] and value of $\Lambda = \pi_e A^2 \eta^2 E/p_E$ was found to be 14.7 instead of 15.

The number of coefficients necessary for sufficient accuracy increases with Λ. For $\Lambda = 14.7$, as we have seen, 30 terms are needed; 23 terms suffice for $\Lambda = 7.3$.

The authors recommend a similar procedure for problems of this type involving circular plates with radial symmetry.

5—Discussion of Results

Curves are given showing the behavior of the deflection w, the slope $-dw/dr = rq/R$, the radial membrane stress p, the circumferential membrane stress p_c, and the radial bending stress σ_b. Actually, these quantities are multiplied by appropriate factors, so chosen that all remain finite as Λ tends to infinity, and are plotted with Λ as variable. Table 2 gives the calculated

TABLE 2

Fig.	2	6	6	6	8	8
Λ	p_0/p_e	$\dfrac{p_{ce}}{p_e\sqrt{\Lambda}}$	$\dfrac{\sigma_{b0}}{p_e\sqrt{\Lambda}}$	$\dfrac{\sigma_{b\max}}{p_e\sqrt{\Lambda}}$	$\dfrac{q_e}{\sqrt{(p_e\Lambda/E)}}$	$\dfrac{w_0}{\gamma h\sqrt{\Lambda}}$
1.00	1.00	1.00	0	0	0	0
1.016	0.959	1.06	1.17	1.17	0.57	....
1.065	0.823	1.21	2.15	2.15	1.09	1.58
1.145	0.634	1.42	2.83	2.83	1.55	....
1.26	0.416	1.67	3.18	3.18	1.93	2.92
1.40	0.195	1.90	3.25	3.25	2.22	....
1.57	0	2.11	3.13	3.13	2.44	3.91
1.78	—0.182	2.30	2.88	....	2.63	....
2.01	—0.326	2.46	2.59	2.87	2.77	4.65
2.49	—0.519	2.64	2.03	2.75	2.91	5.15
3.76	—0.714	2.88	1.14	....	3.08	....
4.08	—0.730	2.92	1.00	....	3.09	....
5.12	—0.758	3.00	0.68	2.58	3.15	6.10
7.31	—0.750	3.09	0.35	....	3.20	....
14.74	—0.673	3.20	0.07	2.49	3.27	6.65
∞	—0.473	3.33	0	2.29	3.33	6.78

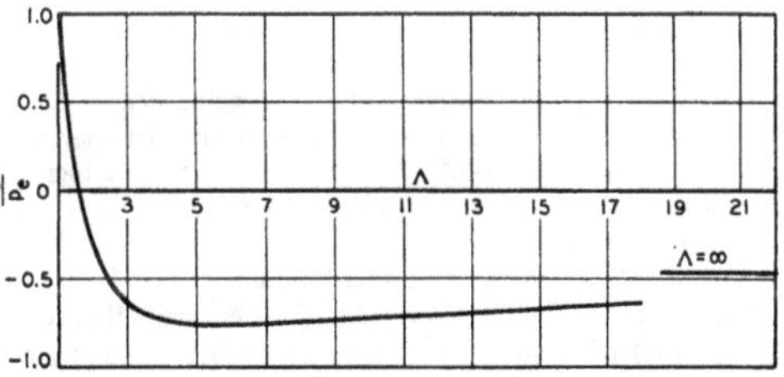

Fig. 2 Radial Membrane Stress at Center

values from which the curves were obtained. Graphs are also given, for various fixed values of Λ, with r/R as variable.

It is convenient to discuss here limit situations for $\Lambda \to \infty$ and plot limit values of all quantities, though the mathematical discussion of these matters is given in the next section.[12]

Radial Membrane Stress. In Fig. 2 the curve for the quantity p_0/p_e is given to show the behavior of the radial membrane stress p_0 at the center of the plate in its dependence on Λ. The plate

[12] The earlier ones were obtained from the perturbation method. For $\Lambda = 1.57$ identical results were obtained using the power-series method.

is in a state of hydrostatic compression for $\Lambda = 1$, at which buckling just begins. With increase of Λ one expects p_0/p_e to decrease because of the buckling and consequent stretch of the middle surface. From the figure it is seen that p_0/p_e decreases rapidly and, what is surprising, eventually becomes negative,

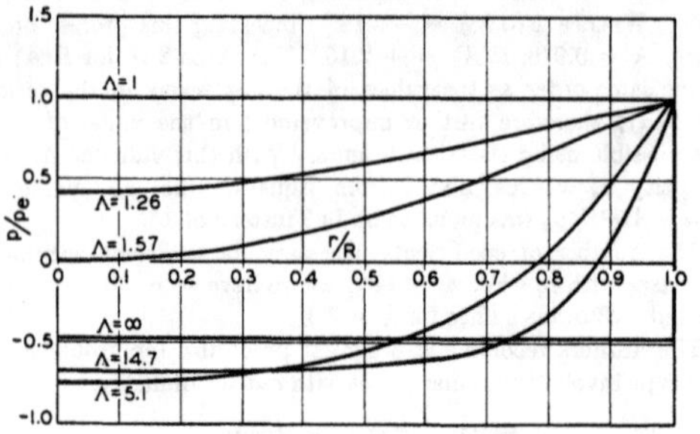

FIG. 3 RADIAL MEMBRANE STRESSES

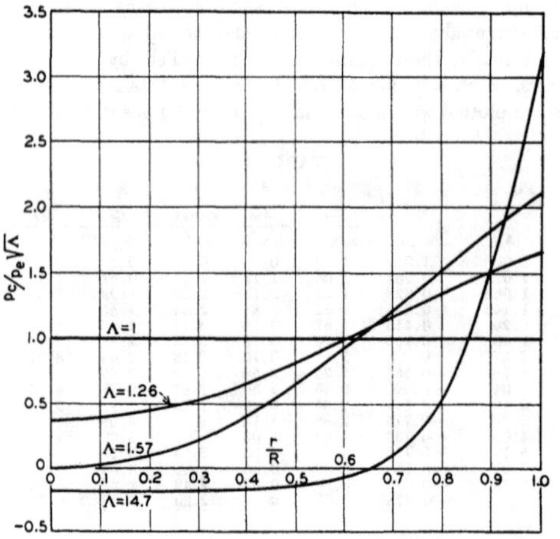

FIG. 4 CIRCUMFERENTIAL MEMBRANE STRESS

i.e., p_0 becomes a tensile stress. The transition occurs for the quite low value $\Lambda = 1.57$. For $\Lambda \sim 5.5$ the quantity p_0/p_e has an absolute minimum (—0.76); from this point on it increases slowly and attains the value —0.67 for $\Lambda = 14.7$. The asymptotic value for $\Lambda \rightarrow \infty$ is —0.47.

Fig. 3 shows curves for p/p_e as a function of r/R for various fixed values of Λ. The main feature of these curves is that for the higher values of Λ this quantity tends to become a negative constant over the central portion of the plate and then to rise rapidly to positive values in a strip near the edge. As will be shown in the next section, the width of this "boundary layer" tends to zero with increasing Λ. The curve marked $\Lambda = \infty$ indicates the limit curve for $\Lambda \rightarrow \infty$.

Maximum Stresses. Fig. 4 depicts the circumferential membrane stress p_c as a function of r/R for various fixed values of Λ. The values of p_c and of the radial membrane stress p are the same at the center of the plate (i.e., $p_{c0} = p_0$) because of symmetry, hence p_{c0} becomes negative beyond $\Lambda = 1.57$. The stress p_c increases much more rapidly than p toward the edge of the plate and for all values of Λ retains its maximum at the edge. These maxima (p_{ce}) increase with Λ in higher degree than the maxima p_e of p: It was hence convenient to plot $p_c/p_e\sqrt{\Lambda}$ rather than p_c/p_e. For high values of Λ the curves indicate that p_c/p_e tends to become constant over an increasingly large portion of the plate. Since $p_{c0} = p_0$, it is clear that p_{c0}/p_e has the limit value —0.47.

Curves for the radial bending stress σ_b (actually $\sigma_b/p_e\sqrt{\Lambda}$) as a function of r/R for various values of Λ are shown by Fig. 5. For $\Lambda = 1$, when buckling just begins, $\sigma_b \equiv 0$. The maximum of σ_b is attained at the center until $\Lambda = 1.57$; up to this point the curves present no striking features. When Λ increases beyond 1.57, however, the point where p_b attains its maximum is no longer at the center, but shifts toward the edge of the plate. When $\Lambda = 2.01$, for example, the maximum is already beyond $r/R = 0.5$; for $\Lambda = 14.7$ the maximum occurs at $r/R = 0.87$. For large values of Λ, $\sigma_b/p_e\sqrt{\Lambda}$ at the center of the plate decreases rapidly and tends to zero as $\Lambda \rightarrow \infty$; it might be mentioned that the same is also true of the quantity σ_b/p_e at the center. With increasing Λ the high values of σ_b tend to be concentrated in a boundary layer.

The variation of $p_{ce}/p_e\sqrt{\Lambda}$, $\sigma_{b0}/p_e\sqrt{\Lambda}$, and $\sigma_b \max/p_e\sqrt{\Lambda}$ with Λ is shown in Fig. 6. The limit values obtained by the asymptotic

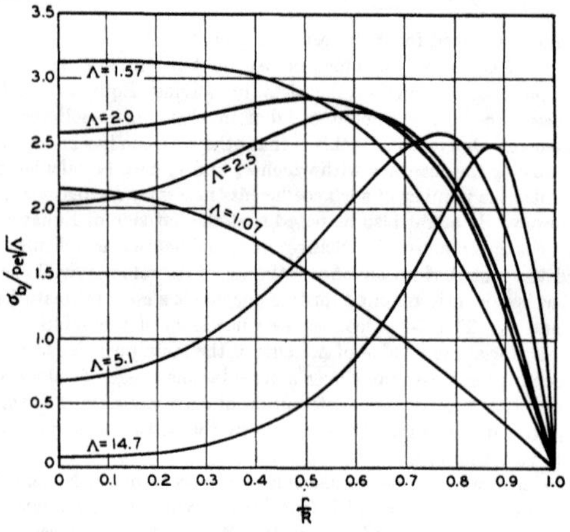

FIG. 5 RADIAL BENDING STRESS

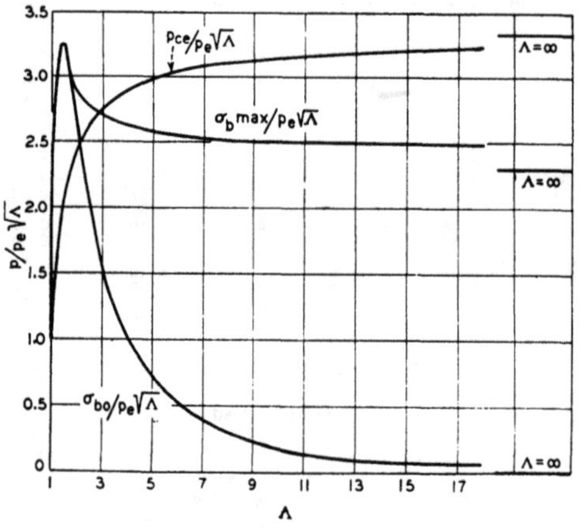

FIG. 6 MAXIMUM STRESSES

theory are also shown and the curves indicate clearly that these quantities really do tend to the limit values.

The circumferential bending stress σ_{bc} is not shown.[13] Its maximum is attained at the center and is the same as that of σ_b at this point up to $\Lambda = 1.57$; beyond this value of Λ, the maximum point also shifts toward the edge but the maximum value is always less than that of σ_b. Asymptotically $\sigma_{bc} = \nu\sigma_b$. Further, the maximum of the radial membrane stress p is always less than that of p_c. Hence, for design purposes, if the maximum-stress theory is used, only the stresses σ_b and p_c need be considered.[14] From $\Lambda = 3$ on, the circumferential membrane stress p_{ce} dominates.[15]

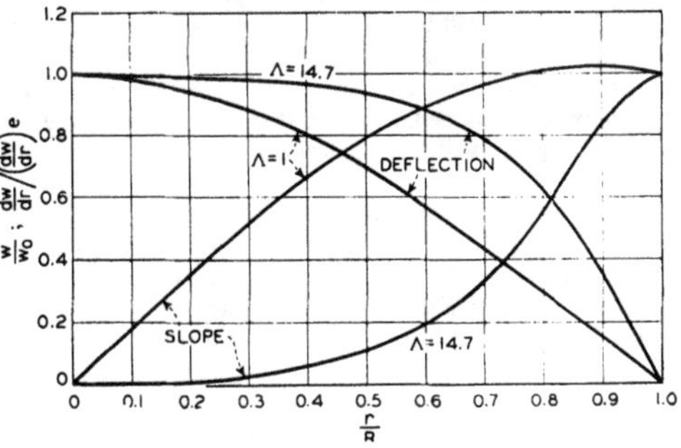

Fig. 7 Slope and Deflection

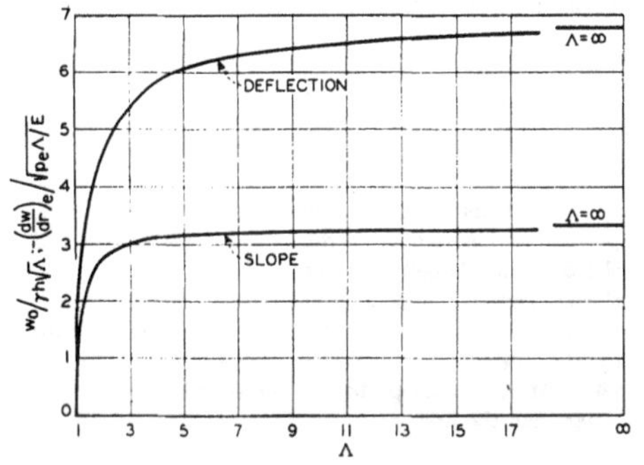

Fig. 8 Slope and Deflection

Slope and Deflection. In Figs. 7 and 8 the deflection w and the slope dw/dr are shown. In Fig. 7 the ratio of w and dw/dr to their maxima, i.e., w/w_0 and $(dw/dr)/(dw/dr)_e = rq/Rq_e$, are given as functions of r/R for the values $\Lambda = 1$ and $\Lambda = 14.7$. There is a quite perceptible flattening of the deflection curve for the higher values of Λ, accompanied by a tendency of the slope to approach zero in the central portion of the plate. Asymptotically the curve for w/w_0 would consist of two segments: a horizontal straight line up to the edge ($r/R = 1$) of the plate and a vertical line from it down to zero. Fig. 8 shows a rapid approach of the quantities $w_0/\gamma h\sqrt{\Lambda}$ and $-(dw/dr)_e/\sqrt{(p_e\Lambda/E)} = q_e/\sqrt{(p_e\Lambda/E)}$ to their asymptotic values.

Asymptotic Range. It is seen from the curves that the solutions from about $\Lambda = 5$ on do not change qualitatively; even quantitatively there is little change. Hence, for the range $\Lambda > 5$, which might be called the "asymptotic range," the solutions are at least roughly approximated by the asymptotic solution.

6—Limit State and Asymptotic Solution

As is seen from the curves, the deflection and stresses are nearly constant over a large portion of the plate when Λ becomes large and abrupt changes in all quantities occur in a narrow strip near the edge of the plate. In addition, the quantities p_0/p_e, $p_{ce}/p_e\sqrt{\Lambda}$, $\sigma_{b_e}/p_e\sqrt{\Lambda}$, etc., appear to approach finite limit values as Λ tends to infinity. It is possible to give a mathematical treatment of this limit process and obtain from it the limit values.

An asymptotic treatment of laterally loaded plates has been given by Hencky following A. Föppl (7, 4). Cf. also (1). It consists in omitting the terms referring to bending in the von Kármán equations (i.e., the left member of Equation [2]). One is tempted to apply an analogous treatment to the present problem, but this would lead to incorrect results. The procedure analogous to Hencky's would be as follows: Since $\Lambda = p_e/4.24\eta^2E$ it is clear that $\Lambda \to \infty$ if the slenderness ratio η tends to zero while p_e is held fixed. Upon allowing $\eta \to 0$ in the differential equations [8] and [9] one obtains

$$E^{-1}\Gamma p = q^2/2, \quad pq = 0 \qquad [29, 30]$$

the only nonsingular solution of which is $q \equiv 0$, $p \equiv$ const. If, then, this constant were identified with the edge value p_e, one would obtain as limit state a hydrostatic compression, which would clearly be wrong since the limit state appears rather to be one of constant tension. See Fig. 3. The error does not lie in the solution $q \equiv 0$, $p \equiv$ const but in the manner of fixing the constant. The sudden change in p from tension to compression that takes place across the boundary layer must be considered; the value of the constant under discussion is to be identified with the limit value of p at the inner edge of the boundary layer. Hence it can be found in this case only by an independent treatment of the boundary layer.[16]

The essential feature in the mathematical treatment of the boundary layer (as is also the case with Prandtl's boundary-layer theory for viscous flow) consists in stretching the scale with increasing Λ in such a way that the width of the boundary layer (measured in the new scale) does not shrink to zero. This can be done by introducing the new variable

$$\beta = (1 - r/R)\lambda, \quad 0 \leq \beta \leq \lambda, \text{ where } \lambda = (p_e/E)^{1/2}/\eta$$

It is necessary also to introduce new dependent variables which remain finite as $\Lambda \to \infty$. Such variables are found to be $P = p/p_e$ and $Q = q/\gamma\sqrt{(p_e/E)}$. Introducing the new variables into the differential Equations [8] and [9] and then allowing λ to tend to infinity, gives the differential equations

$$\frac{d^2P}{d\beta^2} = \frac{1}{2}Q^2, \quad \frac{d^2Q}{d\beta^2} + PQ = 0 \qquad [31, 32]$$

[13] The stress σ_{bc} is given by the formula

$$\sigma_{bc} = 6\gamma\eta E[\nu r q_r + (1 + \nu)q]$$

[14] The relative magnitudes of the stresses are here given for the case $\Lambda = 14.7$. The stresses at the edge ($r/R = 1$) where p_c is maximum, and at r/R where σ_b is maximum, are as follows

r/R	$p/p_e\sqrt{\Lambda}$	$\sigma_b/p_e\sqrt{\Lambda}$	$p_c/p_e\sqrt{\Lambda}$	$\sigma_{bc}/p_e\sqrt{\Lambda}$	$\tau_b/p_e\sqrt{\Lambda}$
1.0	0.3	0	3.2	0.7	1.8
0.87	0	2.5	1.2	1.4	1.2

The quantity τ_b is the horizontal shear stress due to bending. (τ_b is, as one expects, small and it vanishes asymptotically.) It is clear that the stress p should be combined with σ_b and the stress p_c with σ_{bc}. But, in any case, the stress p_c at the edge governs for $\Lambda = 14.7$.

[15] This is, however, not true in the case of the clamped plate, where the maximum of σ_b as well as of p_c is always attained at the edge. The authors have solved the asymptotic problem for this case and have found that in the limit $\sigma_{be}/p_{ce} = 6\gamma = 1.8$; hence the bending stress governs.

[16] Hencky's method is, however, legitimate in his case where neither edge *compression* nor edge moment is prescribed.

The range of the variable β is obviously $0 \leq \beta \leq \infty$, where $\beta = 0$, $\beta = \infty$ correspond to the outer and inner edges of the boundary layer respectively. The boundary conditions become

$$P = 1, \quad dQ/d\beta = 0 \text{ for } \beta = 0$$

and regularity for $\beta = \infty$.

The authors have solved this nonlinear boundary-value problem and proved that its solution gives the limit values correctly (26)[17,18]. From the functions P and Q thus obtained finite limit values were also found for all quantities discussed in the preceding section. In particular, $P(\infty) = -0.47$, which is the limit value for p/p_e at the inner edge of the boundary layer. Hence the constant $-0.47\ p_e$ is the one needed to complete the solution by the method of Hencky. Consequently it is proper to take -0.47 as the limit value for the ratio p_0/p_e, which refers to the value of p/p_e at the center of the plate.

Mathematically there is only one passage to the limit characterized by

$$\Lambda \to \infty \text{ or } p_e/\eta^2E = p_eR^2/\gamma^2h^2E \to \infty \ldots\ldots\ldots [33]$$

For a fixed plate (R and h fixed), this would require $p_e \to \infty$, implying infinitely high stress. If p_e is held fixed, one might consider a series of plates for which the slenderness ratio $\eta \to 0$; for example the radius R might be held fixed and the thickness h allowed to approach zero. In this case the governing stresses would tend to become infinite, as one sees from the limit relation

$$p_{ce} \sim (3.33)p_e\sqrt{\Lambda} = 3.33\, p_e^{2/3}/(4.24)^{1/3}\eta E^{1/3} \to \infty \ldots\ldots [34]$$

However, for a value of Λ as large as one pleases, plates could be designed for which stress and slope would be as small as necessary to insure the validity of the von Kármán equations. In order to see this let η and p_e tend to zero in such a way that $p_e/\eta^2E \to \infty$ while $p_e/\eta E = \sqrt{(4.24\Lambda p_e/E)}$ remains sufficiently small; then it follows from the asymptotic formula $p_{ce} \sim (3.33)p_e\sqrt{\Lambda}$ for the greatest stress and the formula $q_e \sim (3.33)\sqrt{(p_e\Lambda/E)}$ for the greatest slope (see Table 2), that p_{ce} and q_e can be made as small as desired.

A valid objection on physical grounds to the application of our formulas for high values of Λ may well be raised: It is likely that there is a value of Λ for which the buckled state in its turn becomes unstable, and what might be called a "second buckling" sets in; such a second buckling would not, of course, possess the radial symmetry of the first buckled state. Rough calculations indicate, however, that the second buckling will probably not begin until Λ is somewhere in the asymptotic range ($\Lambda > 5$).

If one were to pursue the second buckling into its nonlinear range, it is likely that it would eventually also become instable and a "third buckling" ensue, and so on; that is, more and more wrinkles would appear at the edge of the plate. Various experimenters, working with rectangular plates, have found that the plates do become wrinkled near the edges (2, 3, 8, 12, 13, 18, 19). The authors feel that such wrinkles are to be explained as the result of a buckling of order higher than the first.

It is of interest to indicate the relation[19] between the authors' asymptotic results and the notion of effective width introduced by von Kármán for the rectangular plate which is simply supported on two opposite sides and subjected to a thrust at the other two sides. In spite of the obvious difference between the present case and that of von Kármán, it is nevertheless true that these results throw some light on the latter case and contribute to substantiate certain assumptions on which von Kármán bases the derivation of his formula. These assumptions concerning the stress distribution for high values of the thrust (or for small thickness) are as follows: 1 The stresses will be concentrated in a strip along the simply supported edge; and 2 the membrane stress on an element perpendicular to this strip rather than the bending stress will be the maximum stress. The same properties of the stress distribution are found mathematically for the present case of the simply supported circular plate with radial symmetry. It may be mentioned, however, that the stress distribution in the circular plate will not conform to property 2 if the plate is clamped.[15]

Von Kármán's formula can be considered as a relation between maximum and average value of the membrane stress on an element perpendicular to the boundary strip. In the present case the stress on an element perpendicular to the boundary strip is the circumferential membrane stress p_c. Its average value $\bar{p}_c$ is given by

$$\bar{p}_c = R^{-1}\int_0^R p_c dr = R^{-1}\int_0^R \frac{d}{dr}(rp)dr = p_e \ldots [35]$$

From the asymptotic formula

$$p_{ce}/p_e\sqrt{\Lambda} \to 3.3 \ldots\ldots\ldots\ldots\ldots\ldots [36]$$

there results

$$(\bar{p}_c/\eta^2)/(p_{ce}/\eta^2)^{2/3} \to 0.728 \ldots\ldots\ldots\ldots [37]^{20}$$

In von Kármán's formula the power 1/2 takes the place of the power 2/3. A strong discrepancy was to be expected owing to the different shape of the plate, the different way in which the thrust is applied, and the different wave pattern that is assumed. The authors think it likely that the discrepancy is caused mostly by the different wave pattern. The wave pattern assumed in von Kármán's case probably corresponds to buckling of higher order than the first.

[17] In much the same manner the asymptotic solutions for the buckling of the clamped plate and the plate under edge moment were obtained. It may be of interest to note that the boundary-layer phenomenon occurs in these cases also.

[18] It is possible to apply a perturbation method in the neighborhood of $\lambda = \infty$ by expansions in powers of $1/\lambda$. The foregoing solution would constitute the first step in such a procedure. The authors have carried out the second step also (26).

[19] This relation is also discussed by von Kármán (25)

[20] This number was given incorrectly in (5). See reference (26).

BIBLIOGRAPHY

1 "The Clamped Square Sheet," by D. G. Bourgin, *American Journal of Mathematics*, vol. 61, 1939, p. 418.

2 "Summary of the Present State of Knowledge Regarding Sheet-Metal Construction," by H. L. Cox, Aeronautical Research Committee Report No. 1553, 1933, His Majesty's Stationery Office, London, England, pp. 1–20.

3 "Buckling of Thin Plates in Compression," by H. L. Cox, Aeronautical Research Committee Report 1554, 1933, His Majesty's Stationery Office, London.

4 "Vorlesungen über technische Mechanik," by August Föppl, B. G. Teubner, Leipzig, Germany, 1907, vol. 5, section 24, p. 139.

5 "The Non-Linear Boundary Value Problem of the Buckled Plate," by K. O. Friedrichs and J. J. Stoker, Proceedings of the National Academy of Sciences, vol. 25, 1939, p. 535.

6 "Sur la rigidité de la tôle flambée," by A. Grzedzielski and W. Billewicz, *Sprawozdania Instytut Techniczny Lotnictwa*, vol. 10, 1937, p. 5.

7 "Über den Spannungszustand in kreisrunden Platten mit verschwindender Biegungssteifigkeit," by H. Hencky, *Zeitschrift für Mathematik und Physik*, vol. 63, 1915, p. 311.

"Die Berechnung dünner rechteckiger Platten mit verschwindender Biegungssteifigkeit," *Zeitschrift für angewandte Mathematik und Mechanik*, vol. 1, 1921, pp. 81, 423.

8 "Festigkeitsprobleme im Maschinenbau," by Th. von Kármán, *Enzyklopädie der mathematischen Wissenschaften*, vol. 4, 1910, article 27.8, p. 348.

9 "The Strength of Thin Plates in Compression," by Th. von Kármán, E. E. Sechler, and. L. H. Donnell, Trans. A.S.M.E., vol. 54, 1932, paper APM-54-5.

10 "Die über die Ausbeulgrenze belastete Platte. Energieansatz und Differentialgleichungen," by K. Marguerre, *Zeitschrift für angewandte Mathematik und Mechanik*, vol. 16, 1936, p. 353.

11 "Über die Tragfähigkeit eines längsbelasteten Plattenstreifens nach Überschreiten der Beullast," by K. Marguerre and E. Trefftz, *Zeitschrift für angewandte Mathematik und Mechanik*, vol. 17, 1937, p. 85.

12 "Die mittragende Breite der gedrückten Platte," by K. Marguerre, *Luftfahrtforschung*, vol. 14, 1937, p. 121.

13 "Verhalten eines von Schub- und Druckkräften beanspruchen Plattenstreifens oberhalb der Beulgrenze," by A. Kromm and K. Marguerre, *Luftfahrtforschung*, vol. 14, 1937, p. 627.

14 "Zum Problem der Plattenstabilität," by P. Polubarinova-Kotschina, *Applied Mathematics and Mechanics*, vol. 3, 1936, p. 16.

15 "The Mechanics of Viscous Fluids," by L. Prandtl, "Aerodynamic Theory," vol. 3, Julius Springer, Berlin, Germany, 1935, sec. 13, 14, pp. 80–90.

16 "Applied Elasticity," by John Prescott, Longmans, Green and Co., London, England, 1924, p. 496.

17 "Die Überschreitung der Knickgrenze bei dünnen Platten," by G. Schnadel, Proceedings, Third International Congress of Applied Mechanics, vol. 3, 1930, p. 73.

18 "Strength of Rectangular Flat Plates Under Edge Compression," by Louis Schuman and Goldie Back, National Advisory Committee for Aeronautics, Technical Report No. 356, 1930, Superintendent of Documents, Washington, D. C.

19 "Stress Distribution in Stiffened Panels Under Compression," by E. E. Sechler, *Journal of the Aeronautical Sciences*, vol. 4, 1937, p. 320.

20 "Theory of Elastic Stability," by S. Timoshenko, McGraw-Hill Book Co., Inc., New York, N. Y., First edition, 1936, pp. 321–323, 390–395.

21 "Theory of Elasticity," by S. Timoshenko, McGraw-Hill Book Co., Inc., New York, N. Y., First edition, 1934, chapter 3, section 21.

22 "Bending of Circular Plates With Large Deflection," by Stewart Way, Trans. A.S.M.E., vol. 56, 1934, p. 627.

23 "Buckling and Failure of Thin Rectangular Plates in Compression," by M. Yamamoto and K. Kondo, Tokyo Imperial University, Aeronautical Research Institute, Report No. 119, 1935.

24 "Tragfähigkeit der über die Beulgrenze belasteten Kreisplatte," by K. Federhofer, *Forschung auf dem Gebiet des Ingenieurwesens*, vol. 11, 1940, p. 97.

25 "The Engineer Grapples With Nonlinear Problems," by Th. von Kármán, *Bulletin of the American Mathematical Society*, vol. 46, 1940, p. 631.

26 "The Nonlinear Boundary Value Problem of the Buckled Plate," by K. O. Friedrichs and J. J. Stoker. To be published in the *American Journal of Mathematics*.

[47–2] **On the boundary-value problems of the theory of elasticity and Korn's inequality, Ann. Math., 48 (1947), 441–471.**

Commentary:

In the classical linear theory of elasticity the usual boundary conditions either involve prescribed displacements u_i ("first boundary value problem") or prescribed tractions ("second boundary value problem"), or a combination of these. Proving existence of solutions of the second problem is made difficult by the fact that the potential energy density is a definite quadratic form in the strain components $s_{ik} = \frac{1}{2}(u_{i/k} + u_{k/i})$ but is only semi-definite in the displacement gradients $u_{i/k} = \partial u_i/\partial x_k$ (since rigid motions evoke no stresses). Korn's inequality bridges this gap by asserting that the Hilbert norm of the $u_{i/k}$ can be estimated in terms of that of the s_{ik}, provided the resultant of the $r_{ik} = \frac{1}{2}(u_{i/k} - u_{k/i})$ vanishes. (Equivalently the norm of the $u_{i/k}$ can be estimated in terms of the sum of the norms of the s_{ik} and u_i without restricting the r_{ik}). Friedrichs gave the first accessible proof of this fundamental non-trivial inequality, by reducing it to the solution of the first boundary value problem.

Korn's inequality contains a constant K ("Korn constant"), which depends on the domain. Numerical estimates for K for special types of domains have been given by B. Bernstein and R. Toupin, [1], and J.H. Bramble and L.E. Payne, [2]. The best value of K for a sphere was found by L.E. Payne and H.F. Weinberger, [3].

Different proofs, based on singular integrals, for Korn's inequality and more general coercive inequalities have been given by K.T. Smith, [4] and J. Gobert, [5]. For a proof using functional analysis arguments see G. Duvaut and J.L. Lions, [6].

[1] B. Bernstein and R. Toupin. Korn inequalities for the sphere and circle, Arch. Rat. Mech. Anal., 6 (1960), 51–64.

[2] J.H. Bramble and L.E. Payne. Some inequalities for vector functions with applications in elasticity, Arch. Rat. Mech. Anal., 11 (1962), 16–26.

[3] L.E. Payne and H.F. Weinberger. On Korn's inequality, Arch. Rat. Mech. Anal., 8 (1961), 89–98.

[4] K.T. Smith. Inequalities for formally positive integro-differential forms, Bull. A.M.S., 6 (1961), 368–370.

[5] J. Gobert. Une inegalite fondamentale de la theorie de l'elasticite, Bull. Soc. Roy. Sci. Liege, 31 (1962), 182–191.

[6] G. Duvaut and J.L. Lions. Inequalities in Mechanics and Physics, Springer-Verlag, 1976).

ANNALS OF MATHEMATICS
Vol. 48, No. 2, April, 1947

ON THE BOUNDARY-VALUE PROBLEMS OF THE THEORY OF ELASTICITY AND KORN'S INEQUALITY

BY K. O. FRIEDRICHS

(Received March 21, 1946)

When one attacks the boundary-value and eigen-value problems of elasticity by means of minimum properties one encounters a peculiar difficulty. The strain energy of a deformation $\{u_1, u_2, u_3\}$ considered a function of x_1, x_2, x_3 in a connected domain $\mathfrak{R}$ is given by[1]

$$E(u) = \int_{\mathfrak{R}} \{a s_{\kappa\lambda}^2 + b s_{\kappa\lambda}^2\}\, dx, \qquad dx = dx_1\, dx_2\, dx_3, \tag{1}$$

where

$$s_{kl} = \tfrac{1}{2}(u_{k/l} + u_{l/k}) \tag{2}$$

with

$$u_{k/l} = \partial u_k / \partial x_l\,;$$

a and b are appropriate positive constants. The minimum problem consists in minimizing the total potential energy of which $E(u)$ is the dominant part.

In order to construct the solution from a minimizing sequence one needs several estimates. Such estimates can be obtained once one can establish that the Dirichlet integral

$$D(u) = \int_{\mathfrak{R}} (u_{\kappa/\lambda})^2\, dx \tag{3}$$

is bounded for a minimizing sequence. However, from the fact that the total potential energy approaches its lower limit, one can immediately only conclude that the expression $E(u)$ remains bounded. To derive the boundedness of $D(u)$ from that of $E(u)$ is the peculiar task that one encounters. The key for overcoming this difficulty is the inequality

$$D(u) \leqq K_0 E(u), \tag{4}$$

which holds under appropriate conditions discussed below, K_0 being a constant depending only on the domain $\mathfrak{R}$. The significance of this inequality will become clearer from an equivalent form of it, which in itself offers some interest.

In addition to the symmetric part s_{kl} of the derivative $\partial u_k/\partial x_l$ we introduce the antisymmetric part

$$r_{kl} = \tfrac{1}{2}(u_{kl} - u_{l/k}) \tag{5}$$

so that

$$u_{kl} = s_{kl} + r_{kl}\,. \tag{6}$$

---

[1] In this paper we omit the summation sign when it refers to greek subscripts occurring twice or in squared terms. Thus $s_{\kappa\kappa} = \sum_\kappa s_{\kappa\kappa}$, and $s_{\kappa\lambda}^2 = \sum_{\kappa\lambda} (s_{\kappa\lambda})^2$.

Relations (2) and (5) then imply

(7) $$u_{kl}^2 = s_{kl}^2 + r_{kl}^2 .$$

Further we set

(8) $$S(u) = \int_{\Re} s_{\kappa\lambda}^2 \, dx$$

and

(9) $$R(u) = \int_{\Re} r_{\kappa\lambda}^2 \, dx;$$

relation (7) then yields

(10) $$D(u) = S(u) + R(u).$$

On the other hand we have

(11) $$aS(u) \leqq E(u) \leqq (a + 3b)S(u)$$

as follows from

(12) $$s_{\kappa\kappa}^2 < 3s_{\kappa\lambda}^2 .$$

Inequality (4) is then seen to be equivalent with the inequality

(13) $$R(u) \leqq \mathrm{K}S(u),$$

K_1 being a constant depending only on the domain $\Re$. Clearly (4) follows from (13) by virtue of (11) with $\mathrm{K}_0 \geqq (\mathrm{K} + 1)/a$, and (13) follows from (4) by virtue of (12) with $\mathrm{K} \geqq (a + 3b)\mathrm{K}_0 - 1$. (Note that a $\mathrm{K}_0 \geqq 1$ since $aD(u) = E(u)$ for any gradients $u_k = \partial\varphi/\partial x_k$ of a potential function φ).

Inequality (13) cannot hold without imposing conditions on the functions u, since $S(u) = 0$ for the vectors

(14) $$b^{(12)} = (x_2 , -x_1 , 0), \; b^{(31)} = (-x_3 , 0, x_1), \quad b^{(23)} = (0, x_3 , -x_2),$$

for which $R(u) = 0$. It is immediately seen that the vectors $b^{(kl)}$ and their linear combinations are the only ones for which $S(u) = 0$ while $R(u) \neq 0$.[2] To exclude these vectors from admission we may impose either of the following two conditions.

First case:

(15) $u = 0$ at the boundary $\mathfrak{B}$ of the domain $\Re$.

Second case:

(16) $$\int_{\Re} r_{kl} \, dx = 0, \qquad k, l = 1, 2, 3.$$

[2] Relation $s_{kl} = 0$ for $k, B = 1, 2, 3$, implies $2u_{k/lm} = s_{kl/m} + s_{km/l} - s_{lm/k} = 0$; hence $u_k = c_{k\lambda}x_\lambda$ with constant c_{kl}. From $c_{kl} + c_{lk} = s_{kl} = 0$ we obtain $u_k = c_{12}b_k^{(12)} + c_{31}b_k^{(31)} + c_{23}b_k^{(23)}$.

It is clear that the vectors $b^{(kl)}$ satisfy neither (15) nor (16). Condition (16) could be interpreted as orthogonality to the vectors $b^{(kl)}$ with respect to the quadratic form R.

The basic inequality (13) under either condition (15) or (16) was formulated and proved by A. Korn (1906) [1] and (109) [2]. Korn's proof for the first case is very simple, that for the second case very complicated.[3]

In the present paper a new direct proof for Korn's inequality in the second case is offered;[4] it is based on a number of identities for the solutions of a differential equation associated with the form $S(u)$.

In the second part of the paper we shall first prove the existence of the solution of the first boundary value problem of elasticity, employing the first case of Korn's inequality; secondly we shall deal with the eigen-value problem under natural boundary conditions employing the second case of Korn's inequality. We base our treatment on the minimum properties of the solution, deviating in several respects from standard methods.[5] In particular we employ the mollifiers, introduced in earlier papers of the author [10], [11], as smoothing operators and further we are able to avoid completely any reference to the fundamental solution of the differential equation.

We prefer to treat the N-dimensional rather than the three-dimensional case. The notions introduced so far with reference to the three dimensions apply in obvious manner to N-dimensions.

Part I. Korn's Inequality

1. The Domain and the Admitted Functions

The "admitted" domains $\Re$ for which we shall prove our theorems are bounded[6] open connected regions in the space of N independent variables $x = x_1, \cdots, x_n$, satisfying certain requirements. To these admitted domains belong those whose boundaries have continuous normal vectors, also cubes and other domains with corners or edges.

We denote by R_δ the subdomain of all points in $\Re$ having a distance greater than δ from the boundary $\mathfrak{B}$ of $\Re$.[7] By $\Re - \Re_\delta$ we then denote the

[3] In his note, (see reference at the end of this paper) Korn refers to another paper of his [3], in which he deals at great length with the boundary value problem of elasticity from the point of view of integral equations referring to still earlier papers of his. The author of the present paper has been unable to verify Korn's proof for the second case.

[4] For the case of two dimensions a new proof of Korn's inequality was given in an earlier paper [9]. The present proof, when specialized to two variables is, however, simpler than that in the paper [9].

[5] References to a great number of papers dealing with boundary and characteristic value problems of elasticity from the point of view of integral equations are given in the bibliography.

[6] For the sake of simplicity we restrict ourselves to bounded domains. Modifications in definitions and argument sufficient to handle certain classes of unbounded domains will be indicated.

[7] If $\Re$ is not bounded we add to the definitions of $\Re_\delta$ the requirement that it lies in the sphere $S_\delta : x_\kappa^2 \leqq \delta^{-2}$.

boundary strip consisting of all points in $\mathfrak{R}$ having a distance of at most δ from $\mathfrak{B}$.[8] The nature of the admitted domains $\mathfrak{R}$, which we shall term Ω-*domains*, is characterized first by the existence of a vector $\Omega_k(x)$, $k = 1, \cdots, N$, defined in the domain $\mathfrak{R}$ and on its boundary $\mathfrak{B}$ such that

1) Ω_k is continuous in $\mathfrak{R} + \mathfrak{B}$
2) $\Omega_k = 0$ on $\mathfrak{B}$
3) Ω_k possesses in $\mathfrak{R}$ continuous derivatives

$$\Omega_{k/l} = \partial\Omega_k/\partial x_l.$$

4) The derivatives $\Omega_{k/l}$ are bounded in $\mathfrak{R}$,

$$| \Omega_{k/l} | \leqq C_1 . \tag{1.1}$$

5) For an appropriate connected boundary strip $\mathfrak{R} - \mathfrak{R}_{2v}$

$$\Omega_{\kappa/\kappa} \geqq 1 \text{ in } \mathfrak{R} - \mathfrak{R}_{2v} . \tag{1.2}$$

We secondly require the existence of a number C_2 and to every $\delta > 0, \leqq 2v$, the existence of a function $\mathrm{H} = \mathrm{H}^\delta(x)$ with continuous first and second derivatives $\mathrm{H}_{/l} = \mathrm{H}_l$ and $\mathrm{H}_{/lm} = \mathrm{H}_{lm}$ in $\mathfrak{R}$ and a $\delta' > 0, < \delta$, such that

$$\begin{aligned} \mathrm{H}(x) &= 1 \text{ in } \mathfrak{R}_\delta \\ &= 0 \text{ in } \mathfrak{R} - \mathfrak{R}_{\delta'} , \end{aligned} \tag{1.3}$$

$$0 < \mathrm{H}(x) \leqq 1 \text{ in } \mathfrak{R}, \text{ and}$$

$$| \mathrm{H}_l | \leqq C_2/\delta \text{ in } \mathfrak{R}. \tag{1.4}$$

These requirements express exactly those properties of the domain $\mathfrak{R}$ which are used in proving Korn's inequality. If the boundary $\mathfrak{B}$ of the domain $\mathfrak{R}$ can be imbedded in a set of equidistant surfaces with continuously differentiable normal vector, then the construction of functions Ω and H is very easy. Continuity of the normal vector is, however, sufficient. More specifically we state the

THEOREM: *The domain $\mathfrak{R}$ is an Ω-domain if there exists an inside neighborhood $\mathfrak{S}^\circ$ of its boundary $\mathfrak{B}$ which is a one-to-one image of a spherical shell $\rho_0 \leqq \rho \leqq 1$, ($\rho^2 = \xi_\kappa^2$), of a $[\xi_1, \cdots, \xi_N]$-space given by functions $x = x(\xi)$ which possess continuous first derivatives with non-vanishing Jacobian in $\rho_0 \leqq \rho \leqq 1$.*

The domain $\mathfrak{R}$ is also an Ω-domain if its boundary consists of a finite number of pieces of the character just described.

The proof of this theorem is given in a different paper [17]. In the same paper it is also shown that $\mathfrak{R}$ is an Ω-domain if it possesses a finite number of corners or edges, the neighborhoods of which are images of the corner or edge $0 < y_1, 0 < y_2, \cdots, 0 < y_n, n \leqq N$ by one-to-one transformations $x_k = x_k(y_1, \cdots, y_N)$ with bounded second derivatives and positive Jacobian. Accordingly, a cube or parallelopiped is an Ω-domain.

Of the functions $u = \{u_k(x)\}$ we assume that they possess continuous deriva-

[8] Or being outside of S_δ for unbounded domains $\mathfrak{R}$.

tives $u_{k/l}$ in $\mathfrak{R}$ such that the Dirichlet integral (cf. (3))

$$D(u) = \int_{\mathfrak{R}} u_{\kappa/\lambda}^2 \, dx \tag{1.5}$$

is finite. (Note that summation with respect to K and λ from 1 to N is implied.) The class of such functions will be called $\mathfrak{D}$. By $\mathfrak{D}^{\bullet}$ we denote the subspace of functions $\dot{u}$ to which there is a $\delta > 0$ such that $\dot{u} = 0$ in $\mathfrak{R} - \mathfrak{R}_\delta$; by $\mathfrak{D}^\circ$ we denote the class of all functions u in $\mathfrak{D}$ which can be approximated by functions $\dot{u}$ in $\mathfrak{D}$ such that $D(u - \dot{u})$ is arbitrarily small. The condition that a function u be in $\mathfrak{D}^\circ$ will serve us as a substitute for the condition (2) that u vanishes on $\mathfrak{B}$.

2. Korn's Inequality in the First Case

For functions in the class $\mathfrak{D}^\circ$ Korn's inequality in the form

$$R(u) \leqq S(u) \qquad \text{if } u \text{ in } \mathfrak{D}^\circ \tag{2.1}$$

is immediately established. It follows from the identity

$$\int_{\mathfrak{R}} r_{\kappa\lambda}^2 \, dx + \int_{\mathfrak{R}} s_{rr}^2 \, dx = \int_{\mathfrak{R}} s_{\kappa\lambda}^2 \, dx \tag{2.2}$$

valid for all functions in $\mathfrak{D}^\circ$.[9] To prove this identity we observe that by (6) $s_{\kappa\lambda}^2 - r_{\kappa\lambda}^2 = u_{\kappa/\lambda} u_{\lambda/\kappa}$, whence

$$\int_{\mathfrak{R}} \{s_{\kappa\lambda}^2 - s_{\mu\mu}^2 - r_{\kappa\lambda}^2\} \, dx = \int_{\mathfrak{R}} \{u_{\kappa/\lambda} u_{\kappa/\lambda} - u_{\kappa/\kappa} u_{\lambda/\lambda}\} \, dx.$$

We first assume that u is in $\mathfrak{D}^{\bullet}$ and possess continuous second derivatives. In view of relation (6) we then have

$$u_{\kappa/\lambda} u_{\lambda/\kappa} - u_{\kappa/\kappa} u_{\lambda/\lambda} = (u_\kappa u_{\lambda/\kappa})_\lambda - (u_\kappa u_{\lambda/\lambda})_\kappa .$$[10]

The integral of the last term over $\mathfrak{R}$ vanishes since u vanishes in the boundary strip $\mathfrak{R} - \mathfrak{R}_\delta$. Hence identity (2.1) holds for functions u in $\mathfrak{D}^{\bullet}$ which possess continuous second derivatives. Clearly, any function u in $\mathfrak{D}^{\bullet}$ can be approximated by functions in $\mathfrak{D}^{\bullet}$ possessing continuous second derivatives in such a manner that the first derivatives are approximated uniformly in $\mathfrak{R}$. Therefore (2.2) holds for all u in $\mathfrak{D}^{\bullet}$. Every function in $\mathfrak{D}^\circ$ can by definition be approximated by functions $\dot{u}$ in $\mathfrak{D}^{\bullet}$ such that $D(u - \dot{u}) \to 0$. In view of (7) and (12) it is clear that the terms occurring in (2.2) taken for $\dot{u}$ approach those taken for u. Thus identity (2.2) and inequality (2.1), is proved for all u in $\mathfrak{D}^\circ$. This proof, except for the handling of the boundary condition, was already given by Korn.

Korn's inequality for u in $\mathfrak{D}^\circ$ is already sufficient for the treatment of the first boundary problem of elasticity and, as a matter of fact, the existence of

---

[9] See A. Korn [1] and [2].

[10] Subscripts k, l, ... attached to parentheses indicate differentiation with respect to x_k, x_l, ...

the solution of this problem is needed to establish the second case of Korn's inequality, i.e., under condition (16). Instead of first treating the first boundary problem and then the second case of Korn's inequality we prefer to reverse the systematic order and show first how the second case of Korn's inequality can be handled.

3. Korn's Inequality in the Second Case

We first show that the second case of the inequality,

$$R(u) \leqq K_2 S(u) \tag{3.1}$$

under the condition

$$\int_{\mathfrak{R}} r_{kl}\, dx = 0, \qquad k, l = 1, 2, \cdots, N, \tag{3.2}$$

cf. (16), can be reduced to the first case once the inequality is established for all functions satisfying (3.2) which possess continuous derivatives of every order in $\mathfrak{R}$ and satisfy the differential equation[11]

$$s_{k\lambda/\lambda} = \tfrac{1}{2}(u_{k/\lambda\lambda} + u_{\lambda/\lambda k}) = 0, k = 1, \cdots, N. \tag{3.3}$$

We then speak of the "main" case of Korn's inequality, since the proof of the inequality in this case requires the main effort.

The reduction to the main case is based on the possibiity of splitting every function u into $\mathfrak{D}$ into

$$u = u' + u'' \tag{3.4}$$

such that u'' is in $\mathfrak{D}$, possesses continuous second derivatives in $\mathfrak{R}$, and satisfies equation (3.3) while u' is in $\mathfrak{D}^\circ$.

This possibility of splitting follows immediately from the theorem of the existence of the solution of the first boundary problem, to be proved in section 7, (see footnote[20]), in view of the fact that equation (3.1) is nothing but the differential equation of elastic equilibrium for the case that the elastic constants a, b are $a = 1, b = 0$. At the same time it is shown that relation

$$S(u'', u^\circ) = \int_{\mathfrak{R}} s''_{\kappa\lambda} s_{\kappa\lambda}\, dx = 0 \tag{3.5}$$

holds for every function u° in $\mathfrak{D}^\circ$, whence in particular

$$S(u'', u') = 0. \tag{3.6}$$

Consequently, the relation

$$S(u) = S(u') + S(u'') \tag{3.7}$$

holds, which expresses the fact that splitting of the function u implies splitting of the form $S(u)$. We must further show that also the side condition (3.2) splits.

---

[11] See A. Korn [2] §3.

We first note that relation (3.2) is satisfied for every function $\dot{u}$ in $\mathfrak{D}^\bullet$, as follows immediately by integration of $\dot{r}_{kl}$ over $\mathfrak{R}$. Let now u' be any funtion in $\mathfrak{D}^\circ$, and let $\dot{u}$ in $\mathfrak{D}^\bullet$ be such that $D(u' - \dot{u})$ is arbitrarily small. By virtue of (10) we have for the difference $u^* = u' - u$,

$$\left|\int_{\mathfrak{R}} r_{\kappa\lambda}^* \, dx\right|^2 \leqq R(u^*) \int_{\mathfrak{R}} dx \leqq D(u^*) \int_{\mathfrak{R}} dx \to 0.$$

Therefore (3.2) holds for any function u' in $\mathfrak{D}^\circ$.

If now the function u satisfies (3.2), it follows that the same is true for the component u''. Suppose now that Korn's inequality is valid in the main case:

$$R(u'') \leqq \mathrm{K}_3 S(u''). \tag{3.8}$$

By Schwarz's inequality and by (2.1) and (3.8) we then have

$$R(u) \leqq 2R(u') + 2R(u'') \leqq 2S(u') + 2\mathrm{K}_3 S(u'') \leqq \mathrm{K}_2\{S(u') + S(u'')\};$$

$$\mathrm{K}_2 = 2 \max\{1, K_3\}.$$

By virtue of (3.7) we therefore arrive at (3.1). Thus it is shown that to prove Korn's inequality in the second case it is sufficient to prove the inequality in the main case (3.8).

4. Korn's Inequality in the Main Case

We now assume that u in $\mathfrak{D}$ possesses continuous derivatives of every order and satisfies the equation, cf. (3.3),

$$s_{k\lambda/\lambda} = \tfrac{1}{2}(u_{k/\lambda\lambda} + u_{\lambda/\lambda k}) = 0 \tag{4.01}$$

and condition, cf. (3.2),

$$\int_{\mathfrak{R}} r_{kl} \, dx = 0, \qquad k, l = 1, \cdots, N. \tag{4.02}$$

Our proof of Korn's inequality, cf. (3.8),

$$R(u) \leqq \mathrm{K}_3 S(u) \tag{4.03}$$

for such functions u proceeds in four steps.[12] *First* we shall establish the inequality

$$\int_{\mathfrak{R}-\mathfrak{R}_{2v}} r_{\kappa\lambda}^2 \, dx < C_3 \int_{\mathfrak{R}_{2v}} r_{\kappa\lambda}^2 \, dx + C_4 \int_{\mathfrak{R}} s_{\kappa\lambda}^2 \, dx + C_5 \sqrt{\int_{\mathfrak{R}} s_{\kappa\lambda}^2 \, dx \int_{\mathfrak{R}} r_{\kappa\lambda}^2 \, dx}$$

or, in obvious notation,

$$R(u) - R_{2v}(u) \leqq C_3 R_{2v}(u)\, C_4 S(u) + C_5 \sqrt{S(u) R(u)}. \tag{4.04}$$

[12] Korn in his proof, [2], expresses the functions u_k through potential functions u_k' by $u_k = 4u_k' + \varphi_{/k}$ where φ is such that $\varphi_{/\lambda\lambda} = -2u_{\lambda/\lambda}$. For the potential functions u_k' an expansion with respect to a system of special potential functions is employed.

This inequality allows to estimate the integral of $r_{\kappa\lambda}^2$ extended over the boundary strip $\mathfrak{R} - \mathfrak{R}_{2\nu}$ terms of the integral of $r_{\kappa\lambda}^2$ over the subdomain $\mathfrak{R}_{2\nu}$ and of $S(u)$. To establish inequality (4.04) is the decisive step in the proof. The further three steps are concerned with estimating $R_{2\nu}(u)$ in terms of $S(u)$.

To derive inequality (4.04) we employ a technique based on formal identities. Once the usefulness of such identities for the present purpose has been seen it is not difficult to find all such identities by systematic manipulation. For the sake of brevity we note and verify those identities that serve our purpose. They are

$$(4.05) \qquad (u_{\kappa/\lambda}u_{\lambda/\kappa} - u_{\kappa/\kappa}u_{\lambda/\lambda})_m + 2(u_{\kappa/m}u_{\lambda/\lambda} - u_{\kappa/\lambda}u_{\lambda/m})_\kappa = 0$$

and

$$(4.06) \qquad (u_{\kappa/\kappa}u_{\lambda/\lambda})_m + 2(u_{m/\kappa}u_{\lambda/\lambda} - r_{\kappa\lambda}u_{m/\lambda})_\kappa = 4u_{\lambda/\lambda}s_{m\kappa/\kappa} + 2u_{m/\lambda}s_{\lambda\kappa/\kappa} = 0.$$

Note that identity (4.06) holds because the vector function u is assumed to satisfy the differential equation (4.01); identity (4.05) on the other hand holds for any vector function u. The inequalities are immediately verified by first interchanging κ and λ, then adding the results and finally carrying out the differentiation. We now insert $u_{k/l} = s_{kl} + r_{kl}$ in these identities and pay in particular attention to the terms which are quadratic in r_{kl}. The term $-(r_{\kappa\lambda}^2)_m$ occurs in the first term of (4.05) while the terms $\pm(r_{\kappa\lambda}r_{\lambda m})_\kappa$ occur in the second terms of (4.05) and (4.06) respectively. We eliminate the latter term by adding the two identities and obtain

$$(4.07) \qquad (s_{\kappa\lambda}^2 - r_{\kappa\lambda}^2)_m + 2(2s_{m\kappa}s_{\lambda\lambda} - 2_{m\lambda}s_{\kappa\lambda} - 2s_{m\lambda}r_{\kappa\lambda} + r_{m\lambda}s_{\kappa\lambda})_\kappa = 0.$$

We now multiply identity (4.07) by the function Ω_m defined in section 1, integrate over $\mathfrak{R}$, and apply integration by parts. We then would obtain the relation

$$(4.08) \qquad \int_{\mathfrak{R}} \{\Omega_{\mu/\mu}(r_{\kappa\lambda}^2 - s_{\kappa\lambda}^2) + 2\Omega_{\mu/\kappa}(s_{\mu\lambda}s_{\kappa\lambda} + 2s_{\mu\lambda}r_{\kappa\lambda} - 2s_{\mu\kappa}s_{\lambda\lambda} - s_{\kappa\lambda}r_{\mu\lambda})\}\,dx = 0$$

if we were sure that vanishing of Ω_m at the boundary were sufficient to insure vanishing of the boundary term resulting from integration by parts. This, however, cannot be inferred immediately since the derivatives of u need not be bounded on $\mathfrak{B}$.

To overcome this little difficulty we multiply identity (4.07) by $H^\delta\Omega_m$, where H^δ is the function defined in section 1, which equals 1 in $\mathfrak{R}_\delta$ and 0 outside of $\mathfrak{R}_\delta$. After integration by parts we obtain

$$(4.09) \qquad \int_{\mathfrak{R}} \{(H^\delta\Omega_\mu)_\mu(r_{\kappa\lambda}^2 - s_{\kappa\lambda}^2) + 2(H^\delta\Omega_\mu)_\kappa(s_{\mu\lambda}s_{\kappa\lambda} + 2s_{\mu\lambda}r_{\kappa\lambda} - 2s_{\mu\kappa}s_{\lambda\lambda} - s_{\kappa\lambda}r_{\mu\lambda})\}\,dx = 0.$$

Observing that $|\Omega_m| \leqq NC_1\delta$ in $\mathfrak{R} - \mathfrak{R}_\delta$ by virtue of (1.1), we find from (1.1), (1.3) and (1.4),

$$|(H^\delta\Omega_m)_k| \leqq (1 + NC_2)C_1 = C_6 \text{ in } \mathfrak{R} - \mathfrak{R}_\delta.$$

Hence we obtain from (4.08), (note that $H^\delta = 1$ in R_δ),

$$
(4.10)\quad \begin{aligned} &\left| \int_{\mathfrak{R}_\delta} \{\Omega_{\mu/\mu}(r_{\kappa\lambda}^2 - s_{\kappa\lambda}^2) + 2\Omega_{\mu/\kappa}(s_{\mu\lambda} s_{\kappa\lambda} + 2s_{\mu\lambda} r_{\kappa\lambda} \right. \\ &\qquad \left. - 2s_{\mu\kappa} s_{\lambda\lambda} - s_{\kappa\lambda} r_{\mu\lambda})\} \, dx \right| \leqq 7NC_6 \int_{\mathfrak{R}-\mathfrak{R}_\delta} (u_{\kappa/\lambda})^2 \, dx, \end{aligned}
$$

where Schwarz's inequality, $s_{kl}^2 + r_{kl}^2 = u_{\kappa/l}^2$, and (11) have been used. Letting δ approach zero one obtains (4.08) as desired.

We now write identity (4.08) in the form

$$
(4.11)\quad \begin{aligned} \int_{\mathfrak{R}-\mathfrak{R}_{2v}} \Omega_{\mu/\mu} r_{\kappa\lambda}^2 \, dx &= \int_{\mathfrak{R}_{2v}} \Omega_{\mu/\mu} r_{\kappa\lambda}^2 \, dx + 2 \int_{\mathfrak{R}} (\Omega_{\kappa/\mu} - 2\Omega_{\mu/\kappa}) s_{\mu\lambda} r_{\kappa\lambda} \, dx \\ &\quad + \int_{\mathfrak{R}} \{\Omega_{\mu/\mu} s_{\kappa\lambda}^2 - 2\Omega_{\mu/\kappa} s_{\mu\lambda} s_{\kappa\lambda} + 4\Omega_{\mu/\kappa} s_{\mu\kappa} s_{\lambda\lambda}\} \, dx. \end{aligned}
$$

Making use of the assumed relations (1.1) and (1.2), and applying Schwarz's inequality on the right hand side of (4.11) we immediately obtain inequality (4.04).

The *second step* consists in estimating $\int_{\mathfrak{R}_{2v}} r_{\kappa\lambda}^2 \, dx$ in terms of $\int_{\mathfrak{R}_v} r_{\kappa\lambda/\mu}^2 \, dx$. To this end we employ a slightly weakened form of Poincaré's inequality:[13]

Let v be any function with continuous derivatives $v_{/k} = v_k$ in $\mathfrak{R}$. Then

$$
(4.12)\quad \int_{\mathfrak{R}_{2v}} v^2 \, dx \leqq C_7 \int_{\mathfrak{R}_v} v_\kappa^2 \, dx + C_8 \left(\int_{\mathfrak{C}_0} v \, dx \right)^2,
$$

$\mathfrak{C}_0$ being a cube lying in $\mathfrak{R}_{2v}$, and having a diameter less than v. For the sake of completeness we present the simple proof of this inequality. Let $\mathfrak{C}$ be any cube $x_k^0 < x_k < x_k^0 + a$ in $\mathfrak{R}_v$, then for any points x' and x'' in $\mathfrak{C}$ we have

$$
\begin{aligned} &| v(x_1'', x_2'' \cdots, x_{N-1}'', x_N'') - v(x_1', x_2'', \cdots, x_{N-1}', x_N') |^2 < | v(x_1'', x_2'', \cdots, x_N'') \\ &\quad - v(x_1', x_2'', \cdots, x_N'') | + \cdots + | v(x_1', \cdots, x_{N-1}', x_N'') - v(x_1', \cdots, x_{N-1}', x_N') | \\ &\leqq \left\{ \int_{x_2=x_2'', \cdots, x_N=x_N''} | v_1 | \, dx_1 + \cdots + \int_{x_1=x_1', \cdots, x_{N-1}=x_{N-1}'} | v_N | \, dx_N \right\}^2 \\ &\leqq Na \left\{ \int_{x_2=x_2'', \cdots, x_N=x_N''} | v_1 | \, dx_1 + \cdots + \int_{x_1=x_1', \cdots, x_{N-1}=x_{N-1}'} | v_N | \, dx_N \right\}. \end{aligned}
$$

After integration with respect to x' and x'' over the cube one obtains

---

[13] The strict form of Poincaré's inequality is

$$
\int_{\mathfrak{C}} v^2 \, dx \leqq C \int_{\mathfrak{C}} v_\mu^2 \, dx + \left(\int_{\mathfrak{C}} v \, dx \right)^2 \Big/ \int_{\mathfrak{C}} dx
$$

for connected regions $\mathfrak{C}$ of finite area enjoying some additional properties. Cf. R. Courant and D. Hilbert [12], Vol. II Ch. VII §§3.1, 6.3, 8.

$$2a^N \int_{\mathfrak{C}} v^2\,dx - 2\left(\int_{\mathfrak{C}} v\,dx\right)^2 \leqq Na^{N+2} \int_{\mathfrak{C}} v_\kappa^2\,dx,$$

or

$$\int_{\mathfrak{C}} v^2\,dx < C_9 \int_{\mathfrak{C}} v_\kappa^2\,dx + C_{10}\left(\int_{\mathfrak{C}} v\,dx\right)^2. \tag{4.13}$$

Consider now two adjacent cubes $\mathfrak{C}_0$ and $\mathfrak{C}_1$ (i.e. two cubes having one face in common). We then find easily

$$\left|\int_{\mathfrak{C}_0} v\,dx - \int_{\mathfrak{C}_1} v\,dx\right|^2 \leqq a^{N+2} \int_{\mathfrak{C}_0+\mathfrak{C}_1} v_\kappa^2\,dx, \tag{4.14}$$

hence

$$\left(\int_{\mathfrak{C}_1} v\,dx\right)^2 \leqq 2\left(\int_{\mathfrak{C}_0} v\,dx\right)^2 + 2a^{N+2} \int_{\mathfrak{C}_0+\mathfrak{C}_1} v_\kappa^2\,dx.$$

Consequently, applying (4.13) to $\mathfrak{C}_1$,

$$\int_{\mathfrak{C}_1} v^2\,dx = C_{11} \int_{\mathfrak{C}_0+\mathfrak{C}_1} v_\kappa^2\,dx + 2\left(\int_{\mathfrak{C}_0} v\,dx\right)^2.$$

Similarly, for any cube $\mathfrak{C}_r$ that can be connected by a chain of successively adjacent cubes with $\mathfrak{C}_0$, we obtain

$$\int_{\mathfrak{C}_r} v^2\,dx < C_{12} \int_{\mathfrak{C}_0+\mathfrak{C}_1+\cdots+\mathfrak{C}_r} v_\kappa^2\,dx + C_{13}\left(\int_{\mathfrak{C}_0} v\,dx\right)^2. \tag{4.15}$$

Since we have assumed that the diameter of the cube is less than v, every cube that has one point in common with $\mathfrak{R}_{2v}$ lies completely in $\mathfrak{R}_v$. By virtue of the assumption that $\mathfrak{R}_{2v}$ is connected we can cover $\mathfrak{R}_{2v}$ by a finite number of such cubes which all lie in $\mathfrak{R}_v$. Applying relation (4.15) to all such cubes relation (4.12) follows by addition.

Substituting r_{kl} for v we obtain

$$\int_{\mathfrak{R}_{2v}} r_{\kappa\lambda}^2\,dx < C_7 \int_{\mathfrak{R}_v} r_{\kappa\lambda/\mu}^2\,dx + C_8\left(\int_{\mathfrak{C}_0} r_{\kappa\lambda}\,dx\right)^2. \tag{4.16}$$

The *third step* consists in deriving the inequality

$$\int_{\mathfrak{R}_v} r_{\kappa\lambda}^2\,dx < C_{14} \int_{\mathfrak{R}} s_{\kappa\lambda}^2\,dx + C_{15}\sqrt{\int_{\mathfrak{R}} s_{\kappa\lambda}^2\,dx \int_{\mathfrak{R}} r_{\kappa\lambda}^2\,dx}, \tag{4.17}$$

essentially making use of the condition that u satisfy the differenial equation $s_{k\lambda/\lambda} = 0$. This could be done in several ways. In a rather direct way relation (4.17) follows from the identity

$$\begin{aligned} 4\int_{\mathfrak{R}} \mathrm{H} r_{\kappa\lambda/\mu}^2\,dx &= 2\int_{\mathfrak{R}} \mathrm{H}[u_{\kappa/\lambda\mu}^2 - u_{\kappa/\lambda\mu} u_{\mu/\lambda\kappa}]\,dx \\ &= \int_{\mathfrak{R}} \Big\{\mathrm{H}_{\lambda\lambda}(3u_{\mu/\kappa}^2 + 3u_{\mu/\kappa} u_{\kappa/\mu}) - 2\mathrm{H}_{\kappa\lambda}(2u_{\mu/\kappa} u_{\mu/\lambda} \\ &\quad + 3u_{\mu/\kappa} u_{\lambda/\mu} + u_{\kappa/\mu} u_{\lambda/\mu} - 2(u_{\kappa/\lambda} + u_{\lambda/\kappa}) u_{\mu/\mu}\Big)\,dx \\ &= 2\int_{\mathfrak{R}} \Big\{3\mathrm{H}_{\lambda\lambda} s_{\kappa\mu}^2 - 2\mathrm{H}_{\kappa\lambda}(s_{\kappa\mu} s_{\lambda\mu} - 2s_{\kappa\mu} s_{\mu\mu} - s_{\kappa\mu} r_{\lambda\mu})\Big\}\,dx. \end{aligned} \tag{4.18}$$

Here $H = H^v$ is the function defined in section 1, satisfying (1.3) and (1.4). Without restriction we may assume that H possesses continuous second derivatives. To prove the identity (4.18) one best proceeds backwards and verifies

$$(4.19)\quad \begin{aligned}&(3u_{\mu/\kappa}^2 + 3u_{\mu/\kappa}u_{\kappa/\mu})_{\lambda\lambda} - (4u_{\mu/\kappa}u_{\mu/\lambda} + 6u_{\mu/\kappa}u_{\lambda/\mu} + 2u_{\kappa/\mu}u_{\lambda/\mu} - 4(u_{\kappa/\lambda}\\ &+ u_{\lambda/\kappa})u_{\mu/\mu})_{\kappa\lambda} = 2u_{/\mu\kappa\lambda}^2 - 2u_{\mu/\kappa\lambda}u_{\kappa/\lambda\mu} - 4u_{\mu/\kappa\kappa}(u_{\mu/\lambda\lambda} + u_{\lambda/\lambda\mu}) - 2(u_{\mu/\kappa\kappa}\\ &+ u_{\kappa/\kappa\mu})u_{\lambda/\lambda\mu} - 2u_{\mu/\kappa}(u_{\mu/\lambda\lambda\kappa} + u_{\lambda/\kappa\lambda\mu}) = 2u_{\mu/\kappa\lambda}^2 - 2u_{\lambda/\kappa\lambda}u_{\kappa/\lambda\mu} = 4r_{\kappa\mu/\lambda}^2.\end{aligned}$$

The property of identity (4.18), which is important for our purpose, is that no term quadratic in r_{kl} occurs on the right hand side. It is not obvious before hand whether or not an identity exists for which this is the case, but it is of course possible to answer this question by systematic enumeration of all possible identities of the type of (4.18).

We note that by (1.3) $H^v \geqq 0$ in $\mathfrak{R}$ and $H^v = 1$ in $\mathfrak{R}_v$. The left hand side in (4.18) is therefore greater than $\int_{\mathfrak{R}_{2v}} r_{\kappa\lambda/\mu}^2 dx$. Hence, using Schwarz's inequality, we obtain

$$(4.20)\qquad \int_{\mathfrak{R}_v} r_{\kappa\lambda/\mu}^2\, dx \leqq C_{16}\int_{\mathfrak{R}} s_{\kappa\lambda}^2\, dx + C_{17}\sqrt{\int_{\mathfrak{R}} s_{\kappa\lambda}^2\, dx \int_{\mathfrak{R}} r_{\kappa\lambda}^2\, dx}.$$

Combining this result with (4.16) we find

$$(4.21)\qquad R_{2v}(u) \leqq C_{18}\, s(u) + C_{19}\sqrt{S(u)R(u)} + C_8\left(\int_{\mathfrak{C}_0} r_{\kappa\lambda}\, dx\right)^2$$

as the combined result of the second and third step.

We now combine (4.21) with (4.04). To this end we add $R_{2v}(u)$ on both sides of (4.04) and then insert (4.21). The result is

$$(4.22)\qquad R(u) \leqq C_{20}\, S(u) + C_{21}\sqrt{S(u)R(u)} + C_{22}\left(\int_{\mathfrak{C}_0} r_{\kappa\lambda}\, dx\right)^2,$$

whence, in view of $2C_{21}\sqrt{S(u)R(u)} < C_{21}^2 S(u) + R(u)$,

$$R(u) \leqq (2C_{20} + C_{21})S(u) + 2C_{22}\left(\int_{\mathfrak{C}_0} r_{\kappa\lambda}\, dx\right)^2, \quad \text{or}$$

$$(4.23)\qquad R(u) < C_{23}\, S(u) + C_{23}\left(\int_{\mathfrak{C}_0} r_{\kappa\lambda}\, dx\right)^2,$$

which is the combined result of the first three steps.

In the *fourth step* we make use of the condition

$$(4.02)\qquad \int_{\mathfrak{R}} r_{kl}\, dx = 0, \qquad k, l = 1 \cdots, N,$$

in order to eliminate the last term in (4.23). To this end we observe that for the vectors $b^{(ml)}$, defined by

$$(4.24)\qquad b_l^{(lm)} = x_m\, , b_m^{(lm)} = -x_l\, , b_k^{(lm)} = 0, k \neq l, \neq m,$$

(cf. (14)), we have

$$\tfrac{1}{2}(b_{j/k}^{(lm)} - b_{k/j}^{(lm)}) = 1 \text{ if } j = l, k = m; = -1 \text{ if } j = m, k = l; \quad = 0 \text{ otherwise,} \tag{4.25}$$

$$\tfrac{1}{2}(b_{j/k}^{(lm)} + b_{k/j}^{(lm)}) = 0 \text{ for all } j, k. \tag{4.26}$$

Hence by adding to u a linear combination $c_{\lambda\mu} b^{(\lambda\mu)}$ of these vectors $b^{(lm)}$, with constant $c_{lm} = -c_{ml}$, we can change r_{jk} by an arbitrary constant without changing s_{jk}. Suppose now u satisfies (4.02), then we can find constants $c_{lm} = -c_{ml}$ such that

$$\tilde{u} = u + c_{\lambda\mu} b^{(\lambda\mu)} \tag{4.27}$$

satisfies

$$\int_{\mathfrak{C}_0} \tilde{r}_{kl}\, dx = 0 \qquad \text{for all } k, l. \tag{4.28}$$

Hence by (4.23)

$$R(\tilde{u}) \leqq C_{23} S(\tilde{u}) = C_{23} S(u). \tag{4.29}$$

Now in view of (4.02), and

$$b_{j/k}^{(lm)} - b_{k/j}^{(lm)} = 2 \text{ if } j = l, k = m, \quad = -2 \text{ if } j = m, k = l \quad = 0 \text{ otherwise}$$

we have $R(u, b^{(lm)}) = 0$; hence by (4.27)

$$R(u) \leqq R(u) + R(c_{\lambda\mu} b^{(\lambda\mu)}) = R(\tilde{u}). \tag{4.30}$$

Consequently (4.30) and (4.29) combined yield

$$R(u) \leqq C_{23} S(u) \qquad \text{if (4.02) holds.} \tag{4.31}$$

Thus Korn's inequality (4.03) in the main case is proved.

Part II. The Boundary Value Problems

5. The Extended Function Spaces

Before proving the unique existence of the solutions of the boundary value problems it is advisable to extend the manifold of functions considered so as to include all quadratically integrable functions. Instead of using the notions of Lebesgue integration we prefer to introduce "ideal" functions in the following sense: A sequence $u^\sigma = \{u_k^\sigma\}$ of continuous functions (or vectors) defined in $\Re$ is said to define "function" $u = \{u_k\}$ if

$$H(u^{\sigma\tau}) = \int_{\Re} (u_\kappa^{\sigma\tau})^2\, dx \to 0, \qquad \sigma, \tau \to 0, \tag{5.01}$$

where we use the abbreviation $u^{\sigma\tau} = u^\sigma - u^\tau$. It is readily seen that the manifold $\mathfrak{H}$ of "ideal functions" so defined form a complete linear space, a Hilbert space, comprising the continuous functions. For "functions" u^1, u^2 and u in $\mathfrak{H}$ the bilinear form

$$H(u^1, u^2) = \int_{\mathfrak{R}} u^1_\kappa u^2_\kappa \, dx \tag{5.02}$$

and the quadratic form

$$H(u) = \int_{\mathfrak{R}} u^2_\kappa \, dx \tag{5.03}$$

can be defined such that for the sequence u^σ defining u

$$H(u^\sigma - u) = \int_{\mathfrak{R}} (u^\sigma_\kappa - u_\kappa)^2 \, dx \to 0 \qquad \text{as} \qquad \sigma \to \infty. \tag{5.04}$$

Clearly,

$$H(u) \geqq 0 \tag{5.05}$$

and u^1 and u^2 are said to be "equal" if $H(u^1 - u^2) = 0$.

Suppose a sequence of functions u^σ in $\mathfrak{D}$, i.e. possessing continuous derivatives in $\mathfrak{R}$ (cf. end of section 1), defines an ideal function u in $\mathfrak{H}$, such that also the derivatives $u^\sigma_{/l}$ of u^σ, $l = 1, \cdots, N$ define ideal functions $u_{/l}$ in $\mathfrak{H}$; then we assign the functions $u_{/l}$ to the function u as its derivatives.[14] The manifold of such functions u we denote by $\mathfrak{G}$. Before showing that this assignment is independent of the defining sequence u^σ we note that the relation

$$\int_{\mathfrak{R}} w_\kappa u_{\kappa/l} \, dx = -\int_{\mathfrak{R}} w_{\kappa/l} u_\kappa \, dx \tag{5.06}$$

or

$$H(w, u_{/l}) = -H(w_{/l}, u)$$

holds for all functions w in $\mathfrak{D}^\bullet$, i.e., for functions w with continuous derivatives vanishing identically in a boundary strip. If $u = 0$ we have $H(w, u_{/l}) = 0$; from this relation one can conclude[15] that $H(u_{/l}) = 0$. In view of the definition of equality we thus have found *the fundamental fact that* $u = 0$ *entails* $u_{/l} = 0$.[16] This shows at the same time that the assignment of $u_{/l}$ to u in $\mathfrak{G}$ is independent of the defining sequence and constitutes a linear operator.

For functions u^1, u^2 and u in $\mathfrak{G}$ the bilinear form

$$D(u^1, u^2) = H(u^1_{/\lambda}, u^2_{/\lambda}) = \int_{\mathfrak{R}} u^1_{\kappa/\lambda} u^2_{\kappa/\lambda} \, dx \tag{5.07}$$

---

[14] This is essentially the "strong" derivative in the sense introduces in [10] and [11]; relation (5.06) expresses that it is at the same time a "weak" derivative.

[15] To this end one need only show that the space $\mathfrak{D}^\bullet$ is dense in $\mathfrak{H}$. We shall prove the statement slightly differently in the next section.

[16] Or that $H(u) = 0$ entails $D(u) = 0$.

and the quadratic form

$$D(u) = H(u_{/\lambda}) = \int_{\mathfrak{R}} (u_{\kappa/\lambda})^2 \, dx \tag{5.08}$$

are defined. Relation (5.04) applied to the derivatives of a function u in $\mathfrak{G}$ and its defining sequence u^σ gives

$$D(u^\sigma - u) \to 0 \text{ as } \sigma \to \infty. \tag{5.09}$$

For the treatment of the first boundary problem a submanifold $\mathfrak{G}^\circ$ of $\mathfrak{G}$ is fundamental. To define it we employ the manifold $\mathfrak{D}^\bullet$ of functions having continuous derivatives in $\mathfrak{R}$ and vanishing in a boundary strip, which was introduced at the end of section 1. The ideal functions that result from sequences of functions u^σ in $\mathfrak{D}^\bullet$ such that

$$D(u^{\sigma\tau}) \to 0 \tag{5.10}$$

and

$$H(u^{\sigma\tau}) \to 0 \tag{5.11}$$

then form the manifold $\mathfrak{G}^\circ$. Clearly relations (5.04) and (5.09) hold for u in $\mathfrak{G}^\circ$.

One further easily verifies that *the space* $\mathfrak{G}$ *is complete* in this sense: to every sequence u^σ of functions in $\mathfrak{G}$ for which

$$D(u^{\sigma\tau}) \to 0, \; H(u^{\sigma\tau}) \to 0 \text{ as } \sigma, \tau \to \infty \tag{5.12}$$

there is a function u in $\mathfrak{G}$ such that

$$D(u^\sigma - u) \to 0, \; H(u^\sigma - u) \to 0 \text{ as } \sigma \to \infty. \tag{5.13}$$

If the functions u^σ lie in $\mathfrak{G}^\circ$ the same is true for the limit u, as is easily seen. Thus *the space* $\mathfrak{G}^\circ$ *is also complete.*

Finally we note that to every open subdomain $\mathfrak{R}'$ of $\mathfrak{R}$ we can define by a limit process the forms

$$H'(u^1, u^2) = \int_{\mathfrak{R}'} u^1_\kappa u^2_\kappa \, dx, \qquad H'(u) = \int_{\mathfrak{R}'} u^2_\kappa \, dx \tag{5.14}$$

for functions in $\mathfrak{H}$, and

$$D'(u^1, u^2) = H'(u^2_{/\lambda}, u^2_{/\lambda}), \; D'(u) = H'(u_{/\lambda}) \tag{5.15}$$

for funtions in $\mathfrak{G}$. We state the

LEMMA. *If* $H'(u) = 0$ *for all proper open subdomains* $\mathfrak{R}'$ *of* $\mathfrak{R}$ *then* $u = 0$. To show this we let M be an upper bound for $H(u^\sigma)$ and choose to a given $\epsilon > 0$ a σ such that $H(u^\sigma - u^\tau) \leqq \epsilon^2/36M$ for all $\tau > \sigma$. This implies $| H(u^\sigma) - H(u^\tau) | \leqq \epsilon/3$ and $| H(u) - H(u^\sigma) | \leqq \epsilon/3$. Then we choose a proper open subdomain $\mathfrak{R}'$ of $\mathfrak{R}$ such that $H(u^\sigma) - H'(u^\sigma) \leqq \epsilon/3$. Clearly $H'(u^\sigma - u^\tau) \leqq \epsilon^2/36M$, and $| H'(u^\sigma) - H'(u^\tau) | \leqq \epsilon/3$ whence $| H'(u^\sigma) | = | H'(u) - H'(u^\sigma) | \leqq \epsilon/3$. Therefore $H(u) \leqq \epsilon$ which implies $H(u) = 0$ and $u = 0$.

We feel that it is of considerable advantage to introduce ideal functions. Of course, it would be possible to circumvent introducing them and to remain entirely in the field of continuously differentiable functions; but one would not gain in brevity by such a procedure. For, the same arguments that are used to establish the properties of the ideal functions, such as the possibility of defining the bilinear forms for them, would then have to be applied to the minimizing sequence and to some related sequences. As a whole, the reasoning would become less concise. Introduction of ideal functions, on the other hand, provides a clear arrangement of the arguments, showing which steps are of a formal, and which of a constructive nature.

6. The Mollifiers

As a basic tool we employ a sequence of smoothing integral operators, the "mollifiers". Their kernel is of the form

$$j_\rho(x - y) = \rho^{-N} j\left(\frac{x_1 - y_1}{\rho}\right) \cdots j\left(\frac{x_N - y_N}{\rho}\right), \tag{6.01}$$

where the function $j(\xi)$ of the variable ξ possesses derivatives of every order and satisfies

$$j(\xi) = 0 \text{ for } |\xi| > 1, \tag{6.02}$$

$$\int_{-\infty}^{\infty} j(\xi)\, d\xi = 1. \tag{6.03}$$

As we have proved at some other place,[17] the integral operator J_ρ defined by

$$J_\rho u(x) = \int_{\mathfrak{R}} j_\rho(x - y) u(y)\, dy \tag{6.04}$$

produces from every function in $\mathfrak{H}$ a function with continuous derivatives of every order enjoying the following properties:

Let $\mathfrak{R}'$ be any proper subregion. Let ρ be so small that for every point x in $\mathfrak{R}'$ the points y with $|x_k - y_k| \leqq 2\rho$, $k = 1, \cdots, N$ are in $\mathfrak{R}$. Then $J_\rho u$ *is a function in* $\mathfrak{D}$. Further we have

$$H'(J_\rho u) \leqq H(u) \tag{6.05}$$

and

$$H'(J_\rho u - u) \to 0 \text{ as } \rho \to 0. \tag{6.06}$$

Finally: for functions u in $\mathfrak{G}$, the fundamental identity

$$(J_\rho u)_{/l} = J_\rho u_{/l} \text{ in } \mathfrak{R}' \tag{6.07}$$

holds. We confine ourselves to prove the last statement, (6.07). In view of

$$\partial j_\rho(x' - x)/\partial x_l = \partial j_\rho(x' - x)/\partial x'_l,$$

[17] See [10] and [11].

(6.07) is equivalent with

$$-\int_{\mathfrak{R}} (\partial j_\rho(x' - x)/\partial x_l) u_k(x)\, dx = \int_{\mathfrak{R}'} j_\rho(x' - x) u_{k/l}(x)\, dx$$

for x' in $\mathfrak{R}'$. Introducing the function $j(x) = \{j_m(x)\}$ by

$$j_k(x) = j_\rho(x' - x),\ j_m(x) = 0,\ m \neq k,$$

the latter relation can be written in the form

$$-H(j_{/l}, u) = H(j, u_{/l})$$

Hence the statement (6.07) follows from (5.06) if it is shown that j is in $\mathfrak{D}^{\bullet}$. This is evidently the case for x' in $\mathfrak{R}'$ by virtue of the restriction imposed on ρ in relation to $\mathfrak{R}'$.

Relations (6.07), (6.06) imply

(6.08) $$D'(J_\rho u - u) \to 0 \text{ as } \rho \to 0.$$

We now may supply the proof of the statement made in section 5 that $H(w, u) = 0$ *for all* w *in* $\mathfrak{D}^{\bullet}$ *entails* $u = 0$, which was needed to show that $u = 0$ implies $u_{/l} = 0$. We need only set $w = j$, this function being in $\mathfrak{D}^{\bullet}$ for x' in $\mathfrak{R}'$. Then $H(w, u) = 0$ yields $J_\rho u(x') = 0$, x' in $\mathfrak{R}'$. From relation (6.06) we then conclude $H'(u) = 0$ for all proper subdomains $\mathfrak{R}'$ and this is sufficient to insure $u = 0$ according to the lemma proved in section 5.

7. The First Boundary Problem

In this section we shall prove the *unique existence*[18] of a solution $u = \{u_1, \cdots, u_N\}$ of the differential equation

(7.01) $$a u_{\kappa/\lambda\lambda} + (a + b) u_{\lambda/\lambda\kappa} = a s_{\kappa\lambda/\lambda} + b s_{\lambda\lambda/\kappa} = 0,$$

which at the boundary $\mathfrak{B}$ of the domain $\mathfrak{R}$ assumes the same values as a given function $\bar{u}$ in $\mathfrak{G}$ in the sense that the *difference* $u - \bar{u}$ *is a function* in the manifold $\mathfrak{G}^{\circ}$. This form of the boundary condition implies that u lies in $\mathfrak{G}$.[19] The solution u is further a "real" function having continuous derivatives of every order in the ordinary sense.[20] The constants a and b are to satisfy

(7.02) $$a > 0,\ b > -a/N.$$

We first make a few preliminary remarks. The differential equation (7.01)

[18] For existence proofs by reduction to integral equations see Lauricella [4], Korn [3], [5], Weyl [7], Lichtenstein [8]. A possible treatment employing the minimum property is indicated in Courant and Hilbert [12], Vol. II, ch. VII, §9.7.

[19] We say the system $u = \{u_1, \dots, u_N\}$ lies in any of the functions spaces of section 5 if each component u_k does.

[20] It then follows that the statements made at the beginning of section 3 are correct: u lies in $\mathfrak{D}$ and, if $\bar{u}$ is in $\mathfrak{D}$, $u - \bar{u}$ lies in $\mathfrak{D}^{\circ}$. Hence $\bar{u}$ is split into $u = u' + u''$ with $u' = u - \bar{u}$, $u'' = u$. Relation (3.3) is nothing but relation (7.04) for $u^1 = u''$ and $u^2 = u^{\bullet}$ in $\mathfrak{D}^{\bullet}$.

is connected with the bilinear form

$$(7.03) \qquad E(u^1, u^2) = \int_{\mathfrak{R}} \{a s^1_{\kappa\lambda} s^2_{\kappa\lambda} + b s^1_{\kappa\lambda} s^2_{\kappa\lambda}\}\, dx$$

by virtue of the fact that

$$(7.04) \qquad E(u^1, u^2) = 0$$

if u^1 in $\mathfrak{G}$ has continuous second derivatives and satisfies (7.01), and u^2 is in $\mathfrak{G}^\circ$. Relation (7.04) follows immediately by integration by parts if the function u^2 is in $\mathfrak{D}^\bullet$, vanishing in a boundary strip. By virtue of (5.09) for u^ν in $\mathfrak{D}^\bullet$ it then carries over to u^2 in $\mathfrak{G}^\circ$.

For the quadratic form

$$(7.05) \qquad E(u) = \int_{\mathfrak{R}} \{a(s_{\kappa\lambda})^2 + b(s_{\kappa\kappa})^2\}\, dx$$

we establish successively the following inequalities

$$(7.06) \qquad S(u) = \int_{\mathfrak{R}} (s_{\kappa\lambda})^2\, dx < \Theta E(u)$$

$$(7.07) \qquad D(u) = \int_{\mathfrak{R}} (u_{\kappa/\lambda})^2\, dx \leqq 2S(u)$$

$$(7.08) \qquad H(u) = \int_{\mathfrak{R}} (u_\kappa)^2\, dx \leqq \Pi_1 D(u)$$

all valid for u in $\mathfrak{G}^\circ$. Relation (7.06) follows immediately from (7.02) with $\Theta = 1/a$ if $b > 0$, $\Theta = 1/a + Nb$ if $b < 0$, (cf. (11)). Relation (7.07) combined with (7.06) is *Korn's inequality*, cf. (4) which in the equivalent form (13) or (2.1) was proved for functions u in $\mathfrak{D}^\circ$ in section 2. By virtue of (5.09) it also holds in $\mathfrak{G}^\circ$.

The "basic inequality" (7.08) is easily proved since we have assumed the domain $\mathfrak{R}$ to be bounded.[21] Let $\mathfrak{K}$ be a cube containing $\mathfrak{R}$. For simplicity we assume that $\mathfrak{K}$ be given by $0 < x_k < A$, $k = 1, \cdots, N$. Then any function u in $\mathfrak{D}^\bullet$ can be extended to $\mathfrak{K}$ by setting $u = 0$ in $\mathfrak{K} - \mathfrak{R}$. Evidently we have

$$|u_k(x_1, x_2, \cdots, x_N)|^2 < \left| \int_0^{x_1} u_{k/1}(y, x_2, \cdots, x_N)\, dy \right|^2 \leqq A \int_0^A (u_{k/1})^2\, dy,$$

[21] It would be sufficient to use a restricted form of this inequality in which the left hand integral is extended over a proper subdomain $\mathfrak{R}'$. The inequality would then remain true for unbounded domains $\mathfrak{R}'$ with bounded $\mathfrak{R}'$ provided that the bojndary $\mathfrak{B}$ of $\mathfrak{R}$ contains a section which after suitable coordinate transformation can be written in the form $0 \leqq x_N \leqq f(x_1, \ldots, x_{N-1})$, $0 \leqq x_k \leqq a$, $k = 1, \ldots, N-1$, $b \leqq f \leqq c$, $a, b, c = \text{const.}$ In general, however, the inequality is not valid for unbounded domains. For certain unbounded domains, viz. for those for which the constant lies in the manifold $G^\bullet$, the first boundary problem coalesces with the second one and can be treated as the latter will be treated in section 8. These difficulties encountered with inequality (7.08) are the reason why we have confined ourselves to bounded domains.

whence (7.08) follows by integration with respect to $x_2, \cdots, x_N$. The relation (7.08) thus proved for u in $\mathfrak{D}^\bullet$ carries over to u in $\mathfrak{G}^\circ$ by (5.09) and (5.04).

The enumerated inequalities allow an immediate proof of the *uniqueness* of the solution of the boundary problem. We have to show that a solution u of (7.01) which is in $\mathfrak{G}^\circ$ vanishes. Since relation (7.04) is applicable to $u^1 = u^2 = u$, we have $E(u) = 0$. By virtue of the inequalities (7.06), (7.07), (7.08) we have $H(u) = 0$, whence by definition of equality $u = 0$. Thus the uniqueness is proved.

This uniqueness proof shows clearly the significance of the three inequalities (7.06), (7.07), (7.08). What is needed is only the combined inequality

$$H(u) \leqq \Pi E(u), \tag{7.09}$$

which follows with $\Pi = 2\Theta\Pi_1$. To prove this combined inequality, Korn's inequality is decidedly employed. Korn's inequality will further be used in the existence proof.

To prove the existence of a solution of the boundary problem we first show that among all functions in $\mathfrak{G}$ satisfying the boundary condition

$$u - \bar{u} \text{ is in } \mathfrak{G}^\circ \tag{7.10}$$

there is one that minimizes the integral

$$E(u) = \int_{\Re} \{a s_{\kappa\lambda}^2 + b s_{\kappa\kappa}^2\}\, dx. \tag{7.11}$$

Let E_0 be the greatest lower bound of $E(u)$ such that

$$E(u) \geqq E_0 \qquad \text{for all admitted } u, \tag{7.12}$$

a function in $\mathfrak{G}$ being "admitted" if it satisfies (7.10). Let u^σ be a minimizing sequence, viz. a sequence such that

$$E(u^\sigma) \to E_0 \text{ as } \sigma \to \infty. \tag{7.13}$$

Then we have

$$E(u^{\sigma\tau}) = \tfrac{1}{2}E(u^\sigma) + \tfrac{1}{2}E(u^\tau) - E(\tfrac{1}{2}(u^\sigma + u^\tau)). \tag{7.14}$$

Clearly $\frac{1}{2}(u^\sigma + u^\tau)$ is admitted and hence by (7.14)

$$\tfrac{1}{4}E(u^{\sigma\tau}) \leqq \tfrac{1}{2}E(u^\sigma) + \tfrac{1}{2}E(u^\tau) - E_0. \tag{7.15}$$

From (7.13) it therefore follows that

$$E(u^{\sigma\tau}) \to 0 \text{ as } \sigma, \tau \to \infty.\text{[22]} \tag{7.16}$$

From relation (7.16) we obtain by virtue of Korn's inequality (7.07) together with (7.06)

---

[22] This argument using the convexity of a quadratic form is somewhat simpler than that due to E. E. Levi customarily used for proving the projection theorem in the Hilbert space. Cf. F. Rellich [13], F. Riesz [14].

(7.17) $$D(u^{\sigma\tau}) \to 0 \text{ as } \sigma, \tau \to \infty$$

and from the basic inequality (7.08)

(7.18) $$H(u^{\sigma\tau}) \to 0 \text{ as } \sigma, \tau \to \infty,$$

so that by virtue of the completeness of the space $\mathfrak{S}^{\circ}$ we are sure of the existence of an admitted function u° in $\mathfrak{S}$ with

(7.19) $$D(u^{\sigma} - u^{\circ}) \to 0, \sigma \to \infty.$$

This relation obviously implies $E(u^{\sigma} - u^{\circ}) \to 0$ and hence

(7.20) $$E(u^{\sigma}) \to E(u^{\circ}), \text{ as } \sigma \to \infty.$$

In view of (7.13) we infer

(7.21) $$E(u^{\circ}) = E_0.$$

Therefore the function u° minimizes the functional $E(u)$.

The *existence of the solution* u° *of the minimum problem* has thus been proved completely. As a consequence of the minimum property of u° the first variation vanishes:

(7.22) $$E(u^{\circ}, \delta u) = 0 \qquad \text{for all } \delta u \text{ in } \mathfrak{S}^{\circ},$$

$E(u, v)$ being the bilinear form associated with the quadratic form $E(u)$.

We now proceed to show that u° *possesses derivatives of every order and satisfies the differential equation* (7.01). This will be accomplished on the basis of three lemmas, which we shall formulate and prove in the next section in a form slightly more general than needed for the present purpose, since we shall make use of this more general form for the free boundary problem in section 9.

8. Three Lemmas

LEMMA 1. *Let* u° *be a funttion in* $\mathfrak{S}$, f° *a funttion in* $\mathfrak{H}$ *such that*

(8.01) $$E(u^{\circ}, \delta u) = H(f^{\circ}, \delta u)$$

holds for all δu *in* $\mathfrak{S}^{\circ}$; *then* $u = J_{\rho}u^{\circ}$ *and* $f = J_{\rho}f^{\circ}$ *satisfy the differential equation*

(8.02) $$as_{\kappa\lambda/\lambda} + bs_{\lambda\lambda/\kappa} + f_k = 0, \quad k = 1, \cdots, N.$$

To prove the lemma we choose for δu the set of functions

(8.03) $$\delta u(x) = g_l(x' - x) = g_{lk}(x' - x),$$

with

(8.04) $$g_{ll}(\xi) = j_{\rho}(\xi), \quad g_{lk}(\xi) = 0, \quad k \neq l,$$

where $j_{\rho}(\xi)$ is the mollifier kernel introduced in section 6. We let x' be in a domain $\mathfrak{R}'$ and ρ so small that $g(x' - x) = 0$ when x is in an appropriate boundary strip $\mathfrak{R} - \mathfrak{R}_s$, (cf. section 6). Indicating differentiation with respect to x'_k by $[\cdots]_k$ we derive from (8.01) the relation

$$0 = \int_{\Re} \{as^{\circ}_{\kappa\lambda}(g_{m\kappa/\lambda} + g_{m\lambda/\kappa}) + bs^{\circ}_{\kappa\kappa}g_{m\lambda/\lambda}\}\,dx - \int_{\Re} f^{\circ}_{\kappa}g_{m\kappa}\,dx$$

$$= \int_{\Re} \{au^{\circ}_{\kappa/\lambda}(g_{m\kappa/\lambda} + g_{m\lambda/\kappa}) + bu^{\circ}_{\kappa\kappa}g_{m\lambda/\lambda}\}\,dx$$

$$- \int_{\Re} f^{\circ}_{\kappa}g_{m\kappa}\,dx = -\int_{\Re} u^{\circ}_{\kappa}\{ag_{m\kappa/\lambda\lambda} + (a+b)g_{m\lambda/\lambda\kappa}\}\,dx$$

$$- \int_{\Re} f^{\circ}_{\kappa}g_{m\kappa}\,dx = -a\left[\int_{\Re} u^{\circ}_{\kappa}g_{m\kappa}\,dx\right]_{\lambda\lambda}$$

$$- (a+b)\left[\int_{\Re} u^{\circ}_{\kappa}g_{m\lambda}\,dx\right]_{\lambda\kappa} - \int_{\Re} f^{\circ}_{\kappa}g_{m\kappa}\,dx$$

$$= -a[J_{\rho}u^{\circ}_{m}]_{\lambda\lambda} - (a+b)[J_{\rho}u^{\circ}_{\kappa}]_{m\kappa} - J_{\rho}f^{\circ}_{m}.$$

The result states nothing else but that the functions $u = J_{\rho}u^{\circ}(x)$ and $f = J_{\rho}f^{\circ}(x)$ satisfy equation (8.02). Thus Lemma 1 is proved.

We know that the functions $J_{\rho}u^{\circ}$ and $J_{\rho}f^{\circ}$ approach in the mean the functions u° and f° as $\rho \to 0$.[23] Thus u° can in the mean be approximated by functions with continuous derivatives of every order which satisfy a differential equation whose right member approaches f°. From this fact we shall conclude that u° itself possesses continuous second derivatives and satisfies the differential equation.

This can be done in several ways, for example, by employing the fundamental solution of the differential equation. We shall show, however, that this can be also done without that tool.[24]

We shall formulate this result in two lemmas. Before doing so we introduce the operator ∇ which takes a function u into the function $\nabla u = \{u_{/k}\}$ in $\mathfrak{H}$. If ∇u itself is in the operation $\nabla^2 u$ is defined:[25] we then say u possesses second derivatives $\nabla^2 u$ in $\mathfrak{H}$. Quite generally the expression: "u has derivatives $\nabla^h u$ in $\mathfrak{H}$" can be successively defied. By $(\nabla^h u)^2$ we denote the sum of the squares of all derivatives of order h.

We further introduce the function spaces $\mathfrak{H}'$ and $\mathfrak{G}'$ with respect to a proper subdomain $\Re'$ of $\Re$ in the same maner as $\mathfrak{H}$ and $\mathfrak{G}$ were defined with reference to $\Re$.

Then we formulate

LEMMA 2. *Let f° be a function with derivatives up to the order $h - 1$ in $\mathfrak{H}''$ with reference to every subdomain $\Re''$ of $\Re$. Let u° be a function in $\mathfrak{G}$ such that the functions $u = J_{\rho}u^{\circ}$ and $f = J_{\rho}f^{\circ}$ defined in $\Re'$ (if ρ is sufficiently small) satisfy*

[23] This statement is equivalent to the statement that u°, satisfying the differential equation in the "weak" sense, also satisfies it in the "strong" sense; cf. [10] and [11]; see also the footnote in p. 21.

[24] Our procedure is similar to that employed in [15], pp. 49–51 for existence proofs derived from solutions of finite difference equations.

[25] $\nabla^2 u = \{u_{/kl}\}$ here and not $\nabla^2 u = u_{/kk}$.

the differential equation (8.02). *The* u^0 *possesses derivatives up to the order* h *in* $\mathfrak{H}'$.

To prove the lemma, we let u and f be any functions with continuous derivatives of every order in $\mathfrak{R}$ satisfying the differential equation (8.02)

$$au_{k/\lambda\lambda} + (a + b)u_{\lambda/k\lambda} + f_k = 0.$$

For such functions we then derive the "fundamental" inequality

$$\text{(8.05)} \qquad \int_{\mathfrak{R}_\delta} (u_{\kappa/\lambda\mu})^2 \, dx \leqq C_\delta \int_{\mathfrak{R}_{\delta'}} \{(u_{\kappa/\lambda})^2 + f_\kappa^2 + (f_{\kappa/\kappa})^2\} \, dx,$$

with an appropriate constant $C_\delta > 0$ depending on the subdomains $\mathfrak{R}_\delta$ and $\mathfrak{R}_{\delta'}$. Note that in this inequality a quadratic integral over a subdomain involving second derivatives is estimated by integrals over a larger subdomain involving first, second, and third derivatives.[26]

To prove inequality (8.05) we first note that the relation (8.02) implies

$$\text{(8.06)} \qquad (2a + b)u_{\kappa/\kappa\lambda\lambda} + f_{\kappa/\kappa} = 0.$$

Using this relation we find the identity

$$\text{(8.07)} \qquad \begin{aligned} \int_{\mathfrak{R}} \mathrm{H}\{(2a + b)a^2(u_{\kappa/\lambda\mu})^2 &+ (2a + b)(au_{\kappa/\lambda\lambda} - (a+b)u_{\lambda/\kappa\lambda})f_\kappa \\ &- (a + b)^2 u_{\kappa/\kappa} f_{\lambda/\lambda}\} \, dx \\ = (2a + b) \int_{\mathfrak{R}} \{a^2[\mathrm{H}_{\mu\mu}(u_{\kappa/\lambda})^2 &- \mathrm{H}_{\lambda\mu} u_{\kappa/\lambda} u_{\kappa/\mu}] + \tfrac{1}{2}(a + b)^2 \mathrm{H}_{\lambda\lambda}(u_{\kappa/\kappa})^2\} \, dx, \end{aligned}$$

where $\mathfrak{H} = \mathfrak{H}^\delta$ is a function, defined in section 1, which equals 1 in $\mathfrak{R}_\delta$ and vanishes outside of $\mathfrak{R}_{\delta'}$. Identity (8.07) is readily verified by integration by parts. In fact the right member equals

$$\int_{\mathfrak{R}} \mathrm{H}\{(2a + b)[a^2(u_{\kappa/\lambda\mu})^2 - a^2 u_{\kappa/\lambda\lambda} u_{\kappa/\mu\mu} + (a+ b)^2(u_{\kappa/\kappa\lambda})^2] + (a + b)^2 u_{\kappa/\kappa} u_{\lambda/\lambda\mu\mu}\} \, dx$$

and hence is equal to the left member of (8.07) by (8.02) and (8.06).

We now make use of the fact that by (7.02), $2a + b > a + b > 0$, since $N > 1$ was assumed. Observing further $\mathrm{H} = 1$ in $\mathfrak{B}_\delta$, $\geqq 0$ in $\mathfrak{R}_\delta - \mathfrak{R}_{\delta'}$, $= 0$ in $\mathfrak{R} - \mathfrak{R}_{\delta'}$, and using Schwarz's inequality, we see that identity (8.06) entails inequality (8.05).

Since the derivatives $\nabla^h u$ of u, $h = 1, 2, \cdots$, satisfy the differential equation

$$a\nabla^h u_{k/\lambda\lambda} + (a + b)\nabla^h u_{\lambda/k\lambda} + \nabla^h f_k = 0,$$

a relation similar to (8.07) holds for all derivatives $\nabla^h u$ of u. Thus we obtain

$$\text{(8.08)} \qquad \int_{\mathfrak{R}_\delta} (\nabla^h u)^2 \, dx < C_{\delta,h} \int_{\mathfrak{R}_{\delta h}} \{(\nabla u)^2 + \textstyle\sum_{\eta=0}^{h-1} (\nabla^\eta f)^2\} \, dx,$$

[26] Cf. the "fundamental" inequality in [15] p. 51 footnote.

where δ_h is defined by the recursion $\delta_{h+1} = \delta_h'$.

From relation (6.08) we derive, for $\rho, \sigma \to 0$,

$$D'(J_\rho u^\circ - J_\sigma u^\circ) \to 0, \qquad \text{or}$$

$$H'(\nabla(J_\rho u^\circ - J_\sigma u^\circ)) \to 0 \qquad \text{and, using (6.07)}$$

$$H'(\nabla^\eta(J_\rho f^\circ - J_\sigma f^\circ)) \to 0, \qquad \eta = 0, \cdots, h - 1$$

Employing these relations for $\mathfrak{R}_{\delta_h}$ instead of $\mathfrak{R}'$ and choosing δ such that $\mathfrak{R}_\delta$ contains an arbitrary $\mathfrak{R}'$ we obtain from (8.08), applied to $u = J_\rho u^\circ - J_\sigma u^\circ$, $f = J_\rho f^\circ - J_\sigma f^\circ$,

$$\int_{\mathfrak{R}'} (\nabla^h J_\sigma u^\circ - \nabla_h J_\sigma u^\circ)^2 \, dx \to 0 \qquad \text{as} \quad \rho, \sigma \to 0. \tag{8.09}$$

By virtue of the definition of the spaces $\mathfrak{H}'$ and $\mathfrak{G}'$, relation (8.09) implies the statement of Lemma 2.

LEMMA 3. *Let u° be a function with derivatives up to the order*

$$k > M = \left[\frac{N+1}{2}\right],$$

in $\mathfrak{H}'$, then the funtion u° equals a continuous function with continuous derivatives up to the order $k - M$.

If u° satisfies the conditions of Lemma 2 then u° satisfies the differential equation (8.02) in the proper sense.

Let, as before, $u = J_\rho u$. Then relation (8.09) holds. From this relation we shall conclude that the function $u = J_\rho u^\circ$ converge uniformly in every proper subdomain $\mathfrak{R}'$ to a limit function u^1 possessing continuous derivatives up to the order $k - M$, and that the derivatives of $J_\rho u^\circ$ approach uniformly those of u^1. To this end we derive through integration by parts the relation

$$\int_{\mathfrak{R}} \mathrm{H} r^{-N} x_{\kappa_1} \cdots x_{\kappa_M} \nabla_{\kappa_1} \cdots \nabla_{\kappa_M} u \, dx$$

$$= (-1)^{M-1} \int_{\mathfrak{R}} \nabla_{\kappa_{M-1}} \cdots \nabla_{\kappa_1} \mathrm{H} r^{-N} x_{\kappa_1} \cdots x_{\kappa_M} \nabla_{\kappa_M} u \, dx,$$

assuming the origin to lie in $\mathfrak{R}_{2\delta}$. By virtue of the fact that $\mathrm{H} = 1$ in $\mathfrak{R}_\delta$ and that for every m

$$\nabla_{\kappa_m} \cdots \nabla_{\kappa_1} r^{-N} x_{\kappa_1} \cdots x_{\kappa_m} = 0, \tag{8.10}$$

we have

$$\nabla_{\kappa_{M-1}} \cdots \nabla_{\kappa_1} \mathrm{H} r^{-N} x_{\kappa_1} \cdots x_{\kappa_M} = (m-1)!\, r^{-N} x_{\kappa_M} \text{ in } \mathfrak{R}_\delta .$$

Hence, after another integration by parts we find

$$\begin{aligned} (m-1)!\, S_N u(0) = &- \int_{\mathfrak{R}} \nabla_{\kappa_M} \cdots \nabla_{\kappa_1} \mathrm{H} r^{-N} x_{\kappa_1} \cdots x_{\kappa_M} u \, dx \\ &+ \int_{\mathfrak{R}} \dot{\mathrm{H}} r^{-N} x_{\kappa_1} \cdots x_{\kappa_M} \Delta_{\kappa_1} \cdots \nabla_{\kappa_M} u \, dx. \end{aligned} \tag{8.11}$$

The coefficients of u vanishes in $\mathfrak{R}_\delta$. The coefficient of $\nabla^M u$ is in $\mathfrak{H}$; for $\int_{\mathfrak{R}} \mathrm{H}^2 r^{-2N+2m}\, dx$ is finite since $2M - N - 1 = 2[\frac{1}{2}(N + 1)] - (N + 1) \geqq 0$.

Consequently, we obtain from (8.11) the estimate[27]

$$u^2(0) < C_\delta \int_{\mathfrak{R}_\delta} \{u^2 + (\nabla^M u)^2\}\, dx \tag{8.12}$$

with an appropriate constant C_δ. Clearly, such an estimate holds for every point 0 in $\mathfrak{R}_{2\delta}$. Also for the derivatives $\nabla^k u$ one has

$$(\nabla^{k-M} u(0))^2 < C_{\delta,k} \int_{\mathfrak{R}_\delta} \{(\nabla^{k-M} u)^2 + (\nabla^k u)^2\}\, dx. \tag{8.13}$$

Relation (8.09) for $h = k$ and $h = k - M$ now yields immediately

$$\nabla^{k-M} J_\rho u^\circ \ - \ \nabla^{k-M} J_\sigma u^\circ \ \rightarrow \ 0 \tag{8.14}$$

uniformly in every proper subdomain $\mathfrak{R}'$. Consequently it is true that $J_\rho u^\circ$ approaches uniformly a function u^1 together with all derivatives up to the order $k - M$.

We now need only show that $u^\circ = u^1$.

It is clear that $J_\rho u^1$ approaches u^1 uniformly in $\mathfrak{R}'$, hence

$$J_\rho(u^1 - u^\circ) \rightarrow 0 \text{ as } \rho \rightarrow 0 \text{ as } \rho \rightarrow 0 \text{ in } \mathfrak{R}', \tag{8.15}$$

whence

$$H'(J_\rho u^1 - u^\circ) \rightarrow 0 \qquad \text{as } \rho \rightarrow 0. \tag{8.16}$$

By (6.06) this entails

$$H'(u^1 - u^\circ) = 0. \tag{8.17}$$

According to the lemma at the end of section 5, this implies

$$u^\circ = u^1. \tag{8.18}$$

Clearly, if the function u° satisfies the conditions of Lemma 2, it follows that it satisfies the differential equation (8.02). Thus Lemma 3 is proved.

Applying the three lemmas to the solution u° of the minimum problem in section 7, with $f^\circ = 0$, we find by virtue of (7.22) that this function u° equals a continuous function with continuous derivatives of every order in every proper subdomain $\mathfrak{R}'$, satisfying the differential equation (8.02) = (7.01). Thus the existence of a solution of the first boundary value problem is established.

We should mention that this existence theorem was used in the proof of Korn's inequality in the second case.[28] Thus we have now proved this inequality completely.

---

[27] Cf. Sobolev [16].

[28] Incidentally, we could have arranged the proof of Korn's inequality in the second case such that Lemmas 2 and 3 had not been used.

9. The Free Characteristic Value Problem

A boundary value problem may be termed "free" if in it no boundary values are prescribed for the desired function, but if instead natural boundary conditions (involving derivatives), are to be satisfied. The natural boundary conditions associated with the differential equations (7.01) are

$$a s_{k\lambda} \xi_\lambda + b s_{\lambda\lambda} \xi_k = g_k \text{ at } \mathfrak{B}, \tag{9.01}$$

where $\xi = \{\xi_k\}$ is the unit vector of the external normal at $\mathfrak{B}$ and $g = \{g_k\}$ is a prescribed vector, the "stressvector" acting against the surface $\mathfrak{B}$. The vector g is to satisfy the conditions

$$\int_{\mathfrak{B}} g_k \, dB = 0, \qquad \int_{\mathfrak{B}} (g_k x_l - g_l x_k) \, dB = 0 \tag{9.02}$$

dB being the surface differential on $\mathfrak{B}$; these conditions express that the total force and the total moment resulting from the prescribed stress vector are zero. The solutions u are further to satisfy the conditions

$$\int_{\mathfrak{B}} u_k \, dx = 0, \qquad k = 1, \cdots, N, \tag{9.03}$$

and

$$\int_{\mathfrak{B}} (u_k x_l - u_l x_k) \, dx = 0, \qquad k, l = 1, \cdots, N. \tag{9.04}$$

These conditions exclude the $N(N+1)/2$ functions $u = a^{(m)}$ and $u = b^{(lm)}$ given by

$$a_k^{(l)} = 1 \quad \text{for } l = k, \ = 0 \quad \text{for } l \neq k, \tag{9.05}$$

$$b_k^{(m)} = x_m \text{ for } k = l \neq m, \ = -x_l \text{ for } k = m \neq l, \ = 0 \text{ otherwise.} \tag{9.06}$$

For, (9.03) expresses the orthogonality $H(a^{(k)}, u) = 0$, $k = 1, \cdots, N$, and (9.04) expresses the orthogonality $H(b^{(kl)}, u) = 0$, $k, l = 1, \cdots, N$. The functions $a^{(m)}$ and $b^{(lm)}$ (cf. (4.24) and (4.25)) are solutions of the homogeneous problem, i.e. of the differential equation (7.01) and the boundary conditions (8.01) with $g = 0$. (They correspond to N translations and $N(N-1)/2$ rotations). Without restrictions we may assume that the coordinate system is so placed that

$$H(b^{(lm)}, a^{(k)}) = 0, \qquad \text{for } k, l, m = 1, \cdots, N. \tag{9.07}$$

The *unique existence of a solution* of the problem can then be established along the lines followed in section 7, after the conditions (9.01) and (9.02) have been replaced by different ones (cf. (9.09)) which do not refer to the boundary explicitly and involve only integrals over the domain $\mathfrak{R}$. Korn's inequality in the second, full, case is an essential tool needed to estimate the Dirichlet integral $D(u)$ in terms of the integral $E(u)$.

We shall not carry through the details. Instead we prefer to present a treatment of the *free eigen-value problem.*[29] Here Korn's inequality will essentially be used to show that a discrete set of eigen-values exists which approach ∞.

The problem then is to establish a sequence of values $\lambda_1, \lambda_2, \cdots \to \infty$ and corresponding functions $u^{(1)}, u^{(2)}, \cdots$ in $\mathfrak{S}$ which possess derivatives up to every order and satisfy the differential equation

$$(9.08) \qquad a s_{k\mu/\mu} + b s_{\mu\mu/k} + \lambda u_k = 0, \qquad k = 1, \cdots, N,$$

and instead of (9.01) with $g = 0$ the boundary condition

$$(9.09) \qquad \int_{\mathfrak{R}_\delta} w_\kappa \{a s_{\kappa\mu} H^\delta_\mu + b s_{\mu\mu} H^\delta_\kappa\}\, dx \to 0 \quad \text{as} \quad \delta \to 0,$$

for an arbitrary function w in $\mathfrak{S}$. Here H^δ is the function defined in section 1, which equals 1 in $\mathfrak{R}_\delta$ and equals 0 in $\mathfrak{R} - \mathfrak{R}_{\delta'}$. The expression $a s_{k\mu} H^\delta_\mu + b s_{\mu\mu} H^\delta_k$ may be considered an approximation to the left hand side of (9.01) since the derivatives $\{H^\delta_k\}$ are an approximation to the normal vector $\{\xi_k\}$ except for a factor. Therefore one may consider relation (9.09) as a substitute for (9.01).

One further has to establish the well-known orthogonality relations

$$(9.10) \quad H(u^{(\alpha)}, u^{(\beta)}) = \int_{\mathfrak{R}} u^{(\alpha)}_\kappa u^{(\beta)}_\kappa\, dx = \delta_{\alpha\beta}, = 1 \text{ if } \alpha = \beta, = 0 \text{ if } \alpha \neq \beta,$$

$$(9.11) \quad \begin{aligned} E(u^{(\alpha)}, u^{(\beta)}) &= \int_{\mathfrak{R}} \{a s^{(\alpha)}_{\kappa\mu} s^{(\beta)}_{\kappa\mu} + b s^{(\alpha)}_{\kappa\kappa} s^{(\beta)}_{\mu\mu}\}\, dx \\ &= \lambda_\alpha = \lambda_\beta \text{ if } \alpha = \beta, = 0 \text{ if } \alpha \neq \beta. \end{aligned}$$

We note that the *first* $M = N(N+1)/2$ *eigen-values are zero*, the corresponding eigen-functions being the N constants (9.05) and the $N(N-1)/2$ linear functions (9.06).

For every function u in $\mathfrak{S}$ with continuous second derivatives which satisfies equation (9.08), relation (9.09) is equivalent with

$$(9.12) \qquad E(w, u) - \lambda H(w, u) = 0$$

for arbitrary w in $\mathfrak{S}$. Clearly, (9.12) is equivalent with

$$\int_{\mathfrak{R}} H^\delta \{a w_{\kappa/\mu} s_{\kappa\mu} + b w_{\kappa/\kappa} s_{\mu\mu} + \lambda w_\kappa u_\kappa\}\, dx \to 0 \qquad \text{as } \delta \to 0,$$

and, after integration by parts, this relation is seen to be the same as (9.09) for functions u satisfying (9.08).

We shall employ relation (9.12) to dispose of the casé that the eigen-value is zero. In this case relation (9.12) with $u = 0$ gives $E(u) = 0$ whence $s_{km} = 0$; the only solution of this equation are the translations and rotations (9.05) and (9.06). Thus $\lambda = 0$ is eigen-value with multiplicity $M = N(N+1)/2$.

[29] For treatments of eigen-value problems by reduction to integral equations see A. Korn [1] (for the first boundry condition) and H. Weyl [7].

If there is another eigen-value it is positive, since (9.12) for $w = u$ implies $\lambda H(u, u) = E(u, u) \geqq 0$.

A decisive step in the theory is that the eigen-values are defined in a different manner:[30] λ_n is the g. l. b. of the ratio

$$E(u)/H(u) \tag{9.13}$$

for all functions in $\mathfrak{G}$ for which

$$H(u, u^{(\alpha)}) = 0 \qquad \text{for } \alpha = 1, \cdots, n-1. \tag{9.14}$$

When this definition is adopted it is no longer *a priori* obvious that the $M + 1^{st}$ eigen-value is positive. This fact follows, however, immediately by combining various inequalities. Firstly inequality (7.06)

$$S(u) < \Theta E(u), \tag{9.15}$$

secondly, inequality

$$D(u) < \Gamma S(u) \tag{9.16}$$

under the condition

$$\int_{\mathfrak{R}} r_k \, dx = 0, \tag{9.17}$$

which, with $\Gamma = \mathrm{K}_2 + 1$, is equivalent with Korn's inequality in the second case (3.1) under condition (3.2); finally Poincaré's inequality in the strict form[31]

$$H(u) < \Pi_2 D(u) \tag{9.18}$$

under the condition (9.03). Combining we obtain with $\Pi = \Gamma\Theta\Pi_2$

$$H(u) < \Pi E(u) \tag{9.19}$$

under conditions (9.03) and (9.17). We proceed to show that *the inequality* (9.18) *holds also under the conditions* (9.03) *and* (9.04). To show this we observe that, by adding an appropriate combination b of the vectors (9.06), we can find to every function u satisfying (9.03) and (9.04) a function

$$u^* = u + b, \; b = c_{\mu\nu} b^{(\mu\nu)}$$

which satisfies (9.17) and hence (9.19), $H(u^*) = \Pi E(u^*)$. The addition of the vector b to u does not change s_{kl} i.e. $s^*_{kl} = s_{kl}$; hence $E(u^*) = E(u)$. On the other hand condition (9.04) implies the orthogonality $H(u, b) = c_{\mu\nu} \int_{\mathfrak{R}} u_\kappa b_\kappa^{(\mu\nu)} \, dx$ $= c_{\mu\nu} \int_{\mathfrak{R}} (u_\mu x_\nu - u_\nu x_\mu) \, dx = 0$. Hence $H(u^*) = H(u) + H(b) \geqq H(u)$.

---

[30] See Courant-Hilbert [12] Vol. I Chap. VI, Vol. II, Chap. VII.

[31] See [12] Vol. II, Chap. VII, §§3.1, 6.3, 8. Poincaré's inequality is proved there under certain assumptions on the nature of the domain, which are satisfied for the Ω-domains here considered.

Thus (9.19) for u^* implies that (9.17) also holds for the function u satisfying (9.03) and (9.04).

For the following we need that (9.16) also holds (perhaps with a different constant Γ) if (9.03) and (9.04) are satisfied. To this end we determine the coefficients c_{mn} in $u_* = u + c_{\mu\nu}b^{(\mu\nu)}$ such that u_* satisfies (9.04). In view of (9.07) the function u_* also satisfies (9.03). Since (9.04) is equivalent with $H(b^{(mn)}, u) = 0$ we find $2c_{mn}H(b^{(mn)}, b^{(mn)}) = H(b^{(mn)}, u^*)$, whence $c_{\mu\nu}^2 < C_{29}H(u^*)$ with an appropriate constant C_{29}; since u^* satisfies (9.03) and (9.17) we may apply (9.18) and obtain

$$c_{\mu\nu}^2 \leqq \Pi C_{29}E(u^*) = \Pi C_{29}E(u). \tag{9.20}$$

On the other hand we have

$$D(u) = 2D(u^*) + D(c_{\mu\nu}b^{(\mu\nu)}) \leqq 2D(u^*) + C_{30}c_{\mu\nu}^2,$$

whence by (9.16), (9.15), and (9.20)

$$D(u) \leqq (2\Gamma\Theta + C_{30}\Pi)E(u).$$

In other words relation (9.16) is valid with a different constant Π if (9.03) and (9.04) are satisfied.

Conditions (9.03) and (9.04) express just the orthogonality of the function u to the eigen-functions (9.05), (9.06) with the eigen-value zero; i.e. they are conditions (9.13) for $n = M$. Hence inequality (9.19) under (9.03), (9.04) is equivalent with

$$\lambda_{M+1} > 0, \tag{9.21}$$

this eigen-value being defined as the g. l. b. of (9.13) under condition (9.14).

By similar arguments we shall prove that the *eigen-values increase indefinitely* with n. We cover the domain $\mathfrak{R}$ with a finite number of spheres of radius ϵ such that every point is covered at most 2^N times. We then apply to every intersection $\mathfrak{F}$ of such a sphere with $\mathfrak{R}$ Poincaré's inequality

$$\int_{\mathfrak{F}} u_\kappa^2\, dx \leqq F^{-1}\left(\int_{\mathfrak{F}} u_\kappa\, dx\right)^2 + 2N\epsilon^2 \int_{\mathfrak{F}} (u_{\kappa/\lambda})^2\, dx, \tag{9.22}$$

F being the area of $\mathfrak{F}$, and Korn's inequality in the form

$$\int_{\mathfrak{F}} r_{\kappa\lambda}^2\, dx < F^{-1}\left(\int_{\mathfrak{F}} r_{\kappa\lambda}\, dx\right)^2 + C_{31}\int_{\mathfrak{F}} s_{\kappa\lambda}^2\, dx. \tag{9.23}$$

We can easily find functions v in $\mathfrak{D}^*$ having continuous derivatives and vanishing in a boundary strip, such that

$$\left(\int_{\mathfrak{F}} r_{\kappa\lambda}\, dx\right)^2 - \left(\int_{\mathfrak{F}} v r_{\kappa\lambda}\, dx\right)^2 \leqq \tfrac{1}{2}F\int_{\mathfrak{F}} r_{\kappa\lambda}^2\, dx.$$

We need only choose v as a continuously differentiable function vanishing in a boundary strip of $\mathfrak{F}$ such that $\int_{\mathfrak{F}} (v - 1)^2\, dx$ is sufficiently small.

In view of

$$\left(\int_{\mathfrak{F}} vr_{\kappa\lambda}\,dx\right)^2 = \tfrac{1}{4}\left(\int_{\mathfrak{F}} (v_{/\lambda}u_\kappa - v_{/\kappa}u_\lambda)\,dx\right)^2 < \left(\int_{\mathfrak{F}} v_{/\lambda}u_\kappa\,dx\right)^2$$

we find from (9.23)

$$\frac{1}{2}\int_{\mathfrak{F}} r_{\kappa\lambda}^2\,dx \leqq F^{-1}\left(\int_{\mathfrak{F}} v_{/\mu}u_\kappa\,dx\right)^2 + C_{31}\int_{\mathfrak{F}} s_{\kappa\lambda}^2\,dx.$$

We insert this relation in (9.22), employing $(u_{\kappa/\lambda})^2 = s_{\kappa\lambda}^2 + r_{\kappa\lambda}^2$; finally we add the resulting inequalities and divide by 2^N. Then we obtain an inequality

$$H(u) \leqq \sum_{\eta=1}^{h} \{H(w^{(\eta)}, u)\}^2 + C_{32}\epsilon^2 S(u), \tag{9.24}$$

or

$$H(u) \leqq \sum_{\eta=1}^{h} \{H(w^{(\eta)}, u)\}^2 + C_{33}\epsilon^2 E(u), \tag{9.25}$$

with appropriate functions $w^{(\eta)}$, $\eta = 1, \cdots, h$, depending on ϵ; C_{31}, C_{32}, and C_{33} depending only on $\mathfrak{R}$ and N.

From this inequality (9.25) one concludes (cf. [12] vol. II chap. VII, §4, Theorem 6) that the *sequence of eigen-values defined as minima increases indefinitely*. Further one deduces from this inequality (cf. Rellich's theorem [12] vol. II, chap. VII, §3) the following

LEMMA. *From any sequence of functions u^σ in $\mathfrak{G}$ for which $E(u^\sigma)$ is bounded one can select a subsequence u_*^σ for which $H(u_*^{\sigma\tau}) \to 0$ as $\sigma, \tau \to \infty$.*

We now proceed to establish the existence of an eigen-function $u^{(n)}$, assuming that the existence of the eigen-functions $u^{(1)}, \cdots, u^{(n-1)}$ is already established. We define $u = u^{(n)}$ as a function in $\mathfrak{G}$, which satisfies the condition:

I. $H(u^{(\alpha)}, u^{(n)}) = 0, \qquad \alpha = 1, \cdots, n-1$

II. $H(u^{(n)}) = 1$

III. $E(u^{(n)}) = \lambda_n$

IV. $E(w, u^{(n)}) = \lambda_n H(w, u^{(n)})$ for all w in $\mathfrak{G}$.

For $w = u^{(\alpha)}$, $\alpha = 1, \cdots, n-1$, I to IV entail

V. $E(u^{(\alpha)}, u^{(n)}) = 0, \qquad \alpha = 1, \cdots, n-1.$

Relations I, II, III, V are equivalent with (9.10), (9.11).

VI. The eigen-functions u are complete in this sense: Defining for every function u in $\mathfrak{G}$ the projection $P^{(n)}u$ by

$$P^{(n)}u = \sum_{\alpha=1}^{n} H(u, u^{(\alpha)})u^{(\alpha)} \tag{9.26}$$

we have

$$H(u - P^{(n)}u) \to 0, \tag{9.27}$$

$$E(u - P^{(n)}u) \to 0 \text{ as } n \to \infty.$$

VII. $u^{(n)}$ possesses continuous derivatives of every order in $\mathfrak{R}$.

VIII. $u^{(n)}$ satisfies the differential equation

$$as^{(n)}_{k\mu/\mu} + bs^{(n)}_{\mu\mu/k} + \lambda_n u^{(n)}_k = 0$$

and the boundary condition (9.09).

We take a sequence u^σ of functions satisfying I and II for which $E(u^\sigma) \to \lambda_n$. By virtue of the lemma formulated above it is no restriction to assume that

(9.28) $$H(u^{\sigma\tau}) \to 0 \text{ as } \sigma,\ \tau \to \infty.$$

For any function u satisfying I we evidently have $E(u) \geqq \lambda_n H(u)$; hence

(9.29) $$E(u^\sigma - u^\tau) - \lambda_n H(u^\sigma - u^\tau) \geqq 0$$

(9.30) $$E(u^\sigma + u^\tau) - \lambda_n H(u^\sigma + u^\tau) \geqq 0.$$

The sum of the two left members here is

$$2\{E(u^\sigma) - \lambda_n H(u^\sigma)\} + 2\{E(u^\tau) - \lambda_n H(u^\tau)\},$$

and approaches zero as σ or τ increase indefinitely. Hence the left members in (9.29) and (9.30) approach zero as σ and τ increase indefinitely. In particular

(9.31) $$E(u^{\sigma\tau}) - H(u^{\sigma\tau}) \to 0, \qquad \text{as } \sigma,\ \tau \to \infty.$$

In view of (9.24) we conclude

(9.32) $$E(u^{\sigma\tau}) \to 0, \qquad \text{as } \sigma,\ \tau \to \infty.$$

Since $u^{\sigma\tau}$ satisfies (9.03), (9.04), Korn's inequality (9.16) is applicable, whence, by (9.15),

(9.33) $$D(u^{\sigma\tau}) \to 0, \qquad \text{as } \sigma,\ \tau \to \infty.$$

By virtue of the completeness of the space $\mathfrak{G}$ (see section 5) there exists a function $u^{(n)}$ in $\mathfrak{G}$ such that

(9.34) $$D(u^\sigma - u^{(n)}) \to 0$$

and

(9.35) $$H(u^\sigma - u^{(n)}) \to 0 \text{ as } \sigma \to \infty.$$

Clearly also

(9.36) $$E(u^\sigma - u^{(n)}) \to 0,$$

hence

$$H(u^\sigma) \to H(u^{(n)}), \qquad E(u^\sigma) \to E(u^{(n)})$$

whence

(9.37) $$H(u^{(n)}) = 1, \quad \text{(II)}$$

(9.38) $$E(u^{(n)}) = \lambda_n \quad \text{(III)}$$

Since the functions u^ν of the minimizing sequence satisfy I, it is clear by (9.35) that also u satisfies I. Hence $u^{(n)}$ *solves the minimum problem.*

In a standard way one derives that the first variation of $E(u)/H(u)$ vanishes for $u = u^{(n)}$, and thus finds that relation IV holds for all w' in $\mathfrak{S}$ satisfying

$$(9.39) \qquad H(w', u^{(\alpha)}) = 0, \qquad \alpha = 1, \cdots, n-1.$$

Let w be an arbitrary function in $\mathfrak{S}$; then one can find a function $w' = w + c_1u^{(1)} + \cdots + c_{n-1}\, u^{(n-1)}$ such that (9.37) holds: $c_mH(u^{(m)}) = -H(w, u^{(m)}), m = 1, \cdots, n-1$. In view of I and V for $m < n$ instead of n, one has $H(w, u^{(n)}) = H(w', u^{(n)})$ and $E(w, u^{(n)}) = E(w', u^{(n)})$. Hence IV holds without the restriction (9.39). V then follows as indicated.

To establish property VI we first note that by I and V, $P^{(n)}u$ satisfies $H(u - P^{(n)}u, u^{(\alpha)}) = 0, E(u - P^{(n)}u, u^{(\alpha)}) = 0, \alpha = 1, \cdots, n$. Hence by the minimum property of λ_{n+1}

$$\lambda_{n+1}H(u - P^{(n)}u) \geqq E(u - P^{(n)}u) = E(u) - E(P^{(n)}u) > E(u).$$

Hence $\lambda_{n+1} \to \infty$, as $n \to \infty$, implies (9.27), (cf. [3], Chap. VII, §33). From

$$0 \leqq E(P^{(m)}u - P^{(n)}u) = E(P^{(m)}u) - E(P^{(n)}u) < E(u), m > n,$$

we see that $E(P^{(n)}u)$ increases monotonically and is bounded. Hence (cf. [3], Ch. VII, §33)

$$E(P^{(m)}u - P^{(n)}u) \to 0, \qquad \text{as } m, n \to \infty.$$

Similarly

$$H(P^{(m)}u - P^{(n)}u) \to 0, \qquad \text{as } m, n \to \infty.$$

By virtue of the definition of the space $\mathfrak{S}$ (in section 6) it follows that there is a function u^* in $\mathfrak{S}$ such that

$$H(u^* - P^{(n)}u) \to 0, \qquad E(u^* = P^{(n)}u) \to 0, \qquad n \to \infty.$$

From (9.27) we see that $H(u^* - u) = 0$, hence $u^* = u$.[32] Thus (9.28) is also established.

To establish that $u^{(n)}$ equals a function which possesses derivatives of every order, (VI), and satisfies differential equation and boundary condition, (VII), we may employ the three lemmas of section 8.

Lemma 1 is applicable with $u^\circ = u^{(n)}, f^\circ = \lambda_n u^{(n)}$. Lemma 2 is so far applicable only with $h = 2$ since it is so far not known whether or not $f^\circ = \lambda_n u^{(n)}$ possesses derivatives of higher than first order in $\mathfrak{H}'$ (cf. Sec. 8, p. 33). The statement of Lemma 2 for $h = 2$, however, is just that $u^{(n)}$ does possess second derivatives in $\mathfrak{H}'$. Hence Lemma 2 can now be applied for $h = 3$. In other words Lemma 2 re-enforces itself indefinitely. Consequently Lemma 3 can

[32] This conclusion is less obvious than it may appear. What is actually involved is that $H(u^* - u) = 0$ entails $D(u^* - u) = 0$ and hence $E(u^* - u) = 0$, a conclusion which was formulated in section 5 and proved in section 6.

be applied for any value of h. Thus the properties VI and VII are estabished and the *existence of the eigen-values and functions for the free elastic body is established.*

NEW YORK UNIVERSITY

BIBLIOGRAPHY

[1] A. KORN, *Die Eigenschwingungen eines elastischen Koerpers mit ruhender Oberflaeche.* Akad. der Wissensch., Munich, Math. phys. Kl., Berichte 36 p. 351 (1906).

[2] A. KORN, *Ueber einige Ungleichungen, welche in der Theorie der elastischen und elektrischen Schwingungen eine Rolle spielen.* Bulletin International, Cracovie. Akademija umiejet. *Classe des sciences mathèmatiques et naturels.* (1909) pp. 705–724.

[3] A. KORN, *Solution générale du problème d'èquilibre dans la théorie de l'élasticité, dans le cas òu les effets sont donnés à la surface.* Annales, Toulouse Université, 1908.

For references to the work of Lauriella [4] from 1894 on and [5] Korn up to1906 see Encyklopaedie der mathematischen Wissenschaften, IV Teil 4 (1906), 34) p. 91 and 53), 54), p. 103.

[6] A. KORN, *Ueber die Loesungen des Grundproblems der Elastizitaetstheorie.* Mathematische Annalen Bd. 75, p. 497. (1914).

[7] H. WEYL, *Das asymptotische Vertellungsgesetz der Eigenschwingungen eines beliebig gestalteten elastischen Koepers.* Rendiconti, Circolo Matematico di Palermo. Vol. XXXLX (1915), p. 1.

[8] L. LICHTENSTEIN, *Ueber die erste Randwertaufgabe der Elastizitaetstheorie.* Mathematische Zeitschrift. Bd. 20. p. 21 (1924).

These papers contain references to the works of Fredholm, Marcolongo, Boggio and Korn.

[9] K. FRIEDRICHS, *On certain Inequalities and Characteristic Value Problems for Analytic Functions and for Functions of two Variables.* Trans., Am. Math. Soc., Vol. 41 (1937), p. 321.

[10] K. FRIEDRICHS, *On Differential Operators in Hilbert Spaces.* Am. Jour. Math., Vol. LXI (1939), p. 525.

[11] K. FRIEDRICHS, *The Identity of Weak and Strong Extensions of Differential Operators.* Trans., Am. Math. Soc., Vol. 55 (1944), p. 132.

[12] R. COURANT AND D. HILBERT, Methoden der mathematischen Physik Vol. I, 2nd ed. (1931), Vol. II (1937).

[13] F. RELLICH, *Spektraltheorie in nicht separablen Raeumen.* Math. Annalen, 110 (1934), p. 344.

[14] F. RIESZ, *Zur Theorie des Hilbertschen Raumes.* Acta Literarium ac Scientiarum, sectio Scientarum Mathematicarum. Szeged. Vol. VII (1934), p. 34.

[15] R. COURANT, K. FRIEDRICHS, H. LEWY, *Ueber die partiellen Differenzengleichungen der mathematischen Physik.* Mathematische Annalen, Vol. 100 (1928), p. 32.

[16] S. SOBOLEV, *Sur quelques èvaluations concernant les familles des functions ayant des dérivées à carré intégrable.* Comptes Rendues de l'Académie des Sciences USSR, N. S. L. 279–282 (1936).

[17] K. O. FRIEDRICHS, *An Inequality for Potential Functions,* Am. Jour. Math. Vol. LXVIII (1946), p. 581.

IV
Papers in Asymptotics

Commentaries by Wolfgang Wasow

[46–1] **Singular perturbations of non-linear oscillations, (with W.R. Wasow), Duke Math. J., 13 (1946), 367–381.**

Commentary:

The expression "singular perturbation" which is today quite standard appears for the first time in the title of this paper. In later years neither of the two authors could remember to which of them belongs the credit for coining this terminology, but I believe that it was Friedrichs. The formulation of the problem, the plan of attack and several ingenious details, particularly in the convergence proof, are due to Friedrichs. The particular matrix formulation of the non-degeneracy condition was introduced by me.

For several years Friedrichs had been interested in generalizations of relaxation oscillations of the Van der Pol type to systems of order higher than two. One result of these studies was his paper: "On non-linear vibrations of third order," Lecture Notes [46–1]. He considered our joint paper as a preliminary to one in which the limit of the periodic solution, as the parameter tends to zero, is discontinuous.

A short announcement [3] shows that Norman Levinson carried out that extension in great generality. He published the full proofs, which were exceedingly involved, in [4]. The complexity of those problems may be the reason that there appears to be no later contribution to the theory of relaxation oscillations in more than two dimensions in the mathematical literature. There are several such papers for the two-dimensional case and there also exist investigations of singular perturbations in higher dimensions that have a discontinuous limit, but not with periodicity as the boundary condition.

There are, however, more papers on periodic singular perturbations with a smooth limit than can be listed here. In fact, as the asymptotic solution of the differential equation for fixed initial data is the first part of our proof, references to this paper can be found in many articles that are not concerned with periodicity. A few contributions to problems for periodic solutions are listed below. They have been chosen because they introduce decisively different new methods. In [5] the periodic solution is represented by a certain convergent series, somewhat more complicated than a power series. In [2] the existence of a periodic perturbation is proved by means of differential inequalities and suitably constructed upper and lower functions. The paper [1] is based on the qualitative properties of the flow defined by the differential equation, which then leads to asymptotic expansions for the solutions and ultimately makes possible a study of the stability of the solutions.

[1] Neil Fenichel. Geometric singular perturbation theory for ordinary differential equations, J. Diff. Equations, 31 (1979), 53–98.

[2] F.A. Howes. An application of Nagumo's lemma to some singularly perturbed systems, Int'l. J. Non-Linear Mech., 10 (1975), 315–324.

[3] Norman Levinson. Perturbations of discontinuous solutions of non-linear systems of differential equations, Proc. Nat. Acad. Sci. USA, 33 (1947), 214–218.

[4] Norman Levinson. Perturbations of discontinuous solutions of non-linear systems of differential equations, Acta. Math., 82 (1951), 71–106.

[5] W.R. Wasow. On the construction of periodic solutions of singular perturbation problems, in Contributions to the Theory of Nonlinear Oscillations, Ann. Math. Studies, 20 (1950), Princeton, 313–350.

SINGULAR PERTURBATIONS OF NON-LINEAR OSCILLATIONS

By K. O. Friedrichs and W. R. Wasow

The perturbation method is one of the strongest tools for establishing periodic solutions of systems of ordinary differential equations. Suppose a system of n equations

$$X'_\mu = F_\mu(X_1, \cdots, X_n; \epsilon) \qquad (\mu = 1, \cdots, n)$$

for n functions $X_\mu(t)$ with $X'_\mu = dX_\mu/dt$ is given whose right-hand members depend on a parameter ϵ, and do not involve the independent variable t directly. Suppose that for $\epsilon = 0$ the equations possess a known periodic solution, the "base solution". With the aid of well-known perturbation methods one can then obtain periodic solutions of the equations for sufficiently small values of ϵ in the neighborhood of the base solution, provided the right members of the differential equations depend regularly on ϵ, for $\epsilon = 0$. The first, and decisive step in justifying the perturbation procedure is the theorem of Poincaré that, under appropriate conditions and for sufficiently small values of ϵ, a unique periodic solution exists which, as ϵ approaches zero, tends to the base solution for $\epsilon = 0$. (See [1; 401], [5], [6], [7; Chapters II–IV], [8]).

There exists, however, an important class of perturbation problems which is not covered by Poincaré's investigations. In Poincaré's theory essential use is made of the assumption that the functions $F_\mu(X_1, \cdots, X_n; \epsilon)$ are continuous functions of ϵ in the neighborhood of $\epsilon = 0$. (Poincaré even limits his discussion to functions F_μ which are regular analytic in all variables.) This implies, in particular, that the order of the differential system is the same for $\epsilon = 0$ as for $\epsilon \neq 0$. We shall term such problems *regular*. While such regular perturbation problems are common in the field of non-linear vibrations, (see [3], [5], [6], [8] and literature mentioned there), there exists also an important type of perturbation problems in which the dependence on the parameter ϵ is singular in such a way that the order of the system of differential equations is reduced for $\epsilon = 0$. Usually, such systems have the form of equations (1) below. Problems of this nature we shall call *singular* perturbation problems. They arise in mechanical problems concerned with the oscillations of elastically bound masses, if one of the masses is much smaller than the others; for oscillations in electrical networks, if the inductivity or the capacity of one of the meshes is very small. (It seems, though, that these problems have not been given the attention they deserve. But see [4] where such cases of non-linear electrical oscillations are considered in detail.)

As we shall show, it remains true for such singular problems that, under appropriate conditions and for sufficiently small values of ϵ a unique periodic

Received March 22, 1946.

solution exists, which, as $\epsilon \to 0$, approaches the base solution for $\epsilon = 0$. Our methods are related to those of Poincaré, but the theory is considerably complicated by the fact that most solutions of such singular differential equations depend non-uniformly on the parameter ϵ. In particular, the arguments of §§4 and 5 of the present paper have no equivalent in the reasoning for regular problems.

It should be mentioned that there exists a type of singular perturbation problems, not covered by the results of this paper, in which the periodic base solution is discontinuous. The problems of relaxation oscillations, for example, are of that type. For such relaxation problems even the formal perturbation procedure has not yet been developed. The existence of periodic solutions in the neighborhood of the base solution has so far been established only in a special, though typical relaxation problem with two unknown functions. (See [2]).

1. **Formulation of the problem.** We consider the system of differential equations

$$(1)\qquad \begin{aligned} X_i' &= F_i(X_1, \cdots, X_n) \qquad (i = 1, 2, \cdots, n-1), \\ \epsilon X_n' &= F_n(X_1, \cdots, X_n), \end{aligned}$$

where the X_n denote functions of the variable t and the prime indicates differentiation with respect to t. ϵ is a small parameter. It is assumed that the "limiting" differential equations

$$(2)\qquad \begin{aligned} x_i' &= F_i(x_1, \cdots, x_n), \\ 0 &= F_n(x_1, \cdots, x_n) \end{aligned}$$

have a periodic solution $x_\mu = u_\mu(t)$, $(\mu = 1, 2, \cdots, n)$ with period T, the "base solution". Without loss of generality it can be assumed that $u_\mu(0) = 0$, $(\mu = 1, 2, \cdots, n)$ and that the origin is not a singular point of the curve $x_\mu = u_\mu(t)$, $(\mu = 1, 2, \cdots, n)$.

The functions $F_\mu(X_1, \cdots, X_n)$ as well as their first partial derivatives are supposed to be continuous in the closed domain $\mathcal{E}$ of the $X_1, \cdots, X_n$-space determined by the inequalities

$$(3)\qquad | X_\mu - u_\mu(t) | \leq 2k, \text{ for all } t \qquad (\mu = 1, 2, \cdots, n),$$

where k is a constant.

For the results of this paper the condition

$$(4)\qquad \frac{\partial F_n(u_1, \cdots, u_n)}{\partial u_n} \neq 0 \qquad (0 \leq t \leq T)$$

is essential. For the present it will be assumed that ϵ is positive and $\partial F_n/\partial u_n$ is negative, say

$$\frac{\partial F_n(u_1, \cdots, u_n)}{\partial u_n} \leq -2m < 0 \qquad (0 \leq t \leq T), \tag{4a}$$

where m is a constant. We shall see later that our results remain valid, if ϵ and $\partial F_n/\partial u_n$ have the same sign. If the constant k in (3) is chosen sufficiently small the last inequality implies

$$\frac{\partial F_n(X_1, \cdots, X_n)}{\partial X_n} \leq -m < 0 \text{ in } \mathcal{E}. \tag{4b}$$

In the following, Greek letters μ, ν, ρ, etc., will be used for subscripts when a formula is true for all subscripts from 1 to n. Latin subscripts other than n will indicate statements which are true for subscripts from 1 to $n - 1$ only. We shall frequently write $F_\mu(X)$ for $F_\mu(X_1, X_2, \cdots, X_n)$, etc. The abbreviation

$$\frac{\partial}{\partial X_\nu} F_\mu(X) = F_{\mu\nu}(X)$$

will be used throughout. With this notation the inequality (4b) can be written in the simpler form

$$F_{nn}(X) \leq -m < 0 \text{ in } \mathcal{E}.$$

In order to arrive at a simple statement about the existence and uniqueness of periodic solutions of (1) certain exceptional cases have to be excluded by imposing a *non-degeneracy condition*. This condition assures that the base solution $x_\mu = u_\mu(t)$ of (2) is an *isolated* periodic solution in the sense that all solutions of (2) in its neighborhood deviate from periodicity by terms of the *first* order. We leave a precise formulation of this condition to §2 and proceed to formulate our

MAIN THEOREM. *If the differential equations* (2) *possess a non-degenerate periodic solution* $x_\mu = u_\mu(t)$ *satisfying the condition*

$$F_{nn}(u_\mu) \neq 0, \qquad \text{(condition (b))},$$

then, for every sufficiently small value of ϵ, *say,* $|\epsilon| < \epsilon_0$, *the differential equations* (1) *admit exactly one closed integral curve* $X_\mu = U_\mu(t; \epsilon)$ *with the following properties*: (a) *As* ϵ *approaches* 0, *the functions* U_μ *tend to* u_μ, *the assumed periodic base solution for* $\epsilon = 0$. *The convergence is uniform in* t. (b) $X_\mu = U_\mu(t; \epsilon)$ *possesses a period* $T + \tau(\epsilon)$, *which tends to the period* T *of* $u_\mu(t)$ *as* ϵ *approaches* 0. (c) *The* $U_\mu(t; \epsilon)$ *are continuous functions of* ϵ *for* $|\epsilon| < \epsilon_0$.

2. **The non-degeneracy condition.** Before beginning the proof of the Main Theorem we must supply a precise definition of the condition of non-degeneracy. To this end we first observe that the solutions of (2) and also those of (1) are

not uniquely determined by the curves they represent, since the functions $x_\mu(t)$ and $x_\mu(t + t_1)$, where t_1 is a constant, describe the same curve. The correspondence between the integral curves and the solutions can be made bi-unique by introducing some condition which fixes the position of the point $t = 0$ on each integral curve. We therefore add the condition that the point $t = 0$ on each integral curve of the systems (1) or (2) shall lie in the (hyper) plane $\sum_{i=1}^{n-1} u_i'(0)x_i = 0$. (Note that this assumption is satisfied for the base solution, since we have assumed $u_i(0) = 0$.) The choice of this particular condition is motivated by our wish to obtain a symmetric form for the non-degeneracy condition. In order to show that the integral curves in the neighborhood of the curve $x_\mu = u_\mu(t)$ actually intersect this (hyper) plane, it suffices to prove that the curve $x_\mu = u_\mu(t)$ is not tangent to the plane at the origin, through which it passes by assumption. To that end we first remark that tangency of $x_\mu = u_\mu(t)$ to the plane $\sum_{i=1}^{n-1} u_i'(0)x_i = 0$ would imply $\sum_{i=1}^{n-1} u_i'^2(0) = 0$, i.e., $u_i'(0) = 0$. But, on the other hand, the last of the equations (2) implies the condition $\sum_{\nu=1}^{n} u_\nu'(0)F_{n\nu}(u(0)) = 0$, obtained by substituting $x_\mu = u_\mu(t)$ and differentiating with respect to t. Since $F_{nn}(u) \neq 0$, in consequence of (4), —i.e. condition (b)—it follows also that $u_n'(0) = 0$. But $u_\mu'(0) = 0$ (remember that Greek subscripts refer to all integers from 1 to n) contradicts the assumption that $t = 0$ is a regular point of the curve $x_\mu = u(t)$. Hence, this curve is not tangent to the plane $\sum u_i'(0)x_i = 0$ at $t = 0$. Since we are interested here only in integral curves which approximate the closed curve $x_\mu = u_\mu(t)$ along its whole extent the fact that the restriction $\sum u_i'(0)x_i = 0$ for $t = 0$ excludes integral curves of (2) which do not intersect that plane does not imply any loss of generality.

Consider now the solution of (2) which for $t = 0$ assumes initial values $x_\mu = a_\mu$. By virtue of condition (b) it is possible to eliminate in $\mathcal{E}$ the function x_n from the first $n - 1$ of the equations (2). These represent then a system of $n - 1$ differential equations for the functions $x_i(t)$. The classical existence theorem establishes for sufficiently small t the existence and uniqueness of a solution $x_i = x_i(t, a)$ of (2) assuming preassigned values a_i for $t = 0$, provided these values are so small that the point determined by $x_i = a_i$ and $F_n(a_1, \cdots, a_{n-1}, x_n) = 0$ lies in the interior of $\mathcal{E}$. (We recall once more that by writing a_i, with a Latin subscript, we exclude the number a_n.) The corresponding value of x_n can then be determined from the last of the equations (2) as a function $x_n(t, a)$. For the special values $a_i = 0$ we obtain the periodic solution $x_\mu = u_\mu(t)$.

We now adopt the following

Definition. Let $x_\mu(t, a)$ denote the solution of (2) determined by the initial values $x_i(0, a) = a_i$, subject to the condition $\sum u_i'(0)a_i = 0$. Then the periodic solution $x_\mu = u_\mu(t)$ is called *degenerate*, if for the solution $x_i(t, a)$ of (2) with the initial values $x_i = a_i$ the first total differentials of

$$x_i(T + \tau, a) - a_i \tag{5}$$

with respect to τ and the a_j at $\tau = 0$, $a_j = 0$ are equal to zero for approximate values of the increments da_i and $d\tau$.

Observe that the first differential of $x_n(T + \tau, a) - x_n(0, a)$ also vanishes in this case because the x_i determine x_n by means of the equation $F_n(x) = 0$.

The preceding definition of degeneracy can be formulated as a statement about the solutions of the "*variational differential equations*"

$$\text{(6)} \qquad \begin{aligned} &\text{(a) } v_i' = \sum_\mu F_{i\mu}(u)v_\mu \\ &\text{(b) } 0 = \sum_\mu F_{n\mu}(u)v_\mu . \end{aligned}$$

The significance of these equations is that the derivatives

$$v_\mu = u_\mu'$$

are among their solutions, as seen by differentiating equations (2). Since v_n can be eliminated from (6), in consequence of condition (b), this is a system of $n - 1$ linear homogeneous differential equations for $v_1, v_2, \cdots, v_{n-1}$. Denote by $v_{1j}(t), v_{2j}(t), \cdots, v_{n-1,j}(t)$ the solution of (6) with initial values $v_{ij}(0) = \delta_{ij}$, where δ_{ij} is the identity matrix. Then we have the

LEMMA. *The periodic solution* $x_\mu = u_\mu(t)$ *of* (2) *is degenerate if and only if the determinant*

$$\text{(7)} \qquad \begin{vmatrix} u_1'(0) & v_{11}(T) - 1 & v_{12}(T) & \cdots & v_{1,n-1}(T) \\ u_2'(0) & v_{21}(T) & v_{22}(T) - 1 & \cdots & v_{2,n-1}(T) \\ \cdots & \cdots & \cdots & \cdots & \cdots \\ u_{n-1}'(0) & v_{n-1,1}(T) & v_{n-1,2}(T) & \cdots & v_{n-1,n-1}(T) - 1 \\ 0 & u_1'(0) & u_2'(0) & \cdots & u_{n-1}'(0) \end{vmatrix}$$

is zero.

For the proof we form the total differential of (5) at $\tau = a_j = 0$ and express the condition of degeneracy in the form

$$\text{(8)} \qquad x_i'(T, 0)\, d\tau + \sum_{j=1}^{n-1} \left(\frac{\partial x_i(T, 0)}{\partial a_j} - \delta_{ij} \right) da_j = 0,$$

where $d\tau$, da_j are certain increments, not all zero. But $y_i = \partial x_i(t, a)/\partial a_j$ is that solution of the system of differential equations

$$\text{(9)} \qquad \begin{aligned} y_i' &= \sum_\mu F_{i\mu}(x)y_\mu \\ 0 &= \sum_\mu F_{n\mu}(x)y_\mu \end{aligned}$$

which satisfies the initial conditions $y_i = \delta_{ij}$ for $t = 0$. This is seen by substituting the functions $x_i = x_i(t, a)$ into the system (2) and differentiating the resulting equations with respect to a_j. For $a_j = 0$ the system (9) reduces to (6), hence

$$\frac{\partial x_i(t, 0)}{\partial a_j} = v_{ij}(t).$$

Also, $x_i'(T, 0) = u_i'(T) = u_i'(0)$ so that (8) can be written in the form:

$$u_i'(0)\, d\tau + \sum_j (v_{ij}(T) - \delta_{ij})\, da_j = 0.$$

To this homogeneous system of linear algebraic equations in the quantities $d\tau$, da_j the equation $\sum u_j'(0)\, da_j = 0$ has to be added. The existence of a non-trivial solution of this system is equivalent to the vanishing of the determinant (7).

Remark. By a simple argument, omitted here, it can be shown that the preceding definition of degeneracy is also equivalent to the statement that the characteristic polynomial of the matrix $\{v_{ij}(T)\}$ has 1 as a multiple root.

In the following *it will be assumed that $u_\mu(t)$ is a non-degenerate periodic solution.*

3. **The convergence of X_μ to x_μ.** Before we take up the question of the existence of *periodic* solutions of the system (1) we have to investigate the asymptotic behavior for small ϵ of the solutions $X_\mu(t)$ of (1) with prescribed *constant* initial values $X_\mu(0) = a_\mu$. These solutions will be seen to tend to those solutions $x_\mu(t) = x_\mu(t, a)$ of the system (2) which are determined by the initial conditions $x_i(0) = a_i$; (a_i here represents the values a_μ for $\mu \neq n$). If the $a_i(0)$ are sufficiently small, then, irrespective of the value of a_n, the curve $x_\mu(t)$ will, for $0 \leq t \leq 2T$, lie in the interior of the closed neighborhood $\mathcal{E}$ of the closed curve $u_\mu(t)$, determined by the inequalities $| x_\mu - u_\mu(t) | \leq 2k$. We shall assume throughout that

$$(10) \qquad | a_\mu | \leq \alpha \leq \frac{k}{4},$$

where α is chosen so small that the curves $x_\mu(t) = x_\mu(t, a)$ satisfy the stronger inequalities

$$(11) \qquad | x_\mu(t) - u_\mu(t) | \leq \frac{k}{4} \text{ in } 0 \leq t \leq 2T$$

(k was defined in equation (3)). We note for later use that (11) implies in particular that

$$(12) \qquad | x_n(0) | \leq \frac{k}{4}$$

and therefore, in conjunction with (10), that

$$| a_n - x_n(0) | \leq \frac{k}{2}. \tag{13}$$

From here on throughout the paper ϵ is assumed to be positive unless otherwise stated. In the sequel we shall denote by $M, M_1, M_2, \cdots$, numbers which may depend on m, α and k, but which are independent of the a_μ and ϵ.

THEOREM 1. *Let $X_\mu(t) = X_\mu(t; a, \epsilon)$ be the solution of* (1) *determined by the initial values $X_\mu(0) = a_\mu$ where the a_μ are constants independent of ϵ. (No periodicity is required here.) Let $x_\mu(t) = x_\mu(t, a)$ be the solution of* (2) *determined by the initial values $x_i(0) = a_i$. There exist positive constants ϵ_0 and M such that for $\epsilon \leq \epsilon_0$ one has*

$$\left.\begin{aligned} &| X_i(t) - x_i(t) | < \epsilon M \\ &| X_n(t) - x_n(t) | < \epsilon M + | a_n - x_n(0) | e^{-\epsilon^{-1}mt} \end{aligned}\right\} \text{in } 0 \leq t \leq 2T,$$

where m is the constant introduced in (4a).

The difference in the behavior of the X_i and of X_n which this theorem indicates was to be expected, for $X_n(0) = a_n$, while $x_n(0)$ is the solution of $F_n(a_1, \cdots, a_{n-1}, x_n(0)) = 0$. Except if accidentally $a_n = x_n(0)$, the function $X_n(t)$ will not tend to $x_n(t)$ in an interval including $t = 0$. The convergence of X_n is therefore, in general, non-uniform in intervals including $t = 0$. By leaving the difference $a_n - x_n(0)$ in our estimate we set in evidence the fact that the convergence is uniform up to $t = 0$, if $a_n = x_n(0)$.

For the proof of Theorem 1 we define functions Δx_μ by

$$X_\mu = x_\mu + \Delta x_\mu . \tag{14}$$

The functions Δx_μ, (which depend on ϵ), are solutions of the system of differential equations

$$\begin{aligned} (\Delta x_i)' &= F_i(x + \Delta x) - F_i(x) \\ \epsilon(\Delta x_n)' &= F_n(x + \Delta x) - F_n(x) - \epsilon x_n' \end{aligned} \tag{15}$$

obtained by substituting (14) into (1) and using (2). The Δx_μ satisfy the initial conditions

$$\Delta x_i(0) = 0, \qquad \Delta x_n(0) = a_n - x_n(0). \tag{16}$$

The differential equations (15) can be formally linearized. To that end we introduce the abbreviations

$$J_{\mu\nu} = \int_0^1 F_{\mu\nu}(x + \sigma\Delta x)\, d\sigma. \tag{17}$$

Since

$$F_\mu(x + \Delta x) - F_\mu(x) = \int_0^1 \frac{d}{d\sigma} F_\mu(x + \sigma\Delta x)\, d\sigma = \sum_{j=1}^{n} J_{\mu\nu}\Delta x_\nu$$

the system (15) can be written in the form

$$(18a) \qquad (\Delta x_i)' = \sum_{\nu=1}^{n} J_{i\nu}\Delta x_\nu ,$$

$$(18b) \qquad \epsilon(\Delta x_n)' - J_{nn}\Delta x_n = \sum_{j=1}^{n-1} J_{nj}\cdot \Delta x_j - \epsilon x_n' .$$

The quantities $J_{\mu\nu}$ depend on the unknown functions Δx_μ. But for $\epsilon > 0$ the Δx_μ and therefore the $J_{\mu\nu}$ are continuous functions of t and the a. Integrating (18a) with respect to t and treating (18b) formally as a linear differential equation in Δx_n one obtains for the Δx_μ the system of integral equations

$$(19a) \qquad \Delta x_i = \int_0^t \sum_\nu J_{i\nu}(\tau)\Delta x_\nu(\tau)\, d\tau,$$

$$(19b) \quad \Delta x_n = \epsilon^{-1} \int_0^t \left\{ \sum_j J_{nj}(\tau)\cdot \Delta x_j(\tau) - \epsilon x_n'(\tau)\right\} \exp\left[\epsilon^{-1} \int_\tau^t J_{nn}(\rho)\, d\rho\right] d\tau$$

$$- (a_n - x_n(0)) \exp\left[\epsilon^{-1} \int_0^t J_{nn}(\rho)\, d\rho\right]$$

by means of which the Δx_μ will now be estimated for small ϵ. Let us denote by $\mathfrak{D}$ the closed subdomain of $\mathcal{E}$ defined by the inequalities $| x_\mu - u_\mu(t) | \leq k$. Because of the restriction (10) the point $t = 0$ on the curve X_μ lies in the interior of $\mathfrak{D}$. Hence, X_μ lies in $\mathfrak{D}$ for a certain interval of positive values of t, depending on the value of ϵ. Let $T(\epsilon)$ denote the greatest value of t in the interval $0 \leq t \leq 2T$ such that $X_\mu(t)$ lies in $\mathfrak{D}$. This implies that $X_\mu(T(\epsilon))$ lies on the boundary of $\mathfrak{D}$, if $T(\epsilon) < 2T$. (We shall show later that $T(\epsilon) = 2T$ for sufficiently small ϵ).

In order to arrive at an estimate for the Δx_μ from equations (19) we introduce quantities $L_\mu = L_\mu(t)$ by the equations

$$L_\mu = \max | \Delta x_\mu(\tau)e^{-\lambda\tau} | \qquad (0 \leq \tau \leq t),$$

where λ is a positive constant to be chosen later. L_μ depends on t, the a_ν and ϵ. We have by definition

$$(20) \qquad | \Delta x_\mu | \leq L_\mu e^{\lambda t} \qquad (0 \leq t \leq T(\epsilon)).$$

(This procedure is motivated by the fact that the regular solutions of non-singular differential equations are always smaller than some exponential function. In fact, the usual existence proofs by successive approximations consist essentially in constructing a series for the solution which is dominated by the series for an exponential function. For sufficiently large λ the quantities L_μ can therefore be expected to be small, uniformly in t.) Let M_1 denote an upper bound for the quantities $| F_{\mu\nu} |$ in $\mathcal{E}$. Then the $| J_{\mu\nu} |$ have the same bound as long as $0 \leq t \leq T(\epsilon)$. To see this it suffices to show that the points $x_\mu + \sigma\Delta x_\mu$ occurring in (17) lie in $\mathcal{E}$ for $0 \leq t \leq T(\epsilon)$. This is the case because

$$\begin{aligned} | x_\mu + \sigma\Delta x_\mu - u_\mu | &= | x_\mu + \sigma(X_\mu - u_\mu + u_\mu - x_\mu) - u_\mu | \\ &\leq 2 \, | x_\mu - u_\mu | + | X_\mu - u_\mu | \\ &\leq 2 \cdot \frac{k}{4} + k < 2k \qquad (0 \leq t \leq T(\epsilon)), \end{aligned}$$

by virtue of (14), (11), the definition of $\mathfrak{D}$, $T(\epsilon)$ and $\mathcal{E}$, by (3). We then obtain from (19a) and (20) the inequalities

$$| \Delta x_i | \leq M_1 \sum_i L_i \int_0^t e^{\lambda\tau} \, d\tau + M_1 \int_0^t | \Delta x_n | \, d\tau$$

or

$$| \Delta x_i | < \lambda^{-1} M_1 e^{\lambda t} \sum_i L_i + M_1 \int_0^t | \Delta x_n | \, d\tau. \tag{21}$$

To obtain a similar estimate for Δx_n from (19b) we observe first that formula (4b)—implies the inequality $J_{nn} < -m$ for X_μ in $\mathfrak{D}$, i.e. for $0 \leq t \leq T(\epsilon)$. Let M_2 be a common upper bound of $| x_n' |$ in $0 \leq t \leq 2T$ for all initial values with $| a_i | < \alpha$. That such a bound exists, is clear since x_n' can be calculated from the equation $x_n' = -F_{nn}^{-1}(x) \cdot \sum_i F_{ni}(x) \cdot x_i' = -F_{nn}^{-1}(x) \sum_i F_{ni}(x) F_i(x)$, which follows from (2) by differentiation. The $x_\mu(t)$ lie in $\mathfrak{D}$ for $0 \leq t \leq 2T$ and therefore $| F_{nn} | > m > 0$. Then, using condition (4b) and (13) we find from (19b)

$$\begin{aligned} | \Delta x_n | \leq (e^{\lambda t} M_1 \sum_i L_i + \epsilon M_2) \int_0^t \epsilon^{-1} e^{-\epsilon^{-1} m (t-\tau)} \, d\tau \\ + | a_n - x_n(0) | \, e^{-\epsilon^{-1} m t}, \qquad (0 \leq t \leq T(\epsilon)), \end{aligned}$$

or

$$| \Delta x_n | < \frac{M_1}{m} e^{\lambda t} \sum_i L_i + \epsilon \frac{M_2}{m} + | a_n - x_n(0) | \, e^{-\epsilon^{-1} m t}. \tag{22}$$

Hence

$$\int_0^t | \Delta x_n | \, d\tau < \frac{M_1}{m} \lambda^{-1} e^{\lambda t} \sum_i L_i + \epsilon \frac{2TM_2}{m} + \frac{\epsilon}{m} | a_n - x_n(0) |. \tag{23}$$

Inserting (23) into (21) it follows that

$$| \Delta x_i | < M_3 \lambda^{-1} e^{\lambda t} \sum_i L_i + \epsilon M_4$$

or

$$| \Delta x_i | \, e^{-\lambda t} < M_3 \lambda^{-1} \sum_i L_i + \epsilon M_4 \, ,$$

hence,

$$L_i < M_3 \lambda^{-1} \sum_i L_i + \epsilon M_4 ,$$

and

$$\sum_i L_i < (n-1) M_3 \lambda^{-1} \sum_i L_i + \epsilon(n-1) M_4 .$$

Now choose λ so large that $\lambda^{-1} M_3(n-1) < 1$, say $\lambda = 2(n-1)M_3$. Then the last inequality leads to

$$\sum L_i < \epsilon M_5 . \tag{24}$$

The same inequality holds therefore for each of the non-negative quantities L_i separately. By (20), and since $e^{\lambda t} \leq e^{2\lambda T}$,—$\lambda$ being chosen,—this implies that

$$| \Delta x_i | < \epsilon M_6 \qquad (0 \leq t \leq T(\epsilon)). \tag{25}$$

An analogous inequality for $| \Delta x_n |$ is obtained if (24) is inserted in (22):

$$| \Delta x_n | < \epsilon M_7 + | a_n - x_n(0) | e^{-\epsilon^{-1} m t} \qquad (0 \leq t \leq T(\epsilon)). \tag{26}$$

(25) and (26) establish the statement of Theorem 1 for the interval $0 \leq t \leq T(\epsilon)$. It remains to be shown that $T(\epsilon) = 2T$ for sufficiently small ϵ. To that end we note first that because of (13) the inequalities (25) and (26) imply the existence of an ϵ_0 so small that $\Delta x_\mu < \frac{3}{4}k$ in $0 \leq t \leq T(\epsilon)$ whenever $\epsilon \leq \epsilon_0$. This inequality and (11) leads to

$$| X_\mu - u_\mu | \leq | \Delta x_\mu | + | x_\mu - u_\mu | < k$$

for $0 \leq t \leq T(\epsilon)$ and $\epsilon \leq \epsilon_0$, so that X_μ lies in the *interior* of $\mathfrak{D}$ for $t = T(\epsilon)$. But then it follows from the definition of $T(\epsilon)$ that $T(\epsilon) = 2T$. This completes the proof.

4. **The convergence of $\partial x_\mu / \partial a_\nu$ and of X'_μ.** A statement similar to that of Theorem 1 will now be proved about the derivatives $\partial X_\mu / \partial a_\nu$, where X_μ has the same meaning as in Theorem 1. For each ν these derivatives $Y_\mu = \partial X_\mu / \partial a_\nu$ are solutions of the differential equations

$$Y'_i = \sum_\mu F_{i\mu}(X) Y_\mu , \tag{27a}$$

$$\epsilon Y'_n = \sum_\mu F_{n\mu}(X) Y_\mu \tag{27b}$$

characterized by the initial values $Y_\mu(0) = \delta_{\mu\nu}$. This is a *linear* system—the X_μ are now to be considered as given functions—whose coefficients $F_{\sigma\mu}$ are functions of t and of the $n+1$ parameters ϵ, a_μ.

The derivatives $y_\mu = \partial x_\mu / \partial a_j$ satisfy for each j the linear system (9) whose coefficients depend on t and the a_i only. They satisfy the initial conditions

$y_i(0) = \delta_{ij}$. Observe that $\partial x_\mu/\partial a_n = 0$, since the x_μ do not depend on a_n. To find the initial values of $\partial x_n/\partial a_j$ use the last of equations (9) and write

$$\frac{\partial x_n}{\partial a_j} = -F_{nn}^{-1}(x) \sum_i F_{ni}(x) \frac{\partial x_i}{\partial a_j}$$

whence

$$\left(\frac{\partial x_n}{\partial a_j}\right)_{t=0} = F_{nn}^{-1}(a_i, x_n(0))F_{nj}(a_i, x_n(0)),$$

where $x_n(0)$ is determined from $F_n(a_1, \cdots, a_{n-1}, x_n(0)) = 0$.

THEOREM 2. *As* $\epsilon \to 0$

$$\frac{\partial X_i}{\partial a_\nu} \to \frac{\partial x_i}{\partial a_\nu} \qquad (0 \le t \le 2T)$$

$$\frac{\partial X_n}{\partial a_\nu} \to \frac{\partial x_n}{\partial a_\nu} \qquad (0 < t_0 \le t \le 2T)$$

$$\left|\frac{\partial X_n}{\partial a_\nu}\right| < C \qquad (0 \le t \le 2T).$$

Here t_0 *is arbitrary.* C *is a constant. The convergence is uniform with respect to* t *and the* a.

Observe that Theorem 2 implies that $\partial X_\mu/\partial a_n \to 0$, since $\partial x_\mu/\partial a_n = 0$.

The theorem and its proof differ somewhat from Theorem 1. The main difference is caused by the fact that the coefficients in (27) contain the parameters a_μ and t explicitly.

In order to shorten the notation we continue to write $\partial X_\mu/\partial a_\nu = Y_\mu$, $\partial x_\mu/\partial a_\nu = y_\mu$. The reasoning below will hold for $\nu = 1, 2, \cdots, n$. With this notation the initial values of the functions $\Delta y_\mu = Y_\mu - y_\mu$ are zero for $\mu < n$ and continuous functions of the a_σ independent of ϵ for $\mu = n$. Subtracting the equations (9) from the corresponding equations (27) one obtains for the Δy_μ the differential equations

$$\text{(28a)} \qquad (\Delta y_i)' = \sum_\mu F_{i\mu}(X)\Delta y_\mu + \sum_\mu [F_{i\mu}(X) - F_{i\mu}(x)]y_\mu,$$

$$\text{(28b)} \qquad (\Delta y_n)' = \sum_\mu F_{n\mu}(X)\Delta y_\mu + \sum_\mu [F_{n\mu}(X) - F_{n\mu}(x)]y_\mu - \epsilon y_n'.$$

By Theorem 1, $X_\mu \to x_\mu$ as $\epsilon \to 0$ uniformly with respect to t and the a_μ in every interval $0 < t_0 \le t \le 2T$ and the X_μ are bounded in $0 \le t \le 2T$. Hence the equations (28) can be written

$$(\Delta y_i)' = \sum_\mu F_{i\mu}(X)\Delta y_\mu + \omega_i,$$

$$\epsilon(\Delta y_n)' = \sum_\mu F_{i\mu}(X)\Delta y_\mu + \omega_n,$$

where the ω_μ are functions of t, ϵ and the a_σ which remain bounded in $0 \leq t \leq 2T$ uniformly in the a_σ as ϵ tends to zero and for which $\omega_\mu \to 0$, as $\epsilon \to 0$, uniformly with respect to t and the a_σ in every interval $0 < t_0 \leq t \leq 2T$.

From here on the proof proceeds along the same lines as that of Theorem 1, with only obvious modifications, and a repetition is therefore unnecessary. (Observe for the proof that $\int_0^{2T} | \omega_\mu | \, dt \to 0$, as $\epsilon \to 0$.)

THEOREM 3. *As* $\epsilon \to 0$,

$$X_i' \to x_i', \quad | X_n' | < M_9$$

in $0 < t_0 \leq t \leq 2T$, *uniformly with respect to* t *and the* a_μ.

The uniform convergence of the functions X_i' to x_i' as $\epsilon \to 0$, except at $t = 0$ follows immediately from Theorem 1 and the differential equations (1). For X_n, only the boundedness can be deduced so simply:

$$| X_n' | = | \epsilon^{-1} F_n(X) | = \epsilon^{-1} | F_n(x + \Delta x) | = \epsilon^{-1} | F_n(x + \Delta x) - F_n(x) |$$
$$= \epsilon^{-1} \Big| \sum_\mu J_{n\mu} \cdot \Delta x_\mu \Big| < M_9$$

in $0 < t_0 \leq t \leq 2T$ and for $\epsilon \leq \epsilon_0$ by Theorem 1. It is possible to show that X_n' is not only bounded but tends to x_n' except at $t = 0$. But since this result is not needed here, the proof, which consists in a sharper estimate using equations (19), is omitted.

5. **The existence of periodic solutions of** (1). We are now prepared to prove the Main Theorem as stated in equation (1). For the proof we assume at first that $\epsilon > 0$. A solution of (1) with initial values a_μ will be periodic with period $T + \tau$ if $X_\mu(T + \tau) - a_\mu = 0$. Define functions $P_\mu(\epsilon; \tau, a_\nu)$ by

$$P_\mu \equiv X_\mu(T + \tau) - a_\mu . \tag{29}$$

For $\epsilon > 0$ these functions are continuous and continuously differentiable in all variables. We define

$$(P_\mu)_{\epsilon=0} = x_\mu(T + \tau) - a_\mu ,$$

then the Theorems 1, 2 and 3 imply that the functions P_i are continuous functions of τ, ϵ, a_ν with continuous derivatives with respect to τ and the a_ν in a domain defined by $0 \leq \epsilon \leq \epsilon_0$, $| a_\nu | \leq \alpha$, $| \tau | \leq \tau_1 < T$. The theorems of the preceding sections show that the convergence of P_i to $x_i(T + \tau) - a_i$ is *uniform* with respect to τ and the a_ν. Since the P_i as well as the $x_i(T + \tau) - a_i$ are continuous in τ and a_ν combined, it follows that P_i is continuous in τ, a_ν *and* ϵ *combined.* (Observe that the non-uniform convergence of X_μ to x_μ at $t = 0$ does not interfere with the reasoning here, since we operate in the neighborhood of $t = T$.) P_n is continuous and has continuous derivatives with respect to a_μ

while for the derivatives of P_n with respect to τ only the boundedness has been proved. If to the $n - 1$ equation $P_i = 0$ the condition

$$Q(a) \equiv \sum_i u_i'(0)a_i = 0$$

is added, a system of n equations is obtained. We shall now show that this system can be solved for the n quantities τ, a_i in terms of ϵ and a_n:

The n equations are satisfied for $\epsilon = \tau = a_\mu = 0$, because

$$P_i(0; 0, 0) = u_i(T) = u_i(0) = 0,$$

$$Q(0) = 0.$$

The Jacobian of the system $P_i = 0$, $Q = 0$ is

$$\begin{vmatrix} F_1(X) & \dfrac{\partial X_1}{\partial a_1} - 1 & \dfrac{\partial X_1}{\partial a_2} & \cdots & \dfrac{\partial X_1}{\partial a_{n-1}} \\ F_2(X) & \dfrac{\partial X_2}{\partial a_1} & \dfrac{\partial X_2}{\partial a_2} - 1 & \cdots & \dfrac{\partial X_2}{\partial a_{n-1}} \\ \cdots & \cdots & \cdots & \cdots & \cdots \\ F_{n-1}(X) & \dfrac{\partial X_{n-1}}{\partial a_1} & \dfrac{\partial X_{n-1}}{\partial a_2} & \cdots & \dfrac{\partial X_{n-1}}{\partial a_{n-1}} - 1 \\ 0 & u_1'(0) & u_2'(0) & \cdots & u_{n-1}'(0) \end{vmatrix}.$$

For $\epsilon = 0$, $\partial X_i/\partial a_j$ approaches $\partial x_i/\partial a_j$ by Theorem 2. For $a_\mu = 0$, $\partial x_i/\partial a_j$ equals the quantity v_{ij} introduced in §2. For $\epsilon = a_\mu = \tau = 0$, therefore, the present determinant reduces to the determinant (7), which is different from zero, because we assume the problem to be non-degenerate. By virtue of the theorem on implicit functions we can therefore conclude that the n equations possess for small ϵ unique solutions

$$\tau = \chi(\epsilon; a_n), \qquad a_i = \varphi_i(\epsilon; a_n) \tag{30}$$

which reduces to 0 for $\epsilon = a_n = 0$, further, that the functions χ and φ_i are continuous in ϵ and a_n, and have continuous derivatives with respect to a_n. The latter property follows from the fact that P_i and Q have continuous derivatives with respect to a_n. (Here, as in the following, the fact that the functions are not defined for $\epsilon < 0$ does not preclude the application of the implicit function theorem. The proof of this theorem requires a *full* neighborhood of the given solution only with respect to the variables for which the equations are to be solved, not with respect to the variables in terms of which the solution is to be found.) It remains to be shown that a_n can be chosen so as to satisfy also the equation $P_n = 0$. If the functions (30) are substituted into $P_n = 0$ we obtain an equation

$$\Phi(\epsilon; a_n) = 0$$

for a_n. In order to show that $\partial\Phi/\partial a_n$ does not vanish for $\epsilon = a_n = 0$, we first note that

$$\frac{\partial \Phi}{\partial a_n} = X_n' \frac{\partial \chi}{\partial a_n} + \sum_i \frac{\partial X_n}{\partial a_i} \frac{\partial \varphi_i}{\partial a_n} + \frac{\partial X_n}{\partial a_n} - 1. \tag{31}$$

To evaluate the limits of the quantities $\partial\chi/\partial a_n$, $\partial\varphi_i/\partial a_n$, we observe that they satisfy the equation

$$\begin{gathered} X_i'(T + \tau) \frac{\partial \chi}{\partial a_n} + \sum_k \left(\frac{\partial X_i(T + \tau)}{\partial a_k} - \delta_{ik} \right) \frac{\partial \varphi_k}{\partial a_n} = - \frac{\partial X_i(T + \tau)}{\partial a_n}, \\ \sum_k u_k'(0) \frac{\partial \varphi_k}{\partial a_n} = 0 \end{gathered} \tag{32}$$

obtained by differentiating the equations $P_i = 0$, $\sum u_k'(0) a_k = 0$, with respect to a_n after having substituted the functions (30) for τ and the a_i. The equations (32) constitute a system of linear equations for the quantities $\partial\chi/\partial a_n$ and $\partial\varphi_i/\partial a_n$. For $\epsilon = a_n = 0$ the right members of (32) vanish by Theorem 2, since the x_i are independent of a_n. The determinant of the left members of (32) is a continuous function of a_n and ϵ which, by Theorems 2 and 3, reduces to (7) for $\epsilon = a_n = 0$. Since (7) is a non-vanishing determinant, the quantities $\partial\chi/\partial a_n$, $\partial\varphi_i/\partial a_n$ vanish for $\epsilon = a_n = 0$. From Theorem 3 we know that $X_n'(T + \tau)$ remains bounded as $\epsilon \to 0$, and from Theorem 2 it follows that $\partial X_n(T + \tau)/\partial a_n$ tends to zero, since $\partial x_n/\partial a_n = 0$, while the $\partial X_i/\partial a_n$ remain bounded. Hence we have from (31)

$$\lim_{\epsilon \to 0} \frac{\partial \Phi(\epsilon, a_n)}{\partial a_n} = -1 \neq 0.$$

Using again the implicit functions theorem (this time for one function only) we conclude that the equation $\Phi(\epsilon; a_n) = 0$ can be solved uniquely for small positive ϵ, leading to a continuous function $a_n(\epsilon)$ with $a_n(0) = 0$. If this function is substituted for a_n in (30) we obtain continuous functions $\tau(\epsilon)$, $a_\nu(\epsilon)$ such that $\tau(0) = a_\nu(0) = 0$ and $X_\nu(t, a_\nu(\epsilon)) = U_\mu(t; \epsilon)$ is a periodic solution of (1) with period $T + \tau(\epsilon)$. This establishes parts (b) and (c) of the Main Theorem.

In order to prove part (a) of the Main Theorem note that for these periodic solutions Theorem 1 implies

$$| X_\mu(t, a_i(\epsilon)) - x_\mu(t, a_i(\epsilon)) | < \epsilon M_{10}$$

for $0 < t_1 \leq t \leq t_1 + T + \tau(\epsilon) < 2T$, ($t_1$ a constant). Now write

$$| X_\mu - u_\mu | \leq | X_\mu - x_\mu | + | x_\mu - u_\mu |, \tag{33}$$

where X_μ and x_μ are determined by the initial values $a_r(\epsilon)$ and $a_i(\epsilon)$, respectively. As ϵ tends to zero both terms of the right member of (33) tend to zero uniformly in t in the interval $t_1 \leq t \leq t_1 + T + \tau(\epsilon)$; the first, because of Theorem 1, the second because the initial values $a_i(\epsilon)$ tend to zero, the initial values of u_μ. Hence $X_\mu(t, a_\mu(\epsilon))$ tends to $u_\mu(t)$ uniformly for the values of t indicated above. But since $X_\mu(t, a_\mu(\epsilon))$ is periodic this means that the convergence takes place for *all* t. This completes the proof of the Main Theorem for the case that ϵ is positive, which has been assumed so far. When ϵ is negative, the substitution $t^* = -t$ reduces (1) to a system of differential equations in the variable t^* to which the preceding theory can be applied. This shows also that for the Main Theorem the assumption $F_{nn}(u) < 0$ can be replaced by the weaker one that $F_{nn} \neq 0$, as mentioned at the beginning of this paper.

Bibliography

1. E. V. Appleton and W. Greaves, *On the solution of the representative differential equation of the triode oscillator*, Philosophical Magazine, vol. 45(1923), pp. 401–414.
2. D. A. Flanders and J. J. Stoker, *The Limit Case of Relaxation Oscillations, Studies of Non-linear Vibrations*, New York University, 1946.
3. K. O. Friedrichs and J. J. Stoker, *Forced vibrations of systems with non-linear restoring force*, Quarterly of Applied Mathematics, vol. 1(1943), pp. 97–115.
4. K. O. Friedrichs, *On Non-linear Vibrations of Third Order, Studies of Non-linear Vibrations*, New York University, 1946.
5. N. Kryloff, N. Bogoliuboff, and S. Lefschetz, *Introduction to Non-linear Mechanics*, Princeton, 1943.
6. N. Minorsky, *Introduction to non-linear mechanics, Part* II, *Analytical Methods of non-linear mechanics*, The David Taylor Model Basin, September, 1945.
7. H. Poincaré, *Les Méthodes Nouvelles de la Mécanique Céleste*, vol. 1, Paris, 1892.
8. J. J. Stoker and A. Peters, *Seminar Notes*, New York University, 1943.

New York University.

[55–2] **Asymptotic phenomena in mathematical physics, Bull. A.M.S., 61 (1955), 485–504.**

Commentary:

When Friedrichs was invited to give the 1954 Gibbs Lecture, he had been actively interested in asymptotic problems for more than fifteen years. The comparatively small number of his published papers in this field is no adequate measure of Friedrichs' importance and influence in the development of the mathematical area he describes in his Gibbs Lecture. The flourishing of "Asymptotics" since 1954 is due in large part to Friedrichs' work and teaching before and after the Gibbs Lecture. Actually, at the time of that lecture, Friedrichs had already turned most of his creative interest to other topics. His most important published contribution to the asymptotic theory of differential equations is probably contained in his joint papers with J.J. Stoker, notably [41–2]. Some mathematicians might argue, however, that two works which he never chose to have published in properly printed form have been more influential: Namely, his Lecture Notes [53–1], and his Reports [53–1] and [56–1].

Very likely Friedrichs furthered the development of the asymptotic theory of differential equations at least as much by stimulating his students and colleagues as by his publications, including this Gibbs Lecture. Some of the open questions he mentioned in the lecture have now been settled or substantially clarified by mathematicians whose work is indebted to Friedrichs.

One of these is the problem of "matching" solutions. It had been known, since Prandtl introduced the idea of a boundary layer in flows with low viscosity, that one must introduce "outer" and "inner" solutions obtained by "stretching transformations," and that these solutions must be matched. As far as I know, Friedrichs' work with Stoker is the first example where this process was performed in a mathematically complete way without having to rely on plausibility arguments. The problem in question was the boundary layer effect in the buckling of thin plates, see [39–3], [41–2], and [42–2]. Much progress has been made since then, but no complete mathematical theory exists, not even for linear ordinary differential equations. The rule of thumb which Friedrichs and Stoker took over from boundary layer theory, is that the outer solution at the initial point must have the same value as the inner solution at infinity. It was proved by them to be correct for the plate problem, but it does not suffice for a general theory.

There are many theorems in the older asymptotic theory which prove that frequently the formal series solutions represent true solutions asymptotically, but for many problems of particular interest in physics the methods known in 1954 did not suffice, and in his lecture Friedrichs explicitly points out the need for more work in this direction. Diffraction theory, which deals with the transition from physical optics to geometric optics, and which explains the occurrence of shadow regions was singled out by him as a typical and striking application. The asymptotic theory behind these phenomena is now almost completely understood. The many research workers in this area are too numerous to list here, but most of them have been greatly stimulated in this work by their contact with Friedrichs.

ASYMPTOTIC PHENOMENA IN MATHEMATICAL PHYSICS

K. O. FRIEDRICHS

The problems I intend to speak about belong to the somewhat undefined and disputed region at the border between mathematics and physics. The fields of physics from which these problems originate are rather classical; the mathematical questions involved are also rather classical. That does not mean that these problems belong to the past. On the contrary, they are quite alive today and—I am convinced—they will remain so for some time.

The problems concern what may be called asymptotic phenomena. Instead of explaining in general terms what I mean by asymptotic phenomena, I prefer to single out at first one class of such phenomena: *discontinuities*. A typical discontinuity of the kind I have in mind is the boundary of the *shadow* which appears when a light wave passes an object. Now, the propagation of light is governed by a partial differential equation which has continuous solutions. How then is it possible that a discontinuity arises? Of course, actually there is no sharp discontinuity at the shadow boundary; there is a transition from light to dark which takes place across a very narrow strip along the shadow boundary. Nevertheless, it is remarkable enough that the differential equations of wave motion have solutions which involve such quick transitions—in fact, most differential equations of physics possess such solutions—and it is an interesting task to study those features of these equations which make such quick transitions possible.

Discontinuities and quick transitions occur in various branches of physics. A striking example of a discontinuity is the *shock* in gas motion. Quick transitions occur frequently in situations in which one perhaps would not speak of a discontinuity. A case in point is Prandtl's ingenious conception of the *boundary layer*. This is a narrow layer along the surface of a body, traveling in a fluid, across which the flow velocity changes quickly. Prandtl's observation of this quick transition was the starting point for his theory of fluid resistance. Other cases, closely related to the boundary layer phenomenon, are the so-called *edge effect* in the deformation of elastic plates and shells and the *skin effect* in the flow of electric currents. A number of other such effects will be described in the later parts of this lecture. All

The twenty-eighth Josiah Willard Gibbs lecture delivered at Pittsburgh, Pennsylvania, December 27, 1954 under the auspices of the American Mathematical Society; received by the editors June 24, 1955.

these effects may be regarded as typical asymptotic phenomena.

Without attempting to give a precise definition of this term, I shall simply call asymptotic all those phenomena which show discontinuities, quick transitions, nonuniformities, or other incongruities resulting from approximate description.

In the mathematical treatment of such phenomena, physicists have to a certain degree relied on their intuition, and very effectively so; but they have also developed or employed systematic mathematical procedures.

In such a systematic approach one may develop an appropriate quantity with respect to powers of a parameter, ϵ. This expansion is to be set up in such a way that the quantity is continuous for $\epsilon > 0$ but discontinuous for $\epsilon = 0$. Naturally, a series expansion with this character must have peculiar properties. A most remarkable property is that in general these series *do not converge.*

No doubt, divergent series are very useful; it has even been said that they are more useful than convergent ones. However this may be, if a divergent series is useful it must be meaningful.

The use of a series which does not necessarily converge is a typical instance of a "formal procedure" and I should perhaps say a word about the role of *formal procedures* in mathematical physics.

Those who employ mathematics as a tool have rarely been inhibited by the fear of divergence; they have always been confident that, somehow or other, formal procedures are valid. A mathematician may be inclined to frown on this attitude as a superstition; but, on second thought he will yield and try to show that formal procedures—I mean those used by good physicists—indeed are valid if only the meaning of validity is properly interpreted.

There are numerous instances of justification of formal procedures by re-interpretation. I need only refer to the generalizations of the notions of function and differential operator, which have been very effective, in particular, in recent years. It would certainly be interesting to trace the effect of these generalizations in mathematical physics; but I do not intend to do so.

The present talk will be solely concerned with the formal expansions used in the analysis of asymptotic phenomena. The idea of giving validity to these formal series is classical: essentially it goes back to Poincaré.

Poincaré advanced this idea in his work on ordinary differential equations in 1886. Before that time many formal series solutions of such equations had been developed and it was found that they did not converge—in general. Poincaré proved that these formal series solutions represent *asymptotic expansions of actual solutions*. Thus it became clear in which way formal series solutions may be regarded as "valid."

Let me explain the meaning of the phrase "formal series solution" and its "asymptotic" character in connection with an elementary differential equation, namely the *differential equation of the second order*,

$$\epsilon u'' + au' + bu = 0. \tag{1}$$

Here u is a function of a variable z, which in the present context may just as well be taken as real. Furthermore, a and b are analytic functions of z and ϵ is a parameter. Note that this parameter occurs in such a way that the order of the differential equation is reduced for $\epsilon=0$.

We are interested not in solutions of this differential equation for each fixed value of the parameter ϵ, but in the dependence of such solutions on this parameter, in particular, in the neighborhood of $\epsilon=0$.

The formal series which we shall consider are not simply power series in ϵ; they are rather of the form

$$e^{S(z)/\epsilon} \sum_n \epsilon^n v_n(z), \tag{2}$$

which will be referred to as the "standard" form. One may try to find solutions of the differential equation (1) which admits such a series expansion. To this end one tries to determine the functions S and v_n by inserting this series into the differential equation and setting the coefficient of every power of ϵ equal to zero. For the functions S and v_n one then finds simple differential equations which are easily solved. The equation to be satisfied by the function S, the so-called "characteristic equation," is

$$(S')^2 + aS' = 0. \tag{3}$$

Here S' is the derivative of S. Inserting the functions S and v_n thus found into the series (2) one obtains a *formal series solution* of the differential equation. In general, though, this series does not converge.

A formal series of the type described is said to represent the *asymptotic expansion* of a function $u(z, \epsilon)$ if the remainder of the terms up to the Nth order is of the order $N+1$; precisely, if

$$e^{-S(z)/\epsilon} u(z, \epsilon) = \sum_{n=0}^{N} \epsilon^n v_n(z) + O(\epsilon^{N+1}),$$

uniformly in an appropriate z-interval.

The problem treated by Poincaré was a little different from the one just described since he did not consider expansions with respect to powers of ϵ but with respect to powers of z^{-1}. Nevertheless, it was to be expected that the analogue of what he proved also holds in the case considered here. That is, each formal series of the standard type should be the asymptotic expansion of an actual solution. That this is so for equations of the second order was proved as early as 1899 by Horn. The corresponding general theory for equations of the nth order, developed in 1908 by Birkhoff, initiated an extensive literature in this and related fields.

These results may be used to answer questions concerning the behavior of specific solutions of the differential equation as the parameter ϵ tends to zero. I should like to discuss one such question, which is extremely elementary, but nevertheless leads in a natural way to the boundary layer phenomenon.

Let us *prescribe boundary values* for the solution of our differential equation at two points, $z=0$ and $z=z_1$, and ask how the solution of this boundary value problem behaves as $\epsilon \to 0$. Note that for $\epsilon=0$ the differential equation reduces to an equation of the first order. One may therefore wonder whether the solution of the equation of the second order approaches a solution of the equation of the first order. Now a solution of the first order equation is already determined by one boundary condition; one cannot expect that both conditions will be satisfied in the limit. One *boundary condition*—at least—*will get lost.* The question is, which one?

This question and related questions can easily be answered with the aid of two solutions possessing a standard expansion. The answer is that under appropriate conditions the solution of the boundary value indeed does converge to a solution of the first order equation. This solution assumes one of the two boundary values but not the other one. Which boundary value is lost depends on the sign of a/ϵ. Let us assume that the lost boundary value is the one prescribed at $z=0$.

The process of losing a boundary value takes place through *nonuniform convergence.* If the parameter ϵ is small, the solution will run near the limit solution except in a small segment at the end point $z=0$ where it changes quickly in order, as it were, to retrieve the boundary value about to be lost.

Thus a "quick transition" is found to occur. It must occur since a boundary condition is about to be lost; and this loss in turn is necessary since the order of the differential equation is about to drop. To be sure, the *reduction of the order of a differential equation* combined with the *loss of a condition such as a boundary condition* is the most characteristic mathematical feature of asymptotic phenomena.

The next step in the asymptotic analysis of our problem consists in a detailed *description* of the solution of the boundary value problem *within the transition layer*. To this end we introduce a new independent variable by *stretching* the original variable in an appropriate manner. Specifically, we introduce the ratio

$$\zeta = z/\epsilon$$

as a new variable. We then consider the quantity u as a function of ζ, in addition to ϵ, and ask whether or not this new function approaches a limit as $\epsilon \to 0$. This is indeed the case. The limit process is now uniform even near $z=0$. Therefore the new limit function may serve as an approximate description of the quick change of u in the transition layer.

The new limit function is defined for all $\zeta \geqq 0$; in fact it approaches a definite value as $\zeta \to \infty$. Remarkably enough, this value of u at $\zeta = \infty$ is exactly equal to the value which the limit function in the first "direct" process assumes at $z=0$.

This peculiar phenomenon may at first sight appear a little paradoxical, but actually, it is quite natural. Evidently, any fixed z-neighborhood of the point $z=0$ corresponds to an arbitrarily large part of the ζ-axis if only ϵ is made sufficiently small. It is therefore clear that a connection of the two limit functions must involve the behavior of the direct limit function at $z=0$ and the behavior of the second limit function at infinity. The phenomenon just described will be referred to as "identification phenomenon."

The results discussed in connection with the simple equation of the second order are rather typical and they may frequently serve as a guide in understanding other asymptotic phenomena.

As an example, let us consider Prandtl's *boundary layer theory*. This theory was developed in order to solve the problem of fluid resistance, which had caused great difficulties since the time of d'Alembert. It was known that the resistance is due to the viscosity of the fluid; for, it was known that nonviscous fluids do not exert a force on bodies through it. Still, for fluids with low viscosity the assumption of absence of viscosity led to a very satisfactory description of the flow around the body, although it did not yield a resistance.

Prandtl in 1904 resolved this dilemma by advancing the hypothesis that the effect of viscosity is concentrated in a narrow layer near the surface of the body. On the basis of this hypothesis he was able to give a detailed description of the flow in it. With unfailing intuition he appraised the order of magnitude of the various terms of the governing differential equation and rejected those that he judged to be insignificant. The simplified equations thus obtained could then be solved.

There was never any doubt that the boundary layer theory gave a proper account of physical reality, but its mathematical aspects remained a puzzle for some time. Only when this theory is fitted into the framework of asymptotic analysis, does its mathematical structure become transparent.

Viscous fluid flow—in two dimensions, for simplicity—is governed by a partial differential equation of the fourth order. If the viscosity vanishes, the equation reduces to one of the third order. The expansion of viscous fluid flow in the neighborhood of inviscid fluid flow thus appears as an asymptotic expansion, the viscosity being the parameter. A viscous fluid sticks to the wall; hence two boundary conditions are imposed on the viscous fluid flow: namely the conditions that the tangential and normal velocity components vanish. An inviscid fluid is permitted to slide; hence only one condition is imposed on it. Thus one boundary condition gets lost when the viscosity becomes zero.

It is now clear that, before the boundary condition is lost, a quick change must take place across a thin layer near the boundary. This layer, of course, is Prandtl's boundary layer.

Prandtl's detailed description of the flow in the boundary layer can be re-derived by a stretching procedure similar to the one described above. The new stretched variable must be so chosen that with respect to it the boundary layer does not shrink to zero as the viscosity tends to zero.

The approach to the boundary layer theory outlined leads to a definite clarification of the issue but it does not yield a rigorous justification of this theory. The main reason for this difficulty is the nonlinearity of the problem.

The situation is similar in many other nonlinear asymptotic problems. Methods for approximate solutions of such problems are frequently suggested by the facts discussed in connection with the simple ordinary differential equation of the second order.

The boundary layer effect discussed is not the only form of breakdown of uniform convergence. Such a breakdown may also happen in the interior of the domain. A most remarkable such occurrence, in its mathematical aspects even more striking than the boundary layer phenomenon, is the phenomenon discovered by *Stokes* in 1857.

This phenomenon may be explained in connection with the simple differential equation of the second order (1) discussed before. In doing this it is preferable to assume that the coefficients and the solution are analytic functions of the complex variable z.

It was mentioned above that every formal series solution of the standard type is the asymptotic expansion of an actual solution of equation (1), but it was not stated where this asymptotic expansion is valid. In fact, it may happen that this expansion is valid in only a part of the domain in which the function $u(z, \epsilon)$ is defined. That is, it may happen that the function u admits the standard asymptotic expansion in only a part of the z-plane. In other parts it then will possess a completely different asymptotic expansion. The lines which separate subregions of different expansions are called "Stokes lines." The change of the asymptotic expansion on crossing these lines is the "Stokes phenomenon."

In short one may say: *the Stokes phenomenon* obtains at a line *if the asymptotic expansion of the analytic continuation* of u across this line *is not given by the analytic continuation* of the terms *of the asymptotic expansion.*

If a Stokes phenomenon is present, the leading term of the asymptotic expansion of the solution u changes its character on crossing the Stokes line; one may therefore say that this term is discontinuous across the Stokes line. Suppose the function u stands for a physical quantity and suppose this quantity is approximately described by the leading term of the expansion. If this term is discontinuous, the physical quantity is approximately described as being discontinuous, although actually it is continuous. Thus we have encountered the possibility of describing continuous quantities as discontinuous ones by describing them asymptotically. This possibility is of great significance. To be sure, a large class of *discontinuity phenomena in mathematical physics may be interpreted as Stokes or boundary layer phenomena.*

For an ordinary linear differential equation it is easy to locate lines at which a Stokes phenomenon occurs. One can always find such lines near a "turning point." A *turning point* or *transition point* is a point z at which two roots $S'(z)$ of the characteristic equation (3) coalesce: $S_1'(z) = S_2'(z)$. Stokes phenomena then may occur at certain rays through the turning point. For equations of the type here considered, these rays are curves on which the real parts of $S_1'(z)$ and $S_2'(z)$ agree: $\text{Re } S_1'(z) = \text{Re } S_2'(z)$.

For the differential equation (1), in particular, there are four such rays issuing from a turning point. At which ones of these rays the Stokes phenomenon occurs depends on the solution considered.

Specifically, one can find a solution which possesses a standard asymptotic expansion in the open region R generated by three of the four sectors that are formed by the four rays. It possesses an asymptotic expansion also in the fourth sector, at the two rays bounding the fourth sector, and at the turning point; but all three expansions differ from each other and from the standard expansion in the region R. Clearly, a Stokes phenomenon is present; the two rays bounding the fourth sector are Stokes lines.

Of course, one wants to know how to find these different expansions.

Before indicating how one may attack this "continuation" problem I should mention that such a turning point problem was first treated by Jeffreys in 1923 in connection with a somewhat different differential equation.

One possible approach to a solution of the continuation problem consists in reducing it to the problem of finding the asymptotic expansion *at* the turning point. This problem will be referred to as the "connection problem."

I shall briefly indicate a formal procedure by which this connection problem may be attacked.

In this approach one again employs the "method of stretching." For equation (1), in particular, one introduces

$$\zeta = z/\epsilon^{1/2},$$

instead of z, as new independent variable, assuming the turning point to be at $z=0$. It would not be difficult to motivate the choice of $\epsilon^{1/2}$ as stretching factor instead of ϵ. Again, any fixed neighborhood of the turning point will eventually cover the whole ζ-plane. The quantity u, when considered as a function of ζ, now possesses an expansion with respect to powers of $\epsilon^{1/2}$. The terms of this expansion, defined in the whole ζ-plane, can be determined by identifying their behavior at $\zeta=\infty$ with the behavior of the terms of the direct expansion at $z=0$. In other words, one may employ an identification procedure similar to the one discussed in connection with the boundary value problem. The expansion of the solution u at the turning point can then be found.

As mentioned before, this approach is only a formal procedure; naturally, one will ask: does it yield correct results? For differential equation (1), in particular, it is not too difficult to prove that this is so. Such a proof is not so easy, however, in more complicated cases;

for example if an additional singularity is present, or if the equation is of higher order.

The first decisive step in developing a rigorous turning point theory was taken by Langer in 1934. Langer, in fact, treated at first a problem which is not as simple as the one discussed here. Subsequently, a considerable amount of work on the turning point problem has been done, and is being done today.

There are many interesting problems of *mathematical physics* in which a *turning point* analysis plays a role.

Wentzel, Brillouin, and Kramers in 1926 used asymptotic approximations and turning point considerations in solving eigenvalue problems in quantum theory.

A very remarkable problem which requires a turning point analysis is the problem of the *stability of viscous fluid* and the onset of turbulence. Quite a number of aerodynamicists and mathematicians have worked on this somewhat controversial question. Early theoretical investigations led to the prediction that such an instability should occur under peculiar circumstances. This prediction should perhaps have been believed by aerodynamicists, but it was not generally accepted at first. Eventually, the prediction was confirmed by experiment with surprising accuracy.

The pertinent mathematical situation was definitely clarified only in recent years by Wasow through a rigorous turning point analysis.

The asymptotic phenomena of ordinary differential equations which I have described up to now involve linear equations; of course such phenomena have also been studied in connection with *nonlinear* equations. An interesting problem concerns periodic solutions of a differential equation of the form

$$\epsilon u'' = f(u', u).$$

The question is what happens with these periodic solutions as $\epsilon \to 0$, in particular if the limit equation

$$f(u', u) = 0$$

has no periodic solution. Of course there could be no boundary layer effect in the strict sense since there is no boundary. What happens is that the limit function—if it exists—satisfies the equation $f(u', u) = 0$ except at certain points where the derivative u' has a jump discontinuity.

A problem of this type was first treated by van der Pol, 1927, who in this way explained the occurrence of certain jerky oscillations in electric networks, which he called "relaxation oscillations." Subsequently much work was done on electrical and mechanical oscillations of this kind, as well as on the purely mathematical aspects of the problem. Strong results on asymptotic periodic solutions have been obtained by Levinson since 1942.

Another, rather spectacular, case of a discontinuity which may take place in the interior of the domain and not at a boundary is the gas dynamical *shock*. The shock may also be interpreted as the limit of a quick transition and be treated by an asymptotic analysis which involves the drop of the order of a differential equation. The same may be said about the closely related phenomena of *explosion* or *detonation*.

Let me turn to *partial differential equations* and first consider the *hyperbolic* equation

$$u_{tt} - \Delta u = 0, \tag{4}$$

called the "wave equation." Here u is a function of t, x, y, z and Δ is the Laplacian.

The propagation of electromagnetic and acoustic waves is governed by this equation; but these processes are frequently treated in a different manner, in the manner of geometrical optics. One is led to this second treatment in a natural way by asking for formal solutions of the "standard" form

$$u = e^{S/\epsilon} \sum_n \epsilon^n v_n. \tag{5}$$

Here S and v_n are functions of t, x, y, z.

For these functions simple equations are found and readily solved. These equations are of the first order; the drop in order, so typical for asymptotic problems, is thus apparent. The equation for S, the characteristic equation

$$S_t^2 = (\nabla S)^2, \tag{6}$$

is nonlinear.

The formal series (5) is similar to that used for ordinary differential equations. There is a slight difference, however, since the parameter ϵ entering it does not occur in the differential equation. If a concrete problem is to be solved by using this formal solution, the parameter will have to be identified with one of the data of the problem, such as a wave length or a pulse width.

In accordance with the principle of Poincaré, one expects that there exist actual solutions having these formal solutions as asymptotic expansions. That this is so has apparently not yet been proved. One should think, however, that the available methods of proving the existence of solutions of hyperbolic equations would be strong enough for this purpose. But, the primary interest of the asymptotic theory of the wave equation seems to lie in the asymptotic expansion of the solutions of specific problems.

In many cases one can describe a wave process with the aid of the leading term of the above expansion (5). This description now leads to *geometrical optics*. The function S is the eiconal, and the equation $S=\text{const.}$ describes the motion of a wave front. It has been known for a long time that the transition of wave optics to geometrical optics involves asymptotic expansion; but little attention was paid to the fact that this expansion enables one also to determine the propagation of the amplitude. A systematic exploitation of this possibility was started only about ten years ago by Luneburg.

The most interesting asymptotic phenomenon of wave motion occurs when the eiconal S develops singularities on certain surfaces, called *caustics*. Such a singularity may occur since the differential equation satisfied by S is nonlinear.

Suppose now the wave function u possesses a standard asymptotic expansion on one side of the caustic, then its expansion on the other side will be a different one. In other words, a *Stokes phenomenon appears at the caustic.*

The situation is similar at a *shadow boundary*. The *transition from light to shadow is also a Stokes phenomenon.* For, a shadow boundary is just a line across which the asymptotic expansion changes, in other words, a Stokes line. To determine the asymptotic expansion in the shadow region is an interesting problem which has not yet been solved completely.

In connection with the interpretation of the shadow as a Stokes phenomenon I may perhaps make a general remark about *the role of discontinuities in the description of nature.* On the one hand, discontinuities appear to play a secondary role, namely when they are considered as approximate descriptions of continuous phenomena involving quick transitions. On the other hand, discontinuities play a primary role. For, the experimental description of nature and the theoretical description based on it involves objects with more or less sharp outlines. Therefore, nature could not be described in this way if natural objects did not possess sharp outlines, i.e. discontinuities. In other words, the quantities employed to describe nature could not even be defined if discontinuities did not occur. In this sense, discontinuities appear to play a primary role.

It may be debated whether or not this situation involves a vicious or a nonvicious circle. In any case, one may say that asymptotic description is not just a matter of imperfection, but is an essential element in the mathematical description of nature.

The next subject for discussion naturally would be *elliptic partial differential equations*. Various important mathematical results have been obtained in this field. I need only refer to the classical asymptotic theory of eigenvalues developed by Weyl, Courant, and Carleman.

Instead of discussing these results I prefer to discuss a few problems from *mechanics* which involve elliptic equations.

Let us first turn to problems of *elasticity*. One such problem arises when a thin circular disk is subjected to lateral pressure applied along the edge. The disk will deflect if this pressure is large enough. This problem was investigated by Stoker and myself in 1940. The main question was what happens if the lateral pressure is increased indefinitely or—what is equivalent—if one lets the thickness of the plate shrink to zero. The answer was quite unexpected to us.

The deflection w and the stress function ϕ, considered as functions of x and y, satisfy a pair of differential equations

$$h^2\Delta^2 w = f, \qquad \Delta^2\phi = g$$

in which Δ^2 is the biharmonic operator, and f and g are quadratic functions in the second derivatives of w and ϕ; furthermore, h is the thickness of the plate.

The equations which result when one sets $h=0$ imply a constant distribution of the pressure in the plate. The question then arose, what is the value of this pressure? Is it the value prescribed at the edge?

One really had no right to expect this, since the order of the system of equations drops if one sets $h=0$, and one must face the possibility that at least one boundary condition gets lost. This might be the boundary condition concerning the pressure. If so, there should be a thin boundary layer across which internal and external pressure are connected.

The answer to this question could be derived from a boundary layer analysis of the type described before with the aid of the method of stretching. The answer was that the interior limit pressure indeed is not equal to the external pressure; but in addition it was found that this pressure is negative, that is represents *tension*. Thus tension should prevail over most of the plate in spite of the fact that a compression is applied at its edge.

This result was very surprising and we wondered whether we were not misled by having employed a boundary layer analysis heuristically. However, the validity of this procedure was proved rigorously in this case; in fact, the present case is one of the few involving nonlinear equations in which this was possible.

Quick transitions of stresses and strains at the boundary of a deformed elastic body have been observed in many cases. Instead of a boundary layer phenomenon one then speaks of an "edge effect." A number of edge effects closely related to the one discussed have been treated in the last ten years. The first detailed mathematical analysis of an edge *effect* was given by H. Reissner in 1912 in his theory of shells.

A rather famous edge effect was observed much earlier. This is the effect in the *bending of thin plates with free edges*. The differential equation for the deflection w of such a plate is the biharmonic equation $\Delta^2 w = 0$. Two boundary conditions should be imposed on w on mathematical grounds, but three conditions were strongly favored on physical grounds. A pair of two very peculiar boundary conditions were proposed by Kirchhoff in 1850 and later on justified by Kelvin and Tait by qualitative arguments which essentially involved a boundary layer. But only recently was this problem treated by a consistent asymptotic analysis.

In such an analysis, all significant quantities must first be developed with respect to the thickness of the plate. A stretching technique of the type discussed earlier leads to a description of the stresses in the boundary layer. In this way one finds that Kirchhoff's conditions indeed are correct, but, in addition, one can clearly understand in detail how the third boundary condition gets lost.

Asymptotic approximation of quantities defined in thin layers may lead to strange phenomena. I should like to mention one such phenomenon in connection with a problem in *fluid dynamics;* namely the problem of determining the flow of a layer of fluid over a bottom surface under the influence of gravity. This flow is described by a potential function.

There exists a very effective approximate treatment of such flow based on the assumption that the layer of fluid is very thin. This is the so-called "shallow water theory."

A peculiar feature of this approximation is that in it the motion is governed by a hyperbolic differential equation, while originally it is described by a potential function, the solution of an elliptic equation. Small disturbances, which in the original description would affect the whole flow instantaneously would be propagated with a finite speed according to the *shallow water theory*. How is this possible?

This discrepancy is a typical symptom of asymptotic approximation. The shallow water theory results from the leading term of an expansion with respect to the average thickness of the layer. In deriving this expansion one must stretch the vertical variable y and keep the horizontal variable x. The potential equation then goes over into the equation

$$h^2\phi_{xx} + \phi_{\eta\eta} = 0$$

where $\eta = y/h$. Clearly, for $h=0$ the elliptic character of the differential equation is lost. One may perhaps hesitate to destroy the potential equation and to spoil the advantage of working with potential functions. Still, it is appropriate to do so. It has been suggested to call this procedure the "method of spoiling."

Incidentally, it was recently shown by Hyers and myself that a similar *method of spoiling* is the clue to a rigorous treatment of the "solitary wave," i.e. a steady shallow water wave with a single hump.

One more problem from fluid dynamics should be discussed, Prandtl's theory of the *airfoil of finite span.* If a thin wing of finite span travels through the air, a vortex sheet will develop at the trailing edge. A precise treatment of the resulting airflow offers insurmountable difficulties, but Prandtl gave an approximate treatment derived from rather intuitive arguments. He replaced the wing by a line, called "lifting" line at which the flow is assumed to have an appropriate singularity and made other simplifications. This procedure was strikingly effective in the case of normal flight, i.e. flight in the direction perpendicular to the wing, but the method breaks down when applied to a wing in yaw or a swept back wing.

This difficulty was quite recently overcome by a systematic asymptotic treatment of the problem. The wing was imbedded in a set of wings with the same span and similar cross-section profiles. When the chord ϵ of the profile approaches zero, the wing shrinks to a line, the "lifting line." The potential function describing the airflow past the wing is now developed with respect to ϵ about $\epsilon = 0$. The leading term in this expansion describes exactly Prandtl's approximate flow; it can easily be given explicitly as soon as the circulation around the lifting line is known.

In two dimensional airfoil theory the circulation can be deduced from the shape of the profile by Kutta and Joukowski's theory. In the present treatment the shape of the airfoil has disappeared in the limit. To find the missing circulation this shape must be recovered. That can be done by the method of stretching. The new, stretched variables may be so chosen that the profile remains fixed; but then the length of the span tends to infinity as $\epsilon \rightarrow 0$. With respect to the new variables the flow approaches a new limit. The new limit flow is essentially a two dimensional flow around an airfoil with infinite span. Such flow is well determined if its behavior at infinity is known. Now, infinity in the new variables corresponds to the neighborhood of the lifting line in the original variables. By identifying the behavior of the terms in the direct expansion at the lifting line with the behavior at infinity of the terms of the expansion after stretching one is able to find the missing circulation.

In this way, one can retrieve Prandtl's results for a wing flying in the head-on direction, and, in addition, one can treat *wings in yaw, and swept back wings*, which up to now appeared not to be amenable to an approach employing a lifting line.

In most problems discussed so far quite *similar methods of asymptotic analysis* were employed. These methods led to success in a number of cases, but still their scope is limited. That applies, in particular, to the simple method of stretching which played such a prominent part in our discussion.

Although the scope of the method of stretching is rather limited, the general idea of employing appropriate *transformations of the independent variable*, depending on the parameter, seems to be very fruitful. The importance of this idea, which occurs already in Poincaré's work, was strongly emphasized by Lighthill and various specifically adapted transformations were employed by him and others with remarkable success. One of the goals which may be attained in this way is uniformity. That is, one desires an approximation which is uniformly valid on the boundary and off the boundary in a problem of the boundary layer type, or on the Stokes line and off the Stokes line when a Stokes phenomenon is involved.

There are innumerable other asymptotic problems in mathematical physics more or less related to those discussed. An important field of such questions is concerned with physical processes which do or do not approach a *steady limit* as time goes on indefinitely. In fact, the occurrence of *stability and instability* may be regarded as an asymptotic phenomenon; and the decision between stability and instability therefore requires an asymptotic analysis.

It should not be forgotten that the foundations of *statistical mechanics* originated by Boltzmann and Gibbs abound with asymptotic problems of great significance and great difficulty.

It is not my intention to speak about these various fields. There is only one field of physics in which asymptotic problems occur to which I should like to refer: *quantum theory*.

It would have been tempting to speak about the *quantum theory of fields*. Here physicists have developed formal series expansions with great ingenuity. These series, to say the least, do not converge, and yet, to an amazing degree, they make sense physically. To be sure, to justify these formal procedures is a challenge to the mathematician.

However, I want to confine myself to discussing one or two problems in quantum theory which are well understood mathematically.

The first of these problems concerns the differential equation

$$\epsilon i \frac{d}{dz} \psi = H(z)\psi$$

in which ψ is an element of a *Hilbert space* which depends on the variable z and H is a self-adjoint operator, which is also assumed to depend on z. One is interested in the asymptotic behavior of a solution for small values of ϵ.

The equation is essentially of the same type as the ordinary differential equation considered before, the only difference being that the function ψ is an element of a Hilbert space. For this reason, a few technical difficulties must be overcome; but the idea of asymptotic analysis is exactly the same as for ordinary differential equations.

The interest in this problem arises in connection with the "adiabatic theorem" in quantum theory. The differential equation is the *Schrödinger equation* for the state ψ of the system if z/ϵ is regarded as the time. The operator $H(z)$ is then a Hamiltonian which varies "slowly" if ϵ is small. The *adiabatic theorem* now states: if the state ψ was an eigenstate of H originally, $t=0$ say, it will remain approximately an eigenstate if t increases provided ϵ is small enough.

This theorem now results from the leading term of the asymptotic expansion of the solution ψ. In addition, however, the asymptotic analysis enables one to determine terms of higher order in the expansion and thus to estimate the "probability of nonadiabatic transition."

This is of particular interest in the case in which *two eigenvalues* of the operator *H coalesce* at some time during the process and the question has been asked what happens in such a case. Now, if one looks at the problem as an asymptotic problem, one need only realize that coalescence of eigenvalues corresponds to *a turning point*. A turning point analysis then gives a simple answer.

I should like to *summarize* some of the ideas presented. I have tried to show that a great number of asymptotic problems in mathematical physics have important features in common: in particular the drop of the order of the differential equation and the loss of a continuity or boundary condition. I have tried to show that many of these problems can be attacked by similar methods. These methods involve asymptotic expansion, and the analysis of the regions of nonuniformity by stretching or adjustment of the independent variables combined with an appropriate identification procedure.

Furthermore, as I had mentioned in connection with the problem of the shadow, *asymptotic description* is not only a convenient tool in the mathematical analysis of nature, it *has a* more *fundamental significance.*

This fact is also apparent in the relationship between classical and quantum mechanics; a few words may be said about this relationship.

When wave mechanics was discovered it was immediately recognized that the relationship between wave mechanics and classical mechanics is essentially the same as that between wave optics and geometrical optics. That is to say, classical mechanics results from the leading term in the asymptotic expansion of quantum mechanics and in this sense classical mechanics plays a secondary role. On the other hand, as has been stated frequently in discussions of the foundations of quantum-theory, it is impossible to define and explain the basic notions of quantum mechanics without reference to classical mechanics. In this sense then, classical mechanics plays a primary role.

Thus we meet again the same circular situation which we had discussed in connection with the problem of the shadow. Indeed, *the relationship between classical and quantum mechanics affords a striking illustration of the fundamental role which asymptotic description plays in the mathematical description of nature.*

Selected references

From the vast literature on asymptotic phenomena we shall present only a small selection. Additional references will be found in the publications quoted; but even with these additions the bibliography would be far from complete. No references will be given to work on various subjects which were only slightly touched upon in the lecture.

In particular, no reference will be made to the extensive literature on asymptotic series (e.g. the work of van der Corput) and on methods of asymptotic expansion of integrals and special functions (such as the saddle point method), except the résumé which Stokes gave on the phenomenon named after him,

G. G. Stokes, *On the discontinuity of arbitrary constants that appear as multipliers of semi-convergent series*, Acta Math. vol. 26 (Abel Centenary Volume) (1902) pp. 393–397. Reprinted in *Mathematical and Physical Papers*, vol. V, pp. 283–287.

Ordinary linear differential equations

Among the *classical work on asymptotic expansion* of solutions of ordinary differential equations we mention[1] that of Poincaré (1886), Horn (1899), Birkhoff (1908), Noaillon (1912), Tamarkin (1917), Perron (1918). Among more *recent work* we mention that of Trjitzinski (1934), Turittin (1936), and Hukuhara (1937). For the application of this theory to *eigenvalue problems* see the work of Tamarkin (1917 and 1927) and Birkhoff (1923); for the application to *boundary value problems* see Wasow (1944).

References to this work will be found in most *expository presentations* such as

G. D. Birkhoff and R. E. Langer, *The boundary problems and developments associated with a system of ordinary linear differential equations of the first order*, Proceedings of the American Academy of Arts and Sciences vol. 58 (1923) pp. 51–128. Reprinted in *Collected mathematical papers*, vol. I, pp. 347–424.

M. Hukuhara, *Sur les propriétés asymptotiques des solutions d'un système d'équations différentielles linéaires contenant un parametre*, Kyushu Imperial University, Faculty of Engineering Memoirs, vol. 8, 1936–1940, pp. 249–280.

H. L. Turittin, *Asymptotic expansions of solutions of systems of ordinary linear differential equations containing a parameter*, Contributions to the Theory of Nonlinear Oscillations, vol. II, Annals of Mathematical Studies, no. 29, Princeton University Press, 1952, pp. 81–116.

W. Wasow, *Introduction to the asymptotic theory of ordinary linear differential equations*, Report of the National Bureau of Standards, 1954, pp. 54–56.

For the work of Jeffreys (1923), Langer (1934), Cherry (1949), and others on the *turning point problem* see the

Review of the Literature, *Asymptotic solutions of differential equations with turning points*, Department of Mathematics, California Institute of Technology, Pasadena, 1953.

This is referred to as "Cal. Tech. Review" in the following.

NONLINEAR ORDINARY DIFFERENTIAL EQUATIONS

For a summary of earlier work see

M. H. Dulac, *Points singuliers des equations differentielles*, Memorial des Sciences Mathématique, no. 61, 1934.

For the later work by Hukuhara (1935), Nagumo (1939), Malmquist (1940) and others see

W. Wasow, *Singular perturbations of boundary value problems for nonlinear differential equations of the second order*, Communications on Pure and Applied Mathematics, vol. 9 (1956) no. 1.

For the work of Lighthill (1949) and others on *uniform expansions* see

H. S. Tsien, *The PLK method*, Advances in Applied Mechanics, vol. IV, Academic Press Inc., New York, 1955.

W. Wasow, *On the convergence of an approximation method of M. J. Lighthill*, Journal of Rational Mechanics and Analysis, to appear in 1955.

See also the work by Bromberg and Latta referred to below.

For the work by various authors on "singular perturbations" of solutions of *boundary or initial value problems* and of *periodic solutions* see

[1] The year indicated frequently refers only to the first of a number of publications.

N. Levinson, *Perturbation of discontinuous solutions of nonlinear systems of differential equations*, Acta Math. vol. 82 (1950) pp. 71–106

and also

A. B. Vasilieva, Mat. Sbornik N.S. vol. 31 (1952) pp. 587–644.

Relaxation oscillations

For van der Pol's relaxation oscillations and the asymptotic theory of Haag (1943), Dorodnitsyn (1947), and others see

J. J. Stoker, *Nonlinear vibrations*, New York, Interscience, 1950, Chap. V. 6 and Appendix IV.

Stability

For a treatment of problems of stability see

R. Bellman, *Stability theory of differential equations*, New York, McGraw-Hill, 1953

and the survey article

V. V. Nemitzki, Uspehi Matematičeskih Nauk N.S. vol. 9 (1954) pp. 39–56.

Partial differential equations

The asymptotic theory of the equation

$$\epsilon\Delta u = Lu$$

in which Lu is an expression of the first order was treated by Wasow (1944) and

N. Levinson, Ann. of Math. vol. 51 (1950) pp. 428–445.

Asymptotic properties of the Navier-Stokes equation and related equations were investigated by Lagerstrom, Cole, Latta and others; see

P. A. Lagerstrom and J. D. Cole, *Examples illustrating expansion procedures for the Navier-Stokes equations*, Guggenheim Aeronautical Laboratory, California Institute of Technology, Pasadena, California, May, 1955.

Wave equation and geometrical optics

R. K. Luneburg, Lecture Notes, Brown University, 1944, and New York University, 1947–1948.

F. G. Friedlander, Proc. Cambridge Philos. Soc. vol. 43 (1946) pp. 284–286.

F. G. Friedlander and J. B. Keller, *Asymptotic expansions of solutions of* $(\nabla^2+k^2)u=0$, New York University, Institute of Mathematical Sciences, Division of Electromagnetic Research, Research Report No. EM-67, September, 1954.

Elasticity

For the edge effect in shells and plates see

H. Reissner, *Spannungen in Kugelschalen (Kuppeln)*, Festschrift H. Mueller, Breslau, Leipzig, 1912, pp. 181–193.

O. Blumenthal, Archiv der Mathematik und Physik (3) vol. 19 (1912) pp. 136–174.

E. Reissner, Quarterly of Applied Mathematics vol. 10 (1952) pp. 167–173.

For the Kirchhoff's boundary conditions see a paper by R. Dressler and the author to appear in the Communications on Pure and Applied Mathematics.

For edge effects occurring in nonlinear problems see

K. O. Friedrichs and J. J. Stoker, Amer. J. Math. vol. 63 (1941) pp. 839–888,
E. Bromberg and J. J. Stoker, Quarterly of Applied Mathematics vol. 3 (1945) pp. 246–265

and a paper by E. Bromberg to appear in the Communications.

Fluid dynamics

For the principles of *boundary layer theory* see

L. Prandtl, *The mechanics of viscous fluids*, Aerodynamic Theory, Ed. by W. F. Durand, vol. III G. (1935).
S. Goldstein, *Modern developments in fluid dynamics*, vol. I, Oxford, 1938.

For the *stability of viscous fluid flow* see the survey

W. Tollmien, *Laminare Grenzschichter*, Fiat Review of German Science, 1939–1946, vol. II, Hydro- and Aerodynamics, Wiesbaden, 1948, pp. 21–53.

Also see

C. C. Lin, *Hyarodynamic stability*, Proceedings of the Fifth Symposium in Applied Mathematics, New York, McGraw-Hill, 1954, pp. 1–18.

For the pertinent turning point analysis see the work by Wasow (1948–1952) quoted in the Cal. Tech. Review.
For the *shallow water theory* see the book

J. J. Stoker, *Water waves*, New York, Interscience

to appear.
For the principles of airfoil theory see the *Aerodynamic theory*, vol. I. For the approach described in the lecture see a paper by S. Ciolkowski and the author to appear in the Communications.

Quantum theory

For references to the work by Born and Fock (1926), Kato (1950) on the *adiabatic theorem* see a report by the author issued by the Institute of Mathematical Sciences, New York University.
The asymptotic expansion which leads from quantum mechanics to classical mechanics is clearly described by

G. D. Birkhoff, Bull. Amer. Math. Soc. vol. 39 (1933) pp. 681–700.

New York University

V
Contributions to Magnetohydrodynamics

by Harold Weitzner

K. O. Friedrichs' work in ideal magnetohydrodynamics constitutes an exceedingly small part of his scientific output. In these papers, techniques of applied mathematics in whose development he played a major rôle were employed in the relatively new field of ideal magnetohydrodynamics. Although the works are of limited interest solely in terms of their mathematical content, they provide the logical basis for the exposition of an important subfield of physics. In addition, the explicit results established the fundamental understanding in their respective domains. This work has been the basis for numerous other studies and the offshoots of this work are at the vital center of a growing field.

Report [54–1] reprinted here was actually never published. It was written as a Los Alamos Scientific Lab Report, LAMS-2105, and had wide circulation in that form. A later version with H. Kranzer "Nonlinear Wave Motion," appeared in 1958 as a (Courant Institute of Mathematical Sciences) Report, [58–1]. Three distinct, but closely related, topics are examined. It is shown first that the equations of ideal magnetohydrodynamics form a symmetric hyperbolic system. Thus, one may formulate well-posed problems for the system and study wave propagation problems. The wave normal locus and the characteristic locus for this system are first presented here. In the physics literature, the characteristic locus is frequently called the Friedrichs diagram. He also describes the simple waves for the system. The bulk of the study concerns shocks and contact discontinuities in ideal magnetohydrodynamics and concludes with the analog of the solution of a shock tube, or Riemann problem. An earlier paper of F. de Hoffman and E. Teller [1] initiated a study of magnetohydrodynamic shocks, but Friedrichs' paper laid out the broad fundamentals: the possible types of shocks, properties, entropy changes, and the admissibility of shocks. It would be pointless to cite literature that has grown out of this work. Any paper that discusses shocks in plasmas, whether they occur in laboratory plasmas or in space plasmas, uses the classification scheme and terminology developed here. It is interesting to note, however, that one question not fully resolved in this paper, the admissibility and stability of shocks, was largely resolved in important works by A. I. Akhiezer, G. J. Liubarskii and R. B. Polovin on evolutionary conditions for shock waves, see [2]. Even this work left some residual questions on the stability of a certain class of shocks, resolved much later in the numerical treatment by C.K. Chu, [3], etc., and analytically by L. Sarason [4].

A paper of Friedrichs not republished here [60–2] carried over from elasticity the well-known concept of bifurcation of equilibrium and possible transfer of stability to the field of magnetohydrodynamics. Bifurcated equilibrium were studied at the Courant Institute for many years and their relevance has become apparent in the physics community: [5], [6], [7], [8], and [9]. At the present time some numerical experiments are trying to generate bifurcated states, see for example [10]. Such equilibria are closely related to the "Heliac" device [11] of current interest. This paper also contained a stability analysis of a bifurcated state. At the time of Friedrichs' investigation very few stability analyses had been carried out and certainly none were done in a geometry as complicated as the helically symmetric bifucated state. The full stability properties of this state were finally resolved many years later by T. Yeh [6].

In a later paper [78–1], included here, Friedrichs extended the concept of symmetric hyperbolic systems and examined the question of the proper formulation of relativistic, resistive magnetohydrodynamics first broached by him in [74–2]. This paper effectively contains an open challenge to physicists. The constraints of mathematical well-posedness and relativistic invariance severely limit the possible forms of electrical resistance, or Ohm's law. At this point we do not know whether the physical law indeed has this form or whether we need new mathematical structures. In any case, this paper will lead the way into new topics.

[1] F. de Hoffman and E. Teller. Magneto-hydrodynamic shocks, Phys. Rev., 80 (1950), 692–703.

[2] A.I. Akhiezer, G.J. Liubarskii and R.B. Polovin. The Stability of shock waves in magneto-hydrodynamics, Soviet Physics JETP, 8 (1959), 507–511;
R.B. Polovin. Shock waves in magneto-hydrodynamics, Soviet Physics Uspekhi, 3 (1961), 677–688.

[3] C.K. Chu. Some remarks on the stability of hydromagnetic shock waves, Proc. 18th Symp. Appl. Math., ed. H. Grad, A.M.S., Providence, (1967), 1–16.

[4] L. Sarason. Hydromagnetic switch-on shocks, J. Math. Physics, 6 (1965), 1508–1518.

[5] B. Marder and H. Weitzner. A bifurcation problem in *E*-layer equilibria, Plasma Physics, 12 (1970), 435–445.

[6] T. Yeh. Bifurcation and stability of long-wavelength helical hydromagnetic equilibria, Phys. Fluids, 16 (1973), 516–528.

[7] J. Kappraff. Bifurcated stability of a family of stellarator equilibria, Phys. Fluids, 19 (1976), 675–682.

[8] Y.P. Pao. Non-linear behavior of linearly unstable magnetohydrodynamic modes, Phys. Fluids, 21 (1978), 765–772.

[9] Y.P. Pao and P. Rosenau. Non-linear stabilization of $m=2$MHD kink modes, Comm. Pure Appl. Math., XXXI (1978), 647–658.

[10] F. Bauer, O. Betancourt and P.R. Garabedian. A Computational Method in Plasma Physics, Springer-Verlag, New York, (1978).

[11] H. Furth. How to make an optimal stellarator, Proc. Fourth International Stellarator Workshop, Cape May, New Jersey, Published by Princeton Plasma Physics Lab I (1982).

[54–1] **Nonlinear wave motion in magneto-hydrodynamics. Los Alamos Scientific Lab Reports, LA MS-2105.**
(Listed in Reports).

Corrections

Page 354 – after 2.3, $\nabla_n = n \cdot \nabla$ stands for differentiation in the normal direction and the following replacements are made: $\nabla \to n\delta$, $\delta \to -(u_n \pm c)\,\delta$.
B 1, should read $\pm c\delta H + H\delta u_n - H_n \delta u = 0$.

Page 378 – remove the $\sim$ in equation 9.3

Page 385 – equation 11.6 $\gamma = \dfrac{\mathrm{d}(a^2 \varrho)}{a^2\,\mathrm{d}\varrho}$

Preface

The work presented in this report originated in connection with a research project in magneto-hydrodynamics conducted by Professor H. Grad at the Institute of Mathematical Sciences at New York University; the report was completed while the author was a consultant at Los Alamos Scientific Laboratory.

The report contains an expository analysis of the mathematical structure of magneto-hydrodynamic theory; specifically, of the theory of the interaction of magnetic fields with conducting compressible fluids. One of the major aims is to emphasize the remarkably close parallelism between this theory and ordinary gas dynamics. While hydromagnetic shocks and small disturbance waves have been treated before, it seems that hydromagnetic "simple waves" are here treated for the first time. In a typical example it is shown that - as in gas dynamics - simple one-dimensional flow problems can be solved with the aid of shocks and simple waves.

Throughout the exposition a familiarity with the basic notions of gas dynamics - covered e.g. by the first three chapters of Supersonic Flow and Shock Waves - is presupposed.

The bibliography refers only to those papers which are concerned with the subjects treated. No reference is made to a number of significant contributions to hydromagnetics concerned with subjects - such as astro- and geophysical applications, effect of turbulence, stability questions - which are not covered in this report.

Introduction

In the present exposition magneto-hydrodynamics will be confronted with gas dynamics, i.e. with the dynamics of compressible fluids. It will be shown that the basic equations governing magneto-hydrodynamics - or hydromagnetics for short - have essentially the same mathematical character as those governing gas dynamics and that, consequently, essentially the same mathematical methods can be employed which have proved successful in gas dynamics. This fact will be illustrated by discussing simple problems of one-dimensional wave motion.

Hydromagnetic wave motion was apparently discovered by the astrophysicist Alfvén in 1942 [1]. Alfvén observed that waves of a peculiar type will occur in a conducting fluid in which a magnetic field is present provided the conductivity is sufficiently large and the magnetic field is sufficiently strong. The speed with which such a wave travels depends on the magnetic field H and on the density of the fluid ρ. Assuming the fluid to be incompressible and the small disturbance wave to travel in the direction of the field, Alfvén found that this speed - here denoted by c_{Alf} or by b - is given by the expression

(1) $$b = c_{Alf} = [\mu/\rho]^{1/2}|H|$$

in which μ is the permeability.*

Following Alfvén, various authors have dealt with hydromagnetic problems. The work by de Hoffmann and Teller (1950) [2] on hydromagnetic shocks and the "Studies in Magneto-Hydrodynamics" by Lundquist (1952) [5] may be mentioned. A list of other publications is given in the bibliography.

The interest in hydromagnetic wave motions arose in connection with problems in astrophysics. The gases which compose the sun or interstellar clouds, for example, were regarded as conducting fluids in the presence of magnetic fields. Hence hydromagnetic considerations entered the discussion of the origin of sunspots and of cosmic rays. Naturally, one will ask whether or not hydromagnetic considerations will play a role in fields of physics other than astro- and geophysics. In any case, though, the problem of what happens when a magnetic field interacts with a conducting fluid would seem to be of sufficient interest to justify a mathematical investigation of hydromagnetic theory independently of any specific application.

The theory of hydromagnetics presupposes that the medium to which it is applied possesses a number of appropriate properties. First of all the medium must be conducting so that electric currents may flow.

---

* Throughout this exposition Georgi or mks units are used, cf. Stratton [10].

Thus, the medium may be mercury at room temperature, or a plasma, composed of positive and negative and perhaps neutral particles.

Next, the medium should be such that it may be considered a fluid in the sense of fluid dynamics so that the notions of fluid velocity, density, pressure, entropy and temperature are meaningful. If the fluid is a mixture of various components, such as electrons and positive particles, these fluid dynamical notions are to be understood as appropriate mean values.

The pressure is assumed to be a function of density and entropy and this function is assumed to have properties usually required in gas dynamics. Moreover, we shall assume in our discussion that heat conduction and viscosity may be disregarded. As a consequence various types of gases are excluded from treatment, such as gases in which the mean free path is not small compared with the significant dimensions of the problem.

We also shall assume that the flow velocity is small compared with the speed of light. Furthermore we assume that the (mean) electric charge is negligible, so that the medium is essentially neutral, and that the displacement current may be neglected.*

On the other hand, we do not assume that the flow velocity is small compared with the speed of sound. In other words, we assume the fluid to be compressible.

* It would not be necessary to make these assumptions. A strictly Lorentz-invariant counterpart of the treatment given in this report is reserved for a later publication.

A requirement on which the theory of magneto-hydrodynamics depends essentially is that the electric conductivity of the medium is rather large.* As is done frequently, we shall assume this conductivity to be infinite. Infinite conductivity is a consequence of the basic requirement** of magneto-hydrodynamics, which we now formulate.

Whenever the fluid is at rest at a certain point the electric field should vanish there. Since the flow velocity depends on the velocity of the frame of reference from which it is observed the assumption implies this: Whenever the fluid is at rest at a certain point when observed from a frame of reference moving with an appropriate velocity, the electric field vanishes there. The fact, implied in this assumption, that the electric field depends on the frame of reference is, of course, in agreement with the relativistic invariance - i.e. invariance under Lorentz transformation - of Maxwell's equations. Using this invariance, one may derive the basic formula***

$$E = \mu H \times u \tag{2}$$

for the electric field E in terms of the magnetic field and the flow

---

* For an estimate of the necessary magnitude of the conductivity, see Lundquist [5].

** This requirement could vice versa be derived from that of infinite conductivity if one requires, in addition, that the "Hall effect" may be neglected; cf. [9].

*** If the conductivity were allowed to be finite, a generalized Ohm's law would take the place of (2).

velocity u. Here the symbol x indicates the outer product. Vice versa, of course, formula (2) implies the assumption that the electric field vanishes where the flow velocity vanishes.

It should be mentioned that the magnetic field H is independent of the frame of reference. For, the modifications resulting after Lorentz transformation are negligible since they are of the order of magnitude of the square of the ratio of flow speed to light speed.

It is not our intention to discuss in detail the significance of these various physical assumptions. Our aim is mathematical. We want to describe some of the mathematical consequences of the basic differential equations to be derived from these assumptions.

A remark might be made, though, about the assumption of compressibility. As is well known, compressibility need be taken into account only if the flow speed $|u|$ is comparable with the sound speed

$$a = c_{sound} = [dp/d\rho]^{1/2}. \tag{3}$$

Moreover, our approach will in general be worthwhile only if the Alfvén speed c_{Alf}, see (1), is comparable with the sound speed since otherwise hydromagnetic and compressibility effects could be separated. If c_{Alf} is much less than c_{sound}, which will frequently be the case, it might be possible to separate compressibility effects from hydromagnetic effects by first treating compressibility effects in the absence of hydromagnetic effects and only then considering the latter ones assuming the fluid incompressible.

1. Basic Equations

On the basis of all the assumptions described one can now derive a system of differential equations which governs the flow of the fluid and the changes in the electromagnetic field. We denote by t, x, y, z the time and the coordinates of space; by

$$\nabla = \{\partial/\partial x, \ \partial/\partial y, \ \partial/\partial z\} \tag{1.1}$$

we denote the gradient considered a vector; further we set $\partial/\partial t = \nabla_t$. H is the magnetic field vector; the electric field vector is given by (3) in terms of the flow velocity u. In absence of the displacement current, the current per unit area, J, can be expressed as the curl of H,

$$J = \nabla \times H. \tag{1.2}$$

The pressure $p = p(\rho, S)$ is a given function of density and entropy S.

The two Maxwell equations which do not involve current and charge are retained; they may be written as

$$A_0 \qquad \nabla \cdot H = 0;$$

i.e. div $H = 0$, and

$$A_1 \qquad \nabla_t H + \nabla \times [H \times u] = 0.$$

The second term here is curl E/μ by (3). The force which enters Newton's second law consists of the per unit volume, "Lorentz force" $-\mu H \times J$ and the pressure gradient $-\nabla p$. Using (1.2) we may write this law in the form

$$A_2 \qquad \rho \nabla_t u + \rho(u \cdot \nabla) u + \nabla p + \mu H \times [\nabla \times H] = 0.$$

The continuity equation of fluid dynamics is retained

$$A_3 \qquad \nabla_t \rho + \nabla \cdot (\rho u) = 0.$$

Finally, we add the law that the entropy S per unit mass is carried unchanged by the particles,

$$A_4 \qquad \nabla_t S + (u \cdot \nabla) S = 0.$$

These equations, essentially formulated by Lundquist, 1952, [5], and sometimes referred to as "Lundquist equations", will be the basis of our discussion.

It would be sufficient to require equation A_0 to hold only at an initial time; as is well known, it then follows from A_1 that this equation is satisfied at all times.

Equations A_1 to A_4 are a system of eight non-linear partial differential equations of the first order for the eight quantities $H_x, H_y, H_z, u_x, u_y, u_z, \rho$, and S. The first fact we like to emphasize is that <u>these equations are hyperbolic</u>. Specifically, they belong to the special class of <u>symmetric hyperbolic equations</u>, which is particularly well understood mathematically; see [11]. The equations share this property with Maxwell's equations, with the equations of elasticity, and with the equations of gas dynamics. Just as the latter equations, they are nonlinear with coefficients which involve the dependent but not the independent variables. It is because of the latter property, in addition to the hyperbolic character, that the same methods can be applied to the equations A_1 to A_4 that have proved successful in gas dynamics.

The fact that these equations of magneto-hydrodynamics belong to a class of equations which are most consistent in their mathematical character supports the confidence in the physical consistency of the various assumptions from which they were derived.

As is well known, physical processes governed by hyperbolic equations have the property that <u>disturbances are propagated with finite speed</u>. Thus, in ordinary compressible fluids, disturbances travel with the speed of sound, cf. [12], relative to the motion of the fluid. Hence, also in magneto-hydrodynamics disturbances travel with finite speed. However, in contrast to gas dynamics, there are three such speeds. Accordingly, <u>there are three modes of propagation in each direction</u>. Moreover, these speeds depend on the direction; specifically, on the angle between the direction of propagation and the direction of the magnetic field. This remarkable fact was discovered by Herlofson [3] and by van de Hulst [4], in connection with sinusoidal linearized wave motion.

2. Characteristic Manifolds and Propagation of Disturbances

Characteristic manifolds - three dimensional manifolds in (x,y,z,t)-space - associated with a differential equation may be defined in many different ways; cf. [11, 12]. Instead of giving a precise such definition, it is sufficient in the present context to say that a solution of the differential equation may possess a "small" discontinuity only on such a manifold. We may consider such a manifold as being swept out by surfaces $\mathcal{S} = \mathcal{S}(t)$ in (x,y,z)-space; the motion of these surfaces will then be called "characteristic". If in a process described by a solution of the differential equation a small discontinuity or "disturbance" was present on a surface $\mathcal{S}(t_0)$ at an initial time t_0, the disturbance may at later times be present only on surfaces $\mathcal{S}(t)$ which move characteristically. Such a process will be called a "disturbance wave" or simply a "wave".

We introduce the normal vector n of unit length at each point of the surface $\mathcal{S}(t)$ and characterize the motion of the surface $\mathcal{S}(t)$ by its velocity c_{ch} in the normal direction at each of its points. It is convenient to introduce the normal component

(2.1) $$u_n = (n \cdot u)$$

of the flow velocity at this point and to denote the characteristic velocity there by

(2.2) $$c_{ch} = u_n \pm c.$$

Thus $\pm c$ is the normal component of the characteristic velocity relative to the flow velocity.

In order to find the possible values of c one may first set up the relations between the possible discontinuities δH, δu, $\delta \rho$, δS of the quantities H, u, ρ, S on the surface $\mathcal{S}(t)$. Using the formalism of the theory of characteristics, cf. [11], the following relations are found, in which the symbol

(2.3) $$\nabla_n = n \cdot \nabla$$

stands for differentiation in the normal direction.

B_1 $$\mp c\,\delta H + H\,\delta u_n - H_n\,\delta u = 0,$$

B_2 $$\mp \rho c\,\delta u + a^2 n\,\delta \rho + \mu n(H \cdot \delta H) - \mu H_n\,\delta H = 0,$$

B_3 $$\mp c\,\delta \rho + \rho\,\delta u_n = 0,$$

B_4 $$\mp c\,\delta S = 0.$$

Here a is the speed of sound, given by (3), further

(2.4) $$\delta u_n = n \cdot \delta u$$

and

(2.5) $$H_n = n \cdot H$$

is the normal component of the magnetic field.

The determinant of this system of eight homogeneous equations for the eight quantities δH, δu, $\delta \rho$, δS is found to be

$$(2.6) \qquad \det(B) = \mp \rho c^2(\rho c^2 - \mu H_n^2)\left\{\rho c^4 - (\rho a^2 + \mu H^2)c^2 + a^2 \mu H_n^2\right\}.$$

The characteristic velocities $\mp c$ are obviously the roots of the equation

$$(2.7) \qquad \det(B) = 0.$$

Evidently all eight roots of this equation are real, in accordance with the fact that the equations are hyperbolic.

3. Fast and Slow Disturbance Waves

We shall first discuss the roots of the last factor of det (B). The condition that this factor vanishes can be written in the form

(3.1) $$c^2(\rho c^2 - \mu H^2) = a^2(\rho c^2 - \mu H_n^2)$$

or in the form

(3.2) $$(c^2 - a^2)(\rho c^2 - \mu H_n^2) = c^2\mu(H^2 - H_n^2).$$

The larger and the smaller of the roots $c > 0$ of this equation will be denoted by

$$c_{fast} \text{ and } c_{slow}.$$

From equation (3.2) one immediately deduces the inequalities

(3.3) $$c_{slow} \leq a \leq c_{fast},$$

and

(3.4) $$c_{slow} \leq b_n \leq c_{fast}.$$

Here

(3.5) $$b_n = [\mu H_n^2/\rho]^{1/2},$$

is the Alfvén velocity, cf. (1), with the normal component H_n of the magnetic field instead of the complete magnetic field vector H.

The sound speed a was given by (3). An equality sign can hold in relations (3.3) and (3.4) only if $H = H_n n$, so that the right member of (3.2) vanishes. In this case one of the two speeds equals a, the other equals b_n.

The possible disturbance, i.e. the solutions of equations B_1 to B_4 associated with $c = c_{slow}$ or $c = c_{fast}$ are found to be

$$(3.6)\qquad \begin{aligned} \delta H &= k\rho c^2(H - H_n n), \\ \delta u &= \mp kc(\mu H_n H - \rho c^2 n), \\ \delta\rho &= k\rho(\rho c^2 - \mu H_n^2), \\ \delta S &= 0. \end{aligned}$$

Here k is any number $\neq 0$.

It is to be noted that δH has a tangential direction so that

$$(3.7)\qquad \delta H_n = n\cdot\delta H = 0.$$

The disturbance $\delta\rho$ vanishes when $c = b_n$. The disturbance δu may be written in the form

$$(3.8)\qquad \delta u = \mp(\rho c)^{-1}H_n\,\delta H \pm \rho^{-1}c\,\delta\rho\, n,$$

which is on occasion useful. We also note down the relation

$$(3.9)\qquad \delta(p + \tfrac{1}{2}\mu H^2) = a^2\delta\rho + \mu H\cdot\delta H = k\rho c^2(\rho c^2 - \mu H_n^2),$$

which follows from (3.6) and will prove useful later on.

Particular attention should be paid to the cases where the normal n is parallel or perpendicular to the magnetic field H.

In the first case, $H = H_n n$, one of the roots c agrees with the sound speed a. Unless $a = b_n^2 = b^2$ formulas (3.6) remain valid. They must be modified if $c = b$ since this root agrees with a root of another factor of $\det(B)$.

In the second case, $H_n = 0$, the fast speed is given by the noteworthy formula

$$(3.10) \qquad c_{fast} = \left[\frac{dp}{d\rho} + \mu H^2 \rho^{-1}\right]^{1/2} = \left[a^2 + b^2\right]^{1/2}, \quad H_n = 0.$$

Formulas (3.6) remain valid in this case. The other root

$$(3.11) \qquad c_{slow} = 0, \quad H_n = 0,$$

however, agrees with the root of another factor of $\det(B)$. Hence formulas (3.6) must be modified.

The needed modification of formulas (3.6) will not be described here.

In order to illustrate these facts we shall consider segments of plane wave fronts, fast and slow, which travel in the direction of the normal n after having passed through the origin at the time $t = 0$. At a time $t > 0$ these fronts pass through the points ctn with $c = c_{fast}$ and $c = c_{slow}$. The locus of these points is shown in Figure 1, see end of the report. In this figure we have assumed $c_{Alf} < c_{sound}$, so that the slow speed c_{slow} agrees with the Alfvén speed for $H \parallel n$ since this seems to be the more frequent situation. We also have drawn wave fronts of a third, intermediate, type which we are going to describe now.

4. Transverse Waves and Alfvén Waves

An intermediate wave belongs to the root

$$c = b_n = [\mu H_n^2/\rho]^{1/2} \tag{4.1}$$

of equation (3.1). As seen from (3.4) the speed b_n lies between c_{slow} and c_{fast} except if $H \parallel n$ or $H \perp n$. The possible discontinuities associated with this wave given as the solutions of equations (B), are

$$\begin{aligned} \delta H &= \mp k \rho c H \times n, \\ \delta u &= k \mu H_n H \times n, \\ \delta \rho &= \delta S = 0, \end{aligned} \tag{4.2}$$

with an appropriate constant k. In this case, thus, the disturbances are tangential to the wave front and perpendicular to the magnetic field. Relation

$$\delta H^2 = 2H \delta H = 0 \tag{4.3}$$

implied by (4.2) shows that the magnetic field undergoes a small rotation. This intermediate type of disturbance wave will also be referred to as "transverse wave".

At this place we may interpose a remark about the Alfvén wave in an incompressible fluid. The conditions on the possible disturbances in this case are obtained from conditions (B) by setting $a^2\delta = \delta p$ in B_2 and $\delta\rho = 0$ in B_3. In addition to the double root $c = 0$, one finds that $4c = b_n$ is a double root. The corresponding disturbances are

$$
\begin{aligned}
\delta H &= kH_n n^*, \\
\delta u &= \mp k b_n n^* \\
\delta p &= -k\mu H_n(H\cdot n^*), \\
\delta S &= 0,
\end{aligned}
\tag{4.4}
$$

where n^* is an arbitrary unit vector perpendicular to n, and k is an arbitrary number. Evidently, these possible disturbances form a two-dimensional manifold in accordance with the fact that $c = b_n$ is a double root. Relations (4.4) imply the important relation

$$
\delta(p + \tfrac{1}{2}\mu H^2) = \delta p + \mu H\cdot \delta H = 0,
\tag{4.5}
$$

which expresses the fact that the sum of fluid pressure p and "magnetic pressure" $\frac{1}{2}\mu H^2$ is continuous across the surface $\mathcal{S}(t)$. (The notion of magnetic pressure will be discussed in Section 6.)

One may say that the Alfvén wave* results if the sound speed a and hence the fast disturbance speed c_{fast} become infinite while the slow wave speed c_{slow} coalesces with the Alfvén wave speed $c_{Alf} = b_n$.

---

* We use this term for disturbance waves in any direction in an incompressible medium although Alfvén described only waves traveling in the direction of H.

The fast, slow, and intermediate waves described in Sections 3 and 4 represent the three modes of wave propagation referred to at the end of Section 1. Since each type of wave may move in the direction of n or of $-n$, they correspond to six roots c. The propagation of disturbances associated with the remaining roots $c = 0$ will not be referred to as wave motion.

5. Contact Disturbances

About the remaining roots $c = 0$ only a few remarks need be made. This root has the multiplicity 2 unless $H_n = 0$, in which case the multiplicity is 4. We confine ourselves to the case $H_n \neq 0$. In that case the relations B possess only the solutions

$$\delta H = kn \tag{5.1}$$

$$\delta u = 0, \qquad \delta \rho = 0,$$

$$\delta S = k_1,$$

with arbitrary k and k_1. We may just as well set $k = 0$, or what is equivalent

$$\delta H_n = 0. \tag{5.2}$$

It is consistent to do so; for, as could be shown, this condition is satisfied on every surface $\mathcal{S}(t)$ if it is satisfied initially, on $\mathcal{S}(t_o)$.

The remaining possibility of an entropy disturbance corresponds to a contact discontinuity, or a "contact disturbance" as we prefer to say, since the discontinuity is small.

In striking contrast to the situation in gas dynamics <u>a contact discontinuity in a hydromagnetic fluid does not permit a discontinuity in the tangential component of the velocity, provided</u> $H_n \neq 0$. This remarkable fact will play a considerable role in the discussions of wave motions given in Section 13.

6. Conservation Laws

In this section we shall discuss the conservation form of the Lundquist equations, partly as preparation for the discussion of shocks.

It is customary to say that a system of partial differential equations has "conservation form" if each equation consists of the sum of derivatives of functions of the unknown quantities. The reason for this expression is that the laws of conservation of mass, momentum and energy are of this form. On the other hand the possibility of writing the equations in this form is the condition for the possibility of setting up shock relations.

Equation A_o (see Section 1) evidently has conservation form,

$$C_o \qquad \nabla \cdot H = 0.$$

This same is true of equation A_1, but we prefer to write it in the form

$$C_1 \qquad \nabla_t H + \nabla \cdot uH - \nabla \cdot Hu = 0.$$

Here, and correspondingly in the following, the inner product in $\nabla \cdot uH$ involves only ∇ and u while the differentiation ∇ applies to the product uH. Thus, the x-component of $\nabla \cdot uH$ is

$$(\nabla \cdot uH)_x = \nabla_x(u_x H_x) + \nabla_y(u_y H_x) + \nabla_z(u_z H_x).$$

Similarly, the x-component of $\nabla \cdot Hu$ is

$$(\nabla \cdot Hu)_x = \nabla_x(H_x u_x) + \nabla_y(H_y u_x) + \nabla_z(H_z u_x).$$

Equation A_2 can be written in the form

$$C_2 \qquad \nabla_t \rho u + \nabla \cdot u \rho u + \nabla p + \nabla \tfrac{1}{2}\mu H^2 - \nabla \cdot H \mu H = 0,$$

which expresses the law of conservation of momentum. Suppose the magnetic field H is such that the term $\nabla \cdot H \mu H$ vanishes; - this will be the case under various symmetry conditions, cf. [9]. Then the conservation law C_2 has essentially the same form as in gas dynamics except that the expression $p + \frac{1}{2}\mu H^2$ takes the place of the pressure. It is primarily because of this fact that the term $\frac{1}{2}\mu H^2$ is referred to as "magnetic pressure".

The law of conservation of mass, A_3, is given in conservation form

$$C_3 \qquad \nabla_t \rho + \nabla \cdot \rho u = 0.$$

Equation A_4, however, which describes the transport of entropy, is to be replaced by the law of conservation of energy, which assumes the form

$$C_4 \qquad \nabla_t(\tfrac{1}{2}\rho u^2 + \rho e + \tfrac{1}{2}\mu H^2) + \nabla \cdot u(\tfrac{1}{2}\rho u^2 + \rho e + p + \mu H^2)$$

$$- \nabla \cdot \mu H(H \cdot u) = 0.$$

Here e is the internal energy per unit mass of the fluid, which may be considered as a function of density and entropy and is characterized by the relation

$$de = TdS - pd(\rho^{-1}), \tag{6.1}$$

in which T is the temperature. It is to be noted that the expression $\frac{1}{2}\mu H^2$ in the first term of C_4 plays the role of "magnetic energy per unit volume", while the expression μH^2 in the second term of C_4 plays the role of "magnetic enthalpy per unit volume". It should also be noted that the expression

$$u\mu H^2 - \mu H(H\cdot u)$$

occurring in C_4 may be written in the form

$$u\mu H^2 - \mu H(H\cdot u) = [E \times H] \tag{6.2}$$

by virtue of formula (2). This term is thus recognized as Poynting's energy flux per unit area.

7. Shocks

As is well known, continuous gas dynamical motions will break down at some time if they involve a compression. The same must be expected to happen in hydro-magnetic motions if the compressibility of the fluid cannot be neglected. Mathematically speaking, the solution of the differential equations ceases to exist beyond the time of breakdown. Physically speaking, the phenomenon is no longer governed by the differential equations from that time on. Actually, discontinuities, shocks will appear.

As in gas dynamics one may assume that the quantities on both sides of a shock front are governed by the laws of conservation.

The shock conservation laws can be derived from the conservation form of the differential equations by a formalism which is quite analogous to the formalism by which the equations (B) for the disturbances could be derived from the differential equations. We denote by n the normal vector at any point of the shock front and by

$$[Q] = Q_1 - Q_o$$

the jump of any quantity Q across the shock front; here Q_1 is the value on the side toward which the normal n points and Q_o is the value on the other side. Further we denote by U the velocity of the shock front in the normal direction. The recipe then requires one to replace the symbol ∇_t in equations (C) by $-U[\ldots]$ and the symbol ∇ by $n[\ldots]$. The result is the following set of equations.

D_0 $\qquad [H_n] = 0,$

D_1 $\qquad [(u_n - U)H - H_n u] = 0,$

D_2 $\qquad [(u_n - U)\rho u + (p + \frac{1}{2}\mu H^2)n - \mu H_n H] = 0,$

D_3 $\qquad [(u_n - U)\rho] = 0,$

D_4 $\qquad [(u_n - U)(\frac{1}{2}\rho u^2 + \rho e + \frac{1}{2}\mu H^2) + u_n(p + \frac{1}{2}\mu H^2) - \mu H_n(H \cdot u)] = 0.$

These are the relations which connect the values of the quantities H, u, ρ, S on one side of a shock front, or more generally, of a discontinuity surface, with those on the other side and with the speed U of the surface.

These relations were derived by de Hoffmann and Teller in 1950. These authors set up the conservation laws for shocks directly without relating them to differential equations. Also they derived the equations in a Lorentz invariant form and only afterwards derived equations (D) as the non-relativistic approximation. The direct non-relativistic derivation was given by Lüst [11].

The jump conditions are frequently supplemented by the statement that there is a "sheet current" flowing along the discontinuity surface and that the value of the current J* per unit length is given by the relation

(*) $$J^* = n \times [H]$$

in accordance with relation (1.2). Although this statement is of great significance for the description of the physical phenomena involved, it need not be taken into account in the mathematical analysis of the discontinuity condition.

We will speak of a shock if fluid crosses the front, $u_n - U \neq 0$; in that case we assume the direction of the normal vector n so chosen that fluid crosses in the direction of the normal,

$$(7.1) \qquad u_n - U > 0.$$

If $u_n - U = 0$, so that no fluid crosses, we speak of a contact discontinuity.

We shall present an analysis* of the possible types of shocks and contact discontinuities which is completely analogous to the analysis of disturbance waves given in the preceding sections. Before doing this, however, we shall make an important observation of a general nature.

Gas dynamical shocks involve a rise in pressure and density as follows from the requirement that the entropy should increase across a shock. For hydromagnetic shocks we may state similarly: <u>If the entropy increases across a shock front, pressure and density also increase.</u>**

---

* This analysis differs in significant respects from that given by Helfer [6].

** For parallel and perpendicular shocks, see below, de Hoffman and Teller derive this fact by somewhat more special arguments.

The qualification is necessary; for, as we shall see, there are shocks which do not involve changes in entropy, density and pressure at all. The proof of the statement could be derived from the identity

$$[e + \tilde{p}\rho^{-1}] = -\frac{1}{4}[\rho^{-1}][H]^2, \tag{7.2}$$

in which $\tilde{p}$ is the mean value

$$\tilde{p} = \frac{1}{2}(p_0 + p_1). \tag{7.3}$$

This remarkable identity, first given by Lüst, could be derived by forming a linear combination of relations (D) which is similar to the linear combination which expresses the entropy relation A_4 in terms of the relations (C), cf. Section 6. The left member of (7.2) is the "Hugoniot function" which vanishes for gas dynamical shocks. The right member involves the drop $-[\rho^{-1}]$ in specific volume ρ^{-1} and the square of the jump of the magnetic field. From the known properties of the Hugoniot function [12] one can derive that it has the same sign as $-[\rho^{-1}]$ only if it has the same sign as $[S]$. The statement made above then follows.

From relation (7.2) one can also derive the fact that <u>the increase of entropy across a shock is of the third order in</u> ρ <u>and</u> H.

In order to establish the analogy between the shock relations (D) and relations (B) characterizing disturbances it is convenient to introduce the notion of mean value

$$\tilde{Q} = \frac{1}{2}(Q_0 + Q_1)$$

of any quantity Q and to use the formula

$$[PQ] = \tilde{P}[Q] + [P]\tilde{Q}.$$

Furthermore it is convenient to introduce the specific volume ρ^{-1}

(7.4) $$\tau = \rho^{-1}$$

instead of ρ, and the flux

(7.5) $$m = \rho(u_n - U)$$

instead of u_n. Note that by D_3 the flux is the same on both sides of the shock front. Also H_n may be taken as a constant by D_0. Relations D_1 to D_3 can be written as

E_1 $$m\tilde{\tau}[H] + \tilde{H}[u_n] - H_n[u] = 0,$$

E_2 $$m[u] + [p]n + \mu n\tilde{H}\cdot[H] - \mu H_n[H] = 0,$$

E_3 $$m[\tau] - [u_n] = 0.$$

These equations evidently correspond precisely to equations B_1 to B_3 if one lets

$$m,\ \tilde{\tau},\ \rho^2 a^2,\ H, \text{ correspond to } \mp\rho c,\ \rho^{-1}, -[\tau][p],\ \tilde{H};$$

and lets

$$[u],\ [\tau],\ [H] \text{ correspond to } \delta u,\ -\rho^{-2}\delta,\ \delta H.$$

The analogue of relation B_4 is relation (7.2); it may be disregarded in the present context.

From this analogy, or by direct computation, one finds that the determinant of the system E_1 to E_3 is

$$(7.6) \qquad \det(E) = \tilde{\tau} m\left(\tilde{\tau} m^2 - \mu H_n^2\right)\left(\tilde{\tau} m^4 + (\tilde{\tau}[\tau]^{-1}[p] - \mu \tilde{H}^2)m^2 - [\tau]^{-1}[p]\mu H_n^2\right).$$

Thus it is clear that there are <u>fast, slow, and intermediate shocks</u> and that the relationship between them is the same as that between the corresponding disturbance waves.

The equation

$$(7.7) \qquad \det(E) = 0$$

is an equation for the flux m, but it may just as well be considered an equation for the shock velocity

$$(7.8) \qquad U = \tilde{u}_n - m\tilde{\tau},$$

cf. (7.5).

8. Fast and Slow Shocks

The flux m of a fast or a slow shock satisfies the equation

$$m^2(\tilde{\tau} m^2 - \mu \tilde{H}^2) = -[\tau]^{-1}[p](\tilde{\tau} m^2 - \mu H_n^2), \tag{8.1}$$

which expresses the condition that the last factor of det (E) vanishes; cf. (3.1). In analogy with (3.3) and (3.4) we have

$$m^2_{slow} \leq -[\tau]^{-1}[p] \leq m^2_{fast} \tag{8.2}$$

and

$$m^2_{slow} \leq \tilde{\tau}^{-1} \mu H_n^2 \leq m^2_{fast}. \tag{8.3}$$

The relations between the possible jumps across the shock front are

$$[H] = k_1 m^2(\tilde{H} - H_n n), \tag{8.4}$$

$$[u] = k_1 m(\mu H_n \tilde{H} - \tilde{\tau} m^2 n), \tag{8.5}$$

$$[\tau] = -k_1(\tilde{\tau} m^2 - \mu H_n^2), \tag{8.6}$$

with an arbitrary constant k_1 (instead of $\tilde{\tau} k$) in analogy with (3.6).

The cases of a "parallel" shock, $\tilde{H} = H_n n$, and of a "perpendicular" shock, $H_n = 0$, were discussed in great detail by de Hoffmann and Teller [2]. Here we only mention that in both cases the jump of the velocity u is in the normal direction, as seen from (8.5); therefore, observing the shock from an appropriate frame, the flow velocity is normal on both sides.

For "oblique" shocks, for which $\widetilde{H} \neq H_n n \neq 0$, a frame can be so introduced that the flow velocity u is parallel to the magnetic field H on both sides of the front. This fact, emphasized by de Hoffmann and Teller, can be read off from relation D_1; one need only choose the frame such that $u = H_n^{-1}(u_n - U)H$ on one side; then D_1 implies that the latter relation also holds on the other side. Evidently, in this case the electric field E vanishes on both sides of the front.

Using formulas (8.4) to (8.6) we can easily describe the jump of the absolute value of the magnetic field across the shock front. From formula (8.4) we have

$$[H^2] = 2\widetilde{H}[H] = k_1 m^2(\widetilde{H}^2 - H_n^2);$$

hence, using formula (8.6),

$$[H^2] = -m^2[\tau]\,(\widetilde{\tau} m^2 - \mu H_n^2)^{-1}(\widetilde{H}^2 - H_n^2). \tag{8.7}$$

Since we have assumed the normal n to point in the direction in which the fluid crosses the shock front we have $[\tau] < 0$ according to the statement made preceding formula (7.2). In view of (8.3) we are therefore able to state: The magnetic field strength $|H|$ rises across a fast shock and drops across a slow shock. This fact will prove particularly significant in connection with specific flow problems.

Formula (8.4) implies that the tangential component of the magnetic field jumps in its own direction; i.e. the jump $[H]$ is parallel to H_o. It could be shown that this component retains its direction and changes only its magnitude.

The analysis of shocks as given seems appropriate if one desires to obtain a quick survey of the possible types of shocks and to derive some of their simple properties. In this analysis we have extensively used mean values, involving values on both sides of the front. There are a number of other questions, however, which refer to the behavior of the various quantities on each side separately; to answer such questions our analysis may not be appropriate.

Below we shall make various statements concerning such questions without indicating the way they were derived.

One question that may arise is whether or not a tangential component of the magnetic field may be produced through a shock if it was absent ahead of it, or whether or not it may be wiped out if it was present ahead of it. Shocks through which this happens may be called "complete switch-on or switch-off shocks". Clearly, a complete switch-on shock must be fast or a complete switch-off shock must be slow.

It can be shown that complete switch-on shocks exist only if the Alfvén speed is supersonic ahead of the shock; and even then only if the shock strength (e.g. measured by the pressure ratio p_1/p_o) lies below a critical value. If one lets the shock strength increase, the gain in tangential component of H first increases then decreases, and becomes zero when the critical strength is reached. On the other hand, complete switch-off shocks always exist if the Alfvén speed behind the shock front is subsonic; if this speed is supersonic there, they exist only if the shock strength exceeds a critical value.

Gas dynamical shocks have the property that the normal component of the flow velocity relative to the shock velocity, $u_n - U$, is supersonic ahead of the shock front, i.e. on the side (0) from where the fluid comes, and subsonic behind it. For a hydromagnetic shock we may state: The normal flow velocity relative to a fast shock is greater than the fast disturbance speed ahead of the front and less than it but greater than (or equal to) the transverse disturbance speed behind it,

$$u_n - U > c_{fast} \qquad \text{on side (0), ahead,}$$

$$b_n \leq u_n - U < c_{fast} \qquad \text{on side (1), behind.}$$

The equality sign holds only for complete switch-on shocks. The normal flow velocity relative to a slow shock is less than the slow disturbance speed behind the front and greater than it, but less than (or equal to) the transverse speed ahead of it,

$$u_n - U < c_{slow} \qquad \text{on side (1), behind,}$$

$$b_n \geq u_n - U > c_{slow} \qquad \text{on side (0), ahead.}$$

The equality sign holds only for complete switch-off shocks.

These facts may be illustrated in the following figures, see Figure 2 at the end of the paper, in which the shock is assumed to be stationary, $U = 0$; only those disturbance motions are shown that travel against the flow.

Just as in gas dynamics, a fast or a slow hydromagnetic shock is determined by prescribing all quantities ahead of it and the pressure $p > p^{(o)}$, or the velocity $v_n < v_n^{(o)}$ behind it. It is, however, not possible to prescribe arbitrarily the tangential component of the magnetic field behind the shock front; and where it is possible to do so, the shock may not be uniquely determined. This fact is clearly indicated by the remarks about switch-on and switch-off shocks made above.

9. Transverse Shocks Contact Discontinuities

The root

$$(9.1) \qquad m_{tr} = (\mu H_n^2/\tilde{\tau})^{1/2}$$

of equation det (E) = 0, cf. (7.7), is associated with a type of shock which will be called transverse*. Such transverse shocks correspond to the transverse waves discussed in Section 4. In analogy to expressions (4.2) we find the expressions

$$(9.2) \qquad [H] = km\tilde{\tau}\tilde{H}xn,$$

$$(9.3) \qquad [u] = k\mu H_n \tilde{H}xn,$$

$$(9.4) \qquad [\tau] = 0,$$

$$(9.5) \qquad [S] = 0.$$

The latter relation could be derived from (7.2) and $[\tau] = 0$. Also, $[u_n] = 0$, holds as seen from E_3. Thus, the only quantities

---

* de Hoffmann and Teller call such shocks "symmetrical" and reserve the term "transverse" for small disturbances, across which the magnetic field does not change in lowest order. For reasons indicated below, the term "transverse" would nevertheless seem appropriate, and it seems preferable to the terms "symmetrical" or "symmetrical transverse".

A transverse discontinuity is called a "shock" since fluid crosses the surface; in other respects it is closer to a contact discontinuity, see [13].

that jump across a transverse shock are the tangential components of the magnetic field and of the velocity.

Relation (9.2) implies relation

(9.6) $$[H^2] = 0;$$

in other words, the strength of the magnetic field is unchanged across a transverse shock. The magnetic field therefore rotates in the plane of the shock and the flow velocity undergoes a tangential change parallel to $[H]$. All other quantities remain continuous. The possibility of this occurrence necessitated the qualification made in the statement preceding formula (7.2).

Obviously, as seen from (9.3), a frame can be introduced such that the flow is parallel to the front on both sides. Therefore, the term "transverse" shock, introduced by de Hoffmann and Teller, would seem appropriate.

The possibility of such a transverse shock is, of course, to be understood as a mathematical possibility, referring to the existence of solutions of the equations (E) for the root (9.1). Whether or not such shocks are possible in nature is another question; in fact, it would seem that they are possible only under unusual circumstances; cf. Section 12.

The root $m = 0$ of equation (7.7) corresponds to a discontinuity of the normal component H_n of H. However, this possibility is excluded by condition D_o; cf. the arguments in Section 5, which could be carried over here.

A second root $m = 0$ would have occurred if we had not omitted relation D_4 in changing the system (D) over into the system (E). The corresponding discontinuity would be a proper contact discontinuity involving no flow across the front, i.e.

$$u_n - U = 0.$$

In case $H_n \neq 0$ relations (D) then imply

$$[u] = 0, \quad [H] = 0, \quad [p] = 0.$$

The latter relation does not require $[\rho] = 0$ since $[S] \neq 0$ is compatible with D_4. Because of $[H] = 0$ we have a purely gas dynamical contact discontinuity, in fact a special one, because of $[u] = 0$.

In case $H_n = 0$, on the other hand, we can only conclude

$$[p + \frac{1}{2}\mu H^2] = 0$$

while the tangential components of u and of H may undergo any jumps. In fact, the contact discontinuities coalesce in the case $H_n = 0$.

A contact discontinuity which involves a jump in the tangential flow velocity will be called a "shear flow discontinuity". It is remarkable that in a conducting fluid no shear flow discontinuity can be maintained if the magnetic field has a normal component $H_n \neq 0$ at the front. This fact will be the starting point for our discussion of special flow problems in Section 13.

10. Simple Waves*

Naturally, hydro-magnetic flows most easily accessible to treatment are one-dimensional flows, characterized by the condition that all quantities depend only on one space variable, x say, in addition to the time, and hence are constant on each (y,z)-plane at each time. No restriction need be imposed as to the presence or absence of the y and z-components of the vectors H and u.

In gas dynamics, the simplest types of one-dimensional flows are the "simple waves". These waves are used very effectively as building blocks in constructing solutions of flow problems; primarily because <u>a flow region adjacent to a region of constant state is always a simple wave</u>. A state of flow is referred to as constant in a region if all significant flow quantities are time- and space-independent in the system; (of course the boundaries of such a region need not stay fixed).

Simple waves are also possible in magneto-hydrodynamics. They have essentially the same properties as those in gas dynamics and it appears that, as in gas dynamics, they could be effectively used as building blocks.

A <u>simple wave</u> may be characterized by describing not the motion of a particle but by the motion of a "phase", given by a set of values for the set of quantities H, u, ρ. In a simple wave each phase moves

---

* For a theory of simple waves associated with general systems of differential equations whose coefficients do not involve the independent variables, see P. Lax [13].

with constant velocity; i.e. the values of H, u, ρ are constant on straight lines

$$(10.1) \qquad x = Ut + \xi,$$

in the (x,t) plane. The number U as well as H, u, and ρ may be considered functions of the parameter ξ. Instead of U we introduce the quantity $c \geq 0$ defined by

$$(10.2) \qquad U = u_x \pm c.$$

This quantity c will be recognized as one of the characteristic disturbance speeds; i.e. one of the roots of equation (2.7).

From the characterization given one may derive the following recipe for setting up the equations governing simple waves. In equations A, see Section 1, one should replace ∇ and ∇_t by $\left\{d/d\xi, 0, 0\right\}$ and $-(u_x \pm c)d/d\xi$ or simply by $\left\{d, 0, 0\right\}$ and $-(u_x \pm c)d$, since it does not matter whether or not the phase is considered a function of ξ or of any other parameter. After canceling the terms $-u_x dQ$ and $u_x dQ$, the equations thus obtained are

$F_o \qquad dH_x = 0,$

$F_1 \qquad \mp c\, dH_y + H_y du_x - H_x du_y = 0,$

$\qquad \mp c\, dH_z + H_z du_x - H_x du_z = 0,$

F_2 $$\mp \rho c du_x + dp + \frac{1}{2}\mu dH^2 = 0,$$

$$\mp \rho c du_y - \mu H_x dH_y = 0,$$

$$\mp \rho c du_z - \mu H_x dH_z = 0,$$

F_3 $$\mp c d\rho + \rho du_x = 0,$$

F_4 $$\mp c dS = 0.$$

When the equations F_1 to F_4 are considered linear equations for the differentials dH_y, ..., dS it is clear that the determinant of this system must vanish. One verifies - by direct computation or by comparison with equations (B) in Section 2 - that this determinant is precisely det (B), cf. (2.6). Hence, the speed c must be one of the roots of equation det (B) = 0, cf. (2.7), i.e. one of the characteristic disturbance speeds, of course, considered as a function of the phase, H, u, ρ.

According to which kind of speed c enters, the simple wave is fast slow, or transverse; if $c = 0$ the wave is a "contact layer".

As we shall see in the subsequent sections, the differential equations (F) can be greatly simplified; in the case of a polytropic gas they can even be solved explicitly.

11. Fast and Slow Simple Waves

Equation (3.2) for the fast and slow disturbance speeds may in the present case be written in the form

$$(11.1) \qquad \mu(H_y^2 + H_z^2) = c^{-2}(c^2 - a^2)(\rho c^2 - \mu H_x^2),$$

where, cf. (3),

$$a^2 = dp/d\rho .$$

It is more convenient for the present purpose to introduce the square ratio

$$(11.2) \qquad q = c^2/a^2$$

of disturbance speed to sound speed as independent variable and to try to express all other quantities in terms of q. It is also convenient to introduce the square ratio

$$(11.3) \qquad s = \rho a^2/\mu H_x^2,$$

of sound speed to Alfvén speed as dependent variable. Relation (11.1) then becomes

$$(11.4) \qquad H_y^2 + H_z^2 = (q - 1)(s - q^{-1})H_x^2.$$

From relations $F_{2,x}$ and F_3 one further derives the relation

$$(11.5) \qquad 2(q - 1)ds = \gamma d(q - 1)(s - q^{-1}),$$

where

$$(11.6) \qquad \gamma = a^2 d\rho/dp.$$

For polytropic gases, γ is a constant and relation (11.5) becomes a linear differential equation for s as a function of q, which can be solved explicitly. From relations F_1 and $F_{2,y}$, $F_{2,z}$ one infers that the ratio H_z/H_y is constant so that, except for this constant, H_y and H_z can be determined explicitly by (11.4). Since ρ and a may be regarded as known functions of s, cf. (11.3), also $c = a\sqrt{q}$ may be considered known. The velocity u can now be determined by integrating the differentials

$$(11.7) \qquad du_x = \pm c\,\rho^{-1} d\rho,$$

$$(11.8) \qquad du_y = \mp \rho^{-1} c^{-1} \mu H_x dH_y,$$

$$du_z = \mp \rho^{-1} c^{-1} \mu H_x dH_z.$$

As implied by relation (3.4),

$$(11.9) \qquad \rho c^2 \geq \mu H_x^2 \quad \text{or} \quad s \geq q^{-1} \quad \text{in a fast wave,}$$

$$\rho c^2 \leq \mu H_x^2 \quad \text{or} \quad s \leq q^{-1} \quad \text{in a slow wave;}$$

thus the wave speed c may agree with the Alfvén speed. On the other hand, it could be shown that the wave speed c can never coalesce with the sound speed a; i.e.

(11.10) $$q > 1 \quad \text{in a fast wave,}$$

$$q < 1 \quad \text{in a slow wave.}$$

In a fast wave the compression may tend to infinity, and hence $s \to \infty$, while in a slow wave cavitation may be reached, $s \to 0$.

Of course, a fast wave as well as a slow wave may be a compression wave or a rarefaction wave. The tangential components of the magnetic field $|H_y|$ and $|H_z|$ increase across a fast, decrease across a slow compression wave, but they decrease across a fast, and increase across a slow rarefaction wave.

Finally, we mention that "centered rarefaction waves" exist, as in gas dynamics; they are characterized by the condition that at some initial time, $t = 0$, say, all phases involved are concentrated at the same point, $x = 0$ say. (One then must set $\xi = 0$ in formula (10.1) and use another parameter, instead of ξ, to characterize phases.) If we imagine that a centered wave separates two constant states of the fluid, these states will be adjacent at the time $t = 0$ at the place $x = 0$. In other words, the centered wave then resolves an initial discontinuity.

12. Transverse Waves and Contact Layers

While the ratio of H_z to H_y is constant across fast and slow waves, this is not the case in the transverse waves, which we are going to describe. The speed c is the Alfvén speed in these waves

$$(12.1) \qquad c = (\mu H_x^2/\rho)^{1/2};$$

also H_x, ρ, and S, hence p and u_x are constant across these waves. The tangential magnetic field, however, rotates

$$(12.2) \qquad H_y = G \cos \theta , \quad H_z = G \sin \theta ,$$

with $\theta = \theta(x)$ being any function of the phase. Further,

$$(12.3) \qquad u_y = \alpha_y \mp cH_x^{-1}H_y, \quad u_z = \alpha_z \mp cH_x^{-1}H_z$$

with any numbers α_y, α_z. Thus, <u>the flow in a transverse wave may be considered a shear flow</u>. All particles on the same (y,z)-plane move in the same straight line. This shear flow is evidently a steady flow if it is observed from a frame with respect to which $u_x \pm c = 0$. Thus, <u>there are non-constant steady flows in magneto-hydrodynamics</u> in contrast to gas dynamics.

In order to maintain such a shear flow it would be necessary to have at large (x,y) - distances a mechanism which supplies the velocity u in the proper directions there; it would seem that such a mechanism would not easily occur under natural circumstances and that it would have to be rather artificial.

A transverse wave may connect two constant states with different velocities u and magnetic fields H. One may imagine the layer covered by the wave to be arbitrarily thin; one then may approximate the transition by a discontinuity. This discontinuity would be exactly a transverse shock. Thus, a transverse shock may be considered the limit of transverse simple waves. The remarks about the artificial nature of transverse waves, therefore, apply just as well to transverse shocks.

Only a short remark need be made about "contact layers". There is no flow across such a layer in accordance with $c = 0$; i.e. u_x is constant across it. If $H_x = 0$, all other quantities may vary except that relation

$$(12.4) \qquad p + \frac{1}{2}\mu(H_y^2 + H_z^2) = \text{constant}$$

should hold. If $H_x \neq 0$, however, H, u, and p are constant, only ρ and S may vary. Thus, in contrast to gas dynamics, a shear flow layer across which the tangential flow components u_y, u_z vary, cannot be maintained in a conducting fluid if the magnetic field possesses a normal component $H_x \neq 0$.

13. The Resolution of a Shear Flow Discontinuity

As was explained at the end of Section 9, no shear flow discontinuity can remain unchanged in the presence of a magnetic field provided the normal component of the field does not vanish. If such a discontinuity is present at an initial time a wave motion must result which resolves it. The nature of this wave motion will now be described.

Specifically, we consider the following problem. At an initial time, $t = 0$, the fluid is at rest on one side, $x > 0$, of the plane $x = 0$, while on the other side, $x < 0$, it possesses a constant tangential velocity, $(u_y, u_z) \neq 0$, but no normal velocity, $u_x = 0$. The density is constant and the same on both sides, and a constant magnetic field is present - the same on both sides - with a non-vanishing normal component, $H_x \neq 0$.

The impossibility of maintaining a shear flow discontinuity may be visualized as follows: Instead of a discontinuity, consider a thin shear flow layer across which the tangential flow component, u_y say, varies smoothly from its value, $u_y > 0$ say, on the left hand side to the value zero on the right; for simplicity assume $u_z = 0$ throughout. Also assume $H_y = H_z = 0$ and $H_x > 0$. The basic assumption of magneto-hydrodynamics, embodied in formula (2), now implies that the electric field component E_z does not vanish on the left but does so on the right. Therefore, curl $E \neq 0$ in the layer. From A_1 we then conclude that the magnetic field H, specifically the component $-H_y$, will grow into the layer. Since this component at first remains zero outside of the layer, its curl, and hence the current J, will be different from zero in the layer. Specifically, the component J_z will vary from negative to positive values across the layer. Hence the fluid particles in the layer will experience a force, whose x-component varies from negative to positive values across the layer and hence tends to push the particles away from the layer.

Before describing the details of the resulting wave motion we mention a concrete situation in which such initial shear flow discontinuity may occur; namely, if a jet of conducting fluid shoots into conducting fluid at rest so quickly that the hydro-magnetic adjustments just discussed have not yet developed. A similar case was described by Alfvén [1] in explaining the possible origin of hydro-magnetic waves. Assuming the jet to proceed in the y-direction and to be much wider in the z-direction than in the x-direction, we may approximate it by a constant flow in the y-direction between two parallel planes $x = \text{constant}$. The waves which result from the two interfaces will interact only after some time. Up to this time, therefore, the situation may be described in terms of the wave motion resulting from a single interface.

We maintain that <u>the resolution of the shear flow discontinuity is effected by two fast shocks followed by two slow centered rarefaction waves</u>. After the waves have formed the fluid has acquired the mean tangential velocity while <u>a tangential component of the magnetic field has increased</u>, or developed if none was present originally.

In the (x,t)-diagram given in Figure 3 at the end of the report only the x-component of the particles is indicated.

It may be mentioned, incidentally, without giving supporting arguments, that after the waves coming from the two interfaces have interacted the fluid will come to rest in the jet region while the tangential magnetic field has a further increased value, at least for some time.

In attacking the problem in detail, we have, for simplicity, assumed that all z-components vanish and that at time $t = 0$, H_x is a constant,

$$H_x > 0, \quad \text{and} \quad H_y = 0. \tag{13.1}$$

Further, we have assumed that ρ, p, and u_x are constant, the same on both sides,

$$\rho = \overset{o}{\rho}, \quad p = \overset{o}{p}, \quad u_x = \overset{o}{u}_x. \tag{13.2}$$

For convenience we have assumed that the flow is observed from a frame moving with the mean tangential velocity so that

$$u_y = \overset{o}{u}_y, \quad x < 0, \qquad = -\overset{o}{u}_y, \quad x > 0; \tag{13.3}$$

$\overset{o}{u}_y$ being a positive constant.

The conditions to be satisfied after the passage of the waves simplify because of the symmetry of the problem. They are

$$u_x = 0 \quad \text{at} \quad x = 0 \quad \text{for} \quad t > 0. \tag{13.4$_x$}$$

$$(13.4)_y \qquad u_y = 0.$$

Since H_y is even in x, no condition need be imposed on H_y.

It was mentioned at the end of Section 8 that a shock would be determined if one quantity such as p or u_x is prescribed behind it provided the state in front of it is known. Except for a qualification to be discussed below, the same is true for a simple wave. It is, therefore, natural to expect that two quantities could be prescribed behind a pair of waves provided the state in front is known. It is not obvious, however, whether or not the velocity components u_x, u_y may be prescribed without limitation behind a pair of waves. In the present problem this appears to be the case.

The qualification mentioned above is this: Suppose a piston at one end of a gas filled tube is withdrawn with speed greater than a certain "escape speed", cf. [12]. Then the resulting rarefaction wave will lead to "cavitation". The piston will separate from the gas and a vacuum zone will be formed. The gas at the edge of this zone will move with the escape velocity and not with the piston velocity. For similar reasons condition $(13.4)_x$ must be modified. It should read

$$(13.4)_x^* \qquad \text{Either } u_x = 0 \text{ or } \rho = 0 \text{ at } x = 0 \text{ for } t > 0,$$

thus permitting the presence of a vacuum zone which expands with an appropriate escape speed. Condition $(13.4)_y$, on the other hand, must be retained. In fact, it should hold at the edges of the vacuum zone so that the electric field E vanishes there and in the whole vacuum region.

For the description of the wave motion in detail it is necessary to solve a number of transcendental equations which are expressed in terms of explicitly given integrals derived from the considerations of Section 11. Approximate solutions of these equations have been obtained.

Of particular interest is the magnitude of the tangential component H_y of the magnetic field that has developed at the center $x = 0$ after passage of the waves. It may be described by the formula

(13.5) $$|H_y| = (\kappa \overset{\circ}{\rho}/\mu)^{1/2} |\overset{\circ}{u}_y| ,$$

in which κ in turn depends on $\overset{\circ}{u}_y$. Note that $2\overset{\circ}{u}_y$ is the original shear flow discontinuity.

If this discontinuity approaches zero, κ approaches the value 1;

(13.6) $$\kappa \to 1 \quad \text{as} \quad \overset{\circ}{u}_y \to 0.$$

In fact, the same value of κ would have been obtained if one had assumed the fluid to be incompressible and had described the resolution of the discontinuity with the aid of two Alfvén waves, one traveling in each direction.

If $\overset{\circ}{u}_y \to \infty$, the number κ approaches a value κ_∞ which approximately equals .3.

$$(13.7) \qquad \kappa \to \kappa_0 \sim .3 \text{ as } \overset{\circ}{u}_y \to \infty .$$

Cavitation occurs if $\overset{\circ}{u}_y$ is sufficiently large. It may be worth noticing that in such a case the sound speed eventually becomes less than the Alfvén speed even if it was much larger originally; in such a case therefore the medium could not be considered incompressible.

It is planned to give details of the solution of the problem described and of other one-dimensional wave problems in a later publication.

Although it may well turn out that the problem of one-dimensional waves in a compressible conducting medium are not the most urgent problems of magneto-hydrodynamics that need to be solved, it may be hoped that, as in gas dynamics, the study of such problems may contribute to an understanding of significant hydro-magnetic phenomena.

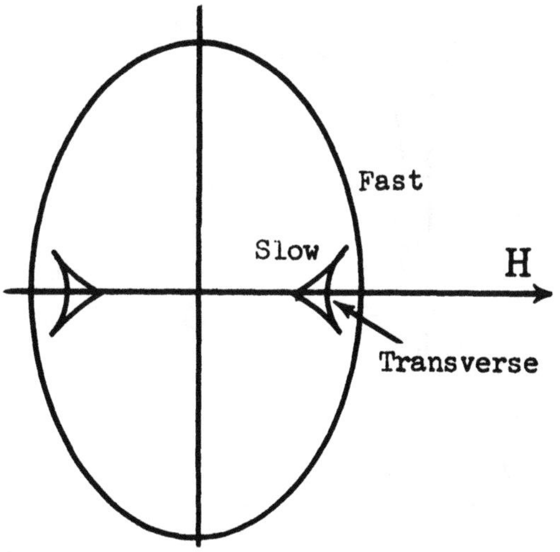

FIGURE 1A

Wave fronts at a time $t > 0$ originating from a point disturbance at the time $t = 0$. They are given by the intersection of the characteristic cone with t = constant and z = constant $(H_z = 0)$.

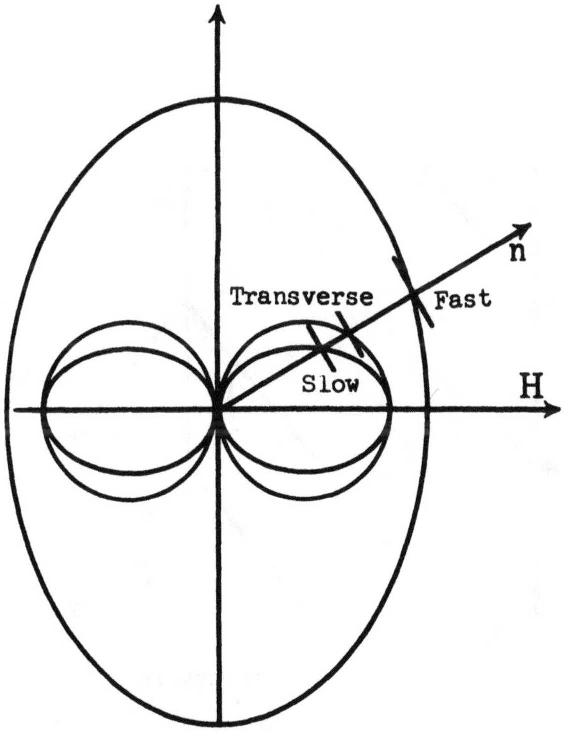

FIGURE 1

The Three Types of Disturbance Waves

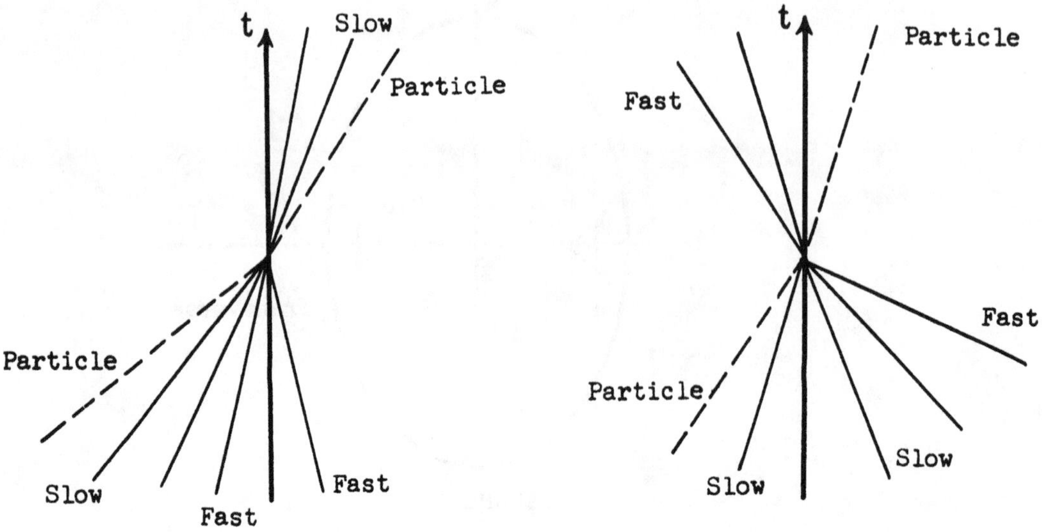

FIGURE 2

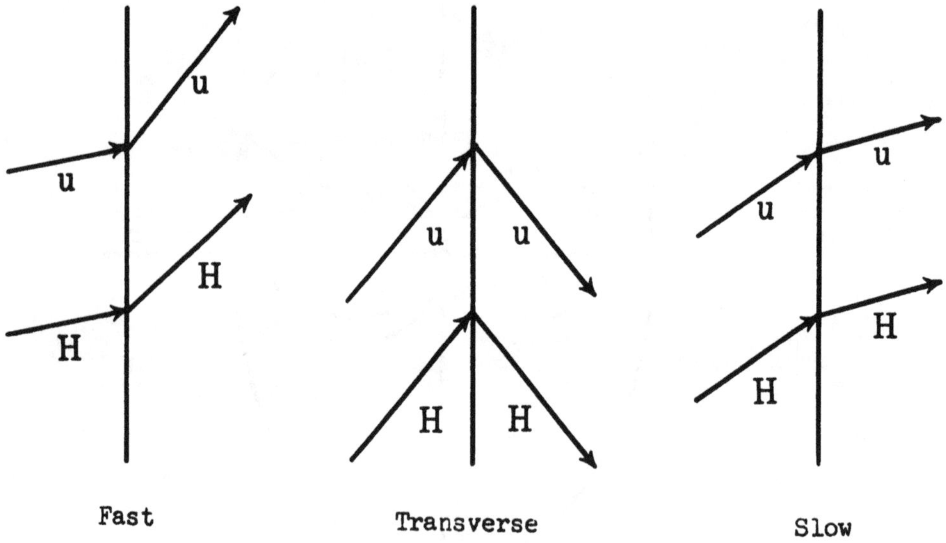

FIGURE 2A

Transitions Through Stationary Shock Fronts

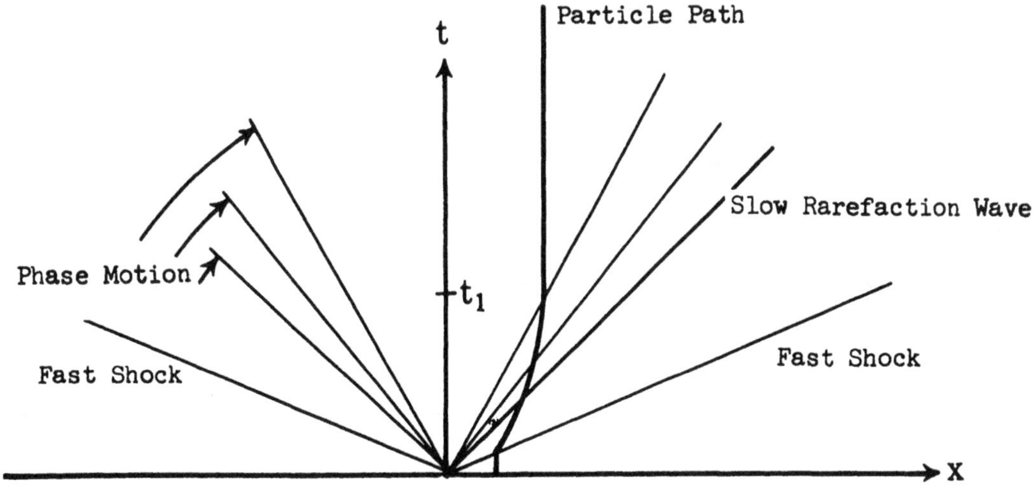

Resolution of a Shear Flow Discontinuity

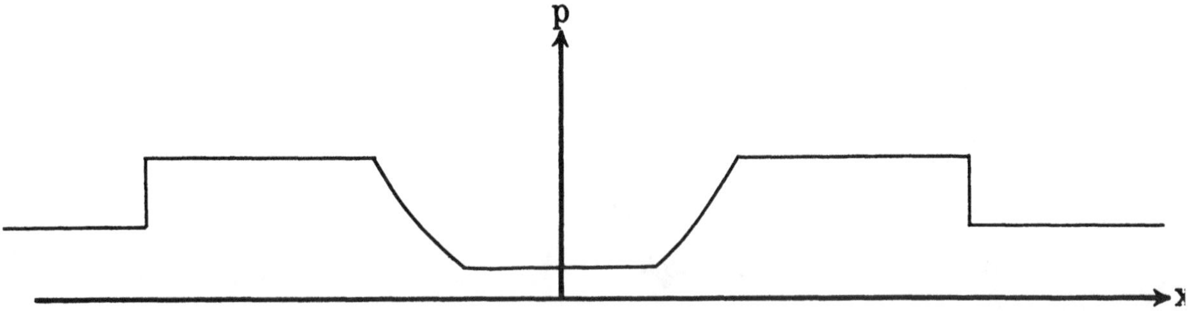

Typical Distribution of Pressure at Time $t_1 > 0$

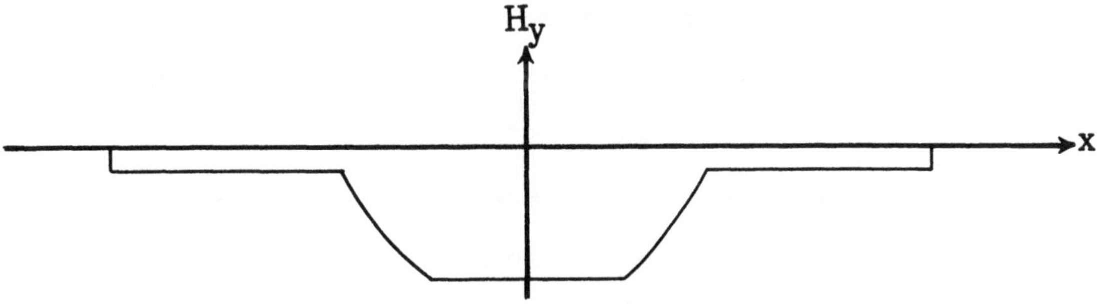

Typical Distribution of Tangential Magnetic Field at Time $t_1 > 0$

FIGURE 3

REFERENCES

[1] H. Alfvén: Cosmical Electrodynamics - Oxford (Clarendon Press) 1950

[2] F. de Hoffmann and E. Teller: Magneto-Hydrodynamic Shocks (Physical Review) November 1950

[3] N. Herlofson: Magneto-Hydrodynamic waves in a compressible fluid conductor (Nature 165, 1020) 1950

[4] H. C. van de Hulst: Interstellar polarization and magneto-hydrodynamic waves in Problems of Cosmical Aerodynamics, p. 46 (Int. Union of Theor. and Appl. Mech. and Int. Astr. Union, Proc. Symposium Paris, 1949) Published by Central Air Documents Office, Dayton, Ohio, 1951

[5] S. Lundquist: Studies in magneto-hydrodynamics (Arkiv för Fysik, Band 5, nr. 15) 1952

[6] L. Helfer: Magneto-Hydrodynamic Shock Waves (The Astrophys. Journal, vol. 117, pp. 177-199) 1953

[7] R. Lüst: Magneto-hydrodynamische Stosswellen in einem Plasma unendlicher Leitfähigkeit (Zeitschrift für Naturforschung, Band 8a, Heft 5, pp. 277-284) 1953

[8] T. G. Cowling: Solar Electrodynamics, The Sun, Chapter 8, Edited by Gerard P. Kuiper, U. of Chicago Press, 1953

[9] H. Grad: Notes on Magneto-Hydrodynamics, mimeographed, Institute of Mathematical Science, New York University, 1954.

Various other references to work on Magneto-hydrodynamics are contained in these papers.

[10] J. A. Stratton: Electromagnetic theory, 1941

[11] R. Courant, D. Hilbert: Methods of Mathematical Physics, Vol. 2, 1954

[12] R. Courant, K. O. Friedrichs: Supersonic Flow and Shock Waves, 1948

[13] P. Lax: Systems of Conservation Laws. To appear in Symposium on Problems in Differential Equations, University of Kansas

[78–1] **Conservation equations and the laws of motion in classical physics, Comm. Pure Appl. Math., XXXI (1978), 123–131.**

Conservation Equations and the Laws of Motion in Classical Physics

K. O. FRIEDRICHS

A Lecture Given at the Quincentenary Celebrations of Uppsala University September 1977

Allow me to say that I greatly appreciate that I have been asked to give a lecture at this unique gathering.

The subject of my lecture involves aspects of pure mathematics, as well as theoretical physics. In its mathematical aspects my talk will be concerned with a certain class of systems of conservation equations. The significance of these equations in physics is that essentially all basic nondegenerate laws of motion in classical physics can be formulated as systems of such equations. The problems of motion governed by these laws can be solved, since the pertinent systems of conservation equations can be reduced to symmetric hyperbolic systems, whose solvability is known.

The conservations equations I shall consider are of the first order. To be sure the basic laws of the motion of continuous media in classical physics are naturally formulated as equations of either the first or the second order; but the latter equations could readily be rewritten as systems of the first order. In this lecture I shall, nevertheless, restrict myself to discussing laws of motion that are naturally formulated as equations of the first order.

These are the laws of fluid dynamics and, more generally, the laws of the interaction of a fluid with an electromagnetic field. The dynamics of this interaction may be referred to as Electro-Magneto-Fluid Dynamics.

There is a limiting case of the latter dynamics, in which no electric field is involved; it is referred to as Magneto-Fluid Dynamics. This very important branch of classical physics was initiated in the years before 1948 here in Sweden by H. Alfvén and was further treated by S. Lundquist and others. Since that time, this theory has undergone a tremendous development. In 1954, when I myself got involved in this field, there were only, as I remember, seven works about this theory that I had to study. Soon afterwards I would have had to read 70 papers about this subject. Now there are perhaps 700 such papers, if not 7,000. Certainly, it is quite exciting for me to have now the opportunity to talk about some aspect of this theory right here in Sweden, where it originated.

Communications on Pure and Applied Mathematics, Vol. XXXI, 123–131 (1978)

I should like to say that my own work in this field is mainly mathematical; but it is not primarily concerned with methods of solving concrete problems, rather it is concerned with the mathematical aspects of a proper formulation of the pertinent differential equations. In the present lecture I shall describe these formulations within the framework of the special theory of relativity. The appropriate non-relativistic formulations can readily be derived from the corresponding relativistic cones.

In the first, mathematical, part of this lecture I shall discuss the pertinent class of conservation equations independently of the laws of physics that can be formulated in terms of such equations. In the second part I shall apply the results to Electro-Magneto-Fluid Dynamics.

Before describing the class of conservation equation to be investigated I must introduce a few notations. There will be m independent variables, denoted by

$$\{x^1, \cdots, x^m\} = \{x^\mu\} = x,$$

and m associated differential operators, denoted by

$$\{\partial_1, \cdots, \partial_m\} = \{\partial_\mu\}.$$

Here $\partial_\mu = \partial/\partial x^\mu$. I also introduce a set of s unknowns, denoted by

$$\{u^1, \cdots, u^s\{ = \{u^\sigma\} = u,$$

which will be regarded as functions $u^\sigma(x)$ of the variables x^μ.

I shall call conservation equation any equation which can be written in the form

$$\partial_\mu q^\mu(u) = g(u),$$

where $q^1, \cdots, q^m$, and g are given functions of u. Summation with respect to μ from 1 to m is implied. Spelled out the equation would be written as

$$\frac{\partial}{\partial u^\sigma} q^\mu(u) \frac{\partial}{\partial x^\mu} u^\sigma = g(u),$$

or simply as

$$\partial_\sigma q^\mu \, \partial_\mu u^\sigma = g,$$

where now also summation with respect to σ from 1 to s is implied.

To justify the name "conservation equation" one may assume that g is 0, and that the q^μ die out sufficiently at infinity. Integration of the equation with respect to $x^1, \cdots, x^{m-1}$ then yields the relation

$$\int \cdots \int q^m(u)\, dx^1 \cdots dx^{m-1} = \text{constant}.$$

Assuming that x^m stands for the time, one can say that the integral on the left is conserved in time.

In the following I shall consider not just one, but a number, r say, of conservation equations involving $r \,.\, m$ functions, $q^\mu_\rho(u)$, and r functions, $g^{(u)}_\rho$, of the s unknowns u^σ. Accordingly, I shall consider a system of r conservation equations of the form

$$\partial_\mu q^\mu_\rho = g_\rho, \qquad \rho = 1, \cdots, r,$$

for s functions $u^\sigma(x)$.

This system of equations is assumed to have certain basic properties. The main property is that the number of equations, r, is greater than the number of unknowns, s:

$$r > s.$$

All subsequent considerations depend on this overdeterminacy.

Clearly, an overdetermined system of equations cannot be expected to have a solution unless the equations depend on each other. Accordingly, I assume that there exists a set of functions of u, $\{y^1, \cdots, y^r\} = \{y^\rho\} = y(u)$, such that the relation

$$\text{I:} \qquad y^\rho\, \partial_\mu q^\mu_\rho = y^\rho g_\rho$$

holds for arbitrary functions u. Summation with respect to ρ from 1 to r and μ from 1 to m is implied.

Relation I will be referred to as the "main dependency relation". In case the excess of the number of equations over the number of unknowns is greater than 1, i.e., if $r - s > 1$, further dependency relations will be assumed to hold.

In addition to the dependency requirements, I shall impose the condition that a certain quadratic form is positive definite. Specifically, this condition is that for suitable numbers $\xi_1, \cdots, \xi_m$ the inequality

$$\text{II:} \qquad -y^\rho \xi_\mu\, \delta^2 q^\mu_\rho > 0,$$

should hold for all unknowns u in a certain neighborhood.

The second variations $\delta^2 q_\rho^\mu$ are assumed to be induced by the variations δu of the unknowns u on which the q_ρ^μ depend. The minus sign is used because of later convenience.

Initial values are to be prescribed on an $(m-1)$-dimensional manifold that, at each of its points, is orthogonal to a covector $\{\xi_1, \cdots, \xi_m\}$ for which inequality II holds. If the number $r-s$, which will be referred to as the "excess number", is greater than one, additional restrictions on the intial values must hold. I shall not describe these restrictions here.

The question now arises how one can show that such an overdetermined system of equations has a solution. As I had mentioned before, that can be done by reducing the system to a symmetric hyperbolic one.

This symmetric hyperbolic system can easily be written down explicitly. To do this one should take the factors $y^\rho = y^\rho$ $(\{u^1, \cdots, u^s\})$, which enter the main dependency relation I, and differentiate them with respect to the unknown u^τ, thus forming the derivatives

$$\partial_\tau y^\rho = \partial y^\rho / \partial u^\tau .$$

Multiplying the basic equations $\partial_\mu q_\rho^\mu = g_\rho$ by these derivatives, one obtains the system of equations

$$\partial_\tau y^\rho \, \partial_\mu q_\rho^\mu = \partial_\tau y^\rho g_\rho , \qquad \tau = 1, \cdots, s,$$

in which summation with respect to ρ is implied. This system of equations, I claim, is symmetric hyperbolic.

To show this one should first spell out the term $\partial_\mu q_\rho^\mu$ and write the equations obtained as

$$\partial_\tau y^\rho \, \partial_\sigma q_\rho^\mu \, \partial_\mu u^\sigma = \partial_\tau y^\rho g_\rho , \qquad \tau = 1, \ldots, s,$$

These equations evidently form a system of s quasilinear differential equations of the first order for the s unknown functions $u^1(x), \cdots, u^s(x)$.

To show that this system is symmetric hyperbolic one must show that the matrix

$$(\partial_\tau y^\rho \, \partial_\sigma q_\rho^\mu) \quad \text{is symmetric in} \quad (\tau, \sigma) \quad \text{for all} \quad \mu,$$

and that the matrix

$$(\xi_\mu \, \partial_\tau y^\rho \, \partial_\sigma q_\rho^\mu) \quad \text{is positive definite.}$$

To this end one should go back to the main dependency relation I, and derive from it that the relation

$$\text{I}': \qquad y^\rho(u)\,\partial_\sigma q_\rho^\mu(u) = 0\,, \qquad \sigma = 1, \cdots, s,\ \mu = 1, \cdots, m\,,$$

holds for all functions u. Differentiating this relation with respect to the unknowns u^τ, considered as independent variables, one obtains the relations

$$\text{I}'': \qquad \partial_\tau y^\rho(u)\,\partial_\sigma q_\rho^\mu(u) \equiv -y^\rho(u)\,\partial_\tau\partial_\sigma q_\rho^\mu(u)\,, \qquad \tau, \sigma = 1, \cdots, s\,.$$

Since evidently the right number here is symmetric in (τ, σ), the same is true of the left number. That is to say, the matrix on the left is indeed symmetric for each μ, as claimed.

Moreover, spelling out the second variations $\delta^2 q^{(u)}$ in relation II and employing relations I″ one can readily verify that the matrix $(\xi_\mu \partial_\tau y^\rho \partial_\sigma q_\rho^\mu)$ is positive definite.

It follows that the system derived is, indeed, symmetric hyperbolic.

The possibility of reducing dependent overdetermined systems of conservation equations to symmetric hyperbolic systems occurred to me late in 1970. When I told my colleague Peter Lax about this he mentioned that he had also used overdetermined systems of conservation equations, but for a different purpose, namely for a general theory of shock discontinuities. (A little later he found out that somewhat earlier C. N. Krushkov had made a similar proposal for a shock theory.) Lax and I combined our results in a common publication; but since then our study of conservation equations has diverged in different directions.

Suppose one has derived a symmetric hyperbolic system from an overdetermined system of conservation equations. Then one can invoke the theory of symmetric hyperbolic systems and assert that the symmetric hyperbolic system here derived has a unique solution that satisfies the prescribed initial conditions.

I should like to mention that in the theory of symmetric hyperbolic equations multiple characteristics need not be excluded and infinite differentiability of the data need not be required.

Once one has shown that the system of the symmetric hyperbolic equations involved has a solution, one must still verify that this solution satisfies the original conservation equations. In case the excess number $r-s$ is 1, this can immediately be done by elementary algebraic arguments. In case $r-s>1$, this can be done with the aid of additional algebraic arguments, provided appropriate $r-s-1$ additional dependency relations, as well as restrictions on the initial data, are satisfied. One can therefore say that the initial problem for our system of conservation equations is well posed.

I shall now proceed to describe in detail the laws of dynamics in classical physics which I have mentioned earlier. The first law that I shall discuss is that of the relativistic dynamics of a compressible fluid.

I shall denote the space-time variables by $\{x^1, x^2, x^3, x^4\} = \{x^\mu\}$ and the relativistic velocity components by $\{w^1, w^2, w^3, w^4\} = \{w^\mu\}$. The components of the associate velocity covector $\{w_1, w_2, w_3, w_4\} = \{w_\mu\}$ should be so taken that the relation

$$w_\mu w^\mu = -c^2$$

holds with c the speed of light. Specifically, these components should be given by $w_{1,2,3} = w^{1,2,3}$ and $w_4 = -c^2 w^4$.

I shall also employ the tensor

$$\Delta^\kappa{}_\lambda \equiv \delta^\kappa_\lambda + c^{-2} w^\kappa w_\lambda\,, \qquad \kappa, \lambda = 1, \cdots, 4\,,$$

which has on occasion been used in the literature. In a rest system characterized by $w^{1,2,3} = 0$, $w^4 = 1$, this tensor reduces to $\Delta^\kappa{}_\lambda = \delta^\kappa_\lambda$, except for $\Delta^4{}_4 = 0$. Therefore, I shall call the tensor $\Delta^\kappa{}_\lambda$ the "spatial delta function".

The velocity vector $\{w^\mu\}$ will be considered as a function of the x^μ. Also the mass density of the fluid, denoted by ρ, will be a function of the x^μ. The law of conservation of mass then has the form

$$\partial_\mu \rho w^\mu = 0\,,$$

The appropriate form of the law of fluid motion is the law of conservation of energy and momentum, which can be written as

$$\partial_k U^\kappa{}_\lambda = 0\,, \qquad \lambda = 1, \cdots, 4\,,$$

where $U^\kappa{}_\lambda$ is the "energy-momentum tensor". This tensor can conveniently be expressed in terms of a function $\mathscr{L}(\rho)$ of the density ρ in the form

$$\Delta^\kappa{}_\lambda \equiv -\mathscr{L}\delta^\kappa_\lambda + \rho\mathscr{L}_\rho\, \Delta^\kappa{}_\lambda\,,$$

where $\mathscr{L}_\rho = \partial\mathscr{L}/\partial\rho$. One then readily verifies the dependency relation

$$\text{I:} \qquad w^\lambda\, \partial_\kappa U^\kappa{}_\lambda + \mathscr{L}_\rho\, \partial_\mu \rho w^\mu \equiv 0\,,$$

The function $\mathscr{L}(\rho)$ will be referred to as a "Lagrangian"—somewhat improperly perhaps. In the usual case this Lagrangian is the function

$$\mathscr{L}(\rho) = \rho c^2 + e(\rho)\,,$$

where ρc^2 is Einstein's rest energy and e is the additional energy, both per unit volume.

The positivity condition II could readily be verified under mild restrictions. I shall not do so.

Evidently, there are five equations in the present problem, but only four unknowns since one of the velocity components, w^4, say, depends on the others. Clearly then, the system of equations is overdetermined, with the excess number $r-s=1$. In forming the symmetric hyperbolic equations by the recipe described earlier one may not, of course, consider w^4 as one of the independent unknowns. Consequently, the symmetric hyperbolic system will not appear in covariant form. But that does not matter since the symmetric hyperbolic system serves only as a technical tool; it is not part of the basic formulation of the laws of nature.

The next problem I want to discuss concerns the interaction of a fluid with an electro-magnetic field. This field will be described by a cotensor $\{B_{\alpha\beta}\}$, $\alpha, \beta = 1, 2, 3, 4$, which is connected with the electric and magnetic field vectors (E_α) and (B_α), $\alpha = 1, \cdots, 4$, through the relations

$$B_{\alpha 4} = E_\alpha, \quad B_{12} = B_3, \quad B_{23} = B_1, \quad B_{31} = B_2 ,$$

Maxwell's first law can now be written as

$$\mathbb{K}_{\kappa\lambda\mu} \equiv \langle \partial_\kappa B_{\lambda\mu} \rangle \equiv \partial_\kappa B_{\lambda\mu} + \partial_\lambda B_{\mu\kappa} + \partial_\mu B_{\kappa\lambda} = 0 ,$$

The second law of Maxwell can be written as

$$\mathbf{J}^\mu = j^\mu ,$$

where $\mathbf{J}^\mu$ is derived from a "Lagrangian" $\mathcal{L}(\rho, \{B_{\alpha\beta}\})$ as

$$\mathbf{J}^\mu \equiv 2\, \partial_\kappa (\partial \mathcal{L} / \partial B_{\mu\kappa}) ,$$

The right number j^μ may be assumed to express Ohm's law in the form

$$j^\mu = \eta w^\mu + \sigma B^{\mu\nu} w_\nu ,$$

where η is the electric charge, per unit volume, carried by the fluid particles, while σ is the "conductivity".

The law of conservation of mass,

$$\mathbf{M} \equiv \partial_\kappa \rho w^\kappa = 0$$

must also be adopted.

The energy-momentum tensor will now have the form

$$U^{\kappa}{}_{\lambda} \equiv -\mathcal{L}\delta^{\kappa}_{\lambda} + \rho\mathcal{L}_{\rho}\,\Delta^{\kappa}{}_{\lambda} + 2(\partial\mathcal{L}/\partial B_{\mu\kappa})B_{\mu\lambda}\,.$$

The associated conservation law,

$$\mathbb{F}_{\lambda} \equiv \partial_{\kappa}U^{\kappa}{}_{\lambda} = f_{\lambda}\,,$$

is assumed to hold with a certain right member, f_{λ}, which could be written down explicitly.

The main dependency relation is

$$w^{\lambda}(\mathbb{F}_{\lambda} - f_{\lambda}) + \mathcal{L}_{\rho}\mathbb{M} + w^{\lambda}(\partial\mathcal{L}/\partial B_{\mu\kappa})\mathbb{K}_{\kappa\lambda\mu} + w^{\lambda}B_{\lambda\mu}(\mathbb{J}^{\mu} - j^{\mu}) \equiv 0\,,$$

as can be verified easily.

Two additional dependency relations will be satisfied:

$$\langle\partial_{\nu}\mathbb{K}_{\kappa\lambda\mu}\rangle \equiv 0\,,$$

in obvious notation, and

$$\partial_{\mu}(\mathbb{J}^{\mu} - j^{\mu}) \equiv 0\,.$$

The latter relation, however, will hold only under suitable restrictions, essentially on the conductivity σ.

The case $\sigma = 0$ fits immediately into our scheme. In case $0 < \sigma < \infty$, additional considerations are needed. In case $\sigma = \infty$, i.e., if the conductivity is infinite, the factor $w^{\lambda}B_{\lambda\mu}$ of σ in the expression for j^{μ} is zero. This fact can be seen to imply that no electric field is present in a rest system. The resultant theory is the relativistic Magneto-Fluid Dynamics. The reduction of the associated conservation equations to symmetric hyperbolic ones described earlier, allows one to show that this theory is well posed. The same is also true of the original, nonrelativistic, theory of Magneto-Fluid Dynamics, developed by Alfvén and Lundquist.

I should like to mention that it is not difficult to reduce the equations of nonrelativistic Magneto-Fluid Dynamics to a symmetric hyperbolic system directly. But to do this for the relativistic equation directly is quite difficult, while this is very easy if one employs the reduction recipé here described.

An extension of the theory of Electro-Magneto-Fluid Dynamics here presented is needed in the case in which the electro magnetic field interacts with a medium in such a way that a polarization field arises. To cover this extension one must allow the Lagrangian $\mathcal{L}$ to depend not only on ρ and the $B_{\alpha\beta}$ but also on the velocity components w^{μ}; that is, one should take

$\mathscr{L} = \mathscr{L}(\rho, \{B_{\alpha\beta}\}, \{w^\mu\})$. The energy momentum tensor must then also be modified. It turns out that this modification is uniquely determined by the requirement that no additional differential equation is involved and that the dependency relations are not changed. This modification of the energy-momentum tensor could easily be written down in several equivalent ways; but I do not want to do that here.

Energy-momentum tensors for polarized fields within the framework of the theory of relativity have been proposed since 1908 by Minkowski, Abraham and many others in the course of time. None of the proposals were quite right, although that of Abraham was very close. The reason that the proposed tensors were not the right ones is that they were not related to a Lagrangian and that no proper attention was paid to the dependency relations.

In all these proposals the question of the symmetry of the energy-momentum tensor $\{U^{\kappa\lambda}\}$ has played a role. Rather artificial symmetrization procedures have on occasion been suggested. Actually, a very simple answer to this question could have been given. Namely, if the Lagrangian function $\mathscr{L}(\rho, \{B_{\alpha\beta}\}, \{w^\mu\})$ is invariant under Lorentz transformation, the tensor $\{U^{\kappa\lambda}\}$, which we have proposed, is symmetric. This can be proved very easily.

Altogether one can say that the use of a Lagrangian is of vital help in the theory presented; but not *via* a variational principle. Instead, the dependency relations for the overdetermined system of equations play the major role.

I should like to mention that the theory here described could be so modified that it covers some problems in which parabolic equations take the place of the hyperbolic ones. It is an open question, however, whether or not this theory can be adapted to more general degenerate problems that result from asymptotic limiting processes or other simplifications.

Bibliography

[1] Friedrichs, K. O., *On the laws of relativistic electro-magneto-fluid dynamics*, Comm. Pure Appl. Math., Vol. 27, 1974, pp. 749–808.

[2] Friedrichs, K. O., *Conservation Equations and the Laws of Motion in Classical Physics*, Research Report 77A, Math. Dept., C.W. Post College, Greenvale, N.Y. 11548, 1977.

Received October, 1977.

$\frac{\partial}{\partial x^\alpha}(P_{\alpha\beta}^{\mu} w^\beta)$. The energy-momentum tensor must then also be modified. It turns out that this modification is uniquely determined by the requirement that no additional differential equation is involved and that the dependency relations are not changed. This modification of the energy-momentum tensor could easily be written down in several equivalent ways, but I do not want to do that here.

Energy-momentum tensors of ponderable fields within the framework of the theory of relativity have been proposed since 1908 by Minkowski, Abraham and many others in the course of time. None of the proposals were quite right, although that of Abraham was very close. One reason that the proposed tensors were not the right ones is that they were not related to a Lagrangian, and that no proper attention was paid to the dependency relations.

In all these proposals the question of the symmetry of the energy-momentum tensor $\{T^{\alpha}{}_{\beta}\}$ has played a role. Rather artificial symmetrization procedures have on occasion been proposed. Actually, a very simple answer to this question could have been given: Namely, if the Lagrangian function $\mathfrak{L}(E, B, w)$ is invariant under Lorentz transformation, the tensor $\{T^{\alpha\beta}\}$, which we have proposed, is symmetric. This can be proved very easily.

Altogether, one can say that the use of a Lagrangian is of vital help in the theory presented, but not the variational principle. Instead, the dependency relations for the over-determined system of equations play the major role.

It should likewise be noted that the theory here described could be so modified that it covers some problems in which parabolic equations take the place of the hyperbolic ones. It is an open question, however, whether or not the theory can be extended to more general degenerate problems that result from approximations, limit processes or other simplifications.

References

[illegible]

[illegible] Vol. [illegible]

[illegible]

(Received [illegible] 197?)

VI
Papers involving Variational Problems

Commentaries by Louis Nirenberg

[29–4] **Ein Verfahren der Variationsrechnung, das Minimum eines Integrals als das Maximum eines anderen Ausdruckes darzustellen, Ges. Wiss. Göttingen, Math.-Phys. Klasse, (1929), 13–20.**

Commentary:

The use of Legendre transforms in transforming a convex variational problem to a dual one is now well known. In this early paper Friedrichs carries out this process for variational expressions for functions of one and two variables. The determination of the boundary conditions for the dual problem is especially interesting.

Ein Verfahren der Variationsrechnung, das Minimum eines Integrals als das Maximum eines anderen Ausdruckes darzustellen.

Von

Kurt Friedrichs, Aachen.

Vorgelegt von R. COURANT in der Sitzung am 7. Dezember 1928.

Bei der numerischen Lösung von Variationsproblemen durch das RITZsche Verfahren ist es wichtig, die Güte der Approximation des Minimalwertes abzuschätzen[1]). E. TREFFTZ[2]) hat ein Verfahren angegeben, — und zwar für das DIRICHLETsche Problem und verwandte Aufgaben — die Lösung von Variationsproblemen so zu approximieren, daß dabei der Minimalwert von unter her angenähert wird. Durch Anwendung des RITZschen und seines Verfahrens gelingt es also, den Minimalwert von beiden Seiten einzugrenzen. Auf anderem Wege und allgemeiner als bei TREFFTZ wird in Folgendem dasselbe Ziel erreicht. Man kann sehr allgemein einem Minimumproblem ein Maximumproblem zuordnen, dessen Maximalwert dem Minimalwert des ursprünglichen gleich ist. Das zugrunde liegende Prinzip ist im wesentlichen eine LEGENDRE-Transformation. Durch eine solche Transformation entsteht z. B. aus dem Variationsproblem für die Lösung der Potentialgleichung mit vorgegebenen Randwerten ein Problem für die konjugierte Potentiafunktion. Durch eine LEGENDRE-Transformation ordnet man ferner in der Elastizitätstheorie dem Prinzip der „virtuellen Verrückungen", das nach CASTIGLIANO benannte Prinzip von der „kleinsten Formänderungsarbeit" zu, in dem nach den Spannungen variiert wird[3]).

---

1) Im allgemeinen liefern die Beweise für die Konvergenz einer Minimalfolge daraus eine Abschätzung für die Annäherung an die Lösungsfunktion selbst.

2) Ein Gegenstück zum RITZschen Verfahren. Verhandlungen des zweiten Internationalen Kongresses für Technische Mechanik, Zürich 1927, S. 131, wo auch an Hand praktisch wichtiger Fälle die numerische Brauchbarkeit nachgeprüft wird. Vgl. auch Konvergenz und Fehlerabschätzung beim RITZschen Verfahren. Math. Annalen, 1923, Bd. 101, wo auch die Approximation der Lösung selbst in einzelnen Punkten auf die Approximation von Minimalwerten zurückgeführt wird.

3) Dies war übrigens der Ausgangspunkt der folgenden Überlegungen. Vgl. RIEMANN, WEBER u. FRANK, Die Differential- u. Integralgleichungen, II, V, § 4, 2. 3.

Um sicher zu gehen, daß das formal neu gebildete Variationsproblem auch wirklich ein Maximum liefert, müssen gewisse starke Definitätsbedingungen gestellt werden, die sich übrigens weitgehend durch Modifikation des Variationsausdruckes erfüllen lassen.

Von Wichtigkeit für die numerische Behandlung könnte auch die folgende Tatsache sein: Sind bei dem einen Problem die Randwerte der Funktion vorgeschrieben, so ist das bei dem zugeordneten Problem nicht der Fall; dort stellen sich die Randbedingungen als „natürliche" ein. Der erste Fall hat den Nachteil, daß man bei der Auswahl der Approximationsfunktionen behindert ist; im zweiten Falle ist die Konvergenz erfahrungsgemäß sehr langsam. Man hat aber nun die Möglichkeit im Einzelfalle zu wählen, welchen Nachteil man mit in Kauf nehmen will.

I. Variationsprobleme in einer Variablen.

Wir beschäftigen uns zunächst mit dem Variationsproblem, das Integral

$$\int_{x_0}^{x_1} F(u', u, x)\, dx \tag{1}$$

zum Minimum zu machen.

Wir verlangen von der gesuchten Funktion $u(x)$, daß sie vorgegebene Randwerte annimmt. Übrigens läßt sich das Folgende ganz ebenso durchführen, wenn Randbedingungen von anderem Typus vorliegen, so daß wir das nicht besonders bemerken.

Daß die Funktion $u'(x)$ die Ableitung von $u(x)$[1]) ist, formulieren wir als eine besondere Gleichung:

$$\frac{du}{dx} = u'. \tag{2}$$

Die Randbedingungen lauten:

(3) $u(x_0) = \bar{u}_0$, $u(x_1) = \bar{u}_1$ oder einfach $u = \bar{u}$ für $x = x_0$, $x = x_1$;

dabei sind $\bar{u}_0$, $\bar{u}_1$ beliebige vorgegebene Werte.

Ferner sei

$$F_{u'u'} > 0, \tag{4}$$

$$\Delta = F_{u'u'} \cdot F_{uu} - F_{uu'}^2 > 0. \tag{5}$$

1) $u(x)$ und $u'(x)$ sollen zwischen oberen und unteren Schranken liegen, die für jedes x in $x_0 \leqq x \leqq x_1$ gegeben seien. $u(x)$ und $u'(x)$ sind also auf einen abgeschlossenen Bereich B beschränkt. In B sei F nach seinen Argumenten zweimal stetig differenzierbar. Auch (4) und (5) sollen in B gelten.

Wir nehmen an, daß das Minimum für eine zweimal stetig differenzierbare ganz innerhalb B liegende Funktion $u(x)$ angenommen wird. Für sie gilt die Eulersche Gleichung

$$\frac{d}{dx} F_{u'} = F_u. \tag{6}$$

Nun führen wir neue Funktionen $p(x)$, $p'(x)$ ein durch die Gleichungen

$$F_{u'} = p, \quad F_u = p', \tag{7}$$

woraus wir wegen (5) u und u' als Funktionen von p und p' berechnen können. Mit ihnen bilden wir einen neuen Integranden:

$$\Phi(p', p, x) = pu' + p'u - F(u', u, x); \tag{8}$$

es gilt dann

$$\Phi_{p'} = u, \quad \Phi_p = u'. \tag{9}$$

Als neues Variationsproblem stellen wir nun

$$\int_{x_0}^{x_1} \Phi\, dx - p\bar{u}\Big|_{x_0}^{x_1} = \text{Min.}\,[1]), \tag{10}$$

wo zur Konkurrenz zugelassen sind alle stetigen Funktionen $p(x)$ und $p'(x)$[2]), ohne Randbedingungen, nur unterworfen der Bedingungsgleichung

$$\frac{dp}{dx} = p'. \tag{11}$$

Wir behaupten, daß gerade die der Lösung $u(x)$ und $u'(x)$ von (1) vermöge (7) zugeordneten Funktionen $p(x)$, $p'(x)$ den Ausdruck (10) zum Minimum machen. Zunächst ist wegen (6) die Gl. (11) erfüllt. Die neue Eulersche Gleichung

$$\frac{d}{dx} \Phi_{p'} = \Phi_p \tag{12}$$

ist wegen (9) nichts anderes als die Bedingungsgleichung (2) und also erfüllt. Die natürliche Randbedingung

$$\Phi_{p'} - \bar{u} = 0 \quad \text{für} \quad x = x_0, \; x = x_1 \tag{13}$$

ist nichts anderes als die Randbedingung (3). Also ist der Ausdruck (10) zunächst stationär. Daß er auch ein Minimum ist, folgt

---

1) Anstatt das Negativum von (10) zum Maximum zu machen.

2) Sie sollen ganz in einem Bereich B^* von derselben Art wie B liegen. B^* soll enthalten sein in dem Bereich der durch Transformation aus B entsteht. p' sei in B^* einmal stetig differenzierbar.

aus der Positivität der zweiten Variation von (10) für alle $p(x)$, $p'(x)$ im Bereich $B^*(x)$. Man errechnet nach (9), (7) unter Beachtung von (4), (5)

$$\Phi_{p'p'} = \frac{1}{\Delta} F_{u'u'} > 0, \tag{14}$$

$$\Phi_{p'p'} \cdot \Phi_{pp} - \Phi^2_{pp'} = \frac{1}{\Delta} > 0. \tag{15}$$

Der Minimalwert des neuen Problemes (10)

$$\int_{x_0}^{x_1} (pu' + p'u - F)\, dx - p\bar{u}\Big|_{x_0}^{x_1} = -\int_{x_0}^{x_1} F\, dx$$

ist überdies gerade dem Minimalwert von (1) entgegengesetzt gleich.

II. Zusätze.

1. Ist die starke Definitheitsbedingung (5) nicht erfüllt, sind aber die größten Werte von $-F_{uu}$ und $|F_{uu'}|$ genügend klein im Verhältnis zum kleinsten Wert von $F_{u'u'}$, so können wir die Bedingung (5) z. B. dadurch befriedigen, daß wir zum Integral (1) noch ein Integral

$$\int_{x_0}^{x_1} \frac{d}{dx} G(u, x)\, dx$$

addieren[1]); indem wir G geeignet — etwa $G \equiv (\alpha x + \beta) . u^2$ — wählen.

2. Ohne daß die Existenz der Lösung von (1) bekannt ist, läßt sich zeigen, daß die Summe der unteren Grenzen von (1) und (10) nicht negativ ist. Es ist nämlich[2]) für beliebige zulässige aber voneinander unabhängige Funktionen $u(x)$, $p(x)$ mit $u' = \frac{du}{dx}$, $p' = \frac{dp}{dx}$

$$\int_{x_0}^{x_1} [F(u) + \Phi(p)]\, dx - pu\Big|_{x_0}^{x_1}$$

$$= \int_{x_0}^{x_1} [F(u) - F(v) + vF_u(v) + v'F_{u'}(v) - uF_u(v) - u'F_{u'}(v)]\, dx,$$

wobei wir $v = \Phi_{p'}$, $v' = \Phi_p$, also $p = F_{u'}(v)$, $p' = F_u(v)$ gesetzt haben. Der Integrand ist nichts anderes als der Rest zweiter

1) Das entspricht der LEGENDREschen Umformung der zweiten Variation.

2) Wir schreiben $F(u)$, $\Phi(p)$ anstatt $F(u', u, x)$, $\Phi(p', p, x)$ usw.

Ordnung der Entwicklung von $F(u)$ nach u', u in der Umgebung von v', v und also wegen der Bedingungen (4), (5) nicht negativ und übrigens nur gleich Null für $u = v$, $u' = v'$, d. h. für die Lösung.

3. Wir betrachten ferner den ausgearteten Fall, daß $F_{uu} = F_{uu'} = 0$ ist, wo wir also

$$F = G(u', x) + k(x) . u \tag{16}$$

setzen können. Hier fordern wir nur (4) $G_{u'u'} > 0$. Die LEGENDRE-Transformation lautet jetzt einfach

$$\begin{gathered} F_{u'} = p, \\ \Phi(p, x) = pu' - G, \qquad \Phi_p = u', \end{gathered} \tag{17}$$

wo wir also u' allein durch p ausdrücken können und umgekehrt. Die aus der Lösung $u(x)$ von (1) so entstehende Funktion $p(x)$ macht den Ausdruck (10) zum Minimum unter allen Funktionen $p(x)$, die noch der Nebenbedingung

$$\frac{dp}{dx} = k \tag{18}$$

genügen. (Im Ausdruck (10) tritt jetzt p' nicht auf). In der Tat erlaubt (18) nur Variationen von der Form $p + \text{const.}$ und es ist die erste Variation von (10) $\int_{x_0}^{x_1} \Phi_p \, dx - \bar{u} \Big|_{x_0}^{x_1} = 0$ wegen $\Phi_p = u'$, während die zweite Variation (4) positiv ist.

4. Unterscheiden wir bei einem Variationsproblem Bedingungsgleichungen: u' ist Ableitung von u, u nimmt gegebene Randwerte an —, und Variationsgleichungen: EULERsche Gleichung, natürliche Randbedingungen —, so ergibt sich aus dem vorangehenden: die Bedingungsgleichungen des ursprünglichen Problems sind die Variationsgleichungen des transformierten und umgekehrt.

5. Methode der LAGRANGE-Faktoren. Die bisher dargestellte Transformation der Variationsprobleme läßt sich noch allgemeiner entwickeln, wenn man von der Methode der LAGRANGE-Faktoren ausgeht. Man ersetzt zunächst das Minimum-Problem (1) mit den Bedingungen (2), (3) durch ein bedingungsloses Variationsproblem: Der Ausdruck

$$I \equiv \int_{x_0}^{x_1} \left[F + p(x) \left(\frac{du}{dx} - u' \right) \right] dx - q(u - \bar{u}) \Big|_{x_0}^{x_1} \tag{19}$$

soll stationär werden. Hierbei sind die von einander unabhängigen Funktionen $u(x)$, $u'(x)$, $p(x)$ und die Werte q_0, q_1 zu variieren. Die erste Variation von I lautet:

$$(20)\quad \delta I = \int_{x_0}^{x_1} \left[\left(F_u - \frac{dp}{dx}\right)\delta u + (F_{u'} - p)\,\delta u' + \left(\frac{du}{dx} - u'\right)\delta p\right] dx$$

$$\left[(q-p)\,\delta u + (u-\bar{u})\,\delta q\right]_{x_0}^{x_1}.$$

Es ergeben sich durch Variation nach p und q: die Gleichungen:

$$(2)\quad \frac{du}{dx} - u' = 0, \qquad (3)\quad u - u = 0 \quad \text{für } x = x_0,\ x = x_1$$

und durch Variation nach u und u':

$$(21)\qquad F_u - \frac{dp}{dx} = 0,$$

$$(22)\qquad F_{u'} - p = 0,$$

$$(23)\qquad q - p = 0.$$

Das Ausgangsproblem entsteht, wenn man (2) und (3) von vornherein als Bedingungen stellt. Dann sind (21), (22) und (23) Variationsgleichungen. Stellt man dagegen diese letzten drei Gleichungen als Bedingungen, so ergeben sich (2) und (3) als Variationsgleichungen. Das ist das transformierte Maximum-Problem. Man hat dabei übrigens zu benutzen, daß durch die Bedingungen (21), (22) die freie Variabilität von $p(x)$ nicht eingeschränkt wird; das ist auch sicher nicht der Fall, sobald u und u' vermöge (21), (22) durch p und $\frac{dp}{dx}$ wie früher angenommen berechenbar sind.

Durch Einsetzen von (21), (22), (23) in (19) erhält man nach partieller Integration den Ausdruck

$$\int_{x_0}^{x_1} [F - u F_u - u' F_{u'}]\, dx + F_{u'} \cdot \bar{u}\Big|_{x_0}^{x_1},$$

der in das Negativum von (10) übergeht, sobald man $p(x)$ als einzige neue Funktion einführt.

Beiläufig sei erwähnt, daß das Variationsproblem der „kanonischen“ Gleichungen (2), (21) aus dem Problem (19) einfach dadurch entsteht, daß man nur die Gleichung (22) als Bedingung stellt (und etwa die Randbedingungen (3)).

Die Methode der LAGRANGE-Multiplikatoren zum Ausgang zu wählen, hat den Vorzug, daß man ohne weiteres übersieht, wie man vorzugehen hat, wenn noch weitere Bedingungsgleichungen beliebiger Art vorliegen.

Anm. bei der Korrektur. Wie Herr COURANT mitteilte, läßt sich der Übergang vom Minimum- zum Maximumproblem rein begrifflich begründen. Das soll an anderer Stelle ausgeführt werden.

III. Variationsprobleme für zwei Variable in Beispielen.

1. Wir betrachten zunächst das Problem

$$(24)\qquad \tfrac{1}{2}\iint_G (u_x^2+u_y^2+au^2)\,dx\,dy = \text{Min.}\qquad u=\bar{u} \text{ auf } \Gamma,\quad a>0.$$

wo G ein Gebiet der (x, y)-Ebene mit dem Rand Γ ist[1]). Es sei $u(x, y)$ die Lösung dieses Problems[2]).

Wir führen neue Funktionen p, q, d ein vermöge der Gleichungen

$$(25)\qquad p=u_x,\quad q=u_y,\quad d=au$$

und betrachten als neuen Integranden

$$pu_x+qu_y+du-\frac{1}{2}(u_x^2+u_y^2+au^2)=\frac{1}{2}\left(\frac{1}{a}d^2+p^2+q^2\right).$$

Wir stellen demgemäß das neue Variationsproblem

$$(26)\qquad \frac{1}{2}\iint_G\left(\frac{1}{a}d^2+p^2+q^2\right)dx\,dy-\int_\Gamma (px_n+qy_n)\bar{u}\,ds = \text{Min.}$$

(x_n, y_n Komponenten der äußeren Normalen s Bogenlänge auf Γ), wobei wir der Eulerschen Gleichung von (19) entsprechend die Gleichung

$$(27)\qquad d=p_x+q_y$$

als einzige Nebenbedingung stellen. Die Eulerschen Gleichungen von (26) lauten

$$(28)\qquad \begin{aligned}\frac{1}{a}(p_x+q_y)_x-p&=0\\ \frac{1}{a}(p_x+q_y)_y-q&=0,\end{aligned}$$

die natürliche Randbedingung

$$(29)\qquad \frac{1}{a}(p_x+q_y)-\bar{u}=0 \text{ auf } \Gamma.$$

Diese Gleichungen besagen, daß die Funktion

$$u=\frac{1}{a}(p_x+q_y)$$

folgendes erfüllt:

$$u_x=p,\quad u_y=q,\quad \Delta u-au=0,\quad u=\bar{u} \text{ auf } \Gamma.$$

---

1) Etwa mit stückweise stetig differenzierbar sich ändernder Tangente.

2) Sie besitzt sicher stetige erste und zweite Ableitungen im Innern von G.

Der Wert des Minimums von (26) ist wieder gleich dem negativen Minimalwert von (24).

2. Das DIRICHLETsche Problem

$$\frac{1}{2}\iint_G (u_x^2+u_y^2)\,dx\,dy = \text{Min.} \quad u = \bar{u} \text{ auf } \Gamma \tag{30}$$

entspricht dem ausgearteten Fall II. 3. Man erhält durch Einführung von $u_x = p$, $u_y = q$, $\Phi(p, q) = \frac{1}{2}(p^2+q^2)$ das Variationsproblem

$$\frac{1}{2}\iint_G (p^2+q^2)\,dx\,dy - \int_\Gamma (p x_n + q y_n)\,\bar{u}\,ds = \text{Min.} \tag{31}$$

unter der Nebenbedingung $p_x + q_y = 0$. Wir befriedigen diese Nebenbedingung durch den Ansatz $p = v_y$, $q = -v_x$ und erhalten nach partieller Integration auf Γ somit das Problem

$$\frac{1}{2}\iint_G (v_x^2+v_y^2)\,dx\,dy + \int_\Gamma v\,\bar{u}_s\,ds = \text{Min.} \tag{32}$$

dessen Lösung v die zu u konjugierte Potentialfunktion ist. Die Gleichung $\Delta v = 0$ und die natürliche Randbedingung $v_n = -\bar{u}_s$ stellt sich von selbst ein.

Ohne die Existenz der Lösung von (30) zu kennen, können wir einsehen, daß die Summen der unteren Grenzen von (30) und (32) nicht negativ ist. Das lehrt die folgende identische Umformung (vgl. II. 2)

$$\iint_G (u_x^2+u_y^2+v_x^2+v_y^2)\,dx\,dy + 2\int_\Gamma v\,u_s\,ds$$
$$= \iint_G [(u_x-v_y)^2+(u_y+v_x)^2]\,dx\,dy.$$

Bei dem „freien" Problem (31) können wir von den Funktionen p und q schon das Erfülltsein der zugehörigen EULERschen Gleichung $p_y - q_x = 0$ verlangen. Wir können dann durch $u_x = p$, $u_y = q$ eine Funktion u einführen und erhalten so das Problem

$$\frac{1}{2}\iint_G (u_x^2+u_y^2)\,dx\,dy - \int_\Gamma u_n\,\bar{u}\,ds = \text{Min.} \tag{33}$$

für alle Funktionen der Nebenbedingung $\Delta u = 0$. Das ist nichts anderes als das Verfahren, welches TREFFTZ angegeben hat. Für numerische Zwecke bleibt es vielleicht zweckmäßig diese Zusatzforderung beizubehalten; sobald es aber möglich ist, ohne Schwierigkeit Approximationsfunktionen anzugeben, die von vornherein den natürlichen Randbedingungen genügen und nicht der EULERschen Gleichung, wird man eine bessere Konvergenz erwarten dürfen, wenn man diese dem RITZschen Verfahren zugrunde legt.

[33–1] **Über ein Minimumproblem für Potentialströmungen mit freiem Rande, Math. Ann., 109 (1933), 60–82.**

Commentary:

This paper studies the problem of symmetric two dimensional compressible flow of water from a nozzle; the shape of the flow after it leaves the nozzle is not known, it forms a free boundary. This has a well known associated variational problem. Friedrichs proves that a stationary point of this variational problem is necessarily a minimum, and it follows that the solution is unique. The proof is very interesting. Friedrichs makes a change of independent variables, introducing the stream function as a new independent variable. The boundary value problem is thus transformed to a complicated looking non-linear one for the vertical component of position – but now the fixed domain is known. Friedrichs then treats the corresponding variational problem.

The corresponding change of variable has been taken up, in a systematic way, only very recently in proving regularity of free boundaries. This paper was, in fact, far ahead of its time.

Über ein Minimumproblem für Potentialströmungen mit freiem Rande.

Von

Kurt Friedrichs in Braunschweig.

Bei den Randwertproblemen der Potentialtheorie, auf welche die wirbelfreien Strömungen mit freien Grenzen führen, bietet die Frage nach der Existenz und Eindeutigkeit ein besonderes Interesse. Deshalb, weil es nicht immer von vornherein selbstverständlich ist, — wie manche Diskussionen gezeigt haben — welches die „richtigen" Randbedingungen sind; d. h. diejenigen, welche Existenz und Eindeutigkeit garantieren.

Zur Klärung solcher Fragen hat sich bisher als Hilfsmittel vor allem die Variationsrechnung bewährt; in der Tat sind auch die typischen Potentialprobleme mit freien Grenzen der Variationsrechnung zugänglich; das soll für eine Reihe von Strömungen, insbesondere für Ausflußströmungen aus einer Düse gezeigt werden. Die entsprechende Randwertaufgabe — allerdings in einer von der physikalischen Frage (I) zunächst etwas abweichenden Form (II) gestellt — entsteht aus der Variation eines einfachen Integrals für die Stromfunktion. Insbesondere ist dabei auch das Integrationsgebiet zu variieren, wodurch diejenige Randbedingung entspringt, die das Druckgleichgewicht am Rande des Strahles ausdrückt[1]).

Darüber hinaus aber zeige ich, daß die Lösung der Randwertaufgabe den Variationsausdruck wirklich zum *Minimum* macht; als unmittelbare Konsequenz dieser Tatsache ergibt sich die *Eindeutigkeit* dieser Lösung.

Bisher war die Eindeutigkeit bei ebenen nach innen konkaven Düsen nur gegenüber infinitesimaler Abänderung bewiesen (Weinstein, Hamel, Weyl [1], [2]), aber auch für das ursprüngliche physikalische Problem (I) (s. unten). Auf diese infinitesimale Eindeutigkeit gründete Weinstein seinen Existenzsatz.

Die *Existenz* der Lösung mit Hilfe des Minimumproblems zu beweisen, steht noch aus.

[1]) Wie mir inzwischen Herr A. Weinstein mitteilt, hat schon Riabouchinsky [1] 1927 solche Variationsprobleme aufgestellt und die erste Variation berechnet, allerdings ohne Bezug auf die Minimumeigenschaft im Falle freier Grenzen.

Ebenso bleibt noch offen, ob die Lösung des Minimumproblems auch *numerisch* approximiert werden kann, womit dann zugleich auch eine numerische Berechnung des Kontraktionskoeffizienten zu gewinnen wäre.

Die hier angegebene Methode ist nicht auf ebene Probleme beschränkt. Es werden neben den ebenen auch die achsensymmetrischen *räumlichen* Ausflußströmungen, die der funktionentheoretischen Behandlung nicht mehr zugänglich sind, auf ein Minimumproblem zurückgeführt.

In einem ersten Anhang wird gezeigt, daß die Minimumeigenschaft unseres ersten Variationsproblems nur eine allgemeine Eigenschaft regulärer Variationsprobleme ausdrückt. In einem zweiten Anhang gewinne ich aus dem Minimumproblem einen neuen Zugang zur Behandlung der oben erwähnten infinitesimalen Eindeutigkeit.

Zunächst seien die behandelten Probleme nur vorläufig formuliert, um ihren endgültigen Ansatz zu motivieren. Die ebene Ausflußströmung einer zur Strömung konkaven Düse (Fig. 1) wird beschrieben erstens durch ein Stromgebiet, das teils vom „Düsenrand" teils vom „Strahlrand" begrenzt wird. Wir beschränken uns dabei auf den Fall der symmetrischen Strömung; wesentlich benutzen wir allerdings nur, daß Aus- und Einströmrichtung (im unendlichen) übereinstimmen und daß der Düsenrand eine eindeutige Projektion auf diese Richtung besitzt. Die Symmetrieachse wählen wir zur x-Achse; dann ist zweitens die Stromfunktion $\psi(x, y)$ für die Strömung kennzeichnend; sie sei Potentialfunktion, Null auf der Achse und gleich Eins auf dem „Rande"; die Quadratsumme ihrer Ableitungen sei konstant v^2 auf dem Strahlrand, wo v^2 die vorgegebene quer über den Strahl konstante Ausströmgeschwindigkeit im Unendlichen ist. Insbesondere ist die Strahlbreite im Unendlichen nicht vorgegeben (Problem I). Wir sind allerdings genötigt, das Problem etwas zu verändern; wir denken uns nämlich die Düse irgendwie mit stetiger Krümmung verlängert; und dann die Strahlbreite im Unendlichen vorgegeben, während die Stelle der Ablösung von der Düse offen gelassen ist (Problem II). Diese Variation der Ablösungsstelle ist wohl kaum physikalisch realisierbar; doch scheint mir das Problem II das mathematische natürlichere zu sein; auch bei den Untersuchungen von Weinstein und Weyl[2]) war das Problem II

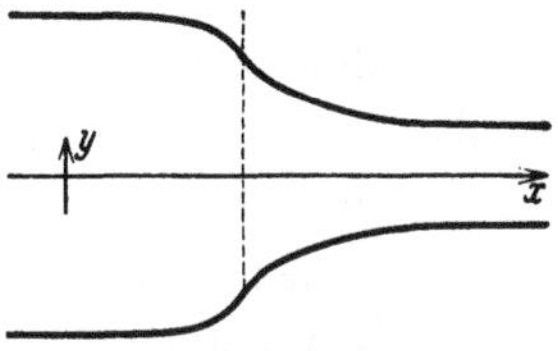

Fig. 1.

[2]) Unser Problem (II) ist dasselbe wie bei Weyl [[2]]. Das ursprüngliche Problem (II) von Weinstein entsteht, indem zusätzlich auch die Ablösungsstelle vorgeschrieben wird.

unmittelbar zugänglich, während die Behandlung von (I) auf die von (II) zurückgeführt wurde. Im Falle einer zur Strömung konvexen Düse ist eine Variation der Ablösungsstelle auch physikalisch denkbar[3]).

Zur konkaven und konvexen Düse gehören zwei Arten von Ablösung; sie unterscheiden sich dadurch, daß in einem Falle die Strömung in der Nähe der Ablösungsstelle die (mit stetiger Krümmung verlängerte) Düse durchsetzt, im anderen Falle nicht; wir sprechen dann von *äußerer* und *innerer* Ablösung. Zunächst soll nur der Fall äußerer Ablösung behandelt werden.

Der Variationsausdruck, der von der richtigen Stromfunktion ψ zum Minimum gemacht wird, lautet nun

$$\iint \{\psi_x^2 + \psi_y^2 + v^2\}\, dx\, dy,$$

über das Stromgebiet integriert; damit das Integral existiert, wird das Stromgebiet im endlichen abgebrochen; der Fall des unendlichen Gebietes erfordert einige Modifikationen, die gesondert behandelt werden.

Von dem Stromgebiet wird nur verlangt, daß sein Rand außerhalb der Düse verläuft und von der Stromfunktion im wesentlichen nur, daß sie auf dem Rande konstant Eins ist. Die Addition der Konstanten v^2 im Integranden[4]) wäre bei festem Gebiet effektlos, hier ist sie wesentlich; durch Gebietsvariation entsteht auf dem Teil des Randes, der sich nicht den Düsenwänden anlegt, d. h. auf dem Strahlrand gerade als natürliche Randbedingung $\psi_x^2 + \psi_y^2 = v^2$, während für die anliegenden Teile Ungleichungen entspringen.

Daß das Minimumproblem überhaupt einen Sinn hat, mag man sich durch folgende Überlegung plausibel machen: Das Minimum von

$$\iint_\Gamma \{\psi_x^2 + \psi_y^2\}\, dx\, dy$$

bei festem Gebiet kann höchstens wachsen, wenn das Gebiet — und damit $\iint_\Gamma v^2\, dx\, dy$ — verkleinert wird; diese Verkleinerung ist ja gleichbedeutend mit der Zusatzbedingung $\psi = 1$ im Differenzgebiet.

Das eigentliche mathematische Problem dieser Arbeit besteht in dem Nachweis, daß die Lösung der Strömungsaufgabe den obigen Ausdruck wirklich zum Minimum macht und also eindeutig ist. Zu dem Zweck

---

[3]) Zur Klärung dieser Möglichkeit hat St. Bergmann [3] als Beispiel eine Kontur konstruiert, bei der sich zwei Strahlbreiten angeben lassen, zu denen auch zwei Ablösungsstellen gehören.

[4]) Übrigens lassen sich auch Wirbelströmungen durch ein solches Variationsproblem kennzeichnen, wenn die Bernoullikonstante $q(\psi)$ als Funktion von ψ vorgegeben und an Stelle von $\frac{v^2}{2}$ eingesetzt wird.

wird die zweite Variation dieses Integrals in einer Weise berechnet, die in Evidenz setzt, daß sie positiv-definit ist. In einer solchen Berechnung der zweiten Variation allein liegen die Schwierigkeiten bei der noch ausstehenden Übertragung auf allgemeine ebene Düsenformen und auf nicht achsensymmetrische räumliche Probleme.

1. Randwert- und Minimum-Problem der endlichen Strömung.

1. 1. *Zulässigkeitsbedingungen.*

Zur Formulierung[5]) des mathematischen Problems [1] ist zunächst anzugeben, welche Gebiete und welche Funktionen in ihm wir „zulassen".

Das abgeschlossene Gebiet Γ mit dem Inneren Γ' sei begrenzt von (Fig. 2)

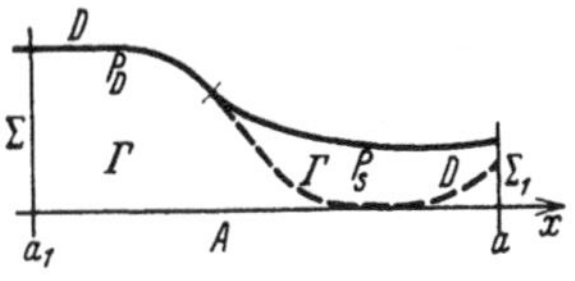

Fig. 2.

der Achse $\quad A:\; y = 0, \qquad a_1 \leqq x \leqq a,$

dem „Rand" $\quad \mathsf{P}:\; y = \Omega(x), \qquad a_1 \leqq x \leqq a,$

den Seitenlinien $\quad \Sigma_1:\; x = a_1, \qquad 0 \leqq y \leqq \Omega(a_1),$

$\quad \Sigma:\; x = a, \qquad 0 \leqq y \leqq \Omega(a).$

Der Rand P sei durch die Düse

$$D:\; y = Y(x), \qquad a_1 \leqq x \leqq a$$

eingeschränkt, vermöge

$$(1) \qquad Y(x) \leqq \Omega(x) \qquad a_1 \leqq x \leqq a.$$

Die Randfunktionen $\Omega(x)$ und $Y(x)$ seien stetig differenzierbar. Wir zerlegen den Rand P, je nachdem ob in (1) die Gleichheit angenommen wird oder nicht, in den *„Düsenrand"* P_D und den *„Strahlrand"* P_S, so daß also gilt:

$$Y(x) = \Omega(x) \quad \text{auf } \mathsf{P}_D,$$
$$Y(x) < \Omega(x) \quad \text{auf } \mathsf{P}_S.$$

Ein Gebiet Γ, das den angegebenen Bedingungen genügt, möge kurz mit $[\Gamma]$ bezeichnet werden.

---

[5]) Wir wollen im folgenden Funktionen und Bereiche durch lateinische oder griechische Buchstaben kennzeichnen, je nachdem ob sie fest oder variabel sind.

Von der in Γ erklärten Funktion $\psi(x, y)$ — dann $[\psi_1]$ genannt — verlangen wir: ψ ist einmal in Γ, zweimal in Γ' stetig differenzierbar; ferner

$$\psi = 1 \quad \text{auf } \mathsf{P}, \tag{2}$$

$$\psi = 0 \quad \text{auf } A, \tag{3}$$

$$\psi_y > 0 \quad \text{in } \Gamma. \tag{4}$$

Gebiete $[\Gamma]$ und Funktionen $[\psi_1]$, die in ihnen erklärt sind, heißen zum Problem [1] *zugelassen.*

Wir haben uns bei der Wahl dieser Bedingungen auf die symmetrische Strömung beschränkt und demgemäß nur eine Hälfte des Strömungsgebietes angesetzt. Die Ungleichung (1) garantiert äußere Ablösung von der Düse. Der Schranke $Y(x)$ brauchen wir keine weitere Bedingung auferlegen, solange wir keinen Existenzsatz formulieren. Das eigentliche Problem der konkaven Düse entsteht, wenn $Y(x)$ in einem Intervall $(a_1, a_3 < a)$ negativ gekrümmt ist, für $x \geqq a_3$ verschwindet.

Übrigens haben wir durch die Forderung der Stetigkeit von $\Omega'(x)$ und $Y'(x)$ den Fall polygonaler Düsen ausgeschlossen. Die in (4) enthaltene Forderung, daß sich die Stromlinien in der Form $y = y(x)$ darstellen lassen, kann zweifellos (auch für den Rand $y = \Omega(x)$) gemildert werden.

1. 2. *Randwertaufgabe.*

Die Lösung der Randwertaufgabe heiße $\psi = \psi^0(x, y)$, das zugehörige Gebiet $\Gamma = \Gamma^0$ mit dem Rand $\mathsf{P} = \mathsf{P}^0$. Es wird dann ψ^0 und Γ^0 — außer durch Zulässigkeit — durch die folgenden Bedingungen gekennzeichnet:

$$\psi_x^0 = 0 \qquad \text{auf } \Sigma_1 \text{ und } \Sigma, \tag{5}$$

$$(\psi_x^0)^2 + (\psi_y^0)^2 = v^2 \qquad \text{auf } \mathsf{P}_S^0, \tag*{(6)$_S$}$$

wo v die vorgeschriebene konstante Strahlrandgeschwindigkeit ist; auf dem festen Rand gelte dagegen die Ungleichung

$$(\psi_x^0)^2 + (\psi_y^0)^2 \leqq v^2 \quad \text{auf } \mathsf{P}_D^0, \tag*{(6)$_D$}$$

im Innern von Γ^0 genüge ψ der Differentialgleichung

$$\psi_{xx} + \psi_{yy} = 0 \qquad \text{in } \Gamma^{0\prime}. \tag{7}$$

An Stelle der Bedingung (5), die senkrechte Ein- und Ausströmung bedeutet, würden andere zu setzen sein, wenn entsprechende Zulässigkeitsbedingungen auf Σ_1 und Σ gestellt sind.

Wesentlich ist — außer $(6)_S$ — die Ungleichung $(6)_D$, die wir — wie ausdrücklich bemerkt sei — als zusätzliche Randbedingung ansehen[6]).

1. 3. *Minimumproblem.*

Die bes· hriebene Randwertaufgabe ist — wie wir zeigen wollen — gleichwertig mit folgender Variationsaufgabe:

Durch Wahl von Funktionen vom Typ $[\psi_1]$ in Gebieten $[\Gamma]$ ist zum Minimum zu machen

$$J[\psi, \Gamma] = \iint_\Gamma \{\psi_x^2 + \psi_y^2 + v^2\}\, dx\, dy.$$

1. 4. *Lösung des Minimumproblems ist Lösung der Randwertaufgabe.*

Satz 1. Ist eine zulässige Funktion $\psi^0(x, y)$ in einem zulässigen Gebiet Γ^0 mit der Randfunktion $\Omega^0(x)$ Lösung des Minimumproblems, so ist ψ^0 in Γ^0 auch Lösung der Randwertaufgabe.

Ich will diesen Satz nur formal begründen.

Ist $\psi^\varepsilon(x, y)$, $\Omega^\varepsilon(x)$ eine Schar zulässiger Funktionen, J^ε der entsprechende Wert von J, so setzen wir

$$\frac{\partial \psi^0}{\partial \varepsilon} = \delta \psi^0, \quad \frac{\partial \Omega^0}{\partial \varepsilon} = \delta \Omega^0, \quad \frac{\partial J^0}{\partial \varepsilon} = \delta J^0.$$

Wir unterstellen, daß für jede zulässige Wahl von $\delta\psi^0$, $\delta\Omega^0$ zugehörige zulässige ψ^ε und Ω^ε zu bilden sind; dabei ist die Ungleichung (1) zu beachten, aus ihr entnehmen wir:

I. Ist $\delta\Omega^0 = 0$ auf P_D^0, so ist eine Schar ψ^ε, Ω^ε für eine Vollumgebung von $\varepsilon = 0$ bildbar. In diesem Falle gilt $\delta J^0 = 0$.

II. Ist $\delta\Omega^0 \geqq 0$ auf P_D^0, so ist eine Schar ψ^ε, Ω^ε für $\varepsilon \geqq 0$ bildbar. Somit folgt hier nur:

$$\delta J^0 \geqq 0.$$

Ferner ist wesentlich zu beachten, daß die Variationen von ψ und Ω auf P voneinander abhängen vermöge $\psi = 1$, der Bedingung (2). Aus ihr entspringt

$$\delta(2) \qquad \delta\psi^0 + \psi_y^0\, \delta\Omega^0 = 0;$$

auf der Achse gilt wegen (3)

$$\delta(3) \qquad \delta\psi^0 = 0 \quad \text{auf} \quad A.$$

Im übrigen setzen wir die Variationen $\delta\psi^0$, $\delta\Omega^0$ als willkürlich an.

---

[6]) Die explizite bekannten Ausflußströmungen bei polygonaler Begrenzung erfüllen sie in einer Umgebung der Ablösungsstelle, wie sich leicht nachprüfen läßt. Bei innerer Ablösung würde in $(6)_D$ das entgegengesetzte Zeichen stehen.

Nunmehr berechnen wir die erste Variation δJ^0

$$\delta J^0 = -2\iint_{\Gamma^0} \{\psi^0_{xx} + \psi^0_{yy}\}\,\delta\psi^0\,dx\,dy + \int_{\mathsf{P}^0} \{\psi^0_y - \psi^0_x \Omega^{0\prime}(x)\}\,\delta\psi^0\,dx$$

$$+ \int_{\mathsf{P}^0} \{(\psi^0_x)^2 + (\psi^0_y)^2 + v^2\}\,\delta\Omega^0\,dx + \left[\int_0^{\Omega(x)} \psi^0_x\,\delta\psi^0\,dy\right]_{a_1}^{a},$$

wobei wir $\delta(3)$ schon berücksichtigt haben. Wir können noch $\Omega^{0\prime}(x)$ vertreiben, indem wir die Gleichung

$$\psi^0_x + \Omega^{0\prime}\,\psi^0_y = 0$$

verwenden, die durch Differentiation aus $\psi(x, \Omega(x)) = 1$ entsteht; indem wir noch $\delta\psi^0$ vermöge $\delta(2)$ durch $\delta\Omega^0$ ausdrücken, entsteht

$$\delta J^0 = -2\iint_{\Gamma^0} \{\psi^0_{xx} + \psi^0_{yy}\}\,\delta\psi^0\,dx\,dy + \int_{P^0} \{v^2 - (\psi^0_x)^2 - (\psi^0_x)^2\}\,\delta\Omega^2\,dx$$

$$+ \left[\int_0^{\Omega(x)} \psi^0_x\,\delta\psi^0\,dy\right]_{a_1}^{a}.$$

Es sei nun zunächst $\delta\Omega^0 = 0$ auf P^0_D, so daß also $\delta J^0 = 0$ gilt; so entspringt aus der Willkürlichkeit von $\delta\psi$ in $\Gamma^{0\prime}$ und auf Σ^0_1, Σ^0 und von $\delta\Omega^0$ auf P^0_S

$$\psi^0_{xx} + \psi^0_{yy} = 0 \text{ in } \Gamma^{0\prime}, \quad v^2 - (\psi^0_x)^2 - (\psi^0_y)^2 = 0 \text{ auf } \mathsf{P}_S$$

$$\psi^0_x = 0 \text{ auf } \Sigma_1,\ \Sigma;\ \text{d. h. (7), (6)}_S\text{, (5).}$$

Ist sodann $\delta\Omega > 0$ auf P^0_D, so bleibt

$$\delta J^0 = \int_{\mathsf{P}^0_D} \{v^2 - (\psi^0_x)^2 - (\psi^0_y)^2\}\,\delta\Omega^0\,dx$$

übrig und aus $\delta J^0 \geqq 0$ folgt $(6)_D$:

$$v^2 - (\psi^0_x)^2 - (\psi^0_y)^2 \geqq 0 \quad \text{auf} \quad \mathsf{P}^0_D.$$

1. 5. *Die Lösung des Randwertproblems löst das Minimumproblem.*

Satz 2: Erfüllt die zulässige Funktion $\psi^0(x, y)$ im zulässigen Gebiet Γ^0 die Bedingungen der Randwertaufgabe, so gilt für jede zulässige Funktion $\psi(x, y)$ im zulässigen Gebiet Γ

$$J[\psi, \Gamma] > J[\psi^0, \Gamma^0],$$

außer wenn $\Gamma = \Gamma^0$ und $\psi = \psi^0$.

Diese Tatsache begründen wir darauf, daß die erste Variation von J, für die Lösung Γ^0, ψ^0 der Randwertaufgabe genommen, nicht negativ ist, während die zweite Variation für jedes zulässige Γ, ψ positiv ist.

Zum Beweis dieses positiven Charakters werden wir Γ^0, ψ^0 und Γ, ψ in eine zulässige Schar Γ^ε, ψ^ε einbetten; das kann auf verschiedene Weise geschehen und nicht in jedem Falle wird die Positivität evident; wir kommen aber zum Ziel, wenn wir diese Einbettung auf dem einfachsten sich darbietenden Wege vollziehen. Wir transformieren das variable Gebiet auf ein festes Gebiet, und zwar, indem wir x und ψ als unabhängige, y als abhängige Veränderliche einführen; wegen der Randbedingungen für ψ (insbesondere $\psi = 1$ auf P) entsteht dann nämlich ein festes Rechteck der (x, ψ)-Ebene. Daß der quadratische Charakter des Integranden und der lineare Charakter der durch Variation entspringenden Gleichungen verloren geht, scheint unvermeidlich zu sein.

Das neue Gebiet G im Bereich der Variablen x, ψ ist das Rechteck

$$G:\quad a_1 \leqq x \leqq a, \quad 0 \leqq \psi \leqq 1;$$

dem Rande P entspricht

$$R:\quad a_1 \leqq x \leqq a, \quad \psi = 1.$$

Im übrigen mögen Gebiet, Kurven, Punkte bei der Abbildung ihre Namen behalten.

Wegen $\psi_y > 0$ in Γ, der Bedingung (4), kann die Gleichung $\psi = \psi(x, y)$ durch eine Funktion $y = y(x, \psi)$ aufgelöst werden. Es ist

$$\psi_y = \frac{1}{y_\psi}, \quad \psi_x = -\frac{y_x}{y_\psi},$$

$$\psi_{xx} = -\left(\frac{y_x}{y_\psi}\right)_x + \left(\frac{y_x}{y_\psi}\right)_\psi \frac{y_x}{y_\psi}, \quad \psi_{yy} = \left(\frac{1}{y_\psi}\right)_\psi \frac{1}{y_\psi}$$

und das Integral (9) geht über in

$$\tilde{J}[y] = \iint\limits_G \left[\frac{y_x^2 + 1}{y_\psi} + v^2 y_\psi\right] dx\, d\psi.$$

Jeder Funktion vom Typ $[\psi_1]$ in einem Gebiet $[\Gamma]$ entspricht eine Funktion $y(x, \psi)$ in G, die zweimal im Inneren, einmal am Rande von G stetig differenzierbar ist und ferner den folgenden Bedingungen genügt:

$(\tilde{1})$ $\qquad Y(x) \leqq y(x, 1)$ auf R.

$(\tilde{3})$ $\qquad y(x, 0) = 0$ auf A.

$(\tilde{4})$ $\qquad y_\psi > 0$ in G.

Eine solche — $[y_1]$ genannte — Funktion $y(x, \psi)$ führt auch stets zu einer Funktion $[\psi_1]$ in einem Gebiet $[\Gamma]$.

Der Lösung der Randwertaufgabe $\psi = \psi^0(x, y)$ in Γ^0 möge die Funktion $y^0(x, \psi)$ entsprechen. Sie genügt den Relationen

$(\tilde{5})$ $\qquad y_x^0 = 0 \qquad$ für $x = a_1$, $x = a$.

$(\tilde{6})_S$, $(\tilde{6})_D$ $\qquad \left(\frac{y_x^0}{y_\psi^0}\right)^2 + \left(\frac{1}{y_\psi^0}\right)^2 - v^2 \begin{cases} = 0 & \text{auf } P_S^0, \\ \leqq 0 & \text{auf } P_D^0. \end{cases}$

(7) $\qquad \tilde{\Delta} y^0 = -\left(\frac{y_x^0}{y_\psi^0}\right)_x + \left(\frac{y_x^0}{y_x^0}\right)_\psi \left(\frac{y_x^0}{y_\psi^0}\right) + \left(\frac{1}{y_\psi^0}\right)_\psi \frac{1}{y_\psi^0} = 0.$

Es sei nun $y^1(x, \psi)$ irgendeine Funktion der Art $[y_1]$. Wir bilden dann

$$\delta y(x, \psi) = y^1(x, \psi) - y^0(x, \psi)$$

und die Schar von Funktionen

$$y^\varepsilon(x, \psi) = y^0(x, \psi) + \varepsilon\, \delta y(x, \psi). \tag{8}$$

Jede dieser Funktionen ist auch vom Typ $[y_1]$ für $0 \leqq \varepsilon \leqq 1$; man beachte insbesondere, daß für solche ε die Ungleichungen (1), (4) erfüllt sind. Wir berechnen nun $\tilde{J}[y^\varepsilon]$ für $\varepsilon = 1$ durch Entwicklung um $\varepsilon = 0$:

$$\tilde{J}[y^1] = \tilde{J}[y^0] + \delta \tilde{J}[y^0] + \int_0^1 (1-\varepsilon)\, \delta^2 \tilde{J}[y^\varepsilon]\, d\varepsilon. \tag{9}$$

Dabei ist

$$\delta \tilde{J}[y^\varepsilon] = \frac{d}{d\varepsilon} \tilde{J}[y^\varepsilon], \quad \delta^2 \tilde{J}[y^\varepsilon] = \frac{d^2}{d\varepsilon^2} \tilde{J}[y^\varepsilon].$$

Wir berechnen zunächst

$$\delta \tilde{J}[y] = \iint_G \left\{ \frac{2 y_x}{y_\psi} \delta y_x - \frac{y_x^2}{y_\psi^2} \delta y_\psi - \frac{1}{y_\psi^2} \delta y_\psi + v^2 \delta y_\psi \right\} dx\, d\psi \tag{10}$$

$$= -2 \iint_G \tilde{\varDelta} y\, \delta y\, dx\, d\psi + \int_R \left\{ -\frac{y_x^2}{y_\psi^2} - \frac{1}{y_\psi^2} + v^2 \right\} \delta y\, dx + 2 \left[\int_0^1 \frac{y_x}{y_\psi} \delta y\, d\psi \right]_{a_1}^{a},$$

wobei der von A herrührende Anteil wegen ($\tilde{3}$) fortfällt.

Für $y = y^0(x, \psi)$ fällt wegen $\tilde{\varDelta} y^0 = 0$ das Flächenintegral fort und ebenso die von $x = a_1$, a und P_S herrührenden Anteile der Randintegrale wegen ($\tilde{5}$), $(\tilde{6})_S$. Auf P_D^0 ist $y^0(x, 1) = Y(x)$ und also wegen ($\tilde{1}$) $\delta y \geqq 0$. Unter Berücksichtigung von $(6)_D$ folgt dann

$$\delta \tilde{J}[y^0] \geqq 0. \tag{11}$$

Sodann berechnet man die zweite Variation

$$\delta^2 \tilde{J}[y] = 2 \iint_G \frac{1}{y_\psi^3} \{(y_\psi \delta y_x - y_x \delta y_\psi)^2 + \delta y_\psi^2\}\, dx\, d\psi. \tag{12}$$

Ohne weiteres nimmt sie eine Gestalt an, die ihren positiven Charakter erkennen läßt.

Es ist also für jede zugelassene Funktion y und jede Variation δy

$$\delta^2 \tilde{J}[y] > 0,$$

außer, wenn $\delta y = 0$ ist. Es ist also insbesondere

$$\int_0^1 (1-\varepsilon)\, \delta^2 \tilde{J}[y^\varepsilon]\, d\varepsilon > 0 \tag{13}$$

außer, wenn $y^1 = y^0$ ist.

Dann aber folgt aus (9), (11), (13) für $\varepsilon = 1$

$$\tilde{J}[y^1] > \tilde{J}[y^0],$$

außer für $y^1 = y^0$ und damit die Behauptung von Satz 2.

1. 6. *Satz von der Eindeutigkeit.*

Aus der Minimumeigenschaft (Satz 2) folgt sofort für Problem [1]:

Satz 3. Es gibt höchstens ein zulässiges Gebiet $\Gamma = \Gamma^0$, zu dem eine zulässige Funktion $\psi = \psi^0(x, y)$ gehört, welche die Randwertaufgabe löst.

2. Unendlicher Strahl und unendliche Düse.

Es macht keine Schwierigkeit, auch den Fall $[1]^\infty$ zu behandeln, daß die Düse und der Strahl sich ins Unendliche erstrecken. Es ist nur zu bedenken, daß das Integral

$$\iint \{\psi_x^2 + \psi_y^2 + v^2\}\, dx\, dy,$$

über ein derartiges unendliches Gebiet erstreckt, nicht existiert.

Es ist aber leicht möglich, solche unendlichen Anteile abzuziehen, die zur Variation nichts beitragen. Man beachte zunächst, daß das Integral $\iint \{\psi_x^2 + (\psi_y - v)^2\}\, dx\, dy$, über Gebiete $[\Gamma]$ erstreckt, sich von dem obigen Integral nur um den Ausdruck $2v \iint \psi_y\, dx\, dy = 2v \int dx$ unterscheidet, der von der Wahl der Funktion ψ wegen (2) und (3) nicht abhängt; andererseits ist beim Abfluß ($x = \infty$) gerade $\psi_y = v$, so daß

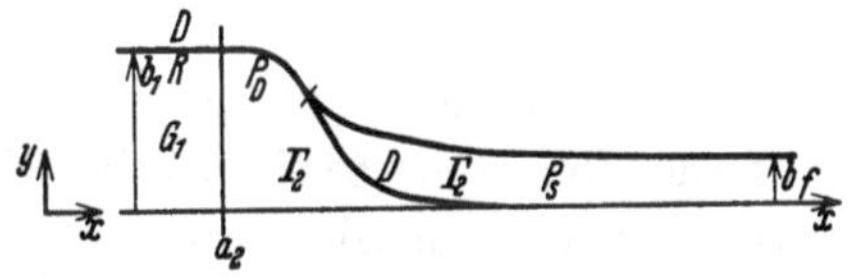

Fig. 3.

dieses Integral nach der Abflußseite ins Unendliche erstreckt werden kann; ist im Zufluß ($x = -\infty$) $\psi_y = v_1$, so wird man für die Zuflußseite das Integral $\iint \{\psi_x^2 + (\psi_y - v_1)^2\}\, dx\, dy$ ansetzen; damit hier die Ersetzung von v durch v_1 auf die Variation keinen Einfluß hat, wird im Integrationsbereich dieses Integrals keine Ablösung zugelassen. So gelangt man zu folgender Formulierung des Problems (Fig. 3).

Das Gebiet Γ und sein Rand P seien bestimmt durch

Γ: $$0 \leqq y \leqq \Omega(x),$$

P: $$y = \Omega(x).$$

Der überall definierten und stetig differenzierbaren Funktion $\Omega(x)$ sei die untere Schranke $y = Y(x)$ gestellt; $Y(x)$ sei stetig differenzierbar und es gelte

$$Y(x) \to b_1, \qquad x \to -\infty.$$

Ferner sei für $\Omega(x)$ die Bedingung

$$(1)_\infty \qquad \Omega(x) \to b, \qquad x \to +\infty$$

gestellt.

Die Zu- und Abflußbreiten b_1 und b sind mit den Zu- und Abflußgeschwindigkeiten v_1 und v gekoppelt vermöge

$$b_1 v_1 = b v = 1.$$

Das Gebiet Γ und der Rand P werde durch eine Gerade $x = a_2$ in einen festen und einen variablen Teil gespalten,

$$\Gamma = G_1 + \Gamma_2, \quad \mathsf{P} = R_1 + \mathsf{P}_2.$$

Für $\Omega(x)$ werden die Bedingungen gestellt:

$$(1)_1 \qquad Y(x) = \Omega(x) \quad \text{für} \quad x \leqq a_2, \text{ d. h. auf } R_1.$$

$$(1)_2 \qquad Y(x) \leqq \Omega(x) \quad \text{für} \quad x \geqq a_2, \text{ d. h. auf } \mathsf{P}_2.$$

Ferner werde P_2 in das „Mundstück" P_D und den Strahlrand P_S geteilt vermöge

$$\begin{array}{ll} \mathsf{P}_D: & y = \Omega(x) = Y(x) \\ \mathsf{P}_S: & y = \Omega(x) > Y(x) \end{array} \Bigg\} \quad x \geqq a_2.$$

Gebiete dieser Art heißen vom Typ $[\Gamma]^\infty$; durch Geraden $x = -a^*$ und $x = a^*$ werden aus ihnen Teilgebiete Γ^* vom Typ $[\Gamma]$ herausgeschnitten.

Von den Funktionen ψ sei verlangt, daß sie in jedem Teilgebiet Γ^* zum Typ $[\psi_1]$ gehören; ferner möge es zwei positive Schranken geben, zwischen denen die Werte der Horizontalgeschwindigkeit ψ_y im ganzen Gebiet Γ liegen, schließlich soll für sie der Variationsausdruck

$$J[\psi, \Gamma] = \iint\limits_{G_1} \{\psi_x^2 + (\psi_y - v_1)^2\}\, dx\, dy + \iint\limits_{\Gamma_2} \{\psi_x^2 + (\psi_y - v)^2\}\, dx\, dy$$

existieren. ψ sei dann als $[\psi_1]^\infty$ bezeichnet.

Funktionen $[\psi_1]^\infty$ in Gebieten $[\Gamma]^\infty$ seien zum Problem $[1]^\infty$ *zugelassen.*

Die *Randwertaufgabe* ist dann durch dieselben Bedingungen wie bei [1] mit Ausnahme von (5) gekennzeichnet.

Das *Minimumproblem* besteht in der Aufgabe, für zulässige ψ und Γ das Integral $J[\psi, \Gamma]$ möglichst klein zu machen.

Die *Sätze* 1, 2, 3 gelten auch hier; sie sind wörtlich zu übertragen.

Die *Beweise* dieser Sätze sind ebenfalls sinngemäß zu übertragen; einige Abänderungen erfordern nur die Schlüsse, die sich an die Formel (9) anschließen.

Es empfiehlt sich, vorerst das Integral $\tilde{J}$ über endliche Rechtecke G^* — durch $-a^* \leqq x \leqq a^*$, $0 \leqq \psi \leqq 1$ gekennzeichnet — zu erstrecken. Wir setzen

$$\tilde{J}^*[y] = \iint\limits_{G_1^*} \left\{\frac{y_x^2 + (1 - v_1 y_\psi)^2}{y_\psi}\right\} dx\, d\psi + \iint\limits_{G_2^*} \left\{\frac{y_x^2 + (1 - v y_\psi)^2}{y_\psi}\right\} dx\, d\psi$$

$$= \iint\limits_{G^*} \left\{\frac{y_x^2 + 1}{y_\psi} + v_2 y_\psi\right\} dx\, d\psi + (v_1^2 - v^2) \iint\limits_{G_1^*} dy\, dx$$

$$- 2 v_1 \iint\limits_{G_1^*} dx\, d\psi - 2 v \iint\limits_{G_2^*} dx\, d\psi.$$

Nun hängen die Gebiete G_1^*, G_2^* der (x, ψ)-Ebene, und auch das Gebiet G_1^* der x, y-Ebene nicht von der Wahl der Funktion $y(x, \psi)$ ab, also auch nicht die drei letzten Integrale. Die erste Variation $\delta \tilde{J}^*[y^0]$ erhält nach partieller Integration, von den Seitenlinien von G^* herrührend, das Zusatzglied

$$Z^* = \left[\int_0^1 \frac{y_x^0}{y_\psi^0} \delta y\, dx\right]_{-a^*}^{a^*},$$

das aber für eine geeignet ins unendliche wachsende Folge a^* verschwindet.

Es ist nämlich $|\delta y| = \left|\int_0^\psi \delta y_\psi\, d\psi\right| \leqq \int_0^1 |\delta y_\psi|\, d\psi$ und also

$$|Z^*| \leqq 2 \left[\int_0^1 \left\{\left(\frac{y_x^0}{y_\psi^0}\right)^2 + \delta y_\psi^2\right\} d\psi\right]_{-a^*}^{a^*};$$

nun bleibt aber — nach Annahme — y_ψ zwischen positiven Schranken und demnach folgt aus der Existenz der Integrale $\tilde{J}[y]$, $\tilde{J}[y^0]$ die Existenz von

$$\iint\limits_G \left\{\left(\frac{y_x^0}{y_\psi^0}\right)^2 + \delta y_\psi^2\right\} dx\, d\psi;$$

also gibt es eine Folge $a^* \to \infty$, für welche $|Z^*| \to 0$ strebt.

Danach folgt genau wie vorher

$$\tilde{J}[y] > \tilde{J}[y^0]$$

außer für $y = y^0$.

3. Andere Begrenzungen.

Es sei kurz angemerkt, daß an Stelle des bisher gewählten Falles der äußeren Ablösung von einer Düse auch allgemeinere Typen von Begrenzungen unserer Behandlung zugänglich sind. An Stelle einer Düse D mögen zwei Grenzen D_i, $D_ä$ vorgegeben sein.

$$D_i\colon \quad y = Y_i(x),$$

$$D_ä\colon \quad y = Y_ä(x).$$

Der Rand $\mathsf{P} : y = \Omega(x)$ möge sich von D_i nach außen, von $D_ä$ nach innen ablösen; diese Forderung ist enthalten in:

$$(1) \qquad Y_i(x) \leqq \Omega(x) \leqq Y_ä(x).$$

Überdies sei der Rand P gezwungen, sich Teilen F_i und $F_ä$ der Grenzen D_i und $D_ä$ anzulegen.

$$(1)_i \qquad \Omega(x) = Y_i(x) \quad \text{auf} \quad F_i,$$

$$(1)_ä \qquad \Omega(x) = Y_ä(x) \quad \text{auf} \quad F_ä.$$

Diejenigen Teile von D_i und $D_ä$, denen sich P ungezwungen anlegt, mögen P_i, $\mathsf{P}_ä$, der freie Rand möge P_S heißen. Dann ist der gesamte Rand P in fünf Teile zerlegt, auf denen gilt:

$$\Omega(x) = Y_i(x) \quad \text{auf} \quad F_i + \mathsf{P}_i,$$

$$\Omega(x) = Y_ä(x) \quad \text{auf} \quad F_ä + \mathsf{P}_ä,$$

$$Y_i(x) < \Omega(x) < Y_ä(x) \quad \text{auf} \quad S.$$

Randwert- und *Minimumproblem*, ebenso wie die *Sätze* 1, 2, 3, lassen sich ohne weiteres übertragen. Zu bemerken ist nur, daß an Stelle der Ungleichung $(6)_D$ die Ungleichungen

$$(6)_i \qquad (\psi_x^0)^2 + (\psi_y^0)^2 \leqq v^2 \quad \text{auf} \quad \mathsf{P}_i^0,$$

$$(6)_ä \qquad (\psi_x^0)^2 + (\psi_y^0)^2 \geqq v^2 \quad \text{auf} \quad \mathsf{P}_ä^0$$

auftreten, während für F_i und $F_ä$ keine weitere Randbedingung entspringt.

Der Fall konvexer Düse mit innerer Ablösung ist für $Y_i(x) = 0$ in diesem allgemeinen Fall mit enthalten.

4. Räumlich achsensymmetrische Strömung.

Die Minimumeigenschaft ermöglicht, auch solche Strahlprobleme anzufassen, die nicht zur Potentialgleichung mit zwei Veränderlichen führen. Das sei für die räumliche achsensymmetrische Ausflußströmung [2] gezeigt.

Es sei die Drehachse zur x-Achse gewählt; als zweite Veränderliche führen wir $y = \frac{1}{2} r^2$ an Stelle des senkrechten Abstandes r von der Drehachse ein. Das zulässige Gebiet Γ der (x, y)-Ebene, aus dem durch Drehung das Strömungsgebiet entsteht, sei vom Typ $[\Gamma]$.

Die Menge, die durch die Kreisscheibe vom Radius r hindurchfließt, sei $2\pi\,\psi$; wir fassen ψ als Funktion $\psi(x, y)$ von x und y im Gebiet Γ auf.

Zulässige ψ — $[\psi_2]$ genannt — sollen dann solche Funktionen $[\psi_1]$ sein, für welche der neue Variationsausdruck

$$J[\psi, \Gamma] = \iint_{\Gamma} \left\{\frac{1}{2y}\psi_x^2 + \psi_y^2 + v^2\right\} dx\,dy$$

existiert.

Der obige Ansatz für $J[\psi, \Gamma]$ motiviert sich daraus, daß nunmehr

$$\psi_y \quad \text{und} \quad -\frac{1}{\sqrt{2y}}\psi_x$$

Axial- und Radialkomponente der Geschwindigkeit sind.

Das *Minimumproblem* lautet: Durch Wahl zulässiger ψ und Γ ist $J[\psi, \Gamma]$ möglichst klein zu machen.

In der *Randwertaufgabe* sind die Bedingungen (6), (7) zu ersetzen durch

$$\text{(6)} \qquad \frac{1}{2y}(\psi_x^0)^2 + (\psi_y^0)^2 - v^2 \begin{cases} = 0 & \text{auf } \mathsf{P}_S^0 \\ \leqq 0 & \text{auf } \mathsf{P}_D^0. \end{cases}$$

$$\text{(7)} \qquad \frac{1}{2y}\psi_{xx}^0 + \psi_{yy}^0 \qquad = 0 \quad \text{in } \Gamma^{0\prime}.$$

Die *Sätze* 1, 2, 3 übertragen sich wörtlich.

Bei der Übertragung der *Beweise* sind neu zu behandeln nur die Randglieder bei der ersten Variation und vor allem der Nachweis, daß die zweite Variation positiv ist.

Zu dem Zweck führen wir zunächst wieder x, ψ als neue Variable, $y(x, \psi)$ als neue Funktion ein. Der Variationsausdruck geht dann über in

$$\tilde{J}[y] = \iint_{G} \left\{\frac{y_x^2}{2\,y\,y_\psi} + \frac{1}{y_\psi}\right\} dx\,d\psi + v^2 \int_{R} y\,dx.$$

Es liegt nun nahe, aus der Lösung $y^0(x, \psi)$ der Randwertaufgabe und einer anderen zulässigen Funktion $y^1(x, \psi)$ die Schar

$$y^0 + \varepsilon(y^1 - y^0)$$

zu bilden und die Variationen zu berechnen; es zeigt sich aber, daß dann der Integrand der zweiten Variation keine positiv-definite Form in δy, δy_x, δy_ψ ist. Merkwürdigerweise gelangt man hier zum Ziel, wenn man

$$\varkappa = \ln y$$

als neue Funktion einführt und diese additiv variiert. Das Integral $\check{J}$ geht dabei über in

$$\check{J}[\varkappa] = \iint\limits_G \left\{\frac{1}{2}\frac{\varkappa_x^2}{\varkappa_\psi} + \frac{e^{-\varkappa}}{\varkappa_\psi}\right\} dx\,d\psi + v^2 \int\limits_R e^{\varkappa}\,d\psi.$$

Die erste Variation wird:

$$\delta \check{J}[\varkappa] = \iint\limits_G \left\{-\frac{\varkappa_x}{\varkappa_\psi}\delta\varkappa_\psi + \frac{1}{2}\left(\frac{\varkappa_x}{\varkappa_\psi}\right)^2 \delta\varkappa_\psi - \frac{e^{-\varkappa}}{\varkappa_\psi}\delta\varkappa - \frac{e^{-\varkappa}}{\varkappa_\psi^2}\delta\varkappa_\psi\right\} dx\,d\psi$$

$$+ v^2 \int\limits_R e^{\varkappa}\,\delta\varkappa\,\delta x.$$

Die zweite Variation nimmt die Gestalt

$$\delta^2 \check{J}[\varkappa] = \iint\limits_G \left\{\begin{array}{l} \dfrac{1}{\varkappa_\psi}\left(\delta\varkappa_x^2 - 2\dfrac{\varkappa_x}{\varkappa_\psi}\delta\varkappa_x\delta\varkappa_\psi + \left(\dfrac{\varkappa_x}{\varkappa_\psi}\right)^2\delta\varkappa_\psi^2\right) \\ + \dfrac{e^{-\varkappa}}{\varkappa_\psi}\left(\delta\varkappa^2 + \dfrac{2}{\varkappa_\psi}\delta\varkappa\,\delta\varkappa_\psi + \dfrac{2}{\varkappa_\psi^2}\delta\varkappa_\psi^2\right) \end{array}\right\} dx\,d\psi$$

$$+ v^2 \int\limits_R e^{\varkappa}\,d\varkappa^2\,dx$$

an; sie ist in der Tat positiv und Null nur für $\delta\varkappa = 0$. Um nun hieraus die Ungleichung

$$\check{J}[\varkappa] > \check{J}[\varkappa^0] \qquad \text{für} \qquad \varkappa \neq \varkappa_0,$$

herzuleiten, wo $\varkappa^0$ die Lösung der Randwertaufgabe, $\varkappa = \varkappa^1$ eine beliebige zugelassene Funktion ist, müssen wir genauer folgendermaßen schließen.

Zunächst ist festzustellen, wie sich $\varkappa, \varkappa_\psi, \varkappa_x$ verhält, wenn ψ bei festem x gegen Null strebt. Wir benutzen dabei die Stetigkeit der Ableitungen ψ_x, ψ_y, die Randbedingung $\psi(x, 0) = 0$ und $\psi_y < 0$; es folgt so die Stetigkeit von $\frac{e^{-\varkappa}}{\varkappa_\psi} = \psi_y = \frac{1}{y_\psi}$, von $e^{\varkappa}\varkappa_\psi = y_\psi$ und von $\frac{\varkappa_x}{\varkappa_\psi} = -\psi_x$; ferner gleichmäßig in x $e^{\varkappa} \to 0$ und $\frac{\varkappa_x}{\varkappa_\psi} \to 0$ für $\psi \to 0$. Aus der Existenz von $\check{J}[\varkappa]$ folgt die Integrierbarkeit von $\frac{\varkappa_x^2}{\varkappa_\psi}$ und also auch von $e^{\varkappa}\varkappa_x^2$.

Eine Funktion $\varkappa$, die diese Eigenschaften besitzt, führt auch immer auf eine Funktion $y(x, \psi)$ bzw. $\psi(x, y)$, die sich auf der Achse so verhält, wie es zur Zulassung nötig ist.

Es folgt danach auch, daß die

$$\varkappa^\varepsilon = \varkappa^0 + \varepsilon\,\delta\varkappa \quad \text{mit} \quad \delta\varkappa = \varkappa^1 - \varkappa^0 \qquad [0 \leqq \varepsilon \leqq 1]$$

zulässig sind, wenn es die $\varkappa^0, \varkappa^1$ sind; es ist dann nämlich

$$y^\varepsilon = y^0 \left(\frac{y^1}{y^0}\right)^\varepsilon$$

und es bleibt y^1/y^0 ebenso wie y^0/y^1 stetig und positiv bei $\psi \to 0$; abgesehen von dem zulässigen Verhalten auf der Achse beachte man noch, daß

$$y^\varepsilon_\psi = (1-\varepsilon)\, y^0_\psi \left(\frac{y^1}{y^0}\right)^\varepsilon + \varepsilon\, y^\varepsilon_\psi \left(\frac{y^0}{y^1}\right)^{1-\varepsilon} > 0$$

ist.

Ferner folgt aus $y^1/y^0 \geqq 1$ auf P^0_D,

$$\delta\varkappa \geqq 0 \quad \text{auf} \quad \mathsf{P}^0_D.$$

Es sei nun — entsprechend dem Vorgehen im zweiten Abschnitt — aus dem Gebiet G durch die Bedingung $\tau \leqq \psi \leqq 1$ das Gebiet G_τ herausgeschnitten; nach partieller Umformung der ersten Variation entsteht auf der Geraden $\psi = \tau$ das Randglied

$$Z^\tau = \int\limits_{\psi=\tau} \left\{\frac{\varkappa^0_x}{\varkappa^0_\psi} + \frac{1}{2}\left(\frac{\varkappa^0_x}{\varkappa^0_\psi}\right)^2 - \left(\frac{e^{-\varkappa_0}}{\varkappa^0_\psi}\right)^2 e^{\varkappa^0}\right\} \delta\varkappa\, dx,$$

dessen Integrand nach oben angegebenen Eigenschaften gleichmäßig in x mit $\psi \to 0$ verschwindet. Alsdann folgert man wie früher $\delta \check{J}[\varkappa^0] \geqq 0$ und wegen $\delta^2 \check{J}[\varkappa] > 0$ auch $\check{J}[\varkappa] > \check{J}[\varkappa^0]$, außer für $\delta\varkappa = 0$.

Auch die achsensymmetrische räumliche Strömung mit *unendlicher Düse* und *unendlichem Strahl* ist als Minimumproblem anzusetzen.

Man hat Gebiete $[\Gamma]^\infty$ zuzulassen, und Funktionen ψ, die in jedem endlichen Teilgebiet Γ^* zum Typ $[\psi_2]$ gehören, für die ψ_y in Γ zwischen positiven Grenzen bleibt und für welche der Variationsausdruck

$$J[\psi, \Gamma] = \iint\limits_{G_1} \left\{\frac{1}{2y}\psi_x^2 + (\psi_y - v_1)^2\right\} dx\, dy + \iint\limits_{\Gamma_2} \left\{\frac{1}{2y}\psi_x^2 + (\psi_y - v^2\right\} dx\, dy$$

existiert. Randwertaufgabe und Sätze 1., 2., 3. übertragen sich. Bei den Beweisen hat man zunächst Integrationsgebiete Γ^* heranzuziehen und in ihnen die Umformungen wie bei [2] zu begründen, alsdann wie bei $[1]^\infty$ den Übergang zu unendlichem Gebiet zu vollziehen.

Anhang.

I. Das Minimumproblem bei beliebigem konvexen Integranden $F(\psi_x, \psi_y)$.

Anhangsweise sei auf einen allgemeinen Satz der Variationsrechnung aufmerksam gemacht, der im wesentlichen folgendes besagt:

Führt ein Variationsproblem, dessen Integrand nur von den Ableitungen der variablen Funktion von zwei Veränderlichen abhängt, bei festem Gebiet zu einem Minimum, so auch, wenn ein Teil des Randes —

auf dem für die variable Funktion ein konstanter Wert vorgeschrieben ist — (bis auf Grenzkurven) freigelassen wird.

Das Strahlproblem [1] erscheint so als ein Sonderfall; und die Minimaleigenschaft hängt nicht an dem linearen Charakter der Differentialgleichung. Offen bleibt noch, ob ein solcher Satz auch für einfache allgemeinere Klassen von Variationsproblemen gilt, deren Integranden noch die Funktion oder die unabhängigen Veränderlichen enthalten dürfen, wie beim Minimumproblem [2] des achsensymmetrischen Strahls.

Sei $F(p,q)$ eine — für alle p,q — zweimal stetig differenzierbare Funktion, für welche die Form der zweiten Variation positiv ist in den $\delta p, \delta q$ für jedes p, q.

Es sei dann das *Minimum* von

$$J[\psi,\Gamma] = \iint_{\Gamma} F(\psi_x, \psi_y)\, dx\, dy$$

gesucht, durch Wahl von Gebieten Γ und von Funktionen $\psi(x,y)$ vom Typ $[\Gamma]$ und $[\psi_1]$.

Die zugehörige *Randwertaufgabe* verlangt für zulässiges $\Gamma = \Gamma^0$ und $\psi = \psi^0$

$$(5) \qquad F_p^0 = 0 \quad \text{auf} \quad \Sigma_1, \Sigma,$$

$$(6) \qquad F^0 - F_p^0\,\psi_x^0 - F_q^0\,\psi_y^0 \begin{cases} = 0 & \text{auf } \mathsf{P}_S^0, \\ \geqq 0 & \text{auf } \mathsf{P}_D^0, \end{cases}$$

$$(7) \qquad \frac{\partial}{\partial x} F_p^0 + \frac{\partial}{\partial y} F_q^0 = 0 \quad \text{in} \quad \Gamma^0.$$

Dabei ist durch Index gekennzeichnet, daß in F, F_p, F_q als Argumente $p = \psi_x^0$ und $q = \psi_y^0$ zu wählen sind.

Es gelten dann in wörtlicher Übertragung die *Sätze* 1, 2, 3, durch die die obige vorläufige Formulierung präzisiert wird.

Bei den *Beweisen* ist nur noch neu zu begründen, daß die zweite Variation von $J[\psi,\Gamma]$ für jedes ψ positiv ist.

Das gelingt, wie beim Problem [1], wenn man $y = y(x,\psi)$ als neue variable Funktion einführt. Das Integral $J[\psi,\Gamma]$ geht dann über in das über das feste Gebiet G erstreckte Integral

$$\tilde{J}[y] = \iint_G F\left(-\frac{y_x}{y_\psi}, \frac{1}{y_\psi}\right) y_\psi\, dx\, d\psi.$$

Nun genügt es, zu zeigen, daß die Form der zweiten Variation des neuen Integranden positiv definit ist. Es handelt sich also um das folgende allgemeine Lemma über eine Funktion von zwei Veränderlichen.

Für die (zweimal stetig differenzierbare) Funktion $F(p, q)$ sei für jedes p, q die Form

$$F''[\delta p, \delta q] = F_{pp}\delta p^2 + 2F_{pq}\delta p\,\delta q + F_{qq}\delta q^2$$

positiv definit; dann ist für die Funktion

$$\Phi(\pi, \varkappa) = \varkappa F\left(-\frac{\pi}{\varkappa}, \frac{1}{\varkappa}\right)$$

ebenfalls die Form

$$\Phi''[\delta\pi, \delta\varkappa] = \Phi_{\pi\pi}\delta\pi^2 + 2\Phi_{\pi\varkappa}\delta\pi\,\delta\varkappa + \Phi_{\varkappa\varkappa}\delta\varkappa^2$$

positiv definit für $\varkappa > 0$.

In der Tat berechnet man mit $p = -\frac{\pi}{\varkappa}$, $q = \frac{1}{\varkappa}$

$$\begin{aligned}\delta\Phi &= \varkappa(F_p\delta p + F_q\delta q) + F\delta\varkappa,\\ \delta^2\Phi &= \varkappa F''[\delta p, \delta q] + F_p(\delta^2 p + 2\delta p\,\delta\varkappa)\\ &\quad + F_q(\delta^2 q + 2\delta q\,\delta\varkappa) + F\delta^2\varkappa.\end{aligned}$$

Für $\delta^2\pi = \delta^2\varkappa = 0$ gilt aber

$$\delta^2\Phi = \Phi''[\delta\pi, \delta\varkappa]$$

und

$$\begin{aligned}\delta^2 p + 2\delta p\,\delta\varkappa &= \delta^2(p\varkappa) = 0,\\ \delta^2 q + 2\delta q\,\delta\varkappa &= \delta^2(q\varkappa) = 0.\end{aligned}$$

Damit folgt

$$\Phi''[\delta\pi, \delta\varkappa] = \varkappa F''[\delta p, \delta q].$$

II. Eindeutigkeit gegenüber infinitesimaler Variation.

Unser Minimumproblem bietet auch einen neuen Zugang zur Untersuchung der infinitesimalen Eindeutigkeit der Strömungsrandwertaufgabe (II), die für Weinstein [1] das entscheidende Hilfsmittel bei der Untersuchung der Einzigkeit und Existenz der Lösung des Problems I bildet (vgl. Einleitung).

Diese infinitesimale Eindeutigkeit wurde von Weinstein [1], Hamel [1] und Weyl [2] unter verschiedenen Bedingungen für die Düsengestalt bewiesen; und zwar auf Grund der Tatsache, daß die Variation der Stromfunktion einen gewissen quadratischen Integralausdruck D zu Null macht, der andererseits mittels Abschätzungen als positiv definit erkannt wird.

Dieser Ausdruck D ist nun gerade die halbe zweite Variation unseres Variationsausdruckes J, für die Lösung genommen. Dieser Zusammenhang legt ein Verfahren nahe, den positiven Charakter auch ohne Abschätzungen zu beweisen. Und zwar ist dieses Verfahren (im wesentlichen) nichts anderes, als ein von Jacobi für solche Zwecke angegebenes Prinzip.

Zunächst seien einige Relationen über die *Strahlrandkrümmung* angemerkt.

Es sei $\psi(x, y)$ eine Funktion, welche auf einer Kurve $\mathsf{P}_S : y = \Omega(x)$ zweimal stetig differenzierbar ist und den Relationen

$$\psi = 1$$

genügt. Dann folgt aus

$$\psi_x dx + \psi_y dy = 0,$$

$$(\sqrt{\psi_x^2 + \psi_y^2})_x dx + (\sqrt{\psi_x^2 + \psi_y^2})_y dy = 0$$

die Existenz eines Faktors $\varkappa(x)$, so daß

(*) $$(\sqrt{\psi_x^2 + \psi_y^2})_x = \varkappa\, \psi_x, \quad (\sqrt{\psi_x^2 + \psi_y^2})_y = \varkappa\, \psi_y$$

wird; ferner gilt, wenn mit $\frac{\partial}{\partial n}$ Differentiation nach einer Normalen bezeichnet wird,

(**) $$\psi_x = \sqrt{\psi_x^2 + \psi_y^2}\, \frac{\partial x}{\partial n}, \quad \psi_y = \sqrt{\psi_x^2 + \psi_y^2}\, \frac{\partial y}{\partial n}$$

und hieraus

$$\frac{\partial}{\partial n} \psi_x = \varkappa\, \psi_x, \quad \frac{\partial}{\partial n} \psi_y = \varkappa\, \psi_y.$$

Wird eine beliebige Geschwindigkeitskomponente $u = \mu\, \psi_y - \nu\, \psi_x$ mit konstantem μ, ν eingeführt, so gilt

$$\frac{\partial u}{\partial n} - \varkappa\, u = 0.$$

Der Faktor $\varkappa$ ist übrigens die Krümmung von P_S.

Um nun den Begriff der *infinitesimalen Variation* zu erklären, gehe man aus von einer Schar $\psi^\alpha(x, y)$ von Lösungen der Randwertaufgabe in Gebieten Γ^α mit der Randfunktion $\Omega^\alpha(x)$; dann bilde man — den oberen Index 0 lassen wir fort —

$$\delta\psi(x, y) = \frac{\partial \psi}{\partial \alpha}(x, y)\Big|^{\alpha = 0}, \quad \delta\Omega(x, y) = \frac{\partial \Omega}{\partial \alpha}(x, y)\Big|^{\alpha = 0}.$$

Aus den Randbedingungen für $\psi^\alpha(x, y)$ erhält man durch Differentiation nach α

$$\delta\psi + \psi_y \delta\Omega = 0 \quad \text{auf} \quad \mathsf{P},$$

$$\delta\psi = 0 \quad \text{auf} \quad \mathsf{P}_D,$$

$$\psi_x \delta\psi_x + \psi_y \delta\psi_y + \tfrac{1}{2}[(\psi_x)^2 + (\psi_y)^2]_y \delta\Omega = 0 \quad \text{auf} \quad \mathsf{P}_S,$$

oder, indem man die Formeln (*), (⁑) verwendet,

$$\frac{\partial}{\partial n}\delta\psi - \varkappa\delta\psi = 0 \quad \text{auf} \quad \mathsf{P}_S.$$

Schließlich ergibt sich für $\delta\psi$ die Potentialgleichung

$$\delta\psi_{xx} + \delta\psi_{yy} = 0 \quad \text{in} \quad \Gamma'.$$

So gelangen wir zu folgender *Formulierung* des Begriffes der *infinitesimalen Variation*, bei der wir uns auf das Problem [1] beschränken.

Es sei ψ in Γ Lösung der Randwertaufgabe des Problems [1]. Überdies soll ψ auch auf der Begrenzung von Γ zweimal stetig differenzierbar sein, außer an den Ablösungsstellen, wo die zweiten Ableitungen von geringerer als erster Ordnung unendlich werden dürfen[7]). Auf dem Strahlrand P'_S sei die Krümmung $\varkappa$ erklärt durch

$$\frac{\partial}{\partial n}\psi_y - \varkappa\psi_y = 0.$$

Es sei $\delta\psi(x,y)$ in Γ eine Funktion mit folgenden Eigenschaften:

$\delta\psi$ ist stetig und stetig differenzierbar in Γ, außer an den Ablösungsstellen, wo $\delta\psi_x, \delta\psi_y$ von geringerer als erster Ordnung unendlich werden können[8]); es sei $\delta\psi$ in Γ' zweimal stetig differenzierbar und erfülle die Bedingungen

$$\begin{aligned} \delta\psi_{xx} + \delta\psi_{yy} &= 0 \quad \text{in} \quad \Gamma', \\ \delta\psi &= 0 \quad \text{auf} \quad \mathsf{P}_D, A, \\ \frac{\partial}{\partial n}\delta\psi - \varkappa\delta\psi &= 0 \quad \text{auf} \quad \mathsf{P}'_S, \\ \delta\psi_x &= 0 \quad \text{auf} \quad \Sigma_1, \Sigma. \end{aligned}$$

[$\frac{\partial}{\partial n}$ Ableitung nach der äußeren Normalen.]

Dann ist $\delta\psi$ eine infinitesimale Variation von ψ in Γ.

Der infinitesimale *Eindeutigkeitssatz* lautet

Satz 4.

$$\delta\psi = 0 \quad \text{in} \quad \Gamma.$$

Der *Beweis* geht mit Weinstein von der Tatsache aus, daß für die infinitesimale Variation $\delta\psi$

$$D[\delta\psi] = \iint\limits_{\Gamma'} \{\delta\psi_x^2 + \delta\psi_y^2\}\, dx\, dy - \int\limits_{\mathsf{P}_S} \varkappa\,\delta\psi^2\, ds = 0$$

---

7) Man darf annehmen, daß diese Bedingungen erfüllt sind, wenn Y'' stetig existiert. Übrigens benutzen wir wesentlich die Existenz der zweiten Ableitungen von ψ nur auf dem Strahlrand P_S.

8) Man könnte zeigen, daß die Ordnung $1/2$ ist, wie das bei Weyl explizite auftritt.

ist, wie aus den Bedingungen für $\delta\psi$ durch bekannte Greensche Umformung folgt; andererseits ist für jede Funktion $\delta\psi$ dieses Integral positiv und nur Null für $\delta\psi = 0$. Das Prinzip von Jacobi, das wir an Stelle der Abschätzungen von Weinstein, Weyl und Hamel zum Beweise hierfür verwenden, lautet — bis auf jeweilige Präzisierung für einzelne Fälle —:

Prinzip von Jacobi. Ein quadratisch homogener Integralausdruck erster Ordnung ist positiv definit, wenn es eine Funktion gibt, die der zugehörigen linearen Differentialgleichung und den zugehörigen natürlichen Randbedingungen genügt und die im Integrationsgebiet nicht verschwindet.

In unserem Fall ist eine solche Funktion leicht zu finden: die Horizontalgeschwindigkeit $u = \psi_y$! In der Tat gilt

$$u_{xx} + u_{yy} = 0 \text{ in } \Gamma', \qquad u_x = 0 \quad \text{auf} \quad \Sigma_1, \Sigma,$$

und auch $\frac{\partial u}{\partial n} - \varkappa u = 0$ auf P_S ist erfüllt. Das ist aber auch von vornherein zu erwarten; denn die Verschiebung der Strömung in y-Richtung stellt eine Variation dar, bei der nur die „festen Randbedingungen" $\psi = 1$ auf P_D und $\psi = 0$ auf A verletzt werden.

Der Beweis des Jacobi-Prinzips — für unseren Fall durchgeführt — verläuft so:

Man führe die Funktion

$$\delta y = -\frac{\delta\psi}{u} \qquad \text{mit} \qquad u = \psi_y$$

ein, die wegen $u > 0$ in Γ stetig und außer bei der Ablösung stetig differenzierbar ist. Alsdann wird

$$\begin{aligned}\delta\psi_x^2 + \delta\psi_y^2 = u^2\{\delta y_x^2 + \delta y_y^2\} - u(u_{xx} + u_{yy})\,\delta y^2 \\ + (u\,u_x\,\delta y^2)_x + (u\,u_y\,\delta y^2)_y.\end{aligned}$$

Nach Integration über Γ entsteht

$$\begin{aligned}D[\delta\psi] = &\iint_\Gamma u^2\{\delta y_x^2 + \delta y_y^2\}\,dx\,dy \\ &- \iint_\Gamma u(u_{xx} + u_{yy})\,\delta y^2\,dx\,dy + \int_{A+\Sigma_1+\Sigma+\mathsf{P}_D} u\frac{\partial u}{\partial n}\,\delta y^2\,ds \\ &+ \int_{\mathsf{P}_S} u\left(\frac{\partial u}{\partial n} - \varkappa u\right)\delta y^2\,ds.\end{aligned}$$

Unter Berücksichtigung von $u_{xx} + u_{yy} = 0$ in Γ', der Randbedingung für δy und u folgt

$$D[\delta\psi] = \iint_{\Gamma} u^2 \{\delta y_x^2 + \delta y_y^2\}\, dx\, dy.$$

So ist der positiv definite Charakter von $D[\delta\psi]$ in Evidenz gesetzt und damit Satz 4 bewiesen.

Die hier gewonnene Form von $D[\delta\psi]$ ermöglicht unmittelbar die Identifizierung mit der halben zweiten Variation unseres Ausdrucks $J[\psi, \Gamma]$ für die Lösung; übrigens war die Vermutung eines solchen Zusammenhanges für mich der Anlaß, diese zweite Variation zu berechnen. Wir gehen von der Form (12)

$$\delta^2 \tilde{J}[y] = 2 \iint_{G} \frac{1}{y_\psi^3} \{(y_\psi \delta y_x - y_x \delta y_\psi)^2 + \delta y_\psi^2\}\, dx\, d\psi$$

aus und führen wieder x und y statt x und ψ ein; die Ausdrücke $y_\psi \delta y_x - y_x \delta y_\psi$ und δy_ψ für die Funktion $\delta y(x, \psi)$ gehen dann in $\frac{1}{\psi_y} \delta y_x$ und $\frac{1}{\psi_y} \delta y_y$ für $\delta y(x, y)$ über; so wird

$$\delta^2 \tilde{J}[y] = 2 \iint_{\Gamma} \psi_y^2 \{\delta y_x^2 + \delta y_y^2\}\, dx\, dy.$$

Wenn also die hier auftretende Variation δy mit der infinitesimalen Variation δy identifiziert wird, so ist auch

$$\delta^2 \tilde{J}[y] = 2 D[\delta y].$$

Zum Schluß sei angemerkt, daß der Satz von der infinitesimalen Eindeutigkeit nicht nur beim unendlichen Gebiet und im räumlichen Falle entsprechend formuliert und bewiesen werden kann, sondern auch in Fällen allgemeineren Gebietes. Man wird dabei, wenn nötig, an Stelle der Horizontalgeschwindigkeit irgendeine andere Geschwindigkeitskomponente als Faktor u anzusetzen haben, die ja auch auf dem Strahlrande die Bedingung $\frac{\partial u}{\partial n} - \varkappa u = 0$ erfüllt.

Bei den Problemen ohne weitere natürliche Randbedingungen, z. B. bei denen mit unendlichem Gebiet, wird dann der Satz von der infinitesimalen Eindeutigkeit die Voraussetzung enthalten: Die Strömung besitzt eine Geschwindigkeitskomponente u, die in Γ' nicht verschwindet.

Die Voraussetzungen von Weinstein, Hamel und Weyl beziehen sich dagegen auf die Gesamtkrümmung der Düse, haben aber unsere Bedingung zur Folge; allerdings verfängt die Methode von Weyl auch noch in einigen Fällen polygonaler Begrenzung, in denen unsere Bedingung nicht erfüllt ist.

VII
Other Papers

[37–1] **On certain inequalities and characteristic value problems for analytic functions and for functions of two variables, Trans. A.M.S., 41 (1937), 321–364.**

Commentary by Peter D. Lax:

This charming paper, crisp and clear, the first one Friedrichs wrote in English, deals with analytic functions $w(z) = u + iv$ defined and square integrable in a domain D bounded by a finite number of curves with continuously turning tangent, except for a finite number of corners. The first result is that there is a constant $\theta < 1$ such that

$$\left| \int_D w^2 \, dxdy \right| \leqq \theta \int_D |w|^2 \, dxdy$$

for all analytic functions whose mean value is zero:

$$\int_D w \, dxdy = 0$$

The second result is:
When the boundary of D has no corners, the quadratic form

i) $$\int_D (u^2 - v^2) \, dxdy$$

is completely continuous with respect to the unit form

ii) $$\int_D (u^2 + v^2) \, dxdy$$

The third result is:
When the boundary of D has corners, then the essential spectrum of the quadratic form i) with respect to the form ii) is contained in the interval $(-M, M)$, where

$$M = \max \frac{\sin \alpha}{\alpha},$$

α the internal angles at the corners. This result is sharp.

Friedrichs notes that the first two results are consequences of the last.

The main step in the proof is an estimate for $\int w^2\, dxdy$ in a sufficiently thin boundary strip. Here Friedrichs uses an auxiliary function j that is approximately antianalytic in the strip i.e. satisfies there $\left|\frac{dj}{dz}\right| < \varepsilon$.

This is an early example, if not the earliest, where nonanalytic functions are used to prove a theorem about analytic functions.

The second part of the paper is about two-dimensional versions of Korn's inequality.

Reprinted from the
TRANSACTIONS OF THE AMERICAN MATHEMATICAL SOCIETY
Vol. 41, No. 3, pp. 321–364
May, 1937

ON CERTAIN INEQUALITIES AND CHARACTERISTIC VALUE PROBLEMS FOR ANALYTIC FUNCTIONS AND FOR FUNCTIONS OF TWO VARIABLES*

BY
KURT FRIEDRICHS

PART I. THE CASE OF ANALYTIC FUNCTIONS

1. INTRODUCTION

In this first part I investigate some properties of the manifold $\mathfrak{F}$ of all analytic functions $u+iv=w(z)$ defined in a bounded open connected domain D of the $(z=x+iy)$-plane for which the integral

$$\iint_D |w|^2 dxdy$$

is finite.†

First I establish the following inequality. There exists a positive constant $\theta<1$ such that, for all functions $w(z)$ which satisfy the additional condition

$$\iint wdxdy = 0,$$

the inequality

$$\left|\iint_D w^2 dxdy\right| \leqq \theta \iint_D |w|^2 dxdy \tag{1.1}$$

is valid.

It will be seen that this inequality is equivalent to

$$\iint_D u^2 dxdy \leqq \Gamma \iint_D v^2 dxdy \tag{1.2}$$

under the additional condition

$$\iint_D udxdy = 0,$$

the constant $\Gamma=(1+\theta)/(1-\theta)$ being greater than 1.

---

* Presented to the Society, October 31, 1936; received by the editors December 6, 1935, and January 27, 1936.

† The same space was investigated in regard to different properties by St. Bergman, Mathematische Annalen, vol. 86 (1922), p. 238, and Berliner Sitzungsberichte, 1927, p. 178; and S. Bochner, Mathematische Zeitschrift, vol. 14(1922), p. 180).

Secondly, I deal with the characteristic value problem for the quadratic form

$$\iint_D w^2 dxdy$$

with respect to the unit-form

$$\iint_D |w|^2 dxdy.$$

I prove the existence of a sequence of characteristic values $\mu_1, \mu_2, \mu_3, \cdots, \mu_n \downarrow 0$, and corresponding characteristic functions $w_1(z), w_2(z), \cdots$ satisfying the conditions

$$\iint_D \overline{w_m(z)} w_n(z) dxdy = \begin{cases} 1, & n = m, \\ 0, & n \neq m, \end{cases}$$

and thus being orthonormal, such that each function $w(z)$ in $\mathfrak{F}$ can be developed in a series

$$w(z) = c_1 w_1(z) + c_2 w_2(z) + c_3 w_3(z) + \cdots$$

converging uniformly in every closed subdomain of D, while the expansions

$$\iint_D |w|^2 dxdy = |c_1|^2 + |c_2|^2 + |c_3|^2 + \cdots$$

$$\iint_D w^2 dxdy = \mu_1 c_1^2 + \mu_2 c_2^2 + \mu_3 c_3^2 + \cdots$$

hold.

The largest characteristic value μ_1 is equal to 1 and the corresponding characteristic function $w_1(z)$ is constant. The inequality stated above expresses nothing but the fact that the second characteristic value μ_2 is less than 1.

The validity of these theorems depends essentially on the nature of the boundary of the domain D. My assumption is that this boundary B consists of a finite number of closed curves having a continuous tangent except at a finite number of corners. Then the inequality holds; but the expansion theorem is valid if and only if there are no corners (except internal cusps). In the case of corners the extreme points of the limit spectrum can be determined.

Let each closed curve of the boundary be represented by a continuous periodic function $z=z(s)$ of a parameter s. Except at a finite number of corners this function shall have a continuous derivative

$$\frac{dz}{ds}=\dot{z}(s),$$

which, at each corner, is continuous on both sides. We can assume

$$|\dot{z}| = \left|\frac{dz}{ds}\right| = 1,$$

so that s is the arc length of the curve, and further that the normal $i\dot{z}$ is directed into the interior of D.

At a corner the argument of $\dot{z}$ has the jump $(1-\omega)\pi$, where $\omega\pi$ is the inner angle of the corner.

We assume $0<\omega\leqq 2$. Thus we exclude external cusps ($\omega=0$); in this case it can happen that not even the inequality holds as we shall show at the end of this part.

2. A basic lemma

We set

$$\mathrm{M} = \max_{\omega}\left|\frac{\sin\omega\pi}{\omega\pi}\right|.$$

Since the case $\omega=0$ has been excluded we have $\mathrm{M}<1$, and $\mathrm{M}=0$ if there are no corners (except internal cusps).

Lemma 1. *Let ϵ be an arbitrary positive number. Then there exists a boundary strip S in D, bounded by the exterior boundary B and an interior boundary B' which consists of a finite number of rectifiable closed curves, such that for every function $w(z)=u(z)+iv(z)$ in $\mathfrak{F}$ the inequality*

$$(2.1)\qquad \left|\iint_S w^2dxdy\right| \leqq (\mathrm{M}+\epsilon)\iint_S |w|^2dxdy + \gamma\int_{B'}|w|^2|dz|$$

holds, γ being a suitable positive constant depending on ϵ. This inequality implies

$$(2.2)\quad (1-\mathrm{M}-\epsilon)\iint_S u^2dxdy \leqq (1+\mathrm{M}+\epsilon)\iint_S v^2dxdy + \gamma\int_{B'}|w|^2|dz|.$$

Without loss of generality we may confine ourselves to the case of a domain D bounded by only one closed curve B.

First we discuss the case of no corners. We may assume $\epsilon<1$. We choose a constant σ in such a way that

$$(2.3)\qquad \left|\frac{z(s')-z(s)}{s'-s}-\dot{z}(s)\right| \leqq \frac{\epsilon}{4} \quad \text{as} \quad |s'-s| \leqq \sigma.$$

We set

$$\Delta z(s) = \frac{z(s+\sigma) - z(s)}{\sigma}, \qquad \Delta \dot{z}(s) = \frac{\dot{z}(s+\sigma) - \dot{z}(s)}{\sigma}.$$

Then we have

$$(2.3)' \qquad | \Delta z(s) - \dot{z}(s) | \leqq \frac{\epsilon}{4},$$

$$\frac{3}{4} \leqq | \Delta z(s) | \leqq \frac{5}{4}.$$

We choose a number $\rho > 0$ in such a way that

$$| s' - s | \leqq \sigma \quad \text{as} \quad | z(s') - z(s) | \leqq \rho$$

and a number $T > 0$ in such a way that

$$T \leqq \frac{2}{5}\rho$$

and

$$(2.4) \qquad T \left| \frac{\Delta z(s') - \Delta z(s)}{s' - s} \right| \leqq \frac{\epsilon}{40} \quad \text{as} \quad | s' - s | \leqq \sigma.$$

Consequently we have

$$(2.4)' \qquad T | \overline{\Delta z} \Delta \dot{z} | \leqq \frac{\epsilon}{32}.$$

Now we introduce a new parameter t, $0 \leqq t \leqq T$, and set

$$z = z(s) + it\Delta z(s).$$

The strip $0 < t < T$ corresponds in a one-to-one way to a certain boundary strip S in D, bounded by the boundary B and an inner curve B' which correspond to $t = 0$ and $t = T$ respectively. To show this we first prove the relation

$$(2.5) \qquad | z' - z - [(s' - s) + i(t' - t)]\dot{z}(s) | \leqq \frac{3}{8}\epsilon | (s' - s) + i(t' - t) |,$$

as $|s' - s| \leqq \sigma$.

In fact, in view of (2.3), (2.4), we have

$$\begin{aligned} | z' - z - [(s' - s) + i(t' - t)]\dot{z}(s) | \\ = | [z(s') - z(s) - (s' - s)\dot{z}(s)] + it'[\Delta z(s') - \Delta z(s)] \\ + i(t' - t)[\Delta z(s) - \dot{z}(s)] | \end{aligned}$$

$$\leqq \frac{\epsilon}{4}|s' - s| + \frac{\epsilon}{40}|s' - s| + \frac{\epsilon}{4}|t' - t|$$

$$\leqq \left| \frac{\epsilon}{4} + \frac{\epsilon}{40} + i\frac{\epsilon}{4} \right| \left| (s' - s) + i(t' - t) \right|.$$

Now let $z' = z$. Then we have

$$\begin{aligned} | z(s') - z(s) | &= | t'\Delta z(s') - t\Delta z(s) | \\ &\leqq t' | \Delta z(s') | + t | \Delta z(s) | \\ &\leqq 2T\frac{5}{4} \leqq \rho. \end{aligned}$$

Therefore $|s'-s| \leqq \sigma$ and relation (2.5) leads to $|(s'-s)+i(t'-t)|(1-\frac{3}{8}\epsilon) \leqq 0.$ Hence $s'=s$, $t'=t$. Thus the one-to-one correspondence of $0<t<T$ and S is proved.

We calculate the Jacobian

$$J = \frac{\partial(x, y)}{\partial(s, t)} = \Re \overline{\Delta z}(\dot{z} + it\Delta\dot{z});$$

from (2.3)′, (2.4)′ we find

$$J \geqq 1 - \frac{\epsilon}{4} - \frac{\epsilon}{32} \geqq \frac{1}{2}. \tag{2.6}$$

In the strip S the parameters s and t can be expressed in terms of x, y, hence in terms of $z = x + iy$, $\bar{z} = x - iy$; and a simple calculation yields the relation

$$2J\frac{\partial}{\partial z} = \overline{\Delta z}\frac{\partial}{\partial s} - i\overline{(\dot{z} + it\Delta\dot{z})}\frac{\partial}{\partial t}.$$

We now introduce the function

$$j(z, \bar{z}) = z(s) - it\Delta z(s)$$

and find

$$J\frac{\partial j}{\partial z} = i\Im(\dot{z}\overline{\Delta z}) + t\Im(\Delta\dot{z}\overline{\Delta z}).$$

From (2.3)′, (2.4)′, (2.6) we get

$$\left| \frac{\partial j}{\partial z} \right| \leqq \epsilon.$$

By a simple calculation, we obtain the identity

$$\iint_{\tau<t<T}\left(1-\frac{\overline{\partial j}}{\partial z}\right)w^2dxdy = \iint_{\tau<t<T}\frac{\partial}{\partial \bar z}[(\overline{z-j})w^2]dxdy$$
$$= \frac{i}{2}\int_{B'}(\overline{z-j})w^2dz$$
$$-\frac{i}{2}\int_{t=\tau}(\overline{z-j})w^2dz$$

for $0<\tau<T$. Since $\int_0^T\int_{t=\tau}|w|^2dsd\tau \leqq 2\iint_S|w|^2dxdy$ is finite, there exists a sequence $\tau\to 0$ for which $\int_{t=\tau}|w|^2ds$ is bounded. If we let τ tend to zero in this way, we have

$$\left|\frac{i}{2}\int_{t=\tau}(\overline{z-j})w^2dz\right| = \left|\int_{t=\tau}t\overline{\Delta z}w^2(\dot z + it\Delta\dot z)ds\right|$$
$$\leqq \tau\left(1+\frac{\epsilon}{2}\right)\int_{t=\tau}|w|^2ds \to 0$$

and therefore the identity

$$\iint_S\left(1-\frac{\overline{\partial j}}{\partial z}\right)w^2dxdy = T\int_{B'}w^2\overline{\Delta z}dz.$$

It yields immediately the inequality

$$\left|\iint_S w^2dxdy\right| \leqq \frac{5}{4}T\int_{B'}|w|^2|dz| + \epsilon\iint_S|w|^2dxdy.$$

3. Case of corners

We now pass on to the case of boundary B having corners z_ν with the angles $\omega_\nu\pi$ ($\nu=1,\cdots,n$). We map the domain D conformally on a domain D^* of the z^*-plane such that the boundary B^* of D^* has a continuous tangent. There exists such a mapping, regular in $D+B$ except at the corners, which behaves at the corners as follows: there is an analytic function

$$\lambda = \lambda(z) \quad \text{with} \quad \lambda(z_\nu) = 0, \quad \lambda'(z_\nu) \neq 0,$$

regular in the neighborhood of $z=z_\nu$, and an analytic function

$$\lambda_* = \lambda_*(z^*) \quad \text{with} \quad \lambda_*(z_\nu^*) = 0, \quad \lambda_*'(z_\nu^*) \neq 0$$

(where z_ν^* corresponds to the corner z_ν), regular in the neighborhood of $z^*=z_\nu^*$, such that

$$\lambda = \lambda_*^{\omega_\nu}.\dagger$$

We shall make use of the fact that there is a constant $\chi > 0$ such that

$$(3.1) \qquad \left| (z^* - z_\nu^*) \frac{d}{dz^*} \log \frac{dz}{dz^*} + (1 - \omega_\nu) \right| \leqq \chi \left| z^* - z_\nu^* \right| + \chi \left| z^* - z_\nu^* \right|^{\omega_\nu}$$

in a certain neighborhood U_ν^* of z_ν^*. We see this by a simple calculation:

$$\frac{dz^*}{dz} = \frac{\lambda_*^{1-\omega_\nu}}{\omega_\nu} \frac{\lambda'}{\lambda_*'},$$

$$\lambda_* \frac{d}{d\lambda_*} \log \frac{dz^*}{dz} = (1 - \omega_\nu) - \lambda_* \frac{\lambda_*''}{(\lambda_*')^2} + \omega_\nu \lambda \frac{\lambda''}{(\lambda')^2}.$$

Since

$$\frac{z^* - z_\nu^*}{\lambda_*} : \frac{dz^*}{d\lambda_*} = 1 + (z^* - z_\nu^*) \vartheta_*(z^*),$$

where $\vartheta_*(z^*)$ is bounded at $z^* = z_\nu^*$, we get

$$(z^* - z_\nu^*) \frac{d}{dz^*} \log \frac{dz^*}{dz} - (1 - \omega_\nu) = (1 - \omega_\nu)(z^* - z_\nu^*) \vartheta_*(z^*) - (z^* - z_\nu^*) \frac{\lambda_*''}{\lambda_*'}$$

$$+ \omega_\nu (z^* - z_\nu^*)^{\omega_\nu} \left(\frac{z^* - z_\nu^*}{\lambda_*} \right)^{1-\omega_\nu} \frac{\lambda_*'}{\lambda'} \frac{\lambda''}{\lambda'}.$$

Since $\lambda_*/(z^* - z_\nu^*) \to \lambda_*'(0) \neq 0$ as $z^* \to z_\nu^*$ and $\lambda'(0) \neq 0$, there is a number $\chi > 0$ and a neighborhood U_ν^* of z_ν^*, where

$$\left| 1 - \omega_\nu \right| \left| \vartheta_* \right| + \left| \frac{\lambda_*''}{\lambda_*'} \right| \leqq \chi, \qquad \omega_\nu \left| \frac{z^* - z_\nu^*}{\lambda^*} \right|^{1-\omega_\nu} \left| \frac{\lambda_*'}{\lambda'} \right| \left| \frac{\lambda''}{\lambda'} \right| \leqq \chi$$

and, thus, (3.1) holds in U_ν^*. We introduce the function

$$\eta = \frac{dz}{dz^*} : \overline{\frac{dz}{dz^*}} \quad \text{with} \quad |\eta| = 1.$$

---

† We can construct such a mapping, e.g., in the following way. We choose a number a_1 in the exterior of $D+B$ such that the function

$$z^{(1)} = \left(\frac{z - z_1}{z - a_1} \right)^{1/\omega_1}$$

maps the domain $D+B$ in a one-to-one way on a domain D^1+B^1 of the $z^{(1)}$-plane; D^1+B^1 has no corner at the point $z^{(1)} = z_1^{(1)}$. In the same way we choose $a_2^{(1)}$ in the exterior of D^1+B^1 and form

$$z^{(2)} = \left(\frac{z^{(1)} - z_2^{(1)}}{z^{(1)} - a_2^{(1)}} \right)^{1/\omega_2},$$

and so on. Then we set

$$z^* = z^{(n)}$$

and take $\lambda_* = z^{(\nu)}$ for the corner $z = z_\nu$.

We observe that relation (3.1) is equivalent to

$$(3.2)\qquad \left|(z^* - z_\nu^*)\frac{\partial\eta}{\partial z^*} + (1-\omega_\nu)\eta\right| \leqq \chi|z^* - z_\nu^*| + \chi|z^* - z_\nu^*|^{\omega_\nu},$$

where η is considered as a function of z^* and $\overline{z^*}$. On setting

$$w^*(z^*) = w(z)\frac{dz}{dz^*}$$

we have

$$\iint_D |w|^2 dxdy = \iint_{D^*} |w^*|^2 dx^*dy^*$$

$$\iint_D w^2 dxdy = \iint_{D^*} (w^*)^2 \bar{\eta}\, dx^*dy^*.$$

4. Continuation

In what follows we omit the sign * and write z, z_ν, D, B, S, U_ν, w instead of z^*, z_ν^*, D^*, B^*, S^*, U_ν^*, w^*.

Let the boundary B be represented by $z=z(s)$, and let s_ν be that value of s for which $z(s_\nu)=z_\nu$; we set $\dot{z}(s_\nu)=\dot{z}_\nu$. We introduce the parameters s and t in the boundary strip as before (§2) and define the function $j(z,\bar{z})$.

$$z = z(s) + it\Delta z(s); \qquad j = z(s) - it\Delta z(s).$$

Let S_ν and B_ν be the domain of all points of S and B respectively for which $|s-s_\nu| \leqq \sigma$, except $z=z_\nu$. We assume $\epsilon<\frac{1}{3}$ and at the same time so small that $B_\nu+S_\nu$ is contained within the neighborhood U_ν of $z=z_\nu$. From (3.2) we get

$$(4.1)\qquad \left|(z-z_\nu)\frac{\partial\eta}{\partial z} + (1-\omega_\nu)\eta\right| \leqq \chi|z-z_\nu| + \chi|z-z_\nu|^{\omega_\nu} \quad \text{in} \quad S_\nu + B_\nu.$$

We take note of the relations

$$(4.2)\qquad \begin{aligned} |(z-z_\nu) - (s-s_\nu+it)\dot{z}_\nu| &\leqq \frac{3}{8}\epsilon|s-s_\nu+it| \\ |(j-z_\nu) - (s-s_\nu-it)\dot{z}_\nu| &\leqq \frac{3}{8}\epsilon|s-s_\nu-it| \end{aligned}$$

which holds for $|s-s_\nu| \leqq \sigma$ and, therefore, in $S_\nu+B_\nu$. They can be derived in the same way as relation (2.5).

We define the functions

$$\mathrm{K}_\nu(z,\bar{z}) = -\frac{z-z_\nu}{\dot{z}_\nu} : \frac{j-z_\nu}{\dot{z}_\nu}$$

$$H_\nu(z, \bar{z}) = -\frac{z - z_\nu}{\dot{z}_\nu} : \frac{\overline{z - z_\nu}}{\dot{z}_\nu}$$

in $S_\nu + B$; we have

(i) $|H_\nu| = 1$,

(ii) $1 - \epsilon \leqq |K_\nu| \leqq 1 + \epsilon$,

(iii) $|H_\nu - K_\nu| \leqq \epsilon$.

In view of (4.2) relation (iii) follows from

$$|H_\nu - K_\nu| = \frac{\left| \overline{\dfrac{z - z_\nu}{\dot{z}_\nu}} - \dfrac{j - z_\nu}{\dot{z}_\nu} \right|}{\left| \dfrac{j - z}{\dot{z}_\nu} \right|} \leqq \frac{\dfrac{3}{4}\epsilon}{1 - \dfrac{3}{8}\epsilon} \leqq \epsilon.$$

The relations (iii) and (i) lead to (ii).

The function K_ν has the value 1 on the lines $s = s_\nu$, $0 < t < T$, and the value -1 on the boundary $t = 0$, $s \neq s_\nu$. Therefore $\log K_\nu$ is defined; we have

$$\log K_\nu = 0 \quad \text{on} \quad s = s_\nu, \quad 0 < t < T;$$
$$\log K_\nu = \mp i\pi \quad \text{on} \quad t = 0, \quad s \lessgtr s_\nu.$$

In view of (4.2) we get

$$|\log K_\nu| \leqq \frac{\dfrac{3}{4}\epsilon}{1 - \dfrac{3}{8}\epsilon} + \left| \log \frac{t + i(s - s_\nu)}{t - i(s - s_\nu)} \right| \leqq \epsilon + \pi.$$

We can define $\log H_\nu$ in such a way that $|\log H_\nu| \leqq 3\epsilon/2$ on the line $s = s_\nu$ and we set

$$K_\nu^{\omega_\nu} = \exp \omega_\nu \log K_\nu, \qquad H_\nu^{\omega_\nu} = \exp \log H_\nu.$$

From (iii) we obtain

(iv) $|K_\nu^{\omega_\nu} - H_\nu^{\omega_\nu}| \leqq 3\epsilon$.

We consider the function

$$P_\nu(z, \bar{z}) = \frac{1}{\omega_\nu} K_\nu^{\omega_\nu} - \frac{\cos \omega_\nu \pi}{\omega_\nu} - \frac{\sin \omega_\nu \pi}{\omega_\nu \pi} \log K_\nu,$$

defined in $S_\nu + B_\nu$; it vanishes on the boundary $t = 0$, $s \neq s_\nu$, and it is bounded. Therefore we have

$$(4.3) \qquad \begin{aligned} (z - z_\nu) P_\nu(z, \bar{z}) &\to 0 \\ (z - z_\nu)^{\omega_\nu} P_\nu(z, \bar{z}) &\to 0 \end{aligned} \qquad \text{as } t \to 0$$

uniformly in s. We calculate the derivative of P_ν with respect to z and find

$$(z - z_\nu)\frac{\partial P_\nu}{\partial z} = \left[K_\nu^{\omega_\nu} - \frac{\sin \omega_\nu \pi}{\omega_\nu \pi}\right]\left(1 + K_\nu \frac{\partial j}{\partial z}\right).$$

According to (ii), (iv) and

$$\left| \frac{\partial j}{\partial z} \right| \leqq \epsilon \tag{4.4}$$

we get

$$\left| (z - z_\nu)\frac{\partial P_\nu}{\partial z} - H_\nu^{\omega_\nu} \right| \leqq \left| \frac{\sin \omega_\nu \pi}{\omega_\nu \pi} \right| + 6\epsilon. \tag{4.5}$$

We choose non-negative functions $\rho_\nu(s)$ which have continuous derivatives, which vanish outside of B_ν, which are equal to 1 for $|s - s_\nu| \leqq \sigma/2$ and for which

$$\sum_\nu \rho_\nu(s) \leqq 1.$$

Here the summation $\sum_\nu$ is extended over all points $z = z_\nu$. We consider the function

$$\Omega(z, \bar{z}) = \sum_\nu \rho_\nu(s) H_\nu^{-\omega_\nu}(z - z_\nu)\eta P_\nu + \left[1 - \sum_\nu \rho_\nu(s)\right]\eta(z - j).$$

Since

$$\rho_\nu(s)(z - z_\nu)P_\nu \to 0, \qquad z - j \to 0 \quad \text{as} \quad t \to 0, \tag{4.6}$$

we have

$$\Omega(z, \bar{z}) \to 0 \quad \text{as} \quad t \to 0.$$

We investigate the derivative $\partial\Omega/\partial z$. First we observe that $\partial\rho_\nu/\partial z = \rho_\nu'(\partial s/\partial z)$ is bounded in S and that $\partial\eta/\partial z$ is bounded in $S - \sum_\nu S_\nu$. Further we calculate

$$\frac{\partial}{\partial z} H_\nu^{-\omega_\nu}(z - z_\nu)\eta = (1 - \omega_\nu)H_\nu^{-\omega_\nu}\eta + H_\nu^{-\omega_\nu}(z - z_\nu)\frac{\partial \eta}{\partial z}.$$

We now use relation (4.1); because of $|H_\nu| = 1$ we get

$$\left| \frac{\partial}{\partial z} H_\nu^{-\omega_\nu}(z - z_\nu)\eta \right| \leqq \chi |z - z_\nu| + \chi |z - z_\nu|^{\omega_\nu}$$

in S_ν. Thus, on account of (4.3) and (4.6) we find

$$\sum_\nu P_\nu \frac{\partial}{\partial z} \rho_\nu H_\nu^{-\omega_\nu}(z - z_\nu)\eta + (z - j)\frac{\partial}{\partial z}\left[1 - \sum_\nu \rho_\nu\right]\eta \to 0 \tag{4.7}$$

as $t \to 0$, uniformly in s.

Consequently we can choose a positive number $T_1 \leqq T$ such that for $0<t\leqq T_1$

$$\left| \frac{\partial \Omega}{\partial z} - \sum_\nu \rho_\nu H_\nu^{-\omega_\nu}(z - z_\nu)\eta \frac{\partial P_\nu}{\partial z} - \left[1 - \sum_\nu \rho_\nu\right]\eta\left(1 - \frac{\partial j}{\partial z}\right)\right| \leqq \epsilon .$$

According to (4.4) and (4.5) this relation leads to

$$\left| \frac{\partial \Omega}{\partial z} - \eta \right| \leqq \sum_\nu \rho_\nu \left| \frac{\sin \omega_\nu \pi}{\omega_\nu \pi} \right| + 7\epsilon \leqq M + \epsilon_1, \tag{4.8}$$

where $\epsilon_1 = 7\epsilon$.

Let S_1 be the strip $0<t<T_1$ and B_1' the curve $t=T_1$; since $\Omega \to 0$ as $t\to 0$ we have the identity

$$\iint_{S_1} w^2 \frac{\partial \bar{\Omega}}{\partial z} dxdy = \frac{i}{2}\int_{B_1'} w^2 \,\bar{\Omega} dz.$$

From this and relation (4.8) we deduce

$$\iint_{S_1} w^2 \bar{\eta} dxdy \leqq \gamma_1 \int_{B_1'} |w|^2 |dz| + (M+\epsilon_1) \iint_{S_1} |w|^2 dxdy$$

where $\gamma_1 = 2 \max |\Omega|$ on B_1'. Since ϵ_1 is arbitrarily small we thus have proved Lemma 1 also for the case where the boundary has corners.

5. Additional lemmas

In proving our theorems we further make use of the following elementary lemmas.

Lemma 2. *Let D' be a closed domain within D. Then for all functions $w(z)$ in $\mathfrak{F}$ and all z' in D' we have*

$$|w(z')|^2 \leqq C_0' \iint_D |w|^2 dxdy \tag{5.1}$$

$$\left| \frac{dw}{dz}(z') \right|^2 \leqq C_1' \iint_D |w|^2 dxdy, \tag{5.2}$$

C_0', C_1' being positive constants depending on D'.

We omit the proof of this well known lemma, which is an immediate consequence of the mean-value theorem and the Schwarz inequality.

Lemma 3. *Let D' be a closed subdomain within D and z_0 a point of D'. Then for all functions $w(z) = u(z) + iv(z)$ for which*

$$u(z_0) = 0$$

the inequality

$$| w(z') |^2 \leqq C' \iint_D v^2 dxdy, \qquad z' \text{ in } D',$$

is valid, C' being a positive constant depending on D' and z_0.

We choose a finite number of points $z_0, z_1, z_2, \cdots, z_j, \cdots, z_k$ in D' and a positive number R such that $|z_j - z_{j-1}| < R$, that every point z' of D' belongs to one of the circles $|z - z_j| < R$ and that the circles $|z - z_j| < 3R$ are in the interior of D. Now we take the relation

$$w(z') - u(z_j) = \frac{1}{2\pi} \int_{|z-z_j|=r} v(z) \frac{z + z' - 2z_j}{z - z'} \frac{dz}{z - z_j}$$

which holds for $|z' - z_j| < r$. We multiply by rdr and integrate with respect to r from $r = 2R$ to $r = 3R$ and apply the Schwarz inequality assuming $|z' - z_j| < R$. Thus we obtain

$$| w(z') - u(z_j) | \leqq 4 \left(\frac{1}{5R^2\pi} \iint_{2R < |z - z_j| < 3R} v^2 dxdy \right)^{1/2} \leqq \frac{2}{R} \left(\iint_D v^2 dxdy \right)^{1/2}.$$

Since $u(z_0) = 0$ and $|z_j - z_{j-1}| < R$, we get for every point z' of D'

$$| w(z') | \leqq (k + 1) \frac{2}{R} \left(\iint_D v^2 dxdy \right)^{1/2}.$$

Herewith we have proved Lemma 3.

6. Fundamental inequalities

We now are ready for the proof of

THEOREM 1. *There exists a positive number Γ such that the inequality*

$$\iint_D u^2 dxdy \leqq \Gamma \iint_D v^2 dxdy$$

holds for all functions $w = u + iv$ in $\mathfrak{F}$ which satisfy the condition

$$\iint_D u dxdy = 0.$$

We first observe that it is sufficient to prove the inequality for all functions $w = u + iv$ in $\mathfrak{F}$ which vanish at a certain point z_0 in D; this fact follows from the relation

$$\iint_D u^2(z)dxdy \leqq \iint_D [u(z) - u(z_0)]^2 dxdy,$$

which is an immediate consequence of $\iint_D u dxdy = 0$. Now we choose a positive number ϵ such that $M+\epsilon<1$; this is possible since $M<1$. By (2.2) of Lemma 1 we have

$$(1 - M - \epsilon)\iint_D u^2 dxdy \leqq \iint_{D-S} u^2 dxdy + (1 + M + \epsilon)\iint_S v^2 dxdy$$
$$+ \gamma \int_{B'} |w|^2 |dz|.$$

We choose a point z_0 of $D-S$ and assume $u(z_0)=0$. Then we apply Lemma 3 to the closed domain $D'=D-S$ and get

$$(1 - M - \epsilon)\iint_D u^2 dxdy$$
$$\leqq \left[C'\iint_{D-S} dxdy + (2 + \epsilon) + C'\int_{B'} |dz|\right]\iint_D v^2 dxdy;$$

thus Theorem 1 is proved.

The inequality of Theorem 1 is equivalent to

$$(\Gamma + 1)\Re\iint_D w^2 dxdy \leqq (\Gamma - 1)\iint_D |w|^2 dxdy,$$

and the additional condition is the same as

$$\Re\iint wdxdy = 0.$$

Now let w be a function in $\mathfrak{F}$ which satisfies the relation

$$\iint_D wdxdy = 0;$$

then we can choose a number η of absolute value 1 such that

$$\Re\iint_D (\eta w)^2 dxdy = \left|\iint_D w^2 dxdy\right|;$$

since $\Re\iint_D \eta w dxdy = 0$ we can apply Theorem 1 to ηw. On setting $\theta = (\Gamma-1)/(\Gamma+1)$ we thus obtain

THEOREM 2. *There exists a constant* $\theta<1$ *such that the inequality*

$$\left|\iint_D w^2 dxdy\right| \leqq \theta \iint_D |w|^2 dxdy$$

holds for all functions in $\mathfrak{F}$ *satisfying the condition*

$$\iint_D w dxdy = 0.$$

7. SPACE $\mathfrak{F}$ AS A HILBERT SPACE

To prove our expansion theorem we start with the observation that the manifold $\mathfrak{F}$ of all functions $w(z)$ for which

$$\iint_D |w|^2 dxdy$$

is finite constitutes a linear metric space. This space is either complex or real depending on whether we allow multiplication by complex or real numbers and define an inner product either by

$$(w_1, w_2) = \iint_D \bar{w}_1 w_2 dxdy$$

or by

$$(w_1, w_2) = \Re \iint_D \bar{w}_1 w_2 dxdy.$$

In either case the modulus

$$(w, w)^{1/2} = \left(\iint_D |w|^2 dxdy\right)^{1/2}$$

has the same value. In order that

$$\Re \iint_D w_1 w_2 dxdy$$

become a symmetric bilinear form we choose the second definition of (w_1, w_2) and thus assume $\mathfrak{F}$ to be a real space. In this case, iw belongs to the space $\mathfrak{F}$ together with w_1; but this function iw is orthogonal to w:

$$(w, iw) = \Re \iint_D \bar{w} iw dxdy = \Re i \iint_D |w|^2 dxdy = 0.$$

Now we establish

LEMMA 4. *The space* $\mathfrak{F}$ *of functions* $w(z)$ *is a Hilbert space.*

First we prove that $\mathfrak{F}$ is complete. Let $w^n(z)$ be a sequence in $\mathfrak{F}$ such that

$$(w^m - w^n, w^m - w^n) = \iint_D |w^m - w^n|^2 dxdy \to 0 \quad \text{as} \quad m, n \to \infty.$$

From the inequality (1) of Lemma 2 we deduce that $w^m(z) - w^n(z)$ converges to zero uniformly in every closed subdomain D' of D. From this fact it follows by a well known procedure that (i) $\iint_D |w|^2 dxdy < \infty$ and therefore $w(z)$ belongs to the space $\mathfrak{F}$; and (ii)

$$\iint_D |w^n - w|^2 dxdy \to 0 \quad \text{as} \quad n \to \infty.$$

$\mathfrak{F}$ is a subspace of the linear space† of all complex-valued functions $k(x, y)$, continuous in D, for which

$$(k \mid k) = \iint_D |k|^2 dxdy < \infty;$$

since, obviously, this space is separable $\mathfrak{F}$ has a like property. Therefore $\mathfrak{F}$ is a Hilbert space.

A sequence of functions $w^1(z), w^2(z), \cdots, w^n(z), \cdots$ in $\mathfrak{F}$ is called *weakly convergent* to a function $w(z)$ in $\mathfrak{F}$,

$$w^n(z) \dot{\to} w(z),$$

if it has the following two properties:

(i) there is a constant $M_0 > 0$ such that

$$\iint_D |w^n|^2 dxdy \leqq M_0;$$

(ii)
$$\iint_D \overline{W}(w^n - w) dxdy \to 0 \quad \text{as} \quad n \to \infty$$

for each function $W(z)$ in $\mathfrak{F}$.

We have the

LEMMA 5. *Let $w^1(z), w^2(z), \cdots, w^n(z)$ be a sequence of functions in $\mathfrak{F}$ for which*

$$\iint_D |w^n|^2 dxdy \leqq M_0$$

and which converges to a function $w_0(z)$ uniformly in the interior. Then $w_0(z)$ belongs to $\mathfrak{F}$ and $w^n(z)$ converges weakly to $w_0(z)$.

† Cf. Part II, §2.

1. From $\iint_{D'}|w^n|^2dxdy\leqq M_0$ we get $\iint_{D'}|w_0|^2dxdy\leqq M_0$ and therefore $w_0(z)$ belongs to $\mathfrak{F}$.

2. Let $W(z)$ be a function of $\mathfrak{F}$. Without loss of generality, we may assume $w_0(z)=0$. To a given number $\epsilon>0$ we choose a subdomain D' such that $\iint_{D-D'}|W|^2dxdy\leqq\epsilon^2$ and a number n_ϵ such that $|w^n(z)|\leqq\epsilon$ in D' for $n\geqq n_\epsilon$. Then we have for $n\geqq n_\epsilon$

$$\left|\iint_D \overline{W}w^n dxdy\right| \leqq \epsilon^2\left[\iint_D dxdy\iint_D |W|^2dxdy + M\right]$$

and therefore $\iint_D\overline{W}w^ndxdy\rightarrow 0$ as $n\rightarrow\infty$.

Lemma 6. *A sequence $\mathfrak{S}$ of functions $w^1(z)$, $w^2(z)$, $\cdots$ in $\mathfrak{F}$ which converges weakly to zero converges to zero uniformly in the interior of D.*

The sequence $\mathfrak{S}$ is equicontinuous in the interior of D; this fact follows from (2) of Lemma 2 and $\iint_D|w^n|^2dxdy\leqq M_0$. Therefore it is sufficient to show that the sequence $\mathfrak{S}$ converges to zero at every point of D, the uniformity being a consequence of the equicontinuity.

Let z_0 be a point of D; let $\mathfrak{S}'$ be a subsequence of $\mathfrak{S}$ which at z_0 converges to a certain value w. Since $\mathfrak{S}'$ is also equicontinuous, it contains a subsequence $\mathfrak{S}''$ which converges uniformly in the interior to a certain function $w_0(z)$. According to Lemma 5, $w_0(z)$ belongs to $\mathfrak{F}$ and $\mathfrak{S}''$ converges weakly to $w_0(z)$. As $\mathfrak{S}''$ also converges weakly to zero, we have $w_0(z)=0$. Therefore $w=w_0(z_0)=0$. Thus $\mathfrak{S}$ converges to zero in every point of D and Lemma 6 is proved.

8. A characteristic value problem

In the present paragraph we discuss the form

$$w_1Vw_2 = \Re\iint_D w_1w_2dxdy$$

where $w_1(z)$ and $w_2(z)$ are functions in $\mathfrak{F}$; we observe that this form is
(i) symmetric,

$$w_1Vw_2 = w_2Vw_1;$$

(ii) bilinear,

$$wV(a_1w_1 + a_2w_2) = a_1(wVw_1) + a_2(wVw_2), \quad (a_1, a_2 \text{ real});$$

(iii) bounded,

$$|wVw| \leqq (w, w).$$

Such a form is called *completely continuous*, if, as $n\to\infty$, $w^n V w^n \to 0$ for every sequence $w^n(z)$ of functions in $\mathfrak{F}$ which converges weakly to zero.

An immediate consequence of Lemmas 4 and 5 is

THEOREM 3. *The form $\Re\iint_D w^2 dxdy$ is completely continuous if the boundary has no corners except internal cusps* ($\omega=2$).

Let $w^n(z)$ be a sequence which converges weakly to zero. To prove the theorem it is sufficient to show that for every number $\epsilon>0$ we can determine a number n_ϵ such that $|\Re\iint_D (w^n)^2 dxdy| \leqq 3M_0\epsilon$ for $n\geqq n_\epsilon$. To do this, we refer to Lemma 1 and the notation there used. Let S be the boundary strip of Lemma 1 corresponding to the given number ϵ, let B' be the inner boundary of S. Then, according to Lemma 6, there is a number n_ϵ such that

$$\iint_{D-S} |w^n|^2 dxdy \leqq \epsilon M_0 \quad \text{and} \quad \gamma\int_{B'} |w^n|^2 |dz| \leqq \epsilon M_0 \quad \text{for} \quad n \geqq n_\epsilon.$$

Now, if we note that M$=0$, in (2.1) of Lemma 1, we have

$$\left|\Re\iint (w^n)^2 dxdy\right| \leqq \left|\iint_{D-S} (w^n)^2 dxdy\right| + \left|\iint_S (w^n)^2 dxdy\right|$$
$$\leqq \epsilon M_0 + 2\epsilon M_0 = 3\epsilon M_0.$$

Now we can apply the general theory of completely continuous forms in Hilbert spaces. This theory shows that there exist two sequences of characteristic values $\mu_1\geqq\mu_2\geqq\mu_3\geqq\cdots\to 0$ and $\mu_{-1}\leqq\mu_{-2}\leqq\cdots\to 0$ and of corresponding characteristic functions $w_1(z), w_2(z), w_3(z), \cdots, w_{-1}(z), w_{-2}(z), \cdots$ satisfying the relations

$$(w_m, w_n) = \begin{cases} 1, & m = n, \\ 0, & m \neq n, \end{cases}$$

with the following properties: Let $w(z)$ be an arbitrary function of $\mathfrak{F}$. On setting

$$a_n = (w_n, w), \qquad n = \pm 1, \pm 2, \pm 3, \cdots$$

we have the relation

$$\mu_n a_n = (w_n V w)$$

and the developments

$$(w, w) = a_1^2 + a_2^2 + \cdots + a_{-1}^2 + a_{-2}^2 + \cdots$$
$$(wVw) = \mu_1 a_1^2 + \mu_2 a_2^2 + \cdots + \mu_{-1}a_{-1}^2 + \mu_{-2}a_{-2}^2 + \cdots.$$

If for any value μ_* a function $w_*(z)=0$ in $\mathfrak{F}$ satisfies the characteristic relation $wVw_* = \mu_*(w, w_*)$ for all w of $\mathfrak{F}$, then μ_* is contained among the values μ_n and the function $w_*(z)$ is a linear combination of the characteristic functions belonging to all $\mu_n = \mu_*$.

We observe that, simultaneously with μ_n and $w_n(z)$, $-\mu_n$ and $iw_n(z)$ also satisfy the characteristic relation. For that reason we can set

$$\mu_{-n} = -\mu_n \quad \text{and} \quad w_{-n}(z) = iw_n(z).$$

Then

$$\iint_D \overline{w_n(z)} w_m(z) dxdy = \begin{cases} 1, & n = m, \\ 0, & n \neq m, \end{cases}$$

and, on setting

$$c_n = a_n - ia_{-n} = \iint_D \overline{w_n(z)} w(z) dxdy,$$

we obtain

$$\iint_D |w|^2 dxdy = |c_1|^2 + |c_2|^2 + \cdots$$

$$\Re \iint_D w^2 dxdy = \Re[\mu_1 c_1^2 + \mu_2 c_2^2 + \cdots].$$

If we take $e^{i\pi/4} w(z)$ instead of $w(z)$, we get

$$\Im \iint_D w^2 dxdy = \Im[\mu_1 c_1^2 + \mu_2 c_2^2 + \cdots].$$

We further observe the relation

$$\iint_D |w - c_1 w_1 - c_2 w_2 - \cdots - c_n w_n|^2 dxdy$$

$$= \iint_D |w|^2 dxdy - |c_1|^2 - |c_2|^2 - \cdots - |c_n|^2 \to 0 \text{ and } n \to \infty,$$

and we deduce from it, according to Lemma 2, that $c_1w_1(z) + c_2w_2(z) + \cdots$ converges to $w(z)$ uniformly in the interior. We may record these results by formulating

THEOREM 4. *If the boundary B has no corners except internal cusps, then there is a sequence of non-negative characteristic values*

$$\mu_1 \geqq \mu_2 \geqq \mu_3 \geqq \cdots \to 0$$

and a sequence of characteristic functions in $\mathfrak{F}$

$$w_1(z),\ w_2(z),\ \cdots$$

orthonormal in the sense

$$\iint_D \bar{w}_n w_m dxdy = \begin{cases} 1, & n = m, \\ 0, & n \neq m, \end{cases}$$

which satisfy the characteristic relation

$$\iint_D w_n w dxdy = \mu_n \iint_D \bar{w}_n w dxdy$$

for every function $w(z)$ *in* $\mathfrak{F}$. *For each function* $w(z)$ *in* $\mathfrak{F}$ *we have an expansion*

$$w(z) = c_1 w_1(z) + c_2 w_2(z) + \cdots$$

converging uniformly in the interior of D *and for the corresponding quadratic forms*

$$\iint_D |w|^2 dxdy = |c_1|^2 + |c_2|^2 + |c_3|^2 + \cdots,$$

$$\iint_D w^2 dxdy = \mu_1 c_1^2 + \mu_2 c_2^2 + \mu_3 c_3^2 + \cdots,$$

where

$$c_n = \iint_D \bar{w}_n w dxdy.$$

Since

$$\left|\iint_D w^2 dxdy\right| \leqq \iint_D |w|^2 dxdy$$

and the equality

$$\iint_D w^2 dxdy = \iint_D |w|^2 dxdy$$

actually holds if and only if w is a real constant, we may state

REMARK 1. *The first characteristic value* $\mu_1 = 1$, *and the first characteristic function* $w(z)$ *is a real constant.*

REMARK 2. *The second characteristic value is less than* 1, $\mu_2 < 1$.

This inequality and the developments of $\iint_D |w|^2 dxdy$ and $\iint_D w^2 dxdy$ confirm Theorem 2, since relation $\iint_D w dxdy = 0$ is equivalent to $c_1 = 0$ and $\mu_n \leqq \mu_2 = \theta$ for $n = 2, 3, 4, \cdots$.

9. CASES OF CIRCLE AND ELLIPSE

For the circle and the ellipse the characteristic values and functions can be given explicitly.

THEOREM 5. *If D is the circle $|z|<1$, then $\mu_2=\mu_3=\cdots=0$.*

For every analytic function $W(z)$ which is regular in $|z|<1$ the mean value theorem gives

$$\iint_D W(z)dxdy = W(0).$$

Since together with $w(z)$ also $w^2(z)$ is analytic in D, we have

$$\iint_D wdxdy = w(0),$$

$$\iint_D w^2dxdy = w^2(0).$$

Consequently $c_1=0$ is equivalent to $w(0)=0$ and for all functions with $c_1=0$ we have

$$\iint_D w^2dxdy = 0.$$

THEOREM 6. *Let D be the ellipse*

$$\frac{x^2}{\cosh^2\sigma}+\frac{y^2}{\sinh^2\sigma}<1,$$

then

$$\mu_n = \frac{n\sinh 2\sigma}{\sinh 2n\sigma}, \qquad n = 1, 2, 3, \cdots$$

and the characteristic functions are the derivatives of the Tchebycheff polynomials,

$$w_n(z) = \rho_n\frac{\sinh(n\operatorname{arc}\cosh z)}{\sinh(\operatorname{arc}\cosh z)} = \rho_n\frac{2^{n-1}}{n}T_n'(z),$$

where

$$\rho_n = \left(\frac{2n}{\pi\sinh 2n\sigma}\right)^{1/2}.$$

These functions $w_1(z)$, $w_2(z)$, $w_3(z)$, $\cdots$ are polynomials of degrees $0, 1, 2, \cdots$. The set of such polynomials is complete in the sense that every function $w(z)$ in $\mathfrak{F}$ can be approximated by a polynomial

$$p_n(z) = c_1^n w_1(z) + c_2^n w_2(z) + \cdots + c_n^n w_n(z),$$

so that

$$\iint_D |w(z) - p_n(z)|^2 dxdy$$

becomes arbitrarily small.† It remains to prove the orthogonality relation and the characteristic relation. To simplify the calculations we transform the ellipse D into a rectangular domain Δ of the $(\zeta = \xi + i\eta)$-plane. We set $z = \cosh \zeta$ and Δ: $0 \leqq \xi < \sigma$, $-\pi < \eta < \pi$. Then we have

$$\iint_D \bar{w}_n w_m dxdy = \rho_n \rho_m \iint_\Delta \cosh n\zeta \cosh m\zeta d\xi d\eta = 0,$$

if $n \neq m$. Also

$$\begin{aligned} \iint_D |w_n|^2 dxdy &= \rho_n^2 \iint_\Delta |\cosh n\zeta|^2 d\xi d\eta \\ &= \rho_n^2 \int_{-\pi}^{\pi}\int_0^\sigma \frac{1}{2} \cosh 2n\xi d\xi d\eta \\ &= \rho_n^2 \pi \frac{1}{2n} \sinh 2n\sigma = 1, \end{aligned}$$

$$\begin{aligned} \iint_D w_n w_m dxdy &= \rho_n \rho_m \iint_\Delta \frac{\sinh n\zeta \sinh m\zeta}{\sinh \zeta} \overline{\sinh \zeta} d\xi d\eta \\ &= \sum_\alpha \iint_\Delta \sinh \alpha\zeta \, \overline{\sinh \zeta} d\xi d\eta, \end{aligned}$$

where α runs through

$$\alpha = m - n + 1,\ m - n + 3, \cdots, m + n - 1, \quad \text{if} \quad m \geqq n.$$

The integral $\iint_\Delta \sinh \alpha\zeta \, \overline{\sinh \zeta} \, d\xi d\eta$ vanishes except when $\alpha = 1$, in which case its value is ρ_1^{-2}; this case occurs only, if $m = n$; therefore we have

$$\iint_D w_n w_m dxdy = \begin{cases} 0, & n \neq m, \\ \rho_n^2/\rho_1^2 = \mu_n, & n = m. \end{cases}$$

10. General case of boundary with corners

Since the form $\Re \iint_D w^2 dxdy$ is symmetric and bounded, the general theory of spectra‡ is applicable also when this form is not completely continuous,

---

† L. Bieberbach, *Zur Theorie und Praxis der konformen Abbildung*, Rendiconti del Circolo Matematico di Palermo, vol. 38 (1914), p. 98.

T. Carleman, *Über die Approximation analytischer Funktionen durch lineare Aggregate von vorgegebenen Potenzen*, Arkiv för Matematik, Astronomi och Fysik, vol. 17, No. 9 (1922).

For a more detailed reference, see J. L. Walsh, *Approximation by Polynomials in the Complex Domain*, Mémorial des Sciences Mathématiques, No. 73, 1935, p. 61.

‡ Cf. M. H. Stone, *Linear Transformations in Hilbert Space and their Applications to Analysis*, American Mathematical Society Colloquium Publications, vol. 15, 1932.

in which case it has a continuous spectrum. We use the notion of limit spectrum (Häufungsspektrum) in the sense of Weyl;† this closed set of values μ consists of all points of the continuous spectrum, of the limit points of the characteristic values and of all characteristic values of infinite order. In every closed interval outside of the limit spectrum there are only a finite number of characteristic values and these have a finite order. The limit spectrum consists of the single point $\mu=0$ if and only if the form is completely continuous. Now we can prove

THEOREM 7. *The l.u.b.* $\bar{\mu}$ *and the g.l.b.* $\underline{\mu}$ *of the limit spectrum of the form* $\Re\iint_D w^2dxdy$ *are precisely* M *and* $-$M *respectively. Therefore Theorem* 4 *is valid, if and only if* $\mathrm{M}=0$; *that is to say, if the boundary has no corner except interior cusps.*

First we show that the limit spectrum is contained in the interval $-\mathrm{M}\leqq\mu\leqq\mathrm{M}$. From Lemma 1 we derive the inequality

$$\Re\iint_D w^2dxdy-\iint_{D-S}|w|^2dxdy-\gamma\int_{B'}|w|^2|dz| \leqq (\mathrm{M}+\epsilon)\iint_D|w|^2dxdy$$

which shows that the l.u.b. of the whole spectrum and therefore also the l.u.b. of the limit spectrum of the form of the left-hand member is not greater than $\mathrm{M}+\epsilon$. Now we refer to the fundamental theorem of Weyl‡ to the effect that the limit spectrum of a form remains unaltered whenever the form is changed by adding a completely continuous form. Now, according to Lemma 6, the forms $\iint_{D-S}|w|^2dxdy$ and $\gamma\int_{B'}|w|^2|dz|$ are completely continuous and therefore the l.u.b. of the limit spectrum of $\Re\iint_D w^2dxdy$ is not greater than M, since ϵ is arbitrarily small. The same reasoning shows that the g.l.b. is not less than $-$M. Thus we have proved that $-\mathrm{M}\leqq\underline{\mu}\leqq\bar{\mu}\leqq\mathrm{M}$.

We now prove that $\bar{\mu}\geqq\mathrm{M}$. According to a remark of H. Weyl‡ this value has the following property: Whenever there is a sequence of functions $w(z)$ in $\mathfrak{F}$ weakly convergent to zero for which

$$\Re\iint_D w^2dxdy:\iint_D|w|^2dxdy\rightarrow\mu,$$

then

$$\mu\leqq\bar{\mu}.$$

† H. Weyl, *Über beschränkte quadratische Formen, deren Differenz vollstetig ist*, Rendiconti del Circolo Matematico di Palermo, vol. 27 (1909), pp. 373–392.

‡ Loc. cit.

Let $z=0$ be a corner of the boundary B with an angle ω and with $|\sin \omega\pi/\omega\pi| = \mathrm{M}$. We set $z=re^{i\vartheta}$. We represent the two branches of B in the neighborhood of $z=0$ by $\vartheta=\vartheta_+(r)$ and $\vartheta=\vartheta_-(r)$ and choose a number $R>0$ such that the domain

$$D_R: \quad \vartheta_-(r) < \vartheta < \vartheta_+(r), \qquad 0 < r < R,$$

is contained in D.

We may assume $\vartheta_+(0)=\omega\pi/2$, $\vartheta_-(0)=-\omega\pi/2$. Then we set $\eta=1$ if $\sin \omega\pi>0$, $\eta=i$ if $\sin \omega\pi<0$ and choose the sequence

$$w_\alpha(z) = \eta(2\alpha)^{1/2}z^{\alpha-1}, \qquad \alpha \to 0.$$

We calculate $\iint_D |w_\alpha|^2 dxdy$ and $\iint_D w_\alpha^2\, dxdy$. On setting

$$\vartheta_+(r) - \vartheta_-(r) = \omega\pi + r\theta(r),$$

$$\int_{\vartheta_-(r)}^{\vartheta_+(r)} e^{(2\alpha-2)i\vartheta} d\vartheta = \sin \omega\pi + r\phi(r),$$

we find that $|\theta(r)|$ and $|\phi(r)|$ are bounded for $0<r<R$. Hence, as $\alpha\to 0$, we have

$$\iint_D | w_\alpha |^2 dxdy = \omega\pi R^{2\alpha} + 2\alpha \int_0^R r^{2\alpha}\theta(r)dr + 2\alpha \iint_{D-D_R} r^{2\alpha-2}dxdy \to \omega\pi,$$

$$\iint_D w_\alpha^2\, dxdy = |\sin \omega\pi|R^{2\alpha} + 2\alpha\eta^2 \int_0^R r^{2\alpha}\phi(r)dr + 2\alpha \iint_{D-D_R} z^{2\alpha-2}dxdy \to |\sin \omega\pi|,$$

and therefore

$$\iint_D w_\alpha^2\, dxdy : \iint_D | w_\alpha |^2 dxdy \to \left| \frac{\sin \omega\pi}{\omega\pi} \right| = \mathrm{M}.$$

According to Lemma 5, the sequence $w_\alpha(z)$ converges weakly to zero, as $\alpha\to 0$. Thus we obtain $\mathrm{M} \leqq \bar{\mu}$. If we take the sequence $w_\alpha(z)=i\eta\ (2\alpha)^{1/2}z^{\alpha-1}$ we get $-\mathrm{M} \geqq \mu$. So we have proved Theorem 7.

Remark. In the case where the boundary B has an external cusp which was excluded hitherto it can happen that the limit spectrum reaches the points $\mu=1$ and $\mu=-1$ and that, consequently, the inequality theorem does not hold.

We assume that the boundary B has an internal cusp at the point $z=0$. On introducing the functions $\vartheta=\vartheta_+(r)$, $\vartheta=\vartheta_-(r)$ and the domain D_R as before we assume that

$$\vartheta_+(r) = \frac{\kappa}{2} r + r^2 \theta_+(r), \qquad \vartheta_-(r) = -\frac{\kappa}{2} r + r^2 \theta_-(r),$$

where $\kappa \neq 0$ and θ_+, θ_- are bounded for $0 < r < R$. Then we choose the sequence

$$w_\alpha(z) = (2\alpha)^{1/2} z^{\alpha - 3/2}.$$

On setting

$$\vartheta_+(r) - \vartheta_-(r) = \kappa r + r^2 \theta(r),$$

$$\int_{\vartheta_-(r)}^{\vartheta_+(r)} e^{(2\alpha-3)i\vartheta} d\vartheta = \kappa r + r^2 \phi(r),$$

we find that $\theta(r)$ and $\phi(r)$ are bounded for $0 < r < R$. Hence, as $\alpha \to 0$, we have

$$\iint_D |w_\alpha|^2 dxdy = \kappa R^{2\alpha} + 2\alpha \int_0^R r^{2\alpha}\theta(r) dr + 2\alpha \iint_{D-D_R} r^{2\alpha-3} dxdy \to \kappa,$$

$$\iint_D |w|^2 dxdy = \kappa R^2 + 2\alpha \int_0^R r^{2\alpha}\phi(r) dr + 2\alpha \iint_{D-D_R} r^{2\alpha-3} dxdy \to \kappa,$$

and, therefore,

$$\iint_D w_\alpha^2 \, dxdy : \iint_D |w_\alpha|^2 dxdy \to 1.$$

Since w_α tends weakly to zero this relation yields $\bar{\mu} = 1$.

Part II. The case of two functions of two variables

This second part is concerned with complex-valued functions $k(x, y)$ of two variables x, y, defined in an open domain D. The manifold of all such functions $k(x, y)$ for which the integral

$$(k \mid k) = \frac{1}{2} \iint_D \left\{ \left| \frac{\partial k}{\partial x} \right|^2 + \left| \frac{\partial k}{\partial y} \right|^2 \right\} dxdy$$

is finite forms a Hilbert space $\mathfrak{K}$.

I consider several subspaces of $\mathfrak{K}$ and investigate their relations. By means of the projectors of these subspaces I can represent the operator which corresponds to the quadratic form treated in Part I.

The inequality and the expansion theorem which I have established for this form can be employed here. Under the same assumption regarding the boundary, I obtain an inequality and expansions for the functions k of the space $\mathfrak{K}$. This inequality statement is that there exists a positive number $\sigma = \sigma_D$ such that

$$\sigma(k \mid k) \leqq \iint_D \left(\Re \frac{\partial k}{\partial z}\right)^2 dxdy + \iint_D \left|\frac{\partial k}{\partial z}\right|^2 dxdy$$

for all functions k in $\mathfrak{R}$ which satisfy the relation $\Im\iint_D(\partial k/\partial z)dxdy=0$. Finally I show that this inequality† plays a decisive part in the theory of equilibrium and vibration of an elastic plate.

1. The relative spectrum of two subspaces

1.1. The smallest angle between two spaces. At first we make some remarks on the relative spectrum of two subspaces of a Hilbert space which we will apply to our question in functional spaces.

Let $\mathfrak{H}$ be a real Hilbert space of elements h with the inner product (h_1, h_2). Let $\mathfrak{F}$ and $\mathfrak{Q}$ be two closed (linear) subspaces of $\mathfrak{H}$ with the projectors F and Q; the spaces of all elements in $\mathfrak{H}$ which are orthogonal to $\mathfrak{F}$ and $\mathfrak{Q}$ respectively are denoted by $\mathfrak{G}$ and $\mathfrak{P}$ with projectors G and P. So we have

$$F+G=1, \qquad FG=0, \qquad P+Q=1, \qquad PQ=0.$$

We denote by $\mathfrak{H}'$, $\mathfrak{F}'$, $\mathfrak{G}'$, $\mathfrak{P}'$, $\mathfrak{Q}'$ the subspaces of $\mathfrak{H}$, $\mathfrak{F}$, $\mathfrak{G}$, $\mathfrak{P}$, $\mathfrak{Q}$ which are orthogonal to the four section spaces $\mathfrak{Q}\mathfrak{F}$, $\mathfrak{P}\mathfrak{F}$, $\mathfrak{Q}\mathfrak{G}$, $\mathfrak{P}\mathfrak{G}$; so we have $\mathfrak{H}'=\mathfrak{P}'\oplus\mathfrak{Q}'=\mathfrak{F}'\oplus\mathfrak{G}'$, $\mathfrak{H}=\mathfrak{H}'\oplus\mathfrak{Q}\mathfrak{F}\oplus\mathfrak{P}\mathfrak{F}\oplus\mathfrak{Q}\mathfrak{G}\oplus\mathfrak{P}\mathfrak{G}$.

We introduce the largest non-negative number $\tau_0\leqq\pi/2$ such that for all elements f in $\mathfrak{F}$, q in $\mathfrak{Q}$ which both are orthogonal to the section $\mathfrak{Q}\mathfrak{F}$ the relation

$$(1.1) \qquad (f, q)^2 \leqq \cos^2 \tau_0(f, f)(q, q)$$

is valid, and we call it the "smallest angle between the spaces $\mathfrak{Q}$ and $\mathfrak{F}$." The inequality (1.1) is equivalent to each of the following four inequalities:

$$(1.2)_Q \qquad (Qf, Qf) \leqq \cos^2 \tau_0(f, f);$$

$$(1.2)_P \qquad (Pf, Pf) \geqq \sin^2 \tau_0(f, f);$$

$$(1.2)_F \qquad (Fq, Fq) \leqq \cos^2 \tau_0(q, q);$$

$$(1.2)_G \qquad (Gq, Gq) \geqq \sin^2 \tau_0(q, q).$$

We show this for $(1.2)_Q$: from (1.1) we get

$$(Qf, Qf)^2 = (f, Qf)^2 \leqq \cos^2 \tau_0(f, f)(Qf, Qf)$$

and thus $(1.2)_Q$; and from $(1.2)_Q$ we get

† It is the analogue of the inequality of A. Korn for functions of three variables. The expansion theorem is related to those of E. and F. Cosserat.

Cf. A. Korn, *Über einige Ungleichungen, welche in der Theorie der elastischen und elektrischen Schwingungen eine Rolle spielen*, Bulletin de l'Académie des Sciences de Cracovie, 1909, vol. 2, pp. 705–724, and literature indicated therein.

$$(f, q)^2 = (Qf, q)^2 \leqq (Qf, Qf)(q, q) \leqq \cos^2 \tau_0 (f, f)(q, q)$$

and thus (1.1).

Remark. Every subspace of finite dimension has a *positive* smallest angle with respect to every other space in which it is not contained.

We have the following

THEOREM 1.1. *If the smallest angle τ_0 between the spaces $\mathfrak{Q}$ and $\mathfrak{F}$ is positive, then there is a constant ρ such that*

$$\rho(h, h) \leqq (h, Ph) + (h, Gh)$$

for all elements h in $\mathfrak{H}$ which are orthogonal to $\mathfrak{Q}\mathfrak{F}$.

For every such element h can be written in the form

$$h = f + q + j,$$

where the elements f in $\mathfrak{F}$, q in $\mathfrak{Q}$, j in $\mathfrak{P}\mathfrak{G}$ are orthogonal to $\mathfrak{Q}\mathfrak{F}$. Then we have

$$\begin{aligned}(h, h) &= (f, f) + 2(f, q) + (q, q) + (j, j)\\ &\leqq (f, f) + 2 \cos \tau_0 [(f, f)(q, q)]^{1/2} + (q, q) + (j, j)\\ &\leqq (1 + \cos \tau_0)[(f, f) + (q, q)] + (j, j)\end{aligned}$$

and

$$\begin{aligned}(h, Ph) + (h, Gh) &= (f, Pf) + (q, Gq) + 2(j, j)\\ &\geqq \sin^2 \tau_0 [(f, f) + (q, q)] + 2(j, j).\end{aligned}$$

Thus Theorem 1.1 is true with $\rho = 1 - \cos \tau_0$.

1.2. **The spectrum of the operator FQF.** Now we discuss the symmetric operator FQF. It transforms every element f in $\mathfrak{F}$ into an element $FQFf$ in $\mathfrak{F}$ and we have

$$0 \leqq (f, FQFf) \leqq (f, f).$$

The operator FQF, considered an operator in $\mathfrak{F}$, has a spectral resolution. For the sake of simplicity we assume that the spectrum is a pure point spectrum. A characteristic value κ is a real number to which there are elements $f_\kappa \neq 0$ in $\mathfrak{F}$ such that

$$FQFf_\kappa = \kappa f_\kappa.$$

The space of all such characteristic elements f_κ may be designated by $\{f_\kappa\}$; the characteristic spaces $\{f_\kappa\}$ for different values κ are orthogonal and all these spaces span the whole space $\mathfrak{F}$: $\sum_\kappa \{f_\kappa\} = \mathfrak{F}$. The characteristic values κ

satisfy the relation $0 \leq \kappa \leq 1$. The characteristic spaces of $\kappa = 1$ and $\kappa = 0$ are the sections of $\mathfrak{Q}$ and $\mathfrak{P}$ respectively with $\mathfrak{F}$:

$$\{f_1\} = \mathfrak{Q}\mathfrak{F}, \quad \{f_0\} = \mathfrak{P}\mathfrak{F}.$$

If we wish to exclude the values $\kappa = 0$ and $\kappa = 1$ we write $\sum'$ instead of $\sum$. So we have $\sum'_\kappa \{f_\kappa\} = \mathfrak{F}'$. Let f_κ be such a characteristic element. Then the element $q_\kappa = Qf_\kappa$ has the properties

$$(q_\kappa, q_\kappa) = (q_\kappa, f_\kappa) = \kappa(f_\kappa, f_\kappa).$$

Therefore, if τ is the angle between q_κ and f_κ we have

$$\kappa = \cos^2 \tau.$$

The g.l.b. of all such numbers $\tau \neq 0$ is the smallest angle τ_0 between the spaces $\mathfrak{Q}$ and $\mathfrak{F}$ and if τ_1 is the l.u.b. of all such numbers $\tau \neq \pi/2$, then $\pi/2 - \tau_1$ is the smallest angle between the spaces $\mathfrak{P}$ and $\mathfrak{F}$. The manifold which is spanned by the characteristic elements f_κ and by $q_\kappa = Qf_\kappa$, provided that $\kappa \neq 0$ and $\kappa \neq 1$, may be denoted by $\{f_\kappa, q_\kappa\}$. The dimension of $\{f_\kappa, q_\kappa\}$ is twice the multiplicity of $\{f_\kappa\}$. The subspaces $\{f_\kappa, q_\kappa\}$ for different values κ are orthogonal to each other, to $\mathfrak{P}\mathfrak{F}$, and to $\mathfrak{Q}\mathfrak{F}$. This fact follows from

$$\kappa'(f_\kappa, f_{\kappa'}) = (f_\kappa, q_{\kappa'}) = (q_\kappa, q_{\kappa'}) = (q_\kappa, f_{\kappa'}) = \kappa(f_\kappa, f_{\kappa'}).$$

These spaces are also orthogonal to the sections $\mathfrak{Q}\mathfrak{G}$ and $\mathfrak{P}\mathfrak{G}$; and we have

THEOREM 1.2. *The spaces $\{f_\kappa, q_\kappa\}$ belonging to all characteristic values $\kappa \neq 0$, $\neq 1$ of the operator FQF in $\mathfrak{F}$ span the whole space $\mathfrak{H}'$: $\mathfrak{H}' = \sum'_\kappa \{f_\kappa, q_\kappa\}$.*

Let h be an element of $\mathfrak{H}'$ which is orthogonal to all manifolds $\{f_\kappa, q_\kappa\}$, $(\kappa \neq 0, \neq 1)$. Then h is orthogonal to all $\{f_\kappa\}$. Since these elements f_κ span the whole space $\mathfrak{F}$, the element h is orthogonal to $\mathfrak{F}$; that means h belongs to $\mathfrak{G}'$. Further, the element h is orthogonal to all elements $q_\kappa = Qf_\kappa$ and, consequently, the element Qh of $\mathfrak{Q}'$ is orthogonal to $\mathfrak{F}'$; that means: Qh belongs to $\mathfrak{G}'$ and, since $\mathfrak{Q}'\mathfrak{G}' = 0$ we have $Qh = 0$. Therefore h belongs to $\mathfrak{P}'$; but, since $\mathfrak{P}'\mathfrak{G}' = 0$, we have $h = 0$. Thus Theorem 1.2 is proved.

1.3. **The spectrum of the operator $aP + bG$.** We investigate the operator

$$aP + bG,$$

where $a \neq 0$ and $b \neq 0$ are given numbers. Since this operator is symmetric and bounded it has a spectral resolution. We can express the characteristic values λ of this operator by the characteristic values $\kappa = \cos^2 \tau$ of the operator FQF in $\mathfrak{F}$. To every such value we determine the solutions λ_κ of the quadratic equation

$$(\lambda_\kappa - a)(\lambda_\kappa - b) = ab\kappa,$$

namely,

$$\lambda_\kappa^+ = \frac{a+b}{2} + \left[\left(\frac{a-b}{2}\right)^2 + ab\kappa\right]^{1/2} = \frac{a+b}{2} + \left[\left(\frac{a+b}{2}\right)^2 - ab\sin^2\tau\right]^{1/2}$$
$$= \frac{a+b}{2} - \left[\left(\frac{a-b}{2}\right)^2 + ab\kappa\right]^{1/2} = \frac{a+b}{2} - \left[\left(\frac{a+b}{2}\right)^2 - ab\sin^2\tau\right]^{1/2}.$$

If $\kappa=0$ we have $\lambda_0^+=\max(a, b)$, $\lambda_0^-=\min(a, b)$; if $\kappa=1$, $\lambda_1^+=\max(a+b, 0)$, $\lambda_1^-=\min(a+b, 0)$. We have the

THEOREM 1.3. *The values $\lambda=\lambda_\kappa^\pm$, for every $\kappa\neq 0, \neq 1$, are characteristic values of the operator $aP+bG$; their characteristic functions are*

$$h_\kappa^\pm = (\lambda_\kappa^\pm - b)f_\kappa - aq_\kappa.$$

The characteristic spaces of the values $\lambda=a$ and $\lambda=b$ are $\mathfrak{P}\mathfrak{F}$ and $\mathfrak{Q}\mathfrak{G}$ respectively and the characteristic spaces of the values $\lambda=a+b$ and $\lambda=0$ are $\mathfrak{P}\mathfrak{G}$ and $\mathfrak{Q}\mathfrak{F}$ respectively. The other characteristic functions span the whole space $\mathfrak{H}'$:

$$\mathfrak{G}' = \sum_\kappa{}'\{h_\kappa^+\} + \sum_\kappa{}'\{h_\kappa^-\}. \tag{1.3}$$

A simple calculation shows that the elements $h_\kappa^\pm$ and the four section spaces are characteristic. Since

$$\lambda_\kappa^+ - \lambda_\kappa^- \neq 0 \text{ if } \kappa \neq 0, \neq 1, \text{ we have}$$
$$f_\kappa = \frac{h^+ - h^-}{\lambda_\kappa^+ - \lambda_\kappa^-}, \qquad q_\kappa = \frac{1}{a}\,\frac{(\lambda_\kappa^+ - b)h_\kappa^+ - (\lambda_\kappa^- - b)h_\kappa^-}{\lambda_\kappa^+ - \lambda_\kappa^-}$$

and

$$\{h_\kappa^+\} \oplus \{h_\kappa^-\} = \{f_\kappa, q_\kappa\}, \qquad \kappa \neq 0 \neq 1.$$

Therefore relation (1.3) follows from Theorem 1.2 and Theorem 1.3 is proved

We remark that the spectrum is contained within the two closed intervals $[\lambda_0^-, \lambda_1^-]$ and $[\lambda_0^+, \lambda_1^+]$, which have a common point only if $a+b=0$ or $a=b$. Every limit point κ_∞ of the κ-spectrum corresponds to two limit points λ_∞^+ and λ_∞^- of the λ-spectrum (except when $a=b$, $\kappa_\infty=0$ or $a+b=0$, $\kappa_\infty=1$).

If the spaces $\mathfrak{Q}$ and $\mathfrak{F}$ have a positive smallest angle $\tau_0>0$, then there is a constant $\rho>0$ such that $\rho(h, h)\leq a(h, Ph)+b(h, Qh)$, if $a>0$, $b>0$ for all $h\perp\mathfrak{Q}\mathfrak{F}$. (The largest possible number ρ is exactly

$$\rho_0 = \frac{a+b}{2} - \left[\left(\frac{a-b}{2}\right)^2 + ab\cos^2\tau_0\right]^{1/2}$$

$$= \frac{ab \sin^2 \tau_0}{\dfrac{a+b}{2} + \left[\left(\dfrac{a-b}{2}\right)^2 + ab \cos^2 \tau_0\right]^{1/2}}.$$

Cf. Theorem 1.1 where $a=b=1$, $\rho_0=1-\cos \tau_0$.)

In the same way we can treat the spectral problem of the operator $aP+bG$ with respect to a different unit-form, for example,

$$\alpha(h_1, Fh_2) + \beta(h_1, Gh_2), \qquad \alpha \geqq 0, \qquad \beta \geqq 0.$$

In this case also every κ corresponds to two characteristic values λ, the solutions of the quadratic equation

$$(\lambda\alpha - a)(\lambda\beta - b) + (\lambda(\beta - \alpha) - b)a\kappa = 0,$$

and the characteristic elements

$$h_\kappa = (\lambda\beta - b)f_\kappa - aq_\kappa.$$

The section spaces $\mathfrak{P}\mathfrak{F}$, $\mathfrak{Q}\mathfrak{G}$, $\mathfrak{P}\mathfrak{G}$, $\mathfrak{Q}\mathfrak{F}$ belong to the characteristic values

$$\frac{a}{\alpha}, \quad \frac{b}{\beta}; \quad \frac{a+b}{\beta}, \quad 0.$$

2. The space $\mathfrak{K}$

2.1. **The space $\mathfrak{R}$ in general.** Let D be an open domain in the z-plane. Let $k=k_x+ik_y$ be a complex-valued function of x and y defined in D and having derivatives

$$\frac{\partial k}{\partial x}, \quad \frac{\partial k}{\partial y}$$

with respect to x and y which are L^2-integrable over any subdomain D^* of D

$$(k \mid k)_{D^*} = \frac{1}{2}\iint_{D^*}\left\{\left|\frac{\partial k}{\partial x}\right|^2 + \left|\frac{\partial k}{\partial y}\right|^2\right\}dxdy < \infty.$$

Considering k as a function of $z=x+iy$ and $\bar{z}=x-iy$ we have the relation

$$(k \mid k)_{D^*} = \iint_{D^*}\left\{\left|\frac{\partial k}{\partial z}\right|^2 + \left|\frac{\partial k}{\partial \bar{z}}\right|^2\right\}dxdy,$$

because of

$$2\frac{\partial}{\partial z} = \frac{\partial}{\partial x} - i\frac{\partial}{\partial y}, \quad 2\frac{\partial}{\partial \bar{z}} = \frac{\partial}{\partial x} + i\frac{\partial}{\partial y}.$$

In case of a multiply-connected domain D we admit many-valued functions provided that they have singled-valued derivatives $\partial k/\partial x$, $\partial k/\partial y$.

We denote by $\mathfrak{R}$ the manifold of all functions $k(x, y)$ of this kind for which the integral

$$(k \mid k) = \iint_D \left\{ \left| \frac{\partial k}{\partial z} \right|^2 + \left| \frac{\partial k}{\partial \bar{z}} \right|^2 \right\} dxdy$$

is finite; $\mathfrak{K}$ is a linear space. Two functions k, k^* of $\mathfrak{K}$ are "equivalent" if $(k^* - k \mid k^* - k) = 0$.† The manifold of all functions of $\mathfrak{K}$ which are equivalent to each other corresponds to one element of the space $\mathfrak{K}$. So, e.g., the function $k(x, y) = \text{const.}$ corresponds to the element zero.

In the space $\mathfrak{K}$ we define the inner product

$$(k_1 \mid k_2) = \Re \iint_D \left\{ \overline{\frac{\partial k_1}{\partial z}} \frac{\partial k_2}{\partial z} + \frac{\partial \bar{k}_1}{\partial z} \frac{\partial k_2}{\partial \bar{z}} \right\} dxdy.$$

With respect to this metric the space $\mathfrak{K}$ is a real one; but to every element $k = k_x + ik_y$ of $\mathfrak{K}$ the operations $\Re k = k_x$, $\Im k = k_y$, $ik = k_y - ik_x$, $\bar{k} = k_x - ik_y$ are feasible. Corresponding to Fischer's form of the theorem of F. Riesz and E. Fischer we have the important fact:

THEOREM 2.1. *The space $\mathfrak{K}$ is complete.*‡

2.2. **The subspace $\mathfrak{F}$.** We introduce the subspace $\mathfrak{F}$ of all elements f in $\mathfrak{R}$ for which $\iint_D |\partial f/\partial \bar{z}|^2 dxdy = 0$; to these elements there correspond functions f with continuous derivatives for which $\partial f/\partial \bar{z} = 0$, that is to say, which are analytic functions of z. In the following we assume the function f in $\mathfrak{F}$ to be analytic. Every function f of this space $\mathfrak{F}$ can also be represented by the derivative $w = df/dz$ and we have

$$(2.1) \qquad (f_1 \mid f_2) = \Re \iint_D \bar{w}_1 w_2 dxdy.$$

Therefore the space $\mathfrak{F}$ can be identified with the space of analytic functions $w(z)$ dealt with in Part I. As we proved there, we have

† For further application we need the obvious

LEMMA 2.1. *Two functions k, k^* are equivalent if*

$$\iint_D |k^* - k|^2 dxdy = 0.$$

‡ For the proof use, e.g., the methods of G. Fubini, *Il principio di minimo e i teoremi di esistenza per i problemi al contorno relativi alle equazioni alle derivate parziali di ordine pari*, Rendiconti del Circolo Matematico di Palermo, vol. 23 (1907), pp. 9–11.

THEOREM 2.2. *The space $\mathfrak{F}$ is complete.*

The space $\overline{\mathfrak{F}}$ of all functions $\bar{f}$ in $\mathfrak{K}$ for which

$$\frac{\partial \bar{f}}{\partial z} = 0$$

consists of the conjugates of all functions f in $\mathfrak{F}$. We have

$$(2.2)\qquad (\bar{f}_1 \mid \bar{f}_2) = \Re \iint_D \frac{d\bar{f}_1}{d\bar{z}} \frac{d\bar{f}_2}{dz} dxdy.$$

From Theorem 2.2 we get:

THEOREM 2.3. *The space $\overline{\mathfrak{F}}$ is complete.*†

Since for every function f_1 in $\mathfrak{F}$, $\bar{f}_2$ in $\overline{\mathfrak{F}}$

$$(f_1 \mid \bar{f}_2) = 0,$$

we have

THEOREM 2.4. *The spaces $\mathfrak{F}$ and $\overline{\mathfrak{F}}$ are orthogonal to each other.*

A function $k(x, y)$ is called a potential function if it has continuous second derivatives satisfying the relation $\partial^2 k/\partial z \partial \bar{z} = 0$. Such a function $k(x, y)$ can be written in the form $k(x, y) = f(z) + \overline{f^*(z)}$, where $f(z), f^*(z)$ are analytic in z; together with k also f and f^* belong to $\mathfrak{F}$. Thus we get

THEOREM 2.5. *The space $\mathfrak{F} \oplus \overline{\mathfrak{F}}$ consists of the potential functions in $\mathfrak{K}$.*

2.3. **The subspace $\mathfrak{G}$.** We introduce the subspace $\dot{\mathfrak{G}}$ of elements g in $\mathfrak{K}$ which are equivalent to functions g in $\mathfrak{K}$ which vanish identically in a boundary strip. We denote by $\mathfrak{G}$ the closure of $\dot{\mathfrak{G}}$. That is to say, $\mathfrak{G}$ consists of all functions g in $\mathfrak{K}$ for which there are functions $\dot{g}$ in $\dot{\mathfrak{G}}$ such that $(\dot{g} - g \mid \dot{g} - g)$ is arbitrarily small.‡ This definition contains the

THEOREM 2.6. *The space $\mathfrak{G}$ is complete.*

We establish the following basic identity:

$$(2.3)\qquad (k \mid g) = 2\Re \iint_D \frac{\overline{\partial k}}{\partial z} \frac{\partial g}{\partial z} dxdy = 2\Re \iint_D \frac{\partial \bar{k}}{\partial z} \frac{\partial g}{\partial \bar{z}} dxdy$$

for all k in $\mathfrak{K}$, g in $\mathfrak{G}$.

---

† This theorem is related to that of S. Zaremba. Cf. S. Zaremba, *Sur un problème toujours possible, comprenant, à titre de cas particulier, le problème de Dirichlet et celui de Neumann*, Journal de Mathématiques, sér. 9, vol. 6 (1927), pp. 127–163; O. Nikodym, *Sur un théorème de M. S. Zaremba concernant les fonctions harmoniques*, Journal de Mathématiques, sér. 9, vol. 12 (1933), pp. 95–109, and *Sur le principe du minimum*, Mathematica Cluj, vol. 9 (1936), p. 123.

‡ Under simple assumptions regarding the boundary it would be possible to give a direct definition of $\mathfrak{G}$ by a boundary condition.

To prove it let $\dot{g}$ be a function of $\dot{\mathfrak{G}}$ which vanishes in a boundary strip S. In D we take a subdomain D^* with rectifiable boundary B^* contained in S. If k is in $\mathfrak{K}$, there exists a function $k^*(x, y)$ which is defined in D^*+B^*, has continuous second derivatives, and approximates k in the sense that $(k^*-k|k^*-k)_{D^*}$ is small. The equations

$$\iint_{D^*} \frac{\overline{\partial k^*}}{\partial z} \frac{\partial \dot{g}}{\partial z} dxdy - \iint_{D^*} \frac{\partial \overline{k^*}}{\partial z} \frac{\partial \dot{g}}{\partial \bar{z}} dxdy$$

$$= \frac{i}{2} \int_{B^*} \frac{\overline{\partial k^*}}{\partial \bar{z}} \dot{g} d\bar{z} + \frac{i}{2} \int_{B^*} \frac{\partial \overline{k^*}}{\partial z} \dot{g} d\bar{z} = 0$$

imply that, for all functions k in $\mathfrak{K}$,

$$\iint_{D} \frac{\overline{\partial k}}{\partial z} \frac{\partial \dot{g}}{\partial z} dxdy - \iint_{D} \frac{\partial \bar{k}}{\partial z} \frac{\partial \dot{g}}{\partial \bar{z}} dxdy$$

$$= \iint_{D^*} \frac{\overline{\partial k}}{\partial z} \frac{\partial \dot{g}}{\partial z} dxdy - \iint_{D^*} \frac{\partial \bar{k}}{\partial z} \frac{\partial \dot{g}}{\partial \bar{z}} dxdy = 0.$$

Since $\dot{\mathfrak{G}}$ is dense in $\mathfrak{G}$ we conclude that (2.3) also holds.

From this identity we immediately get

THEOREM 2.7. *The space* $\mathfrak{G}$ *is orthogonal to* $\mathfrak{F}$ *and to* $\overline{\mathfrak{F}}$: $\mathfrak{G} \perp \mathfrak{F}$, $\mathfrak{G} \perp \overline{\mathfrak{F}}$.

Further we prove the decisive

THEOREM 2.8. *The spaces* $\mathfrak{F}$, $\overline{\mathfrak{F}}$, $\mathfrak{G}$ *span the whole space* $\mathfrak{K}$: $\mathfrak{F} \oplus \overline{\mathfrak{F}} \oplus \mathfrak{G} = \mathfrak{H}$.

We construct† the following function $\dot{g}$ in $\dot{\mathfrak{G}}$:

Let $|z-z_0| \leqq R$ be a circle within D, z' a point within this circle and $|z-z'| \leqq r$ a circle in the interior of $|z-z_0| \leqq R$. Then we put

$$\dot{g}(x, y) = \begin{cases} 0 & \text{for } |z - z_0| \geqq R; \\ \dfrac{1}{\pi} \log \left| \dfrac{R(z - z')}{R^2 - \overline{(z' - z_0)}(z - z_0)} \right| & \text{for } |z - z_0| \leqq R, \\ & \quad \text{but } |z - z'| \geqq r; \\ \dfrac{1}{\pi} \log \left| \dfrac{Rr}{R^2 - \overline{(z' - z_0)}(z - z_0)} \right| & \text{for } |z - z'| \leqq r. \end{cases}$$

Let k be a function orthogonal to $\mathfrak{G}$. Then from $(ig|k)=0$ and $(g|k)=0$ we

† For the following reasoning cf. G. Fubini, loc. cit., p. 10, §§6, 7, and R. Courant, *Über direkte Methoden, bei Variations- und Randwertproblemen*, Jahresbericht der Deutschen mathematiker vereinigung, vol. 34 (1925), pp. 107, 108.

have $\iint_D (dk/dz)(d\bar{g}/d\bar{z})dxdy=0$, according to (2.3). Hence we obtain by integration by parts

$$\frac{1}{2\pi i}\int_{|z-z'|=r} k(x, y)\frac{dz}{z-z'} = \frac{1}{2\pi i}\int_{|z-z_0|=R} k(x, y)\frac{R^2-|z'-z_0|^2}{|z-z'|^2}\frac{dz}{z-z_0}.$$

Consequently the mean value at the left-hand side is independent of r and is a potential function $k(x', y')$ in x', y'. The function $k^*(x, y)-k(x, y)$ has the property that

$$\frac{1}{2\pi i}\int_{|z-z'|=r}(k^*-k)\frac{dz}{z-z'} = 0$$

and, consequently,

$$\iint_{|z-z'|\leqq r}(k^*-k)dxdy = 0.$$

Therefore† $\iint_D |k^*-k|^2 dxdy=0$ and k is equivalent to the potential function k^* (cf. Lemma 2.1). According to Theorem 2.5 the element k belongs to $\mathfrak{F}\oplus\overline{\mathfrak{F}}$. Thus $\mathfrak{K}=\mathfrak{F}\oplus\overline{\mathfrak{F}}\oplus\mathfrak{G}$ as we wished to prove.

We denote by F, $\overline{F}$, G the orthogonal projectors which belong to the closed subspaces $\mathfrak{F}$, $\overline{\mathfrak{F}}$, $\mathfrak{G}$; the projections of a function k of $\mathfrak{K}$ on these spaces are Fk, $\overline{F}k$, Gk respectively. Theorem 2.8 gives the relation

$$F+\overline{F}+G = 1.$$

3. The space $\mathfrak{H}$

3.1. **The metric (,).** In this section we deal with the space

$$\mathfrak{H} = \mathfrak{F} + \mathfrak{G}$$

consisting of all elements h of $\mathfrak{K}$ which are orthogonal to $\overline{\mathfrak{F}}$. We employ the abbreviation

$$(k_1, k_2) = \Re\iint_D \frac{\overline{\partial k_1}}{\partial z}\frac{\partial k_2}{\partial z}dxdy$$

$$(k_1; k_2) = \Re\iint_D \frac{\partial \bar{k}_1}{\partial z}\frac{\partial k_2}{\partial \bar{z}}dxdy$$

for all k in $\mathfrak{K}$, so that

† We make use of the

LEMMA 2.2. *Whenever an L^2-integrable function $\phi(x, y)$ has the property that*

$$\frac{1}{r^2\pi}\iint_{|z'-z|\leqq r}\phi(x', y')dx'dy' = 0$$

for every circle $|z-z'|\leqq r$ within D, then $\iint_D|\phi|^2dxdy=0$.

$$(k_1, k_2) + (k_1; k_2) = (k_1 | k_2).$$

The inner product (h_1, h_2) defines a new metric (,) in the space $\mathfrak{H}$, since the form (h, h) is positive definite and vanishes for h in $\mathfrak{H}$ only if $h=0$. An immediate consequence of identities (2.1) and (2.3) is

THEOREM 3.1. *The spaces $\mathfrak{F}$ and $\mathfrak{G}$ are orthogonal with respect to the metric* (,).

Further we use

LEMMA 3.1. *For elements h in $\mathfrak{H}$ the inequality*

$$\frac{1}{2}(h | h) \leqq (h, h) \leqq (h | h)$$

is valid.

From the identities (2.2) and (2.3), on writing $h=Fh+Gh=f+g$, we get in fact the relation

$$\frac{1}{2}(h | h) = \frac{1}{2}(f | f) + \frac{1}{2}(g | g) = \frac{1}{2}(f, f) + (g, g)$$
$$\leqq (f, f) + (g, g) = (h, h) \leqq (h, h) + (h; h) = (h | h).$$

From this inequality we deduce

LEMMA 3.2. *A subspace of $\mathfrak{H}$ is complete with respect to the metric* (|) *if and only if it is complete with respect to the metric* (,).

Thus we obtain

THEOREM 3.2. *The spaces $\mathfrak{F}$ and $\mathfrak{G}$ are closed with respect to* (,).

In the following part of §3 the terms "orthogonal" and "closed" refer to the metric (,) only.

The projectors F and G for the spaces $\mathfrak{F}$ and $\mathfrak{G}$ satisfy the relation

$$F + G = 1 \quad \text{in} \quad \mathfrak{H}.$$

They belong to the form (,) − (;) and (;) in the sense that

$$(h^*, Fh) = (h^*, h) - (h^*; h), \qquad (h^*, Gh) = (h^*; h)$$

for h^*, h in $\mathfrak{H}$, in accordance with (2.3).

3.2. **The spaces $\mathfrak{P}$ and $\mathfrak{Q}$.** We introduce the symmetric forms

$$h_1Ph_2 = \frac{1}{4}\iint_D \left(\overline{\frac{\partial h_1}{\partial z}} + \frac{\partial h_1}{\partial z}\right)\left(\frac{\partial h_2}{\partial z} + \overline{\frac{\partial h_2}{\partial z}}\right) dxdy = \iint_D \Re\frac{\partial h_1}{\partial z}\,\Re\frac{\partial h_2}{\partial z}\, dxdy$$

$$h_1\mathbf{Q}h_2 = \frac{1}{4}\iint_D\left(\overline{\frac{\partial h_1}{\partial z}} - \frac{\partial h_1}{\partial z}\right)\left(\frac{\partial h_2}{\partial z} - \overline{\frac{\partial h_2}{\partial z}}\right)dxdy = \iint_D \Im\frac{\partial h_1}{\partial z}\,\Im\frac{\partial h_2}{\partial z}\,dxdy$$

defined for h_1, h_2 in $\mathfrak{H}$. We have

$$0 \leqq h\mathbf{P}h \leqq (h, h), \qquad 0 \leqq h\mathbf{Q}h \leqq (h, h)$$

and

$$\mathbf{P} + \mathbf{Q} = (,).$$

The forms $\mathbf{P}$ and $\mathbf{Q}$ correspond to bounded symmetric operators P, Q such that

$$h^*\mathbf{P}h = (h^*, Ph); \qquad h^*\mathbf{Q}h = (h^*, Qh)$$

for h^*, h in $\mathfrak{H}$. P and Q satisfy the relation

$$P + Q = 1.$$

By $\mathfrak{P}$ and $\mathfrak{Q}$ we denote the subspaces of all the elements p and q in $\mathfrak{H}$ for which

$$Qp = 0 \quad \text{and} \quad Pq = 0,$$

respectively.

The functions

$$p = p_x + ip_y \text{ in } \mathfrak{P} \quad \text{and} \quad q = q_x + iq_y \text{ in } \mathfrak{Q}$$

can be characterized as well by

$$p\mathbf{Q}p = \iint_D\left(\Im\frac{\partial p}{\partial z}\right)^2 dxdy = \frac{1}{4}\iint_D\left(\frac{\partial p_y}{\partial x} - \frac{\partial p_x}{\partial y}\right)^2 dxdy = 0$$

and

$$qPq = \iint_D\left(\Re\frac{\partial q}{\partial z}\right)^2 dxdy = \frac{1}{4}\iint_D\left(\frac{\partial q_x}{\partial x} + \frac{\partial q_y}{\partial y}\right)^2 dxdy = 0$$

respectively. We have

THEOREM 3.3. *The spaces $\mathfrak{P}$ and $\mathfrak{Q}$ are closed.*

Since p in $\mathfrak{P}$, q in $\mathfrak{Q}$

$$(p, q) = p\mathbf{P}q + p\mathbf{Q}q = (p, Pq) + (Qp, q) = 0$$

we have

THEOREM 3.4. *The spaces $\mathfrak{P}$ and $\mathfrak{Q}$ are orthogonal*: $\mathfrak{P} \perp \mathfrak{Q}$.

We now prove the basic

THEOREM 3.5. *The spaces $\mathfrak{P}$ and $\mathfrak{Q}$ span the whole space* $\mathfrak{H}$: $\mathfrak{P}\oplus\mathfrak{Q}=\mathfrak{H}$.

Let h be a function in $\mathfrak{H}$ which is orthogonal to $\mathfrak{P}$. Then we have

$$pPh = pPh + pQh = (p, h) = 0. \tag{3.1}$$

We take the function

$$k(x, y) = \begin{cases} z - z_0 & \text{in} \quad |z - z_0| \leqq R, \\ \dfrac{R^2}{\overline{z - z_0}} & \text{in} \quad |z - z_0| \geqq R, \end{cases}$$

which belongs to $\mathfrak{K}$ and has the property

$$\frac{\partial k}{\partial z} = 1 \quad \text{in} \quad |z - z_0| < R, \qquad = 0 \quad \text{in} \quad |z - z_0| > R.$$

The projection of this function on the space $\mathfrak{H}$: $p = k - \overline{F}k$ also has the property

$$\frac{\partial p}{\partial z} = 1 \quad \text{in} \quad |z - z_0| < R, \qquad = 0 \quad \text{in} \quad |z - z_0| > R$$

and, therefore, belongs to $\mathfrak{P}$. By inserting this function p into the relation (3.1) we get

$$\mathfrak{R} \iint_{|z-z_0|<R} \frac{\partial h}{\partial z} dxdy = 0.$$

Since the circle $|z-z_0| < R$ was arbitrary within D, we deduce (cf. Lemma 2.2)

$$\iint_D \left(\mathfrak{R} \frac{\partial h}{\partial z}\right)^2 dxdy = 0.$$

Hence h belongs to $\mathfrak{Q}$ and consequently $\mathfrak{P} \oplus \mathfrak{Q} = \mathfrak{H}$.

From Theorem 3.5 we see that the elements Qh belong to $\mathfrak{Q}$ because they are orthogonal to $\mathfrak{P}$: $(p, Qh) = (Qp, h) = 0$. Thus we have $PQ = 0$ or $Q^2 = Q$; in the same way we find $P^2 = P$. Therefore we have

THEOREM 3.6. *The operators P and Q are the projectors of the spaces $\mathfrak{P}$ and $\mathfrak{Q}$ respectively.*

Hence we see that the theory of §1 is applicable.

3.3. **The section spaces.** Before going into detail we investigate the section spaces $\mathfrak{P}\mathfrak{F}$, $\mathfrak{Q}\mathfrak{F}$, $\mathfrak{P}\mathfrak{G}$, $\mathfrak{Q}\mathfrak{G}$.

We have

$$\mathfrak{P}\mathfrak{F} = \{z\}, \qquad \mathfrak{Q}\mathfrak{F} = \{iz\},$$

for $f=z+\text{const.}$ and $f=iz+\text{const.}$ are the only functions of $\mathfrak{F}$ for which df/dz is real and imaginary respectively. The spaces $\mathfrak{P}\mathfrak{G}$ and $\mathfrak{Q}\mathfrak{G}$ consist of all functions of $\mathfrak{P}$ and $\mathfrak{Q}$, respectively, which are constant at the boundary in the approximate sense of the definition of $\mathfrak{G}$ (cf. §2). Let $\mathfrak{F}'$, $\mathfrak{G}'$, $\mathfrak{P}'$, $\mathfrak{Q}'$ be the subspaces of all functions in $\mathfrak{F}$, $\mathfrak{G}$, $\mathfrak{P}$, $\mathfrak{Q}$ which are orthogonal to the section spaces; then we have $\mathfrak{H}'=\mathfrak{F}'\oplus\mathfrak{G}'=\mathfrak{P}'\oplus\mathfrak{Q}'$. We establish the following theorem, but we shall not make use of it.

THEOREM 3.7. *The elements h of $\mathfrak{H}'$ are equivalent to functions of the form*

$$h = z\overline{\phi_1(z)} + \phi_2(z) + \overline{\phi_0(z)},$$

where $\phi_1(z)$, $\phi_2(z)$, and $\phi_0(z)$ are analytic in z.

To prove this we take a circle $|z-z_0|\leqq 2R$ in D; we choose a real function $\phi(x, y)$ which is four times continuously differentiable and which is $=1$ in $|z-z_0|\leqq R$, $=0$ in $|z-z_0|\geqq 2R$. Then we choose a number $r<R$ and a point z' of $|z'-z_0|<R-r$ and set

$$h^0 = \begin{cases} \dfrac{z-z'}{r^2} & \text{in } \quad |z-z'|\leqq r, \\ \dfrac{\partial}{\partial\bar{z}}\phi\log|z-z'|^2 & \text{in } \quad |z-z'|\geqq r. \end{cases}$$

This function h^0 belongs to $\mathfrak{P}\mathfrak{G}$ and ih^0 belongs to $\mathfrak{Q}\mathfrak{G}$. Therefore $(h^0,h)=(ih^0, h)=0$. This implies

$$\frac{1}{r^2\pi}\iint_{|z-z'|\leqq r}\frac{\partial h}{\partial z}dxdy = -\frac{1}{\pi}\iint_{R\leqq|z-z_0|\leqq 2R}\frac{\partial^2}{\partial z\partial\bar{z}}\phi\log|z-z'|^2\frac{\partial h}{\partial z}dxdy.$$

The integral on the right is a potential function in x', y', which is independent of r and can be written in the form $\overline{\phi_1(z')}+\phi_2'(z')$, where $\phi_1(z)$ and $\phi_2(z)$ are analytic in z. On setting

$$\phi^* = z\overline{\phi_1(z)} + \phi_2(z)$$

we get

$$\iint_{|z-z'|\leqq r}\frac{\partial}{\partial z}(h-\phi^*)dxdy = 0$$

and (cf. Lemma 2.2)

$$\iint_D\left|\frac{\partial(h-\phi^*)}{\partial z}\right|^2 dxdy = 0.$$

Therefore $\overline{h-\phi^*}$ is equivalent to an analytic function $\phi_0(z)$, and h itself is equivalent to $z\,\overline{\phi_1(z)}+\phi_2(z)+\phi_0(z)$.

3.4. **The operator** M. In §1 we investigated the operator FQF in $\mathfrak{F}$. In its place, we now consider the bounded symmetric operator

$$M = F(P - Q)F = F - 2FQF,$$

which takes every element f of $\mathfrak{F}$ into the element Mf in $\mathfrak{F}$. This operator M in $\mathfrak{F}$ belongs to the form

$$f_1\boldsymbol{M}f_2 = f_1\boldsymbol{P}f_2 - f_1\boldsymbol{Q}f_2 = \Re\iint_D \frac{df_1}{dz}\,\frac{df_2}{dz}\,dxdy$$

in the sense that

$$f_1\boldsymbol{M}f_2 = (f_1, Mf_2).$$

Now, this form M is exactly the form dealt with in Part I if $w=df/dz$; and thus we obtain a very simple representation of the operator of this form. On the other hand we can use the properties of this form established in Part I to investigate the new forms $\boldsymbol{P}$ and $\boldsymbol{Q}$.

4. The form E

4.1. **The inequality.** In this section we assume, as in Part I, that the boundary B of the domain D consists of a finite number of curves $z=z(s)$ with continuous tangent $z(s)$ except at a finite number of corners where $\dot{z}(s)$ is continuous on each side and arc $\dot{z}(s)$ jumps by less than π. Then we get (in the sense of §1)

THEOREM 4.1. *The spaces $\mathfrak{F}$ and $\mathfrak{Q}$ (or $\mathfrak{P}$) have a positive smallest angle.*

Under our assumption on the boundary we can deduce from Theorem 1, Part I that there is a constant $\theta<1$ such that for all functions f in $\mathfrak{F}$ satisfying the relation

$$\Re\iint_D \frac{df}{dz}\,dxdy = 0$$

the inequality

$$f\boldsymbol{M}f = \Re\iint_D \left(\frac{df}{dz}\right)^2 dxdy \leqq \theta\iint_D \left|\frac{df}{dz}\right|^2 dxdy$$

is valid. Since the form $\boldsymbol{M}$ at the left-hand side belongs to the operator $M=1-2FQF$ in $\mathfrak{F}$ we get, on setting if instead of f, the inequality

$$-(f,f) + 2(f,Qf) \leqq \theta(f,f)$$

or

$$(f, Qf) \leqq \frac{1+\theta}{2}(f, f),$$

for all functions f in $\mathfrak{F}$ which satisfy the relation

$$(iz, f) = 0$$

and which are therefore orthogonal to $\mathfrak{Q}\mathfrak{F} = \{iz\}$. (In the same way we get $(f, Pf) \leqq (1+\theta)/2\, (f, f)$ under $(z, f) = 0$.) Therefore $\cos^2 \tau_0 \leqq (1+\theta)/2 < 1$ and the smallest angle $\tau_0 < \pi/2$ is positive as we stated in Theorem 4.1.

We introduce the form

$$h_1Eh_2 = a(h_1, Ph_2) + b(h_1, Gh_2) = a(h_1Ph_2) + b(h_1;\, h_2)$$

$$\equiv a \iint_D \Re \frac{\partial h_1}{\partial z} \Re \frac{\partial h_2}{\partial z} dxdy + b\Re \iint_D \frac{\partial \bar{h}_1}{\partial z} \frac{\partial h_2}{\partial \bar{z}} dxdy$$

for h in $\mathfrak{H}$. We assume a and b to be positive. Applying Theorem 1.1 of §1 we deduce immediately from Theorem 4.1 the

THEOREM 4.2. *There is a positive constant* ρ *such that*

$$\rho(h, h) \leqq (hEh)$$

or

$$\rho \iint_D \left|\frac{\partial h}{\partial z}\right|^2 dxdy \leqq a \iint_D \left|\Re \frac{\partial h}{\partial z}\right|^2 dxdy + b \iint_D \left|\frac{\partial h}{\partial \bar{z}}\right|^2 dxdy$$

for all functions h *in* $\mathfrak{K}$ *which satisfy the condition*

$$(iz, h) = \Im \iint_D \frac{\partial h}{\partial z} dxdy = 0.$$

We now consider the form

$$k_1Ek_2 = a \iint_D \Re \frac{\partial k_1}{\partial z} \Re \frac{\partial k_2}{\partial z} dxdy + b\Re \iint_D \frac{\partial \bar{k}_1}{\partial z} \frac{\partial k_2}{\partial \bar{z}} dxdy$$

for k in $\mathfrak{R}$; $a>0$, $b>0$. On setting $k = (F+G)k + \bar{F}k = h + \bar{f}$, we get

$$k_1Ek_2 = h_1Eh_2 + b(\bar{f}_1 \mid \bar{f}_2)$$

$$(k_1 \mid k_2) = (h_1 \mid h_2) + (\bar{f}_1 \mid \bar{f}_2)$$

$$\leqq 2(h_1, h_2) + (\bar{f}_1 \mid \bar{f}_2)$$

(cf. Lemma 3.1), and

$$(iz \mid k) = (iz, h).$$

Therefore we obtain from Theorem 4.2

THEOREM 4.3. *There is a positive constant* σ *such that*

$$\sigma(k \mid k) \leqq (kEk),$$

or

$$\sigma \iint_D \left|\frac{\partial k}{\partial z}\right|^2 dxdy + \sigma \iint_D \left|\frac{\partial k}{\partial \bar{z}}\right|^2 dxdy \leqq a \iint_D \left(\Re \frac{\partial k}{\partial z}\right)^2 + b \iint_D \left|\frac{\partial k}{\partial \bar{z}}\right|^2 dxdy,$$

for all functions k *in* $\mathfrak{K}$ *which satisfy the condition*

$$(iz \mid k) = \Im \iint_D \frac{\partial k}{\partial z} dxdy = 0.$$

4.2. **The spectral resolution.** It is very simple to give the spectral resolution of the form E with respect to the unit-form (,) or ($|$) if we assume that the boundary B has no corners.

First we remark that, according to Theorem 4 of Part I, the operator $M = 1 - 2FQF$ in $\mathfrak{F}$ has a pure point spectrum of values

$$\mu_1 = 1 > \mu_2 \geqq \mu_3 \geqq \cdots \rightarrow 0; \qquad \mu_{-n} = -\mu_n.$$

Therefore the operator FQF in $\mathfrak{F}$ also has a pure point spectrum of values

$$\kappa_n = \frac{1 - \mu_n}{2} \rightarrow \frac{1}{2}, \qquad n = \pm 1, \pm 2, \pm 3, \cdots.$$

The characteristic functions of FQF and M are the same $f_n(z)$; $f_{-n}(z) = if_n(z)$. $f_1(z) = az$, $f_{-1}(z) = iaz$; a = real constant.

We now observe that the form E in $\mathfrak{H}$ belongs to the operator $aP + bG$ with respect to the unit-form (,). Therefore we can apply Theorem 1.3 of §1. We find that this operator has a pure point spectrum of values λ_n^+, λ_n^-, $n = \pm 1, \pm 2, \cdots$, which are the solutions of the quadratic equation

$$(\lambda_n - a)(\lambda_n - b) = ab \frac{1 - \mu_n}{2}.$$

For $n = \pm 1$ we get the characteristic values $\lambda = 0, a, b, a+b$; their characteristic spaces are

$$\mathfrak{Q}\mathfrak{F} = \{iz\}, \qquad \mathfrak{P}\mathfrak{F} = \{z\}, \qquad \mathfrak{Q}\mathfrak{G}, \qquad \mathfrak{P}\mathfrak{G}.$$

The characteristic functions of λ_n for $n = \pm 2, \pm 3, \cdots$ are

$$h_n = (\lambda_n - b)f_n - aQf_n.$$

They belong to the space $\mathfrak{H}'$ and span it.

The set of the characteristic values λ_n^+, λ_n^- has the two limit points

$$\lambda_\infty^+ = \frac{a+b}{2} + \frac{1}{2}[a^2+b^2]^{1/2}, \qquad \lambda_\infty^- = \frac{a+b}{2} - \frac{1}{2}[a^2+b^2]^{1/2}.$$

To give the spectral resolution of the form $k_1 E k_2$ with respect to the unit-form $(|)$ we observe the relations

$$k_1 E k_2 = h_1 E h_2 + b(\bar{f}_1 \mid \bar{f}_2),$$

$$(k_1 \mid k_2) = (h_1, F h_2) + 2(h_1, G h_2) + (\bar{f}_1 \mid \bar{f}_2).$$

Therefore the space $\bar{\mathfrak{F}}$ is characteristic with value b. The other characteristic functions belong to $\mathfrak{H}$; to find them we apply the remarks at the end of §1 and get the following:

The values $\lambda=0$, a, $b/2$, $(a+b)/2$, b are characteristic with spaces $\mathfrak{Q}\mathfrak{F}=\{iz\}$, $\mathfrak{P}\mathfrak{F}=\{z\}$, $\mathfrak{Q}\mathfrak{G}$, $\mathfrak{P}\mathfrak{G}$, $\mathfrak{F}$. The other characteristic values λ_n^+, λ_n^-, $n=\pm 2, \pm 3, \cdots$, are the solutions of the equation

$$(\lambda_n - a)(2\lambda_n - b) + (\lambda_n - b)a\frac{1-\mu_n}{2};$$

they have two limit points λ_∞^+, λ_∞^-, solutions of

$$(\lambda_\infty - a)(2\lambda_\infty - b) + (\lambda_\infty - b)\frac{a}{2} = 0.$$

Their characteristic functions belong to the space $\mathfrak{H}'$ and span it.

5. Application to the theory of the elastic plate

Finally we outline the application of the inequality of Theorem 4.2 to the theory of an elastic plate. Let us imagine an elastic plate spread out over the domain D. We assume that the boundary B of D consists of a finite number of curves with a continuous tangent except at a finite number of corners as in §4. Let $k(x, y)$ be the displacement transforming every point z of D into the point $z+k$. Then the potential energy arising from this deformation† is

† The theory of *transversal* displacements also depends on the form kEk provided that k is the gradient $k=\partial j/\partial z$ of a real function $j(x, y)$. The equilibrium problem has been treated several times with the help of variational methods. For the first time by G. Fubini, loc. cit., and by W. Ritz, Journal für die reine und angewandte Mathematik, vol. 135 (1909). For the vibrations see W. Ritz, Annalen der Physik (1909). For both see K. Friedrichs, Mathematische Annalen, vol. 98 (1927), pp. 206–247. For these problems, where $k=\partial j/\partial z$ belongs to $\mathfrak{P}+\bar{\mathfrak{F}}$, our inequality (Theorem 4.2) need not be used if $b\neq 0$.

$$kEk = a\int\!\!\int_D \left(\Re\frac{\partial k}{\partial z}\right)^2 dxdy + b\int\!\!\int_D \left|\frac{\partial k}{\partial z}\right|^2 dxdy,$$

where a and b are positive constants.†

We restrict k to the subspace $\mathfrak{K}_0$ of one-valued functions in $\mathfrak{K}$. In $\mathfrak{K}_0$ we introduce the form

$$k_1Hk_2 = \Re\int\!\!\int_D \bar{k}_1k_2dxdy.$$

Then the kinetic energy arising from the vibration $ke^{i\omega t}$ is proportional to

$$kHk = \int\!\!\int_D |k|^2dxdy.$$

Let the complex-valued function $\phi(x, y)$ be the density of a force applied over the interior of D; the potential energy of this force is

$$-kH\phi = -\Re\int\!\!\int_D \bar{k}\phi dxdy.$$

There are different cases according as the displacement k has to satisfy boundary conditions or not.

First we take the case (1) of no displacement at the boundary B. Then the displacement g belongs to the subspace $\mathfrak{G}$.

The problem of equilibrium (1) is: to find a displacement $\underline{g}$ in $\mathfrak{G}$ such that

$$2g'E\underline{g} - g'H\phi = 0$$

for all g' in $\mathfrak{G}$.

The theory of vibrations (1) requires the simultaneous spectral resolution of the forms gEg and gHg.

Now, for g in $\mathfrak{G}$ we have the relation

$$gEg \geqq b\int\!\!\int_D \left|\frac{\partial g}{\partial \bar{z}}\right|^2 dxdy = \frac{b}{2}(g \mid g).$$

Since the form H in $\mathfrak{G}$ is bounded and completely continuous with respect to $(\mid)$, it has like properties with respect to E. Therefore no difficulty arises in solving the two problems.

In case (2) there is no boundary condition for the displacement (and no force working at the boundary).

† It is

$$a = \frac{m+1}{m-1}G = \frac{2m}{m-1}E, \qquad b = G = \frac{2m}{m-1}E$$

where E, G, m are the moduls of elasticity.

The question of *equilibrium* (2) is: to find a displacement $\underline{k}$ in $\mathfrak{K}_0$ such that

$$2k'E\underline{k} - k'H\phi = 0$$

for all k' in $\mathfrak{K}'$. Here we must assume $1H\phi=0$, $iH\phi=0$, that is, $\int\int_D \phi dx=0$ and $izH\phi=\mathfrak{J}\int\int_D \bar{z}\phi dxdy=0$ as we shall see.

The theory of *vibrations* (2) requires the simultaneous spectral resolution of the forms kEk and kHk for k in $\mathfrak{K}_0$.

In the case (2) the energy kEk vanishes for a pure translation $k=$const. and a pure rotation $k=iz$. To exclude this we take the accessory conditions

$$(5.1) \qquad 1Hk = 0, \qquad iHk = 0 \qquad \text{or} \qquad \int\int_D kdxdy = 0$$

$$(5.2) \qquad izHk = 0 \qquad \text{or} \qquad \mathfrak{J}\int\int_D \bar{z}kdxdy = 0.$$

By $\mathfrak{K}_1$ or $\mathfrak{K}_2$ we denote the space of all functions k in $\mathfrak{K}_0$ satisfying (5.1) or (5.1) and (5.2) respectively. In $\mathfrak{K}_1$ and $\mathfrak{K}_2$ we use the unit-form $(|)$.

We employ the inequality of Poincaré:†

There is a positive constant π such that for all k in $\mathfrak{R}_1$

$$kHk \leqq \pi(k \mid k).$$

So the form H is bounded in $\mathfrak{R}_1$. Therefore the spaces $\mathfrak{K}_1$ and $\mathfrak{K}_2$ are closed with respect to $(|)$.

Further we note†

The form H is completely continuous in $\mathfrak{K}_1$, with respect to $(|)$.

Now we prove

THEOREM 5.1. *There is a positive constant ϵ such that for all functions k in $\mathfrak{K}_2$*

$$\epsilon(k \mid k) \leqq (kEk).$$

Since the space $\{iz\}$ has the dimension 1 it has a positive smallest angle $\tau_0 \leqq \pi/2$ with respect to the space $\mathfrak{K}_2$ and the section $\{iz\}\mathfrak{K}_2=0$. Let aiz be the projection of k into $\{iz\}$; then we have (cf. $(1.2)_P$) for k in $\mathfrak{K}_2$

$$(k - aiz \mid k - aiz) \geqq \sin^2 \tau_0(k \mid k).$$

Now, $k-aiz$ being orthogonal to $\{iz\}$, Theorem 4.3 gives a constant $\sigma>0$ such that

† Cf. e.g. K. Friedrichs, *Spektraltheorie halbbeschränkter Operatoren und Anwendung auf die spektral Zerlegung von Differentialoperatoren*. II. Mathematische Annalen, vol. 109 (1934), pp. 705–707.

$$\sigma(k - aiz \mid k - aiz) \leqq (k - aiz)\boldsymbol{E}(k - aiz).$$

But we have

$$(k - aiz)\boldsymbol{E}(k - aiz) = (k\boldsymbol{E}k).$$

Thus Theorem 5.1 is proved.

In consequence of Theorem 5.1 we can take $\boldsymbol{E}$ as unit-form in the space $\mathfrak{K}_2$, and we are sure that every subspace of $\mathfrak{K}_2$ which is complete with respect to $(\,|\,)$ is complete with respect to $\boldsymbol{E}$; especially $\mathfrak{K}_2$ itself is closed as to $(\,|\,)$; further that every form in $\mathfrak{K}_2$ which is bounded or completely continuous with respect to $(\,|\,)$ has a like property with respect to $\boldsymbol{E}$.

Now, since $k\boldsymbol{H}\phi$ is a bounded linear form, it is a well known fact that there is a function $\underline{k}$ in $\mathfrak{R}_2$ such that

$$2k'\boldsymbol{E}\underline{k} - k'\boldsymbol{H}\phi = 0$$

for all k' in $\mathfrak{R}_2$; but since we have assumed $(a+ib)\boldsymbol{H}\phi=0$, $iz\boldsymbol{H}\phi=0$, this relation holds for all k in $\mathfrak{R}_0$. This function $\underline{k}$ is the solution of the *equilibrium problem* (2).

The form $\boldsymbol{H}$ is completely continuous with respect to $\boldsymbol{E}$. From this we get the solution of the *vibration problem* (2):

There is a sequence of values $\eta_1 \leqq \eta_2 \leqq \eta_3 \leqq \cdots \to \infty$ *and of functions* $k_1, k_2, k_3, \cdots$ *in* $\mathfrak{R}_2$ *orthogonal with respect to* $\boldsymbol{H}$ *and* $\boldsymbol{E}$ *such that every function* k *in* $\mathfrak{R}_2$ *can be developed into a series*

$$k = a_1k_1 + a_2k_2 + a_3k_3 + \cdots$$

by real coefficients $a_1, a_2, a_3, \cdots$ *in the sense that*

$$k\boldsymbol{H}k = a_1^2 + a_2^2 + a_3^2 + \cdots$$
$$k\boldsymbol{E}k = \eta_1a_1^2 + \eta_2a_2^2 + \eta_3a_3^2 + \cdots.$$

It is beyond our purpose to discuss the nature of the function $\underline{k}$ and of these characteristic functions in detail.

Braunschweig, Germany

[54–2] **The existence of solitary waves, (With D.H. Hyers), Comm. Pure Appl. Math., VII (1954), 517–550.**

Commentary by Louis Nirenberg:

This beautiful paper gives a rigorous solution of the existence of a solitary wave in two dimensional irrotational flow of a fluid of finite height over a horizontal bottom, and which has constant velocity at infinity. When observed from a frame moving with constant speed the flow appears stationary and the shape of the top of the fluid is that of a solitary wave.

Beginning with Boussines*q*, approximate solutions had been described. In this paper the existence of a solution with speed close to critical speed is proved as a convergent expansion in a small parameter, the amplitude, roughly the maximal height of the wave, minus its height at infinity. (M.A. Lavrentiev had earlier obtained such solutions as limits of periodic waves, [1]). One has to solve Laplace equations in the domain with given boundary conditions on the bottom and on the (unknown) top, where the boundary condition is nonlinear.

The shallow water approximation is taken as a first approximation, and the authors take the daring step of stretching the dependent and the independent variables in a suitable way. This has the effect of ruining the Laplace equation. Using power series expansions one has to solve a series of difficult linear problems. This is done with the aid of a fundamental solution due to F. John and a corresponding integral equation.

The paper is very technical but it is a remarkable tour de force; to my mind, one of Friedrichs' most striking papers. As in his 1933 paper on free boundary flow, it illustrates Friedrichs' daring in making a change of variable which appears to ruin a beautiful looking equation.

Recently C.J. Amick and J.F. Toland, [2], by a quite different approach, obtained solitary waves of finite (i.e. not small) amplitude.

[1] Lavrentiev, M.A., On the theory of long waves, Akad. Nauk Ukrain. R.S.R., Zbornik Prac. Inst. Mat. V., (1946), 8, (1947), 13–69; A.M.S. Translation 102 (1954), 3–50.

[2] Amick, D.J. and Toland, J.F., On solitary water-waves of finite amplitude, Arch. Rat. Mech. Anal. 76 (1981), 9–95.

The Existence of Solitary Waves*

By K. O. FRIEDRICHS and D. H. HYERS

1. *Introduction*

A solitary wave is a permanent two-dimensional irrotational flow of a fluid which has a finite altitude over a horizontal bottom and which is at rest at infinity. The qualification "permanent" indicates that the flow appears to be steady when observed with an appropriate velocity. The problem is to establish that such a flow is possible assuming that the only force aside from pressure forces acting on the fluid is gravity.

Approximate descriptions of such solitary waves were given independently by Boussinesq (1871), Lord Rayleigh (1876), and more systematically, by Levi-Cevita (1911), see [1,2,3]. A refinement of these approximations was given in 1926 by A. Weinstein [5].

These approximate descriptions indicate that solitary waves exist only when the wave velocity U—or the velocity of the observer to whom the flow appears steady—is supercritical,

$$U > (gh)^{1/2}.$$

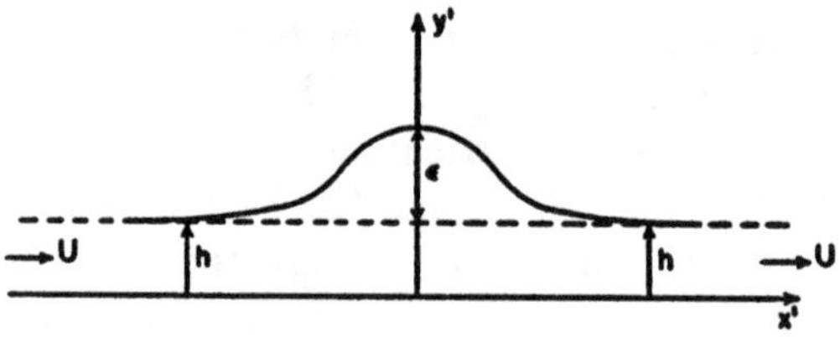

FIG. 1. Solitary wave.

Here h is the altitude—or depth—of the fluid at infinity, assumed to be the same at both ends, and g is the acceleration of gravity. It is further indicated that to every value of the ratio $U(gh)^{-1/2}$ or of the "reduced depth"

$$\gamma = ghU^{-2}$$

a single symmetric solitary wave exists except for translations.

A new approximate treatment of solitary waves was given by J. Keller [9] in connection with an expansion procedure for flow problems of which the first term corresponds to the "shallow water" theory. Specifically, the solitary wave

*This paper represents results obtained under Contract Nonr-285 (06), Office of Naval Research.

could be found only by including terms of second order and performing the expansion near $\gamma = 1$.

Recently, Lavrentiev [7] has given a proof of the existence of permanent solitary waves in which these waves are considered as limits of periodic waves with indefinitely increasing wave length. In the present paper we prove the existence of solitary waves in a direct manner.

The existence of periodic permanent waves of infinite depth was proved by Levi-Civita [3]. He employed an expansion with respect to powers of an appropriate quantity, essentially the elevation of the crest over the trough. As independent variables he used essentially the potential and the stream function of the steady wave motion, the reason being that the domain of these quantities is known a priori. Similarly, Struik [4] proved the existence of periodic permanent waves of finite depth. An attempt to prove the existence of solitary waves in a similar manner would fail, however; the solutions of the equations which would arise succesively in such an expansion procedure would lead only to the parallel flow with constant velocity. This occurrence is avoided in Lavrentiev's procedure of approaching solitary waves by periodic ones.

The failure of a direct expansion procedure is due to a non-uniformity in the convergence of the set of solitary waves to the parallel flow when the elevations of their crests approach zero. Denoting this elevation by ϵ, the top surface is in first approximation given by

$$y = h + \epsilon \operatorname{sech}^2 (3\, \epsilon/4\, h^3)^{1/2} x$$

when the abscissa x runs in the horizontal direction as the ordinate y runs upwards. Evidently, the terms in the expansion of this function with respect to powers of ϵ become infinite as $| x |$ tends to infinity.

This difficulty is easily remedied by introducing $\epsilon^{1/2}x$, or an equivalent quantity, as independent variable. In other words, the horizontal independent variable is subjected to a stretching, dependent on the parameter ϵ, while the vertical independent variable is left untouched. In doing so one destroys the harmonic character of the functions used to describe the flow. Although for each positive value of ϵ the functions are still equivalent to potential functions—inasmuch as they can be transformed into such functions by stretching—this is no longer the case for the limit function approached as $\epsilon \to 0$. Thus one may say that a set of harmonic functions is developed in the neighborhood of a non-harmonic function. Such a procedure seems to be indispensable in a direct approach to the treatment of solitary waves.

A stretching of the horizontal variable[1] is an essential feature of the shallow water theory. The only difference between our approach and the shallow water

[1]Similar stretchings play an essential role in the theory of elasticity in the transition from plates to beams and from elastic bodies to plates. They also occur in certain variants of the boundary layer theory.

approach lies in the use of different independent variables. Except for this difference, we may say that our result establishes the validity of the shallow water approach in a special case. When we speak of the shallow water approach we mean the expansion described by Keller [9] and not simply the shallow water theory of first order in which solitary waves do not occur, since they are essentially of second order.

In Section 2 the problem is described mathematically and the new independent variables (the velocity potential φ and stream function ψ) and dependent variables θ, τ (essentially the angle of the velocity and logarithm of the magnitude of the velocity) are introduced. In place of the reduced depth $\gamma = gh/U^2$, the parameter $a = (-\frac{1}{3}\log\gamma)^{1/2}$ is always used.

An expansion procedure is explained in Section 3 after an appropriate stretching $\hat{\varphi} = a\varphi$ of the independent variable φ as described above. Section 4 is devoted to Green's function for a related linear problem. Using this function, integral equations are set up for the problem in Section 5, and then transformed by means of the transformations $\hat{\varphi} = a\varphi$, $\hat{\theta} = a^{-3}\theta$, $\hat{\tau} = a^{-2}\tau$ into the form (5.18). In order to satisfy the boundary conditions at infinity, it turns out that a certain auxiliary integral condition (5.19) must be satisfied. On the basis of this condition, the system of integral equations (5.18) is replaced by a modified system (5.23). Section 5 is concluded by obtaining a solution of equations (5.23) and (5.19) for the special case $a = 0$.

In Section 6 certain function spaces B_1, B_2 and $B = B_1 \times B_2$ are defined, and it is shown that the system (5.23) may be regarded as an equation $G(\omega, a) = 0$ where $\omega = \{\hat{\theta}, \hat{\tau}\}$ and G are elements of B. It is also proved that G and its Fréchet differential $\delta G(\omega, a; \delta\omega)$ are uniformly continuous (Lemma 6.4).

Section 7 is devoted to an analysis of the variational equation $\delta G(\omega, a; \delta\omega) = \delta\zeta$ corresponding to the equation $G(\omega, a) = 0$, and to a proof of the existence of a two parameter family of solutions $\omega = \omega(a, c, \hat{\varphi})$ of the latter equation. Finally it is shown that for each sufficiently small $a > 0$, the parameter $c = c(a)$ may be chosen so that $\omega = \{\hat{\theta}, \hat{\tau}\}$ also satisfies the auxiliary equation (5.19) or (7.1). This establishes the existence of a symmetric solitary wave for values of the parameter $\gamma = gh/U^2$ which are less than one and sufficiently close to one.

2. *Description of the Problem*

Let U be the velocity of propagation of the wave, and let horizontal and vertical axes be chosen moving with the wave so that the vertical y'-axis passes through the crest C of the wave with the origin taken at the bottom. The asymptotic depth (as $x' \to \infty$) will be denoted by h. With this choice of axes the motion may be regarded as a steady, two-dimensional and irrotational flow, so that we may introduce the velocity potential φ', the stream function ψ' and the complex quantities $z' = x' + iy'$, $\chi' = \varphi' + i\psi'$, $w' = u' - iv'$, where $w' = d\chi'/dz'$. The corresponding dimensionless quantities will be denoted without primes: $z = z'/h$, $w = w'/U$, $\chi = \chi'/hU$, so that $d\chi/dz = w$. Further we introduce the "reduced altitude"

$$\gamma = gh/U^2 \tag{2.1}$$

as the essential dimensionless parameter of the problem.

Along the free surface streamline l, the Bernoulli law gives us the boundary condition

$$\tfrac{1}{2} \mid w \mid^2 + \gamma y = \text{const.} \qquad \text{along } l, \tag{2.2}$$

in the terms of the dimensionless variables. We must also have

$$\begin{aligned} \psi &= 0 \text{ on } y = 0, \qquad \psi = 1 \text{ on } l, \\ w &\to U \text{ as } \mid x \mid \to \infty, \qquad \mathcal{I}m\, w = 0 \text{ on } y = 0. \end{aligned} \tag{2.3}$$

The complex potential function $\chi(z)$ will map the region occupied by the fluid in the z-plane onto the strip $0 < \psi < 1$ in the χ-plane. Taking χ as independent variable we wish to find an analytic function $w(\chi)$ defined in this strip and satisfying the appropriate boundary conditions.

To simplify the form of the boundary condition (2.2) we follow Levi-Civita [3] and introduce instead of w the complex dependent variable $\theta + i\lambda$, where

$$w = \exp\{-i(\theta + i\lambda)\}$$

so that $\mid w \mid = e^\lambda$ and $\theta = \arg \bar{w}$ is the inclination of the velocity vector. Condition (2.2) may now be written in the form

$$\frac{\partial \theta}{\partial \psi} = \gamma e^{-3\lambda} \sin \theta. \tag{2.4}$$

We also introduce the parameter a, given by

$$e^{-3a^2} = \gamma = gh/U^2, \tag{2.5}$$

which will be used throughout instead of γ. It will be assumed that a is real, i.e. that

$$\gamma = gh/U^2 < 1. \tag{2.6}$$

Finally, we put

$$\tau = \lambda + a^2, \tag{2.7}$$

so that condition (2.4) becomes

$$\frac{\partial \theta}{\partial \psi} = e^{-3\tau} \sin \theta \quad \text{along} \quad \psi = 1. \tag{2.8}$$

The other boundary conditions are

(2.9) $$\theta = 0 \quad \text{along} \quad \psi = 0,$$

(2.10) $$\text{as} \quad \varphi \to \pm\infty, \qquad \theta \to 0 \quad \text{and} \quad \tau \to a^2.$$

Our problem is this: for a given value of a find a function $\omega = \omega(\chi) = \theta + i\tau$, analytic in the strip $0 < \psi < 1$, continuous along $\psi = 0$ and $\psi = 1$ and satisfying the boundary conditions (2.8), (2.9) and (2.10). Clearly, such a function $\omega = \omega(\chi)$ leads to a solution of the original problem of finding an appropriate region in the z-plane and an analytic function $\chi = \chi(z)$ defined in it which satisfies conditions (2.2) and (2.3). Having proved the existence of a function $\omega(\chi)$ as described, we may say that we have proved the existence of a solitary wave.

3. *An Approximate Solution*

Although approximate solutions have been obtained in different ways by several authors, as noted in the introduction, we give here another procedure, which not only gives the usual first approximation but also in principle leads to higher order approximations, and will later form the basis of our existence proof.

As explained in the introduction, the essential step which makes this approximation procedure possible consists in an appropriate stretching of the horizontal independent variable φ, leaving the vertical independent variable ψ unchanged. Using the parameter a defined by (2.5), we introduce specifically the new independent variable

$$\phi = a\varphi.$$

Then we consider the dependent variables θ and τ as functions of ϕ, ψ and a and expand them in powers[2] of a:

(3.1) $$\tau = a^2\tau_1(\phi, \psi) + a^4\tau_2(\phi, \psi) + \cdots, \qquad \theta = a^3\theta_1(\phi, \psi) + a^5\theta_2(\phi, \psi) + \cdots.$$

In terms of ϕ and ψ the Cauchy-Riemann equations for θ, τ become

(3.2) $$\theta_\psi = -a\tau_\phi\,; \qquad \tau_\psi = a\theta_\phi\,.$$

If we substitute the expansions (3.1) into (3.2) and equate coefficients of the first two lowest powers of a we get

$$\theta_{1\psi} = -\tau_{1\phi}\,; \qquad \tau_{1\psi} = 0; \qquad \theta_{2\psi} = -\tau_{2\phi}\,; \qquad \tau_{2\psi} = \theta_{1\phi}\,.$$

As is seen from the second equation, the function τ_1 is independent of ψ,

[2]It can be shown that the coefficients of the powers of a omitted may be taken as zero without loss of generality.

$\tau_1 = \tau_1(\phi)$. On integrating the other equations and using the boundary condition $\theta = 0$ for $\psi = 0$ we obtain the relations

$$\theta_1 = -\psi\tau_1'(\phi); \qquad \tau_2 = -\tfrac{1}{2}\psi^2\tau''(\phi) + j(\phi),$$
(3.3)
$$\theta_2 = \tfrac{1}{6}\psi^3\tau_1'''(\phi) - \psi j'(\phi),$$

where $j(\phi)$ is a function of ϕ alone, to be determined from the later steps.

When the expansions (3.1) are substituted into the boundary condition (2.8) and the coefficients of a^3 and of a^5 on both sides of the result are equated, one has

$$\theta_{1\psi}(\phi, 1) = \theta_1(\phi, 1),$$
$$\theta_{2\psi}(\phi, 1) = \theta_2(\phi, 1) - 3\tau_1\cdot\theta_1(\phi, 1).$$

From (3.3) we deduce that the first equation is automatically satisfied and that the second will be satisfied if and only if $\tau_1(\phi)$ satisfies the ordinary differential equation

$$\tau_1''' = 9\tau_1\tau_1' \,. \tag{3.4}$$

In order to specify a solution of this third order equation, three conditions are necessary. We take these:

$$\tau_1'(0) = 0, \qquad \tau_1(\infty) = 1, \qquad \tau_1''(\infty) = 0. \tag{3.5}$$

The first condition should be satisfied since $\theta = 0$ at the crest, while the second and third result from the condition that $\tau \to a^2$ as $\varphi \to \infty$, which we satisfy by taking $\tau_1(\infty) = 1$, $\tau_n(\infty) = 0$, $n > 1$.

The differential equation (3.4) may be integrated by quadratures and the solution satisfying (3.5) is

$$\tau_1 = 1 - 3\,\mathrm{sech}^2\,\tfrac{3}{2}\phi. \tag{3.6}$$

Within terms of higher order in a one easily derives from this result the expressions

$$y = 1 + 3a^2\,\mathrm{sech}^2\,3ax/2,$$
(3.7)
$$u = 1 - 3a^2\,\mathrm{sech}^2\,3ax/2$$

for the wave profile and the velocity along it.

The expression for the wave profile is that given by Boussinesq, Rayleigh, and Keller. The expression for the velocity agrees with that of Keller, except that instead of the factor $3a^2$ he has the factor $3a^2 \exp\{3a^2/2\}\,(1 + 3a^2/2)^{-1}$. As Keller remarks, Boussinesq and Rayleigh give $u = 1$ for the velocity.

Before a comparison with the formula given by Weinstein could be made, it would be necessary to determine the terms τ_2 and θ_2 in the expansion (3.1). This we do not carry out here.

4. *Green's Function*

The principal boundary condition (2.8) may be written in the form

$$\theta_\psi - \theta = e^{-3\tau} \sin\theta - \theta \quad \text{along} \quad \psi = 1. \tag{4.1}$$

In order to arrive at a formulation of our problem in terms of integral equations we shall need Green's function for the strip $0 < \psi < 1$ with the boundary conditions

$$\begin{aligned} \theta_\psi - \theta &= f(\varphi) \quad \text{along} \quad \psi = 1, \\ \theta &= 0 \quad\quad\ \text{along} \quad \psi = 0. \end{aligned} \tag{4.2}$$

This function $J(\chi \mid \chi') = H + iK$ will be specified in the next lemma. It will be of the form[3]

$$J(\chi \mid \chi') = -\frac{1}{2\pi} \log(\chi - \chi') + \rho(\chi, \chi') \tag{4.3}$$

where $\rho(\chi, \chi')$ is a regular analytic function of χ for χ and χ' in the strip $0 < \psi < 1$; its real part H will satisfy the conditions

$$H_\psi - H = 0 \quad \text{on} \quad \psi = 1, \qquad 0 \le \psi' < 1, \tag{4.4}$$

$$H = 0 \quad \text{on} \quad \psi = 0, \qquad 0 < \psi \le 1. \tag{4.5}$$

A proof of the following lemma is outlined in the appendix.

LEMMA 4.1. *Let C denote the contour consisting of the real axis in the χ-plane with an indentation around the origin into the negative half-plane.*

For $0 \le \psi < \psi' \le 1$, the function $J(\chi \mid \chi')$ can then be described[4] *by the formula*

$$J(\chi \mid \chi') = \frac{1}{2\pi i} \int_C \frac{\sin \mu(\chi - \varphi')}{\mu} \cdot \frac{\mu \cosh \mu(1 - \psi') - \sinh \mu(1 - \psi')}{\mu \cosh \mu - \sinh \mu} \, d\mu. \tag{4.6}$$

This function can be extended to the rest of the domain $0 \le \psi \le 1$, $0 \le \psi' \le 1$ in such a way as to satisfy (4.3), (4.4), (4.5).

It will be necessary to split the function J into two parts of which one, J_1, approaches zero as $|\varphi - \varphi'| \to \infty$, so that the other, J_0, describes the behavior of J for $|\varphi - \varphi'|$ large. This split is accomplished according to

LEMMA 4.2. *For $\psi' = 1$ we have the decomposition*

$$J(\chi \mid \varphi' + i) = J_0(\chi \mid \varphi' + i) + J_1(\chi \mid \varphi' + i), \tag{4.7}$$

---

[3]These conditions do not specify J uniquely, for example we can always add a purely imaginary constant to J and get another function satisfying the conditions.

[4]The description of Green's function given here is similar to the description of related Green's functions given by F. John [10].

where

$$J_0(\chi \mid \varphi' + i) = \tfrac{3}{4} is[\tfrac{1}{5} + (\chi - \varphi')^2],$$

$$J_1 = \frac{s}{2\pi} \int_c \frac{\exp\{-is\mu(\chi - \varphi')\}}{\mu \cosh \mu - \sinh \mu} \, d\mu,$$

$$s = \operatorname{sgn}(\varphi - \varphi') = \begin{cases} 1 & \text{if } \varphi - \varphi' > 0 \\ -1 & \text{if } \varphi - \varphi' < 0. \end{cases}$$

When $| \varphi - \varphi' | \to \infty$, $J_1 \to 0$.

The decomposition (4.7) may be established by setting $\psi' = 1$ in (4.6), subtracting J_1, expressing the result as a contour integral around a closed path containing the origin, and evaluating the latter integral by finding the residue at the origin.

LEMMA 4.3. *Let* $f(\varphi)$ *be continuous on* $-\infty < \varphi < \infty$ *and let the integral* $\int_{-\infty}^{\infty} \varphi^2 \mid f(\varphi) \mid d\varphi$ *exist. Then the function*

$$\omega(\chi) = \theta + i\tau = \int_{-\infty}^{\infty} J(\chi \mid \varphi' + i) f(\varphi') \, d\varphi' \tag{4.8}$$

is analytic in the open strip $0 < \psi < 1$, *continuous in the closed strip* $0 \leq \psi \leq 1$, *and* θ *satisfies the boundary conditions* (4.2) *in the sense that*

$$\lim_{\substack{\psi \to 1 \\ \psi < 1}} \{\theta_\psi - \theta\} = f(\varphi), \tag{4.9}$$

$$\theta = 0 \quad \text{for} \quad \psi = 0. \tag{4.10}$$

The proof of Lemma 4.3 is also given in the appendix.

5. *Integral Equations of the Problem*

For $\psi = \psi' = 1$ we write $J(\varphi, \varphi')$ instead of $J(\varphi + i \mid \varphi' + i)$ and define the linear operators Ω, $\Omega(\psi)$ by

$$\begin{aligned} \Omega f &= \int_{-\infty}^{\infty} J(\varphi, \varphi') f(\varphi') \, d\varphi', \\ \Omega(\psi) f &= \int_{-\infty}^{\infty} J(\chi \mid \varphi' + i) f(\varphi') \, d\varphi', \end{aligned} \tag{5.1}$$

f being real-valued and continuous; clearly, $\Omega f = \Omega(1) f$. We set

$$F(\theta, \tau) = e^{-3\tau} \sin \theta - \theta$$

in accordance with (4.1). We shall show that a suitably chosen solution of the integral equation

$$\theta + i\tau = \Omega F(\theta, \tau) + ia^2 \tag{5.2}$$

for the functions θ and τ on the boundary $\psi = 1$ will yield a solution of our problem. By Lemma 4.3, a solution $\theta + i\tau$ of equation (5.2) is an analytic function $\omega(\chi)$ whose boundary values along $\psi = 1$ and $\psi = 0$ satisfy conditions (2.8) and (2.9).

So far nothing has been said regarding the conditions (2.10) at infinity, and we now devote our attention to these.

Corresponding to the decomposition $J = J_0 + J_1$ of Lemma 4.2 we have the decomposition of the operator Ω:

$$\Omega = \Omega_0 + \Omega_1\,, \tag{5.3}$$

in which

$$\Omega_0 = \int_{-\infty}^{\infty} J_0(\varphi, \varphi')f(\varphi')\,d\varphi', \tag{5.4}$$

$$\Omega_1 = \int_{-\infty}^{\infty} J_1(\varphi, \varphi')f(\varphi')\,d\varphi', \tag{5.5}$$

with similar formulas for $\Omega(\psi)$.

As long as $f \to 0$ as $|\varphi| \to \infty$ strongly enough so that f is absolutely integrable on $(-\infty, \infty)$, it can be shown that $\Omega_1(\psi)f \to 0$ as $|\varphi| \to \infty$. However such is obviously not the case for $\Omega_0(\psi)f$, unless we impose much more severe restrictions on f, since J_0 is a quadratic polynomial in φ and φ'. We next investigate the nature of these restrictions.

By (4.7) we have

$$J_0(\chi \mid \varphi' + i) = \frac{3}{4}\,is(\varphi - \varphi')^2 - \frac{3}{2}\,s\psi(\varphi - \varphi') - \frac{3}{5}\,is\left(\frac{5}{4}\,\psi^2 - \frac{1}{4}\right),$$

so that

$$\Omega_0(\psi)f = -\psi S_1 f + iS_0 f - i\left(\frac{5}{4}\,\psi^2 - \frac{1}{4}\right)S_2 f \tag{5.6}$$

where

$$\begin{aligned} S_0 f &= \frac{3}{4}\int_{-\infty}^{\infty} \operatorname{sgn}(\varphi - \varphi')(\varphi - \varphi')^2 f(\varphi')\,d\varphi', \\ S_1 f &= \frac{3}{2}\int_{-\infty}^{\infty} |\varphi - \varphi'|\, f(\varphi')\,d\varphi', \\ S_2 f &= \frac{3}{5}\int_{-\infty}^{\infty} \operatorname{sgn}(\varphi - \varphi')f(\varphi')\,d\varphi'. \end{aligned} \tag{5.7}$$

LEMMA 5.1. *Let f be an odd continuous function which vanishes at ∞ strongly enough so that $\int_{-\infty}^{\infty} \varphi^2 \mid f(\varphi) \mid d\varphi$ exists. Then if $\int_{-\infty}^{\infty} \varphi f(\varphi) d\varphi = 0$, the relation $\Omega_0(\psi) f \to 0$ holds as $\mid \varphi \mid \to \infty$, uniformly for $0 \leq \psi \leq 1$. Moreover,*

$$(5.8)\qquad \begin{aligned} S_0 f &= -3 \int_{\varphi}^{\infty} d\varphi' \int_{\varphi'}^{\infty} d\varphi'' \int_{\varphi''}^{\infty} f(\varphi''')\, d\varphi''', \\ S_1 f &= 3 \int_{\varphi}^{\infty} d\varphi' \int_{\varphi'}^{\infty} f(\varphi'')\, d\varphi'', \\ S_2 f &= -\frac{6}{5} \int_{\varphi}^{\infty} f(\varphi')\, d\varphi'. \end{aligned}$$

Proof: $$\frac{5}{3} S_2 f = \int_{-\infty}^{\varphi} f(\varphi')\, d\varphi' - \int_{\varphi}^{\infty} f(\varphi')\, d\varphi' = -2 \int_{\varphi}^{\infty} f(\varphi')\, d\varphi'$$

since f is odd. Clearly $S_2 f \to 0$ as $\mid \varphi \mid \to \infty$.

$$\frac{2}{3} S_1 f = \int_{-\infty}^{\varphi} (\varphi - \varphi') f(\varphi')\, d\varphi' - \int_{\varphi}^{\infty} (\varphi - \varphi') f(\varphi')\, d\varphi'$$

$$= -2 \int_{\varphi}^{\infty} (\varphi - \varphi') f(\varphi')\, d\varphi',$$

since $\int_{-\infty}^{\infty} \varphi' f(\varphi')\, d\varphi' = 0$ and f is odd.

Hence $\lim_{\mid \varphi \mid \to \infty} S_1 f = 0$, and by integration by parts we get

$$S_1 f = 3 \int_{\varphi}^{\infty} d\varphi' \int_{\varphi'}^{\infty} f(\varphi'')\, d\varphi''.$$

The assertions regarding $S_0 f$ are proved in a similar fashion.

From now on, we shall always assume that *the wave is symmetric*, i.e. that θ is an odd function of φ and that τ is an even function of φ. Then we may restrict outselves to the consideration of odd functions f, since $F(\theta, \tau) = e^{-3\tau} \sin \theta - \theta$ is now odd as a function of φ.

With this understanding, suppose that we had a solution $\theta(\varphi)$, $\tau(\varphi)$ of (5.2) satisfying the auxiliary condition

$$\int_{-\infty}^{\infty} \varphi f(\varphi)\, d\varphi = 0,$$

where $f(\varphi) = F(\theta(\varphi), \tau(\varphi))$ tends to zero as $\mid \varphi \mid \to \infty$ strongly enough so that $\varphi^2 f(\varphi)$ is absolutely integrable on $-\infty < \varphi < \infty$. Then by Lemma 4.3 the function

$$\omega(\chi) = \theta(\chi) + i\tau(\chi) = \Omega(\psi) f + i a^2,$$

having $\theta(\varphi) + i\tau(\varphi)$ as its boundary values on $\psi = 1$, would be analytic for $0 < \psi < 1$ and would satisfy the boundary conditions (2.8), (2.9) on $\psi = 0$ and $\psi = 1$. By Lemma 5.1, $\Omega_0(\psi) f \to 0$ as $\mid \varphi \mid \to \infty$, and it can also be shown (by the same methods used in proving Lemma 6.2 below) that under our hypotheses $\Omega_1(\psi)f \to 0$ as $|\varphi| \to \infty$. Hence $\omega(\chi)$ would also satisfy the condition (2.10) at infinity, and therefore would furnish a solution to our problem. The rest of the paper is devoted to finding such a solution $\theta(\varphi)$, $\tau(\varphi)$ of (5.2).

A decisive step in our procedure is that we make the same *change of independent variable*,

$$\hat{\varphi} = a\varphi, \tag{5.9}$$

that we made in Section 3. In accordance with the developments of Section 3 we also put

$$\theta(\varphi, 1) = a^3\hat{\theta}(\hat{\varphi}); \qquad \tau(\varphi, 1) = a^2\hat{\tau}(\hat{\varphi}). \tag{5.10}$$

The function

$$F(\theta, \tau) = e^{-3\tau} \sin \theta - \theta = e^{-3a^2\hat{\tau}} \sin a^3\hat{\theta} - a^3\hat{\theta}$$

contains a^5 as a factor; therefore we set

$$\hat{F}(\hat{\theta}, \hat{\tau}, a) = a^{-5}F(a^3\hat{\theta}, a^2\hat{\tau}). \tag{5.11}$$

For later use, we note that $\hat{F}$ may be written in the form

$$\hat{F}(\hat{\theta}, \hat{\tau}, a) = -3\hat{\tau}\hat{\theta} + a^2F_1(\hat{\theta}, \hat{\tau}, a), \tag{5.12}$$

where $F_1(\hat{\theta}, \hat{\tau}, a)$ is an entire function of each of its arguments having $\hat{\theta}$ as a factor. Writing $\hat{f}(\hat{\varphi}) = a^{-5}f(\varphi)$, in accordance with (5.11), we have by (5.9):

$$S_1 f = \frac{3}{2}\int_{-\infty}^{\infty} |\varphi - \varphi'| f(\varphi')\, d\varphi' = \frac{3}{2}a^3 \int_{-\infty}^{\infty} |\hat{\varphi} - \hat{\varphi}'| \hat{f}(\hat{\varphi}')\, d\hat{\varphi}' = a^3 S_1 \hat{f}.$$

Similarly $S_0 f = a^2 S_0 \hat{f}$, $S_2 f = a^4 S_2 \hat{f}$, so that

$$\Omega_0 f = -a^3 S_1 \hat{f} + ia^2 S_0 \hat{f} - ia^4 S_2 \hat{f}. \tag{5.13}$$

We also set

$$J_1^a(\hat{\varphi}, \hat{\varphi}') = \frac{s}{2\pi}\int_c \frac{\exp\{-si\mu a^{-1}(\hat{\varphi} - \hat{\varphi}')\}e^{a\mu}\, d\mu}{\mu \cosh \mu - \sinh \mu} \tag{5.14}$$

with $s = \operatorname{sgn} a(\hat{\varphi} - \hat{\varphi}')$, and

$$\Omega_1^a \hat{f} = \int_{-\infty}^{\infty} J_1^a(\hat{\varphi}, \hat{\varphi}')\hat{f}(\hat{\varphi}')\, d\hat{\varphi}'. \tag{5.15}$$

It follows that

$$\Omega_1 f = a^4 \Omega_1^a \hat{f}. \tag{5.16}$$

From now on it will often be convenient to separate the operator Ω_1^a into its real and imaginary parts and write

$$\Omega_1^a = \Theta_1^a + iT_1^a\,. \tag{5.17}$$

In place of the complex equation (5.2) we now have the system of two real non-linear integral equations

(5.18)
$$\hat{\theta}(\hat{\varphi}) = [-S_1 + a\Theta_1^a]F(\hat{\theta}, \hat{\tau}, a),$$
$$\hat{\tau}(\hat{\varphi}) = 1 + [S_0 - a^2S_2 + a^2T_1^a]F(\hat{\theta}, \hat{\tau}, a)$$

for the determination of the functions $\hat{\theta}$, $\hat{\tau}$. We seek a solution $\hat{\theta}$, $\hat{\tau}$ of equation (5.18) which also satisfies the condition

(5.19)
$$\int_{-\infty}^{\infty} \hat{\varphi}F(\hat{\theta}, \hat{\tau}, a)\, d\hat{\varphi} = 0$$

as well as the boundary conditions at ∞:

(5.20)
$$\hat{\theta} \to 0, \qquad \hat{\tau} \to 1 \quad \text{as} \quad |\hat{\varphi}| \to \infty,$$

which result from (2.10).

Let $\eta(\hat{\varphi})$ be a function such that $\int_{-\infty}^{\infty} \hat{\varphi}\eta(\hat{\varphi})d\hat{\varphi} = 1$. Specifically, we can and will always take

(5.21)
$$\eta(\hat{\varphi}) = \frac{9}{2}\frac{\sinh 3\hat{\varphi}}{(1 + \cosh 3\hat{\varphi})^2} = \frac{1}{4}\tau_1'(\hat{\varphi}),$$

where $\tau_1(\hat{\varphi})$ is given by (3.6). Put

(5.22)
$$E(\hat{\theta}, \hat{\tau}, a) = F(\hat{\theta}, \hat{\tau}, a) - \eta(\hat{\varphi}) \int_{-\infty}^{\infty} \hat{\varphi}'F(\hat{\theta}(\hat{\varphi}), \hat{\tau}(\hat{\varphi}'), a)\, d\hat{\varphi}'.$$

Then obviously $\int_{-\infty}^{\infty} \hat{\varphi}E(\hat{\varphi}, \hat{\tau}, a)d\hat{\varphi} = 0$.

Instead of dealing directly with the system (5.18) we shall try to solve the modified system

(5.23)
$$\hat{\theta}(\hat{\varphi}) = [-S_1 + a\Theta_1^a]E(\hat{\theta}, \hat{\tau}, a),$$
$$\hat{\tau}(\hat{\varphi}) = 1 + [S_0 - a^2S_2 + a^2T_1^a]E(\hat{\theta}, \hat{\tau}, a).$$

Then finally we shall try to choose a solution satisfying (5.19) as well. Such a solution will satisfy all our requirements. It will turn out that, for each value of a, equations (5.23) possess a one-parametric manifold of solutions. We shall be able to determine the parameter as a function of a such that condition (5.19) is fulfilled.

We first investigate the solution for $a = 0$. By (5.12) and (5.22) we have

(5.24)
$$E(\hat{\theta}, \hat{\tau}, 0) = -3\left(\hat{\tau}\hat{\theta} - \eta \int_{-\infty}^{\infty} \hat{\tau}(\hat{\varphi}')\hat{\theta}(\hat{\varphi}')\hat{\varphi}'\, d\hat{\varphi}'\right).$$

For $a = 0$, the system (5.23) reduces to

(5.25)
$$\hat{\theta}(\hat{\varphi}) = -S_1E(\hat{\theta}, \hat{\tau}, 0),$$
$$\hat{\tau}(\hat{\varphi}) = 1 + S_0E(\hat{\theta}, \hat{\tau}, 0).$$

From lemma (5.1) it follows that $S_1 \hat{f} = (S_0 \hat{f})'$ and $(S_0 \hat{f})''' = 3\hat{f}$ (the primes denote differentiation with respect to $\hat{\varphi}$). Hence from (5.25) we get

$$\hat{\theta} = -\hat{\tau}', \qquad \hat{\tau}''' = 3E(\hat{\theta}, \hat{\tau}, 0),$$

or

$$\hat{\tau}''' = 9\hat{\tau}\hat{\tau}' + 9k\eta , \tag{5.26}$$

where

$$k = \int_{-\infty}^{\infty} \hat{\tau}\hat{\theta}\hat{\varphi}\, d\hat{\varphi} = -\int_{-\infty}^{\infty} \hat{\tau}\hat{\tau}'\hat{\varphi}\, d\hat{\varphi}.$$

Now in Section 3 we found that the function $\tau_1 = 1 - 3 \operatorname{sech}^2 \frac{3}{2}\hat{\varphi}$ is a solution of the third order equation $\tau_1''' = 9\tau_1\tau_1'$ satisfying the conditions $\tau_1(\infty) = 1$, $\tau_1'(0) = \tau_1'(\infty) = 0$, $\tau_1''(\infty) = 0$. Using integration by parts we find

$$k_1 = -\int_{-\infty}^{\infty} \tau_1\tau_1'\hat{\varphi}\, d\hat{\varphi} = \int_0^{\infty} (\tau_1^2 - 1)\, d\hat{\varphi}$$

for this solution. By integrating the equation $\tau_1''' = 9\tau_1\tau_1'$ and observing that $\tau_1(\infty) = 1$, we get $\tau_1'' = \frac{9}{2}(\tau_1^2 - 1)$. Hence, by integrating from 0 to ∞ we obtain

$$k_1 = \int_0^{\infty} (\tau_1^2 - 1)\, d\hat{\varphi} = \frac{2}{9}\{\tau_1'(\infty) - \tau_1'(0)\} = 0.$$

Therefore $\hat{\tau} = \tau_1(\hat{\varphi})$ is a solution of (5.26), and it follows that

$$\begin{aligned} \hat{\theta} &= -\tau_1'(\hat{\varphi}), \\ \hat{\tau} &= \tau_1(\hat{\varphi}) \end{aligned} \tag{5.27}$$

is a solution of (5.25) which, as we have just seen, also satisfies the auxiliary condition (5.19).

6. *The Function Spaces. Continuity and Differentiability*

Before we can solve equation (5.23) for $a > 0$ we must introduce appropriate function spaces. Although we shall be dealing exclusively with the new variables $\hat{\varphi}$, $\hat{\theta}$, $\hat{\tau}$ (see (5.9), (5.10)) we shall for simplicity drop the circumflexes. We also write $\tau = t + 1$.

Let B_1 denote the class of all those odd continuous functions $\theta(\varphi)$ defined on $-\infty < \varphi < \infty$ for which $e^{2\varphi}\theta(\varphi)$ remains bounded for all $\varphi \geq 0$, and define the norm $\| \theta \|$ as

$$\| \theta \| = \sup_{0 \leq \varphi < \infty} e^{2\varphi} \mid \theta(\varphi) \mid.$$

Similarly, let B_2 denote the class of all those even continuous functions $t(\varphi)$ for which $e^{2\varphi}t(\varphi)$ remains bounded for $\varphi \geq 0$ and put

$$\| t \| = \sup_{0 \leq \varphi < \infty} e^{2\varphi} \mid t(\varphi) \mid .$$

Then with the indicated norms, the classes B_1 and B_2 are Banach spaces, as is easily verified. We shall also often use the product space $B = B_1 \times B_2$ consisting of all pairs

$$\omega = \{\theta, t\} \quad \text{with} \quad \theta \in B_1, \quad t \in B_2$$

and with the norm

$$\| \omega \| = (\| \theta \|^2 + \| t \|^2)^{1/2}.$$

In this section we shall show that the system (5.23) may be regarded as an equation of the form $G(\omega, a) = 0$ in which ω and G belong to the Banach space B and then investigate the continuity and differentiability of G.

The proof will be broken up into a series of lemmas.

LEMMA 6.1. *Let $f \in B_1$ satisfy $\int_{-\infty}^{\infty} \varphi f(\varphi)\, d\varphi = 0$. Then $S_0 f \in B_2$, $S_1 f \in B_1$, $S_2 f \in B_2$ and*

$$\| S_0 f \| \leq \tfrac{3}{8} \| f \|, \qquad \| S_1 f \| \leq \tfrac{3}{4} \| f \|, \qquad \| S_2 f \| \leq \tfrac{3}{5} \| f \|.$$

Lemma 6.1 follows immediately from Lemma 5.1, on using the inequality $\mid f(\varphi) \mid \leq e^{-2\varphi} \| f \|$ in formulas (5.8).

LEMMA 6.2. *If $f \in B_1$ and $0 \leq a \leq \pi/4$, then*

$$\| \Omega_1^a f \| \leq a \| f \|,$$

where $\| \Omega_1^a f \|^2 = \| \Theta_1^a f \|^2 + \| T_1^a f \|^2$.

Proof: We set

$$j(\varphi, a) = \Omega_1^a f = \int_{-\infty}^{\infty} J_1^a(\varphi, \varphi') f(\varphi')\, d\varphi'.$$

We need only consider the case $\varphi > 0$, since $\Theta_1^a f$ is odd and $T_1^a f$ is even in φ. By (5.14) and (5.15),

$$j(\varphi, a) = \frac{1}{2\pi} \int_{-\infty}^{\varphi} f(\varphi')\, d\varphi' \int_C \frac{\exp\{-i\mu(\varphi - \varphi')/a\} e^{\mu}}{\mu \cosh \mu - \sinh \mu}\, d\mu$$
$$- \frac{1}{2\pi} \int_{\varphi}^{\infty} f(\varphi')\, d\varphi' \int_c \frac{\exp\{-i\mu(\varphi' - \varphi)/a\} e^{-\mu}}{\mu \cosh \mu - \sinh \mu}\, d\mu.$$

On the basis of the fact that $e^{\mu}(\mu \cosh \mu - \sinh \mu)^{-1}$ is monotonically decreasing for large $\mid \mu \mid$, one can show that the order of integration can be changed, and hence we have

$$j(\varphi, a) = \frac{1}{2\pi} \int_C \frac{e^\mu \exp\{-i\mu\varphi/a\}}{\mu \cosh \mu - \sinh \mu} d\mu \int_{-\infty}^{\varphi} \exp\{-i\mu\varphi'/a\} f(\varphi')\, d\varphi'$$

$$- \frac{1}{2\pi} \int_C \frac{e^{-\mu} \exp\{i\mu\varphi/a\}}{\mu \cosh \mu - \sinh \mu} d\mu \int_{\varphi}^{\infty} \exp\{-i\mu\varphi'/a\} f(\varphi')\, d\varphi'.$$

Using Euler's formula for the exponential and the oddness of $f(\varphi)$, we can rewrite the integral with limits from $-\infty$ to φ in terms of integrals with limits from 0 to φ and from 0 to ∞. The result is

$$j(\varphi, a) = j_1(\varphi, a) + j_2(\varphi, a), \tag{6.1}$$

where

$$j_1(\varphi, a) = -\frac{1}{\pi} \int_C \frac{\cosh(\mu - i\mu\varphi/a)}{\mu \cosh \mu - \sinh \mu} d\mu \int_{\varphi}^{\infty} \exp\{-i\mu\varphi'/a\} f(\varphi')\, d\varphi',$$

(6.2)

$$j_2(\varphi, a) = \frac{i}{\pi} \int_C \frac{e^\mu \cdot \exp\{-i\mu\varphi/a\}}{\mu \cosh \mu - \sinh \mu} d\mu \int_0^{\varphi} \sin \frac{\mu}{a} \varphi' f(\varphi')\, d\varphi'.$$

The integrals j_1 and j_2 can be estimated by deforming the contour C into a large rectangle in the lower half $\mathscr{I}m\mu < 0$ of the μ-plane, expressing the integrals as a sum of residues and estimating this sum. We illustrate the procedure in the case of $j_2(\varphi, a)$, which is the most sensitive case, the argument for j_1 being similar.

Writing $\mu = \sigma - i\lambda$, where we assume $\lambda \geq 0$, we have

$$\left| \int_0^{\varphi} \sin \frac{\mu}{a} \varphi' f(\varphi')\, d\varphi' \right|$$

$$\leq \frac{1}{2} \| f \| \int_0^{\varphi} \left\{ \exp\left\{\left(\frac{\lambda}{a} - 2\right)\varphi'\right\} + \exp\left\{-\left(\frac{\lambda}{a} + 2\right)\varphi'\right\}\right\} d\varphi'$$

$$\leq \frac{1}{2} \| f \| \left\{ \frac{\exp\left\{\left(\frac{\lambda}{a} - 2\right)\varphi\right\} - 1}{\frac{\lambda}{a} - 2} + \frac{1 - \exp\left\{-\left(\frac{\lambda}{a} + 2\right)\varphi\right\}}{\frac{\lambda}{a} + 2} \right\}.$$

Setting

$$B\left(\varphi, \frac{\mu}{a}\right) = \exp\{-i\mu\varphi/a\} \int_0^{\varphi} \sin \frac{\mu}{a} \varphi' f(\varphi')\, d\varphi', \tag{6.3}$$

we obtain the inequality

$$\left| B\left(\varphi, \frac{\mu}{a}\right) \right| \leq \frac{1}{2} \| f \| e^{-2\varphi} \left\{ \frac{1 - \exp\left\{-\left(\frac{\lambda}{a} - 2\right)\varphi\right\}}{\frac{\lambda}{a} - 2} + \frac{\exp\left\{-\left(\frac{\lambda}{a} + 2\right)\varphi\right\}}{\frac{\lambda}{a} + 2} \right\}. \tag{6.4}$$

The contour integral to be estimated is

$$j_2(\varphi, a) = \frac{i}{\pi} \int_C \frac{B(\varphi, \mu/a)e^{\mu}}{\mu \cosh \mu - \sinh \mu} d\mu. \tag{6.5}$$

In the plane of $\mu = \sigma + i\nu$, let C_n denote the rectangular contour consisting of the real segment $-\infty < \sigma \leq -n$, $\nu = 0$, followed by the vertical segment $\sigma = -n$, $0 > \nu \geq -n\pi$, followed by the horizontal segment $-n < \sigma \leq n$, $\nu = -n\pi$, followed by the vertical segment $\sigma = n$, $-n\pi < \nu \leq 0$, followed by the real segment $n < \sigma < \infty$, $\nu = 0$. Here $n = 1, 2, 3, \cdots$.

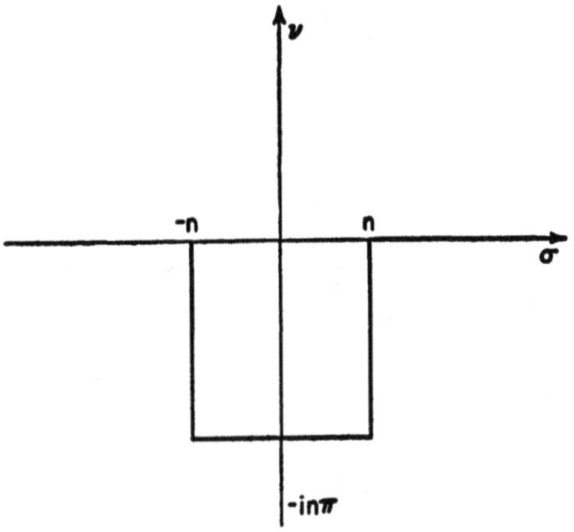

FIG. 2. Contour C_n.

Then the original indented contour C may be replaced by the contour C_n in (6.5) providing we take account of the contributions from the residues at the poles at $\mu_1, \mu_2, \cdots$ of the integrand which lie in the half-plane $\mathscr{I}m\, \mu < 0$ within the contour C_n. These poles are all simple, and are located at the roots $\mu_m = -i\lambda_m$ of the equation $\tanh \mu = \mu$, so we have

$$m\pi < \lambda_m < m\pi + \pi/2, \qquad m = 1, 2, 3, \cdots. \tag{6.6}$$

It is easily seen that the contributions to the integral

$$\int_{C_n} \frac{B(\varphi, \mu/a)e^{\mu}}{\mu \cosh \mu - \sinh \mu} d\mu$$

from the horizontal segments of C_n tend to zero as $n \to \infty$. By integrating the estimate (6.4) for $B(\varphi, \mu/a)$ and using the inequality $| e^{\mu}(\mu \cosh \mu - \sinh \mu)^{-1} | \leq 5/(n - 4)$ which holds along the vertical segments, one can establish that the contributions from the vertical segments also tend to zero as $n \to \infty$. Thus in the limit we obtain

$$j_2(\varphi, a) = 2\pi i \sum_{m=1}^{\infty} R_m , \tag{6.7}$$

where R_m is the residue of the integrand at the pole $\mu = \mu_m = -i\lambda_m$.

Since $\tanh \mu_m = \mu_m$, we get

$$\begin{aligned} R_m &= \frac{i}{\pi} \frac{B\left(\frac{\mu_m}{a}, \varphi\right)(\cosh \mu_m + \sinh \mu_m)}{\mu_m \sinh \mu_m} \\ &= \frac{i}{\pi} B\left(\frac{\mu_m}{a}, \varphi\right) \cdot \frac{1 + \mu_m}{\mu_m^2}. \end{aligned} \tag{6.8}$$

By (6.4) we have

$$\left| B\left(\frac{\mu_m}{a}, \varphi\right) \right| \leq \frac{a}{2} \| f \| e^{-2\varphi} \frac{1 - \exp\left\{-\left(\frac{\lambda_m}{a} - 2\right)\varphi\right\}}{\lambda_m - 2a} + \frac{\exp\left\{-\left(\frac{\lambda_m}{a} + 2\right)\varphi\right\}}{\lambda_m + 2a}.$$

By the assumptions of the lemma, $a \leq \pi/4$ and by (6.6), $\lambda_m > \pi$ for $m = 1, 2, 3, \cdots$. Hence $\lambda_m - 2a > \lambda_m - \frac{1}{2}\lambda_m = \frac{1}{2}\lambda_m$, so that

$$\begin{aligned} | 2\pi i R_m | &\leq a \| f \| e^{-2\varphi}\left(\frac{2}{\lambda_m} + \frac{1}{\lambda_m}\right) \frac{1 + \lambda_m}{\lambda_m^2} \\ &\leq 3a \| f \| e^{-2\varphi} \frac{1}{\lambda_m^2}\left(1 + \frac{1}{\lambda_m}\right) \\ &\leq \frac{4a \| f \| e^{-2\varphi}}{\lambda_m^2} < \frac{4a \| f \| e^{-2\varphi}}{m^2\pi^2} \end{aligned}$$

by (6.6). Thus

$$| j_2(\varphi, a) | < \sum_{m=1}^{\infty} | 2\pi i R_m | < \frac{4a}{\pi^2} \| f \| e^{-2\varphi} \sum_{1}^{\infty} \frac{1}{m^2},$$

or on substituting the known sum $\pi^2/6$ of the series,

$$| j_2(\varphi, a) | < \frac{2a}{3} \| f \| e^{-2\varphi}.$$

By similar methods it can be shown that

$$| j_1(\varphi, a) | \leq \frac{8}{27} a \| f \| e^{-2\varphi},$$

so that

$$\| \Omega_1^a f \| \leq \| j_1(\varphi, a) \| + \| j_2(\varphi, a) \| < a \| f \|,$$

which is the required inequality of Lemma 6.2.

Using similar methods, one can also demonstrate the next lemma.

LEMMA 6.3. *For* $0 \leq a \leq a' \leq \pi/4$, *the inequality*

$$|| aj(\varphi, a') - aj(\varphi, a) || < 3 \, || f || \, a'(a' - a),$$

holds, where $j(\varphi, a) = \Omega_1^a f$, *and* $f \in B_1$.

Proof: As in the preceding lemma, we illustrate the procedure in the case of $j_2(\varphi, a)$ only, the argument for $j_1(\varphi, a)$ being similar. Referring to formulas (6.7), (6.8) and (6.3) we see that we need estimates for the differences

$$\Delta_m = B(\varphi, \mu_m/a') - B(\varphi, \mu_m/a).$$

Setting $\mu = -i\lambda$ and using (6.3) we obtain

$$B(\varphi, \mu/a) = \frac{1}{2i} \int_0^{\varphi} \left\{ \exp\left\{-\frac{\lambda}{a}(\varphi - \varphi')\right\} - \exp\left\{-\frac{\lambda}{a}(\varphi + \varphi')\right\}\right\} f(\varphi')\, d\varphi'.$$

Now assume $0 < a < a' < \pi/4$. Then

$$\Delta = \Delta' - \Delta''$$

where

$$\Delta' = \frac{1}{2i} \int_0^{\varphi} \left\{ \exp\left\{-\frac{\lambda}{a'}(\varphi - \varphi')\right\} - \exp\left\{-\frac{\lambda}{a}(\varphi - \varphi')\right\}\right\} f(\varphi')\, d\varphi',$$

$$\Delta'' = \frac{1}{2i} \int_0^{\varphi} \left\{ \exp\left\{-\frac{\lambda}{a'}(\varphi + \varphi')\right\} - \exp\left\{-\frac{\lambda}{a}(\varphi + \varphi')\right\}\right\} f(\varphi')\, d\varphi'.$$

Using the law of the mean on the difference of the exponentials and recalling that $f \in B_1$, we obtain

$$| \Delta' | \leq || f || \frac{\lambda}{2} \left(\frac{a' - a}{aa'}\right) \int_0^{\varphi} (\varphi - \varphi') \exp\left\{-\frac{\lambda}{a'}(\varphi - \varphi')\right\} e^{-2\varphi'}\, d\varphi'.$$

Assuming $\lambda > 4a'$, we obtain the inequality

$$a \, | \Delta' | \leq || f || \frac{\lambda}{2} a'(a' - a) e^{-2\varphi}/(\lambda - 2a')^2,$$

which implies

$$a \, || \Delta' || \leq \frac{2 \, || f || \, a'(a' - a)}{\lambda}. \tag{6.9}$$

In a similar fashion one can show that the inequality

$$a \, || \Delta'' || \leq \frac{|| f || \, a'(a' - a)}{\lambda} \tag{6.10}$$

also holds for $\lambda > 4a'$. Hence, using (6.7), (6.8), (6.9), (6.10) and (6.6) we find that

$$a \,||\, j_2(\varphi, a') - j_2(\varphi, a) \,||\, \leq 2a \sum_{m=1}^{\infty} ||\, \Delta_m \,||\, \frac{1 + \lambda_m}{\lambda_m^2}$$

$$\leq 8 \,||\, f \,||\, a'(a' - a) \sum_{m=1}^{\infty} \lambda_m^{-2}$$

$$\leq \frac{4}{3} \,||\, f \,||\, a'(a' - a).$$

By analogous methods the inequality

$$a \,||\, j_1(\varphi, a') - j_1(\varphi, a) \,||\, \leq 5/3 \,||\, f \,||\, a'(a' - a)$$

may be demonstrated, and since $j(\varphi, a) = j_1(\varphi, a) + j_2(\varphi, a)$, the lemma is proved.

Returning to the equations (5.23), we drop the circumflexes, write $\tau = t + 1$ and put

$$(6.11) \qquad \begin{aligned} G_1(\omega, a) &= \theta + [S_1 - a\Theta_1^a]E(\theta, t + 1, a), \\ G_2(\omega, a) &= t - [S_0 - a^2 S_2 + a^2 T_1^a]E(\theta, t + 1, a), \\ G(\omega, a) &= \{G_1(\omega, a), G_2(\omega, a)\}. \end{aligned}$$

Then the system (5.23) may be written in the abbreviated form

$$(6.12) \qquad G(\omega, a) = 0.$$

LEMMA 6.4. *For each a in the interval $0 \leq a \leq \pi/4$ the function $G(\omega, a)$ defined by* (6.11) *is a transformation taking the space B into intself. For each a in the interval and each $\omega \in B$, the Fréchet differential $\delta G(\omega, a; \delta\omega)$ exists. The function $G(\omega, a)$ and its differential $\delta G(\omega, a; \delta\omega)$ are each continuous jointly in ω, a and uniformly with respect to ω, a, $\delta\omega$ when these variables range over bounded sets.*

Proof: Let $\omega = \{\theta, t\}$ range over an arbitrary but fixed bounded sphere $M \subset B$, while a is always confined to the interval $\mathcal{J} : 0 \leq a \leq \pi/4$. Since the function $F(\theta, \tau, a)$ has the form

$$F(\theta, \tau, a) = a^{-5}(e^{-3a^2\tau} \sin a^3\theta - a^3\theta),$$

it is clear that F is an entire function of each of its arguments and that F may be factored in the form θ times another entire function. Thus with $\tau = t + 1$, it follows that $F \in B_1$ and that F remains bounded when ω and a vary over M and $\mathcal{J}$, respectively.

Since $\eta \in B_1$ by (5.21), the function

$$E(\theta, \tau, a) = F(\theta, \tau, a) - \eta(\varphi) \int_{-\infty}^{\infty} \varphi' F(\theta, \tau, a) d\varphi'$$

also belongs to B_1, and remains bounded as ω and a vary over their respective ranges:

$$(6.13) \qquad ||\, E(\theta, \tau, a) \,||\, < c \quad \text{for} \quad \omega \in M, \quad a \in \mathcal{J}.$$

From equation (6.11) and Lemmas 6.1, 6.2 it follows that $G(\omega, a) \in B$. To prove the continuity property for G_1, note that

$$G_1(\omega', a') - G_1(\omega, a)$$
$$= \theta' - \theta + [S_1 - a'\Theta_1^{a'}]E(\theta', \tau', a') - [S_1 - a\Theta_1^a]E(\theta, \tau, a)$$
$$(6.14) \qquad = (\theta' - \theta) + S_1[E(\theta', \tau', a') - E(\theta, \tau, a)]$$
$$+ (a' - a)\Theta_1^{a'}E(\theta', \tau', a') + a\{\Theta_1^{a'}E(\theta', \tau', a') - \Theta_1^a E(\theta', \tau', a')\}$$
$$+ a\Theta_1^a[E(\theta', \tau', a') - E(\theta, \tau, a)].$$

From the nature of the function E it is easily seen that $E(\theta, \tau, a)$, regarded as an element of B_1, is uniformly continuous jointly in its three arguments, for $\omega \in M$ and $a \in \mathcal{I}$. Hence, on applying Lemmas 6.1 and 6.2 with $f(\varphi) = E(\theta', \tau', a) - E(\theta, \tau, a)$ we see that the norms of the first, second and last terms of (6.14) can be made arbitrarily small, when $\| \theta' - \theta \|$, $\| \tau - \tau' \|$, $| a' - a |$ are sufficiently small. From inequality (6.13) and Lemma 6.3 it follows that the same is true of the remaining terms and indeed that $G_1(\omega, a)$ is uniformly continuous. A similar argument shows that $G_2(\omega, a)$ and hence $G(\omega, a)$ is uniformly continuous. Using subscripts to denote partial derivatives and setting

$$(6.15) \qquad \delta E(\theta, \tau, a; \delta\omega) = F_\theta\, \delta\theta + F_\tau\, \delta t - \eta \int_{-\infty}^{\infty} \varphi'\{F_\theta\, \delta\theta + F_\tau\, \delta t\}\, d\varphi',$$

it is easily shown by standard methods that $G(\omega, a)$ has the Fréchet differential $\delta G = \{\delta G_1, \delta G_2\}$ where

$$(6.16) \qquad \begin{aligned} \delta G_1(\omega, a; \delta\omega) &= \delta\theta + [S_1 - a\Theta_1^a]\, \delta E(\theta, \tau, a; \delta\omega), \\ \delta G_2(\omega, a; \delta\omega) &= \delta t - [S_0 - a^2 S_2 + a^2 T_1^a]\, \delta E(\theta, \tau, a; \delta\omega). \end{aligned}$$

The uniform continuity of the differential is now proved in the same way as for G itself.

7. *The Existence Theorem*

We are looking for a solution $\omega = \{\theta, t\}$ of the equation (6.12), $G(\omega, a) = 0$, which also satisfies the auxiliary condition

$$(7.1) \qquad \Gamma = \int_{-\infty}^{\infty} \varphi F(\theta, \tau, a)\, d\varphi = 0.$$

For $a = 0$ we have already found (in Section 5) a solution $\omega = \omega_1$ satisfying both conditions, namely $\omega_1 = \{\theta_1, t_1\}$ where

$$(7.2) \qquad t_1(\varphi) = -3\,\mathrm{sech}^2\, \tfrac{3}{2}\varphi, \qquad \theta_1(\varphi) = -t_1'(\varphi), \qquad \tau_1(\varphi) = 1 + t_1(\varphi).$$

Since t_1 and θ_1 each behave like a constant times $e^{-3\varphi}$ as $\varphi \to \infty$ it is clear that $\omega_1 \in B$. In Section 6 we saw that the transformation $G(\omega, a)$ is uniformly continuous and differentiable with respect to ω.

The equation $G(\omega, a) = 0$ could be written in the form $\omega = H(\omega, a)$ in a natural way, see (5.23), and it might be thought that iterations could be applied directly. Such iterations, however, would not converge since the transformation $H(\omega, a)$ is not contracting. If they did, the solution of the equation $G(\omega, a) = 0$ would be uniquely determined and one could not expect that the auxiliary condition (7.1) could be satisfied. We therefore must proceed differently.

In order to show how to continue our solution from $a = 0$ to positive values of a, we first investigate the "variational equation"

$$\delta G(\omega_1, 0; \delta\omega) = \delta\zeta, \tag{7.3}$$

where $\delta\zeta = \{\delta\rho, \delta\sigma\}$ denotes an arbitrary element of B. Setting $a = 0$ in (6.16) we have

$$\delta G_1(\omega_1, 0; \delta\omega) = \delta\theta + S_1\, \delta E(\theta_1, \tau_1, 0; \delta\omega),$$

$$\delta G_2(\omega_1, 0; \delta\omega) = \delta t - S_0\, \delta E(\theta_1, \tau_1, 0; \delta\omega),$$

and by (5.24) and (6.15),

$$\delta E(\theta_1, \tau_1, 0; \delta\omega) = -3\{\tau_1\, \delta\theta + \theta_1\, \delta t - \eta L(\tau_1\, \delta\theta + \theta_1\, \delta t)\},$$

where $L(f)$ denotes the functional

$$L(f) = \int_{-\infty}^{\infty} \varphi f(\varphi)\, d\varphi = 2\int_0^{\infty} \varphi f(\varphi)\, d\varphi, \qquad f \in B_1. \tag{7.4}$$

Thus, on using the representations (5.8) of Lemma 5.1, the variational equation (7.3) takes the form

$$\begin{aligned} &\delta\theta - 9\int_{\varphi}^{\infty} d\varphi' \int_{\varphi'}^{\infty} \{\tau_1\, \delta\theta + \theta_1\, \delta t - \eta L(\tau_1\, \delta\theta + \theta_1\, \delta t)\}\, d\varphi'' = \delta\rho, \\ &\delta t - 9\int_{\varphi}^{\infty} d\varphi' \int_{\varphi'}^{\infty} d\varphi'' \int_{\varphi''}^{\infty} \{\tau_1\, \delta\theta + \theta_1\, \delta t - \eta L(\tau_1\, \delta\theta + \theta_1\, \delta t)\}\, d\varphi''' = \delta\sigma, \end{aligned} \tag{7.5}$$

in which $\delta\rho$, $\delta\sigma$ are arbitrary elements of B_1 and B_2, respectively.

Now the system (7.5) can be replaced by a single ordinary differential equation. For, if we put $y = \delta t - \delta\sigma$, we obtain the relation $-y' = \delta\theta - \delta\rho$ from (7.5). Also

$$\begin{aligned} \tau_1\, \delta\theta + \theta_1\, \delta t &= -\tau_1 y' + \theta_1 y + \tau_1\, \delta\rho + \theta_1\, \delta\sigma \\ &= -(\tau_1 y)' + \tau_1\, \delta\rho + \theta_1\, \delta\sigma, \end{aligned}$$

and by differentiation of the second equation of (7.5) we obtain

$$y''' - 9(\tau_1 y)' + 9\eta L[(\tau_1 y)'] + f(\varphi) = 0, \tag{7.6}$$

where the primes now denote differentiation with respect to φ and where

$$f(\varphi) = 9\{\tau_1\, \delta\rho + \theta_1\, \delta\sigma - \eta L(\tau_1\, \delta\rho + \theta_1\, \delta\sigma)\} \tag{7.7}$$

is to be regarded as a known function. From the form of f it is obvious that

$$L(f) = \int_{-\infty}^{\infty} \varphi f(\varphi)\, d\varphi = 0, \tag{7.8}$$

so that f is not completely arbitrary in B_1. From (7.5) we have $y(\infty) = y''(\infty) = 0$, and by (5.21) and (7.2), $\eta(\varphi) = \frac{1}{4} t_1'(\varphi)$. Hence, integrating (7.6) we get

$$y'' - 9\tau_1 y + ct_1 = g(\varphi), \tag{7.9}$$

where

$$c = \frac{9}{4} L[(\tau_1 y)'], \tag{7.10}$$

$$g(\varphi) = \int_{\varphi}^{\infty} f(\varphi')\, d\varphi'. \tag{7.11}$$

From (7.8) it follows by integration by parts that

$$\int_0^{\infty} g(\varphi)\, d\varphi = 0. \tag{7.12}$$

Solutions of equation (7.9) satisfying appropriate boundary conditions lead to solutions of the variational equation (7.3). In order to investigate the solutions of (7.9) we first prove:

LEMMA 7.1. *For $h \in B_2$ the inhomogeneous equation*

$$y'' - 9\tau_1 y = h(\varphi) \tag{7.13}$$

has a unique solution y satisfying the boundary conditions

$$y'(0) = 0, \qquad y(\infty) = 0, \quad \text{and} \quad y \in B_2, \qquad y' \in B_1. \tag{7.14}$$

Proof: We consider the homogeneous equation $y'' - 9\tau_1 y = 0$, where

$$\tau_1 = 1 - 3 \operatorname{sech}^2 \tfrac{3}{2}\varphi;$$

it is easy to verify by direct substitution that this equation has the solution

$$v = \tfrac{2}{3} \operatorname{sech}^2 \tfrac{3}{2}\varphi \tanh \tfrac{3}{2}\varphi.$$

Clearly $v(\varphi)$ behaves like $4e^{-3\varphi}/3$ as $\varphi \to \infty$ and $v'(0) = 1$. Let $u(\varphi)$ be the solution of the same homogeneous equation satisfying the initial conditions $u(0) = 1$, $u'(0) = 0$. Then the Wronskian is $uv' - u'v = 1$, so that u and v are linearly independent. It is easily shown (u can also be explicitly calculated by a known device) that $u \sim Ae^{3\varphi}$ as $\varphi \to \infty$ where $A = \text{const.} \neq 0$. Thus there is no solution of the homogeneous equation satisfying the boundary conditions (7.14).

Since the Wronskian of u and v is 1, the solution of (7.13) satisfying the boundary conditions (7.14) is given by

$$y = u(\varphi) \int_{\varphi}^{\infty} v(\varphi')h(\varphi')\, d\varphi' + v(\varphi) \int_{0}^{\varphi} u(\varphi')h(\varphi')\, d\varphi', \tag{7.15}$$

as can be verified by direct substitution. By the properties of u and v just discussed and by the hypothesis of the lemma, there exist constants A_1 and B_1 such that

$$| u(\varphi) | \leq A_1 e^{3\varphi}, \qquad | v(\varphi) | \leq B_1 e^{-3\varphi}, \qquad | h(\varphi) | \leq || h || e^{-2\varphi}.$$

Hence

$$| y | \leq A_1 e^{3\varphi} \int_{\varphi}^{\infty} B_1 e^{-3\varphi'} || h || e^{-2\varphi'}\, d\varphi' + B_1 e^{-3\varphi} \int_{0}^{\varphi} A_1 || h || e^{\varphi'}\, d\varphi'$$

$$\leq \frac{6}{5} A_1 B_1 || h || e^{-2\varphi}.$$

Since y is an even function, it follows that $y \varepsilon B_2$. By differentiation of (7.15) we obtain

$$y'(\varphi) = u'(\varphi) \int_{\varphi}^{\infty} v(\varphi')h(\varphi')\, d\varphi' + v'(\varphi) \int_{0}^{\varphi} u(\varphi')h(\varphi')\, d\varphi'. \tag{7.16}$$

It can be shown that $u'(\varphi) \sim C_1 e^{3\varphi}$, $v'(\varphi) \sim C_2 e^{-3\varphi}$ as $\varphi \to \infty$. By an argument similar to the above for u and v we find that $y' \varepsilon B_1$.

Lemma 7.2. *Denote the right member of* (7.15) *by* $M[h]$, *so that* $M[h]$ *is a linear bounded operator on* B_2 *to* B_1. *Then for any arbitrary value of the parameter* c, *the solution in* B_2 *of* (7.9) *is given by*

$$y = M[g] - cM[t_1]. \tag{7.17}$$

Proof: In view of the proof of Lemma 7.1 the only thing to be verified is that for an arbitrary value of c, (7.10) always holds. Since $y \varepsilon B_2$ and $y' \varepsilon B_1$ and since y evidently satisfies (7.9), we get by integrating the latter equation from $\varphi = 0$ to $\varphi = \infty$ and using (7.12),

$$-9 \int_0^{\infty} \tau_1 y\, d\varphi + c \int_0^{\infty} t_1(\varphi)\, d\varphi = 0.$$

By (5.21), using integration by parts

$$\int_0^{\infty} t_1(\varphi)\, d\varphi = -\int_0^{\infty} \varphi t_1'(\varphi)\, d\varphi = -4 \int_0^{\infty} \varphi\eta(\varphi)\, d\varphi = -2.$$

Again, by integration by parts:

$$\int_0^{\infty} \tau_1 y\, d\varphi = -\int_0^{\infty} \varphi(\tau_1 y)'\, d\varphi = -\frac{1}{2} L[(\tau_1 y)'].$$

Thus we always have $c = (9/4)\, L[(\tau_1 y)']$, so that (7.10) always holds.

Summarizing our results in terms of equation (7.3), we now have:

LEMMA 7.3. *Set* $u_1(\varphi) = M[t_1]$ *and* $\nu_1(\varphi) = \{-u_1'(\varphi), u_1(\varphi)\}$. *Then there exists a linear bounded operator K on B to B such that for each real number c and each* $\delta\zeta \in B$

$$\delta\omega = K[\delta\zeta] - c\nu_1$$

is a solution of the variational equation

$$\delta G(\omega_1, 0; \delta\omega) = \delta\zeta.$$

Proof: Replacing y by $\delta t - \delta\sigma$, y' by $-\delta\theta + \delta\rho$ we see that

$$\delta t = \delta\sigma + M[g] - cu_1,$$

$$\delta\theta = \delta\rho - M[g]' + cu_1'$$

is a solution of (7.5), where g is given by (7.11) and (7.7). Since g depends linearly (and continuously) on $\delta\rho$ and $\delta\sigma$, the result follows.

Returning to the equation $G(\omega, a) = 0$, (6.12), we note that it is equivalent to

$$0 = G(\omega_1, 0) - G(\omega, a),$$

since ω_1 is the solution for $a = 0$. Introducing the variable $\xi = \omega - \omega_1$ in place of ω and adding $\delta G(\omega_1, 0; \xi)$ to both sides we have

$$\delta G(\omega_1, 0; \xi) = \delta G(\omega_1, 0; \xi) - G(\omega_1 + \xi, a) + G(\omega_1, a) + G(\omega_1, 0) - G(\omega_1, a).$$

Thus the equation $G(\omega, a) = 0$ is equivalent to

$$\delta G(\omega_1, 0; \xi) = R(\xi, a) + H(a), \tag{7.18}$$

where

$$R(\xi, a) = \delta G(\omega_1, 0; \xi) - G(\omega_1 + \xi, a) + G(\omega_1, a), \tag{7.19}$$

$$H(a) = G(\omega_1, 0) - G(\omega_1, a).$$

LEMMA 7.4. *Given* $\epsilon > 0$, *there exist positive numbers* α *and* β *such that*

$$\| R(\zeta, a) - R(\xi, a) \| \leq \epsilon \| \zeta - \xi \| \tag{7.20}$$

when $0 \leq a < \alpha$, $\| \xi \| < \beta$, $\| \zeta \| < \beta$, *and*

$$\| H(a) \| < \epsilon \tag{7.21}$$

when $0 \leq a < \alpha$.

Proof:

$$\begin{aligned} R(\zeta, a) &- R(\xi, a) \\ &= \delta G(\omega_1, 0; \zeta - \xi) - G(\omega_1 + \zeta, a) + G(\omega_1 + \xi, a) \\ &= \delta G(\omega_1, 0; \zeta - \xi) - \delta G(\omega_1 + \xi, a; \zeta - \xi) \\ &\quad + \delta G(\omega_1 + \xi, a; \zeta - \xi) - G(\omega_1 + \zeta, a) + G(\omega_1 + \xi, a) \\ &= \delta G(\omega_1, 0; \zeta - \xi) - \delta G(\omega_1 + \xi, a; \zeta - \xi) \\ &\quad + \int_0^1 \{\delta G(\omega_1 + \xi, a; \zeta - \xi) - \delta G(\omega_1 + \xi + s(\zeta - \xi), a; \zeta - \xi)\}\, ds. \end{aligned}$$

The Lipschitz condition (7.20) now follows from the fact that $\delta G(\omega, a\, ;\, \delta\omega)$ is continuous in ω, a, uniformly with respect to ω, a, $\delta\omega$, which was proved in Lemma 6.4. The continuity condition (7.21) also follows from the same lemma.

Applying Lemma 7.3 to equation (7.18), we find that the equation $G(\omega, a) = 0$ is equivalent to

$$\xi = KH(a) + KR(\xi, a) - c\nu_1\,. \tag{7.22}$$

When $a = 0$ and $c = 0$ we have the solution $\xi = 0$. Since K is a bounded operator it follows from Lemma 7.4 that the right member of (7.22) satisfies a Lipschitz condition in ξ with the Lipschitz constant less than one, provided that $a > 0$ is sufficiently small. Thus if a and $|c|$ are sufficiently small, equation (7.22) has a solution $\xi(a, c)$, which can be deduced[5] by the method of iteration.

This is *the one-parametric family of solutions of equations* (5.23) equivalent to equation (6.12) which was indicated in Section 5, the new parameter being c.

We next have to show that for each sufficiently small $a > 0$, a value of c can be found such that the auxiliary condition (5.19) (or (7.1)) is satisfied.

We observe[5] that $\xi(a, c)$ is continuous and that it has a continuous partial derivative ξ_c in the neighborhood of $a = 0, c = 0$. In fact ξ_c satisfies the equation

$$\xi_c = K[\delta G(\omega_1, 0; \xi_c) - \delta G(\omega_1 + \xi, a; \xi_c)] - \nu_1\,, \tag{7.23}$$

and its existence may also be demonstrated by iteration. In particular we note that

$$\xi_c(0, 0) = -\nu_1\,. \tag{7.24}$$

Corresponding to the solution $\xi(a, c)$ of (7.22) we have the solution $\omega(a, c) = \omega_1 + \xi(a, c)$ of the equation $G(\omega, a) = 0$. The question is, can we choose the constant c, so far arbitrary, so as to satisfy the auxiliary condition (7.1)? With $\omega(a, c) = \{\theta(a, c, \varphi), t(a, c, \varphi)\}$, set

$$\Gamma(a, c) = \int_{-\infty}^{\infty} F(\theta(a, c, \varphi), 1 + t(a, c, \varphi), a)\varphi\, d\varphi,$$

[5] See for example Hildebrandt and Graves [11].

so that our auxiliary equation (7.1) reduces to the "bifurcation equation"

$$\Gamma(a, c) = 0 \tag{7.25}$$

for the parameter c.

It was shown that $\omega_c = \xi_c$ exists and is continuous. Using this, it is easily demonstrated that differentiation under the integral sign with respect to the parameter c is permissible, so that

$$\Gamma_c(a, c) = \int_{-\infty}^{\infty} \{F_\theta \theta_c + F_\tau t_c\} \varphi \, d\varphi,$$

and $\Gamma_c(a, c)$ is continuous at $a = 0, c = 0$. Now for $a = c = 0$ we have $\theta = \theta_1$, $t = t_1$, $F_\theta = -3\tau_1$, $F_\tau = -3\theta_1$ where $\tau_1 = 1 + t_1$. Hence

$$\Gamma_c(0, 0) = -3 \int_{-\infty}^{\infty} \{\tau_1 \theta_c + \theta_1 t_c\} \varphi \, d\varphi.$$

By (7.24) and the definition of ν_1 as given in Lemma 7.3 it follows that

$$t_c = -u_1(\varphi), \qquad \theta_c = u_1'(\varphi),$$

where, according to Lemma 7.2, $u_1(\varphi)$ satisfies the equations

$$u_1'' - 9\tau_1 u_1 = t_1(\varphi); \qquad u_1'(0) = u_1'(\infty) = 0. \tag{7.26}$$

Since by (7.2), $\theta_1 = -\tau_1'$, we get

$$\Gamma_c(0, 0) = -6 \int_0^{\infty} (\tau_1 u_1' + \tau_1' u_1) \varphi \, d\varphi = -6 \int_0^{\infty} (\tau_1 u_1)' \varphi \, d\varphi = 6 \int_0^{\infty} \tau_1 u_1 \, d\varphi.$$

Integration of the first equation of (7.26) from $\varphi = 0$ to $\varphi = \infty$ and use of the boundary conditions gives

$$-9 \int_0^{\infty} \tau_1 u_1 \, d\varphi = \int_0^{\infty} t_1(\varphi) \, d\varphi = -2.$$

Hence $\Gamma_c(0, 0) = 4/3 \neq 0$.

Now $\Gamma(a, c)$ is continuous with a continuous partial derivative Γ_c in the neighborhood of $a = 0, c = 0$, and since ω_1 satisfies (7.1) for $a = 0$ it follows that $\Gamma(0, 0) = 0$. Therefore by the implicit function theorem there exists a continuous solution $c = c(a)$ of the bifurcation equation (7.25) for a positive and sufficiently small. Thus the function

$$\omega = \omega(a) = \omega(a, c(a))$$

is a solution of the equation (6.12), $G(\omega, a) = 0$, for all sufficiently small positive values of a, which also satisfies the auxiliary condition (7.1). This function $\omega(a)$ satisfies all our requirements and yields the boundary values $\theta(\varphi) = \theta(\varphi, 1)$, $\tau(\varphi) = \tau(\varphi, 1)$ of the harmonic functions $\theta(\varphi, \psi)$, $\tau(\varphi, \psi)$ along the upper boundary $\psi = 1$ in the χ-plane. In view of the remarks made at the end of Section 2 we may summarize these results in the form of the

EXISTENCE THEOREM. *When the reduced altitude $\gamma = gh/U^2$ is less than one, but differs from one by a sufficiently small amount, a symmetric solitary wave exists.*

Appendix. *Properties of Green's Function*

Proof of Lemma 4.1: Set

$$N(\mu, \psi') = \frac{\mu \cosh \mu(1 - \psi') - \sinh \mu(1 - \psi')}{\mu \cosh \mu - \sinh \mu}; \tag{8.1}$$

then the formula (4.6) for $J(\chi \mid \chi')$ becomes

$$J(\chi \mid \chi') = \frac{1}{2\pi i} \int_C \frac{\sin \mu(\chi - \varphi')}{\mu} N(\mu, \psi')\, d\mu, \tag{8.2}$$

for $0 \leq \psi < \psi' \leq 1$.

In the neighborhood of $\mu = 0$ the integrand admits the expansion

$$\frac{\sin \mu(\chi - \varphi')}{\mu} N(\mu, \psi') = \frac{3\psi'(\chi - \varphi')}{\mu^2} + O[1]. \tag{8.3}$$

For $\mu > 2$, on the other hand,

$$\left| \frac{\sin \mu(\chi - \varphi')}{\mu} [N(\mu, \psi') - e^{-\mu\psi'}] \right| \leq \frac{3\sqrt{2}}{\mu} \exp\{-\mu(2 - \psi - \psi')\}. \tag{8.3'}$$

Let S_δ denote the semi-circle of radius δ with center at the origin in the half-plane $\mathcal{I}m\, \mu < 0$. Then $H(\chi \mid \chi') = \mathcal{R}e\, J(\chi \mid \chi')$ is given by

$$H(\chi \mid \chi') = \mathcal{R}e \left\{ \frac{1}{2\pi i} \int_{S_\delta} \frac{\sin \mu(\chi - \varphi')}{\mu} N(\mu, \psi')\, d\mu \right\}$$
$$+ \frac{1}{\pi} \int_\delta^\infty \frac{\cos \mu(\varphi - \varphi') \sinh \mu\psi}{\mu} N(\mu, \psi')\, d\mu.$$

For any $\delta > 0$, the second integral obviously vanishes for $\psi = 0$, $\psi' > 0$. We infer from (8.3) that the value of the first integral for $\psi = 0$ will differ by less than any given $\epsilon > 0$ from

$$\mathcal{R}e \left\{ \frac{1}{2\pi i} \int_{S_\delta} \frac{3\psi'(\varphi - \varphi')}{\mu^2}\, d\mu \right\} = 0$$

when δ is chosen sufficiently small. Hence $H = 0$ for $\psi = 0$ and $\psi' > 0$, so that the boundary condition (4.5) is verified.

By differentiation under the integral sign, which is easy to justify, we obtain $H_{\psi'} - H = 0$ along $\psi' = 1$, for $\psi < 1$. Hence to verify (4.4) it is sufficient to show that $H(\chi \mid \chi')$ is symmetric in its two arguments. Before proving the symmetry, however, we must show how the integral J defined by (8.2) may be extended to the range $\psi \geq \psi'$.

According to Frullani's integral formula we have

$$
(8.4)\qquad
\begin{aligned}
2i\int_0^\infty \frac{\sin\mu(\chi-\varphi')e^{-\mu\psi'}}{\mu}\,d\mu &= \int_0^\infty \frac{\exp\{i\mu(\chi-\bar{\chi}')\}-\exp\{i\mu(\chi'-\chi)\}}{\mu}\,d\mu \\
&= \log\frac{\chi'-\chi}{\chi-\bar{\chi}'},
\end{aligned}
$$

for $0 \leq \psi < \psi' \leq 1$, where the bar denotes the complex conjugate. Hence

$$
\begin{aligned}
J = \frac{1}{2\pi i}\int_C \sin\mu(\chi-\varphi')N(\mu,\psi')\,\frac{d\mu}{\mu} \\
-\frac{1}{\pi i}\int_0^\infty \sin\mu(\chi-\varphi')e^{-\mu\psi'}\,\frac{d\mu}{\mu} - \frac{1}{2\pi}\log\frac{\chi'-\chi}{\chi-\bar{\chi}'}
\end{aligned}
$$

for $0 \leq \psi < \psi' \leq 1$, since the last two terms add up to zero. Let the sum of the first two terms be denoted by $Q(\chi \mid \chi')$. Then we have

$$
(8.5)\qquad J(\chi \mid \chi') = -\frac{1}{2\pi}\log\frac{\chi'-\chi}{\chi-\bar{\chi}'} + Q(\chi \mid \chi'),
$$

where

$$
(8.6)\qquad
\begin{aligned}
Q(\chi \mid \chi') = \frac{1}{2\pi i}\int_{S_\delta} \sin\mu(\chi-\varphi')N(\mu,\psi')\,\frac{d\mu}{\mu} - \frac{1}{\pi i}\int_0^\delta \sin\mu(\chi-\varphi')e^{-\mu\psi'}\,\frac{d\mu}{\mu} \\
+\frac{1}{\pi i}\int_\delta^\infty \sin\mu(\chi-\varphi')[N(\mu,\psi')-e^{-\mu\psi'}]\,\frac{d\mu}{\mu}.
\end{aligned}
$$

Because of (8.3′) the last integral is absolutely convergent for $0 \leq \psi + \psi' < 2$ and uniformly convergent for $0 \leq \psi + \psi' \leq 2\psi_0$, where $0 < \psi_0 < 1$. It follows that the right member of (8.6) is analytic in χ for all values of ψ, ψ' satisfying $0 < \psi + \psi' < 2$, so that $Q(\chi \mid \chi')$ and hence by means of (8.5), also $J(\chi \mid \chi')$, have been extended to the entire domain $0 < \psi < 1$, $0 < \psi' < 1$. Moreover, we have also shown that J satisfies (4.3).

To demonstrate the symmetry of $H = \mathcal{R}e J$, we observe that the real part of the last integral in the right member of (8.6) is

$$
\frac{2}{\pi}\int_\delta^\infty \frac{\cos\mu(\varphi-\varphi')\sinh\mu\psi\sinh\mu\psi'(\mu+1)\,d\mu}{\mu[(\mu-1)e^{2\mu}+\mu+1]};
$$

this expression is obviously symmetric in $\chi = \varphi + i\psi$ and $\chi' = \varphi' + i\psi'$ for every $\delta > 0$. Using (8.3) we see that for small δ, the real part of the sum of the first two integrals of (8.6) is

$$
\mathcal{R}e\left\{-\frac{1}{2\pi i}\frac{3\psi'(\chi-\varphi')}{\mu}\Bigg|_{\mu=-\delta}^{\mu=\delta}\right\} + O[\delta] = -\frac{3}{\pi\delta}\psi\psi' + O[\delta].
$$

Hence it follows from (8.5) and (8.6) that

$$H(\chi \mid \chi') - H(\chi' \mid \chi) = O[\delta],$$

and since δ can be made arbitrarily small, that H is symmetric in χ and χ' for $0 \leq \psi + \psi' < 2$.

This completes the proof of Lemma 4.1, except for the extension of $J(\chi \mid \chi')$ to points where both ψ and ψ' are equal to one. This extension will follow from proposition I below, and the discussion preceding it.

Proof of Lemma 4.3: As we have already shown, $J(\chi \mid \varphi' + i)$ is an analytic function of χ in the open strip $0 < \psi < 1$. Moreover, by Lemma 4.2, $|J(\chi \mid \varphi' + i)|$ behaves like $3\varphi'^2/4$ as $|\varphi'| \to \infty$. Hence, since $\varphi^2 f(\varphi)$ is assumed to be absolutely integrable on $-\infty < \varphi < \infty$, it follows that the integral in the right member of (4.8) converges absolutely and uniformly for χ in the neighborhood of a point in the open strip. Hence $\omega = \theta + i\tau$ is analytic in χ for $0 < \psi < 1$.

The difficult part of the proof is the verification that $\theta(\chi)$ satisfies (4.9) along the upper boundary $\psi = 1$. This involves differentiating with respect to ψ and evaluating the result for $\psi' = 1$ and $\psi \to 1$. In doing this, the last integral in the right member of (8.6) will give trouble if we try to differentiate under the integral sign; therefore we alter its form, separating out its "singular part." To do this we observe first that the function $N(\mu, \psi')$ defined by (8.1) can be written as

$$N(\mu, \psi') = e^{-\mu\psi'} + e^{-(2-\psi')\mu} + \frac{e^{-\mu(1-\psi')}[1 - \frac{1}{2}(\mu + 1)(e^{-2\mu} + e^{-2\mu\psi'})}{\mu \cosh \mu - \sinh \mu}. \tag{8.7}$$

Now the integral

$$\int_0^\infty \frac{\sin \mu(\chi - \varphi')}{\mu} e^{-(2-\psi')\mu}\, d\mu$$

can be evaluated explicitly in terms of logarithms by replacing ψ' by $2 - \psi'$ in (8.4). Hence, on carrying out this evaluation, substituting (8.7) into the last integral of (8.6), and using (8.5), we obtain

$$\begin{aligned} J(\chi \mid \chi') = &-\frac{1}{2\pi} \log \frac{\chi' - \chi}{\chi - \overline{\chi}'} - \frac{1}{2\pi} \log \frac{\varphi - \varphi' + i(2 - \psi - \psi')}{\varphi' - \varphi + i(2 + \psi - \psi')} \\ &+ Q_1(\chi \mid \chi') + Q_2(\chi \mid \chi'), \end{aligned} \tag{8.8}$$

where we have set

$$\begin{aligned} Q_1(\chi \mid \chi') = &\frac{1}{2\pi i} \int_{S_\delta} \frac{\sin \mu(\chi - \varphi')}{\mu} N(\mu, \psi')\, d\mu \\ &- \frac{1}{\pi i} \int_0^\delta \frac{\sin \mu(\chi - \varphi')}{\mu} [e^{-\mu\psi'} + e^{-\mu(2-\psi')}]\, d\mu, \end{aligned} \tag{8.9}$$

$$Q_2(\chi \mid \chi') = \frac{1}{\pi i} \int_\delta^\infty \frac{\sin \mu(\chi - \varphi')[1 - \frac{1}{2}(\mu + 1)(e^{-2\mu} + e^{-2\mu\psi'})]e^{-(1-\psi')\mu}}{\mu^2 \cosh \mu - \mu \sinh \mu}\, d\mu. \tag{8.10}$$

Differentiating (8.8) with respect to ψ and taking real parts we obtain

$$
\begin{aligned}
&H_\psi(\chi \mid \chi') \\
&= -\frac{1}{2\pi}\frac{\psi - \psi'}{(\varphi - \varphi')^2 + (\psi - \psi')^2} - \frac{1}{2\pi}\frac{\psi + \psi' - 2}{(\varphi - \varphi')^2 + (\psi' + \psi - 2)^2} \\
&\quad + \frac{1}{2\pi}\frac{\psi + \psi'}{(\varphi - \varphi')^2 + (\psi + \psi')^2} + \frac{1}{2\pi}\frac{\psi - \psi' + 2}{(\varphi - \varphi')^2 + (\psi - \psi' + 2)^2} \\
&\quad + \Re e\, Q_{1\psi} + H_{2\psi},
\end{aligned} \tag{8.11}
$$

where

$$
\begin{aligned}
&H_{2\psi}(\chi \mid \chi') = \Re e\, Q_{2\psi} \\
&= \frac{1}{\pi}\int_\delta^\infty \frac{\cos \mu(\varphi - \varphi') \cosh \mu\psi[1 - \frac{1}{2}(\mu + 1)(e^{-2\mu} + e^{-2\psi'\mu})]e^{-(1-\psi')\mu}}{\mu \cosh \mu - \sinh \mu}\, d\mu.
\end{aligned} \tag{8.12}
$$

The formulas (8.8) – (8.12) certainly hold for $\psi + \psi' < 2$. Now when $\varphi - \varphi' \neq 0$ and $\psi = \psi' = 1$, we observe that the right members of all these formulas are well defined, and we *define* the functions $J = H + iK$ and H_ψ by means of these formulas, for $\psi = \psi' = 1$ and $\varphi - \varphi' \neq 0$. With this understanding we shall next establish the following proposition:

I. *$J(\chi \mid \chi')$ and $H_\psi(\chi \mid \chi')$ are each continuous jointly in χ and χ' at a pair of distinct points $\chi = \varphi + i$, $\chi' = \varphi' + i$, on the upper boundary.*

The only non-obvious part of the proof of this proposition concerns the integral $H_{2\psi}$ given by (8.12). The only portion of this integral which does not converge absolutely and uniformly near $\varphi = \varphi'$, $\psi = 1$ is the integral

$$
L(\varphi, \varphi', \psi, \psi') = \int_\delta^\infty \frac{\cos \mu(\varphi - \varphi') \cosh \mu\psi e^{-(1-\psi')\mu}}{\mu \cosh \mu - \sinh \mu}\, d\mu.
$$

It is easily verified that for $\psi \leq 1$, $\psi' \leq 1$, the coefficient of the cosine in the integrand of L is a positive, steadily decreasing function of μ for $\mu \geq \delta$. It is evidently also an increasing function of ψ and of ψ'. From these facts it follows readily (e.g. by Chartier's test for convergence) that the integral L converges uniformly with respect to the four parameters $\varphi, \varphi', \psi, \psi'$ for $|\varphi - \varphi'| \geq \epsilon$ and $\psi \leq 1$, $\psi' \leq 1$, where ϵ is any given positive number. Hence L is continuous in all four arguments for $|\varphi - \varphi'| \geq \epsilon$, $\psi = 1$, $\psi' = 1$ and thus proposition I is established.

To complete the proof of Lemma 4.3, we must prove three more propositions in addition to proposition I.

II. *Let $\varphi^2 f(\varphi)$ be absolutely integrable on $-\infty < \varphi < \infty$. Then, for a given $\epsilon > 0$, the integrals*

$$\int_{|\varphi'-\varphi|\geq\epsilon} J(\chi \mid \varphi' + i)f(\varphi')\,d\varphi', \qquad \int_{|\varphi'-\varphi|\geq\epsilon} H_\psi(\chi \mid \varphi' + i)f(\varphi')\,d\varphi'$$

converge uniformly with respect to ψ *for* $0 \leq \psi \leq 1$.

Proof: Setting $\psi' = 1$ in (8.11) and (8.12) we have

$$\begin{aligned} H_\psi(\chi \mid \varphi' + i) &= -\frac{1}{\pi}\frac{\psi - 1}{(\varphi - \varphi')^2 + (\psi - 1)^2} + \frac{1}{\pi}\frac{\psi + 1}{(\varphi - \varphi')^2 + (\psi + 1)^2} \\ &\quad + \Re e\, Q_{1\psi}(\chi \mid \varphi' + i) + H_{2\psi}(\chi \mid \varphi' + i), \end{aligned} \tag{8.13}$$

$$H_{2\psi}(\chi \mid \phi' + i) = \frac{1}{\pi}\int_\delta^\infty \frac{\cos\mu(\phi - \phi')\cosh\mu\psi[1 - (\mu + 1)e^{-2\mu}]}{\mu\cosh\mu - \sinh\mu}\,d\mu. \tag{8.14}$$

Clearly, the first two terms in the right member of (8.13) tend to zero as $|\varphi - \varphi'| \to \infty$, and therefore are bounded for $|\varphi' - \varphi| \geq \epsilon$ and $0 \leq \psi \leq 1$. It is not difficult to show that the same statements are true for $H_{2\psi}$, since the integral in (8.14) converges uniformly for $0 \leq \psi \leq 1$, $|\varphi - \varphi'| \geq \epsilon$.

We now examine the function

$$\begin{aligned} Q_{1\psi}(\chi \mid \varphi' + i) &= \frac{1}{2\pi}\int_{S_\delta} \frac{\mu\cos\mu(\chi - \varphi')}{\mu\cosh\mu - \sinh\mu}\,d\mu \\ &\quad - \frac{2}{\pi}\int_0^\delta \cos\mu(\chi - \varphi')e^{-\mu}\,d\mu. \end{aligned} \tag{8.15}$$

The second integral in the right member of (8.15) is clearly bounded for $0 \leq \psi \leq 1$, $|\varphi - \varphi'| \geq \epsilon$. The first integral may be written as

$$\begin{aligned} \int_{S_\delta} \frac{\mu\cos\mu(\chi - \varphi')}{\mu\cosh\mu - \sinh\mu}\,d\mu &= \int_{S_\delta} \frac{\mu\exp\{\mp i\mu(\chi - \varphi')\}}{\mu\cosh\mu - \sinh\mu}\,d\mu \\ &\quad \pm i\int_{S_\delta} \frac{\mu\sin\mu(\chi - \varphi')}{\mu\cosh\mu - \sinh\mu}\,d\mu, \end{aligned} \tag{8.16}$$

where we take the upper or lower sign according as $\varphi - \varphi' \gtrless 0$. In the first integral of the right member of (8.16), put $\mu = \sigma + i\nu$, where $\nu < 0$. Then

$$|\exp\{\mp i\mu(\chi - \varphi')\}| = \exp\{\nu \mid \varphi - \varphi' \mid\}\cdot\exp\{\sigma\psi\}.$$

Hence this integral is bounded for $|\varphi - \varphi'| \geq \epsilon$, $0 \leq \psi \leq 1$. The second integral of the right member of (8.16) has an odd integrand, and thus can be written as half of the integral extended over a circle with center at the origin. Hence

$$\int_{S_\delta} \frac{\mu\sin\mu(\chi - \varphi')}{\mu\cosh\mu - \sinh\mu}\,d\mu = \pi i R_0 = 3\pi i(\chi - \varphi'),$$

where $R_0 = 3(\chi - \varphi')$ is the residue at the origin. It follows that

$$H_\psi(\chi \mid \varphi' + i) = -\tfrac{3}{2}\,|\varphi - \varphi'| + b(\varphi - \varphi', \psi)$$

where $b(\varphi - \varphi', \psi)$ is continuous and bounded for $0 \leq \psi \leq 1$, $|\varphi - \varphi'| \geq \epsilon$, and indeed tends to zero as $|\varphi - \varphi'| \to \infty$. Therefore, in view of the hypothesis made on the function $f(\varphi)$, the integral

$$\int_{|\varphi'-\varphi|\geq\epsilon} H_\psi(\chi \mid \varphi' + i)f(\varphi')\, d\varphi'$$

converges uniformly with respect to ψ on $0 \leq \psi \leq 1$.

The argument regarding the integral involving J is similar, except that the asymptotic form of J is quadratic in $\varphi - \varphi'$ and is given as the function J_0 of Lemma 4.2.

From propositions I and II together with formula (8.8) it follows readily that $\theta + i\tau = \int_{-\infty}^{\infty} J(\chi \mid \varphi' + i)\, f(\varphi')d\varphi'$ is continuous up to the boundaries $\psi = 0$ and $\psi = 1$.

III. *In the neighborhood of* $\varphi' = \varphi, \psi = 1$ *we have*

$$H_\psi(\chi \mid \varphi' + i) = -\frac{1}{\pi}\frac{\psi - 1}{(\varphi - \varphi')^2 + (\psi - 1)^2} + M(\varphi - \varphi', \psi),$$

where

$$\left| \int_{\varphi-\epsilon}^{\varphi+\epsilon} M(\varphi - \varphi', \psi)f(\varphi')\, d\varphi' \right| < m(\epsilon),$$

and where $m(\epsilon) \to 0$ as $\epsilon \to 0$.

Proof: According to (8.13), $M(\varphi - \varphi', \psi)$ consists of three terms. The only term which is not continuous for $\varphi' = \varphi$ and $\psi = 1$ is $H_{2\psi}$, which is given by (8.14). Instead of $H_{2\psi}$ itself, we may deal with the complex function

$$Q_{2\psi}(\chi \mid \varphi' + i) = \frac{1}{\pi}\int_\delta^\infty \frac{\cos \mu(\chi - \varphi')[1 - (\mu + 1)e^{-2\mu}}{\mu \cosh \mu - \sinh \mu}\, d\mu \tag{8.17}$$

of which $H_{2\psi}$ is the real part. We need be concerned only with the integral

$$I_2 = \frac{1}{\pi}\int_\delta^\infty \frac{\cos \mu(\chi - \varphi')}{\mu \cosh \mu - \sinh \mu}\, d\mu, \tag{8.18}$$

since the rest of the integral $Q_{2\psi}$ converges absolutely and uniformly in the neighborhood of $\varphi' = \varphi$, $\psi = 1$. Integrating the right member of (8.18) by parts we obtain

$$I_2 = -\frac{1}{\chi - \varphi'}\frac{\sin \delta(\chi - \varphi')}{\delta \cosh \delta - \sinh \delta} + \frac{1}{\chi - \varphi'}\int_\delta^\infty \frac{\sin \mu(\chi - \varphi')\mu \sinh \mu}{(\mu \cosh \mu - \sinh \mu)^2}\, d\mu.$$

As $\mu \to \infty$ the coefficient of the sine function in the integrand of the last integral behaves like $2\mu^{-1}e^{-\mu}$. Indeed, a simple calculation shows that

$$\frac{\sin \mu(\chi - \varphi')\mu \sinh \mu}{(\mu \cosh \mu - \sinh \mu)^2} = \frac{2 \sin \mu(\chi - \varphi')e^{-\mu}}{\mu} + O\left[\frac{1}{\mu^2}\right],$$

where the bound for the last term is independent of φ, φ' and ψ, in the neighborhood of $\varphi = \varphi'$, $\psi = 1$. Hence, on using (8.4) with $\psi' = 1$, we find that

$$J_2 = -\frac{i}{\chi - \varphi'} \log \frac{\varphi' - \varphi + i(1 - \psi)}{\varphi - \varphi' + i(1 + \psi)} + \beta(\varphi - \varphi', \psi),$$

where $\beta(\xi, \psi)$ is continuous in the neighborhood of $\xi = 0$, $\psi = 1$. Thus the function $M(\varphi - \varphi', \psi)$ has only a logarithmic singularity at $\varphi' = \varphi$, $\psi = 1$, and proposition III holds.

IV. *Let φ be any fixed real number and let*

$$\theta(\chi) = \int_{-\infty}^{\infty} H(\chi \mid \varphi' + i) f(\varphi')\, d\varphi'. \tag{8.19}$$

Then as $\psi \to 1$ $(\psi < 1)$,

$$\theta_\psi(\varphi + i\psi) - \theta(\varphi + i\psi) \to f(\varphi),$$

provided that $f(\varphi)$ satisfies the hypotheses of Lemma 4.3.

Proof: For a given $\eta > 0$, select $\epsilon > 0$ so small that $| f(\varphi') - f(\varphi) | < \eta$ for $| \varphi' - \varphi | \leq \epsilon$, and $m(\epsilon) < \eta$(cf. proposition III). Then for $\psi < 1$ we have

$$\begin{aligned}\theta_\psi(\chi) - \theta(\chi) = &\int_{|\varphi' - \varphi| \geq \epsilon} \{H_\psi(\chi \mid \varphi' + i) - H(\chi \mid \varphi' + i)\} f(\varphi')\, d\varphi' \\ &+ \int_{\varphi-\epsilon}^{\varphi+\epsilon} H_\psi(\chi \mid \varphi' + i) f(\varphi')\, d\varphi' - \int_{\varphi-\epsilon}^{\varphi+\epsilon} H(\chi \mid \varphi' + i) f(\varphi')\, d\varphi'.\end{aligned} \tag{8.20}$$

Now the first integral on the right side of (8.20) tends to zero as $\psi \to 1$. For, by Lemma 4.1 we know that $H_\psi(\varphi + i, \varphi' + i\psi') = H(\varphi + i, \varphi' + i\psi')$ for $\psi' < 1$, and by proposition I this equality holds in the limit as $\psi' \to 1$ for $| \varphi' - \varphi | \geq \epsilon$, so that $H_\psi(\varphi + i, \varphi' + i) - H(\varphi + i, \varphi' + i) = 0$. Again by proposition I,

$$\begin{aligned}&\lim_{\psi \to 1} \{H_\psi(\varphi + i, \varphi' + 1) - H(\varphi + i\psi, \varphi' + i)\} \\ &\qquad = H_\psi(\varphi + i, \varphi' + i) - H(\varphi + i, \varphi' + i) = 0\end{aligned}$$

for $| \varphi - \varphi' | \geq \epsilon$. Also, by proposition II, we may take limits inside the integral sign so that

$$\lim_{\psi \to 1} \int_{|\varphi - \varphi'| \geq \epsilon} \{H_\psi(\chi \mid \varphi' + i) - H(\chi \mid \varphi + i)\} f(\varphi')\, d\varphi' = 0.$$

In the case of the second integral on the right side of (8.20) we use proposition III and get

$$\begin{aligned}&\left| \int_{\varphi-\epsilon}^{\varphi+\epsilon} H_\psi(\chi \mid \varphi' + i) f(\varphi')\, d\varphi' - f(\varphi) \right| \\ &\qquad \leq \left| -\frac{1}{\pi} \int_{\varphi-\epsilon}^{\varphi+\epsilon} \frac{(\psi - 1) f(\varphi')\, d\varphi'}{(\varphi - \varphi')^2 + (\psi - 1)^2} - f(\varphi) \right| \\ &\qquad\qquad + \left| \int_{\varphi-\epsilon}^{\varphi+\epsilon} M(\varphi - \varphi', \psi) f(\varphi')\, d\varphi' \right|.\end{aligned}$$

Now the integral

$$-\frac{1}{\pi}\int_{\varphi-\epsilon}^{\varphi+\epsilon}\frac{(\psi-1)}{(\varphi-\varphi')^2+(\psi-1)^2}\,d\varphi' = -\frac{2}{\pi}\arctan\frac{\epsilon}{\psi-1}$$

tends to one monotonically as $\psi \to 1$ $(\psi < 1)$.

It follows that when $1 - \psi$ is positive and sufficiently small, we have

$$\left|\int_{\varphi-\epsilon}^{\varphi+\epsilon} H_\psi(\chi \mid \varphi' + i)f(\varphi')\,d\varphi' - f(\varphi)\right| \le 2\eta + m(\epsilon) < 3\eta.$$

As for the last integral on the right side of (8.20), it is easily seen on the basis of formulas (8.8), (8.9) and (8.10) that it also tends to zero with ϵ. Thus proposition IV holds.

It is clear that $\theta(\chi) = 0$ for $\psi = 0$ since, by Lemma 4.1, $H(\chi \mid \varphi' + i) = 0$ for $\psi = 0$. Hence $\theta(\chi)$ satisfies the boundary conditions (4.9) (4.10) and the proof of Lemma 4.3 is complete.

Bibliography

[1] Boussinesq, J., *Théorie de l'intumescence liquide appelée onde solitaire où de translation se propagant dans un canal rectangulaire*, Comptes rendus de l'académie des sciences, Institut de France, 1871, pp. 755ff.

[2] Lord Rayleigh, *On waves*, Philosophical Magazine (5), V. 1, 1876, pp. 257–279; *On periodical irrotational waves at the surface of deep water*, ibid. (6), V. 32, pp. 381–389.

[3] Levi-Cività, T., *Sulle onde progressive di tipo permànente*, Rendiconti della accademia dei Lincei (5), 16^{II}, 1907, pp. 776–790; *Détermination rigoureuse des ondes permanentes d'ampleur finie*, Mathematische Annalen, V. 93, 1925, pp. 264–314.

[4] Struik, D. J., *Détermination rigoureuse des ondes irrotationelles périodiques dans un canal à profondeur finie*, Mathematische Annalen, V. 95, 1926, pp. 595–634.

[5] Weinstein, A., *Sur la vitesse de propagation de l'onde solitaire*, Rendiconti della accademia dei Lincei (6), V. 3, 1926, pp. 463 ff.

[6] Lamb, H., *Hydrodynamics*, 6th edition, 1945, pp. 417–426. (Contains additional references.)

[7] Lavrentiev, M., *On the theory of long waves*, Akad. Nauk Ukrain. R. S. R., Zbornik Prac. Inst. Mat. V. 1946, No. 8, 1947, pp. 13–69 (in Ukranian).

[8] J. J. Stoker, *The formation of breakers and bores*: Appendix by K. O. Friedrichs, *On the derivation of the shallow water theory*, Communications on Pure and Applied Mathematics, V. 1, 1948, pp. 81–85.

[9] Keller, J. B., *The solitary wave and periodic waves in shallow water*, Communications on Pure and Applied Mathematics, V. 1, 1948, pp. 323–329.

[10] John, F., *On the motion of floating bodies, II: Appendix*, Communications on Pure and Applied Mathematics, V. 3, 1950, pp. 92–100.

[11] Hildebrandt, T. H., and Graves, L. M., *Implicit Functions and their differentials in general analysis*, Transactions of the American Mathematical Society, V. 29, 1927, pp. 127–153.

[55–1] **Differential forms on Riemannian manifolds, Comm. Pure Appl. Math., VIII (1955), 551–590.**

Commentary by Louis Nirenberg:

After Friedrichs' beautiful proof of regularity of generalized (weak) solutions of elliptic equations (this technique is now considered the standard one), in a Hilbert space framework, see [53–4], he applied the method in this paper to prove the regularity of "harmonic" differential forms on a manifold. In order to prove that the weak solutions, obtained via a Hilbert space argument, are "strong" he employs, in addition to the usual invariant "Dirichlet integral", a non-invariant form for all coefficients of the differential forms. This conforms with his frequent viewpoint that one should be willing to break invariance.

The proof involves a natural extension of the classical inequality of Korn in elasticity, to forms which are orthogonal to a certain subspace which, in this case, is infinite dimensional. It is a very interesting paper, especially in its treatment of a manifold with boundary. The techniques have had much influence in later work.

COMMUNICATIONS ON PURE AND APPLIED MATHEMATICS, VOL. VIII, 551–590 (1955)

Differential Forms on Riemannian Manifolds*

By K. O. FRIEDRICHS

The aim of the present paper is to show that various tools which proved to be useful in the theory of linear elliptic differential equations can be put to good use in the theory of differential forms on Riemannian manifolds.

Among these tools are integral inequalities which involve what we shall call the "complete" Dirichlet integral in contrast to the integral

$$\int dv * dv + \delta v * \delta v$$

called "Dirichlet integral" by workers in the theory of differential forms.

First we shall treat problems which are concerned with notions of Hilbert space theory. In treating these problems we shall use only tools of this theory and we shall not invoke the theory of integral equations, employed by Hodge, Kodaira, Bidal and de Rham. We also shall not employ the elegant heat equation method of Milgram and Rosenbloom.

The complete Dirichlet integral was already used by Gaffney for closed manifolds; we shall exploit this notion even more fully, in particular, in treating manifolds with a smooth boundary. Our treatment of manifolds with boundaries differs from that by Spencer and Duff, but uses observations made by Conner.

Our method is close to, though different from, that developed independently by Morrey.

We shall not work with the operator $d\delta + \delta d$, employed by the authors mentioned (except Hodge), but with the operators d and δ directly. In this respect our approach was influenced by the work of Weyl in his paper on "The Method of Orthogonal Projection in Potential Theory".

The emphasis in the present paper is on the methods. Still some results appear to be new; also, they are derived under conditions on the metric tensor, which are weaker than those required by other authors except Morrey. The degree of strength of the results thus obtained proved decisive in our approach of manifolds with boundaries.

We should like to explain our approach first in the case of vector fields in a three-dimensional Euclidean domain $\mathfrak{M}$, i.e. in the case considered by Weyl in 1940. One of the problems to be solved is the unique decomposition of any square-integrable vector $v_0 = v_0(x_1, x_2, x_3)$ as the sum

$$v_0 = v_a + h + v_s \tag{1}$$

*The work for this paper was supported by the Office of Naval Research, United States Navy, under contract N6ori-201 T.O.No. 1.

of a gradient, a curl and a harmonic vector. In other words, v_a and v_s are assumed to be of the form

$$v_a = \text{grad}\,\phi, \qquad v_s = \text{curl}\,w, \tag{2}$$

ϕ being a scalar field and w being a vector field; h should satisfy the equations

$$\text{curl}\,h = 0, \qquad \text{div}\,h = 0.$$

Appropriate boundary conditions must be imposed: either conditions of the first kind,

$$(3)_1 \qquad \phi = 0, \qquad h_t = 0 \quad \text{at } \mathcal{B}$$

or conditions of the second kind,

$$(3)_2 \qquad w_t = 0, \qquad h_n = 0 \quad \text{at } \mathcal{B},$$

where the subscripts t and n denote tangential and normal components at the boundary $\mathcal{B}$ of $\mathfrak{M}$.

By orthogonal projection one easily obtains a "weak" decomposition of the form (1) with vectors v_a and v_s which are not necessarily of the form (2) but are known to be limits, in the L^2 sense, of a sequence of vectors of the form grad $\phi^{(\sigma)}$ and curl $w^{(\sigma)}$, $\sigma = 1, 2, \cdots$, as $\sigma \to \infty$. This is sufficient to show that these vectors satisfy the equations curl $v_a = 0$, div $v_s = 0$, in the "weak" sense, i.e. that

$$\int_{\mathfrak{M}} v_a \,\text{curl}\, u \, dx = 0, \qquad \int_{\mathfrak{M}} v_s \,\text{grad}\, \psi \, dx = 0$$

holds for all vectors u and scalars ψ which vanish identically near the boundary of $\mathfrak{M}$. The decomposition (1) may be called "strong" if the vectors v_a and v_s have the form (2), in which the operators curl and div are being understood in the "strong" sense. To establish this strong character of the decomposition one must exhibit the scalar ϕ and the vector w entering (2).

To construct ϕ one can show that the scalars $\phi^{(\sigma)}$ converge to a scalar ϕ with grad $\phi = v_a$. To this end one may rely on the well-known inequality

$$\int_{\mathfrak{M}} \phi^2 \, dx \leqq \int_{\mathfrak{M}} (\text{grad}\,\phi)^2 \, dx, \qquad dx = dx_1 \, dx_2 \, dx_3 , \tag{4}$$

which holds with the boundary condition of the first kind; if ϕ is orthogonal to a constant it is valid without a boundary condition.

In order to find a vector w such that curl $w = v_s$, one applies the weak decomposition to the vectors $w^{(\sigma)}$ of which curl $w^{(\sigma)} \to v_s$ is known. The component $w_s^{(\sigma)}$ of $w^{(\sigma)}$ satisfies the relations div $w_s^{(\sigma)} = 0$ and curl $w_s^{(\sigma)} \to v_s$, the operators div and curl being understood in the weak sense.

Now one can state that as $\sigma \to \infty$ the sequence $w_s^{(\sigma)}$ converges to a limit w_s with the desired properties.

To establish this fact, one may apply to the vector $v = w_s^{(\sigma)} - w_s^{(\tau)}$ the "main inequality"

$$H(v) \leqq CG(v), \qquad C = \text{const.} \tag{5}$$

in which

$$H(v) = \int_{\mathfrak{M}} v^2 \, dx \tag{6}$$

and

$$G(v) = \int_{\mathfrak{M}} [(\text{curl } v)^2 + (\text{div } v)^2] \, dx. \tag{7}$$

Inequality (5) holds if the vector v admits curl and div in the weak sense, satisfies a boundary condition either of the first or of the second kind, and is orthogonal to all harmonic vectors which satisfy the same boundary condition.

In order to prove this statement one should show that a form v which admits curl and div in the weak sense, and satisfies a boundary condition, possesses strong first derivatives and satisfies the "auxiliary inequality"

$$D_c(v) \leqq C_0 G(v) + GH(v), \tag{8}$$

in which $D_c(v)$ is the complete Dirichlet integral

$$D_c(v) = \int_{\mathfrak{M}} \sum_{\alpha = \beta = 1}^{3} (\partial v_\alpha / \partial x_\beta)^2 \, dx. \tag{9}$$

(A short proof of this inequality for a special domain $\mathfrak{M}$ will be given in the appendix to the introduction.)

The use of this "complete" Dirichlet integral is essential for our procedure. The advantage of working with it stems from the fact that it involves all first derivatives of all components of v and not just the two linear combinations curl v and div v.

A well-known consequence of the presence of all first derivatives in D_c is the *complete continuity* of the form $H(v)$ with respect to $D_c(v) + H(v)$ as unit form. Using the auxiliary inequality (8) one now may conclude that the form $H(v)$ is completely continuous with respect to $G(v) + H(v)$ as unit form.

The complete continuity of H, relative to $G + H$, implies that the eigenvalues of H, relative to $G + H$, are discrete and have a finite multiplicity. If there are harmonic vectors (satisfying the boundary condition) they are evidently eigen vectors with the eigenvalue $\lambda = 1$. Hence the space of these harmonic vectors h is finite dimensional. Evidently, there is no eigenvalue > 1, and the discreteness of the eigenvalues implies that there is a largest eigenvalue $\mu < 1$. (Clearly, μ is the second or first eigenvalue according as $\lambda = 1$ is or is not an eigenvalue.) Hence

$$H(v) \leqq \mu[G(v) + H(v)] \qquad \text{for } v \perp h.$$

Thus the main inequality (5), with $C = \mu/(1 - \mu)$, is found.

Now, as was said earlier, the vector w entering relation (2) can be constructed with the aid of this inequality. Thus the strong decomposition is established.

The approach outlined can be entirely extended to skew-symmetric tensors or—what is equivalent—to (exterior) differential forms. Moreover, the reasoning can be extended to the case in which the manifold $\mathfrak{M}$ is of any dimension and carries any Riemannian metric.

The operators d and δ, which transform a form v into its differential dv and codifferential δv, replace the operators curl and div. The integral

$$G(v) = \int [dv * dv + \delta v * \delta v] \tag{10}$$

takes the place of the integral of $(\text{curl } v)^2 + (\text{div } v)^2$. The expression $G(v)$ was employed by Hodge in his theory; it was called the "Dirichlet integral" by him. In our approach we shall employ, in addition to the expression $G(v)$, an integral $D_c(v)$ which involves the sum of the squares of *all* derivatives of *all* components of the form v with respect to coordinate systems defined in the various patches which cover the manifold. This expression $D_c(v)$ will be called the "complete" Dirichlet integral.

It is true that the complete Dirichlet integral is not invariant; but this defect is quite irrelevant. It would be possible to introduce an invariant complete Dirichlet integral by employing the complete covariant derivative of the skew-tensor, and not only its skew-symmetric part; by adding an appropriate term one could even set up an integral for which the differential operator $d\delta + \delta d$ is the Euler-Lagrange derivative—as it is for the integral $G(v)$—but it is not at all necessary to do this.

We shall first assume the manifold to be closed.

The analogue of the auxiliary inequality (8) for closed manifolds was derived by Gaffney, who assumed the components of v to have second derivatives. He used this result to prove the complete continuity of the operator $[d\delta + \delta d]^{-1}$.

We shall exploit Gaffney's inequality further. Using the theorem of the identity of the strong and the weak extension of the pair of operators d and δ—or rather a slight extension of it—we shall prove:

A) Whenever a square-integrable form admits both operators d and δ in the weak sense, the components of this form possess "strong" derivatives, so that $D_c(v)$ is finite.

The desired decomposition can now be obtained by arguments similar to those described for a Euclidean manifold. By orthogonal projection one finds that every square-integrable form v permits the "weak" decomposition

$$v = v_a + h + v_s, \tag{11}$$

given by Kodaira, in which v_a and v_s are limits of forms given as du^σ and δw^σ respectively, while h is a harmonic form

$$dh = 0, \qquad \delta h = 0. \tag{12}$$

If v admits both d and δ weakly, it follows from statement **A** that D_e is finite for each of the three components. Hence v admits d and δ strongly.

Since the form $H(v)$ is completely continuous with respect to $D_e(v) + H(v)$ we conclude, as before, that it is completely continuous with respect to $G(v) + H(v)$. It follows that the space of harmonic forms is finite dimensional and furthermore that the main inequality holds for forms orthogonal to harmonic fields. It is then not difficult to prove that the decomposition (1) is "strong", i.e. that the forms v_a and v_s entering the weak decomposition (1) are of the form

(13) $$v_a = du_s, \qquad v_s = \delta w_a,$$

with d and δ understood in the strong sense.

The results described were obtained, without employing the operator $d\delta + \delta d$, by studying the operators of the first order d and δ alone. This appears to yield a more direct approach to the theory than the study of the compound operator $d\delta + \delta d$ of the second order. In a natural way we shall obtain results about the operators $d\delta$ and δd. In a sense, therefore, our method has points of contact with Hodge's original method.

It has been suggested that one of the advantages of employing the operator $d\delta + \delta d$ is that the theory of differential forms results simply as an application of the well-established theory of differential equations of the second order. One may, however, just as well offer an argument in the opposite direction, namely, that certain points of view from which the theory of differential equations should be attacked have become clear only through ideas coming from topology.

One should observe that the results described are obtained by arguments which remain strictly within the framework of Hilbert space theory combined with a few rather unspecific considerations involving differential operators of the first order. It is not necessary to invoke the theory of integral equations and to employ fundamental solutions and related functions.

Of course, there are problems in the theory of differential forms whose solution requires tools which are foreign to Hilbert space theory simply because the notions involved in these problems lie outside of it. Such questions are treated in Part II. In this connection we mention the following result: **B**) If both the strong differential dv and strong codifferential δv of a form v are bounded and measurable, the form v itself is continuous.

This result makes it easy to overcome a difficulty pointed out by Weyl: if $v_a = du$ is the differential of a form u with continuous derivatives, its integral over a cycle $\mathfrak{z}$ obviously vanishes:

(14) $$\int_{\mathfrak{z}} v_a = 0;$$

however, it is not obvious that this is so if v_a is the L^2-limit of such differentials even if this limit is continuous.

Still, this is the case, as was shown by Weyl for three-dimensional Euclidean domains. To prove that this is so in general one may use an argument which is somewhat more direct than that employed by Weyl. One represents the limit form v_a as a strong differential $v_a = du_s$ of a form u_s, which is possible by virtue of the strong decomposition, cf. (13); then one may make use of the fact, implied by statement **B**, that the form u_s is continuous.

An advantage of our approach to the theory of differential forms, which should be emphasized, is that only rather weak conditions on the metric tensor are required: namely, that the components $g^{\alpha\beta}$ satisfy a Lipschitz condition[1].

That the theory for closed manifolds was developed under such weak conditions has proved decisive for our approach to the theory of manifolds with smooth boundaries. As was observed by Conner, this theory can be reduced to that of forms on manifolds without boundary by the simple device of doubling, provided the derivatives of the components $g^{\alpha\beta}$ need not be assumed to be continuous. Since our approach is developed without such an assumption, the method of doubling is applicable.

These results, described in Part III, do not cover those obtained by Spencer and Duff, and Conner because of the smoothness condition we impose on the boundary.

Among the statements proved there is the inequality

$$D_c(v) \leqq CG(v), \tag{15}$$

which holds for all forms v which are orthogonal to all harmonic fields, ie., to all forms $\overset{0}{v}$ which satisfy the equations $d\overset{0}{v} = 0$, $\delta\overset{0}{v} = 0$, but not necessarily any boundary conditions. Also no boundary condition need be imposed on v.

The attempt at proving this inequality was the origin of this paper. Inequality (15) is of interest independently of the theory of differential forms. It is the counterpart of Korn's inequality [17C] and the remarkable inequalities recently discovered by Aronszajn [18].

In Korn's inequality, which originated in the theory of elasticity and refers to a Euclidean space, the integrand of the form $G(v)$ is a non-negative quadratic form in the derivatives of v which vanishes for a finite dimensional manifold of vectors $\overset{0}{v}$; the inequality holds if v is orthogonal to all vectors $\overset{0}{v}$.

The situation is similar for Aronszajn's inequalities, in which v is a scalar, and D_c and G involve derivatives of higher than first order of v.

A remarkable feature of inequality (15), in which it differs from Korn's and Aronszajn's inequalities, is that the manifold of forms $\overset{0}{v}$ for which $G(\overset{0}{v}) = 0$, i.e., the manifold of harmonic fields, is infinitely dimensional.

Appendix to the Introduction

In this appendix we shall present a proof of the auxiliary inequality (8) for a special domain of the Euclidean space. Also we shall make some general remarks about our derivation of the main inequality (5).

---

[1]The same conditions are required by Morrey [13].

We first prove a weakened form of inequality (8), [17B]. It involves a proper subdomain $\mathfrak{M}'$ of the domain $\mathfrak{M}$ and reads

$$(8)' \qquad D'_c(v) \leqq C_0 G(v) + CH(v)$$

in which $D'_c(v)$ is the complete Dirichlet integral (9) extended over the subdomain $\mathfrak{M}'$.

We make use of the identity

$$\sum_{\alpha,\beta} (\partial v_\alpha/\partial x_\beta)^2 = (\text{curl } v)^2 + (\text{div } v)^2 + \sum_\alpha \partial(v_\alpha \, \partial v_\beta/\partial x_\beta)/\partial x_\alpha - \partial(v_\alpha \, \partial v_\beta/\partial x_\alpha)/\partial x_\beta .$$

Then we select a function $\omega(x_1, x_2, x_3)$ with continuous first derivatives which equals 1 in $\mathfrak{M}'$, is $\leqq 1$ in $\mathfrak{M}$, and vanishes identically in a neighborhood of the boundary of $\mathfrak{M}$. We multiply the preceding identity by ω^2, integrate over $\mathfrak{M}$ and apply integration by parts on the last term. Thus we obtain

$$\begin{aligned}\int_{\mathfrak{M}} \omega^2 \sum_{\alpha,\beta} (\partial v_\alpha/\partial x_\beta)^2 \, dx &= \int_{\mathfrak{M}} \omega^2[(\text{curl } v)^2 + (\text{div } v)^2] \, dx \\ &\quad - 2\int_{\mathfrak{M}} \omega \sum_{\alpha,\beta} [(\partial\omega/\partial x_\alpha) v_\alpha \, \partial v_\beta/\partial x_\beta - (\partial\omega/\partial x_\beta) v_\alpha \, \partial v_\beta/\partial x_\alpha] \, dx \\ &\leqq \int_{\mathfrak{M}} \omega^2[(\text{curl } v)^2 + (\text{div } v)^2] \, dx + 8\epsilon^{-1} \int_{\mathfrak{M}} (\text{grad } \omega)^2 v^2 \, dx \\ &\quad + \epsilon \int_{\mathfrak{M}} \omega^2 \sum_{\alpha,\beta} (\partial v_\alpha/\partial x_\beta)^2 \, dx.\end{aligned}$$

Here $\epsilon > 0$ is arbitrary. Choosing $\epsilon < 1$ and subtracting the last term from the left member we obtain

$$(16) \qquad \begin{aligned}&(1 - \epsilon) \int_{\mathfrak{M}} \omega^2 \sum_{\alpha,\beta} (\partial v_\alpha/\partial x_\beta)^2 \\ &\leqq \int_{\mathfrak{M}} \omega^2[(\text{curl } v)^2 + (\text{div } v)^2] \, dx + 8\epsilon^{-1} \int_{\mathfrak{M}} (\text{grad } \omega)^2 v^2 \, dx\end{aligned}$$

and hence

$$(17) \qquad \begin{aligned}&\int_{\mathfrak{M}'} \sum_{\alpha,\beta} (\partial v_\alpha/\partial x_\beta)^2 \, dx \\ &\leqq C_0 \int_{\mathfrak{M}} [(\text{curl } v)^2 + (\text{div } v)^2] \, dx + C \int_{\mathfrak{M}} v^2 \, dx,\end{aligned}$$

i.e. inequality (8)′.

If the vector v is assumed only to admit the operators curl and div in the weak sense one can prove inequality (8)′ by first approximating v by vectors v^σ with continuous second derivatives, using (8)′ for the vectors v^σ, and then passing to the limit $\sigma \to \infty$.

We now proceed to prove the original inequality (8) for the case that the manifold $\mathfrak{M}$ is a parallelopiped $| x_1 | \leqq a_1$, $| x_2 | \leqq a_2$, $| x_3 | \leqq a_3$ in three dimensional Euclidean space. We assume the vector v to have continuous first derivatives up to the boundary and to satisfy the second boundary condition

$$v_n = 0$$

at the boundary of $\mathfrak{M}$.

We can reduce inequality (8) to inequality (8)′ in the following manner:

We reflect the domain $\mathfrak{M}$ first in the faces $x_1 = \pm a_1$, then the three resulting domains in the faces $x_2 = \pm a_2$, finally the nine domains obtained in $x_3 = \pm a_3$. The resulting domain $\hat{\mathfrak{M}}$ consists of $\mathfrak{M}$ and 26 replicas on $\mathfrak{M}$. We extend the function $v(x)$ into $\hat{\mathfrak{M}}$ by defining v at an image point A obtained by reflection of a point A' in a face as the vector obtained from the vector v at A' by changing the sign of its component perpendicular to this face. One then verifies that the vector v thus defined in $\hat{\mathfrak{M}}$ is continuous and has piecewise continuous first derivatives.

Furthermore, one observes that div v has the same values at the image point and the antecedent, while the vector curl v at the image point is obtained from $-$curl v at the antecedent by changing the sign of the normal component. In any case the integrals of v^2, $(\text{curl } v)^2$, $(\text{div } v)^2$ and $\sum_{\alpha,\beta} (\partial v_\alpha / \partial x_\beta)^2$ over $\hat{\mathfrak{M}}$ are 27 times the integrals of these quantities over $\mathfrak{M}$.

Inequality (8)′ applied to $\hat{\mathfrak{M}}$ then immediately yields the auxiliary inequality (8) for $\mathfrak{M}$.

From the arguments given earlier it now follows that the main inequality (5)

$$\int_{\mathfrak{M}} v^2 \, dx \leqq C \int_{\mathfrak{M}} [(\text{curl } v)^2 + (\text{div } v)^2] \, dx$$

holds for all functions v which admit curl and div in the weak sense. Orthogonality to harmonic vectors satisfying the boundary condition need not be imposed since there are no such vectors in the present case.

Inequality (5)—under the assumption div $v = 0$—was proved by Weyl employing an expansion of v in a Fourier series. It may perhaps be noted that Weyl considered it unlikely that this inequality could be approached in an elementary way, using direct estimates. Our derivation of the main inequality (5) tends to confirm this opinion. Although our proof of the auxiliary inequality is elementary, our proof of the main inequality is not. While we do not employ a complete series expansion, we use the existence of a second or first eigenvalue, which in turn is derived from the most important ingredient of the series expansion, namely the complete continuity of the form, $H(v)$, with respect to the form, $G(v) + H(v)$.

I. HILBERT SPACE OF FORMS ON CLOSED MANIFOLDS

1. *Basic Notions. Decomposition*

The Riemannian m-dimensional manifold $\mathfrak{M}$ consisting of points X is assumed to be covered by patches $\mathcal{P}$, one-to-one images of open spherical regions $x_1^2 + \cdots + x_m^2 < 1$ of an m-dimensional Euclidean space. To each patch $\mathcal{P}$ we assign a "core" $\mathcal{R}$, the image of the closed subsphere $x_1^2 + \cdots + x_m^2 \leqq R^2 < 1$. Each point X of $\mathfrak{M}$ is assumed to lie in at least one such core and there should be an integer $\iota > 0$ such that each point X lies in at most ι patches. The identification transformation of the two coordinate systems in regions covered by two patches is assumed to have Lipschitz continuous first derivatives; of course, the Jacobian should be bounded away from zero. A partition of unity—after Bochner—is supposed to exist. By this we mean here that it should be possible to assign to every patch $\mathcal{P}_\sigma$ a function $\overset{\sigma}{\eta}(X)$ with bounded first derivatives which has the properties

$$0 \leqq \overset{\sigma}{\eta}(X) \leqq 1, \tag{1.1}$$

$\overset{\sigma}{\eta}(X) = 0$ for X outside $\mathcal{R}_\sigma$, and

$$\sum_\sigma \overset{\sigma}{\eta}(X) = 1. \tag{1.2}$$

To every scalar function $\phi(X)$, we assign the function

$$\overset{\sigma}{\phi}(\overset{\sigma}{x}) = \phi(X)$$

of the coordinates $\overset{\sigma}{x}$ of every point X in the patch $\mathcal{P}_\sigma$. If this function $\overset{\sigma}{\phi}$ possesses a derivative with respect to $\overset{\sigma}{x}$, we write

$$\partial\overset{\sigma}{\phi}/\partial\overset{\sigma}{x}_\kappa = \nabla_\kappa \overset{\sigma}{\phi}. \tag{1.3}$$

A differential form v of rank p is an entity assigned to the points X of $\mathfrak{M}$ which, with reference to each patch, can be expressed as

$$v = \overset{\sigma}{v}_{\kappa_1 \cdots \kappa_p} \, d\overset{\sigma}{x}^{\kappa_p} \cdots d\overset{\sigma}{x}^{\kappa_1}, \tag{1.4}$$

in the sense of Cartan, in terms of the components of a skew-symmetric tensor.

If the components of the form v with respect to each patch are continuous we say: the form v lies in the space $\mathfrak{C} = \underset{p}{\mathfrak{C}}$. If the components of v are Lipschitz-continuous, we say: v lies in the space $\underset{p}{\mathfrak{C}_1}$. Such a form possesses (almost everywhere) bounded (measurable) first derivatives, as is well known.

For v in $\underset{p}{\mathfrak{C}_1}$ we may introduce the—not covariant—derivative

$$\nabla_\kappa v = \nabla_\kappa v_{\kappa_1 \cdots \kappa_p} \, dx^{\kappa_p} \cdots dx^{\kappa_1}. \tag{1.5}$$

On the left hand side we have not indicated a reference to the patch $\mathcal{P}_0$ although, strictly speaking, this would be necessary. We shall always omit such a reference in the following when it is implied by the context.

The differential dv of the form v in $\underset{p}{\mathfrak{C}_1}$ can be given in terms of the asymmetric part of the components $\nabla_{\kappa_{p+1}} v_{\kappa_1 \cdots \kappa_p}$ of $\nabla_\kappa v$ as

$$(1.6) \qquad dv = (p+1)\mathrm{Asy}_{p+1}[\nabla_{\kappa_{p+1}} v_{\kappa_1 \cdots \kappa_p}]\, dx^{\kappa_{p+1}} \cdots dx^{\kappa_1}$$

in obvious notation.

In each patch a positive-definite metric tensor $g^{\kappa\lambda}$—or rather $\underset{\sigma}{g}^{\kappa\lambda}$—is supposed to be defined; we assume $g^{\kappa\lambda}$ to be Lipschitz continuous. We set

$$(1.7) \qquad \gamma = \pm[\det g^{\kappa\lambda}]^{-1/2};$$

the sign is to be so chosen that

$$(1.8) \qquad \int \cdots \int_{\mathcal{P}} f(x)\gamma(x)\, dx^1 \cdots dx^n > 0$$

whenever $f(x) > 0$. We find it convenient to introduce the invariant

$$(1.9) \qquad \begin{aligned} \vartheta G v &= \vartheta_{\kappa_1 \cdots \kappa_p} g^{\kappa_1\lambda_1} \cdots g^{\kappa_p\lambda_p} v_{\lambda_1 \cdots \lambda_p}, \qquad && p > 0, \\ &= \vartheta v, && p = 0, \end{aligned}$$

associated with two forms v and ϑ of rank p, and to employ the customary notation

$$(1.10) \qquad \int \vartheta * v = \frac{1}{p!} \sum_\sigma \int \cdots \int_{\mathcal{P}_\sigma} \eta_\sigma \vartheta G v \gamma\, dx^1 \cdots dx^n$$

for the integral of this invariant over the manifold $\mathfrak{M}$.

The adjoint or co-differential δv of a form v in $\underset{p}{\mathfrak{C}_1}$ is a form of rank $p - 1$ uniquely defined by the condition that the "adjointness identity"

$$(1.11) \qquad \int du * v = -\int u * \delta v$$

should hold for all forms u in $\underset{p-1}{\mathfrak{C}_1}$. In terms of components the δv may be written as

$$(1.12) \qquad \begin{aligned} \delta v = \gamma^{-1} & g_{\kappa_1\lambda_1} \cdots g_{\kappa_{p-1}\lambda_{p-1}} \\ & \cdot \nabla_\nu(\gamma g^{\lambda_1\mu_1} \cdots g^{\lambda_{p-1}\mu_{p-1}} g^{\mu\nu} v_{\mu_1 \cdots \mu_{p-1}})\, dx^{\kappa_{p-1}} \cdots dx^{\kappa_1} \end{aligned}$$

in obvious notation.

It should be noted that identity (1.11) is equivalent with the set of identities

$$\int \eta\, du * v + \int \eta u * \delta v$$

(1.11)′

$$= -\frac{1}{p!}\int \cdots \int_{\mathcal{P}} \nabla_{\kappa_p} \eta u_{\kappa_1 \cdots \kappa_{p-1}} g^{\kappa_1 \lambda_1} \cdots g^{\kappa_p \lambda_p} v_{\kappa_1 \cdots \kappa_p} \gamma\, dx^1 \cdots dx^n$$

for all patches $\mathcal{P} = \mathcal{P}_\sigma$.

By $\underset{p}{\mathfrak{B}}$ we denote the *Hilbert space* of all forms with quadratically integrable components. We use the form

(1.13) $$(\vartheta, v)_0 = \int \vartheta * v$$

as inner product and the expression

$(1.13)_0$ $$\| v \|_0 = | (v, v)_0 |^{1/2} = \left[\int v * v \right]^{1/2}$$

as norm. Of course, we set $v = 0$ if $\| v \| = 0$.

2. *Extensions and Decompositions*

It is necessary to extend the operators d and δ to appropriate subspaces of the Hilbert space $\underset{p}{\mathfrak{B}}$.

Before doing this we introduce a convenient convention. As much as possible, we want to avoid showing the rank p explicitly as a subscript; we therefore write simply

$$\underset{p}{\mathfrak{B}} = \mathfrak{B};$$

if any subscript, 0 say, is attached to $\mathfrak{B}$ we write $\underset{p}{\mathfrak{B}_0} = \mathfrak{B}_0$. Frequently, both spaces $\underset{p}{\mathfrak{B}}$ and $\underset{p-1}{\mathfrak{B}}$ or $\underset{p}{\mathfrak{B}}$ and $\underset{p+1}{\mathfrak{B}}$ will enter the discussion. In such cases we shall write

$$\underset{p-1}{\mathfrak{B}} = \mathfrak{U} \qquad \text{and} \qquad \underset{p+1}{\mathfrak{B}} = \mathfrak{W}$$

for $p > 0$ and $p < n$ respectively. On occasion we shall set

$$\underset{p+1}{\mathfrak{U}} = \mathfrak{B} \qquad \text{and} \qquad \underset{p-1}{\mathfrak{W}} = \mathfrak{B}.$$

We now introduce the space $\mathfrak{B}_d$ of all those forms v in $\mathfrak{B}$ for which there is a form w in $\mathfrak{W}$ such that the relation

(2.1) $$\int \mathfrak{w} * w = -\int \delta \mathfrak{w} * v$$

holds for all forms $\mathfrak{w}$ in $\underset{p+1}{\mathfrak{C}_1}$, $0 \leq p < n$.

Furthermore we introduce the space $\mathfrak{V}_\delta$ of all forms v in $\mathfrak{V}$ for which there is a form u in $\mathfrak{U}$ such that the relation

$$\int \tilde{u} * u = -\int d\tilde{u} * v \tag{2.2}$$

holds for all $\tilde{u}$ in $\underset{p-1}{\mathfrak{C}_1}$, $0 < p \leqq n$.

Clearly, the forms w and u are uniquely assigned to the form v; hence we may define operators d and δ by

$$dv = w, \qquad \delta v = u \tag{2.3}$$

for v in $\mathfrak{V}_d$ and $\mathfrak{V}_\delta$ respectively. These operators are then the "weak" extensions of the operators d and δ defined in $\mathfrak{C}_1$. Evidently, *the spaces* $\mathfrak{V}_d$ *and* $\mathfrak{V}_\delta$ *are closed.*

It may be noted incidentally that the space $\mathfrak{V}_d$ is actually independent of the choice of the metric tensor $g^{\alpha\beta}$, although this tensor plays a part in the definition of $\mathfrak{V}_d$. Also; the space of tensor densities with components $\gamma g^{\kappa_1\lambda_1} \cdots g^{\kappa_p\lambda_p} v_{\lambda_1 \cdots \lambda_p}$ formed with tensors v from $\mathfrak{V}_\delta$ is independent of the metric.

We proceed to show that the weak extensions of the operators d and δ are equal to their "strong" extensions. This means that the operators d and δ defined in $\mathfrak{V}_d$ and $\mathfrak{V}_\delta$ are the closures of the operators defined in $\mathfrak{C}_1$. It is true, decomposition theorems could be formulated without using this fact, nevertheless we shall make use of this fact now since we shall need it later in any case. We first formulate

LEMMA 2.1. *If the form* v *is in* $\mathfrak{V}_d$, *then* ηv *is also in* $\mathfrak{V}_d$ *and*

$$(\eta v) = \eta\, dv + v \wedge d\eta. \tag{2.4}$$

Here $v \wedge d\eta$ is $(p + 1)$ times the form whose component $(v \wedge d\eta)_{\kappa_1 \cdots \kappa_{p+1}}$ is the skew-symmetric part of $v_{\kappa_1 \cdots \kappa_p}\, \eta/\kappa_{p+1}$. Similarly if the form v is in $\mathfrak{V}_\delta$, ηv is also in $\mathfrak{V}_\delta$ and

$$\delta(\eta v) = \eta\, \delta v + vG\, d\eta. \tag{2.5}$$

Here $vGd\eta$ is the form with the components

$$(vG\, d\eta)_{\kappa_1 \cdots \kappa_{p-1}} = v_{\kappa_1 \cdots \kappa_{p-1}\kappa}\, g^{\kappa\lambda} \nabla_\lambda \eta.$$

To prove the lemma we first note that identities (2.4) and (2.5) hold for v in $\underset{p}{\mathfrak{C}_1}$. Using (2.5) for $\tilde{w}$ in $\underset{p+1}{\mathfrak{C}_1}$ we verify relation (2.1) for ηv and $\eta dv + v \wedge d\eta$ in place of v and w and using (2.4) for $\tilde{u}$ in $\underset{p-1}{\mathfrak{C}_1}$ relation (2.2) for ηv and $\eta \delta v + vGd\eta$ in place of v and u. Thus the statement of the lemma follows.

Next we formulate

LEMMA 2.2. *To every form* v *in* $\mathfrak{V}$ *there exists a sequence of forms* v^α *in* $\mathfrak{C}_1$ *with the following two properties*: 1) *relation*

(2.6) $$\| v^{\alpha} - v \|_0 \to 0$$

holds as $\alpha \to \infty$, 2) *if* v *belongs to* $\mathfrak{B}_{\alpha}$, *relation*

(2.7) $$\| dv^{\alpha} - dv \|_0 \to 0$$

holds and if v *belongs to* $\mathfrak{B}_{\delta}$, *relation*

(2.8) $$\| \delta v^{\alpha} - \delta v \|_0 \to 0$$

holds.

If v *belongs to both* $\mathfrak{B}_d$ *and* $\mathfrak{B}_{\delta}$, *both relations,* (2.6) *and* (2.7), *hold with the same sequence* v^{α}.

Let us first assume that the form v vanishes outside of the core of a patch. The statement of the lemma is then implied by the theorem of the identity of the weak and the strong extension of differential operators of the first order [17_A], applied to the differential operator which transforms the form v into the pair of forms $\{dv, \delta v\}$. Any form v in $\mathfrak{B}_d$ or $\mathfrak{B}_{\delta}$ will now be written in the form

$$v = \sum_{\sigma} \overset{\sigma}{\eta} v.$$

By Lemma 2.1 the forms $\overset{\sigma}{\eta} v$ are, respectively, also in $\mathfrak{B}_d$ or $\mathfrak{B}_{\delta}$. Hence the statement holds for each form $\overset{\sigma}{\eta} v$. With the aid of the approximating forms $(\overset{\sigma}{\eta} v)^{\alpha}$ one then forms the approximating form

$$v^{\alpha} = \sum_{\sigma} (\overset{\sigma}{\eta} v)^{\alpha}.$$

Clearly, relations (2.7) and (2.8) hold for the form v^{α} if they hold for $(\overset{\sigma}{\eta} v)^{\alpha}$. Thus the statement of Lemma 2.2 follows.

It should be noted that the differential operator $\{d, \delta\}$ has non-constant coefficients. If only one of the two statements (2.7) or (2.8) is to be proved, non-constancy of the coefficients can be avoided.

The statement that the operators d in $\mathfrak{B}_d$ and δ in $\mathfrak{B}_{\delta}$ are the closures of the operators d and δ in $\mathfrak{C}_1$ is an immediate consequence of Lemma 2.2. Another consequence is that relation

(2.9) $$(\tilde{w}, dv) = -(\delta \tilde{w}, v)$$

holds for v in $\mathfrak{B}_d$ and $\tilde{w}$ in $\mathfrak{W}_{\delta}$ and, what is equivalent, that relation

(2.10) $$(\tilde{u}, \delta v) = -(d\tilde{u}, v)$$

holds for v in $\mathfrak{B}_{\delta}$ and $\tilde{u}$ in $\mathfrak{U}_d$.

The spaces of all forms v in $\mathfrak{B}_d$ with $dv = 0$ or in $\mathfrak{B}_{\delta}$ with $\delta v = 0$ will be denoted by

$$\mathfrak{B}_{d-0} \quad \text{and} \quad \mathfrak{B}_{\delta-0},$$

respectively. The spaces of all forms $v = du$ with u in $\mathfrak{U}_d$ or $v = \delta w$ with w in $\mathfrak{W}_\delta$ will be denoted[2] by

$$d\mathfrak{U} \quad \text{and} \quad \delta\mathfrak{W},$$

respectively. The closures[3] of these spaces will be denoted by

$$\overline{d\mathfrak{U}} \quad \text{and} \quad \overline{\delta\mathfrak{W}}.$$

Relations (2.9) and (2.10) imply that the spaces $\mathfrak{V}_{d=0}$ and $\mathfrak{V}_{\delta=0}$ are orthogonal to the spaces $\overline{\delta\mathfrak{W}}$ and $\overline{d\mathfrak{U}}$, respectively:

$$\mathfrak{V}_{d=0} \perp \overline{\delta\mathfrak{W}}, \qquad \mathfrak{V}_{\delta=0} \perp \overline{d\mathfrak{U}}. \tag{2.11}$$

From the definitions of $\mathfrak{V}_d$ and $\mathfrak{V}_\delta$, on the other hand, we infer that every form $\perp \overline{\delta\mathfrak{W}}$ or $\perp \overline{d\mathfrak{U}}$ lies in $\mathfrak{V}_{d=0}$ and $\mathfrak{V}_{\delta=0}$, respectively. The projection theorem of the Hilbert space theory then yields the "decomposition theorems":

THEOREM 2.1.

$$\mathfrak{V} = \overline{d\mathfrak{U}} \oplus \mathfrak{V}_{\delta=0}, \tag{2.12}$$

$$\mathfrak{V} = \mathfrak{V}_{d=0} \oplus \overline{\delta\mathfrak{W}}. \tag{2.13}$$

Next we state that spaces $\overline{d\mathfrak{U}}$ and $\overline{\delta\mathfrak{W}}$ are perpendicular:

$$\overline{d\mathfrak{U}} \perp \overline{\delta\mathfrak{W}}. \tag{2.14}$$

To prove this statement we consider a form w in $\underset{p+1}{\mathfrak{C}_1}$ which vanishes outside of the core $\mathcal{R}$ of a patch $\mathcal{P}$ on a form u where components with respect to $\mathcal{P}$ have continuous second derivatives. Since for such forms

$$ddu = 0,$$

we have

$$\int \delta w * du = 0. \tag{2.15}$$

Now, every form u in $\underset{p-1}{\mathfrak{C}_1^{p-1}}$ can be approximated by forms u^α with second derivatives in $\mathcal{P}$ in such a way that

$$\int_{\mathfrak{R}} (u^\alpha - u) * (u^\alpha - u) \to 0, \qquad \int_{\mathfrak{R}} (du^\alpha - du) * (du^\alpha - du) \to 0$$

as $\alpha \to \infty$. Incidentally, this elementary fact could be derived as a by-product when proving the identity of the strong and the weak extension referred to in

[2]It would perhaps be more consistent to denote these spaces by $d\mathfrak{U}_\delta$ and $\delta\mathfrak{W}_\delta$ since they are the ranges of $\mathfrak{U}_d$ and $\mathfrak{W}_\delta$ under d and δ respectively. The notation adopted is more convenient; it is not likely to lead to misunderstandings.

[3]Later on, at the end of Section 5, we shall show that the spaces $d\mathfrak{U}$ and $\delta\mathfrak{W}$ are closed: $\overline{d\mathfrak{U}} = d\mathfrak{U}$, $\overline{\delta\mathfrak{W}} = \delta\mathfrak{W}$.

the proof of Lemma 2.2. Observing this fact we realize that relation (2.15) holds for every u in $\underset{p-1}{\mathfrak{C}_1}$.

Now let w be any form in $\underset{p+1}{\mathfrak{C}_1}$; then ηw vanishes outside of the core $\mathcal{R}$ of $\mathcal{P}$. Hence relation (2.15) holds for ηw in place of w. Summation with respect to all patches yields formula (2.15) for all w in $\underset{p-1}{\mathfrak{C}_1}$.

By virtue of Lemma 2.2 this formula holds for any u in $\mathfrak{U}_d$ and w in $\mathfrak{W}_\delta$. Closure then yields formula (2.14).

Since a form $\perp \overline{d\mathfrak{U}}$ lies in $\mathfrak{B}_{\delta-0}$ and a form $\perp \overline{\delta\mathfrak{W}}$ lies in $\mathfrak{B}_{d-0}$, relation (2.14) yields the statement

$$\overline{d\mathfrak{U}} \subset \mathfrak{B}_{d-0}, \qquad \overline{\delta\mathfrak{W}} \subset \mathfrak{B}_{\delta-0}. \tag{2.16}$$

This statement is evidently the extension of the identity

$$ddu = 0,$$

which holds for forms u which have second derivatives, and of the corresponding identity

$$\delta\delta w = 0.$$

We now introduce "harmonic forms" as forms which belong to both $\mathfrak{B}_{d-0}$ and $\mathfrak{B}_{\delta-0}$. The space of these harmonic fields will be denoted by $\mathfrak{H}$:

$$\mathfrak{H} = \mathfrak{B}_{d-0} \cap \mathfrak{B}_{\delta-0}. \tag{2.17}$$

Combining formula (2.13) with Theorem 2.1 we then obtain the "weak decomposition theorem",

THEOREM 2.2.

$$\mathfrak{B} = \overline{d\mathfrak{U}} \oplus \mathfrak{H} \oplus \overline{\delta\mathfrak{W}}. \tag{2.18}$$

Except for a somewhat different definition of the spaces involved, Theorem 2.2 is the decomposition theorem of Kodaira and the analogue of the decomposition theorem by Weyl for three-dimensional Euclidean manifolds with boundaries.

3. *The Space* $\mathfrak{B}_d \cap \mathfrak{B}_\delta$

In this section we shall describe a few properties of the intersection $\mathfrak{B}_d \cap \mathfrak{B}_\delta$ of the spaces $\mathfrak{B}_d$ and $\mathfrak{B}_\delta$. In this space, $\mathfrak{B}_d \cap \mathfrak{B}_\delta$, we introduce the inner product

$$(\phi, v)_1 = \int d\phi * dv + \int \delta\phi * \delta v + \int \phi * v \tag{3.1}$$

and the norm

$$\| v \|_1 = [(v, v)_1]^{1/2}. \tag{3.2}$$

As an immediate consequence of the closure of the spaces $\mathfrak{B}_d$ and of $\mathfrak{B}_\delta$, *the space* $\mathfrak{B}_d \cap \mathfrak{B}_\delta$ *is closed with respect to the norm* $\| v \|_1$.

We now introduce the intersection spaces

$$\overline{d\mathfrak{U}}_\delta = \overline{d\mathfrak{U}} \cap \mathfrak{B}_\delta , \qquad \overline{\delta\mathfrak{W}}_d = \overline{\delta\mathfrak{W}} \cap \mathfrak{B}_d : \tag{3.3}$$

Since $\overline{d\mathfrak{U}}_\delta \subset \overline{d\mathfrak{U}} \subset \mathfrak{B}_{d-0} \subset \mathfrak{B}_d$ and $\overline{\delta\mathfrak{W}}_d \subset \overline{\delta\mathfrak{W}} \subset \mathfrak{B}_{\delta-0} \subset \mathfrak{B}_\delta$, by (2.16), we have

$$\overline{d\mathfrak{U}}_\delta \subset \mathfrak{B}_d \cap \mathfrak{B}_d , \qquad \overline{\delta\mathfrak{W}}_d \subset \mathfrak{B}_d \cap \mathfrak{B}_\delta . \tag{3.4}$$

Also, of course,

$$\mathfrak{H} \subset \mathfrak{B}_d \cap \mathfrak{B}_\delta . \tag{3.5}$$

Using (2.14) and (2.11) we infer that the spaces $\overline{d\mathfrak{U}}_\delta$, $\overline{\delta\mathfrak{W}}_d$, $\mathfrak{H}$ are orthogonal to each other. From the decomposition Theorem 2.2 we then obtain the "modified decomposition theorem",

THEOREM 3.1. *The space* $\mathfrak{B}_d \cap \mathfrak{B}_\delta$ *admits the decomposition*

$$\mathfrak{B}_d \cap \mathfrak{B}_\delta = \overline{d\mathfrak{U}}_\delta \oplus \mathfrak{H} \oplus \overline{\delta\mathfrak{W}}_d .$$

Using Theorem 3.1 we infer

THEOREM 3.2. *Each of the spaces* $\overline{d\mathfrak{U}}_\delta$, $\overline{\delta\mathfrak{W}}_d$, $\mathfrak{H}$, *is closed with respect to the norm* $\| v \|_1$.

4. *Strongly Differentiable Forms. Strong Decomposition*

A form v in $\mathfrak{B}$ will be called "strongly differentiable" if its components $\overset{\sigma}{v}_{\kappa_1 \cdots \kappa_p}$ in each patch possess strong derivatives $\nabla_\kappa \overset{\sigma}{v}_{\kappa_1 \cdots \kappa_p}$[4]. By this we mean that to every component $\overset{\sigma}{v}_{\kappa_1 \cdots \kappa_p}$ (with reference to the patch) there exist functions $\nabla_\kappa \overset{\sigma}{v}_{\kappa_1 \cdots \kappa_p}$, for $\kappa = 1, \cdots , n$, quadratically integrable in every core $\mathcal{R}_\sigma$ and a sequence of functions $\nabla_\kappa \overset{\sigma(\alpha)}{v}_{\kappa_1 \cdots \kappa_p}$ for $\alpha \to 0$, which possess continuous derivatives $\nabla_\kappa \overset{\sigma(\alpha)}{v}_{\kappa_1 \cdots \kappa_p}$ in $\mathcal{R}_\sigma$ and for which, as $\alpha \to 0$,

$$\int \cdots \int_{\mathcal{R}_\sigma} | \overset{\sigma(\alpha)}{v}_{\kappa_1 \cdots \kappa_p} - \overset{\sigma}{v}_{\kappa_1 \cdots \kappa_p} |^2 \, dx^1 \cdots dx^n \to 0 \tag{4.1}$$

and

$$\int \cdots \int_{\mathcal{R}_\sigma} | \nabla_\kappa \overset{\sigma(\alpha)}{v}_{\kappa_1 \cdots \kappa_p} - \nabla_\kappa \overset{\sigma}{v}_{\kappa_1 \cdots \kappa_p} |^2 \, dx^1 \cdots dx^n \to 0, \quad \kappa = 1, \cdots , n, \tag{4.1$_1$}$$

for every core $\mathcal{R}_\sigma$ of the patch $\mathcal{P}_\sigma$.

The form whose components are the derivatives $\nabla_\kappa \overset{\sigma}{v}_{\kappa_1 \cdots \kappa_p}$ may conveniently be denoted by $\nabla_\kappa v$

$$\nabla_\kappa v = \nabla_\kappa \overset{\sigma}{v}_{\kappa_1 \cdots \kappa_p} \, dx^{\kappa_p} \cdots dx^{\kappa_1}; \tag{4.2}$$

---

[4]We shall not make use of the fact that the strong derivative is almost everywhere the actual derivative.

but this form is defined only in the core $\mathcal{R}_0$. The system of forms $\nabla_\kappa v$ will be denoted by

(4.3) $$\nabla v = \{\nabla_\kappa v\} = \{\nabla_1 v \cdots \nabla_n v\}.$$

Using the tensor

(4.4) $$G = \underset{p}{G} = \{g^{\kappa_1\lambda_1} \cdots g^{\kappa_p\lambda_p}\}$$

and the notation

$$vGv = v_{\kappa_1\cdots\kappa_p} g^{\kappa_1\lambda_1} \cdots g^{\kappa_p\lambda_p} v_{\lambda_1\cdots\lambda_p}$$

of (1.9), we introduce the expressions

(4.5) $$\| v \|_{\mathcal{R}} = \left[\frac{1}{p!}\int_{\mathcal{R}} vGv\gamma\, dx^1 \cdots dx^n\right]^{1/2},$$

(4.6) $$\| \nabla v \|_{\mathcal{R}} = \left[\frac{1}{p!}\int_{\mathcal{R}} \nabla_\kappa v g^{\kappa\lambda} G \nabla_\lambda v\gamma\, dx^1 \cdots dx^n\right]^{1/2}$$

Relations (4.1) and $(4.1)_1$ are obviously equivalent to the relations

(4.7) $$\| v^{(\alpha)} - v \|_{\mathcal{R}} \to 0,$$

(4.8) $$\| \nabla v^{(\alpha)} - \nabla v \|_{\mathcal{R}} \to 0, \qquad \alpha \to \infty.$$

The space of all strongly differentiable forms will be denoted by

$$\mathfrak{B}_1 .$$

As a consequence of well known facts we have

LEMMA 4.1. *The space $\mathfrak{B}_1$ is complete with respect to the norm*[5]

$$\| v \| + \sum_\sigma \| \nabla v \|_{\mathcal{R}_\sigma} .$$

[5]The expression $D_c(v) = \sum_\sigma \| \nabla v \|^2_{\mathcal{R}_\sigma}$ is the "complete" Dirichlet integral. In its place we could have worked with the invariant expression

$$\| \tilde{\nabla} v \|^2 + C \| v \|^2$$

as square of the norm, in which $\tilde{\nabla} v$ is the covariant derivative of v,

$$\tilde{\nabla}_\kappa v = \nabla_\kappa v - \Gamma^\sigma_\kappa \wedge v_\sigma$$

and C is sufficiently large. Here $\Gamma^\sigma_\kappa = \{\Gamma^\sigma_{\kappa\alpha}\}$, $\Gamma^\sigma_{\kappa\alpha}$ being the Christoffel symbol, $v_\sigma = \{v_{\alpha_1\cdots\alpha_{p-1}\sigma}\}$, and

$$\| \tilde{\nabla} v \|^2 = \frac{1}{p!}\int \cdots \int \tilde{\nabla}_\kappa v g^{\kappa\lambda} G \tilde{\nabla}_\lambda v\gamma\, dx^n \cdots dx^1.$$

The Euler-Lagrange derivative of the integral $\| \tilde{\nabla} v \|^2$ is proportional to the expression

$$(d\delta + \delta d)v + \frac{1}{p} g^{\kappa\lambda} R^\sigma_{\kappa\lambda} \wedge v_\sigma - \frac{2}{p} g^{\kappa\tau} R^\sigma_\kappa \wedge v_{\sigma\tau},$$

in which

$$R^\sigma_{\kappa\lambda} = \{R^\sigma_{\kappa\lambda\alpha}\}, \qquad R^\sigma_\kappa = \{R^\sigma_{\kappa\alpha_1\alpha_2}\},$$

$R^\sigma_{\kappa\lambda\mu}$ being the Riemannian curvature tensor. Evidently, by adding to $\| \tilde{\nabla} v \|^2$ an appropriate term, which contains the tensor R but not the derivatives of v, one can form an expression whose Euler-Lagrange derivative is exactly $(d\delta + \delta d)v$.

Clearly, the operator d is applicable to forms v in $\mathfrak{B}_1$ since each derivative contributing to the component of dv is defined, cf. (1.6). Similarly, one observes that the operator δ is applicable to v in $\mathfrak{B}_1$ since all those terms are defined which result when the differentiation with respect to x^κ, indicated in (1.12), is carried out on the factors of the product of which the components of γGv consist. One also verifies that the identities (1.11)′ hold for each patch $\mathcal{P}_\sigma$. Hence, by virtue of the invariance of dw and δv, the adjointness identity (1.11) holds for v in $\mathfrak{B}_1$ and w in $\mathfrak{W}_1$. It follows that the forms in $\mathfrak{B}_1$ belong to $\mathfrak{B}_d$ as well as to $\mathfrak{B}_\delta$,

$$\mathfrak{B}_1 \subset \mathfrak{B}_d \cap \mathfrak{B}_\delta, \tag{4.9}$$

and that *the operations d and δ can be expressed in terms of strong derivatives with respect to variables $x^1, \cdots, x^n$.*

The *main statement* to be made in the present section is that the converse of relation (4.9) also holds:

$$\mathfrak{B}_1 \supset \mathfrak{B}_d \cap \mathfrak{B}_\delta. \tag{4.10}$$

That is, whenever a form v in $\mathfrak{B}$ admits both the operation d as well as δ (in the weak sense), its components with respect to any patch have strong derivatives with respect to each coordinate x^κ, $\kappa = 1, \cdots, n$. Moreover, the approximating components can be taken as those of approximating functions which are defined on the whole manifold.

We combine the statements (4.9) and (4.10) in

THEOREM 4.1. $\mathfrak{B}_1 = \mathfrak{B}_d \cap \mathfrak{B}_\delta$.

Statement (4.10) is essentially a special case of a "differentiability theorem" proved elsewhere [17D]. Still, it seems desirable to present here the major steps of its proof. To this end we first select a patch $\mathcal{P}$, a core $\mathcal{R}$ of $\mathcal{P}$, then another core $\mathcal{Q}$ of $\mathcal{P}$ which contains $\mathcal{R}$. We select a non-negative function $\omega(x)$ with derivatives of any order, which equals 1 in $\mathcal{R}$ and 0 outside. Finally, we introduce the space $\mathfrak{B}_1(\mathcal{Q})$ of all forms v defined in $\mathcal{Q}$ whose components have strong derivatives, square integrable over $\mathcal{Q}$. We then formulate

LEMMA 4.2. *There are constants $C_0 > 0$ and $C_1 > 0$ such that for all forms v in $\mathfrak{B}_1(\mathcal{Q})$ the inequality*

$$\| \nabla v \|_{\mathcal{R}} \leqq C_0 \| dv \|_{\mathcal{Q}} + C_0 \| \delta v \|_{\mathcal{Q}} + C_1 \| v \|_{\mathcal{Q}} \tag{4.11}$$

holds, in which the terms on the right hand side are to be understood in obvious analogy with the convention (4.5).

This inequality (4.11) was derived and used by M. Gaffney [7]; we shall refer to it as "Gaffney's inequality". (In case of a Euclidean metric a similar inequality was proved earlier [17B].) For the sake of completeness we present a proof which differs from Gaffney's proof in a minor detail, namely, we do not assume that the patches are sufficiently small.

It is convenient to assign to the form v of the rank p the form v_λ of rank

$p - 1$ by "freezing" the last subscript $\lambda_p = \lambda$; i.e., if $v_{\lambda_1 \ldots \lambda_p}$ are the components of v with respect to a coordinate system, then the components of v_λ are

$$v_{\lambda_1 \cdots \lambda_{p-1}\lambda} .$$

Similarly, we may freeze two components. Then the identity

$$\nabla_\nu v_\lambda = (dv)_{\lambda\nu} + (d(v_\nu))_\lambda , \tag{4.12}$$

holds as readily verified. Here $\nabla_\nu v_\lambda$ is the form whose components are the derivatives with respect to x^ν of the form v_λ, cf. (4.3).

Next we employ the tensor $\underset{p-1}{G}$, defined by (4.4), and verify the relation

$$\nabla_\mu v_\kappa \underset{p-1}{G} \nabla_\nu v_\lambda = (p + 1)^{-1}(dv)_{\kappa\mu} \underset{p-1}{G} (dv)_{\lambda\nu} + \nabla_\mu v_\kappa \underset{p-1}{G} (d(v_\nu))_\lambda . \tag{4.13}$$

We multiply this relation by $g^{\kappa\lambda}g^{\mu\nu}$ and carry out summation with respect to $\kappa, \lambda, \mu, \nu$. By virtue of the skew symmetry of the components of v, the resulting formula can be written in the form

$$\nabla_\mu v g^{\mu\nu} \underset{p}{G} \nabla_\nu v = (p + 1)^{-1} dv \underset{p+1}{G} dv + p\nabla_\mu v_\kappa g^{\kappa\lambda} g^{\mu\nu} \underset{p-1}{G} \nabla_\lambda v_\nu . \tag{4.14}$$

Next we choose a function ω with bounded derivatives for which

$$\begin{aligned} \omega &\equiv 1 \quad \text{in } \mathfrak{R}, \\ \omega &\equiv 0 \quad \text{outside of } \mathfrak{Q} \end{aligned} \tag{4.15}$$

and multiply formula (4.14) by $\omega^2\gamma$. It is easy to verify that the result can be written in the form

$$\begin{aligned} \omega^2\gamma\nabla_\mu v g^{\mu\nu} \underset{p}{G} \nabla_\nu v &= (p + 1)^{-1}\omega^2\gamma \, dv \underset{p+1}{G} dv \\ &\quad + p\nabla_\mu[\omega^2\gamma v g^{\kappa\lambda} g^{\mu\nu} \underset{p-1}{G} \nabla_\lambda v_\nu] - p\nabla_\lambda[\omega^2\gamma v g^{\kappa\lambda} g^{\mu\nu} \underset{p-1}{G} \nabla_\mu v_\nu] \\ &\quad + p\omega^2\gamma \, \delta v \underset{p-1}{G} \delta v \\ &\quad + vP^\rho\omega\nabla_\rho v + vQv, \end{aligned} \tag{4.16}$$

where P^ρ and Q are matrices which are polynomials in the functions $g^{\alpha\beta}$, γ, ω and their first derivatives. Integration over the core $\mathfrak{Q}$ and application of Schwarz's inequality yields the relation

$$(1 - \epsilon)[|| \omega\nabla v ||_{\mathfrak{R}}']^2 \leqq || \omega \, dv ||^2 + || \omega \, \delta v ||^2 + C[|| v ||_{\mathfrak{Q}}]^2, \tag{4.17}$$

where ϵ is an arbitrary positive number < 1 and C is an appropriate constant. Using (4.15) we obtain inequality (4.11).

We now derive the statement (4.10) from Gaffney's inequality by using Lemma 2.2. Let v^α be the sequence whose existence is stated in this lemma.

We apply inequality (4.11) to the differences $v^{\alpha\beta} = v^\alpha - v^\beta$. For each patch $\mathcal{P}$ we find

$$\| v^{\alpha\beta} \|_{\mathfrak{R}} \to 0, \qquad \| \nabla v^{\alpha\beta} \|_{\mathfrak{R}} \to 0, \qquad \text{as } \alpha, \beta \to \infty .$$

The statement (4.10) is then a consequence of Lemma 4.1. Thus Theorem 4.1 is proved.

We now formulate

THEOREM 4.2. $\mathfrak{H} \subset \mathfrak{B}_1$.

It is an immediate application of Theorem 4.1 as inferred from the definition $\mathfrak{H} = \mathfrak{B}_{d=0} \cap \mathfrak{B}_{\delta=0}$ of $\mathfrak{H}$, combined with (3.5). Theorem 4.2 states that *any harmonic field possesses strong derivatives.*

From the differentiability theorem for solutions of elliptic differential equation—which can be proved by extending the arguments used in proving Theorem 4.1, cf. [17D]—it would follow that a harmonic field is continuous and possesses derivatives of any desired order provided the metric tensor $g^{\kappa\mu}$ and the identification transformation satisfy sufficiently strong differentiability conditions. In Section 6 a related fact will be derived under weaker conditions.

It is convenient to characterize the intersection of any subspace of $\mathfrak{B}$ with $\mathfrak{B}_1$ by a subscript 1. Then Theorem 4.2 may also be expressed in the form

$$\mathfrak{H} = \mathfrak{H}_1 . \tag{4.18}$$

Furthermore, the spaces $\overline{d\mathfrak{U}_s}$ and $\overline{\delta\mathfrak{W}_d}$, defined by (3.3) and further characterized by (3.4), can be expressed as

$$\overline{d\mathfrak{U}_s} = \overline{d\mathfrak{U}_1} , \qquad \overline{\delta\mathfrak{W}_d} = \overline{\delta\mathfrak{W}_1} \tag{4.19}$$

by virtue of Theorem 4.1. Combining this theorem with the modified decomposition Theorem 3.2, we obtain the "strong decomposition theorem",

THEOREM 4.3.

$$\mathfrak{B}_1 = \overline{d\mathfrak{U}_1} \oplus \mathfrak{H} \oplus \overline{\delta\mathfrak{W}_1} . \tag{4.20}$$

In words it may be stated as follows: *If a form possesses strong derivatives then its projections into $\overline{d\mathfrak{U}}$, $\mathfrak{H}$, and $\overline{\delta\mathfrak{W}}$ possess strong derivatives.*

In the formulation and use of this theorem, our approach to the theory of differential forms differs definitely from previous approaches.

5. *Complete Continuity. Connection Theorem*

In this section we shall show first that the space $\mathfrak{H}$ of harmonic forms has a finite dimension, and secondly that the operators d and δ map dense subspaces of the spaces $\overline{\delta\mathfrak{B}_d}$ and $\overline{d\mathfrak{B}_1}$ linearly in a one-to-one way into the spaces $\overline{d\mathfrak{U}}$ and $\overline{\delta\mathfrak{W}}$, respectively. These facts will follow from the *complete continuity* of the form

$$(v, v)_0 = \int v * v \tag{5.1}$$

with respect to the form

$$(v, v)_1 = \int dv * dv + \int \delta v * \delta v + \int v * v, \tag{5.2}$$

cf. (3.1), as unit form. This complete continuity is equivalent with the statement of

THEOREM 5.1. *There exists a sequence of forms $j^{(\rho)}$, $\rho = 1, 2, \cdots$, in $\mathfrak{B}_1$ and to every $\epsilon > 0$ an integer r_ϵ such that the inequality*

$$\| v \|_0^2 \leqq \epsilon \| v \|_1^2 + \sum_{\rho=1}^{r_\epsilon} (j^{(\rho)}, v)^2 \tag{5.3}$$

holds for every form v in $\mathfrak{B}_1$.

The subsequent arguments, used to prove the complete continuity of $\| v \|_0$ with respect to $\| v \|_1$, are closely related to the arguments used by Rellich in proving his Selection Theorem [15]. This selection theorem together with inequality (4.11) was used by Gaffney to prove the complete continuity of Green's operator, i.e. the inverse of the operator $d\delta + \delta d$.

Inequality (5.3) will be derived as a consequence of

LEMMA 5.1. *To each patch $\mathcal{P}$ there exists a sequence g^μ, $\mu = 1, 2, \cdots$, of forms in $\mathfrak{B}_1(\mathcal{Q})$ and to every $\beta > 0$ an integer m_β such that the inequality*

$$\| v \|_{\mathcal{R}}^2 \leqq \beta \| \nabla v \|_{\mathcal{R}}^2 + \sum_{\mu=1}^{m_\beta} (v, g^\mu)^2 \tag{5.4}$$

holds.

The latter inequality in turn is an immediate consequence of the corresponding inequality for v in $\mathfrak{C}_1$ proved in Courant-Hilbert, Vol. II [16] with the aid of Poincaré's inequality.

Inserting inequality (4.11) in (5.4) and using the fact that $\eta = 0$ outside of $\mathcal{R}$ we obtain the inequality

$$\int_{\mathcal{P}} \eta v * v \leqq C_0 \int_{\mathcal{Q}} dv * dv + C_0 \int_{\mathcal{Q}} \delta v * \delta v$$
$$+ C_1 \int_{\mathcal{Q}} v * v + \sum_{\mu=1}^{m_\delta} \left| \int v * g^{(\mu)} \right|^2.$$

We add these inequalities for all patches $\mathcal{P} = \mathcal{P}_0$ and use the obvious inequality

$$\sum_\sigma \int_{\mathcal{P}_\sigma} v * v \leqq \iota \int_{\mathcal{P}} v * v$$

involving the maximum number ι of patches which may cover any point, cf.

Section 1. Setting

$$\hat{C}_0 = \max_\sigma \overset{\sigma}{C}_0 , \qquad \hat{C}_1 = \max_\sigma \overset{\sigma}{C}_1 ,$$

we find

$$(5.5) \qquad \| v \|_0^2 \leqq \beta_\iota C_0[\| dv \|_0^2 + \| \delta v \|_0^2] + \beta_\iota C_1 \| v \|_0^2 + \sum_\sigma \sum_{\mu=1}^{\overset{\sigma}{m}_\beta} (v, \overset{\sigma}{g}\mu)^2.$$

We choose $\beta < 1/\hat{C}_1$, and set

$$\epsilon = \beta_\iota C_0/1 - \beta_\iota \hat{C}_1 .$$

Furthermore we take $r_\epsilon = \sum_\sigma \overset{\sigma}{m}_\beta$ and rearrange the double sequence of forms $[1 - \delta_\iota \hat{C}_1]^{-1/2} g_\mu^{(\sigma)}$ as a single sequence of forms $j^{(\rho)}$. Then inequality (5.5) leads to inequality (5.3).

As a first consequence of Theorem 5.1 we state

THEOREM 5.2. $\dim \mathfrak{H} < \infty$.

In other words, *the space of harmonic forms has a finite dimension.*

In fact, we may state specifically

$$\dim \mathfrak{H} \leqq r_\epsilon$$

for any $\epsilon > 0$. For, otherwise the space $\mathfrak{H}$, which by Theorem 4.2 is contained in $\mathfrak{B}_1$, would contain a form $v \neq 0$ perpendicular to $j^{(1)}, \cdots , j^{(r_\epsilon)}$. Since $dv = \delta v = 0$ for a harmonic form, inequality (5.3) would yield $\| v \|_0 = 0$ in contradiction to $v \neq 0$.

As another consequence of Theorem 5.1 we shall establish

THEOREM 5.3. *There is a positive number κ such that the inequality*

$$(5.6) \qquad \int v * v \leqq \kappa \int dv * dv + \kappa \int \delta v * \delta v$$

holds for all v in the space $\overline{d\mathfrak{U}_s} \oplus \overline{\delta\mathfrak{W}_d}$.

Note that by (4.18) and (4.19) the space $d\mathfrak{U}_s \oplus \delta\mathfrak{W}_s$ consists of exactly those forms v in $\mathfrak{B}_1$ which are orthogonal to $\mathfrak{H}$ with respect to either $(v, v)_1$ or $(v, v)_0$. Inequality (5.6), to be proved, can evidently be written in the form

$$(5.7) \qquad (1 + \kappa)(v, v)_0 \leqq \kappa(v, v)_1 \qquad \text{for } v \perp \mathfrak{H}.$$

Therefore, it expresses the fact that, except for the value $\lambda = 1$, the spectrum of the form $(v, v)_0$ with respect to the unit form $(v, v)_1$ is bounded by the value $\kappa[1 + \kappa]^{-1} < 1$, and thus stays away from the value 1. Note that $\lambda = 1$ is a possible eigenvalue of the form $(v, v)_0$, attained exactly for the harmonic forms.

The statement expressed by inequality (5.7) is an immediate consequence of the complete continuity of the form $(v, v)_0$ with respect to $(v, v)_1$, expressed by inequality (5.3). This complete continuity insures that the form $(v, v)_0$ has a complete discrete spectrum of eigenvalues of finite multiplicity approaching zero. However, we do not use this completeness for our purposes. We need only two consequences of the complete continuity. Theorem 5.2, which states that the eigenvalue $\lambda_1 = 1$ has a finite multiplicity, is the first such consequence. The other one is the fact that the form $(v, v)_0$ has a second eigenvalue, $\lambda_2 < 1$.

In proving the latter fact we first make sure that the form $(v, v)_1$ may serve as unit form for the Hilbert space $\mathfrak{B}_1$. This is guaranteed by Theorem 3.1. Then we can use standard arguments, cf. Courant-Hilbert [16]. These arguments yield that the ratio

$$(v, v)_0/(v, v)_1$$

taken for all functions v in $\mathfrak{B}_1$ and $\perp \mathfrak{H}$ assumes a maximum λ_2 for a certain form v_2, which satisfies the relation

$$(v_2, v_2)_0 = \lambda_2(v_2, v_2)_1$$

or

$$\lambda_2 \int dv_2 * dv_2 + \lambda_2 \int \delta v_2 * \delta v_2 = (1 - \lambda_2) \int v_2 * v_2$$

This relation shows that $\lambda_2 = 1$ would imply $dv_2 = \delta v_2 = 0$ in contradiction to $v_2 \perp \mathfrak{H}$. Hence $\lambda_2 < 1$.

Inequalities (5.7) and (5.6) then hold with

$$\kappa = \lambda_2[1 - \lambda_2]^{-1}.$$

As an immediate consequence of Theorem 5.3 combined with inequality (4.11), we mention

LEMMA 5.2 *There exists a constant $C > 0$ such that for all forms v in $\overline{d\mathfrak{U}_s} \oplus \overline{\delta\mathfrak{W}_d}$ the inequality*

$$\sum_{\sigma} \| \nabla v \|^2_{\mathfrak{R}_\sigma} \leqq C \int dv * dv + C \int \delta v * \delta v \tag{5.9}$$

holds.

Another important consequence of inequality (5.6) is that the forms $\| dv \|_0^2 = \int dv * dv$ and $\| \delta v \|_0^2 = \int \delta v * \delta v$ may serve as unit forms in the spaces $d\mathfrak{U}_1$ and $\delta\mathfrak{W}_1$.

We shall not formulate this fact separately; we shall use it implicitly in the proof of the main result of Part I to which we shall refer as the "Connection Theorem":

THEOREM 5.4. *To every form w in $\overline{d\mathfrak{B}}$ there is assigned in a unique way a form v in $\overline{\delta\mathfrak{W}_1}$ such that*

$$w = dv.$$

To every form u in $\overline{\delta\mathfrak{W}}_1$ there is assigned in a unique way a form v in $\overline{d\mathfrak{U}}_1$ such that

$$u = \delta v.$$

To prove this theorem we take a sequence of forms v^ν in $\mathfrak{V}$ such that $dv^\nu \to w$ and let v_a^ν be the projection of v^ν into the space $\overline{\delta\mathfrak{W}}$. From (2.11) and (2.13) we conclude that $d(v_a^\nu - v^\nu) = 0$. Since v^ν is in $\mathfrak{V}_d$, the same is true of v_a^ν and hence v_a^ν is in $\overline{\delta\mathfrak{W}}_d = \overline{\delta\mathfrak{W}}_1$, cf. (4.19).

Evidently, $dv_a^\nu = dv^\nu \to w$. Consequently, $d(v_a^\nu - v_a^\mu) \to 0$ as $\nu, \mu \to \infty$. Since $\delta(v_a^\nu - v_a^\mu) = 0$, inequality (5.6) yields $v_a^\nu - v_a^\mu \to 0$. Hence there exists a form v_a in $\overline{\delta\mathfrak{W}}$ such that $v_a^\nu \to v_a$. Clearly, v_a is in v_d, hence in $\overline{\delta\mathfrak{W}}_d = \overline{\delta\mathfrak{W}}_1$ with $dv_a = w$.

The second part of the theorem follows similarly. As an evident Corollary of Theorem 5.4 we state

THEOREM 5.5. *The spaces $d\mathfrak{V}$ and $\delta\mathfrak{V}$, and hence also $d\mathfrak{U}$ and $\delta\mathfrak{W}$, are closed: $d\mathfrak{V} = \overline{d\mathfrak{V}}$, $\delta\mathfrak{V} = \overline{\delta\mathfrak{V}}$, or*

$$d\mathfrak{U} = \overline{d\mathfrak{U}}, \qquad \delta\mathfrak{W} = \overline{\delta\mathfrak{W}}. \tag{5.10}$$

Therefore we may omit the bar from now on.

II. CONTINUOUS FORMS ON CLOSED MANIFOLDS

6. *Properties of Forms Involving Continuity*

The results which we have developed so far refer to forms with quadratically integrable components. From them one can derive statements involving forms with continuous components, as we shall show in this section. Among these statements will be a variant of the Full Decomposition Theorem of Bidal and de Rham. We first formulate[6]

THEOREM 6.1. *A form v in $\mathfrak{V}_1$ for which dv and δv are bounded and measurable forms is itself continuous.*

By saying that v *is* continuous we mean that it differs from a form v_c with continuous components by a form of vanishing norm: $\| v - v_c \| = 0$. The same remark applies to the measurable boundedness of dv and δv.

The statement of Theorem 6.1 can be derived from well known properties of integral operators with singular, but integrable, kernels.

We select a patch $\mathcal{P}$ and employ the function $\omega(x)$, introduced in Section 4, which has the properties (4.15).

The value of any function $\phi(x)$ at any point $\overset{0}{x}$ in $\mathcal{R}$ will be denoted by

$$\phi(\overset{0}{x}) = \overset{0}{\phi}; \tag{6.1}$$

[6]The idea to use such a theorem was suggested by L. Bers.

in particular we set

$$g_{\kappa\lambda}(\overset{0}{x}) = \overset{0}{g}_{\kappa\lambda} . \tag{6.2}$$

We also use the abbreviation

$$x^\kappa - \overset{0}{x}{}^\kappa = \underset{0}{x}{}^\kappa \tag{6.3}$$

for any points x in $\mathcal{P}$, $\overset{0}{x}$ in $\mathcal{R}$. We then introduce the—non-geodetic—distance $\underset{0}{r}$ by

$$\underset{0}{r}{}^2 = \overset{0}{g}_{\kappa\lambda}\underset{0}{x}{}^\kappa\underset{0}{x}{}^\lambda. \tag{6.4}$$

For any function $\phi(x)$ with continuous first derivatives we have

$$\Omega\overset{0}{\phi} = -\int_{\mathcal{P}} \underset{0}{r}{}^{-m}\underset{0}{x}{}^\lambda\nabla_\lambda(\omega\phi)\overset{0}{\gamma}\,dx \tag{6.5}$$

where Ω is the Euclidean content of the unit sphere $|\underset{0}{x}| = 1$ in the m-dimensional space.

Now let v be a form in $\underset{p}{\mathfrak{C}}_1$, i.e. one whose components have continuous first derivatives. The value of this form at the point $\overset{0}{x}$ can then be expressed in terms of its derivatives as follows:

$$\overset{0}{v} = -\Omega^{-1}\int_{\mathcal{P}} \underset{0}{r}{}^{-m}\underset{0}{x}{}^\mu\nabla_\mu(\omega v)\overset{0}{\gamma}\,dx.$$

Here we have used the convention (4.3). Carrying out the differentiation with respect to x^μ we obtain

$$\overset{0}{v} = -\Omega^{-1}\int_{\mathcal{P}} \underset{0}{r}{}^{-m}\underset{0}{x}{}^\mu\omega\nabla_\mu v\,\overset{0}{\gamma}\,dx + \overset{0}{A}v,$$

where A stands for the integral operator which transforms the form v into a form whose value at the point $\overset{0}{x}$ is

$$\overset{0}{A}v = -\Omega^{-1}\int_{\mathcal{P}} \underset{0}{r}{}^{-m}\underset{0}{x}{}^\lambda(\nabla_\mu\omega)v\,\overset{0}{\gamma}\,dx. \tag{6.6}$$

Next we express $\nabla_\beta v$ in terms of dv and $d(v_\beta)$ through formula (4.12):

$$\nabla_\beta v = (dv)_\beta + d(v_\beta).$$

Here $d(v_\beta)$ is the differential of the form v_β . Recall that v_β is the form of rank $p - 1$ obtained from the form v of rank p by freezing the subscript β.

We then set

$$v = Av + B\,dv + Zv, \tag{6.7}$$

defining the integral operators B and Z by

$$\overset{0}{B}dv = -\Omega^{-1}\int_{\mathcal{P}} \underset{0}{r}^{-m}\underset{0}{x}^{\beta}\omega(dv)_{\beta}\overset{0}{\gamma}\,dx, \tag{6.8}$$

$$\overset{0}{Z}v = -\Omega^{-1}\int_{\mathcal{P}} \underset{0}{r}^{-m}\underset{0}{x}^{\beta}\omega d(v_{\beta})\overset{0}{\gamma}\,dx. \tag{6.9}$$

Further integrations by parts will make it possible to express the operator Z in terms of an operator which acts on δv and operators which act on v.

Before doing this it is convenient to adopt various conventions. First we introduce the form y of rank 1 with the components

$$y_{\alpha} = \overset{0}{g}_{\alpha\beta}\underset{0}{x}^{\beta}\underset{0}{r}^{-m} = \tfrac{1}{2}\partial\underset{0}{r}^{2}/\partial\underset{0}{x}^{\alpha} \tag{6.10}$$

and write formula (6.9) in the form

$$\overset{0}{Z}v = -\Omega^{-1}\int_{\mathcal{P}} \omega\overset{0}{\gamma}\,d(v_{\beta})\overset{0}{g}^{\beta\mu}y_{\mu}\,dx. \tag{6.11}$$

Then we make use of the relation

$$\int_{\mathcal{P}}(\nabla_{\alpha}\psi^{\mu})y_{\mu}\,dx - \int_{\mathcal{P}}(\nabla_{\mu}\chi^{\mu})y_{\alpha}\,dx = \int_{\mathcal{P}}[\chi^{\mu} - \psi^{\mu}]\nabla_{\mu}y_{\alpha}\,dx, \tag{6.12}$$

which holds for any pair of vectors $\psi^{\mu}(x)$, $\chi^{\mu}(x)$ which have continuous derivatives, vanish outside of $\mathcal{R}$, and agree at $x = \overset{0}{x}$:

$$\overset{0}{\psi}{}^{\mu} = \overset{0}{\chi}{}^{\mu}. \tag{6.13}$$

To derive relation (6.12) one has to integrate by parts, use identity

$$\nabla_{\alpha}y_{\mu} = \nabla_{\mu}y_{\alpha}\,,$$

and verify that the resulting additional terms

$$-m^{-1}\overset{0}{g}_{\alpha\mu}\overset{0}{\psi}{}^{\mu} \qquad \text{and} \qquad -m^{-1}\overset{0}{g}_{\alpha\mu}\overset{0}{\chi}{}^{\mu}$$

cancel by (6.13).

We apply relation (6.12) to the forms

$$\psi^{\mu} = \omega\overset{0}{\gamma}\overset{0}{g}{}^{\mu\beta}v_{\beta}$$

and

$$\chi^{\mu} = \omega\underset{p-1}{\overset{0}{G}}{}^{-1}\gamma(\underset{p}{G}v)^{\mu}$$

which evidently satisfy (6.13). Here $G = \underset{p-1}{G}$, cf. (4.4), and $G^{-1} = \underset{p-1}{G}{}^{-1}$ is the inverse of G.

In the following—as in Lemma 2.1—we shall use the notation $v \wedge q$ for the exterior product of two forms v and q with the components $v_{\alpha_1 \cdots \alpha_{p-1}}$ and q_{α_1}; the components of $v \wedge q$ then are the skew-symmetric parts of $v_{\alpha_1 \cdots \alpha_{p-1}} q_{\alpha_p}$. With this convention we may write Zv in the form

$$Zv = E\,\delta v + Fv + Gv \tag{6.14}$$

where

$$\overset{0}{E}\,\delta v = -p\Omega^{-1} \int_{\mathcal{P}} \omega(\overset{0}{\underset{p-1}{G}}{}^{-1}\gamma \underset{p-1}{G}\,\delta v) \wedge y\,dx, \tag{6.15}$$

$$\overset{0}{F}v = -p\Omega^{-1} \int_{\mathcal{P}} \omega \overset{0}{\underset{p-1}{G}}{}^{-1}(\{\gamma \underset{p}{G} - \overset{0}{\gamma} \overset{0}{\underset{p}{G}}\}v)^{\mu} \wedge \nabla_{\mu} y\,dx \tag{6.16}$$

and

$$\overset{0}{G}v = p\Omega^{-1} \int_{\mathcal{P}} (\overset{0}{\gamma}[v_{\beta} \wedge d\omega]\overset{0}{g}{}^{\beta\mu} y_{\mu}\,dx) - p\Omega^{-1} \int_{\mathcal{P}} (\overset{0}{\underset{p-1}{G}}{}^{-1}\gamma(\underset{p}{G}v)^{\mu}\nabla_{\mu}\omega) \wedge y\,dx. \tag{6.17}$$

Setting

$$L = A + G + F, \tag{6.18}$$

we find that the value $\overset{0}{v}$ of $v(x)$ for $x = \overset{0}{x}$ satisfies the relation

$$\overset{0}{v} = \overset{0}{B}\,dv + \overset{0}{E}\,\delta v + \overset{0}{L}v \qquad \text{if } \overset{0}{x} \text{ is in } \mathcal{R}. \tag{6.19}$$

The kernels of the integral operators B, E, L are continuous in x and $\overset{0}{x}$ except for $x = \overset{0}{x}$; also, these kernels are absolutely integrable with respect to x and the integral of their absolute values is bounded independently of $\overset{0}{x}$. For the kernels B to G this follows from the fact that they are of the order $O(\underset{0}{r}^{-m+1})$ at $x = \overset{0}{x}$; the kernel of F is also of the order $O(\underset{0}{r}^{-m+1})$ since the components $g^{\kappa\lambda}(x)$ satisfy a uniform Lipschitz condition.

It is known that integral operators with these properties transform bounded measurable functions into continuous functions. Moreover, it is known that these operators can be extended to L^2-integrable functions defined in the core $\mathcal{Q} \supset \mathcal{R}$; they then produce functions which are L^2-integrable over $\mathcal{Q}$.

We conclude that for a function v in $\mathfrak{B}_1$ with bounded measurable dv and δv, relation (6.18) holds in $\mathcal{R}$ and that the function

$$B\,dv + E\,\delta v$$

is continuous in $\mathcal{Q}$.

The operator L will not necessarily transform L^2-functions into continuous ones, but a finite power of it will do so, as is well known. In fact, since the $g^{\kappa\lambda}(x)$ satisfy a Lipschitz condition, the operator L^k has this property if $k > m/2$, since the kernel of L^k, which is of the order $O(r^{-m+k})$, is then itself quadratically integrable.

Since relation (6.18) is proved only for x in $\mathcal{R}$, iterations of it cannot be carried out without preparations. We consider a nested sequence of sub-patches $\mathcal{P}_0 \subset \mathcal{P}_1 \subset \cdots \subset \mathcal{P}_k = \mathcal{R} \subset \mathcal{P}_{k+1} = \mathcal{Q}$, the images of spheres $|x| \leqq R_0, R_1, \cdots, R_{k+1}$, with $0 < R_0 < R_1 < \cdots < R_{k+1}$. To each of these sub-patches $\mathcal{P}_\kappa$ we assign a continuously differentiable function $\omega(x) = \omega_\kappa(x)$ which vanishes outside of $\mathcal{P}_{\kappa+1}$ and equals 1 in $\mathcal{P}_\kappa$. The operators B, E, L formed with the aid of the function $\omega_\kappa(x)$ will be denoted by B_κ, E_κ, L_κ. From (6.18) we infer that relation

$$v = B_\kappa \, dv + E_\kappa \, \delta v + L_\kappa v \tag{6.20}$$

holds for $v = \overset{0}{v} = v(\overset{0}{x})$ if the point $\overset{0}{x}$ lies in $\mathcal{P}_\kappa$; evidently, this relation involves the values of v, dv, δv only in the sub-patch $\mathcal{P}_{\kappa+1}$. By successive use of (6.20) one finds that the relation

$$\begin{aligned} v = B_0 \, dv + E_0 \, \delta v + L_0(B_1 \, dv + E_1 \, \delta v) \\ + \cdots + L_0 L_1 \cdots L_{k-1}(B_k \, dv + E_k \, \delta v + L_k v) \end{aligned} \tag{6.21}$$

holds for $\overset{0}{x}$ in $\mathcal{P}_0$.

According to what was said before, the function $L_0 \cdots L_{\kappa-1} (B_\kappa \, dv + E_\kappa \, \delta v)$ is continuous in $\mathcal{P}_0$ if v is a square integrable function and $L_0 \cdots L_k$, just as L^k, transforms an L^2-function into a continuous one.
Hence v is continuous in $\mathcal{P}_0$.

Since the choice of the patch $\mathcal{P}_0$ was arbitrary, Theorem 6.1 is proved.

As a first consequence of Theorem 6.1 we note

THEOREM 6.2. *Harmonic fields are continuous,*

$$\mathfrak{H} \subset \mathfrak{C}.$$

Evidently, because of $dv = \delta v = 0$, the assumptions of Theorem 6.1 are satisfied.

Next we formulate

THEOREM 6.3. *If the form v in $\mathfrak{V}_1$ possesses bounded measurable differentials dv and δv, its projections into the spaces $d\mathfrak{V}$, $\mathfrak{H}$, and $\delta\mathfrak{V}$, are continuous.*

To prove this theorem, let v_z, h, v_a be the projections of v into $d\mathfrak{V}$, $\mathfrak{H}$, $\delta\mathfrak{V}$ according to Theorem 2.2. Since $dh = dv_a = 0$, the component v_z is in $\mathfrak{V}_d$ with $dv_z = dv$. Since $\delta v_z = 0$, Theorem 6.1 is applicable and yields that v_z is continuous. Similarly we have v_a in $\mathfrak{C}$.

Theorem 6.3, combined with Theorem 2.2, is a variant of the full decomposition theorem of Bidal and de Rham. Our conditions on the coefficients $g^{\kappa\lambda}(x)$ are weaker than those made by these and other authors. On the other hand, they make stronger statements about the nature of the components v_z and v_a.

A certain strengthening is afforded by

THEOREM 6.4. *A continuous form w in $d\mathfrak{B}$ is the differential dv of a continuous form v in $\delta\mathfrak{W}$. A continuous form v in $\delta\mathfrak{B}$ is the ∞-differential of a continuous form w in $d\mathfrak{U}$.*

To prove this theorem we rely on Theorem 5.4. Accordingly, there is a form v in $\delta\mathfrak{W}$ with $w = dv$. Since $v = 0$ by (2.13), and $dv = w$ is assumed continuous, Theorem 6.1 is applicable to v. It yields the first statement.

The second statement follows similarly.

7. *Periods*

In this section we are concerned with the periods of integrals of closed forms, i.e. the values of the integrals of such forms over cycles. In particular, we want to establish the equivalence of Hodge's and de Rham's first and second theorems with the aid of the statements of Section 4.

We denote the integral of a form of rank p over a cycle $\mathfrak{z}$ of dimension p by

$$\int_{\mathfrak{z}} v.$$

We need not explain here what is meant by this expression; it is sufficient to note that $\int_{\mathfrak{z}} v$ is defined for forms in $\mathfrak{C}$, is linear, bounded, and vanishes if v is the differential $v = du$ of a form u in $\underset{p-1}{\mathfrak{C}_1}$,

$$\int_{\mathfrak{z}} du = 0. \tag{7.1}$$

The properties named are the only ones that will be used.

We first formulate

THEOREM 7.1. *If the form v in du is in $\mathfrak{C}$, the periods of its integrals vanish,*

$$\int_{\mathfrak{z}} v = 0. \tag{7.2}$$

If v were known to be of the form du with u in $\mathfrak{C}_1$, the statement would be implied by (7.1). By definition, v can be approximated by functions $v^{(\nu)} = du^{(\nu)}$ with $u^{(\nu)}$ in $\mathfrak{C}_1$ such that $\| v^{(\nu)} - v \| \to 0$ as $\nu \to \infty$; but it is a priori not excluded[7] that $\int_{\mathfrak{z}} v^{(\nu)}$ does not converge to $\int_{\mathfrak{z}} v$, even if v is continuous.

We employ Theorem 6.4. According to it there is a continuous form u in $\delta\mathfrak{B}$ such that $v = du$.

We shall approximate this form u by forms in $\mathfrak{C}_1$ whose differentials approximate du. To this end we write

$$u = \sum_{\sigma} \overset{\sigma}{\eta}\, u, \tag{7.3}$$

[7]This difficulty was clearly pointed out by H. Weyl [2].

where the functions $\eta = \overset{0}{\eta}$ constitute a partition of unity, cf. Section 1. According to Lemma 2.1 the form ηu is in $\mathfrak{B}_d$ and has the differential

$$d(\eta\, u) = \eta\, du + u \wedge d\eta. \tag{7.4}$$

The continuity of u and of du now imply that $d(\eta\, u)$ *is also continuous.* In fact, *it is at this place that Theorem 6.4 is used essentially.*

We now make use of

LEMMA 7.1. *Let $\hat{u}$ be a form in $\mathfrak{U}_d$ which together with $d\hat{u}$ is in $\mathfrak{C}$ and vanishes outside of $\mathfrak{Q}$. Then there exists a sequence of forms u^α in $\mathfrak{C}_1$ such that, as $\alpha \to \infty$,*

$$u^\alpha \Rightarrow \hat{u}, \qquad du^\alpha \Rightarrow du \tag{7.5}$$

uniformly in $\mathcal{P}$.

The statement follows from the identity of the weak and strong extensions of the operator d defined with respect to the maximum norm instead of the Hilbert norm. Without reference to this fact it results immediately by using mollifiers; cf. [17A].

From relation (7.5) and the properties of the period formulated in the beginning of this section, we deduce relation

$$\int_s du^\alpha \to \int_s d\hat{u} \qquad \text{as } \alpha \to \infty.$$

Since

$$\int_s du^\alpha = 0$$

we have

$$\int_s d\hat{u} = 0. \tag{7.6}$$

We now apply this result to the function $\hat{u} = \eta u$, which is permitted since $d(\eta\, u)$ is continuous as stated above. Then we obtain

$$\int_s d(\eta\, \hat{u}) = 0. \tag{7.7}$$

Summation with respect to patches $\mathcal{P}_\sigma$ then yields

$$\sum_\sigma \int_s d(\overset{\sigma}{\eta}\, u) = 0$$

or, by (7.4),

$$\int_s \sum_\sigma \overset{\sigma}{\eta}\, d\overset{\prime}{u} + \int_s u \wedge \sum_\sigma d\overset{\sigma}{\eta} = 0. \tag{7.8}$$

Now, $\sum_\sigma \overset{0}{\eta} = 1$ by definition, and hence $\sum_\sigma d\overset{0}{\eta} = 0$. Therefore relation (7.8) yields relation

(7.9) $$\int_{\mathfrak{z}} du = 0$$

which is equivalent to relation (7.2) which was to be proved.

It should be noted that the argument given here yields Theorem 7.1 without recourse to Theorem 6.4, whenever one can choose $\eta \equiv 1$ throughout the manifold. For, Theorem 6.4 was used solely to establish the continuity of $u \wedge d\eta$ in formula (7.4). Note that it is possible[8] to choose $\eta \equiv 1$ if the manifold itself is Euclidean, as was assumed by H. Weyl.

Theorem 7.1 now allows one immediately to *derive Hodge's first theorem from de Rham's first theorem.* By the latter we mean the statement that to every p-dimensional cocycle there exists a closed form v in $\mathfrak{C}_1$, i.e. one with $dv = 0$, whose periods $\int_{\mathfrak{z}} v$ over cycles $\mathfrak{z}$ are just the values which the cocycle (considered a linear assignment of numbers to chains) assigns to the cycle $\mathfrak{z}$. Hodge's theorem states that these forms may even be taken to be harmonic.

Indeed, the form v given by de Rham's theorem may, by Theorem 2.2, be split as

$$v = v_2 + h$$

into a form in $d\mathfrak{B}$ and a harmonic form. Now, v is continuous and h is continuous, by Theorem 6.2; hence v_2 is continuous. Therefore Theorem 7.1 is applicable; it yields

$$\int_{\mathfrak{z}} h = \int_{\mathfrak{z}} v,$$

and thus the statement of Hodge's theorem.

The *second theorem of de Rham*, which states that a closed continuous form is exact if its periods vanish,—or rather a variant of it—now *follows from the second theorem of Hodge*, which states that a harmonic form vanishes if all its periods vanish.

For, the closed form v in $\mathfrak{C}$ can be split into $v = v_a + h$ with v_a in $d\mathfrak{B} \cap \mathfrak{C}$ and h harmonic. By Theorem 7.1 the periods of h are those of v; hence they vanish. By Hodge's second theorem $h = 0$, hence v in $d\mathfrak{B}$. By Theorem 6.4 $v = du$ with u in $\mathfrak{C}$. This statement is a variant of de Rham's second theorem.

III. FORMS ON MANIFOLDS WITH BOUNDARIES

8. *Boundaries and Boundary Conditions*

In describing manifolds with boundaries we shall employ a covering by patches, but, in addition to the "full" patches described in Section 1, we shall introduce "boundary patches", the images of hemispheres

$$x_1^2 + \cdots + x_m^2 < 1, \qquad x_m \geq 0.$$

[8]This was observed by M. Gaffney, according to oral communication. Similar arguments were used by Hausner [10].

If a point X with $x_m = 0$ in a boundary patch $\mathcal{P}$ is contained in another patch, this other patch is assumed to be a boundary patch also and the point X is assumed to have the m-th coordinate $x_m^{(1)} = 0$ with respect to $\mathcal{P}^{(1)}$. Also, every sufficiently small neighborhood of this point X in one patch should lie in the other patch.

A point X with $x_m = 0$ in one patch will be called a boundary point. All boundary points form the boundary, $\mathcal{B}$.

All other properties of patches discussed in Section 1 are to be carried over to boundary patches literally.

On the components of the metric tensor $g^{\alpha\beta}$ we impose the condition

$$g^{\alpha m} = 0, \qquad g_{\alpha m} = 0, \qquad \alpha \neq m, \qquad \text{on } \mathcal{B}. \tag{8.1}$$

This condition is no restriction since it can be achieved by an appropriate coordinate transformation[9], but we prefer to impose it from the outset.

To a form in $\mathfrak{C}$, i.e. one with continuous components, we may assign tangential and normal components at the boundary, namely the components which do not, or do, have m as one of the subscripts. Evidently, this distinction is independent of the choice of the coordinate system.

We now *impose on all forms in* $\mathfrak{C}$ *and* $\mathfrak{C}_1$ *either of two conditions.* The first of these conditions is *that the tangential components of the form should vanish.* This condition will be referred to as the *first boundary condition* and the resulting spaces of forms will be characterized by a front-superscript "1". The forms will be called "of first kind".

The second boundary condition which may be imposed on the forms of $\mathfrak{C}$ *and* $\mathfrak{C}_1$ *is that their normal components vanish on the boundary*; the resulting spaces of forms will be characterized by a front-superscript "2". The forms will be called "of the second kind".

Actually, the two conditions are equivalent inasmuch as they interchange

---

[9]To this end it is sufficient to introduce new coordinates $\hat{x}_1, \ldots, \hat{x}_m$ in each boundary patch such that $x_m \equiv \hat{x}_m$ and $\hat{g}^{\alpha m} = 0$ for $\hat{x}_m = 0$, $\alpha \neq m$. Each new coordinate $\hat{x}_\alpha$ is then to be determined as a function of $x_1, \ldots, x_m$ with Lipschitz continuous first derivatives such that $\hat{x}_\alpha = x_\alpha$ for $x_m = 0$ and

$$\partial x_\alpha / \partial x_m = f_\alpha \qquad \text{for } x_m = 0,$$

where

$$f_\alpha = f_\alpha(x_1, \cdots, x_{m-1}) = -g^{m\alpha}/g^{mm}(x_1, \cdots, x_{m-1}).$$

The functions

$$\hat{x}_\alpha = x_\alpha + x_m \int \cdots \int f_\alpha(x + x_m \xi) j(\xi)\, d\xi_1 \cdots d\xi_{m-1}$$

have this property if the functions $j(\xi_1, \ldots, \xi_{m-1})$ are chosen as "mollifier kernels", i.e. as functions with continuous derivatives (of second order at least) which vanish outside a neighborhood of $\xi = 0$ and for which

$$\int \cdots \int j(\xi)\, d\xi_1 \cdots d\xi_{m-1} = 1.$$

roles if forms of rank p are interpreted as forms of rank $m - p$, as is easily verified, making use of condition (8.1).

We now may define the spaces ${}^1\mathfrak{B}_d$, ${}^1\mathfrak{B}_s$, ${}^1\mathfrak{B}_1$ etc. of forms of the first kind, and the spaces ${}^2\mathfrak{B}_d$, ${}^2\mathfrak{B}_s$, ${}^2\mathfrak{B}_1$ etc. of forms of the second kind, in literally the same way as the spaces $\mathfrak{B}_d$, $\mathfrak{B}_s$, etc. were defined in Part I, the only difference being that the functions in $\mathfrak{C}$ and $\mathfrak{C}_1$ entering these definitions are restricted by the first and second boundary conditions, respectively.

We maintain that then *all lemmas and theorems stated in Part I hold literally as they are formulated there.*

The proofs of these statements, however, do not carry over without modification. We shall not describe such modifications. Instead, we shall show that the statements now made can be deduced from those made in Part I by the simple device of *doubling the manifold.*

We consider a replica $\hat{\mathfrak{M}}$ of the manifold $\mathfrak{M}$ and denote by $\hat{X}$ the point on $\mathfrak{M}$ which corresponds to the point X on $\hat{\mathfrak{M}}$. On $\hat{\mathfrak{M}}$ we define a metric tensor by setting

$$g^{\alpha\beta}(x) = g^{\alpha\beta}(\hat{x}). \tag{8.2}$$

We now consider the pair of manifolds $\mathfrak{M} + \hat{\mathfrak{M}}$ as constituting a single manifold without boundary of the type discussed in Section 1. In doing this we must regard a pair of corresponding boundary patches, each being the image of the same hemisphere, as constituting a single ordinary patch, the image of a full sphere. To this end it is necessary to let one of the two boundary patches be the image of the complementary hemisphere, $x_m \leqq 0$, simply by substituting $-x_m$ for x_m. This change has the effect of changing the signs of all tensor components which have m as subscript an odd number of times.

Accordingly, $g^{\alpha m}$, $\alpha \neq m$, changes sign, but $g^{\alpha m}(x)$ remains continuous by virtue of condition (6.1). In fact, we have imposed condition (6.1) in order to achieve this continuity.

It is important to note that, in general, the derivatives of the metric tensor will not be continuous across the boundary of $\mathfrak{M}$. For, the relative curvature of the boundary, which depends on the first derivatives of $g^{\alpha\beta}$, is covariant and therefore cannot be annihilated by coordinate transformation.

As was first observed by Conner [12], *continuous forms on $\mathfrak{M}$ can be continued to continuous forms on $\mathfrak{M} + \hat{\mathfrak{M}}$ by setting*

$$v(X) = -v(\hat{X}) \tag{^{1}8.3}$$

if v satisfies the first boundary condition, and

$$v(X) = v(\hat{X}) \tag{^{2}8.3}$$

if v satisfies the second boundary condition.

For forms of the first kind the continuity across the boundary is evident for the tangential components since these components are required to vanish on $\mathfrak{B}$. For the normal components this might at first sight not appear to be the case—in general—since these components need not vanish on $\mathfrak{B}$ and have opposite values at corresponding points of the two replicas. However, since the normal components change sign when the two hemispheres are combined into a full sphere, the normal components are seen to be continuous anyway. The continuity of extended forms of the second kind is verified in a similar manner.

Forms v on $\mathfrak{M} + \hat{\mathfrak{M}}$ resulting by ${}^1(8.3)$ and ${}^2(8.3)$ from a form v on $\mathfrak{M}$ will be called "odd" and "even" respectively.

Clearly, the integral

$$\int \bar{v} * v$$

extended over $\mathfrak{M} + \hat{\mathfrak{M}}$ is twice the integral extended over $\mathfrak{M}$ if both v and $\bar{v}$ are either odd or even. Evidently, odd and even forms are perpendicular to each other.

Consider the spaces $\mathfrak{B}_d$ and $\mathfrak{B}_\delta$ defined with reference to $\mathfrak{M} + \hat{\mathfrak{M}}$ but restricted to forms of the first or second kind. In the definitions of these spaces arbitrary functions $\tilde{w}$ and $\tilde{u}$ in $\underset{p+1}{\mathfrak{C}_1}$ and $\underset{p-1}{\mathfrak{C}_1}$ enter; but only the odd or even parts of these forms enter the integrals (2.1) and (2.2). Thus it is seen that *the spaces* $\mathfrak{B}_d$ *and* $\mathfrak{B}_\delta$ *with reference to* $\mathfrak{M} + \hat{\mathfrak{M}}$ *lead to the spaces* ${}^1\mathfrak{B}_d$ *and* ${}^1\mathfrak{B}_\delta$ *or* ${}^2\mathfrak{B}_d$ *and* ${}^2\mathfrak{B}_\delta$ *when restricted to forms of the first or second kind.*

Clearly then, *all the spaces introduced in Part I defined with reference to* $\mathfrak{M} + \hat{\mathfrak{M}}$ *reduce to those of the first and second kind if restricted to odd or even forms.*

Moreover we state that *all the theorems and lemmas stated in Part I with reference to* $\mathfrak{M} + \hat{\mathfrak{M}}$ *but restricted to odd and even forms lead to the corresponding theorems and lemmas for forms of the first and second kind.* In order to verify this statement we need only verify that the conditions on the metric tensor $g^{\alpha\beta}$ adopted in Part I hold for the metric tensor defined in $\mathfrak{M} + \hat{\mathfrak{M}}$. Now, this is the case since we have required of the functions $g^{\alpha\beta}$ only that they are Lipschitz-continuous. The method of doubling would not be applicable, in general, if continuous derivatives of the $g^{\alpha\beta}$ had been required; for, as mentioned above, continuity of the first derivatives of the $g^{\alpha\beta}$ cannot always be enforced in the process of doubling. It is thus clear why it was essential to have developed the theory under rather weak conditions on the metric.

9. *Periods of First and Second Kind*

In a manifold with boundary we distinguish the ordinary Betti group, which refers to ordinary cycles, from the Betti group modulo the boundary $\mathfrak{B}$, which refers to cycles mod $\mathfrak{B}$. A cycle mod $\mathfrak{B}$ is a chain whose boundary lies on $\mathfrak{B}$ and it is said to be homologous zero mod $\mathfrak{B}$ if it is homologous a chain on $\mathfrak{B}$.

The periods of forms with reference to ordinary cycles and cycles mod $\mathfrak{B}$ will be referred to as ordinary periods and periods mod $\mathfrak{B}$ respectively.

As was observed by Conner [12], the relationship between these two Betti groups and that of $\mathfrak{M} + \hat{\mathfrak{M}}$ can be described as follows:

Let ${}^2\mathfrak{z}$ stand for ordinary nonbounding cycles in $\mathfrak{M}$ and ${}^1\mathfrak{z}$ for the cycles in $\mathfrak{M}$ mod $\mathfrak{B}$ which are nonbounding mod $\mathfrak{B}$. Then the nonbounding cycles of $\mathfrak{M} + \hat{\mathfrak{M}}$ are exactly represented as linear combinations of the cycles

$$ {}^2\mathfrak{z} + {}^2\hat{\mathfrak{z}} \quad \text{and} \quad {}^1\mathfrak{z} - {}^1\hat{\mathfrak{z}}. $$

In this way bases for Betti groups of the cycles ${}^1\mathfrak{z}$ and ${}^2\mathfrak{z}$ lead to a base for the Betti group of $\mathfrak{M} + \hat{\mathfrak{M}}$.

We now maintain that the statements of Section 7—including de Rham's and Hodge's two theorems—remain valid *for forms of the first kind with respect to cycles mod* $\mathfrak{B}$, *and for forms of the second kind with respect to the ordinary cycles.* The ordinary periods will therefore also be called periods of the second kind, and those mod $\mathfrak{B}$ of the first kind.

Theorem 7.1 carries over immediately to forms of the second kind. To prove relation (7.2) for a form v of the first kind, let $\hat{\mathfrak{z}}$ be the replica of the cycle $\mathfrak{z}$ mod $\mathfrak{B}$. Then $\mathfrak{z} - \hat{\mathfrak{z}}$ is an ordinary cycle in $\mathfrak{B} + \hat{\mathfrak{B}}$. Now, Theorem 7.1 is applicable to the odd extension of v. Hence we have

$$ 0 = \int_{\mathfrak{z}-\hat{\mathfrak{z}}} v = \int_{\mathfrak{z}} v - \int_{\hat{\mathfrak{z}}} \hat{v} = 2 \int_{\mathfrak{z}} v. $$

Thus the statement of Theorem 7.1 holds for forms of the first kind with respect to cycles mod $\mathfrak{B}$.

To establish *Hodge's first theorem* for forms of the second kind, assign to the cycles ${}^2\mathfrak{z} + {}^2\hat{\mathfrak{z}}$ twice the value assigned to ${}^2\mathfrak{z}$ and the value zero to the cycles ${}^1\mathfrak{z} - {}^1\hat{\mathfrak{z}}$. Hodge's theorem then yields the existence of a harmonic form in $\mathfrak{M} + \hat{\mathfrak{M}}$ whose periods have these values. Split this form into an odd and an even component. Evidently, the periods of the odd part over ${}^1\mathfrak{z} - {}^1\hat{\mathfrak{z}}$ and ${}^2\mathfrak{z} + {}^2\hat{\mathfrak{z}}$ vanish. Hence the odd part itself vanishes by Hodge's second theorem. The even part leads to a form of the second kind in $\mathfrak{M}$ whose periods assume the prescribed values on the cycles ${}^2\mathfrak{z}$.

In the same way Hodge's first theorem follows for forms of the first kind.

The validity of Hodge's second theorem for forms of the first or the second kind is inferred by similar arguments.

Thus it is clear that forms of the first and second kind are uniquely determined by their period of the first and second kind respectively.

It should be noted that to each form v of rank p a "dual" form $*v$ of rank $m - p$ can be assigned in such a way that the operators d and δ interchange their roles (except for sign). The dual form $*v$ is of the second and first kind if v is of the first and second kind, as is immediately verified by using condition (8.1). If v is of the first or second kind, the period of second or first kind of the dual form $*v$ will be called the "dual period of the second or first kind".

Clearly, Hodge's theorems imply that a harmonic form of one kind is uniquely characterized by the dual periods of the other kind.

10. *Relationship Between Forms of First and Second Kind*

Of course, the Hilbert spaces ${}^1\mathfrak{B}$ and ${}^2\mathfrak{B}$ of forms of first or second kind are identical,

$${}^1\mathfrak{B} = {}^2\mathfrak{B} = \mathfrak{B},$$

since these spaces were defined without reference to the boundary condition.

The relationship between the forms of the first and second kind is essentially expressed by the fact that in the definitions of the spaces ${}^2d\mathfrak{U}$ and ${}^1\delta\mathfrak{W}$ the second and first boundary conditions may be omitted. Similarly, in the definitions of the spaces ${}^2\mathfrak{B}_d$ and ${}^1\mathfrak{B}_\delta$ the functions w and u need not be required to satisfy the second and first boundary conditions respectively. In obvious notation we may express these facts in

THEOREM 10.1. ${}^2d\mathfrak{U} = d\mathfrak{U}$, ${}^1\delta\mathfrak{W} = \delta\mathfrak{W}$

and

THEOREM 10.2. ${}^2\mathfrak{B}_d = \mathfrak{B}_d$, ${}^1\mathfrak{B}_\delta = \mathfrak{B}_\delta$.

Of course the statement implies

$${}^2\mathfrak{B}_{d-0} = \mathfrak{B}_{d-0}\,, \qquad {}^1\mathfrak{B}_{\delta-0} = \mathfrak{B}_{\delta-0}\,.$$

To verify these statements we shall prove

LEMMA 10.1. *Every form v in $\mathfrak{C}_1$ can be approximated by forms ${}^2v^\tau$ in ${}^2\mathfrak{C}$ such that*

$$\| {}^2v^\tau - v \| \to 0, \qquad \| d\, {}^2v^\tau - dv \| \to 0 \qquad \text{as } \tau \to \infty$$

and by forms ${}^1v^\tau$ in ${}^1\mathfrak{C}_1$ such that

$$\| {}^1v^\tau - v \| \to 0, \qquad \| \delta\, {}^1v^\tau - \delta v \| \to 0 \qquad \text{as } \tau \to 0.$$

It is sufficient to prove the first statement. The second statement then follows by interpreting δu for a form of rank p as $d\hat{u}$ for an appropriate form of rank $m - p$. In doing this, condition (8.1) will have to be used.

We write v in the form

$$v = \sum_\sigma \overset{\sigma}{\eta} v$$

and consider a boundary patch $\mathcal{P}_\sigma$. In it we introduce a function $\zeta_\tau(x_m)$ with derivatives of any order which vanishes in a neighborhood $0 \leqq x_m \leqq \tau$ and equals one for $x_m \geqq \tau$. We consider the operator Z_τ which multiplies every normal component of $\overset{\sigma}{\eta} v$ by $\zeta_\tau(x_m)$ and leaves the tangential components unchanged. Evidently

$$\| Z_\tau \eta v - \eta v \| \to 0 \qquad \text{as } \tau \to 0.$$

Now we observe that in forming $d(Z_\tau \eta v)$ the derivative of $Z_\tau(x_m)$ with respect to x_m does not enter; for, in forming dv a normal component of v is never differentiated with respect to x_m. Consequently

$$|| d(Z_\tau \eta v) - d\eta v || \to 0 \qquad \text{as } \tau \to 0.$$

Defining the operator $\overset{\bullet}{Z}_\tau$ by $\overset{\bullet}{Z}_\tau = 1$ for ordinary patches we may define the operator Z_τ by

$$Z_\tau v = \sum_{\bullet} \overset{\bullet}{Z}_\tau \overset{\bullet}{\eta} v$$

in obvious notation. The statement of the lemma then follows with $v^\tau = Z_\tau v$.

The statements of Theorems 10.1 and 10.2 are now immediate consequences of Lemma 10.1.

These two theorems have evidently the following corollaries:

COROLLARY 10.1.
$$^1d\mathfrak{U} \subset {}^2d\mathfrak{U},$$
$$^2\delta\mathfrak{W} \subset {}^1\delta\mathfrak{W}.$$

COROLLARY 10.2.
$$^1\mathfrak{B}_d \subset {}^2\mathfrak{B}_d,$$
$$^2\mathfrak{B}_\delta \subset {}^1\mathfrak{B}_\delta ;$$

Corollary 10.1 implies that in defining the spaces $^1d\mathfrak{U}$ and $^2\delta\mathfrak{W}$ one may just as well require the approximating functions to satisfy both boundary conditions instead of only one. In fact, one may even require them to be identically zero in a neighborhood of the boundary. This would correspond to Weyl's definition of these spaces for a Euclidean manifold.

In the following we shall always employ Theorems 10.1 and 10.2 without explicitly referring to them.

First we note that these theorems enable one to combine the decompositions of forms of the first and second kind. From the decomposition

$$\mathfrak{B} = {}^1d\mathfrak{U} \oplus \mathfrak{B}_{\delta-0} = \mathfrak{B}_{d-0} \oplus {}^2\delta\mathfrak{W},$$

see Theorem 2.1, combined with

$$^1d\mathfrak{U} \subset \mathfrak{B}_{d-0}, \qquad {}^2\delta\mathfrak{W} \subset \mathfrak{B}_{\delta-0},$$

see (2.16), we obtain

THEOREM 10.3.
$$\mathfrak{B} = {}^1d\mathfrak{U} \oplus \mathfrak{H} \oplus {}^2\delta\mathfrak{W},$$

in which

$$\mathfrak{H} = \mathfrak{B}_{\delta-0} \cap \mathfrak{B}_{d-0}$$

is the space of all harmonic forms.

As seen from relations (2.13), $(2.16)_1$ for $^1\mathfrak{B}$ and (2.12), $(2.16)_2$ for $^2\mathfrak{B}$, the space $\mathfrak{H}$ in turn admits the following decompositions:

THEOREM 10.4.
$$\mathfrak{H} = {}^1\mathfrak{H} \oplus \mathfrak{B}_{d-0} \cap \delta\mathfrak{W},$$
$$\mathfrak{H} = \mathfrak{B}_{\delta-0} \cap d\mathfrak{U} \oplus {}^2\mathfrak{H},$$

where

$$ {}^1\mathfrak{H} = {}^1\mathfrak{B}_{d-0} \cap \mathfrak{B}_{\delta-0} \quad \textit{and} \quad {}^2\mathfrak{H} = {}^2\mathfrak{B}_{\delta-0} \cap \mathfrak{B}_{d-0}\,. $$

Under certain circumstances these two decompositions can be combined.

We first mention that the ordinary cycles ${}^2\mathfrak{z}$ are obviously contained among the cycles ${}^1\mathfrak{z}$ mod $\mathfrak{B}$. The periods ${}^2\pi$ of the second kind are therefore contained among those of the first kind; of course the periods ${}^2\pi$ over cycles which are homologous to a cycle on $\mathfrak{B}$ necessarily vanish for forms of the first kind. Now, it could happen that all periods ${}^2\pi$ of the second kind vanish for the form v. For this case we can state

THEOREM 10.5. *Harmonic forms of the first kind, i.e. forms v in ${}^1\mathfrak{H}$ for which all periods of the second kind vanish, belong to $d\mathfrak{B}$.*

By Theorem 5.4 such a form is the differential of a form in $\delta\mathfrak{B}$. To prove the theorem we note that v is in ${}^1\mathfrak{B}_{d-0}$, hence in $\mathfrak{B}_{d-0} = {}^2\mathfrak{B}_{d-0}$. Therefore we may regard v as a form of the second kind. We then may split v into a component in ${}^2d\mathfrak{B}$ and a component in ${}^2\mathfrak{H}$. By Theorem 7.1 the periods ${}^2\pi$ of v are those of its component in ${}^2\mathfrak{H}$. Since these periods vanish this component vanishes by Hodge's second theorem. In other words v is in ${}^2d\mathfrak{B} = d\mathfrak{B}$. Thus Theorem 10.5 is proved.

We observe that it may happen that the periods ${}^2\pi$ of the second kind vanish *for all* forms v in ${}^1\mathfrak{H}$. This does happen, for example, if the manifold $\mathfrak{R}$ can be imbedded in an m-dimensional Euclidean space, since then all cycles are homologous to cycles on $\mathfrak{B}$. For this case we may state

THEOREM 10.6. *Suppose that all cycles of the second kind are homologous to cycles on $\mathfrak{B}$. Then*

$$ \mathfrak{H} = {}^1\mathfrak{H} \oplus d\mathfrak{U} \cap \delta\mathfrak{W} \oplus {}^2\mathfrak{H}. $$

and

$$ {}^1\mathfrak{H} \subset d\mathfrak{U}, \qquad {}^2\mathfrak{H} \subset \delta\mathfrak{W}. $$

The statement results if we first derive ${}^1\mathfrak{H} \subset d\mathfrak{U}$ from Theorem 10.5 and then use Theorem 10.4.

Finally we consider the space $\mathfrak{B}_1$ of all forms in $\mathfrak{B}$ which possess strong derivatives in the sense explained in Section 2, i.e. which can be approximated by forms v in $\mathfrak{C}_1$ such that for each patch, including boundary patches, the derivatives of the components of v approach limits with respect to the L_2-norm. For forms in the space $\mathfrak{B}_1$ we then can state

THEOREM 10.7. *The projections of a form in $\mathfrak{B}_1$ into ${}^1d\mathfrak{U}$ and ${}^2\delta\mathfrak{W}$ belong to ${}^1d\mathfrak{U}_\delta$ and ${}^2\delta\mathfrak{W}_d$, respectively. Vice versa, these spaces lie in $\mathfrak{B}_1$.*

The spaces ${}^1d\mathfrak{U}_\delta$ and ${}^1\delta\mathfrak{W}_d$ were introduced in Section 3 for manifolds without boundaries. For manifolds with boundaries they may be defined by

$$ {}^1d\mathfrak{U}_\delta = {}^1d\mathfrak{U} \cap \mathfrak{B}_\delta\,, \qquad {}^2\delta\mathfrak{W}_d = {}^2\delta\mathfrak{W}_\delta \cap \mathfrak{B}_d\,. $$

Now, let v be a form in $\mathfrak{B}_1$ and let ${}^1v$ be its projection into ${}^1d\mathfrak{U}$; then $v - {}^1v$ belongs to ${}^1\mathfrak{B}_{s-0} \subset {}^1\mathfrak{B}_s = \mathfrak{B}_s$. Since v evidently belongs to $\mathfrak{B}_s$, the same is true of ${}^1v$. Similarly the statement follows for the projection ${}^2v$ into ${}^2\delta\mathfrak{W}$. The last part of Theorem 10.7 is obvious.

11. *An Inequality*

We consider the space $\mathfrak{B} \ominus \mathfrak{H}$ of all quadratically integrable forms which are orthogonal to all harmonic forms and the space $\mathfrak{B}_1 \ominus \mathfrak{H}$ of all forms which have strong first derivatives and at the same time are orthogonal to all harmonic forms. We then state

THEOREM 11.1. *There is a constant C such that*

$$\sum_{\bullet} \| \nabla v \|^2_{\mathfrak{A}_\bullet} \leqq C \int dv * dv + \delta v * \delta v \tag{11.1}$$

for all forms v in the space $\mathfrak{B}_1 \ominus \mathfrak{H}$.

The meaning of the term $\| \nabla v \|_{\mathfrak{A}}$ is given by (4.6). The left member of (11.1) is essentially the integral of the sum of the squares of all first derivatives of v; i.e., it is essentially the non-invariant complete Dirichlet integral.

Inequality (11.1) was proved for manifolds without boundary in Section 5, Lemma 5.2.

Now let $\mathfrak{M}$ be a manifold with boundary. By Lemma 5.2 inequality (11.1) holds for forms in either

$${}^1d\mathfrak{U}_s \oplus {}^1\delta\mathfrak{W}_d \qquad \text{or} \qquad {}^2d\mathfrak{U}_s \oplus {}^2\delta\mathfrak{W}_d\,,$$

hence a fortiori for forms in either

$${}^1d\mathfrak{U}_s \qquad \text{or} \qquad {}^2\delta\mathfrak{W}_d\,.$$

Now let v be a form in $\mathfrak{B}_1 \ominus \mathfrak{H}$, and ${}^1v$, ${}^2v$ its projections into ${}^1d\mathfrak{B}$ and ${}^2\delta\mathfrak{B}$. Since v is orthogonal to $\mathfrak{H}$ we have

$$v = {}^1v + {}^2v,$$

by Theorem 10.3. From Theorem 10.7 we infer that ${}^1v$ and ${}^2v$ are respectively in ${}^1d\mathfrak{B}_s$ and ${}^2\delta\mathfrak{B}_d$. Hence inequality (11.1) holds for ${}^1v$ and for ${}^2v$ with constants ${}^1C$ or ${}^2C$. It then follows that the inequality holds for $v = {}^1v + {}^2v$ with $C = 2 \max \{{}^1C, {}^2C\}$.

As was mentioned in the introduction the proof of this inequality was the original aim of the present investigation.

Bibliography

[1] Hodge, W. V. D., *The Theory and Applications of Harmonic Integrals*, Cambridge University Press, 1941.

[2] Weyl, H., *The method of orthogonal projection in potential theory*, Duke Math. J., Vol. 7, 1940, pp. 411–444.

[3] de Rham, G., *Sur la théorie des formes différentielles harmoniques*, Ann. Univ. Grenoble, Vol. 22, 1946, pp. 135–152.

[4] Bidal, P., and de Rham, G., *Les formes différentielles harmoniques*, Comment. Math. Helv., Vol. 19, 1946, pp. 1–49.

[5] Kodaira, K., *Harmonic fields in Riemannian manifolds*, Ann. of Math., Vol. 50, 1949, pp. 587–665.

[6] Milgram, A. N., and Rosenbloom, P. C., *Harmonic forms and heat conduction. I: Closed Riemannian manifolds*, Proc. Nat. Acad. Sci. U.S.A., Vol. 37, No. 3, 1951, pp. 180–184. *Heat conduction on Riemannian manifolds. II: Heat distribution on complexes and approximation theory*, Ibid., Vol. 37, No. 7, 1951, pp. 435–438.

[7] Gaffney, M., *The harmonic operator for exterior differential forms*, Dissertation, University of Chicago, March 1951, and Proc. Nat. Acad. Sci. U.S.A., Vol. 37, 1951, pp. 48–50.

[8] Duff, G. F. D., *Boundary value problems associated with the tensor Laplace equation*, Canadian J. Math., Vol. 5, 1952, pp. 57–80. *Differential forms in manifolds with boundary*, Ann. of Math., Vol. 56, 1952, pp. 115–127.

[9] Duff, G. F. D., and Spencer, D. C., *Harmonic tensors on Riemannian manifolds with boundary*, Ann. of Math., Vol. 56, 1952, pp. 128–156.

[10] Hausner, M., *Dirichlet's Principle and Generalized Boundary Values*, Ann. of Math., Vol. 57, No. 3, 1953, pp. 475–489.

[11] Duff, G. F. D., *A tensor boundary value problem of mixed type*, Canadian J. Math., Vol. 6, 1954. *Various classes of harmonic forms*, (to appear).

[12] Conner, P. E., *The Green's and Neumann's problems for differential forms on Riemannian manifolds*, (to appear).

[13] Morrey, C. B., Jr., and Eells, J., Jr., *A variational method in the theory of harmonic integrals, I*, to appear in Ann. of Math.

[14] Morrey, C. B., Jr., *A variational method in the theory of harmonic integrals II*, Air Force Report, Institute for Advanced Study.

[15] Rellich, F., *Ein Satz über mittlere Konvergenz*, Göttingen Ges. Wiss. Nachr. Math.-Phys. Kl., No. 4, 1930, pp. 30–35.

[16] Courant, R., and Hilbert, D., *Methoden der mathematischen Physik*, Vol. II, Chap. 7, Springer, Berlin, 1937.

[17] Friedrichs, K. O.,
A. *The identity of weak and strong extensions of differential operators*, Trans. Amer. Math. Soc., Vol. 55, No. 1, 1944, pp. 132–151.
B. *A theorem of Lichtenstein*, Duke Math. J., Vol. 14, No. 1, 1947, pp. 67–82.
C. *On the boundary-value problem of the theory of elasticity and Korn's inequality*, Ann. of Math., Vol. 48, No. 2, 1947, pp. 441–471.
D. *On the differentiability of the solutions of linear elliptic differential equations*, Comm. Pure Appl. Math., Vol. 6, No. 3, 1953, pp. 299–325.

[18] Aronszajn, N., *Coercive quadratic integro-differential forms*, Abstract 67^b, Bull. Amer. Math. Soc., Vol. 60, No. 6, 1954, p. 533.

Received May 1, 1955.

[27–1] **Eine invariante Formulierung des Newtonschen Gravitationsgesetzes und des Grenzüberganges vom Einsteinschen zum Newtonschen Gesetz, Math. Ann., 98 (1927), 566–575.**

This was Friedrichs' first published work and is described in D. Isaacson's remarks, p. 602.

Eine invariante Formulierung des Newtonschen Gravitationsgesetzes und des Grenzüberganges vom Einsteinschen zum Newtonschen Gesetz.

Von

Kurt Friedrichs in Göttingen.

Das Einsteinsche Gravitationsgesetz kann nur dann Anspruch auf Gültigkeit erheben, wenn es das Newtonsche Gesetz als Grenzfall enthält. Daß dies in der Tat der Fall ist, hat man auf verschiedene Weisen gezeigt, die alle im wesentlichen auf folgendes herauskommen. Man legte ein spezielles Koordinatensystem zugrunde, nämlich dasjenige, in dem das Newtonsche Gesetz seine gewöhnliche Gestalt annimmt, und machte für die Einsteinschen Gravitationspotentiale g_{ik} in erster Näherung den Ansatz

$$g_{11} = g_{22} = g_{33} = 1, \qquad g_{44} = A - 2U, \qquad g_{ik} = 0 \qquad (i \neq k).$$ [1]

In erster Näherung bedeutet dabei, daß noch Zusatzglieder auftreten, die mit wachsendem A gegen Null gehen. Die geodätischen Linien der durch das obige g_{ik}-Feld bestimmten Metrik gehen bei unbeschränkt wachsendem A [2] gegen die raum-zeitlichen Kurven einer Bewegung im Newtonschen Gravitationsfeld mit dem Gravitationspotential U.

Im folgenden soll dieser Zusammenhang dadurch wiedergegeben werden, daß auch für das Newtonsche Gesetz ebenso wie für das Einsteinsche eine gegenüber raum-zeitlichen Transformationen invariante Formulierung zugrunde gelegt wird. Eine solche Darstellung führt dann zu einer invarianten Formulierung eines Grenzüberganges, durch den aus Einsteinschen Gravitationsfeldern das allgemeinste Newtonsche Gravitationsfeld entsteht. Daß es für das Newtonsche Gesetz eine invariante Darstellung geben muß, ist von vornherein klar. Es scheint aber bisher niemals eine solche angegeben

---

[1]) Dies gilt auch für den kugelsymmetrischen Fall, wo man das Feld g_{ik} explizite kennt.

[2]) Es läßt sich $\sqrt{A}$ mit der Lichtgeschwindigkeit identifizieren.

worden zu sein. Insbesondere ist die Bemerkung zu vermissen, daß ebenso wie beim Einsteinschen auch beim Newtonschen Gravitationsfeld im materiefreien Raum der verjüngte Krümmungstensor verschwindet.

1. Das Newtonsche Gesetz in der alten Form.

Das Newtonsche Gesetz besagt, daß die Beschleunigung eines „Probekörpers", der nur der Gravitation unterliegt, nicht von seiner Natur und von seinem Bewegungszustand abhängt, sondern allein durch das „Gravitationsfeld" bestimmt ist[3]), das in bestimmter Weise von der anziehenden Materie abhängt. Hierbei ist aber ein bestimmtes Koordinatensystem, das der Trägheitsbewegung, vorausgesetzt.

Mathematisch können wir dies Gesetz in folgenden drei Stufen formulieren. Wir verstehen dabei unter x_1, x_2, x_3 die drei räumlichen Koordinaten, unter t die Zeit.

I. Die Gravitation bestimmt an jeder Stelle zu jeder Zeit einen Vektor p^1, p^2, p^3, die „Feldstärke", so daß in einem gewissen „ausgezeichneten" System — wir brauchen nicht zu untersuchen, wie wir zu ihm kommen — für die Bahn $x_i(t)$ $(i = 1, 2, 3)$ eines Probekörpers die Gleichung

$$\frac{d^2 x_i}{dt^2} = p^i \qquad (i = 1, 2, 3)$$

gilt.

II. Die Feldstärke p^i läßt sich aus einem Potential ableiten, d. h. es gibt einen Skalar U, so daß

$$p^i = -\frac{\partial U}{\partial x_i} \qquad (i = 1, 2, 3)$$

ist.

III. Gegenüber diesem Gesetz, das mehr die inneren Eigenschaften des Gravitationsfeldes charakterisiert, betrifft das dritte den Zusammenhang des Feldes mit der felderzeugenden Materie. Wir greifen nur heraus, daß im materiefreien Gebiet die Divergenz der Feldstärke verschwindet.

$$\frac{\partial p^1}{\partial x_1} + \frac{\partial p^2}{\partial x_2} + \frac{\partial p^3}{\partial x_3} = 0 \quad \text{oder} \quad \Delta U = \frac{\partial^2 U}{\partial x_1^2} + \frac{\partial^2 U}{\partial x_2^2} + \frac{\partial^2 U}{\partial x_3^2} = 0\,.$$

2. Das Newtonsche Gesetz in invarianter Form.

Wir wollen diese drei Gesetze jetzt für ein beliebiges System von raum-zeitlichen Koordinaten x_1, x_2, x_3, x_4 formulieren. Das *Gesetz I* lautet

---

[3]) Eigentlich ist diese Eigenschaft eine Definition der Gravitation, durch die sie aus anderen Naturkräften herausgehoben wird.

bekanntlich invariant folgendermaßen: Die Gravitation bestimmt ein Feld von Pseudotensoren Γ^i_{kl}, so daß bei geeigneter Wahl des Parameters t

(1) $$\frac{d^2 x_i}{dt^2} + \Gamma^i_{\alpha\beta} \frac{dx_\alpha}{dt} \frac{dx_\beta}{dt} = 0 \qquad (i = 1, 2, 3, 4)$$

gilt[4]). In unserem ausgezeichneten System wird $t = x_4$,

$$\Gamma^i_{44} = -p^i \qquad (i = 1, 2, 3)$$

und sonst

$$\Gamma^i_{kl} = 0.$$

Wir wollen nun zunächst *Gesetz III* invariant formulieren. Wir rechnen zu dem Zweck die „kleine" Krümmung F_{ik} von Γ^i_{kl} aus und erhalten:

$$F_{44} = -\left(\frac{\partial p^1}{\partial x_1} + \frac{\partial p^2}{\partial x_2} + \frac{\partial p^3}{\partial x_3}\right),$$

$$F_{ik} = 0 \text{ in den anderen Fällen.}$$

Wir finden somit, daß im leeren Raum der verjüngte Krümmungstensor verschwindet.

(2) $$F_{ik}[\Gamma] = 0.$$

Wir bemerken nun, daß das Newtonsche Gesetz (im Gegensatz zu dem Eindruck, den man aus den üblichen Darstellungen gewinnt) in den Anteilen I und III vollkommen mit dem Einsteinschen Gravitationsgesetz übereinstimmt. Der Unterschied liegt allein im *Anteil II.* Das Einsteinsche Gesetz stellt hier die einfache Forderung, daß sich das Γ-Feld als Christoffelsymbol eines Tensorfeldes g_{ik} darstellen läßt.

(3) $$\Gamma^i_{kl} = \frac{1}{2} g^{i\alpha}\left(\frac{\partial g_{k\alpha}}{\partial x_l} + \frac{\partial g_{l\alpha}}{\partial x_k} - \frac{\partial g_{kl}}{\partial x_\alpha}\right),$$

wofür wir auch die Abkürzung

$$\Gamma^i_{kl} = \{g\}^i_{kl}$$

---

[4]) Wollen wir uns von der Notwendigkeit, den Parameter t geeignet zu wählen, freimachen, so schreiben wir „parameterinvariant"

$$\begin{vmatrix} \frac{dx_i}{dt} & \frac{d^2 x_i}{dt^2} + \Gamma^i_{\alpha\beta} \frac{dx_\alpha}{dt} \frac{dx_\beta}{dt} \\ \frac{dx_k}{dt} & \frac{d^2 x_k}{dt} + \Gamma^k_{\alpha\beta} \frac{dx_\alpha}{dt} \frac{dx_\beta}{dt} \end{vmatrix} = 0.$$

Durch das Verschwinden dieses Ausdrucks ist der Pseudotensor $\Gamma^i_{\alpha\beta}$ nicht eindeutig bestimmt. Die Gravitation liefert also unmittelbar nur die Schar der „projektiv gleichwertigen" Γ-Felder, unter denen es eins gibt, für das die folgenden Gesetze II und III gelten. Ob dies ausgezeichnete Γ-Feld durch diese Forderung eindeutig bestimmt ist, oder ob dazu der Zusammenhang mit anderen Naturkräften (Elektrizität) herangezogen werden muß, braucht hier nicht erörtert zu werden.

verwenden wollen; wir merken noch an, daß diese Beziehung mit der Gleichung

$$\frac{\partial g_{ik}}{\partial x_l} - g_{i\alpha}\Gamma^{\alpha}_{kl} - g_{k\alpha}\Gamma^{\alpha}_{il} = 0 \tag{4}$$

äquivalent ist.

Das Newtonsche Γ-Feld läßt sich, wie man leicht einsieht, nicht in dieser Weise aus einem Tensorfeld g_{ik} ableiten. Betrachten wir nun die Werte, die der Newtonsche Γ-Tensor in einem ausgezeichneten Koordinatensystem annimmt, so erscheint es als natürlich, in diesem Koordinatensystem den Ansatz

$$\Gamma^i_{kl} = P^{i\alpha}_{kl}\frac{\partial U}{\partial x_\alpha}$$

zu machen, wo P^{im}_{kl} sich als Tensor transformieren und in unserem Koordinatensystem die Werte $P^{ii}_{44} = 1$ für $i \neq 4$ und sonst $P^{im}_{kl} = 0$ annehmen soll. Es ist zweckmäßig dies durch die Darstellung

$$P^{im}_{kl} = -\eta_{kl}h^{im}$$

zum Ausdruck zu bringen, wobei der Tensor h^{ik} dadurch bestimmt ist, daß er in unserem System die Werte

$$h^{11} = h^{22} = h^{33} = 1 \quad \text{und sonst} \quad h^{ik} = 0$$

annimmt, während für den Tensor η_{ik} die Beziehungen

$$\eta_{44} = -1 \quad \text{und sonst} \quad \eta_{ik} = 0$$

gelten. Der Pseudotensor

$$\Delta^i_{kl} = \Gamma^i_{kl} + \eta_{kl}h^{i\alpha}\frac{\partial U}{\partial x_\alpha} \tag{5}$$

verschwindet also in unserem ausgezeichneten Koordinatensystem überall.

Nun ist es notwendig und in der Tat möglich, für die Größen Δ^i_{kl}, η_{ik}, h^{ik} derartige invariante, also von unserem speziellen Koordinatensystem unabhängige Bestimmungen anzugeben, daß durch die Gleichung (5) das Gesetz II invariant dargestellt werden kann.

Offenbar ist der Pseudotensor Δ^i_{kl} durch das Verschwinden seiner Krümmung gekennzeichnet,

$$F^i_{klm}[\Delta] = 0. \tag{6}$$

Der symmetrische Tensor h^{ik} hat den Rang 3 und den Realitätscharakter (1, 1, 1, 0); während η_{ik} auch symetrisch ist, den Rang 1 und den Charakter (0, 0, 0, −1) besitzt. Diese beiden Tensoren sind durch die Beziehung

$$\eta_{i\alpha}h^{\alpha k} = 0 \tag{7}$$

verknüpft. Ferner verschwinden die mit dem Pseudotensor Δ^i_{kl} gebildeten kovarianten Ableitungen von h^{ik} und η_{ik}:

$$(8)\qquad \frac{\partial h^{ik}}{\partial x_l} + h^{ia}\Delta^k_{al} + h^{ka}\Delta^i_{al} = 0,$$

$$\frac{\partial \eta_{ik}}{\partial x_l} - \eta_{ia}\Delta^a_{kl} - \eta_{ka}\Delta^a_{il} = 0.$$

Hierfür können wir auch, wie wir unter Beachtung von (5) und (7) finden, die Gleichungen

$$(9)\qquad \frac{\partial h^{ik}}{\partial x_l} + h^{ia}\Gamma^k_{al} + h^{ka}\Gamma^i_{al} = 0,$$

$$\frac{\partial \eta_{ik}}{\partial x_l} - \eta_{ia}\Gamma^a_{kl} - \eta_{ka}\Gamma^a_{il} = 0$$

setzen.

Wir behaupten nun, daß der *zweite Anteil* des Newtonschen Gesetzes sich folgendermaßen *invariant formulieren* läßt:

Es gibt einen Skalar U, symmetrische Tensoren η_{ik}, h^{ik} von dem angegebenen Realitätscharakter und einen Pseudotensor Δ^i_{kl}, von der durch die eben angegebenen Bedingungen (6), (7), (8) charakterisierten Art, so daß sich der Tensor Γ^i_{kl} der Gravitationsfeldstärke in der Form:

$$(10)\qquad \Gamma^i_{kl} = \Delta^i_{kl} - \eta_{kl} h^{ia} \frac{\partial U}{\partial x_a}$$

darstellen läßt.

Diese Darstellung braucht nicht die einzig mögliche invariante Darstellung zu sein[5]); doch sie erweist sich dadurch naturgemäß, daß man den Pseudotensor Δ^i_{kl} als Trägheitsfeld deuten kann, während die Tensoren η_{ik} und h^{ik} die Zeit- und Raummetrik wiedergeben.

Wir müssen zeigen, *daß jedes Γ-Feld, das sich in der angegebenen Weise darstellen läßt, auch ein Newtonsches Feld ist,* wenn es noch der Bedingung III (2) genügt. Wir brauchen zu dem Zweck nur ein ausgezeichnetes Koordinatensystem nachzuweisen, d. h. ein System, in dem

---

[5]) So sind z. B. zu einem gegebenen Gravitationsfeld Γ^i_{kl} die Größen Δ^i_{kl}, η_{ik}, h^{ik}, U nicht eindeutig bestimmt. Diese Mehrdeutigkeit läßt sich jedoch leicht diskutieren. Es zeigt sich, daß der Tensor η_{ik} bis auf einen konstanten Faktor und damit auch *die Aufspaltung in Raum und Zeit* eindeutig bestimmt ist. Der Tensor h^{ik} läßt lineare Transformationen der drei räumlichen Koordinaten zu; natürlich muß die Funktion U entsprechend mittransformiert werden; schließlich kann zur Funktion U noch eine in den drei räumlichen Koordinaten lineare Funktion addiert werden, wenn eine entsprechende Transformation des Pseudotensors Δ^i_{kl} zugleich vorgenommen wird. Dies sind die einzigen Unbestimmtheiten, vorausgesetzt, daß die Krümmung $F^i_{klm}[\Gamma]$ nicht verschwindet.

die Größen Δ^i_{kl}, η_{ik}, h^{ik} gerade die Werte annehmen, von denen wir ausgegangen sind. Das ist in der Tat leicht möglich. Wegen (6) gibt es sicher ein System, in dem Δ^i_{kl} verschwindet. Daran ändert auch eine lineare Transformation nichts, die wir so wählen können, daß h^{ik} in einem Punkte die Werte $h^{11} = h^{22} = h^{33} = 1$ sonst $h^{ik} = 0$ annimmt. Die Beziehung (7) zeigt dann, daß in diesem Punkte $\eta_{ik} = 0$ ist außer für $i = k = 4$. Durch eine konstante Dehnung der Koordinate x_4 erreichen wir, daß $\eta_{44} = -1$ wird. Aus den Gleichungen (8) entstehen wegen $\Delta^i_{kl} = 0$ die Gleichungen

$$\frac{\partial h^{ik}}{\partial x_l} = 0; \quad \frac{\partial \eta_{ik}}{\partial x_l} = 0,$$

welche besagen, daß die Tensoren h^{ik} und η_{ik} konstant sind und überall die gewünschten Werte annehmen.

Schließlich merken wir noch an, daß der *dritte Anteil* des Newtonschen Gesetzes (2) $F_{ik}[\Gamma] = 0$ sich unter Benutzung der Größen U, h^{ik}, η_{ik}, Δ^i_{kl} auf die Gleichung

$$(11) \qquad h^{\alpha\beta}\left(\frac{\partial U^2}{\partial x_\alpha \partial x_\beta} - \Delta^\gamma_{\alpha\beta}\frac{\partial U}{\partial x_\gamma}\right) = 0$$

zurückführen läßt, wie man durch Einsetzen des Wertes (10) von Γ^i_{kl} in (2) erkennt. Diese Gleichung (11) ist eine der möglichen invarianten Darstellungen der dreidimensionalen Laplaceschen Gleichung $\Delta U = 0$, in die sie in einem ausgezeichneten Koordinatensystem übergeht.

3. Der Übergang vom Einsteinschen zum Newtonschen Gesetz.

Nun ist es nicht schwer, auch eine invariante Formulierung des Grenzüberganges vom Einsteinschen zum Newtonschen Gesetz zu finden. Vorerst bemerken wir, daß umgekehrt aus dem Newtonschen das Einsteinsche Gesetz einfach dadurch entsteht, daß man von den verschiedenen Bestandteilen des Gesetzes II nur die Gleichung

$$(9)_2 \qquad \frac{\partial \eta_{ik}}{\partial x_l} - \eta_{k\alpha}\Gamma^\alpha_{il} - \eta_{i\alpha}\Gamma^\alpha_{kl} = 0$$

beibehält, dafür aber verlangt, daß der Tensor η_{ik} nicht mehr ausgeartet ist. Genauer können wir uns das so vorstellen, daß wir die Tensoren h^{ik} und η_{ik} durch nicht ausgeartete Tensoren g^{ik} und γ_{ik} vom Charakter $(1, 1, 1, -1)$ ersetzt denken, die sich von ihnen nur wenig, etwa um Glieder von der Größenordnung des Parameters ε, unterscheiden, und die an Stelle von $\eta_{i\alpha}h^{\alpha k} = 0$ der Relation

$$\gamma_{i\alpha}g^{\alpha k} = \varepsilon\delta_i^k \qquad (\delta_i^i = 1, \quad \delta_i^k = 0, \quad i \neq k)$$

genügen. Wir wollen an Stelle von γ_{ik} den Tensor $g_{ik} = \frac{1}{\varepsilon} \gamma_{ik}$ einführen, so daß g_{ik} und g^{ik} durch die Relation

$$g_{i\alpha} g^{\alpha k} = \delta_i^k \tag{12}$$

verbunden sind. Genügt dann der Tensor g_{ik} der Gleichung

$$\frac{\partial g_{i\alpha}}{\partial x_l} - g_{i\alpha} \Gamma^{\alpha}_{kl} - g_{k\alpha} \Gamma^{\alpha}_{il} = 0, \tag{13}$$

so genügt von selbst der Tensor g^{ik} der Gleichung

$$\frac{\partial g^{ik}}{\partial x_l} + g^{i\alpha} \Gamma^{k}_{\alpha l} + g^{k\alpha} \Gamma^{i}_{\alpha l} = 0, \tag{14}$$

dem Analogon von $(9)_1$.

Allerdings wird dann der Pseudotensor

$$\Delta^i_{kl} = \Gamma^i_{kl} + \varepsilon g_{kl} g^{i\alpha} \frac{\partial U}{\partial x_\alpha}$$

im allgemeinen nicht mehr in einem geeigneten System zum Verschwinden gebracht werden können; doch können wir annehmen, daß er sich bei kleinem ε beliebig wenig von einem solchen Pseudotensor unterscheidet.

Diese Vorbetrachtungen führen uns zu der Erwartung, daß durch folgenden Grenzübergang das Einsteinsche in das Newtonsche Gesetz übergeht [6]).

Wir nehmen an, wir haben eine Folge von Einsteinschen Gravitationsfeldern vor uns, die in folgender Weise gegen ein Grenzfeld konvergieren.

1. Der Tensor g^{ik} des Einsteinschen Feldes geht gegen einen ausgearteten Tensor h^{ik} vom Rang 3 und vom Charakter $(1, 1, 1, 0)$, während die Determinante $|g^{ik}|$ überall von derselben Größenordnung, etwa wie eine Konstante ε, gegen Null geht.

2. Es gibt einen Skalar U, so daß der Pseudotensor

$$\{g\}^i_{kl} + \varepsilon g_{kl} g^{i\alpha} \frac{\partial U}{\partial x_\alpha}$$

gegen einen Pseudotensor Δ^i_{kl} konvergiert [7]), der in einem speziellen System verschwindet, d. h. für den $F^i_{klm}[\Delta] = 0$ ist [8]).

Wenn dies erfüllt ist, so, behaupten wir, konvergieren die Pseudo-

[6]) Einige wesentliche Vereinfachungen verdanke ich hier Herrn Prof. Bernays.

[7]) Man kann auch z. B. verlangen, daß es einen Skalar U gibt, so daß das Christoffelsymbol einer zu g_{ik} konformen Metrik $g_{ik} e^{-2\varepsilon U}$, nämlich $\{g \cdot e^{-2\varepsilon U}\}^i_{kl}$ gegen einen solchen Pseudotensor Δ^i_{kl} geht.

[8]) Da der Tensor g_{ik} nicht beschränkt ist, enthält diese Forderung eigentlich zwei Anteile; vgl. Nr. 4.

tensoren $\{g\}^i_{kl}$ des Einsteinschen Feldes gegen einen Pseudotensor Γ^i_{kl}, der ein Newtonsches Feld darstellt mit U als Gravitationspotential.

Um dies einzusehen, bemerken wir, daß der Tensor εg_{ik} gegen einen Tensor η_{ik} konvergiert, wie wir aus der Darstellung $\varepsilon g_{ik} = \varepsilon \frac{G_{ik}}{|g^{ik}|}$ erkennen, wo die Größen G_{ik} die Unterdeterminanten der Größen g^{ik} sind. Da der Tensor h^{ik} den Rang 3 hat, können nicht alle Unterdeterminanten G_{ik} gegen Null gehen; also kann auch nirgends der ganze Tensor η_{ik} verschwinden. Aus der Relation $g_{ia} g^{ak} = \delta^k_i$ wird $\eta_{ia} h^{ak} = 0$. Hieraus schließen wir, daß η_{ik} den Rang 1 hat, und aus den Charakteren von g_{ik} und h^{ik} folgern wir, daß η_{ik} den Charakter $(0, 0, 0, -1)$ besitzt. Nunmehr erkennen wir aus der Forderung 2, daß der Pseudotensor $\{g\}^i_{kl}$ gegen den Pseudotensor

$$\Gamma^i_{kl} = \Delta^i_{kl} - \eta_{kl} h^{ia} \frac{\partial U}{\partial x_a}$$

konvergiert. Die Relationen (9) für η_{ik} und h^{ik} erhalten wir in der Grenze aus den entsprechenden (identischen) Relationen für g_{ik} und g^{ik} (vgl. (3), (4)). Für den Pseudotensor Δ^i_{kl} gilt nach Annahme $F^i_{klm}[\Delta] = 0$, und die Gleichung (2) $F_{ik}[\Gamma] = 0$ bleibt bestehen. Damit ist der Nachweis erbracht, daß der Grenzpseudotensor Γ^i_{kl} ein Newtonsches Feld darstellt.

4. Darstellung des Grenzüberganges in einem ausgezeichneten Koordinatensystem.

Nachdem wir einen Grenzübergang vom Einsteinschen zum Newtonschen Gesetz invariant formuliert haben, wollen wir uns klarmachen, wie dieser Grenzübergang in einem ausgezeichneten Koordinatensystem aussieht. Wir nehmen dabei der Einfachheit halber an, daß sich die Komponenten des Einsteinschen Feldes nach dem Parameter ε in der Umgebung des Newtonschen Feldes entwickeln lassen. Wir setzen also

$$(15) \qquad \begin{aligned} g^{ik} &= h^{ik} + \varepsilon H^{ik} + \dots \\ \varepsilon g_{ik} &= \eta_{ik} + \varepsilon \mathsf{H}_{ik} + \dots \end{aligned}$$

und denken uns diese Entwicklungen in den Ausdruck $\{g\}^i_{kl} + \varepsilon g_{kl} g^{ia} \frac{\partial U}{\partial x_a}$ eingesetzt. Das erste Glied dieses Ausdruckes wird

$$\frac{1}{\varepsilon} \frac{1}{2} h^{ia} \left\{ \frac{\partial \eta_{ka}}{\partial x_l} + \frac{\partial \eta_{la}}{\partial x_k} - \frac{\partial \eta_{kl}}{\partial x_a} \right\} = 0$$

wegen $(8)_2$ und (7). Da nun das von ε freie Glied gleich Δ^i_{kl} sein muß, so erhalten wir die Gleichungen

$$(16)\qquad \begin{aligned} &\frac{1}{2}\, h^{i\alpha}\left\{\frac{\partial \mathsf{H}_{\alpha k}}{\partial x_l}+\frac{\partial \mathsf{H}_{\alpha l}}{\partial x_k}-\frac{\partial \mathsf{H}_{k l}}{\partial x_\alpha}\right\}+\frac{1}{2} H^{i\alpha}\left\{\frac{\partial \eta_{\alpha k}}{\partial x_l}+\frac{\partial \eta_{\alpha l}}{\partial x_k}-\frac{\partial \eta_{k l}}{\partial x_\alpha}\right\} \\ &\qquad + \eta_{kl}\, h^{i\alpha}\, \frac{\partial U}{\partial x_\alpha} = \Delta^i_{kl}. \end{aligned}$$

Aus der Beziehung $g_{i\alpha} g^{\alpha k} = \delta^k_i$ erhalten wir als das von ε freie Glied die Relation:

$$(17)\qquad h^{i\alpha}\, \mathsf{H}_{\alpha k} + H^{i\alpha}\, \eta_{\alpha k} = \delta^i_k .$$

Die beiden Gleichungen (16) und (17) dienen uns zur Bestimmung von H_{ik} und H^{ik}. Gehen wir nun auf ein ausgezeichnetes Koordinatensystem über, so erhalten wir aus den Gleichungen (17) die Relationen:

$$\begin{aligned} \mathsf{H}_{ii} &= 1 \quad [i \neq 4], & \mathsf{H}_{ik} &= 0 \quad [i, k \neq 4], \\ \mathsf{H}_{i4} &= H^{i4} \quad [i \neq 4], & H^{44} &= 1, \end{aligned}$$

und aus den Gleichungen (16) entstehen dann für $k = 4$, $i \neq 4$ die Beziehungen

$$\frac{\partial \mathsf{H}_{i4}}{\partial x_l} - \frac{\partial \mathsf{H}_{l4}}{\partial x_i} = 0 \quad [i, l \neq 4], \qquad \frac{\partial \mathsf{H}_{i4}}{\partial x_4} - \frac{\partial\left(\frac{1}{2}\,\mathsf{H}_{44} + U\right)}{\partial x_i} = 0,$$

während die anderen Gleichungen aus (16) von selbst erfüllt sind. Wir erhalten somit für g_{ik} und g^{ik} in zweiter Näherung die Darstellungen

$$g_{ik} \approx \begin{pmatrix} 1 & 0 & 0 & \frac{\partial P}{\partial x_1} \\ 0 & 1 & 0 & \frac{\partial P}{\partial x_2} \\ 0 & 0 & 1 & \frac{\partial P}{\partial x_3} \\ \frac{\partial P}{\partial x_1} & \frac{\partial P}{\partial x_2} & \frac{\partial P}{\partial x_3} & -\frac{1}{\varepsilon} - 2\left(U - \frac{\partial P}{\partial x_4}\right) \end{pmatrix},$$

$$g^{ik} \approx \begin{pmatrix} 1+\varepsilon H^{11} & \varepsilon H^{12} & \varepsilon H^{13} & \varepsilon \frac{\partial P}{\partial x_1} \\ \varepsilon H^{21} & 1+\varepsilon H^{22} & \varepsilon H^{23} & \varepsilon \frac{\partial P}{\partial x_2} \\ \varepsilon H^{31} & \varepsilon H^{32} & 1+\varepsilon H^{33} & \varepsilon \frac{\partial P}{\partial x_3} \\ \varepsilon \frac{\partial P}{\partial x_1} & \varepsilon \frac{\partial P}{\partial x_2} & \varepsilon \frac{\partial P}{\partial x_3} & \varepsilon \end{pmatrix},$$

womit wir, abgesehen von dem Auftreten der Größe P[9]), auf die bekannte Näherungsdarstellung eines Einsteinschen Feldes in der Umgebung eines Newtonschen Feldes zurückkommen, wenn wir $\frac{1}{\varepsilon}$ mit dem Quadrat der Lichtgeschwindigkeit identifizieren[10]).

---

[9]) Alle die Maßbestimmungen, die sich nur in P unterscheiden, besitzen also in erster Näherung dieselben Cristoffelsymbole $\{g\}^{i}_{kl}$.

[10]) Es wäre noch wichtig zu entscheiden, ob ein Einsteinsches Feld $\{g\}^{i}_{kl}$, das gegen ein Newtonsches Feld Γ^{i}_{kl} geht, auch stets in der hier angegebenen Weise konvergiert. Es ist nur gelungen zu zeigen, daß das sicher dann der Fall ist, wenn man z. B. noch fordert, daß das Grenzfeld der Größen g^{ik} mindestens den Rang 3 hat.

(Eingegangen am 16. 11. 1926.)

[80–1] **Von Neumann's Hilbert space theory and partial differential equations, SIAM Review, 22 (1980), 486–493.**

In 1979 Friedrichs delivered the von Neumann lecture to the Society for Industrial and Applied Mathematics and it is reprinted here.

SIAM REVIEW
Vol. 22, No. 4, October 1980

VON NEUMANN'S HILBERT SPACE THEORY AND PARTIAL DIFFERENTIAL EQUATIONS*

K. O. FRIEDRICHS†

In this von Neumann Lecture I shall relate my early encounters with von Neumann, describe the part of my work that was strongly influenced by his work, and briefly discuss subsequent work I performed, some years ago, that is related to the earlier work.

I first met John von Neumann in Göttingen in 1927 or 1926 in a seminar run by Max Born which was concerned with the new quantum theory. After a seminar session most participants walked up a hill to a garden restaurant. On one of these occasions Born asked von Neumann about his interest in science. Von Neumann then told us that his father wanted him to become a mathematician since he had shown very early a gift for mathematics. But he himself wanted to become a banker, as his father was. Finally, they settled on chemistry.

Von Neumann studied chemistry, not in his native Budapest, Hungary, but in Zürich, Switzerland. There he was influenced by the mathematician Hermann Weyl, who at that time was interested in mathematical logic. So von Neumann wrote a paper in this field which turned out to be rather important. At the same time Hilbert was also concerned with mathematical logic, and so von Neumann went to Göttingen to work with him. In Göttingen he learned about the new quantum theory and participated in the seminar on that field.

Von Neumann's interest in quantum theory led him to write a book about the mathematical foundations of this field. The book appeared in 1932; I reviewed it for the Zentralblatt. This book contained introductory mathematical sections, and sections involving quantum theory. At present I think that the mathematical sections, which were based on papers that had appeared in 1929, were the more significant ones. They were rather abstract.

We, in the group around Courant, were quite suspicious about the significance of such abstract work. Nevertheless, when in late 1930 I studied these abstract papers, I was dumbfounded. In fact, I had just handed to Courant for publication a manuscript on spectral theory. I asked him to return the manuscript. I then rewrote the paper in von Neumann's abstract language. That was the origin of a substantial part of my later work.

Before describing this work I should like to go back some years. When the basic formula of the new quantum theory, $pq - qp = -ih$, was set up by Heisenberg, Born, and Jordan, some mathematicians in Göttingen claimed, somewhat sneeringly, that such a formula could not be valid. They claimed that they could prove this by using Hilbert's theory of infinite matrices, since p and q were such matrices.

An infinite matrix is given by a system of real numbers $M_{\alpha\beta}$ with α and β infinite sequences $\alpha = 1, 2, 3, \cdots, \beta = 1, 2, 3, \cdots$. Such a matrix is called bounded if for any sequence of numbers a_α, b_β the finiteness of the sums $\sum_\alpha |a_\alpha|^2$ and $\sum_\beta |b_\beta|^2$ implies the finiteness of the sum

$$\sum_{\alpha,\beta} a_2 M_{\alpha\beta} b_\beta.$$

* 1979 John von Neumann Lecture presented at the SIAM 1979 Spring Meeting, Toronto, Canada, June 11–13, 1979.

† Courant Institute of Mathematical Sciences, New York University, 251 Mercer Street, New York, NY 10012.

Now the matrices Hilbert had dealt with in his theory were such bounded matrices.

The mistake those mathematicians had made was that they tacitly assumed that the matrices p and q were bounded. They were not. Already Born had observed this, but he thought that the same rules would be valid for unbounded matrices as for bounded ones. That was not so.

Some work on unbounded matrices had been done before, but it was not very satisfactory. It was von Neumann who completely cleared up how to handle unbounded infinite matrices.

For bounded matrices it is possible to introduce any other system of coordinates $\{a_\alpha\}$ and $\{b_\beta\}$; but, as von Neumann showed, this is not at all arbitrarily possible for unbounded matrices. He avoided the resulting awkwardness by introducing very general extensions of the spaces employed by Hilbert, the "general Hilbert spaces." In a general Hilbert space bounded and unbounded "operators" are defined quite generally. Such operators may be represented by infinite matrices of a discrete, continuous, or mixed kind. It is the theory of such general Hilbert spaces that had such a strong effect on me.

A certain aspect of von Neumann's theory as it related to concretely defined incomplete Hilbert spaces was particularly significant for me. This aspect was the possibility of taking sequences of operators and square integrable functions defined in such an incomplete Hilbert space, and then defining new operators and functions in the completed Hilbert space by appropriate limiting processes. All my work involving Hilbert space theory was based on this possibility. The resulting abstract operators and square integrable functions I have referred to as "ideal."

By virtue of the possibility of such extensions it is unnecessary to employ Lebesgue theory.

At about the same time von Neumann's work affected me, it also influenced Marshall Stone, who then wrote his monumental treatise. Stone's work and my work, however, proceeded in different directions.

The part of my work that was most strongly influenced by von Neumann's work concerned spectral theory and partial differential equations. I shall not here describe my earlier papers on spectral theory and on elliptic equations; instead, I shall discuss later work on partial differential equations. In it I have restricted myself to linear equations and also to systems of the first order. The latter restriction is convenient and not severe, since equations of higher order can—in general—be reduced to systems of the first order.

A system of partial differential equations of the first order can be written in the form

$$A^{\mu}_{\alpha\beta}\frac{\partial}{\partial x^{\mu}}u^{\beta}+B_{\alpha\beta}u^{\beta}=f_{\alpha},\qquad \alpha,\beta=1,2,\cdots,a,\quad \mu=1,2,\cdots,m.$$

Of course, summation with respect to μ and β is implied. The unknowns u^{β}, as well as the coefficients $A^{\mu}_{\alpha\beta}$, $B_{\alpha\beta}$, f_{α}, are functions of the independent variables $x^1, x^2, \cdots, x^m$. Thus the number m is the dimension of the underlying space.

The matrix

$$\left[A^{\mu}_{\alpha\beta}\frac{\partial}{\partial x^{\mu}}+B_{\alpha\beta}\right]$$

may be said to represent the operator which acts on the system $\{u^{\beta}\}$ of the unknowns $u^1, \cdots, u^a$. The differentiation $\partial/\partial x^{\mu}$ will in the following simply be written as ∂_{μ}.

I have always restricted myself to dealing with symmetric equations, i.e., with equations for which the relation

$$A^{\mu}_{\alpha\beta}=A^{\mu}_{\beta\alpha}\quad \text{holds for } \alpha,\beta=1,2,\cdots,a,\quad \mu=1,2,\cdots,m.$$

The reason for this restriction was that almost all nondegenerate partial differential equations of classical physics appear either naturally as symmetric equations, or can be reduced to symmetric equations. Also, fortunately, it is much easier to solve symmetric equations then nonsymmetric ones.

A symmetric equation may be elliptic, hyperbolic, or of a more general type which I shall mention later on. For these equations it is always assumed that a certain quadratic form is positive-definite.

The first problem is to show that, under appropriate boundary or initial conditions, a unique solution of the equation exists. This turns out to be true provided the unknown function $u(x)$ is allowed to be an ideal function of the kind referred to above.

In briefly describing my method of proving that such solutions exist, I shall restrict myself to the initial problem of symmetric hyperbolic equations, that is, to equations

$$(A^{\mu}_{\alpha\beta}\,\partial_\mu + B_{\alpha\beta})u^{\beta} = f_{\alpha},$$

for which the quadratic form $\zeta_\mu A^{\mu}_{\alpha\beta}$, with suitable ζ, is positive-definite and for which initial data are prescribed. For other types of equations I have used essentially the same approach.

In proving the unique existence of the solution of a symmetric hyperbolic equation I used a multistep procedure. First, I introduced two kinds of derivatives of ideal functions, the "weak" and the "strong" ones. Next I introduced the notions of ideal "weak" and "strong" solutions of the equation. Actually, these functions are so far only candidates for being solutions. I then showed that a weak solution exists, and that a strong solution is unique. Employing a sequence of certain smoothing operators, which I called mollifiers, I could finally show that the weak and the strong solutions are the same functions and hence are actual solutions. Thus, the unique existence of the solution of the equation was established.

Note that distribution theory is not needed in these considerations.

Since the existence statement obtained involves abstract, nonexplicit entities, this result is not completely satisfactory. In particular, it is insufficient if one wants to compute the solution. This shortcoming may seem to be very severe, but it is not. It can be remedied completely if the data involved in the formulation of the problem have continuous ordinary derivatives of a sufficiently high order. In that case the solution itself and its first derivatives are proper, continuous functions. Under such circumstances computation should be possible. The order of differentiability needed for this purpose is finite. But the minimal value of this order depends on the dimension m of the space of independent variables.

Certainly, it would be possible to obtain this result in one step, rather than two. That single step would, however, be very involved. Also, the procedure would then depend on the dimension, m, of the space. Still, some people prefer such a one-step procedure. My preference has been to establish first a minimum result under minimal assumptions and then to strengthen this result according to need. But that is a matter of taste.

The equations treated by the method indicated are linear. The initial problem for a general nonlinear symmetric hyperbolic system was treated in later years by others.[1] For hyperbolic equations that are not symmetric, very original treatments have been developed by many mathematicians. However, in such cases infinite differentiability of the data is in general assumed. For problems of mathematical physics one likes to avoid the assumption of infinite differentiability, especially when one wants to compute. Still,

[1] In 1933 I had prepared a finished manuscript for a very special case of nonlinear hyperbolic equations, which I never published.

in certain (more or less degenerate) cases of mathematical physics infinite differentiability may be needed to handle the problem.

I mentioned earlier that certain symmetric equations can be handled that need not be elliptic or hyperbolic. These are equations which are symmetric and satisfy a certain positive-definiteness condition which involves also the lower order terms $B_{\alpha\beta}$. This condition is much less restrictive than the positive-definiteness conditions for elliptic and hyperbolic equations. Equations that satisfy such a weaker condition I have called "accretive."[2] It is vital for my treatment of accretive equations that they form systems of the first order and are not equations of the second order. For these accretive equations one can prove the existence of the solution under properly chosen boundary conditions.

Accretive equations cover a large class of equations that are elliptic in one region and hyperbolic in another region. In proving the existence of the solution of such equations, and even for a numerical solution, one need not pay attention to whether or not the equation is elliptic, or hyperbolic, or neither.

Mostly, however, people do not deal with systems of the first order. Rather, they employ second order equations, such as the Tricomi equation and its extensions (which are equivalent to some accretive equations). They then handle the elliptic and hyperbolic pieces separately, by different methods. Such a treatment has its advantages. My approach to accretive equations, on the other hand, shows that, in principle, separate treatment should not be necessary.

The theory of accretive equations so far discussed refers to linear equations. The existence of the solution of nonlinear accretive equations, in the case in which the domain covers the whole space, was proved by a student of mine.

I should like to mention another problem involving partial differential equations, although it has no visible relation to applied mathematics. This is the problem of differential forms on Riemannian manifolds. The formulation of this problem is covariant. Earlier workers on this problem tried to handle it by tools that were also covariant, but they ran into trouble. Actually, there is no reason why the tools should also be covariant. I myself (and Morrey independently at the same time) used tools, such as certain positive-definite quadratic forms, which were not covariant. Then there was no trouble and the problem could be solved.

There is another area, not involving partial differential equations, in which von Neumann has done basic work which has influenced me. More precisely, I should say that this work influenced Courant and me. That is the area of shock theory.

Courant and I had some knowledge of the theory of shocks since we knew of Riemann's work and had learned more about shocks in a seminar run by the aerodynamicist Prandtl in Göttingen. In 1943 Courant had just started a seminar on shocks at New York University, when he heard that von Neumann in Washington was also working on shock theory. While our knowledge concerned stationary shocks, von Neumann's work concerned mainly moving shocks. It was based on work of French aerodynamicists. Of course, von Neumann had added a great number of important original features to this theory. Courant and I then worked on both aspects of shock theory. A substantial part of the book on *Supersonic Flow and Shock Waves* was the result of this work.

[2] The name I had originally given such equations was "symmetric positive." While reading proofs of my paper I learned that R. Phillips had written a largely overlapping paper on "dissipative" hyperbolic systems. Since my matrix ($A^{\mu}_{\alpha\beta}$) was the negative of his I suggested in a footnote that my equations should perhaps be called "accretive." This term was taken up by others and subjected to one or more redefinitions. Now I am no longer sure what this term means.

Before describing my last item, I should like to tell a story. In 1954 I was concerned with equations of nonrelativistic magneto-fluid dynamics. I observed that these equations can be reduced to symmetric hyperbolic ones. Also, I formulated the equations of relativistic magneto-fluid dynamics. Fifteen years later I was asked to give a lecture in Lille, France, at a conference on nonrelativistic and relativistic magneto-fluid dynamics. I planned to present the reduction to a symmetric hyperbolic system, but found, to my dismay, that I had not done that for the relativistic case. So I still had to do that. It was very tough. After a considerable effort I found a not very satisfactory way of carrying out this reduction.

At 11 o'clock on the night before the lecture I found a somewhat better way, which I then presented, although I was still not satisfied. Eventually, late one night some years later, I did find a very simple method. I found this simple method because I was annoyed about something; in fact, I was annoyed about myself.

Essentially, all the basic equations of motion of classical physics are naturally given as conservation equations and very much original work had been done about conservation equations. I myself, however, worked only on symmetric equations and I was annoyed that I did not see the right connection between these two types of equations. Finally, that night, I saw it. Through the connection which I then found the reduction of the pertinent conservation equations to symmetric hyperbolic ones becomes nearly trivial.

To illustrate the reduction procedure I shall first present the equations of relativistic electro-magneto-fluid dynamics. These equations involve the space-time variables $x^{1,2,3,4} = \{x^\mu\}$. Differentiation with respect to x^μ will be denoted by ∂_μ as before. The four components of the relativistic velocity $w^{1,2,3,4}$ are dependent through the relation $\sum_\mu w_\mu w^\mu = -c^2$, where c is the speed of light and $w_4 = -c^2 w^4$, $w_{1,2,3} = w^{1,2,3}$, $w^4 > 0$.

The fluid mass density will be denoted by $\omega(x)$ and the electromagnetic field tensor will be denoted by $\{B_{\kappa\lambda}(x)\} = -\{B_{\lambda\kappa}(x)\}$. I employ a Lagrangian L which is a function of ω and B. In terms of these quantities the energy—momentum tensor—can be given as

$$U^\kappa_\lambda = -L\,\delta^\kappa_\lambda + \omega\frac{\partial L}{\partial\omega}(\delta^\kappa_\lambda + c^{-2} w^\kappa w_\lambda) + 2\frac{\partial L}{\partial B_{\mu\kappa}} B_{\mu\lambda}.$$

There are 10 unknown functions of the variables (x^1, x^2, x^3, x^4) to be determined. They are ω, B_{12}, B_{13}, B_{14}, B_{23}, B_{34}, B_{42}, and w^1, w^2, w^3. The unknown w^4 will be regarded as a function of the w^1, w^2, w^3. To simplify the discussion, the right members of the equations will not be written down unless they are zero. The equations then are given as

(1) $\partial_\kappa U^\kappa_\lambda = \cdots,\qquad \lambda = 1, \cdots, 4,$

(2) $\partial_\kappa(\partial L/\partial B_{\mu\kappa}) = \cdots,\qquad \mu = 1, \cdots, 4,$

(3) $\partial_\nu B_{\lambda\mu} + \partial_\lambda B_{\mu\nu} + \partial_\mu B_{\nu\lambda} = 0,\qquad \lambda, \mu, \nu = 1, \cdots, 4,$

(4) $\partial_\kappa \omega w^\kappa = 0.$

Summation with respect to κ from 1 to 4 is implied.

The system of these "basic" equations is to be solved for the 10 unknowns under suitable initial conditions.

The system of basic equations has two vital properties. First of all, these equations are obviously conservation equations. Secondly, the number of equations is greater than the number, 10, of unknowns. Since the number of distinct equations (3) is 4, there are effectively 13 equations altogether. Thus, the system is overdetermined. It is because of this overdetermining, together with its conservation character, that the system of equations can be handled in a relatively simple way.

Of course, the basic equations must depend on each other. There must exist 3 dependency relations. One of these relations is formed by multiplying successively the basic equations by suitable factors and then adding the products. The factors to be taken are

$$w^\lambda, \qquad 2B_{\lambda\mu}w^\lambda, \qquad (\partial L/\partial B_{\mu\nu})w^\lambda, \qquad (\partial L/\partial \omega).$$

Addition of the products yields zero, as is readily verified. Clearly, then, these factors produce a dependency relation.

The other two dependency relations are also readily determined. The existence of the three dependency relations is, of course, vital for the existence of a common solution of the 13 equations.

Clearly, before one can even think of solving this system of 13 equations, one must first select 10 out of the 13 equations, or rather 10 linear combinations of the 13 equations, in such a way that the resulting subsystem of 10 equations for 10 unknowns can be solved. Such a subsystem can be solved if it is symmetric hyperbolic. So, one will try to make it symmetric hyperbolic by proper selection of the 10 combinations. How to do that is not obvious, but simple. It will be shown somewhat later in connection with the general theory of overdetermined system of conservation equations.

After one has succeeded in selecting a symmetric hyperbolic subsystem and has found the solution of this subsystem, one must still show that all 13 equations are satisfied. I shall not discuss here how to do this.

I shall now briefly describe the general theory of overdetermined systems of conservation equations. Such a system involves m independent variables $x^1, x^2, \cdots, x^m$. It also involves s unknown functions $u^\sigma(x)$, where σ runs from 1 to s and x stands for the system $x^1, \cdots, x^m$. The equations for the unknowns have the form $\partial_\mu q^\mu_\rho(u) = \cdots$. Here, u stands for the system $\{u^\sigma(x)\}$ of functions $u^\sigma(x)$. The terms q^μ_ρ, with $\mu = 1, \cdots, m$ and $\rho = 1, \cdots, r$, are given functions $q^\mu_\rho(u)$ of u.

It is vital to assume that the number of equations r is greater than the number of unknowns, s; that is, $r > s$. The system of equations is then overdetermined.

Clearly, these equations must depend on each other. One of the relations expressing such a dependence is assumed to be of the form

$$y^\rho\, \partial_\mu q^\mu_\rho = \cdots,$$

in which the factors y^ρ are suitable functions $y^\rho(u)$ of the u^σ.

The derivatives of the factors y^ρ with respect to $u^1, \cdots, u^s$ denoted by $\partial_{u^1}y^\rho, \cdots, \partial_{u^s}y^\rho$ play an important role in the formulation of the symmetric hyperbolic equations to be derived. To derive these equations one should multiply the differential equations $\partial_\mu q^\mu_\rho = \cdots$ by the derivatives $\partial_{u^\tau}y^\rho$ and then carry out summation with respect to ρ. One then obtains the system of equations

$$(\partial_{u^\tau}y^\rho)\, \partial_\mu q^\mu_\rho = \cdots, \qquad \tau = 1, \cdots, s.$$

Evidently these derived equations are satisfied if the original equations $\partial_\mu q^\mu_\rho = \cdots$ are satisfied. Note that there are exactly as many derived equations as unknowns, namely s. I now claim that the system of these s derived equations is symmetric hyperbolic.

More explicitly, this claim means that the system of equations

$$(\partial_{u^\tau}y^\rho\, \partial_{u^\sigma}q^\mu_\rho)\, \partial_\mu u^\sigma = \cdots, \qquad \tau, \sigma = 1, \cdots, s,$$

which results when one spells out the derived equations, is symmetric hyperbolic.

This statement is easily proved. First we verify that the relations

$$y^{\rho}\,\partial_{u^{\sigma}} q^{\mu}_{\rho} = 0$$

hold for $\sigma = 1, \cdots, s$ and $\mu = 1, \cdots, m$. This can be done by a simple argument. Differentiating these relations with respect to u^{τ} for $\tau = 1, \cdots, s$, one will obtain the system of relations

$$\partial_{u^{\tau}} y^{\rho}\,\partial_{u^{\sigma}} q^{\mu} + y^{\rho}\,\partial_{u^{\tau}}\,\partial_{u^{\sigma}} q^{\mu}_{\rho} = 0, \qquad \tau, \sigma = 1, \cdots, s.$$

Note that the two terms on the left may be regarded as matrices in (τ, σ) for each value of μ. Since the second matrix is symmetric in σ and τ, the same is true of the first matrix. Now, this first matrix is exactly the coefficient matrix of the spelled out derived system. Hence this coefficient matrix is symmetric, and our claim that the derived system is symmetric is therefore proved. The derived system is even symmetric hyperbolic, provided that a suitable quadratic form is positive-definite.

Suppose one has proved that the symmetric hyperbolic subsystem has a solution. One must still show that all equations of our system are satisfied provided suitable additional $r - s - 1$ dependency relations hold. This can be done by simple considerations combined with elementary algebraic arguments.

The theory of overdetermined systems of conservation equations here summarized is of a rather general significance since, as was said earlier, it covers essentially all nondegenerate equations of motion in classical physics.

I end my presentation with a question:

For the systems of equations I have discussed here, the symmetry feature was a derived property. Now, in many branches of physics, in particular in the various branches of quantum theory, symmetries play a fundamental role, but all these symmetries—as it seems to me—are assumed and not derived. I now wonder whether or not in quantum theory, or other branches of physics, symmetries can also be derived from the overdeterminancy of basic conservation equations?

REFERENCES

RICHARD COURANT AND KURT O. FRIEDRICHS, *Supersonic Flow and Shock Waves*, Interscience, New York, 1948.

S. HAHN-GOLDBERG, *Generalized linear and quasilinear accretive systems of partial differential equations*, Comm. Partial Differential Equations, 2, Part 1 (1977), pp. 109–164.

KURT O. FRIEDRICHS, *Spektraltheorie halbbeschränkter Operatoren*, Math. Ann., 109 (1934), pp. 465–487; 110 (1935), pp. 777 ff.

———, *On differential operators in Hilbert spaces*, Amer. J. Math., 61 (1939), pp. 523–544.

———, *On the differentiability of the solutions of linear elliptic differential equations*, Comm. Pure Appl. Math., 6 (1953), pp. 299–326.

———, *Symmetric hyperbolic linear differential equations*, Comm. Pure Appl. Math., 7 (1954), pp. 345–392.

———, *Differential forms on Riemannian manifolds*, Comm. Pure Appl. Math., 8 (1955), pp. 551–590.

———, *Symmetric positive linear differential equations*, Comm. Pure Appl. Math., 11 (1958), pp. 333–418.

———, *On the laws of relativistic electro-magneto-fluid dynamics*, Comm. Pure Appl. Math., 27 (1974), pp. 749–808.

———, *Conservation equations and the laws of motion in classical physics*, Comm. Pure Appl. Math., 30 (1978), pp. 123–131.

K. O. FRIEDRICHS AND P. D. LAX, *Systems of conservation equations with a convex extension*, Proc. Nat. Acad. Sci. U.S.A., 68 (1971), pp. 1686–1688.

C. B. MORREY, JR. AND J. EELLS, JR., *A variational method in the theory of harmonic integrals*, I., Ann. of Math., 63 (1956), pp. 91–128.

R. S. PHILLIPS, *Dissipative hyperbolic systems*, Trans. Amer. Math. Soc., 86 (1957), pp. 109–173.

MARSHALL H. STONE, *Linear Transformations in Hilbert Space and their Applications to Analysis*, American Mathematical Society, New York, 1932.

JOHN VON NEUMANN, *Zur Theorie der unbeschränkten Matrizen*, J. Reine Angew. Math., 161 (1929), pp. 208 ff.

———, *Allgemeine Eigenwerttheorie Hermitescher Funktional operatoren*, Math. Ann., 102 (1929), pp. 50 ff.

———, *Über adjungierte Funktional operatoren*, Ann. of Math., 33 (1932), pp. 299 ff.

———, *Mathematische Grundlagen der Quantentheorie*, Julius Springer, Berlin, 1932.

[81–1] Unobserved observables and unobserved causality, Comm. Pure Appl. Math., XXXIV (1981), 273–283.

This was Friedrichs' last published work and is described in D. Isaacson's remarks, p. 603.

Unobserved Observables and Unobserved Causality*

KURT O. FRIEDRICHS

The subject of this talk is a question of physics; however, I am a mathematician. Nevertheless, I have always been interested in various fields of mathematical physics and in particular in quantum physics. Since my talk involves quantum physics I should like to indicate how I became interested in Quantum Theory. Essentially I became involved in this question in Göttingen as a student and then as Courant's assistant.

Some time early in 1926, Heisenberg gave a lecture to faculty and assistants about his new theory and about Born's and Jordan's extension. It was fascinating; I remember that one of the statements Heisenberg emphasized was that his theory referred only to quantities that could be determined by observation and that the location of a particle was not such a quantity. But James Frank, the experimental physicist, objected; he saw no reason why it should not be possible to determine the position of a particle. He was right, but probably for the wrong reason. Had he been asked whether or not position and momentum could be observed simultaneously, he might have said that that also should be possible.

Of course, all this was cleared up some years later, when Heisenberg and Bohr developed their interpretation of quantum theory, the "Copenhagen interpretation". But even in later years discussions about the meaning of quantum theory have been going on all the time. What I intend to say in this lecture may perhaps be regarded as one of these discussions.

Before starting I would like to talk a little more about my relationship to quantum theory.

In the course of years I had a number of attacks of quantumitis, or rather quantumosis. One such attack occurred in 1930 when I learned about von Neumann's Hilbert space theory, which he had developed in order to give quantum theory a mathematical—or I should perhaps say—rational basis. Von Neumann wrote a book about quantum theory which appeared in 1932, and which I then reviewed favorably. However, now I am somewhat disappointed in that book. Still, von Neumann's Hilbert space theory remains of basic significance.

I had another attack of quantumitis in 1948, when I dealt with perturbation theory. In 1950 and later on I became involved with quantum theory of fields three more times. I ran my head three times against the wall. But the wall did not crack; only a few chips came off. In the meantime the wall had cracked considerably by the effort of others.

*Lecture given at the Courant Institute and at Northwestern University, 1980.

Communications on Pure and Applied Mathematics, Vol. XXXIV, 273–283 (1981)

My last involvement, the one that led to the considerations I shall present today, had started already in 1948. I had a certain idea then; but I let it slide. A few years ago I once again took up that idea. This led to a publication in the journal *Foundations of Physics*.[1] In preparing the present talk I realized that I could simplify and improve some of the arguments of that paper.

In the discussions I shall present now I do not mean to give a new interpretation of quantum mechanics. I only want to describe a notion which can be added to the standard notions of quantum mechanics. This addition need not be adopted. But, if one does adopt it, definite simplifications of descriptions in quantum mechanics are possible. In particular, the principle of causality is then valid in some sense.

The additional notion is that of "unobserved observable". The main aim of employing it is not to resolve difficulties but rather to make a puzzling situation harmless. This puzzling situation involves physical objects to which certain observables are assigned. They may represent positions, momenta, and the energy of particles and they in turn may be represented by infinite matrices or operators. These observables are, in general, non-commuting entities. Observation of such an observable will yield a real number, which is one of a set of such numbers, called eigenvalues of the observable. For simplicity I shall also say that such an eigenvalue is the "value" of the observable found by observation.

The State Paradox

Suppose the values of the observables in a complete system of commuting observables have been determined by observation; then one says—customarily—that the *state* of the object has been determined. Now, if one makes a new observation of the object immediately afterwards—or even later—that is, if one makes the observation of another complete system of observables of that object, one will find that the state of the object has instantly collapsed and that it instantly appears in a new state. This collapse and reemergence of states is naturally felt to be quite a paradox which has upset many physicists.

This paradoxical situation is commonly illustrated by the example of "Schrödinger's cat". The cat sits in a closed cage which also contains a vessel with cyanic acid. Particles from a radioactive substance will be allowed to enter the cage. If such a particle hits the container and breaks it, the acid released will kill the cat. By opening a window and looking into the cage one can find out whether or not the cat is dead. But, already before doing this one can determine

[1] *Remarks on the Notion of State in Quantum Mechanics*, Foundations of Physics, Vol. 9, Nos. 7/8, 1979, pp. 515–524. Some remarks made in footnotes of that paper and alluded to here are not repeated in the present version.

the probability of the cat's state. This probability may, for example, be that the cat is in a state of 60% death and 40% life. After opening the window one will, however, find that the cat is either dead or alive. If it is found to be dead, one must say that the state of the cat is 100% death. Thus, just opening the window has changed the cat from being 60% dead to 100% dead; that is, opening the window has killed the cat. Such a statement is not just paradoxical; it is preposterous.

Various attempts have been made at resolving the paradox that the state of the object changes instantly if an observation is made. I do not mean just the case of the cat, rather I mean the general paradox that a state changes instantly if an observation is made.

One rather common attempt at solving the paradox consists in claiming that the new state is actually produced just by observing the object. That, I should say, is an evasion. In the case of the cat, it would mean that by opening the window and looking through, the observer has killed the cat.

There is, however, a serious problem involving the relationship between the observer and the object. To clarify this relationship it would be necessary to go into an analysis of the problem of the action and reaction between conscious beings and the world of realities. I do not intend to enter here into the philosophical depth needed for such an analysis. All I want to do now is to present a —not very deep—description of the apparent paradox.

In order to present my attitude I must first write down some elementary notions of quantum mechanics. For simplicity I assume that the physical object in question is such that only two independent noncommuting observables, A and B, are assigned to this object. These observables may, for example, represent position and momentum of a particle. Possible values of A and B, that is, eigenvalues of the observables A and B, will be denoted by a and b.

Suppose one has observed the observable A and has found the value of a for it. Then one does not know the value of the observable B; but one knows something about B. Namely, one knows that if one makes an observation of B right after having observed A—or later—the value b will be found with a definite probability. I denote this probability by

$$P = P\begin{pmatrix} A & \big| & B \\ a & \big| & b \end{pmatrix}.$$

Thus, having found the value a by observing A, one will find the value b with the probability P if observation of B is made either immediately afterwards (or later) if no other observation is made in between.

Actually one knows still a little more, namely the "probability amplitudes", which I denote by

$$\left\langle \begin{matrix} A \\ a \end{matrix} \middle| \begin{matrix} B \\ b \end{matrix} \right\rangle.$$

The probability amplitude is related to the probability $P\binom{A}{a}\big|\binom{B}{b}$ by the relation

$$P\left(\begin{array}{c|c} A & B \\ a & b \end{array}\right) = \left|\left\langle\begin{array}{c|c} A & B \\ a & b \end{array}\right\rangle\right|^2.$$

The value of a probability amplitude, which is a complex number, can be determined, except for a factor of absolute value 1, by methods of the theory of Hilbert spaces.

It is rather customary to say that the knowledge of the probability amplitude gives one the knowledge of the *state* of the object. One also says that the probability amplitude represents the state of the object after one has observed a complete set of commuting observables for it. The state represented by the probability amplitude is frequently called the *physical state* of the object.

Very often, however, the notion of "state" is conceived in quite a different, rather natural, way without reference to the "physical state". Quite often such a natural notion of state is—more or less subconsciously—identified with the physical state, mainly because it is thought that there does not exist any other kind of state. That identification produces quite a confusion. The clarification of this confusion is one of my aims.

Unobserved Observables

To give a natural definition of "state", quite independently of the notion of physical state, one may be inclined to say that the state of an object comprises all characteristics of the object. I shall refer to the state so defined as the *full state*. It is essentially the same as the "Laplace state" which plays a role in the principles of classical mechanics.

Certainly, according to the principles of quantum theory, the full state cannot be determined completely by observation. For, if some characteristics of the object have been determined, a further determination may destroy the knowledge previously obtained. For this reason, some—or most—quantum theorists do not consider what I call *full state* as a part of quantum mechanics. They might even forbid others to use this notion in quantum theory. Nevertheless, I shall use it—but not loosely. Rather I shall use a precise definition of the "full state".

I shall define the "forbidden" notion of full state in a specific manner and shall call the state so defined *intrinsic*. I shall then show that some definite statements can be made about the intrinsic state in spite of the fact that it cannot be determined completely.

In order to define the notion of intrinsic state one must relate this notion to observations. One might be tempted to characterize the intrinsic state at a certain time by the outcomes of the observations made at that time of all observables associated with the object; but, of course, this attempt runs into trouble. For, if the value of one observable is determined, the values of those other observables that do not commute with it will in general change due to the interfering action of the observation instrument. If no observation is made, there is no interference; but then one does not know what the outcome would have been. *Still, one may ask what the outcome of an observation would have been, even though the observation was not made.*

Perhaps it will be said that a physicist should not ask such a question. That may be the case if he were acting as a quantum physicist; but as a human being the physicist may—and often will—ask what the outcome would have been in spite of the fact that he cannot find out what the answer is. He might be able to do so if he were endowed with the faculty of "extra physical perception"; however, I do not suppose that there exists anybody with that faculty.

The remarks offered in the following for consideration are based on the willingness to speak of the value an observation of an observable would have shown, even if the observation was not made. I shall present statements about unobserved values of the observables without in any way intefering with the generally accepted statements of quantum mechanics.

I should mention that values of unobserved observables are on occasion referred to in the literature; but the consequences that I shall derive have not been formulated, as far as I know.

The value of an observable that would have been found if it had been observed at a certain time will be referred to as the *intrinsic value* at that time. If the value was found by actual observation, it will be called its *physical value*; otherwise the value might be called "extra physical". I prefer to call it *ultraphysical*. (Such a value might also be called "hidden"; but I rather do not use this term in the present context in order to distinguish my hideout from other hideouts.)

Since intrinsic values are defined independently of each other, we can speak of the totality of the intrinsic values of all observables associated with the object at a certain time. The values of some of the observables taken from a complete set of commuting such observables may have been determined by observation and therefore are physical values. The intrinsic values of all other observables are then ultraphysical. In any case I regard the totality of the intrinsic values of all observables associated with the object at a certain time as representing the *intrinsic state* of the object at that time.

One may perhaps say that the definition of the term *intrinsic state* as employed here evades the issue since it refers to the outcomes of observations which have been made—or have not been made—from the outside, as it were, and that it does not refer to what the object really is in itself and thus does not really describe the Laplace state.

To a certain degree an internal description of "state" was given by Heisenberg, who characterized a physical object as being endowed with the potentiality —or tendency—to respond to observations with definite outcomes. However, I do not want to use such ideas here.

I am not concerned with the question of the meaning of physical reality; my intention is not even to make statements of physical facts. My aim is strictly formalistic.

Still, the formalism to be presented may be helpful in discussions about the proper interpretations of quantum theory, mainly in that it would offer greater freedom of expression since it allows one to speak of unobserved values. It is certainly not claimed that this formalism yields an explanation of the deeper meaning of quantum theory.

I shall now state some circumstances which involve the concept of intrinsic value. Employing this notion one can describe—to a certain degree—what happens to the unobserved observables during an observation process. One may regard the observation instrument as a physical object with which a number of observables are associated. Of course, the instrument and its preparation will have to be so chosen that the interaction energy will be effective essentially only during a short time interval. One can say that the intrinsic values of the instrument's observables as well as the intrinsic values of the original object change during the interaction process, and go over into new intrinsic values. There is no need to assume a "spontaneous creation" of the new values. The impossibility of describing the effect of the "disturbance" completely is simply reduced to the impossibility of finding out what the intrinsic state of the pair of objects was originally.

In the discussion of the disturbance phenomenon just given, ultra-physical statements were made inasmuch as one has talked of intrinsic values. Certainly it is very tempting to make such statements and certainly it sometimes happens that a physicist makes such a statement inadvertently. To be sure, with some care this can be avoided. Quantum theory can be described without referring to ultraphysical states and values. If this is done, however, one should take a strictly positivistic attitude; in particular, one should not regard the collapse of a physical state as paradoxical.

On the other hand, it is very convenient to employ ultraphysical entities in describing quantum physical situations. Our propositions may serve to legalize such a usage.

A few remarks will be made about the implications of the attitude taken here that the full description of the state of an object is not given by its "physical" state. This physical state is represented by a probability amplitude and, according to this attitude, refers to an actual observation made by an observer. Hence, if one admits only physical states as meaningful notions of state, the mysterious jump from one physical state to another must be regarded simply as a jump of the observer's knowledge.

It has been said by various physicists that one does not need an observer to determine the outcome of a measuring process, and that it would be sufficient to store the observation on a recording device. But that is not quite so. To be sure, if one took a strictly positivistic attitude, one would be forced to say that the recorder is in a state which can be characterized by the probability amplitudes that can be deduced from the observation made before. But then one must make an observation to find out what the recorded information is. If an observer says that this information was already on the recorder before he had looked at it—as many physicists would say—then he makes an ultraphysical statement.

It was an essential assumption underlying the remarks made here that the probability amplitude does not represent the "full" state of the physical object, but only supplies all the information an observer can get, however not all that the observer would have to know before he could describe the full state. That is, the observer must resign himself to accepting as a "fact of quantum physics" that he is not allowed to know all that he may feel he is entitled to know; however, he is allowed to talk about the unknown facts, provided he is willing to make ultraphysical statements.

Causality and Jumps

The primary aim in the following discussion is to describe a relation between intrinsic (or physical) values of certain observables, which represents the principle of causality in some sense. In deriving such a relation two basic notions of quantum mechanics will be used: that of a function of observables and that of the probability of the outcome of an observation.

The observables in question may be represented by strictly selfadjoint operators in Hilbert space. Such an observable, A say, is associated with a set of real numbers, $a, \cdots$. These numbers are the eigenvalues of the operator. The set of these eigenvalues forms the spectrum of the observable. Any of the eigenvalues may be the outcome of an observation. This outcome is then called the "value" that the observable is found to have.

The spectrum of an observable need not be discrete; it may be continuous or partly continuous. In such a case the formulas employed should be understood symbolically.

As was said earlier, the probability $P(^A_a \mid ^B_b)$ that the outcome of the observation of the observable B, after the value a was found for the observable A, is the absolute square of the probability amplitude $\langle ^A_a \mid ^B_b \rangle$. For this amplitude the rule

$$\Sigma_b \left\langle \begin{matrix} A \\ a_0 \end{matrix} \middle| \begin{matrix} B \\ b \end{matrix} \right\rangle \left\langle \begin{matrix} B \\ b \end{matrix} \middle| \begin{matrix} A \\ a' \end{matrix} \right\rangle = \left\langle \begin{matrix} A \\ a_0 \end{matrix} \middle| \begin{matrix} A \\ a' \end{matrix} \right\rangle = \delta(a_0, a')$$

holds.

Next I discuss the notion of function of one or more observables. A function, $f(a)$, of a single observable A may be represented by the corresponding function of the operator that represents the observable A. Functions of two observables,

A and B say, may be represented by the real and imaginary parts of products

$$f_1(A)\, g_1(B)\, f_2(A)\, g_2(B) \cdots .$$

Here the observables A and B are not assumed to commute. More general functions of A and B can be formed from such simple products by suitable limiting processes, as elaborated in Hilbert space theory. Such general functions will be denoted by $f(A, B)$. These functions are assumed to be selfadjoint unless otherwise stated.

It is important for the subsequent considerations that functions of observables can themselves be regarded as observables. How to set up an observation procedure to determine the value of an observable $f(A, B)$ is a question of physics that may be difficult to answer. Still, the present considerations depend essentially on accepting that the functions $f(A, B)$ are observables.

Actually in what follows it is not necessary to employ the widest possible class of functions of A and B which can be regarded as observables. It is sufficient that it contain functions $f(A)$, $g(B)$, of single variables, polynomials such as $AB + BA$, and $i(AB - BA)$, and, in particular, those functions of A and B that represent the energy of the object as well as certain functions that in some way involve the energy.

The energy of the object is represented by an appropriate function of A and B. This function will be denoted by

$$\hbar V = \hbar V(A, B), \text{ where } \hbar \text{ is Dirac's factor.}$$

The functions of the energy needed are the real and imaginary parts of the two functions

$$U^{\pm\tau} = \exp\{\pm i\tau V\},$$

where τ is a real number which stands for the time. The functions $U^{\pm\tau}$ are not selfadjoint, but $U^{\tau} + U^{-\tau}$ and $i(U^{\tau} - U^{-\tau})$ are.

The functions $U^{\pm\tau}$ enter the law of dynamics for the physical object under consideration. Here I restrict myself to nonrelativistic quantum mechanics. Specifically I employ the global law of dynamics.

In formulating this law one must use observables associated with definite times. The observable A associated with the time $t = \tau$ will be denoted by $A^{t=\tau}$, that associated with the time $t = 0$ is then denoted by $A^{t=0}$.

The global law of dynamics can now be given in the form

$$A^{t=\tau} = U^{\tau} A^{t=0} U^{-\tau}$$

with $U^{\tau} = \exp\{i\tau V\}$, where $\hbar V$ is the energy. It is to be noted that the term $A^{t=\tau}$ is selfadjoint although the terms $U^{\pm\tau}$ are not. Clearly, the observables $A^{t=0}$ and

$A^{t=\tau}$ have the same spectrum. Their common eigenvalues will be denoted by a, $a', \cdots$.

The global law allows one to determine the probability that the observation of the observables A at the time $t = 0$ will yield the value a', say, if observation of this observable at the time $t = 0$ has given the value a_0. Specifically this probability is the absolute square of the probability amplitude

$$\left\langle \begin{matrix} A^{t=0} \\ a_0 \end{matrix} \middle| \begin{matrix} A^{t=\tau} \\ a \end{matrix} \right\rangle.$$

This amplitude will play a vital role in the subsequent considerations.

I introduce now a new set of observables. These observables are associated with the time $t = 0$ and depend on a parameter, σ say; they are given by

$$A_\sigma^{t=0} = U^{-\sigma} A_0^{t=0} U^{\sigma},$$

where $A_0 = A$. Now I take $\sigma = -\tau$, obtaining the relation

$$A_{-\tau}^{t=0} = U^{\tau} A_0^{t=0} U^{-\tau}.$$

I also set $A = A_0$ in the law of dynamics, writing this law as

$$A_0^{t=\tau} = U^{\tau} A_0^{t=0} U^{-\tau}.$$

Comparing the last two relations one is led to the relation

$$A_{-\tau}^{t=0} = A_0^{t=\tau}.$$

This relation says that the observable $A_{-\tau}$, when observed at the time $t = 0$, is the same as the observable A_0, when observed at the time $t = \tau$. But, this relation does not imply that the values of the two observables that would be found by observation are the same. Still these values are the same after all.[2]

To prove this statement one may combine the probability amplitude $\langle \begin{smallmatrix} A^{t=0} \\ a \end{smallmatrix} | \begin{smallmatrix} A^{t=\tau} \\ a' \end{smallmatrix} \rangle$, now written as $\langle \begin{smallmatrix} A_0^{t=0} \\ a \end{smallmatrix} | \begin{smallmatrix} A_0^{t=\tau} \\ a' \end{smallmatrix} \rangle$, with the amplitude $\langle \begin{smallmatrix} A_{-\tau}^{t=0} \\ a_0 \end{smallmatrix} | \begin{smallmatrix} A_0^{t=0} \\ a \end{smallmatrix} \rangle$, obtaining the amplitude

$$a \left\langle \begin{matrix} A_{-\tau}^{t=0} \\ a_0 \end{matrix} \middle| \begin{matrix} A_0^{t=0} \\ a \end{matrix} \right\rangle \left\langle \begin{matrix} A_0^{t=0} \\ a \end{matrix} \middle| \begin{matrix} A_0^{t=\tau} \\ a' \end{matrix} \right\rangle = \left\langle \begin{matrix} A_{-\tau}^{t=0} \\ a_0 \end{matrix} \middle| \begin{matrix} A_0^{t=\tau} \\ a' \end{matrix} \right\rangle,$$

by virtue of the rule mentioned earlier. Using the relation $A_{-\tau}^{t=0} = A_0^{t=\tau}$, one can

[2]The proof of this statement was not given in the earlier publication.

write the amplitude above as

$$\left\langle \begin{matrix} A_{-\tau}^{t=0} \\ a_0 \end{matrix} \middle| \begin{matrix} A_0^{t=\tau} \\ a' \end{matrix} \right\rangle = \left\langle \begin{matrix} A_0^{t=\tau} \\ a \end{matrix} \middle| \begin{matrix} A_0^{t=\tau} \\ a' \end{matrix} \right\rangle .$$

The value of the latter amplitude is 1 if $a_0 = a'$ and zero if $a_0 \neq a'$, again by virtue of the rule mentioned earlier.

Forming the probabilities associated with the two terms by taking the absolute squares one obtains the result

$$P\left(\begin{matrix} A_{-\tau}^{t=0} \\ a_0 \end{matrix} \middle| \begin{matrix} A_0^{t=\tau} \\ a' \end{matrix} \right) = \begin{cases} 1 & \text{if } a_0 = a', \\ 0 & \text{otherwise.} \end{cases}$$

This statement implies that the relation $a' = a_0$ holds with probability 1. It follows that the values of the observables $A_{-\tau}^{t=0}$ and $A_0^{t=\tau}$ are the same, as was claimed.

The statement just made implies that one can make a prediction with certainty; namely the prediction that the value of the observable A_0, when observed at the time $t = \tau$, is the same as the value that the observable $A_{-\tau}$ was found to have when it was observd at the time $t = 0$. One may also say: If one wants to know what the value of the observable A_0 is at the time $t = \tau$, one need only observe what the value of the observable $A_{-\tau}$ is at the time $t = 0$. I now maintain that a prediction can be made even if the observable $A_{-\tau}^{t=0}$ has not been observed.

To this end one should note that the prediction that the value of A_0 at the time $t = \tau$ will be "a" if the value of "a" was found for $A_{-\tau}$ at the time $t = 0$ is made quite independently of the manner in which the observation of $A_{-\tau}^{t=0}$ was carried out. Thus, this prediction is not affected by the disturbance produced by the process of observing $a_{-\tau}^{t=0}$. Of course, this is also the case if the observation does not produce any disturbance. This does happen, of course, if no observation is made. Surely in that case we may conclude that the value "a" will be found for $A_0^{t=\tau}$ if "a" is the value of $A_{-\tau}^{t=0}$. But now the value "a" is an intrinsic one. That is to say, *the intrinsic values of* $A_0^{t=\tau}$ *and* $A_{-\tau}^{t=0}$ *are the same*. That this is true is our *main proposition*.

The main proposition just formulated implies that there is a one-to-one mapping of the set of the intrinsic values which our observables have at the time $t = 0$ onto the set of intrinsic values which these observables have at some later time. Evidently, this implies that the intrinsic state of our object at the later time is uniquely determined by the intrinsic state at the initial time. But no relationship between the later value of the observable and its earlier value is implied. In this restricted sense the principle of causality is valid.

In concluding I should like to make a remark about jumps of states. It may happen that the intrinsic values of the observable $A_\sigma^{t=0}$ at values $\sigma < -\tau$ and

$\sigma > -\tau$ of the parameter σ are of such a nature that they approach different limits if the parameter σ approaches the value $\sigma = -\tau$ from two different sides. In that case the intrinsic value of the observable A_σ at the time $t = 0$ jumps when the parameter σ crosses the value $-\tau$. From the main proposition one now may deduce that the intrinsic value of the observable A_0 at the time $t = \tau$ also jumps, namely, when the time t passes through the time τ. One then may say that the jump of the observable A_0 at the time $t = \tau$ was preordained, inasmuch as it was already present in the values of the observable A_σ for $\sigma = -\tau$ at the time $t = 0$. In this sense there is no mystery in quantum jumps.

Received July, 1980.

VIII
Contributions to Quantum Mechanics

David Isaacson

Adapted from a talk at the memorial gathering for Professor Friedrichs, January 1983.

Friedrichs contributed strongly to quantum theory and more generally to mathematical physics through his research and teaching. I will describe briefly some of that research and neglect the enormous influence his lectures, writing, and conversations had on the subject.

Friedrichs' work on spectral representation, spectral resolution, perturbation theory for continuous spectra, scattering theory, functional integration, quantum field theory and the foundations of quantum mechanics is marked by several general features. The problems that he chose to study were basic to science as well as mathematics. At the time he began his work the problems were ill posed, and the areas he worked in were frequently in a state of confusion. He took great pains to define the logical structure of these problems. He selected concrete examples that had most of the important features of these problems, which he then solved in detail with highly original methods. Finally, he took great pains in making his solutions look easy and natural, and in pointing out where further difficulties lay.

It is interesting that Friedrichs wrote his first paper [27–1] in mathematical physics while he was still a graduate student. It was in general relativity. Characteristically, this paper made clear the logical significance of Einstein's general covariance postulate. By showing that Newton's laws could be written in a generally covariant form Friedrichs pointed out clearly that the postulate was a statement about the nature of the gravitational force law and not an invariance principle. This is discussed at length in S. Weinberg's recent book [1] with appropriate reference given to Friedrich's paper.

Modern quantum theory began in 1926 with Schrödinger's publication of his wave equation [2]. As far as I can tell, Friedrichs' first contributions to quantum theory were as a participant in a 1927 or 1926 seminar in Göttingen run by Max Born. Von Neumann, who had come to Göttingen to study with Hilbert, also participated in this seminar. One of Hilbert's goals was to axiomatize physics. At this time, he was attempting to axiomatize quantum mechanics. Von Neumann produced a mathematically sound axiomatization of quantum mechanics that he summarized in 1932 in his now famous book [3], which Friedrichs reviewed for the *Zentralblatt*. In Friedrichs' words, from his von Neumann lecture, [80-1].

"This book contained introductory mathematical sections, and sections involving quantum theory. At present, I think that the mathematical sections which were based on papers that had appeared in 1929, were the more significant ones. They were rather abstract.

We, in the group around Courant, were quite suspicious about the significance of such abstract work. Nevertheless, when in late 1930 I studied these abstract papers, I was dumbfounded. In fact, I had just handed to Courant for publication a manuscript on spectral theory. I asked him to return the manuscript. I then rewrote the paper in Von Neumann's abstract language. That was the origin of a substantial part of my later work."

The work that Friedrichs was referring to appeared first in 1934 in a series of papers on the spectral properties of semi-bounded operators, see Kato's commentary, p. 9. This contains Friedrichs' most quoted result in the mathematics of quantum mechanics, namely, his famous theorem for extending semi-bounded symmetric operators, now universally known as the Friedrichs' extension. It is a basic tool in mathematical physics for various reasons. Primarily, according to quantum mechanics, the observable values of the energy of a stable conservative system should be given by the spectra of a semi-bounded self-adjoint operator called the "quantum mechanical Hamiltonian." The existence of such a self-adjoint operator follows immediately from Friedrichs' theorem by establishing that on a dense set the operator in question is symmetric and semibounded. Some well known examples where Friedrichs' theorem applies are the harmonic oscillator, the hyrogen atom, and boson quantum field Hamiltonians.

Friedrichs was ahead of his time with his next contribution. This was his paper [38–1] in which he developed a method for treating perturbations of operators with continuous spectra. This is precisely the type of perturbation problem that arose in quantum mechanical scattering. Thus, when Heisenberg introduced his scattering matrix [4] and Moller introduced his wave operators [5] in 1943 and 1946, respectively, Friedrichs had already developed a mathematical method for their study. His original perturbation method would now be called a time independent approach to scattering theory.

The other method used to study scattering theory is now called the time dependent method. It was also explored and made clear by Friedrichs in [48–8] in which he established the connection between Moller's approach and his earlier work on perturbation theory.

In his later paper, he showed that Moller's wave operators were essentially his intertwining operators and that Heisenberg's scattering matrix was given simply in terms of his Γ operator. He considerably clarified the sense in which these scattering objects existed. His method and equations have been greatly extended to cover problems with Schrödinger operators and quantum field Hamiltonians. Discussions of some of these extensions can be found in the modern books of Dunford and Schwartz [6], Kato [7], Simon [8], and in Friedrichs' classic monograph, Books [65–3].

Friedrichs next contributed to quantum field theory with a pioneering set of articles published in the early 50's. These articles were collected and published in 1953 as *mathematical aspects of the quantum theory of fields*, Books [53–1]. This book contains an extensive treatment of linear quantum field problems. In particular, the book contains explicit examples of inequivalent representations of the canonical commutation relations. Friedrichs called these representations myriotic and amyriotic fields. The book contains the definition of an integral over a Hilbert space which is used to give the Schrodinger representation of the canonical commutation relations for the free boson field. This work on functional integration was studied by Friedrichs and Shapiro at greater depth, see Lecture Notes [57–1]. The book also contains a careful discussion of scattering for boson fields under the influence of linear homogeneous forces.

This book was especially important in that it isolated and gave precise definitions to many of the basic notions used by physicists in quantum field theory. It is even more remarkable when we recall the amount of confusion present in the physics literature of the time. Remember that the formal work on renormalized perturbation theory by the physicists Feynman, Dyson, and Schwinger [9] was completed in 1949. Friedrichs' first paper in the series appeared in 1950. Friedrichs was in his early fifties when this book was published.

I believe that Friedrichs' most important contribution to quantum physics was his *Perturbation of Spectra in Hilbert Space*, Books [65–3]. It was an expanded version of lectures he gave at Boulder, Colorado in 1960. The first two chapters of this work of art contain Friedrichs' description of his approach to perturbation and scattering problems. Chapter III contains his formal perturbation approach to nonlinear quantum field theory. He explains clearly what some of the main difficulties are in constructing quantum fields that require renormalization. In particular, for a restricted class of perturbations, he outlines the construction of a renormalized Hamiltonian, intertwining operators, and a scattering matrix. The methods used are generalizations of his methods for studying the perturbation of continuous spectra.

Extensions of this work were given by Jack Schwartz [10], Raphael Hoegh-Krohn [11], and James Glimm [12]. In particular the first construction of a nonliner boson quantum field theory by J. Glimm and A. Jaffe [13], [14] made significant use of Friedrichs' diagrams, formulas, and ideas. If Glimm and Jaffe are regarded as fathers of the new branch of mathematics called constructive quantum field theory, then Friedrichs is certainly a grandfather.

Friedrichs' most recent contribution to quantum mechanics was his 1981 paper, Unobserved Observables and Unobserved Causality, [81–1]. This paper is a contribution to one of the most fascinating controversies of 20th century science, the debate between Niels Bohr and Albert Einstein over the completeness of quantum mechanics. Einstein felt that the laws of nature should be deterministic or causal, and for this reason he feld the quantum mechanical notion of the state of a particle was incomplete.

Friedrichs introduced a new notion which he called the "intrinsic state" of a particle and he argued that for this intrinsic state, causality is valid, but in accordance with quantum theory, not verifiable.

We may all speculate on what Einstein and Bohr would have thought of this notion of Friedrichs. However, there is no doubt in my mind that a slight rephrasing of Einstein's words of praise for his friend, Max Planck, apply also to Friedrichs.

"In the temple of science are many mansions, and various indeed are they that dwell therein. Many take to science out of a joyful sense of superior intellectual power; science is their own special sport to which they look for vivid experience and the satisfaction of ambition; many others are to be found in the temple who have offered the products of their brains on this altar for purely utilitarian purposes. Were an angel of the Lord to come and drive all the people belonging to these two categories out of the temple, the assemblage would be seriously depleted, but there would still be some men, of both present and past times, left inside." Our Friedrichs was one of them, and that is why we loved him.

[1] S. Weinberg. Gravitation and Cosmology, John Wiley, New York, 1972.

[2] E. Schrödinger. Vier Vorlesungen über Wellenmechanik, Royal Institution, London, 1928, Berlin, Springer, 1928. English translation by J. F. Shearer and W. M. Deans, Selected Papers on Wave Mechanics, Blackie & Son, Ltd., London, 1928.

[3] J. von Neumann. Mathematische Grundlagen der Quantenmechanik, Berlin, Springer, 1932. English translation by Robert T. Beyer, Mathematical Foundations of Quantum Mechanics, Princeton University Press, Princeton, New Jersey, 1955.

[4] W. Heisenberg. Die "beobachtbaren Großen" in der Theorie der Elementarteilchen I, Zs. f. Phys. 120 (1943), 513; II, 120 (1943), 673.

[5] C. Moller. General properties of the characteristic matrix in the theory of elementary particles, II, Danske Vid. Selsk. Mat.-Fys. Medd. 22 (1946), 19.

[6] N. Dunford and J. T. Schwartz. Linear Operators, III, Interscience Publishers, New York, 1971.

[7] T. Kato. Perturbation Theory for Linear Operators, Springer-Verlag, New York, 1976.

[8] B. Simon. Quantum Mechanics for Hamiltonians Defined as Quadratic Forms, Princeton Univ. Press, New Jersey, 1971.

[9] J. Schwinger, Selected papers on Quantum Electrodynamics, Dover, New York, 1958.

[10] J. T. Schwartz. Mass renormalization and spectral shifts, New York University Physics Dept., New York, 1960.

[11] R. Hoegh-Krohn. Partly gentle perturbation with application to perturbation by annihilation-creation operators, Proc. Nat. Acad. Sci. USA, 56 (1967), 2187.

[12] J. Glimm. Models for Quantum Field Theory, Local Quantum Theory, ed. by R. Jost, Academic Press, New York, 1969.

[13] J. Glimm and A. Jaffe. A $\lambda(\phi^4)_2$ quantum field theory without cutoffs, I, Phys. Rev., 176 (1968), 1945.

[14] J. Glimm and A. Jaffe. Quantum Physics: A Functional Integral Point of view, Springer-Verlag, New York, 1981.

Honorary doctoral degrees

Technische Hochschule, Aachen	24 June 1971
Uppsala Universitet, Uppsala, 500th Anniversary	30 September 1977
Technische Universität Carolo Wilhelmina zu Braunschweig	27 September 1980
Columbia University	13 May 1981
New York University	30 October 1981

Awards

Guggenheim Fellow	1962
First recipient of Applied Mathematics and Numerical Analysis Award given at the National Academy of Sciences of the U.S.A. in Washington, D.C.	24 April 1972
Humboldt Award	1974
The National Medal of Science for 1976 awarded by President Carter in Washington, D.C. "... for bringing the powers of modern mathematics to bear on problems in physics, fluid dynamics and elasticity."	22 November 1977

Acknowledgements

Birkhäuser Boston thanks the original publishers of the papers of Kurt Otto Friedrichs for granting permission to reprint the following papers:

[34–1] Reprinted from *Math. Ann. 109, 110*, © 1934, 1935 by Springer-Verlag.
[34–2] Reprinted from *Math. Ann. 110*, © 1934 by Springer-Verlag.
[35–1] Reprinted from *Math. Ann. 112*, © 1935 by Springer-Verlag.
[38–1] Reprinted from *Math. Ann. 115*, © 1938 by Springer-Verlag.
[48–1] Reprinted from *R. Courant Anniversary Volume*, © 1948 by John Wiley & Sons, Inc.
[48–8] Reprinted from *Comm. Pure Appl. Math. I* © 1948 by John Wiley & Sons, Inc.
[66–2] Reprinted from *Perturbation Theory and Its Application in Quantum Mechanics*, © 1976 by Mathematics Research Center, University of Wisconsin.
[41–2] Reprinted from *Amer. J. Math. LXIII*, © 1941 by The Johns Hopkins University Press.
[42–2] Reprinted from *J. Appl. Mech. 9*, © 1942 by The American Society of Mechanical Engineers.
[46–1] Reprinted from *Duke Math. J. 13*, © 1946 by Duke University Press.
[55–2] Reprinted from *Bull. A.M.S. 61*, © 1955 by American Mathematical Society.
[54–1] Reprinted from *LAMS 2105 Physics*, © 1954, 1957 by Los Alamos National Laboratory and the University of California.
[78–1] Reprinted from *Comm. Pure Appl. Math. XXXI*, © 1978 by John Wiley & Sons, Inc.
[29–4] Reprinted from *Math.-Phys. Klasse*, © 1929 by Georg Olms Verlag AG.
[33–1] Reprinted from *Math. Ann. 109*, © 1933 by Springer-Verlag.
[37–1] Reprinted from *Trans. A.M.S. 41*, © 1937 by American Mathematical Society.
[54–2] Reprinted from *Comm. Pure Appl. Math. VII*, © 1954 by John Wiley & Sons, Inc.
[55–1] Reprinted from *Comm. Pure Appl. Math. VIII*, © 1955 by John Wiley & Sons, Inc.
[27–1] Reprinted from *Math. Ann. 98*, © 1927 by Springer-Verlag.
[80–1) Reprinted from *SIAM Review 22*, © 1980 by the Society for Industrial and Applied Mathematics.
[81–1] Reprinted from *Comm. Pure Appl. Math. XXXIV*, © 1981 by John Wiley & Sons, Inc.

Printed in the USA
CPSIA information can be obtained
at www.ICGtesting.com
LVHW070342201123
764347LV00011B/1001

9 780817 632694